XXIst International Conference on Integrable Systems and Quantum Symmetries

(ISQS21)

AA002738

Journal of Physics: Conference Series Volume 474

Prague, Czech Republic
12-16 June 2013

ISBN: 978-1-62993-451-8
ISSN: 1742-6588

Printed from e-media with permission by:

Curran Associates, Inc.
57 Morehouse Lane
Red Hook, NY 12571

Some format issues inherent in the e-media version may also appear in this print version.

Printed by Curran Associates, Inc. (2014)

For permission requests, please contact the Institute of Physics
at the address below.

Institute of Physics
Dirac House, Temple Back
Bristol BS1 6BE UK

Phone: 44 1 17 929 7481
Fax: 44 1 17 920 0979

techtracking@iop.org

Additional copies of this publication are available from:

Curran Associates, Inc.
57 Morehouse Lane
Red Hook, NY 12571 USA
Phone: 845-758-0400
Fax: 845-758-2634
Email: curran@proceedings.com
Web: www.proceedings.com

TABLE OF CONTENTS

Journal of Physics: Conference Series **474** (2013) 012001 doi:10.1088/1742-6596/474/1/012001

Type-II defects in the super-Liouville theory

A R Aguirre

Instituto de Física Teórica – IFT/UNESP, Rua Dr. Bento Teobaldo Ferraz 271, Bloco II, 01140-070, São Paulo, Brazil

E-mail: aleroagu@ift.unesp.br

Abstract. The introduction of type-II defects is discussed under the Lagrangian formalism and Lax representation for the $N = 1$ super-Liouville model. We derive a new kind of super-Bäcklund transformation for the model and show explicitly the conservation of the modified energy and momentum, as well as supercharge.

1. Introduction

The study of two-dimensional integrable field theories in the presence of defects or impurities has evolved into a rich subject in recent years from both classical and quantum points of view [1]-[14]. The Lagrangian formalism was initially introduced in [3] to describe integrable defect conditions through a suitable local Lagrangian density located at a fixed point. The fields on either side of the defect only interact with each other at the boundary, which characterizes the type-I defects. Several types of bosonic field theories [3, 4] allow this kind of defects preserving modified charges after including some defect contributions. Their integrability can be ensured by using the well-known inverse scattering method formalism where the defect conditions corresponding to frozen Bäcklund transformations turn to be encoded in the defect matrix [5].

Soon after, a generalization of the original Lagrangian description was proposed in [6], by allowing additional degrees of freedom associated with the defect itself, which is now called type-II defects. This alternative framework was analyzed in the cases of the sine/sinh-Gordon, massive free field, Liouville and Tzitzéica-Bullough-Dodd models in [6, 7]. For the supersymmetric extensions of sine-Gordon model [8, 9], and for the massive Thirring models [10, 11], those additional degrees of freedom had already appeared naturally. An important feature of this kind of defects turns to be the fact of having characteristics resembling a pair of fused type-I defects.

On the other hand, it was also proposed in [12] a fully algebraic approach involving the classical r-matrix structure, as well as a modified transition matrix to describe the Liouville-integrable defects, which allowed mainly to show the involutivity of the modified charges in several integrable models [12, 13, 14].

In this paper we will provide a Lagrangian description for the type-II defects within the $N = 1$ super-Liouville model. We will propose a generalization of the super-Bäcklund transformation for the super-Liouville equation given in [15], by including a chiral superfield in the formulation. We will also derive the defect matrices for the Liouville and super-Liouville models and show explicitly the conservation of the modified momentum, energy and supercharge. We also show that the conformal invariance is guaranteed after introducing the type-II defects in the models, which implies they are indeed topological defects.

2. Type-II defect Liouville field theory

In this section, we will review the type-II defects in the Liouville field theory by using the Lagrangian framework, and present also the Lax formalism.

2.1. Lagrangian description

Let us introduce a defect located at $x = 0$, with $\phi_1(x,t)$ be a Liouville field on the left side of it, $\phi_2(x,t)$ be also a Liouville field on the right side of it, and $\Lambda_0(t)$ a boundary field associated with the defect itself. Then, we start with the following Lagrangian density,

$$\mathcal{L} = \theta(-x)\mathcal{L}_1 + \theta(x)\mathcal{L}_2 + \delta(x)\mathcal{L}_D, \tag{1}$$

with

$$\mathcal{L}_p = (\partial_x\phi_p)^2 - (\partial_t\phi_p)^2 + V_p, \qquad p = 1, 2, \tag{2}$$

$$\mathcal{L}_D = (\phi_2\partial_t\phi_1 - \phi_1\partial_t\phi_2) - \Lambda_0\partial_t(\phi_1 - \phi_2) + (\phi_1 - \phi_2)\partial_t\Lambda_0 + B_0(\phi_1, \phi_2, \Lambda_0), \tag{3}$$

where the Liouville potential is given by $V_p = \mu^2 e^{2\phi_p}$ and the defect potential B_0 can be decomposed into $B_0 = B_0^+(\phi_+ - \Lambda_0, \phi_-) + B_0^-(\phi_-, \Lambda_0)$, after introducing the variables $\phi_\pm = \phi_1 \pm \phi_2$ [6]. Here μ is a scale parameter sometimes called cosmological constant. It is not difficult to show that the modified energy $E + B_0$ is conserved, where E denotes the canonical energy. On the other hand, by requesting a conserved modified momentum, the defect potentials $B_0^\pm$ have to satisfy the non-linear relation,

$$\left(\frac{\partial B_0^+}{\partial\phi_-}\right)\left(\frac{\partial B_0^-}{\partial\Lambda_0}\right) - \left(\frac{\partial B_0^+}{\partial\Lambda_0}\right)\left(\frac{\partial B_0^-}{\partial\phi_-}\right) = (V_1 - V_2), \tag{4}$$

and can be adequately written as

$$B_0^+ = -2i\mu\beta^2 e^{(\phi_+ - \Lambda_0)}, \qquad B_0^- = \frac{i\mu}{\beta^2} e^{\Lambda_0}(\cosh\phi_- + \kappa), \tag{5}$$

where κ is an arbitrary parameter. Besides the respective bulk field equations, namely

$$\partial_x^2\phi_p - \partial_t^2\phi_p = \mu^2 e^{2\phi_p}, \qquad p = 1, 2, \tag{6}$$

we also obtain the defect conditions at $x = 0$,

$$\partial\phi_+ + 2\partial_t\Lambda_0 = -\frac{i\mu}{\beta^2} e^{\Lambda_0}\sinh\phi_-, \tag{7}$$

$$\bar{\partial}\phi_- = 2i\mu\beta^2 e^{(\phi_+ - \Lambda_0)}, \tag{8}$$

$$\partial\phi_- = -\frac{i\mu}{\beta^2} e^{\Lambda_0}(\cosh\phi_- + \kappa), \tag{9}$$

where the following light-cone notation has been used, $z = (x - t)/2$, $\bar{z} = (x + t)/2$, $\partial \equiv \frac{\partial}{\partial z} = \partial_x - \partial_t$, $\bar{\partial} \equiv \frac{\partial}{\partial\bar{z}} = \partial_x + \partial_t$. If these type-II defect conditions hold for every $x \in \mathbb{R}$, we get

$$\partial(\phi_+ - \Lambda_0) = -\frac{i\mu}{\beta^2} e^{\Lambda_0}\sinh\phi_-, \tag{10}$$

$$\bar{\partial}\Lambda_0 = 0, \tag{11}$$

$$\bar{\partial}\phi_- = 2i\mu\beta^2 e^{(\phi_+ - \Lambda_0)}, \tag{12}$$

$$\partial\phi_- = -\frac{i\mu}{\beta^2} e^{\Lambda_0}(\cosh\phi_- + \kappa), \tag{13}$$

Journal of Physics: Conference Series **474** (2013) 012001 doi:10.1088/1742-6596/474/1/012001

the type-II Bäcklund transformations for Liouville equation which couples an auxiliary holomorphic field $\Lambda_0 = \Lambda_0(z)$ to the Liouville fields. From the above equations we can also find an anti-holomorphic functional, namely,

$$\partial \left[e^{-(\phi_+ - \Lambda_0)} \left(\cosh \phi_- + \kappa \right) \right] = 0. \tag{14}$$

Now, the modified conserved momentum is given explictly by $\mathcal{P} = P + \left[B_0^+ - B_0^- \right]_{x=0}$, where P denotes the canonical momentum. It is worth also noting that when $\Lambda_0 = 0$, the type-I defect first proposed in [3] is recovered, which can be described by the defect conditions,

$$\partial \phi_+ \;=\; -\frac{i\mu}{\beta^2} \sinh \phi_-, \qquad \bar{\partial}\phi_- \;=\; 2i\mu\beta^2 \, e^{\phi_+}. \tag{15}$$

In the next subsections, we will present a discussion about the conformal symmetry in the present of the defects, and the defect matrix which guarantees the existence of higher-order modified conserved quantities.

2.2. Defect conformal symmetry

In this subsection we discuss if the defect conditions introduced in subsection (2.1) indeed respect the conformal symmetry of the original bulk Liouville theory. The defect is called *conformal* if the energy-momentum tensor flow is continuos across the defect, namely,

$$\left[T^{(1)}(z) - \overline{T}^{(1)}(\bar{z}) \right]_{x=0} \;=\; \left[T^{(2)}(z) - \overline{T}^{(2)}(\bar{z}) \right]_{x=0}, \tag{16}$$

where $T^{(p)}(z)$ and $\overline{T}^{(p)}(\bar{z})$ for $p = 1, 2$, are the respective holomorphic and anti-holomorphic components of the stress tensor, and are given by [16]

$$T^{(p)} = (\partial \phi_p)^2 - \partial^2 \phi_p, \qquad \overline{T}^{(p)} = (\bar{\partial}\phi_p)^2 - \bar{\partial}^2 \phi_p. \tag{17}$$

From the equations of motion, we can show that the conservation laws $\bar{\partial} T = \partial \overline{T} = 0$ are properly satisfied. Now by using the type-II defect conditions, one can directly show that the holomorphic and anti-holomorphic part of the stress-tensor are continuous across the defect individually, namely

$$T^{(1)}\Big|_{x=0} \;=\; T^{(2)}\Big|_{x=0}, \tag{18}$$

$$\overline{T}^{(1)}\Big|_{x=0} \;=\; \overline{T}^{(2)}\Big|_{x=0}. \tag{19}$$

Therefore the type-II integrable defect described by the conditions belong to the class of *purely transmitting defects*, or sometimes called *topological defects* [17], which is one of the extremal solutions for the gluing condition (16). For the type-I integrable defect ($\Lambda_0 = 0$), we find that only the second condition (19) is still satisfied, but the first one (18) differs by a total time-derivative term, namely,

$$T^{(1)}\Big|_{x=0} \;=\; T^{(2)}\Big|_{x=0} + \partial_t \left[2\partial\phi_- + \frac{2i\mu}{\beta^2} \left(\cosh \phi_- + \kappa \right) \right]_{x=0}. \tag{20}$$

Then, the type-I defect is integrable but not conformal, and the Λ_0 field can be thought of as responsible for recovering the conformal invariance of the original theory after introducing a defect. It is also worth noting that an alternative type-I defect can be considered for the Liouville theory which is derived just by exchanging $\partial \leftrightarrow \bar{\partial}$ in condition (15), and that satisfies the first gluing condition (18) but not the second one (19). Then, it is quite natural understand the type-II integrable defect for Liouville theory also as being the result of fusing (in the sense of [18]) these two different kinds of type-I defects, as was already suggested in the sine-Gordon case [6] (see also [19] for fusing defects in the $a_r^{(1)}$ Toda models).

2.3. Defect matrix

Let us define the following auxiliary linear system:

$$\partial\Psi = A(\lambda)\Psi, \qquad \bar{\partial}\Psi = \bar{A}(\lambda)\Psi, \tag{21}$$

where Ψ is a two-dimensional column vector, λ is a spectral parameter, and A and $\bar{A}$ are the Lax connections given by,

$$A(\lambda) = \begin{bmatrix} -\frac{\partial\phi}{2} & -\lambda\mu\, e^{\phi} \\ 0 & \frac{\partial\phi}{2} \end{bmatrix}, \qquad \bar{A}(\lambda) = \begin{bmatrix} \frac{\bar{\partial}\phi}{2} & 0 \\ -\frac{\mu}{\lambda}\, e^{\phi} & -\frac{\bar{\partial}\phi}{2} \end{bmatrix}. \tag{22}$$

Then, the Liouville field equation can be derived as a compatibility condition, or in other words from the zero curvature condition or Lax-Zhakharov-Shabat equation, namely,

$$\bar{\partial}A - \partial\bar{A} + \left[A, \bar{A}\right] = 0. \tag{23}$$

Let us now consider the defect matrix $K(\lambda)$ connecting two different auxiliary problems, $\Psi_1 = K(\lambda)\Psi_2$, by using the following ansatz,

$$K_{ij}(\lambda) = \alpha_{ij} + \lambda^{-1}\beta_{ij} + \lambda^{-2}\gamma_{ij}, \tag{24}$$

which satisfies the differential equations,

$$\partial K = A_1 K - K A_2, \qquad \bar{\partial}K = \bar{A}_1 K - K\bar{A}_2. \tag{25}$$

There are two kind of solutions for the K-matrix. The first one involves the set $\{\alpha_{11}, \alpha_{22}, \beta_{21}, \beta_{12}, \gamma_{11}, \gamma_{22}\}$. Then, we get

$$\alpha_{11} = a_{11}\, e^{\frac{\phi_-}{2}}, \qquad \alpha_{22} = a_{11} e^{-\frac{\phi_-}{2}}, \qquad \gamma_{11} = c_{11}\, e^{-\frac{\phi_-}{2}}, \qquad \gamma_{22} = c_{11}\, e^{\frac{\phi_-}{2}}, \tag{26}$$

$$\beta_{12} = -2i\beta^2 c_{11}\, e^{(\frac{\phi_+}{2} - \Lambda_0)}, \qquad \beta_{21} = \frac{ia_{11}}{\beta^2} e^{-(\frac{\phi_+}{2} - \Lambda_0)} \left(\cosh\phi_- + \kappa\right), \tag{27}$$

where a_{11} and c_{11} are two arbitrary constants. Then, the defect matrix takes the following form,

$$K = \begin{bmatrix} a_{11}\, e^{\frac{\phi_-}{2}} + \frac{c_{11}}{\lambda^2}\, e^{-\frac{\phi_-}{2}} & -\frac{2i\beta^2}{\lambda} c_{11}\, e^{(\frac{\phi_+}{2} - \Lambda_0)} \\ \frac{ia_{11}}{\lambda\beta^2}\, e^{-(\frac{\phi_+}{2} - \Lambda_0)} \left(\cosh\phi_- + \kappa\right) & a_{11}\, e^{-\frac{\phi_-}{2}} + \frac{c_{11}}{\lambda^2} e^{\frac{\phi_-}{2}} \end{bmatrix}. \tag{28}$$

The other solution involves the following set of components $\{\alpha_{12}, \beta_{11}, \beta_{22}, \gamma_{21}\}$. In this case we find,

$$\alpha_{12} = a_{12}\, e^{(\frac{\phi_+}{2} - \Lambda_0)}, \qquad \beta_{11} = \beta_{22} = b_{11} \cosh\left(\frac{\phi_-}{2}\right), \tag{29}$$

$$\gamma_{21} = c_{21} e^{-(\frac{\phi_+}{2} - \Lambda_0)} \left(\cosh\phi_- + \kappa\right), \tag{30}$$

where $a_{12} = -i\beta^2 b_{11}$, $c_{21} = \frac{ib_{11}}{2\beta^2}$. Then, the defect matrix can be written as,

$$K' = \begin{bmatrix} \frac{b_{11}}{\lambda} \cosh\left(\frac{\phi_-}{2}\right) & -i\beta^2 b_{11}\, e^{(\frac{\phi_+}{2} - \Lambda_0)} \\ \frac{ib_{11}}{2\beta^2\lambda^2}\, e^{-(\frac{\phi_+}{2} - \Lambda_0)} \left(\cosh\phi_- + \kappa\right) & \frac{b_{11}}{\lambda} \cosh\left(\frac{\phi_-}{2}\right) \end{bmatrix}. \tag{31}$$

We remark that the existence of the defect matrix provides a sufficient condition to show that integrability is preserved after introducing the type-II defects described by the conditions (7)–(9), since the defect contributions for an infinite number of modified conserved quantities can be recursively computed from the entries of the defect matrix as it was already shown explicitly in the sine-Gordon, Tzitzeica-Bullough-Dodd [7] and Thirring models [11].

Journal of Physics: Conference Series **474** (2013) 012001 doi:10.1088/1742-6596/474/1/012001

3. Super-Liouville theory with defects

In this section, we propose a supersymmetric extension of the type-II defect Lagrangian density described in (1)–(3) for the $N = 1$ supersymmetric Liouville field theory.

3.1. Lagrangian description

Let us consider the supersymmetric Liouville field theory with type-II defects decribed by the following Lagrangian density,

$$\mathcal{L} = \theta(-x)\mathcal{L}_1 + \theta(x)\mathcal{L}_2 + \delta(x)\mathcal{L}_D, \tag{32}$$

with

$$\mathcal{L}_p = (\partial_x\phi_p)^2 - (\partial_t\phi_p)^2 + \bar{\psi}_p(\partial_x - \partial_t)\bar{\psi}_p + \psi_p(\partial_t + \partial_x)\psi_p + \mu^2 e^{2\phi_p} + 2i\mu e^{\phi_p}\bar{\psi}_p\psi_p, \tag{33}$$

and the defect Lagrangian can be written as $\mathcal{L}_D = \mathcal{L}_b + \mathcal{L}_f$, with

$$\begin{aligned}\mathcal{L}_b &= \phi_2\partial_t\phi_1 - \phi_1\partial_t\phi_2 - \Lambda_0\partial_t(\phi_1 - \phi_2) + (\phi_1 - \phi_2)\partial_t\Lambda_0 + \mathcal{B}_0^+ + \mathcal{B}_0^-, && (34)\\ \mathcal{L}_f &= \bar{\psi}_1\bar{\psi}_2 - \psi_1\psi_2 + (\psi_1 - \psi_2)\Lambda_1 - if_1\partial_t f_1 + \mathcal{B}_1^+ + \mathcal{B}_1^- + \mathcal{B}_{\Lambda_1}, && (35)\end{aligned}$$

where the defect potentials are given by,

$$\mathcal{B}_0^+ = -2i\mu\beta^2 e^{(\phi_1+\phi_2-\Lambda_0)}, \qquad \mathcal{B}_0^- = \frac{i\mu}{\beta^2}e^{\Lambda_0}\left[\cosh(\phi_1 - \phi_2) + \kappa\right], \tag{36}$$

$$\mathcal{B}_1^+ = \sqrt{\mu}\beta\, e^{\frac{(\phi_1+\phi_2-\Lambda_0)}{2}}(\bar{\psi}_1 + \bar{\psi}_2)f_1, \qquad \mathcal{B}_1^- = \frac{\sqrt{\mu}}{\beta}e^{\frac{\Lambda_0}{2}}\cosh\left(\frac{\phi_1 - \phi_2}{2}\right)(\psi_1 - \psi_2)f_1, \tag{37}$$

$$\mathcal{B}_{\Lambda_1} = \frac{\sqrt{\mu}}{\beta}e^{\frac{\Lambda_0}{2}}\sinh\left(\frac{\phi_1 - \phi_2}{2}\right)\Lambda_1 f_1. \tag{38}$$

Here, besides the bosonic field Λ_0 the defect potentials also depend on two more auxiliary fermionic fields Λ_1 and f_1. Then, the bulk field equations are

$$\begin{aligned}\partial\bar{\partial}\phi_p &= \mu^2 e^{2\phi_p} + i\mu e^{\phi}\bar{\psi}_p\psi_p, && (39)\\ \bar{\partial}\psi_p &= i\mu e^{\phi_p}\bar{\psi}_p, && (40)\\ \partial\bar{\psi}_p &= -i\mu e^{\phi_p}\psi_p, \qquad p = 1, 2, && (41)\end{aligned}$$

and the defects conditions at $x = 0$ are given as follows,

$$\begin{aligned}\partial_x\phi_1 - \partial_t(\phi_2 - \Lambda_0) = {}& i\mu\beta^2 e^{(\phi_1+\phi_2-\Lambda_0)} - \frac{i\mu}{2\beta^2}e^{\Lambda_0}\sinh(\phi_1 - \phi_2) - \frac{\beta\sqrt{\mu}}{4}e^{\frac{(\phi_1+\phi_2-\Lambda_0)}{2}}(\bar{\psi}_1 + \bar{\psi}_2)f_1\\ & -\frac{\sqrt{\mu}}{4\beta}e^{\frac{\Lambda_0}{2}}\left[\sinh\left(\frac{\phi_1 - \phi_2}{2}\right)(\psi_1 - \psi_2) + \cosh\left(\frac{\phi_1 - \phi_2}{2}\right)\Lambda_1\right]f_1, \end{aligned}\tag{42}$$

$$\begin{aligned}\partial_x\phi_2 - \partial_t(\phi_1 - \Lambda_0) = {}& -i\mu\beta^2 e^{(\phi_1+\phi_2-\Lambda_0)} - \frac{i\mu}{2\beta^2}e^{\Lambda_0}\sinh(\phi_1 - \phi_2) + \frac{\beta\sqrt{\mu}}{4}e^{\frac{(\phi_1+\phi_2-\Lambda_0)}{2}}(\bar{\psi}_1 + \bar{\psi}_2)f_1\\ & -\frac{\sqrt{\mu}}{4\beta}e^{\frac{\Lambda_0}{2}}\left[\sinh\left(\frac{\phi_1 - \phi_2}{2}\right)(\psi_1 - \psi_2) + \cosh\left(\frac{\phi_1 - \phi_2}{2}\right)\Lambda_1\right]f_1, \end{aligned}\tag{43}$$

$$\partial_t(\phi_1-\phi_2) = i\mu\beta^2\, e^{(\phi_1+\phi_2-\Lambda_0)} + \frac{i\mu}{2\beta^2}\, e^{\Lambda_0}\,(\cosh(\phi_1-\phi_2)+\kappa) - \frac{\beta\sqrt{\mu}}{4} e^{\frac{(\phi_1+\phi_2-\Lambda_0)}{2}}(\bar{\psi}_1+\bar{\psi}_2)f_1$$
$$+\frac{\sqrt{\mu}}{4\beta}\, e^{\frac{\Lambda_0}{2}}\left[\sinh\left(\frac{\phi_1-\phi_2}{2}\right)(\psi_1-\psi_2)+\cosh\left(\frac{\phi_1-\phi_2}{2}\right)\Lambda_1\right] f_1, \tag{44}$$

$$\psi_1+\psi_2 = \Lambda_1 + \frac{\sqrt{\mu}}{\beta}\, e^{\frac{\Lambda_0}{2}}\cosh\left(\frac{\phi_1-\phi_2}{2}\right) f_1, \tag{45}$$

$$\psi_1-\psi_2 = \frac{\sqrt{\mu}}{\beta}\, e^{\frac{\Lambda_0}{2}}\sinh\left(\frac{\phi_1-\phi_2}{2}\right) f_1, \tag{46}$$

$$\bar{\psi}_1-\bar{\psi}_2 = \sqrt{\mu}\beta\, e^{\frac{(\phi_1+\phi_2-\Lambda_0)}{2}} f_1, \tag{47}$$

$$\partial_t f_1 = \frac{i\sqrt{\mu}}{2\beta} e^{\frac{\Lambda_0}{2}}\left[\cosh\left(\frac{\phi_1-\phi_2}{2}\right)(\psi_1-\psi_2)+\sinh\left(\frac{\phi_1-\phi_2}{2}\right)\Lambda_1\right]$$
$$+\frac{i\sqrt{\mu}\beta}{2} e^{\frac{(\phi_1+\phi_2-\Lambda_0)}{2}}(\bar{\psi}_1+\bar{\psi}_2). \tag{48}$$

These equations agree with the type-II super-Bäcklund transformation for the super-Liouville model derived in the Appendix B, and can be seen has a straightforward extension of the already known Bäcklund transfomation proposed in [15]. Then, these defect equations are invariant under the supersymmetry transformations,

$$\delta\phi_p = \varepsilon\,\psi_p + \bar{\varepsilon}\,\bar{\psi}_p, \qquad \delta\Lambda_0 = \varepsilon\,\Lambda_1, \tag{49}$$

$$\delta\psi_p = -\varepsilon\,\partial\phi_p - i\mu\,\bar{\varepsilon}\,e^{\phi_p}, \qquad \delta\Lambda_1 = -\varepsilon\partial\Lambda_0, \tag{50}$$

$$\delta\bar{\psi}_p = -\bar{\varepsilon}\,\bar{\partial}\phi_p + i\mu\,\varepsilon\,e^{\phi_p}, \qquad \delta f_1 = \frac{2i\varepsilon\sqrt{\mu}}{\beta}\, e^{\frac{\Lambda_0}{2}}\sinh\left(\frac{\phi_1-\phi_2}{2}\right) - 2i\sqrt{\mu}\beta\,\bar{\varepsilon}\,e^{\frac{(\phi_1+\phi_2-\Lambda_0)}{2}}, \tag{51}$$

where the parameter κ in (44) must be set to -1. Now, we notice that it is possible to eliminate the Lagrange multiplier Λ_1 and simplify the defect conditions as follows,

$$\partial_x\phi_1 - \partial_t(\phi_2-\Lambda_0) = i\mu\beta^2\, e^{(\phi_1+\phi_2-\Lambda_0)} - \frac{i\mu}{2\beta^2}\, e^{\Lambda_0}\sinh(\phi_1-\phi_2) - \frac{\beta\sqrt{\mu}}{4} e^{\frac{(\phi_1+\phi_2-\Lambda_0)}{2}}(\bar{\psi}_1+\bar{\psi}_2)f_1$$
$$-\frac{\sqrt{\mu}}{4\beta}\, e^{\frac{\Lambda_0}{2}}\cosh\left(\frac{\phi_1-\phi_2}{2}\right)(\psi_1+\psi_2)f_1, \tag{52}$$

$$\partial_x\phi_2 - \partial_t(\phi_1-\Lambda_0) = -i\mu\beta^2\, e^{(\phi_1+\phi_2-\Lambda_0)} - \frac{i\mu}{2\beta^2}\, e^{\Lambda_0}\sinh(\phi_1-\phi_2) + \frac{\beta\sqrt{\mu}}{4} e^{\frac{(\phi_1+\phi_2-\Lambda_0)}{2}}(\bar{\psi}_1+\bar{\psi}_2)f_1$$
$$-\frac{\sqrt{\mu}}{4\beta}\, e^{\frac{\Lambda_0}{2}}\cosh\left(\frac{\phi_1-\phi_2}{2}\right)(\psi_1+\psi_2)f_1, \tag{53}$$

$$\partial_t(\phi_1-\phi_2) = i\mu\beta^2\, e^{(\phi_1+\phi_2-\Lambda_0)} + \frac{i\mu}{\beta^2}\, e^{\Lambda_0}\sinh\left(\frac{\phi_1-\phi_2}{2}\right)$$
$$-\frac{\sqrt{\mu}}{4}\left[\beta\, e^{\frac{(\phi_1+\phi_2-\Lambda_0)}{2}}(\bar{\psi}_1+\bar{\psi}_2) - \frac{1}{\beta} e^{\frac{\Lambda_0}{2}}\sinh\left(\frac{\phi_1-\phi_2}{2}\right)(\psi_1+\psi_2)\right] f_1, \tag{54}$$

$$\psi_1-\psi_2 = \frac{\sqrt{\mu}}{\beta}\, e^{\frac{\Lambda_0}{2}}\sinh\left(\frac{\phi_1-\phi_2}{2}\right) f_1, \tag{55}$$

$$\bar{\psi}_1-\bar{\psi}_2 = \sqrt{\mu}\beta\, e^{\frac{(\phi_1+\phi_2-\Lambda_0)}{2}} f_1, \tag{56}$$

$$\partial_t f_1 = \frac{i\sqrt{\mu}\beta}{2} e^{\frac{(\phi_1+\phi_2-\Lambda_0)}{2}}(\bar{\psi}_1+\bar{\psi}_2) + \frac{i\sqrt{\mu}}{2\beta} e^{\frac{\Lambda_0}{2}}\sinh\left(\frac{\phi_1-\phi_2}{2}\right)(\psi_1+\psi_2). \tag{57}$$

These relations can be derived from an alternative defect Lagrangian density $\mathcal{L}'_D = \mathcal{L}_b + \mathcal{L}'_f$, with

$$\begin{aligned}\mathcal{L}'_f &= \psi_1\psi_2 + \bar{\psi}_1\bar{\psi}_2 - if_1\partial_t f_1 + \frac{\sqrt{\mu}}{\beta}e^{\frac{\Lambda_0}{2}}\sinh\left(\frac{\phi_1-\phi_2}{2}\right)(\psi_1+\psi_2)f_1 \\ &+\sqrt{\mu}\beta\, e^{\frac{(\phi_1+\phi_2-\Lambda_0)}{2}}(\bar{\psi}_1+\bar{\psi}_2)f_1. \end{aligned} \tag{58}$$

From these Lagrangian density we derive the corresponding canonical momentum, which is given by

$$P = \int_{-\infty}^{0} dx\Big[2(\partial_t\phi_1)(\partial_x\phi_1) + \bar{\psi}_1\partial_x\bar{\psi}_1 - \psi_1\partial_x\psi_1\Big] + \int_0^{\infty} dx\Big[2(\partial_t\phi_2)(\partial_x\phi_2) + \bar{\psi}_2\partial_x\bar{\psi}_2 - \psi_2\partial_x\psi_2\Big]. \tag{59}$$

By taking its time derivative, we get

$$\begin{aligned}\frac{dP}{dt} &= \Big[(\partial_x\phi_1)^2 + (\partial_t\phi_1)^2 + \bar{\psi}_1\partial_t\bar{\psi}_1 - \psi_1\partial_t\psi_1 - \mu^2e^{2\phi_1} - 2i\mu e^{\phi_1}\bar{\psi}_1\psi_1\Big]_{x=0} \\ &-\Big[(\partial_x\phi_2)^2 + (\partial_t\phi_2)^2 + \bar{\psi}_2\partial_t\bar{\psi}_2 - \psi_2\partial_t\psi_2 - \mu^2e^{2\phi_2} - 2i\mu e^{\phi_2}\bar{\psi}_2\psi_2\Big]_{x=0}. \end{aligned} \tag{60}$$

Now, by using the defect conditions (52)–(57) the right-hand-side of (60) becomes a total time-derivative and then the modified conserved momentum is given by the combination,

$$\begin{aligned}\mathcal{P} &= P + (\bar{\psi}_1\bar{\psi}_2 - \psi_1\psi_2) - \left[\frac{2i\mu}{\beta^2}e^{\Lambda_0}\sinh\left(\frac{\phi_1-\phi_2}{2}\right) + 2i\mu\beta^2e^{(\phi_1+\phi_2-\Lambda_0)}\right] \\ &+\left[\frac{\sqrt{\mu}}{\beta}e^{\frac{\Lambda_0}{2}}\sinh\left(\frac{\phi_1-\phi_2}{2}\right)(\psi_1+\psi_2) - \sqrt{\mu}\beta\, e^{\frac{(\phi_1+\phi_2-\Lambda_0)}{2}}(\bar{\psi}_1+\bar{\psi}_2)\right]f_1. \end{aligned} \tag{61}$$

Now, for the energy

$$\begin{aligned}E &= \int_{-\infty}^{0} dx\Big[(\partial_x\phi_1)^2 + (\partial_t\phi_1)^2 + \bar{\psi}_1\partial_x\bar{\psi}_1 + \psi_1\partial_x\psi_1 + \mu^2e^{2\phi_1} + 2i\mu e^{\phi_1}\bar{\psi}_1\psi_1\Big] \\ &+\int_0^{\infty} dx\Big[(\partial_x\phi_2)^2 + (\partial_t\phi_2)^2 + \bar{\psi}_2\partial_x\bar{\psi}_2 + \psi_2\partial_x\psi_2 + \mu^2e^{2\phi_2} + 2i\mu e^{\phi_2}\bar{\psi}_2\psi_2\Big], \end{aligned} \tag{62}$$

the modified conserved energy is respectively given by,

$$\begin{aligned}\mathcal{E} &= E + (\bar{\psi}_1\bar{\psi}_2 + \psi_1\psi_2) + \left[\frac{2i\mu}{\beta^2}e^{\Lambda_0}\sinh\left(\frac{\phi_1-\phi_2}{2}\right) - 2i\mu\beta^2e^{(\phi_1+\phi_2-\Lambda_0)}\right] \\ &+\left[\frac{\sqrt{\mu}}{\beta}e^{\frac{\Lambda_0}{2}}\sinh\left(\frac{\phi_1-\phi_2}{2}\right)(\psi_1+\psi_2) + \sqrt{\mu}\beta\, e^{\frac{(\phi_1+\phi_2-\Lambda_0)}{2}}(\bar{\psi}_1+\bar{\psi}_2)\right]f_1. \end{aligned} \tag{63}$$

Besides being integrable, the defect theory is also invariant under the supersymmetry transformations, the associated conserved supercharges are given by,

$$Q_\varepsilon = -\int_{-\infty}^{\infty} dx\left(\psi\partial\phi + i\mu e^{\phi}\bar{\psi}\right), \qquad \bar{Q}_{\bar{\varepsilon}} = \int_{-\infty}^{\infty} dx\left(\bar{\psi}\bar{\partial}\phi - i\mu e^{\phi}\psi\right). \tag{64}$$

Then, by considering now the defect conditions we found that the modified conserved supercharges are given by,

$$\mathcal{Q} = Q_\varepsilon - \frac{2\sqrt{\mu}}{\beta}e^{\frac{\Lambda_0}{2}}\sinh\left(\frac{\phi_1-\phi_2}{2}\right)f_1, \tag{65}$$

$$\bar{\mathcal{Q}} = \bar{Q}_{\bar{\varepsilon}} - 2\sqrt{\mu}\beta\, e^{\frac{(\phi_1+\phi_2-\Lambda_0)}{2}}f_1. \tag{66}$$

Here it is worth pointing out that this is the first time that a type-II defect is encompassed within both supersymmetric and conformal model. Presumably, super-extensions of type-II integrable defects for some affine Toda models could be found, for instance for super sinh-Gordon model.

3.2. Defect superconformal symmetry

The defect is called *superconformal* if both the stress tensor and the supercurrents, which are the generators of the supersymmetry transformations, are continuous across the defect. These conditions are written as

$$\left[T^{(1)}(z) - \overline{T}^{(1)}(\bar{z})\right]_{x=0} = \left[T^{(2)}(z) - \overline{T}^{(2)}(\bar{z})\right]_{x=0}, \tag{67}$$

$$\left[J^{(1)}(z) - \overline{J}^{(1)}(\bar{z})\right]_{x=0} = \left[J^{(2)}(z) - \overline{J}^{(2)}(\bar{z})\right]_{x=0}, \tag{68}$$

where the holomorphic and antiholomorphic components of the stress tensor and supercurrents are given by [16],

$$T^{(p)} = (\partial\phi_p)^2 - \partial^2\phi_p + \psi_p\partial\psi_p, \qquad \overline{T}^{(p)} = (\bar{\partial}\phi_p)^2 - \bar{\partial}^2\phi_p + \bar{\psi}_p\bar{\partial}\bar{\psi}_p, \tag{69}$$

$$J^{(p)} = \psi_p\partial\phi_p - \partial\psi_p, \qquad \bar{J}^{(p)} = \bar{\psi}_p\bar{\partial}\phi_p - \bar{\partial}\bar{\psi}_p, \tag{70}$$

for $p = 1, 2$. From the equations of motion, we can show again that the conservation laws $\bar{\partial}T = \partial\overline{T} = \bar{\partial}J = \partial\bar{J} = 0$ are indeed satisfied. Now, after some quite straightforward computations, we conclude that the type-II defect introduced in the previous subsection for the $N = 1$ super-Liouville model is indeed a topological defect, i.e. each holomorphic and anti-holomorphic part of the corresponding currents satisfy the following constraints,

$$T^{(1)}\big|_{x=0} = T^{(2)}\big|_{x=0}, \qquad J^{(1)}\big|_{x=0} = J^{(2)}\big|_{x=0}, \tag{71}$$

$$\overline{T}^{(1)}\big|_{x=0} = \overline{T}^{(2)}\big|_{x=0}, \qquad \bar{J}^{(1)}\big|_{x=0} = \bar{J}^{(2)}\big|_{x=0}. \tag{72}$$

Therefore, we have found that the $N = 1$ super-Liouville model with type-II defect introduced by the Lagrangian density (32) preserves not only supersymmetry but also the conformal symmetry.

3.3. Lax representation

Let us first describe the super-Liouville equation (B.2) as a compatibility condition of the following two linear systems of differential equations,

$$D\mathcal{V} = \mathcal{A}\mathcal{V}, \qquad \bar{D}\mathcal{V} = \bar{\mathcal{A}}\mathcal{V}, \tag{73}$$

where D and $\bar{D}$ are the superderivatives introduced in Appendix B, $\mathcal{V}(x, \theta; \lambda)$ is a vector-valued superfield whose components are the bosonic superfields $\mathcal{V}_1$ and $\mathcal{V}_2$, and the fermionic superfield $\mathcal{V}_3$, and λ is the spectral parameter. The super-Lax connections $\mathcal{A}$ and $\bar{\mathcal{A}}$ are 3×3 graded matrices that can be written in the following form,

$$\mathcal{A} = -\frac{1}{2}(D\Phi)\,\mathbf{H} + \sqrt{\lambda\mu}\exp\left(\frac{\Phi}{2}\right)\mathbf{F}^+, \tag{74}$$

$$\bar{\mathcal{A}} = \frac{1}{2}(\bar{D}\Phi)\,\mathbf{H} - i\sqrt{\frac{\mu}{\lambda}}\exp\left(\frac{\Phi}{2}\right)\mathbf{F}^-, \tag{75}$$

where Φ is a bosonic superfield given by

$$\Phi = \phi + i\theta\bar{\psi} + i\bar{\theta}\psi + i\bar{\theta}\theta F, \tag{76}$$

and $\{\mathbf{H}, \mathbf{F}^{\pm}\}$ are generators of the $osp(1,2)$ Lie superalgebra (see Appendix A). Then, the zero-curvature condition in the superspace, namely

$$\bar{D}\mathcal{A} + D\bar{\mathcal{A}} - \{\bar{\mathcal{A}}, \mathcal{A}\} = 0, \tag{77}$$

allows us to recover the super-Liouville equation (B.2). Now, if we consider the θ-expansions of $\mathcal{V}, \mathcal{A}$ and $\bar{\mathcal{A}}$ respectively, we can find directly the bosonic Lax operator for the supersymmetric Liouville model. Using (B.1), we obtain

$$\partial\Psi = A\Psi, \qquad \bar{\partial}\Psi = \bar{A}\Psi, \tag{78}$$

where,

$$A = -\frac{1}{2}(\partial\phi)\mathbf{H} - \lambda\mu\, e^{\phi}\mathbf{E}^{+} + i\sqrt{\lambda\mu}\,\psi\, e^{\phi/2}\mathbf{F}^{+}, \tag{79}$$

$$\bar{A} = \frac{1}{2}(\bar{\partial}\phi)\mathbf{H} - \frac{\mu}{\lambda}\, e^{\phi}\mathbf{E}^{-} + \sqrt{\frac{\mu}{\lambda}}\,\bar{\psi}\, e^{\phi/2}\mathbf{F}^{-}. \tag{80}$$

It is worth noting that if the fermionic fields vanish we recover the Lax connections for the Liouville model given in the linear problem in eq. (22).

3.4. Defect super-matrix

Now we are interested in deriving the defect matrix for the super-Liouville theory. Let us consider the graded matrix $\mathcal{K}$ connecting two different configurations, namely $\Psi_1 = \mathcal{K}(\lambda)\Psi_2$, satisfying the following equations

$$\partial\mathcal{K} = A_1\mathcal{K} - \mathcal{K}A_2, \qquad \bar{\partial}\mathcal{K} = \bar{A}_1\mathcal{K} - \mathcal{K}\bar{A}_2, \tag{81}$$

where A and $\bar{A}$ are given in eqs. (79) and (80). Let us consider the following ansatz for the λ-expansion of $\mathcal{K}$,

$$\mathcal{K}_{ij} = \alpha_{ij} + \beta_{ij}\,\lambda^{-1/2} + \lambda^{+1/2}\,\delta_{ij}. \tag{82}$$

Then, by solving the differential equations (81), we get

$$\mathcal{K} = \left[\begin{array}{cc|c} \frac{b_{11}}{\lambda^{1/2}}\, e^{-\frac{\phi_-}{2}} + d_{11}\lambda^{1/2}\, e^{\frac{\phi_-}{2}} & d_{12}\lambda^{1/2}\, e^{(\frac{\phi_+}{2}-\Lambda_0)} & -\beta b_{11}\, e^{\frac{(\phi_2-\Lambda_0)}{2}} f_1 \\ \frac{2id_{11}}{\lambda^{1/2}\beta^2}\, e^{(\Lambda_0-\frac{\phi_+}{2})}\sinh^2\left(\frac{\phi_-}{2}\right) & \frac{b_{11}}{\lambda^{1/2}}\, e^{\frac{\phi_-}{2}} + d_{11}\lambda^{1/2}\, e^{-\frac{\phi_-}{2}} & \frac{id_{11}}{\beta}\, e^{\frac{(\Lambda_0-\phi_1)}{2}}\sinh\left(\frac{\phi_-}{2}\right) f_1 \\ \hline -\frac{id_{11}}{\beta}\, e^{\frac{(\Lambda_0-\phi_2)}{2}}\sinh\left(\frac{\phi_-}{2}\right) f_1 & \beta b_{11}\, e^{\frac{(\phi_1-\Lambda_0)}{2}} f_1 & \frac{b_{11}}{\lambda^{1/2}} + d_{11}\lambda^{1/2} \end{array}\right], \tag{83}$$

where b_{11} and d_{11} are two arbitrary constants. Then, this $\mathcal{K}$-matrix can be thought of as generating the type-II defect conditions for the $N = 1$ super-Liouville. In addition, besides the defect contributions to the modified conserved energy, momentum and supercharges, explicitly contributions for higher order modified conserved quantities could be also derived following [11].

Acknowledgments

I am very grateful to the organisers of the XXIst International Conference ISQS21 for the oportunity to present these ideas, and to J.F. Gomes and A.H. Zimerman for valuable discussions. I would like also to thank FAPESP São Paulo Research Foundation for financial support under the PD Fellowship 2012/13866-3.

Appendix A. The $osp(1,2)$ Lie superalgebra

The $osp(1,2)$ Lie superalgebra contains three bosonic generators $\mathbf{H}, \mathbf{E}^{\pm}$ which correspond to the $sl(2)$ Lie algebra, and two fermionic generators $\mathbf{F}^{\pm}$. The three-dimensional matrix representation is given below

$$\mathbf{H} = \left[\begin{array}{cc|c} 1 & 0 & 0 \\ 0 & -1 & 0 \\ \hline 0 & 0 & 0 \end{array}\right], \quad \mathbf{E}^{+} = \left[\begin{array}{cc|c} 0 & 1 & 0 \\ 0 & 0 & 0 \\ \hline 0 & 0 & 0 \end{array}\right], \quad \mathbf{E}^{-} = \left[\begin{array}{cc|c} 0 & 0 & 0 \\ 1 & 0 & 0 \\ \hline 0 & 0 & 0 \end{array}\right] \tag{A.1}$$

$$\mathbf{F}^{+} = \left[\begin{array}{cc|c} 0 & 0 & 1 \\ 0 & 0 & 0 \\ \hline 0 & 1 & 0 \end{array}\right], \quad \mathbf{F}^{-} = \left[\begin{array}{cc|c} 0 & 0 & 0 \\ 0 & 0 & -1 \\ \hline 1 & 0 & 0 \end{array}\right]. \tag{A.2}$$

Usually this superalgebra is labelled $B(0,1)$ and describe the simplest example of a superconformal Toda theory based on a contragredient Lie superalgebra [20, 21]. The (anti) commutation relations are given by,

$$\left[\mathbf{H}, \mathbf{E}^{\pm}\right] = \pm 2\mathbf{E}^{\pm}, \qquad \left[\mathbf{H}, \mathbf{F}^{\pm}\right] = \pm \mathbf{F}^{\pm}, \tag{A.3}$$

$$\left[\mathbf{E}^{+}, \mathbf{E}^{-}\right] = \left\{\mathbf{F}^{+}, \mathbf{F}^{-}\right\} = \mathbf{H} \tag{A.4}$$

$$\left[\mathbf{E}^{+}, \mathbf{F}^{-}\right] = -\mathbf{F}^{+}, \qquad \left[\mathbf{E}^{-}, \mathbf{F}^{+}\right] = -\mathbf{F}^{-}, \tag{A.5}$$

$$\left\{\mathbf{F}^{+}, \mathbf{F}^{+}\right\} = 2\mathbf{E}^{+}, \qquad \left\{\mathbf{F}^{-}, \mathbf{F}^{-}\right\} = -2\mathbf{E}^{-}. \tag{A.6}$$

Appendix B. Type-II super-Bäcklund transformations

Let us consider a bosonic superfield $\Phi = \phi + i\theta\bar{\psi} + i\bar{\theta}\psi + i\bar{\theta}\theta F$, and the superderivatives,

$$D = \frac{\partial}{\partial\theta} + \theta\partial, \qquad \bar{D} = \frac{\partial}{\partial\bar{\theta}} + \bar{\theta}\bar{\partial}, \quad D^2 = \partial, \quad \bar{D}^2 = \bar{\partial}, \qquad D\bar{D} = -\bar{D}D. \tag{B.1}$$

Then, the $N=1$ supersymmetric Liouville equation,

$$D\bar{D}\Phi = -i\mu e^{\Phi}, \tag{B.2}$$

can be derived from the action,

$$S = \int d^2z\, d\bar{\theta}d\theta \left[D\Phi\bar{D}\Phi - 2i\mu e^{\Phi}\right] = \int d^2z \Big[\partial\phi\bar{\partial}\phi + \psi\bar{\partial}\psi + \bar{\psi}\partial\bar{\psi} + \mu^2 e^{2\phi} + 2i\mu e^{\phi}\bar{\psi}\psi\Big]. \tag{B.3}$$

By extending the auxiliary field Λ_0 in (11) to a chiral superfield, namely $\Lambda = \Lambda_0 + i\theta\Lambda_1$, satisfying $\bar{D}\Lambda = 0$, we propose a generalization of the super-Bäcklund transformation for the super-Liouville equation [15], as follows

$$D(\Phi_+ - \Lambda) = \frac{i\sqrt{\mu}}{\beta}\Xi e^{\frac{\Lambda}{2}}\cosh\Big(\frac{\Phi_-}{2}\Big), \qquad \bar{D}\Lambda = 0, \tag{B.4}$$

$$D\Phi_- = \frac{i\sqrt{\mu}}{\beta}\Xi e^{\frac{\Lambda}{2}}\sinh\Big(\frac{\Phi_-}{2}\Big), \qquad \bar{D}\Phi_- = i\sqrt{\mu}\beta\,\Xi\exp\Big(\frac{\Phi_+ - \Lambda}{2}\Big), \tag{B.5}$$

where the fermionic superfield $\Xi = f_1 + \theta b_1 + \bar{\theta} b_2 + \bar{\theta}\theta f_2$, satisfies

$$D\Xi = -\frac{2\sqrt{\mu}}{\beta}e^{\frac{\Lambda}{2}}\sinh\Big(\frac{\Phi_-}{2}\Big), \qquad \bar{D}\Xi = 2\sqrt{\mu}\beta\exp\Big(\frac{\Phi_+ - \Lambda}{2}\Big). \tag{B.6}$$

In components, these equations take the following form,

$$\partial(\phi_+ - \Lambda_0) = -\frac{i\mu}{\beta^2} e^{\Lambda_0} \sinh\phi_- - \frac{\sqrt{\mu}}{2\beta} e^{\frac{\Lambda_0}{2}} \sinh\left(\frac{\phi_-}{2}\right)\psi_- f_1 - \frac{\sqrt{\mu}}{2\beta} e^{\frac{\Lambda_0}{2}} \cosh\left(\frac{\phi_-}{2}\right)\Lambda_1 f_1, \tag{B.7}$$

$$\psi_+ - \Lambda_1 = \frac{\sqrt{\mu}}{\beta} e^{\frac{\Lambda_0}{2}} \cosh\left(\frac{\phi_-}{2}\right) f_1, \tag{B.8}$$

$$F_+ = -\frac{\sqrt{\mu}}{\beta} e^{\frac{\Lambda_0}{2}} \left[b_2 \cosh\left(\frac{\phi_-}{2}\right) + \frac{i}{2}\sinh\left(\frac{\phi_-}{2}\right)\bar{\psi}_- f_1\right], \tag{B.9}$$

$$\begin{aligned}\partial\bar{\psi}_+ &= \frac{\sqrt{\mu}}{2\beta} e^{\frac{\Lambda_0}{2}} \left[i\sinh\left(\frac{\phi_-}{2}\right)\left(b_1\bar{\psi}_- + \frac{i}{2}(\Lambda_1 f_1)\bar{\psi}_- - b_2\psi_- - F_- f_1\right)\right. \\ &\quad \left. - \cosh\left(\frac{\phi_-}{2}\right)\left(2f_2 + i\Lambda_1 b_2 + \frac{1}{2}\bar{\psi}_-\psi_- f_1\right)\right],\end{aligned} \tag{B.10}$$

$$\bar{\psi}_- = \sqrt{\mu}\beta e^{\frac{(\phi_+ - \Lambda_0)}{2}} f_1, \tag{B.11}$$

$$\begin{aligned}F_- &= \sqrt{\mu}\beta e^{\frac{(\phi_+ - \Lambda_0)}{2}} \left[b_1 + \frac{i}{2}(\psi_+ - \Lambda_1) f_1\right], \\ &= -\frac{i\sqrt{\mu}}{\beta} e^{\frac{\Lambda_0}{2}} \left[b_2 \sinh\left(\frac{\phi_-}{2}\right) + \frac{i}{2}\cosh\left(\frac{\phi_-}{2}\right)\bar{\psi}_- f_1\right],\end{aligned} \tag{B.12}$$

$$\bar{\partial}\phi_- = i\sqrt{\mu}\beta e^{\frac{(\phi_+ - \Lambda_0)}{2}} \left[b_2 + \frac{i}{2}\bar{\psi}_+ f_1\right], \tag{B.13}$$

$$\bar{\partial}\psi_- = \frac{\sqrt{\mu}\beta}{2} e^{\frac{(\phi_+ - \Lambda_0)}{2}} \left[2f_2 - ib_1\bar{\psi}_+ + ib_2(\psi_+ - \Lambda_1) + \left(iF_+ + \frac{1}{2}\bar{\psi}_+(\psi_+ - \Lambda_1)\right) f_1\right], \tag{B.14}$$

$$b_1 = -\frac{2\sqrt{\mu}}{\beta} e^{\frac{\Lambda_0}{2}} \sinh\left(\frac{\phi_-}{2}\right), \tag{B.15}$$

$$f_2 = \frac{i\sqrt{\mu}}{\beta} e^{\frac{\Lambda_0}{2}} \cosh\left(\frac{\phi_-}{2}\right)\bar{\psi}_- = i\sqrt{\mu}\beta e^{\frac{(\phi_+ - \Lambda_0)}{2}}(\psi_+ - \Lambda_1), \tag{B.16}$$

$$\partial f_1 = -\frac{i\sqrt{\mu}}{\beta} e^{\frac{\Lambda_0}{2}} \left[\cosh\left(\frac{\phi_-}{2}\right)\psi_- + \sinh\left(\frac{\phi_-}{2}\right)\Lambda_1\right], \tag{B.17}$$

$$\partial b_2 = \frac{\sqrt{\mu}}{\beta} e^{\frac{\Lambda_0}{2}} \left[i\cosh\left(\frac{\phi_-}{2}\right) F_- + \frac{1}{2}\sinh\left(\frac{\phi_-}{2}\right)\bar{\psi}_-\psi_- + \frac{1}{2}\cosh\left(\frac{\phi_-}{2}\right)\bar{\psi}_-\Lambda_1\right], \tag{B.18}$$

$$b_2 = 2\sqrt{\mu}\beta\, e^{\frac{(\phi_+ - \Lambda_0)}{2}}, \tag{B.19}$$

$$\bar{\partial} f_1 = i\sqrt{\mu}\beta\, e^{\frac{(\phi_+ - \Lambda_0)}{2}}\bar{\psi}_+, \tag{B.20}$$

$$\bar{\partial} b_1 = \sqrt{\mu}\beta\, e^{\frac{(\phi_+ - \Lambda_0)}{2}} \left[iF_+ + \frac{1}{2}\bar{\psi}_+(\psi_+ - \Lambda_1)\right], \tag{B.21}$$

$$\psi_- = \frac{\sqrt{\mu}}{\beta} e^{\frac{\Lambda_0}{2}} \sinh\left(\frac{\phi_-}{2}\right) f_1, \tag{B.22}$$

$$\partial\phi_- = \frac{i\sqrt{\mu}}{\beta} e^{\frac{\Lambda_0}{2}} \left[b_1 \sinh\left(\frac{\phi_-}{2}\right) + \frac{i}{2}\sinh\left(\frac{\phi_-}{2}\right)\Lambda_1 f_1 + \frac{i}{2}\cosh\left(\frac{\phi_-}{2}\right)\psi_- f_1\right], \tag{B.23}$$

$$\begin{aligned}\partial\bar{\psi}_- &= \frac{i\sqrt{\mu}}{4\beta} e^{\frac{\Lambda_0}{2}} \left[\cosh\left(\frac{\phi_-}{2}\right)(\Lambda_1 f_1)\bar{\psi}_- + \sinh\left(\frac{\phi_-}{2}\right)(\bar{\psi}_-\psi_-) f_1 + 2i\cosh\left(\frac{\phi_-}{2}\right) F_- f_1\right. \\ &\quad \left. -2i\cosh\left(\frac{\phi_-}{2}\right) b_1\bar{\psi}_- + 2i\cosh\left(\frac{\phi_-}{2}\right) b_2\psi_- + 2i\sinh\left(\frac{\phi_-}{2}\right) b_2\Lambda_1 + 4f_2\sinh\left(\frac{\phi_-}{2}\right)\right]\end{aligned} \tag{B.24}$$

where we have denoted

$$\phi_\pm = \phi_1 \pm \phi_2, \quad \psi_\pm = \psi_1 \pm \psi_2, \quad \bar{\psi}_\pm = \bar{\psi}_1 \pm \bar{\psi}_2, \quad F_\pm = F_1 \pm F_2. \tag{B.25}$$

We notice that these equations can be simplify as follows,

$$\partial(\phi_+ - \Lambda_0) = -\frac{i\mu}{\beta^2} e^{\Lambda_0} \sinh \phi_- - \frac{\sqrt{\mu}}{2\beta} e^{\frac{\Lambda_0}{2}} \cosh\left(\frac{\phi_-}{2}\right)\psi_+ f_1, \tag{B.26}$$

$$\partial\phi_- = -\frac{2i\mu}{\beta^2} e^{\Lambda_0} \sinh^2\left(\frac{\phi_-}{2}\right) - \frac{\sqrt{\mu}}{2\beta} e^{\frac{\Lambda_0}{2}} \sinh\left(\frac{\phi_-}{2}\right)\psi_+ f_1, \tag{B.27}$$

$$\bar{\partial}\phi_- = 2i\mu\beta^2 e^{(\phi_+ - \Lambda_0)} - \frac{\beta\sqrt{\mu}}{2} e^{\frac{(\phi_+ - \Lambda_0)}{2}} \bar{\psi}_+ f_1, \tag{B.28}$$

$$\psi_- = \frac{\sqrt{\mu}}{\beta} e^{\frac{\Lambda_0}{2}} \sinh\left(\frac{\phi_-}{2}\right) f_1, \tag{B.29}$$

$$\bar{\psi}_- = \sqrt{\mu}\beta\, e^{\frac{(\phi_+ - \Lambda_0)}{2}} f_1, \tag{B.30}$$

$$\partial f_1 = -\frac{i\sqrt{\mu}}{\beta} e^{\frac{\Lambda_0}{2}} \sinh\left(\frac{\phi_-}{2}\right)\psi_+, \tag{B.31}$$

$$\bar{\partial} f_1 = i\sqrt{\mu}\beta\, e^{\frac{(\phi_+ - \Lambda_0)}{2}} \bar{\psi}_+, \tag{B.32}$$

$$\bar{\partial}\Lambda_0 = \bar{\partial}\Lambda_1 = 0, \tag{B.33}$$

after eliminating the auxiliary fermionic field Λ_1.

References

[1] Delfino G, Mussardo G and Simonetti P *Nucl. Phys.* B **432** (1994) 518, hep-th/9409076; *Phys. Lett.* B **328** (1994) 123, hep-th/9403049.
[2] Saleur H, *Lectures on non-perturbative fied theory and quantum impurity problems I*, cond-mat/9812110; Saleur H *Lectures on non-perturbative fied theory and quantum impurity problems II*, cond-mat/0007309.
[3] Bowcock P, Corrigan E and Zambon C *Int. J. Mod. Phys.* A **19S2** (2004) 82, hep-th/0305022; *JHEP* **0401** (2004) 056, hep-th/0401020.
[4] Corrigan E and Zambon C, *J. Phys.* A **37** (2004) L471, hep-th/0407199; *Nonlinearity* **19** (2006) 1447, nli.SI/0512038.
[5] Caudrelier V, *IJGMMP* **vol.5 No. 7** (2008) 1085-108.
[6] Corrigan E and Zambon C, *J. Phys.* A **42** (2009) 475203, hep-th/0908.3126.
[7] Aguirre A R, Araujo T R, Gomes J F and Zimerman A H *JHEP* **12** (2011) 056, nlin/1110.1589.
[8] Gomes J F, Ymai L H, and Zimerman A H *J. Phys. A : Math. Gen.* **39** (2006) 7471, hep-th/0601014.
[9] Gomes J F, Ymai L H, and Zimerman A H *JHEP* **03** (2008) 001.
[10] Aguirre A R, Gomes J F, Ymai L H, and Zimerman A H *Proceedings of Science*, **PoS(ISFTG)031** (2009), arXiv:0910.2888v2 [nlin.SI];
Aguirre A R, Gomes J F, Ymai L H, and Zimerman A H *JHEP* **02** (2011) 017, nlin/1012.1537.
[11] Aguirre A R *J. Phys. A: Math. Theor.* **45** (2012) 205205, arXiv:1111.5249.
[12] Avan J and Doikou A *JHEP* **01** (2012) 040, arXiv:1110.4728 [hep-th].
[13] Avan J and Doikou A *JHEP* **11** (2012) 008, arXiv:1205.1661;
Avan J and Doikou A 2013 *J. Phys.: Conf. Series* **411** 012003, arXiv:1302.2456.
[14] Doikou A and Karaiskos N *Nucl.Phys.* B **867** (2013) 872, arXiv:1207.5503.
[15] Chaichian M and Kulish P *Phys. Lett* B **78** (1978) 413.
[16] Liao H C and Mansfield P *Nucl.Phys.* B **344** (1990) 696.
[17] Bachas C and Gaberdiel M R *JHEP* **0411** (2004) 065, hep-th/0411067.
[18] Corrigan E and Zambon C *J. Phys. A: Math. Theor.* **43** (2010) 345201, arXiv:1006.0939.
[19] Robertson C *Folding defect affine Toda field theories*, arXiv:1304.3129.
[20] Kac V G *Adv Math.* **30** (1978) 85.
[21] Olshanetsky M A *Commun. Math. Phys.* **88** (1983) 63.

Journal of Physics: Conference Series **474** (2013) 012002 doi:10.1088/1742-6596/474/1/012002

On $D = 4$ Stationary Black Holes

L. Andrianopoli, R. D'Auria, A. Gallerati and M. Trigiante

Laboratory of Theoretical Physics, DISAT, Politecnico di Torino, Corso Duca degli Abruzzi 24, 10129 Turin, Italy

E-mail: rdauria@polito.it, laura.andrianopoli@polito.it, antonio.gallerati@polito.it, mario.trigiante@polito.it

Abstract. We review some recent results concerning non-extremal and extremal stationary, asymptotically flat black hole solutions in extended $D = 4$ supergravities, and their properties with respect to the global symmetries of the theory. More specifically we refer to the effective three-dimensional description of these solutions and their classification within orbits with respect to the action of the global symmetry group, illustrating, for single-center solutions, the general mathematical relation between the orbits of non-extremal and extremal black holes.

1. Introduction

The effective three-dimensional description [1] of (asymptotically flat) stationary solutions to $D = 4$ supergravity theories have provided a valuable tool for their classification [2, 3, 4, 5, 6, 7, 8, 9, 10]. This approach consists in describing this kind of solutions as solutions to an effective $D = 3$ Euclidean sigma-model which is formally obtained by reducing the $D = 4$ theory along the time direction and dualizing the vector fields into scalars. The main advantage of such a description, with respect to the $D = 4$ one, is that it makes a larger global symmetry group G manifest. The set of all stationary solutions to the original four-dimensional theory is indeed invariant with respect to the global symmetry group G of the Euclidean three-dimensional sigma-model, which is the isometry group of its target space. Being G larger that the analogous symmetry group of the parent $D = 4$ model, transformations can be used to generate new solutions from known ones which were not available in four dimensions. This *solution-generating technique* has been first used in order to construct non-extremal, rotating, electrically charged black hole solutions coupled to scalar fields [2, 11] and, more recently, found application in the context of *subtracted geometry* [12, 13, 14].

Stationary, asymptotically flat, black holes can therefore be conveniently classified in orbits of with respect to the action of G. We shall restrict ourselves here to the single-center case. General features of the solution like its rotation and extremality (related to the temperature) are in particular associated with invariants of G. As far as the rotational property of the black hole is concerned, this statement was proven in [15] by defining a matrix Q_ψ which, just like the Noether charge matrix Q, lies in the Lie algebra $\mathfrak{g}$ of G, and which vanishes if and only if the solution is static. In terms of Q and Q_ψ the regularity condition for the black hole solution was written in a G-invariant way. The matrix Q_ψ allows to easily infer how the angular momentum $\mathscr{J}$ transforms under G. These tools were then applied in [16] in order to define the general algebraic procedure for connecting the orbit of non-extremal solutions to those of extremal ones. In particular, as far as extremal under-rotating and static black holes are concerned, this mechanism makes use of *singular Harrison transformations* and generalizes previous results in the literature, related to specific electric-magnetic frames. We shall review this analysis below.

2. The $D = 3$ description of stationary solutions

Our original setting is $D = 4$ extended (i.e. $\mathcal{N} > 1$), ungauged supergravity, whose bosonic sector consists in n_s scalar fields $\phi^r(x)$, n_v vector fields $A^\Lambda_\mu(x)$, $\Lambda = 0, \dots, n_v - 1$, and the graviton $g_{\mu\nu}(x)$, which are described by the following Lagrangian [1]:

$$\mathcal{L}_4 = e\left(\frac{R}{2} - \frac{1}{2} G_{rs}(\phi^t)\,\partial_\mu\phi^r\,\partial^\mu\phi^s + \frac{1}{4} I_{\Lambda\Sigma}(\phi^r)\,F^\Lambda_{\mu\nu}\,F^{\Sigma\,\mu\nu} + \frac{1}{8\,e} R_{\Lambda\Sigma}(\phi^r)\,\epsilon^{\mu\nu\rho\sigma}\,F^\Lambda_{\mu\nu}\,F^\Sigma_{\rho\sigma}\right). \tag{1}$$

A distinctive feature of supergravity models is that the scalar fields are described by a non-linear sigma-model, namely they are coordinates of a Riemannian target-space $\mathcal{M}^{(4)}_{scal}$, with positive-definite metric $G_{rs}(\phi)$. We shall restrict our analysis to scalar manifolds which are homogeneous symmetric, namely have the general form $\mathcal{M}^{(4)}_{scal} = G_4/H_4$, where G_4 is a semi-simple, non-compact Lie group and H_4 its maximal compact subgroup. Notice, moreover, that the scalar fields are non-minimally coupled to the vector fields through the (negative-definite) matrix $I_{\Lambda\Sigma}(\phi^r)$ (which generalizes the inverse squared-coupling constants) and the matrix $R_{\Lambda\Sigma}(\phi^r)$ (generalizing the theta-angle). The space-time metric of a stationary solution has the general form:

$$ds^2 = -e^{2U}\,(dt + \omega_i\,dx^i)^2 + e^{-2U}\,g_{ij}\,dx^i\,dx^j\,, \tag{2}$$

where $i, j = 1, 2, 3$ label the spatial coordinates and U, ω_i, g_{ij} are all functions of x^i. Since the seminal work by Breitenlohner, Gibbons and Maison [1], it is known that stationary solutions to the $D = 4$ theory are solutions to an effective sigma model defined on an Euclidean three-dimensional space, formally obtained by first reducing the four-dimensional model along the time direction, and then dualizing the $D = 3$ vector fields into scalars. The scalar fields of this effective model $\phi^I(x^i)$ are $n = 2 + n_s + 2n_v$ and comprise, besides the original four-dimensional scalars $\phi^r(x^i)$, the warp function U, the scalar a dual to the vector $\omega_i(x^j)$ and the $2n_v$ scalars $\mathcal{Z}^M = (\mathcal{Z}^\Lambda,\, \mathcal{Z}_\Lambda)$. The precise relation between the scalars a, $\mathcal{Z}^M$ and the four-dimensional fields is:

$$\begin{aligned} A^\Lambda &= A^\Lambda_0(dt + \omega) + A^\Lambda_{(3)}\,, \quad A^\Lambda_{(3)} \equiv A^\Lambda_i dx^i\,, & (3)\\ \mathbb{F}^M &= \begin{pmatrix} F^\Lambda_{\mu\nu} \\ G_{\Lambda\,\mu\nu} \end{pmatrix} \frac{dx^\mu \wedge dx^\nu}{2} = d\mathcal{Z}^M \wedge (dt + \omega) + e^{-2U}\mathbb{C}^{MN}\mathcal{M}_{(4)NP}{}^{*3}d\mathcal{Z}^P\,, & (4)\\ da &= -e^{4U}\,{}^{*3}d\omega - \mathcal{Z}^T\mathbb{C}d\mathcal{Z}\,, & (5) \end{aligned}$$

where $\mathbb{C}_{MN}$ is the $(2n_v) \times (2n_v)$ symplectic invariant, antisymmetric matrix, $\omega = \omega_i\,dx^i$, $*_3$ is the Hodge operation in the $D = 3$ Euclidean space, $F^\Lambda_{\mu\nu}$ are the vector field strengths and $G_{\Lambda\,\mu\nu}$ their magnetic-duals. The symmetric, symplectic negative-definite matrix $\mathcal{M}_{(4)NP}$ is built out of $I = (I_{\Lambda\Sigma})$, $R = (R_{\Lambda\Sigma})$ as follows:

$$\mathcal{M}_{(4)MN} = \begin{pmatrix} (I + R\,I^{-1}\,R)_{\Lambda\Sigma} & -(R\,I^{-1})_\Lambda{}^\Gamma \\ -(I^{-1}\,R)^\Delta{}_\Sigma & I^{-1\,\Delta\Gamma} \end{pmatrix}. \tag{6}$$

The effective sigma-model Lagrangian has the general form:

$$\mathcal{L}_3 = e_3\left(\frac{R_3}{2} - \frac{1}{2} G_{IJ}(\phi^K)\,\partial_i\phi^I\,\partial^i\phi^J\right), \tag{7}$$

[1] Here we adopt the notations and conventions of [15, 16] (in particular we use the "mostly plus" convention and $8\pi G = c = \hbar = 1$).

where $R_3 = R[g_{ij}]$ and $e_3 = \sqrt{\det(g_{ij})}$. The scalars $\phi^I(x^i)$ span a new manifold $\mathcal{M}_{scal}$, which is homogeneous symmetric if and only if $\mathcal{M}^{(4)}_{scal}$ is. Therefore, in our analysis, it can then be expressed as $\mathcal{M}_{scal} = G/H$, where G is the (semisimple) isometry group and H is a maximal subgroup of G. This target manifold is pseudo-Riemannian and its metric $G_{IJ}(\phi^K)$ reads:

$$\frac{1}{2}G_{IJ}(\phi^K)\,\partial_i\phi^I\,\partial^i\phi^J = \partial_i U\partial^i U + \frac{1}{2}G_{rs}\,\partial_i\phi^r\,\partial^i\phi^s + \frac{1}{2}\mathrm{e}^{-2U}\,\partial_i\mathcal{Z}^T\,\mathcal{M}_{(4)}\,\partial^i\mathcal{Z} + \\ + \frac{1}{4}\mathrm{e}^{-4U}\,(\partial_i a + \mathcal{Z}^T\mathbb{C}\partial_i\mathcal{Z})(\partial^i a + \mathcal{Z}^T\mathbb{C}\partial^i\mathcal{Z})\,, \quad (8)$$

its negative-signature directions being related to the scalar fields $\mathcal{Z}^M$. This feature defines H as a suitable non-compact real form of the maximal compact subgroup of G.

Spherically-symmetric, asymptotically -flat black hole solutions are described in this setting by geodesics $\phi^I = \phi^I(\tau)$ on $\mathcal{M}_{scal}$, in which the affine parameter $\tau \leq 0$ is a harmonic function of the radial coordinate r. Let us first review the geodesic description of static solutions and then move to the description of more general ones. A geodesic is uniquely defined by its initial

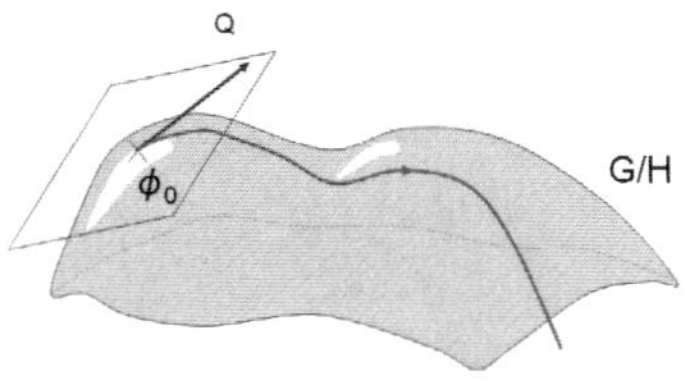

Figure 1. A geodesic on $\mathcal{M}_{scal}$ defined by its initial data.

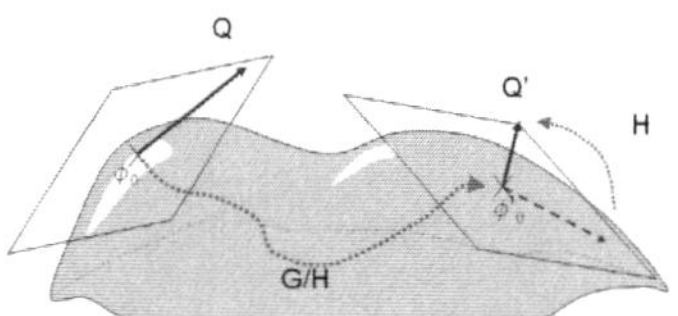

Figure 2. Action of G on a geodesic.

data: an initial point $\phi_0 = (\phi_0^I)$ at radial infinity, corresponding to the limit $\tau \to 0^-$, and an "initial velocity vector" Q in the tangent space $T_{\phi_0}(\mathcal{M}_{scal})$ to the manifold in ϕ_0, see Fig. 1. The global symmetry group of the effective three-dimensional theory is the isometry group G, whose action on a geodesic (ϕ_0, Q) can be described as follows: Using G/H we can freely move the initial point ϕ_0 over the manifold, to a new point ϕ_0' with "velocity vector" Q'; then, for a fixed initial point ϕ_0', we can still act on the initial velocity vector Q' using the stability group H of ϕ_0', acting on $T_{\phi_0'}(\mathcal{M}_{scal})$, see Fig. 2. Since the action of G/H on ϕ_0 is transitive, we can always fix ϕ_0 to coincide with the origin O (defined by the vanishing values of all the scalars) and then classify the orbits of the geodesics under the action of G (i.e. in maximal sets of solutions connected through the action of G) in terms of the orbits of the velocity vector $Q \in T_O(\mathcal{M}_{scal})$ under the action of H.

As we consider more general stationary solutions, they are clearly no longer described by geodesics on the scalar manifold, since the scalars ϕ^I will also depend on the angular coordinates. We shall focus our attention on axisymmetric, single-center black holes, for which the scalar fields will be functions of r and θ only: $\phi^I = \phi^I(r, \theta)$. Nevertheless, the solution will still be characterized by a unique point ϕ_0 at radial infinity ("initial point"):

$$\phi_0^I = \lim_{r\to\infty} \phi^I(r,\theta)\,, \quad (9)$$

and by an "initial velocity" vector Q. By the same token, we can fix $\phi_0 \equiv O$ and classify these solutions according to the action of H on $Q \in T_O(\mathcal{M}_{scal})$.

This tangent space at the origin is isomorphic to subspace $\mathfrak{K}$ (the *coset space*) of the Lie algebra $\mathfrak{g}$ of G, complement to the Lie algebra $\mathfrak{H}$ of H: $T_O(\mathcal{M}_{scal}) \sim \mathfrak{K}$. The space $\mathfrak{K}$, in turn, is the carrier of a representation of H with respect to its adjoint action: $H^{-1}\,\mathfrak{K}\,H \subset \mathfrak{K}$, which realizes the action of the isotropy group H on the tangent space. Therefore the "initial velocity" vector Q should be viewed as a matrix in $\mathfrak{K}$. In fact, Q is the Noether charge matrix of the solution:

$$Q = \frac{1}{4\pi}\int_{S_2} {}^{*3}J\,, \tag{10}$$

$J = J_i\,dx^i$ being the Noether current. The explicit form of J is given by the standard theory of sigma models on coset manifolds:

$$J_i \equiv \frac{1}{2}\partial_i\phi^I\,\mathcal{M}^{-1}\partial_I\mathcal{M}\,, \tag{11}$$

where $\mathcal{M}(\phi^I) = \mathbb{L}(\phi^I)\eta\mathbb{L}(\phi^I)^\dagger$ is an H-invariant symmetric matrix built out of the representative $\mathbb{L}(\phi^I)$ of G/H at the point ϕ^I. The involution defining the subalgebra $\mathfrak{H}$ of H in $\mathfrak{g}$ is defined in terms of an H-invariant real metric η as follows: $h \in \mathfrak{H} \;\Leftrightarrow\; \eta\,h^\dagger\,\eta = -h$ (here we work in some matrix representation of G). The components of Q along a suitable basis $\{K_A\} = \{K_0,\, K_\bullet,\, K_r,\, K_\Lambda,\, K^\Lambda\}$ of $\mathfrak{K} \sim T_O(\mathcal{M}_{scal})$ are the physical quantities characterizing the solution at spatial infinity, namely the ADM mass M_{ADM}, the NUT-charge n_{NUT}, the $D=4$ scalar charges Σ_r and the electric and magnetic charges q_Λ, p^Λ:

$$Q \propto M_{ADM}\,K_0 + \Sigma^r\,K_r + n_{NUT}\,K_\bullet + p^\Lambda K_\Lambda + q_\Lambda K^\Lambda \in \mathfrak{K}\,.$$

Notice that the angular momentum $\mathcal{J}$ does not appear in this expansion.

2.1. Angular Momentum and Duality

In [15] we posed the question: Can we describe, just as all the other physical quantities, the angular momentum as a suitable component of some characteristic vector Q_ψ in the tangent space? In other words: Can we describe the angular momentum in terms of quantities which are intrinsic to the $D=3$ sigma-model? The answer was given by defining Q_ψ as the following $\mathfrak{K}$-valued matrix:

$$Q_\psi = -\frac{3}{4\pi}\int_{S_2^\infty}\psi_{[i}J_{j]}\,dx^i\wedge dx^j = \frac{3}{8\pi}\int_{S_2^\infty} g_{\varphi\varphi}\,J_\theta d\theta d\varphi \propto \mathcal{J}\,K_\bullet + \cdots \in \mathfrak{K}\,, \tag{12}$$

$\psi = \partial_\varphi$ being the angular Killing vector of the axisymmetric solution. Q_ψ is a matrix in the coset space which describes the rotation of the solution. This quantity, as we review below, is important since it allows to easily derive the action of G on $\mathcal{J}$.

2.2. Action of G on a Solution

Being G the global symmetry of the $D=3$ model, if $\phi^I(x^i)$ is a solution, the configuration $\phi'^I(x^i)$, obtained by acting on it by means of an element g of G, is still a solution. Let us denote the g-transformed $\phi' = (\phi'^I)$ of the point $\phi = (\phi^I)$ by $\phi' = g\star\phi$. The (non-linear) relation between ϕ'^I and ϕ^I follows from basic coset-space geometry:

$$\forall g\in G\ :\ \mathcal{M}(g\star\phi) = D[g]\,\mathcal{M}(\phi)\,D[g]^\dagger\,, \tag{13}$$

where $D[g]$ is the matrix associated with g in the chosen representation of G. Equation (13) can be used to devise a *solution generating technique* [2, 11] in order to construct new stationary

Journal of Physics: Conference Series **474** (2013) 012002 doi:10.1088/1742-6596/474/1/012002

solutions from known ones. In particular, from eq.s (11), (10) and (12), it follows that, under a global symmetry transformation g, J, Q, Q_ψ transform through the adjoint action:

$$J \to (D[g]^{-1})^\dagger \, J \, D[g]^\dagger \; ; \; Q \to (D[g]^{-1})^\dagger \, Q \, D[g]^\dagger \; ; \; Q_\psi \to (D[g]^{-1})^\dagger \, Q_\psi \, D[g]^\dagger \, . \tag{14}$$

Recall now that, having fixed $\phi_0 = O$, G was broken to H, so that the action of G on a solution amounts to an action of H on the tangent space quantities: Q, Q_ψ. The above properties allow us to infer the physical features of the transformed solution (including the angular momentum, by virtue of (12)), without having to directly solve the matrix equation (13).

Moreover, in light of eq.s (14), we can characterize the rotational property of a black hole in the effective $D = 3$ description by the following G-invariant statement:

$$\text{Static solution} \;\Leftrightarrow\; Q_\psi = 0 \, . \tag{15}$$

The presence of a non-vanishing Q_ψ is a characteristic of the G-orbits of rotating solutions and therefore one cannot *generate* rotation on a static $D = 4$ solution using G !

Using Q and Q_ψ one can recast the regularity condition of a rotating solution in a G-invariant form. Consider the Kerr-Newman solution to Einstein-Maxwell theory:

$$ds^2 = \frac{\tilde{\Delta}}{\rho^2}(dt + \omega)^2 - \frac{\rho^2}{\tilde{\Delta}}\left(\frac{\tilde{\Delta}}{\Delta}dr^2 + \tilde{\Delta}d\theta^2 + \Delta \sin^2\theta d\varphi^2\right) \tag{16}$$

where

$$\begin{aligned}
\Delta &= (r - m)^2 - \mathbf{c}^2 \, , \\
\mathbf{c}^2 &= m^2 - \frac{1}{2}(q^2 + p^2) - \alpha^2 \, , \\
\tilde{\Delta} &= \Delta - \alpha^2 \sin^2\theta \, , \\
\rho^2 &= r^2 + \alpha^2 \cos^2\theta \, , \\
\omega &= \alpha \sin^2\theta \, \frac{\rho^2 - \tilde{\Delta}}{\tilde{\Delta}} \, d\varphi \, ,
\end{aligned} \tag{17}$$

where m, q, p are the mass and the electric and magnetic charges while the angular momentum is given by $\mathscr{J} = m\,\alpha$. The vector potential A^0 is given by

$$A^0 = (-q\,r + p\,\alpha\,\cos\theta)\,\frac{dt}{\rho^2} + +[-p\,(\alpha^2 + r^2)\cos\theta + q\,\alpha\,r\,\sin^2\theta]\,\frac{d\varphi}{\rho^2} \, . \tag{18}$$

This solution is regular provided the following condition is satisfied:

$$m^2 - \frac{1}{2}(q^2 + p^2) \geq \alpha^2 \, . \tag{19}$$

In this case Q and Q_ψ are diagonalizable matrices and one finds that:

$$\frac{k}{2}\,\mathrm{Tr}(Q^2) = m^2 - \frac{p^2 + q^2}{2} \quad ; \quad \mathrm{Tr}(Q_\psi^2) = \frac{\mathscr{J}^2}{m^2}\,\mathrm{Tr}(Q^2) \, ,$$

where k is a representation-dependent constant. Using the above relations we can rewrite the regularity condition (19) in the following form:

$$\frac{k}{2}\,\mathrm{Tr}(Q^2) \geq \frac{\mathrm{Tr}(Q_\psi^2)}{\mathrm{Tr}(Q^2)} \, , \tag{20}$$

which, in light of the transformation properties (14), is manifestly G-invariant. This means that it is satisfied by all the solutions in the same G orbit as the regular KN one. Equality in eq. (19), or (20), holds for the *extremal* solutions (i.e, with vanishing Hawking temperature).

3. Extremal Limits

In [16] we addressed the problem of defining a general mechanism for connecting the orbit of non-extremal Kerr (or Kerr-Newman) solutions to the orbits of the known extremal ones. The regularity bound (20) can be saturated while keeping both sides non-vanishing (e.g. Q and Q_ψ diagonalizable just as in the Kerr-Newman case). The resulting solution exhibits an ergo-sphere and is dubbed *extremal over-rotating*. The bound can alternatively be saturated in a non-trivial way, by letting both sides of the inequality vanish separately. The resulting solutions can either be *extremal static* (see for instance [17] and references therein) or *extremal under-rotating* (i.e. rotating with no ergo-sphere) [18, 19, 20, 21, 22].

These limits to extremal (under-rotating or static) solutions have been considered in the literature in specific contexts: Heterotic theory [11, 23]; Kaluza-Klein supergravity [18, 19]. In [16] we defined a general geometric prescription for connecting the non-extremal Kerr-orbit to the extremal static or under-rotating ones, in a way which is frame-independent (i.e. does not depend on the particular string theory and compactification yielding the four-dimensional supergravity). This procedure makes use of *singular Harrison transformations* by means of which an Inönü–Wigner contraction on the matrices Q and Q_ψ is implemented, resulting in the nilpotent matrices $Q^{(0)}$ and $Q_\psi^{(0)}$ associated with extremal static or under-rotating black holes.

Harrison transformations [1] are H-transformations which play a special role in the solution generating techniques: They are not present among the global symmetries of the $D=4$ theory and have the distinctive property of switching on electric or magnetic charges when acting on neutral solutions (like the Kerr or Schwarzshild ones). Their generators $(\mathbb{J}_M) = (\mathbb{J}_\Lambda,\, \mathbb{J}^\Lambda)$ in $\mathfrak{H}$ are in one-to-one correspondence with the electric and magnetic charges $(\mathcal{P}^M) = (p^\Lambda,\, q_\Lambda)$ and are non-compact (i.e. are represented, in a suitable basis, by hermitian matrices). The space $\mathrm{Span}(\mathbb{J}_M)$ generated by $\{\mathbb{J}_M\}$ is the coset space of the symmetric manifold H/H_c, H_c being the maximal compact subgroup of H, and thus it is the carrier of a representation of H_c (the same representation in which the charges $\mathcal{P}^M$ transform with respect to H_c). More specifically $H_c = \mathrm{U}(1)_E \times H_4$, where $\mathrm{U}(1)_E$ is an Ehlers transformation, and H_4 is the maximal compact subgroup of G_4.

In [16] we considered the *maximal abelian subalgebra* (MASA) of the space $\mathrm{Span}(\mathbb{J}_M)$. This is a subspace whose generators $\mathbb{J}^{(N)} = \{\mathcal{J}_\ell\}_{\ell=1,\dots,p}$ are defined by the *normal form* of the electric and magnetic charges, i.e. the minimal subset of charges into which the charges of the most general solution can be rotated by means of H_c. Its dimension p is therefore just the *rank of the coset* H/H_c. In the maximal supergravity, for example, $p = \mathrm{rank}\left(\frac{\mathrm{SO}^*(16)}{\mathrm{U}(8)}\right) = 4$, the same being true for the half-maximal theory, $p = \mathrm{rank}\left(\frac{\mathrm{SO}(6,2)\times\mathrm{SO}(2,6+n)}{\mathrm{SO}(2)^2\times\mathrm{SO}(6)\times\mathrm{SO}(6+n)}\right) = 4$, and for the $\mathcal{N}=2$ symmetric models with rank-3 scalar manifold in $D=4$ (for this class of theories, $p =$ rank $+1$). The simplest representative of the latter class of models is the *STU* one, which is a consistent truncation of all the others, besides being a truncation of the maximal and half-maximal theories. Therefore its space $\mathbb{J}^{(N)}$ is contained in the spaces of Harrison generators of all the above mentioned symmetric models. As a consequence of this, for the sake of simplicity, we can restrict ourselves to the simplest *STU* model since the G-orbits of non-extremal and extremal regular solutions to the broad class of symmetric models mentioned above have a representative in the common *STU* truncation. As for the restricted number of $\mathcal{N}=2$ symmetric models for which the rank of $\mathcal{M}_{scal}^{(4)}$ is less than 3, the following discussion has a straightforward generalization. Depending on the symplectic frame, i.e. on the higher-dimensional origin of the four-dimensional theory, this normal form can consist of different kinds of charges. If the $D=4$ supergravity originates from a dimensional reduction of a $D=5$ theory on a circle, one normal form is $\{p^0,\, q_i\}$ and another is $\{q_0,\, p^i\}$, $i = 1, \dots, p-1$.

The procedure devised in [16] consists in first acting on the Kerr solution, given by eq.s (16),(17),(18) with $p = q = 0$, by means of a Harrison transformation generated by $\mathbb{J}^{(N)}$, of the form:

$$\mathcal{O} = \exp\left(\sum_{\ell=0}^{p-1} \log(\beta_\ell)\mathcal{J}_\ell\right) . \tag{21}$$

The resulting solution is a non-extremal, rotating one, coupled to scalar fields, with charges in the normal form. In the Heterotic model where the normal form consists of two electric and two magnetic charges, this solution was first constructed in [2]. This construction reviewed here applies to any symplectic frame. If we denote by $Q^{(K)}$, $Q_\psi^{(K)}$ the matrices Q, Q_ψ pertaining to the Kerr solution

$$Q^{(K)} = 2\, m_K\, H_0\ ,\ \ Q_\psi^{(K)} = 2\, \mathscr{J}_K\, K_\bullet\, , \tag{22}$$

m_K, $\mathscr{J}_K$ being the mass and the angular momentum of the solution, the corresponding matrices for the transformed one are readily computed from (14):

$$Q' = \mathcal{O}^{-1\dagger}\, Q^{(K)}\, \mathcal{O}^\dagger\ ;\ \ Q'_\psi = \mathcal{O}^{-1\dagger}\, Q_\psi^{(K)}\, \mathcal{O}^\dagger\, . \tag{23}$$

We refer to [16] for the precise dependence of the mass, angular momentum and charges of the new solution on m_K, $\mathscr{J}_K$ and β_ℓ. The next step is to rescale the Harrison parameters $\beta_\ell > 0$ and the original angular momentum $\mathscr{J}_K$ as follows:

$$\beta_\ell = (m_K)^{\sigma_\ell}\, \alpha_\ell\ ;\ \ \mathscr{J}_K = m_K^2\, \Omega\, , \tag{24}$$

where $\sigma_\ell = \pm 1$, and send $m_K \to 0$ while keeping α_ℓ and Ω fixed. Although this is clearly a singular limit for the Harrison transformation, the resulting limiting solution is well defined. The normal-form charges in the limit will only depend on α_ℓ and on the signs σ_ℓ and the effect on Q' and Q'_ψ is, as anticipated above, an Inönü–Wigner contraction yielding nilpotent matrices:

$$\{Q',\, Q'_\psi\} \ \stackrel{m_K \to 0}{\longrightarrow}\ \{Q^{(0)},\, Q_\psi^{(0)}\}\ \text{nilpotent}\, . \tag{25}$$

As a consequence of this $\mathrm{Tr}(Q'^2)$ as well as $\mathrm{Tr}(Q'^2_\psi)$ will both vanish in this limit. However the latter quantity vanishes "faster" than the former and this guarantees that both sides of the regularity bound (20) vanish separately. The resulting solution is therefore extremal. Let us illustrate this in some more detail.

Four-dimensional black hole solutions can be classified in orbits with respect to the global symmetry group G_4 in $D = 4$ (which are clearly smaller than the G-orbits in the $D = 3$ description, being G larger than G_4). This group acts on the $D = 4$ fields as a generalized electric-magnetic duality, which respect to which the electric and magnetic charges transform in some characteristic symplectic representation. Among the G-invariant quantities labeling these orbits, the most important is the *quartic invariant* $I_4(p^\Lambda, q_\Lambda)$ which only depends on the electric and magnetic charges and totally characterizes the geometry near the horizon of the extremal static solutions, by virtue of the *attractor mechanism* [24, 25][2]. Being the horizon area proportional to $\sqrt{|I_4|}$, regular extremal solutions can either have $I_4 > 0$ or $I_4 < 0$. In the whole class of supergravities mentioned above, which have the STU one as a consistent truncation, the former solutions can either be supersymmetric (BPS) or non-supersymmetric, while the latter can only be non-supersymmetric. The sign of I_4 in the limiting solution depends on the signs σ_ℓ. We distinguish between two relevant cases.

[2] For a classification of extremal solutions with respect to G_4 see [26].

Case 1. *Only* for the choices of σ_ℓ yielding $I_4 < 0$, we have a residual angular momentum given by:

$$\mathscr{J} = \frac{\Omega}{2}\sqrt{|I_4(p,q)|} \neq 0\,. \tag{26}$$

It is apparent, from the above expression, that the angular momentum, though not G-invariant, is invariant, as expected, under the action of the global symmetry group G_4 in four-dimensions. Indeed $I_4(p,q)$ is G_4-invariant and Ω, pertaining to a Kerr solution, is G_4-invariant as well. The resulting solution is a non-BPS extremal under-rotating black hole. It is important to emphasize here that, in this limit, we find the most general solution of this class *modulo* action of G (generating solution with respect to G).

We also find that the *degree of nilpotency* of $Q_\psi^{(0)}$ is one unit less than that of $Q^{(0)}$. This explains why $\mathrm{Tr}(Q_\psi'^2)$ vanishes faster than $\mathrm{Tr}(Q'^2)$, as mentioned above.

Case 2. For choices of σ_ℓ yielding $I_4 > 0$, we have no residual rotation, i.e. $Q_\psi^{(0)} = 0$, and the limiting solution is either BPS or extremal non-BPS *static*. $Q^{(0)}$ is nilpotent and non-vanishing. Also in this case we find the generating solutions of these classes of black holes with respect to G.

Singular non-extremal limit. There is an other kind of limit we can consider, which consists in keeping $\mathscr{J}_K/m_K$, instead of $\mathscr{J}_K/m_K^2$, fixed. In this case we only have finite limits for $I_4 > 0$. The matrices $Q^{(0)}$ and $Q_\psi^{(0)}$ of the limiting solution are still both nipotent, though they have now the same degree. As a consequence of this, while in the limit the left hand side of (20) vanishes, the right hand side stays finite. Therefore the resulting solution, violating the regularity bound (20), is *singular.* This class of black holes comprises the rotating BPS ones studied, for instance, in [27].

4. Conclusions

We have reviewed a general mechanism for retrieving, in the effective $D = 3$ description, all the extremal limits, including the (nilpotent) G-orbits of regular single-center extremal static and under-rotating solutions from that of non-extremal regular solutions (the Kerr solution). The procedure does not depend on the particular symplectic frame in which the electric and magnetic charges of the four-dimensional supergravity are defined. A valuable tool in this analysis was the matrix Q_ψ, first introduced in [15], which encodes the rotation-properties of the black hole and allows to infer the transformation properties of the angular momentum under G without having to explicitly derive the new solution.

References

[1] Breitenlohner P Maison D and Gibbons G W 1988 *Commun. Math. Phys.* **120** 295
[2] Cvetic M and Youm D 1996 *Nucl. Phys.* B **472** 249
[3] Gunaydin M Neitzke A Pioline B and Waldron A 2006 *Phys. Rev.* D **73** 084019;
[4] Gaiotto D Li W W Padi M 2007 JHEP **12** 093
[5] Bergshoeff E Chemissany W Ploegh A Trigiante M and Van Riet T 2009 *Nucl. Phys.* B **812** 343
[6] Bossard G Nicolai H Stelle K S 2009 *JHEP* **0907** 003
[7] Kim S S Lindman Hornlund J Palmkvist J and Virmani A 2010 *JHEP* **1008** 072
[8] Fre P Sorin A S and Trigiante M 2012 *JHEP* **1204** 015
[9] Bossard G 2012 JHEP **1205** 113
[10] Chemissany W Giaccone P Ruggeri D and Trigiante M 2012 *Nucl. Phys.* B **863** 260
[11] Cvetic M and Youm D 1996 *Phys. Rev.* D **54** 2612
[12] Bertini S Cacciatori S L and Klemm D 2012 *Phys. Rev.* D **85** 064018
[13] Cvetic M and Larsen F 2012 *JHEP* **1209** 076

[14] Virmani A 2012 *JHEP* **1207** 086
[15] Andrianopoli L D'Auria R Giaccone P and Trigiante M 2012 *JHEP* **1212** 078
[16] Andrianopoli L D'Auria R Gallerati A and Trigiante M 2013 *JHEP* **1305** 071
[17] Andrianopoli L D'Auria R Ferrara S and Trigiante M 2008 *Lect. Notes Phys.* **737** 661
[18] Rasheed D 1995 *Nucl. Phys.* B **454** 379 [hep-th/9505038];
[19] Larsen F 2000 *Nucl. Phys.* B **575** 211
[20] Goldstein K and Katmadas S 2009 *JHEP* **0905** 058
[21] Bena I Dall'Agata G Giusto S Ruef C and Warner N P 2009 *JHEP* **0906** 015
[22] Bena I Giusto S Ruef C and Warner N P 2009 *JHEP* **0911** 032
[23] Astefanesei D Goldstein K Jena R P Sen A and Trivedi S P 2006 *JHEP* **0610** 058
[24] Ferrara S Kallosh R 1996 *Phys. Rev.* **D54**, 1514-1524
[25] Ferrara S Kallosh R Strominger A 1995 *Phys. Rev.* **D52** 5412-5416
[26] Bellucci S Ferrara S Gunaydin M Marrani A 2006 *Int. J. Mod. Phys.* **A21** 5043-5098
[27] Behrndt K Lüst D and Sabra W A 1998 *Nucl. Phys.* B **510** 264

Journal of Physics: Conference Series **474** (2013) 012003 doi:10.1088/1742-6596/474/1/012003

Configuration space approach to nonrelativistic conformal dynamics

K. Andrzejewski , J. Gonera, A. Kijanka-Dec
Department of Theoretical Physics and Computer Science,
University of Łódź,
Pomorska 149/153, 90-236 Łódź, Poland
E-mail: jgonera@uni.lodz.pl

Abstract. Lagrangian point of view on conformally invariant nonrelativistic dynamics is presented.

1. Introduction

We present here the review of the results obtained in our paper [1]. Our aim is to discuss the configuration space point of view of classical dynamical systems invariant under the action of the so-called N-conformal nonrelativistic symmetries. We start with very brief discussion of the orbit method [2] which provides the Hamiltonian description of dynamical systems exhibiting some symmetry. Then we remind the structure of N-conformal Galilean algebras and construct the dynamical systems invariant under the transitive action of these algebras. In the following chapter it is shown that the canonical transformations generated by the action of symmetry algebras can be rewritten as point transformation. We provide also the integrals of motion implied by Noether theory. Last section is devoted to some conclusion.

2. The orbit method

The main idea of orbit method can be described as follows:
Given a Lie group G we are looking for all phase manifolds such that:

- G acts as a group of canonical transformations (i.e. it leaves the symplectic form invariant),
- G acts transitively.

The solution of this problem is provided by the orbit method. Namely, one can show that all symplectic manifolds with the symplectic structure invariant under the transitive action of G are coadjoint orbits equipped with the Kirillov form.

The resulting Poisson structure can be described as follows. Let $\{\hat{X}_i\}$ be a basis in L with the commutation rules

$$[\hat{X}_i, \hat{X}_j] = ic^k_{ij}\hat{X}_k \tag{1}$$

Denote by $\{\xi_i\}$ the coordinates in $\tilde{L}$ (dual to L) corresponding to the basis $\{\tilde{\hat{X}}^i\}$. Then the Poisson structure implied by Kirillov symplectic form reads

$$\{\xi_i, \xi_j\} = c^k_{ij}\xi_k. \tag{2}$$

Journal of Physics: Conference Series **474** (2013) 012003 doi:10.1088/1742-6596/474/1/012003

With the phase space at our disposal we can define the dynamical system by selecting some function on coadjoint orbit as a Hamiltonian H. We assume that H is an element of L (in general one could assume that H belongs to the universal enveloping algebra). We see that, in general, H does not commute with the remaining generators (in fact, it cannot commute with all generators; if it commuted, the dynamics, due to transitivity assumption, would be trivial). As a result, some of the integrals of motion implied by the symmetry group G should be explicitly time-dependent (vide Galilei group).

3. N-conformal Galilean algebras g_N

The N-conformal nonrelativistic algebras g_N have been introduced by J.Negro, M.A. del Olmo, A. Rodrigez-Marco [3]. They are defined by the following commutation rules:

$$[J_j, J_k] = i\epsilon_{jkl}J_l, \quad [D,H] = iH, \quad [D,K] = -iK, \quad [K,H] = 2iD,$$

$$[D_s, C_j^a] = iC_j^a, \quad [J^a, C_j^b] = i\epsilon_{abd}C_j^d, \quad [H, C_j^a] = -ijC_{j-1}^a, \tag{3}$$

$$[D, C_j^a] = i(\tfrac{N}{2} - j)C_j^a, \quad [K, C_j^a] = i(N-j)C_{j+1}^a,$$

$a, b, d = 1, 2, 3 \quad j, k, l = 0, 1, ..., N$, all remaining commutators being vanishing.

In what follows we skip D_s. The resulting algebra has the following structure $\tilde{g}_N = (su(2) \oplus sl(2,\mathcal{R})) \uplus \{C_i^a\}$;
$\{C_i^a\}$ span spin one representation of $su(2)$ and "spin" $\frac{N}{2}$ representation of $sl(2,\mathcal{R})$.
In some cases the algebras g_N admit central extensions. Namely, this is the case for N odd; then the relevant commutation rule reads

$$[C_j^a, C_k^b] = i\delta^{ab}\delta^{N,j+k}(-1)^{\frac{k-j+1}{2}}k!j!M \tag{4}$$

$a, b = 1, 2, 3 \quad j, k = 0, 1, ..., N$ with M being the new central element.
There is one exception: in two-dimensional space the central extension is allowed for all N.
In what follows we will be dealing with centrally extended $\tilde{g}_N$ (i.e. N is odd because we assume three space dimensions).

In order to find the G_N-invariant systems we use the orbit method. The construction is highly simplified due to existence of central extension. It appears then that the orbit is uniquely defined by the choice of $sl(2,R)$ and $su(2)$ coadjoint orbits. The variables parameterizing these orbits we will refer to as internal variables. The remaining variables can be chosen in such a way that they form the Darboux coordinates. Let $\vec{c_j}$ be the coordinates on coadjoint orbits related to the generators $\vec{C}$ while χ, s are the internal coordinates corresponding to $sl(2,R)$ and $su(2)$ respectively.
We define

$$\vec{c} = \begin{cases} (-1)^{j-\frac{N-1}{2}}\vec{p}_j, & j = 0, ..., \frac{N-1}{2} \\ mj!\vec{q}_{N-j}, & j = \frac{N+1}{2}, ..., N. \end{cases} \tag{5}$$

Journal of Physics: Conference Series **474** (2013) 012003 doi:10.1088/1742-6596/474/1/012003

Then [4], [5]

$$
\begin{aligned}
h &= \chi^0 - \chi^1 + \frac{1}{2m}\vec{p}_{\frac{N-1}{2}}\vec{p}_{\frac{N-1}{2}} + \sum_{k=1}^{\frac{N-1}{2}} \vec{q}_k\vec{p}_{k-1},\\
d &= \chi^2 + \sum_{k=0}^{\frac{N-1}{2}} (\frac{N}{2} - k)\vec{q}_k\vec{p}_k,\\
k &= \chi^0 + \chi^1 + \frac{m}{2}\left(\frac{N+1}{2}\right)^2 \vec{q}_{\frac{N-1}{2}}\vec{q}_{\frac{N-1}{2}} - \sum_{k=0}^{\frac{N-3}{2}} (N-k)(k+1)\vec{q}_k\vec{p}_{k+1},\\
\vec{j} &= \vec{s} + \sum_{k=0}^{\frac{N-1}{2}} \vec{q}_k \times \vec{p}_k;
\end{aligned}
\tag{6}
$$

here m is the value of central charge M.
The relevant Poisson brackets read off from Kirillov form are

$$
\begin{aligned}
&\{q_k^a, p_l^b\} = \delta^{ab}\delta_{kl},\\
&\{s^a, s^b\} = \epsilon_{abc}s^c, \quad \vec{s} = const.\\
&\{\chi^\mu, \chi^\nu\} = \epsilon^{\mu\nu}{}_{\alpha}\chi^\alpha, \quad g_{\mu\nu}\chi^\mu\chi^\nu = const.\\
&k = 0, 1, \ldots, N, \quad \mu, \nu = 0, 1, 2, \quad a, b = 1, 2, 3
\end{aligned}
\tag{7}
$$

(in the last equation raising and lowering indices is performed with the help of $g_{\mu\nu} = diag(+--)$)
Summarizing, we arrive at the following structure. There are "external" variables $\vec{q}_k$, $\vec{p}_k$, $k = 0, \ldots, \frac{N-1}{2}$ obeying the canonical equations of motion given by the Hamiltonian

$$
H = \frac{1}{2m}\vec{p}_{\frac{N-1}{2}} \cdot \vec{p}_{\frac{N-1}{2}} + \sum_{k=1}^{\frac{N-1}{2}} \vec{q}_k \cdot \vec{p}_{k-1} \tag{8}
$$

In addition, there are "internal" variables χ^μ and s_a related to $sl(2,\mathcal{R})$ and $su(2)$ subalgebras. The dynamics of χ's is governed by the Hamiltonian

$$
H = \chi^0 - \chi^1. \tag{9}
$$

while the spin variables are constants of motion.
One can easily show [6] that the $sl(2,\mathcal{R})$ dynamics is described by the conformal mechanics of Alfaro, Fubini and Furlan [6] In what follows we select trivial orbit for $su(2)$ and $sl(2,\mathcal{R})$, i.e. there are no internal variables.
We see that (7) is the Ostrogradski Hamiltonian for the system described by higher-derivative Lagrangian:

$$
L = \frac{m}{2}\left(\frac{d^{\frac{N+1}{2}}\vec{q}}{dt^{\frac{N+1}{2}}}\right)^2. \tag{10}
$$

This can be easily seen by writing out the canonical equations of motion

$$\begin{aligned}&\dot{\vec{q}}_k=\vec{q}_{k+1},\quad k=0,\ldots,\frac{N-3}{2};\quad \dot{\vec{q}}_{\frac{N-1}{2}}=\frac{1}{m}\vec{p}_{\frac{N-1}{2}}\\&\dot{\vec{p}}_k=-\vec{p}_{k-1},\quad k=1,\ldots,\frac{N-1}{2};\quad \dot{\vec{p}}_0=0,\end{aligned}\tag{11}$$

which, for the basic variable $\vec{q}=\vec{q}_0$, imply $\vec{q}^{(N+1)}=0$.
So, in the particular case of trivial $SU(2)$ and $SL(2,R)$ orbits, $\tilde{g}_N$ is the symmetry algebra of higher derivative free theory [4].

4. Integrals of motion and canonical transformation

It is a matter of standard procedure to obtain the Noether charges corresponding to the generators of Lie algebra. They read [4], [5]

$$\begin{aligned}&h=h(t),\\&d=d(t)-th(t),\\&k=k(t)-2td(t)+t^2h(t),\\&\vec{j}=\vec{j}(t),\\&\vec{c}_j=\begin{cases}(-1)^{j-\frac{N-1}{2}}\sum_{k=0}^{j}\frac{j!}{(j-k)!}t^{j-k}\vec{p}_k, & 0\leq j\leq\frac{N-1}{2},\\(-1)^{j-\frac{N-1}{2}}\sum_{k=0}^{\frac{N-1}{2}}\frac{j!}{(j-k)!}t^{j-k}\vec{p}_k+\\\quad m\sum_{k=\frac{N+1}{2}}^{j}(-1)^{j-k}\frac{j!}{(j-k)!}t^{j-k}\vec{q}_{N-k}, & \frac{N+1}{2}\leq j\leq N.\end{cases}\end{aligned}\tag{12}$$

where $d(t)$, $h(t)$, $k(t)$ and $\vec{j}(t)$ are given by eqs. (5) with χ^μ and $\vec{s}=0$. These charges generate the canonical transformations representing the action of N-conformal Galilean group on Hamiltonian level. Computing systematically the infinitesimal action of all generators ($\delta_0=G,f$) we find:

- for $\vec{c}_j$:

$$\begin{aligned}&\delta_0\vec{q}_n=\{\sum_j\vec{x}_j\vec{c}_j,\vec{q}_n\}=\sum_{k=0}^{N-n}(-1)^{k+n-\frac{N+1}{2}}\frac{(k+n)!}{k!}t^k\vec{x}_{k+n},\\&\delta_0\vec{p}_n=\{\sum_j\vec{x}_j\vec{c}_j,\vec{p}_n\}=m\sum_{k=0}^{n}(-1)^k\frac{(k+N-n)!}{k!}t^k\vec{x}_{k+N-n},\end{aligned}\tag{13}$$

- for h :

$$\begin{aligned}&\delta_0\vec{q}_n=\{\tau h,\vec{q}_n\}=-\tau\left(\frac{1}{m}\delta_{\frac{N-1}{2},n}\vec{p}_{\frac{N-1}{2}}+(1-\delta_{\frac{N-1}{2},n})\vec{q}_{n+1}\right),\\&\delta_0\vec{p}_n=\{\tau h,\vec{p}_n\}=\tau(1-\delta_{n0})\vec{p}_{n-1},\end{aligned}\tag{14}$$

- for d :

$$\begin{aligned}&\delta_0\vec{q}_n=\{\lambda d,\vec{q}_n\}=\lambda\big(-(\frac{N}{2}-n)\vec{q}_n+t(\frac{1}{m}\vec{p}_{\frac{N-1}{2}}\delta_{\frac{N-1}{2},n}+\\&\quad(1-\delta_{\frac{N-1}{2},n})\vec{q}_{n+1})\big),\\&\delta_0\vec{p}_n=\{\lambda d,\vec{p}_n\}=\lambda\left(\left(\frac{N}{2}-n\right)\vec{p}_n-t(1-\delta_{n0})\vec{p}_{n-1}\right),\end{aligned}\tag{15}$$

- for k :

$$\begin{aligned}
\delta_0\vec{q}_n &= \{ck, \vec{q}_n\} = c\Big((1-\delta_{n0})n(N-n+1)\vec{q}_{n-1}+ \\
&\quad 2t(\frac{N}{2}-n)\vec{q}_n - t^2(\frac{1}{m}\delta_{n,\frac{N-1}{2}}\vec{p}_{\frac{N-1}{2}} + (1-\delta_{n,\frac{N-1}{2}})\vec{q}_{n+1})\Big), \\
\delta_0\vec{p}_n &= \{ck, \vec{p}_n\} = c\Big(m\delta_{n\frac{N-1}{2}}(\frac{N+1}{2})^2\vec{q}_{\frac{N-1}{2}} - \\
&\quad (1-\delta_{n\frac{N-1}{2}})(N-n)(n+1)\vec{p}_{n+1} - 2t(\frac{N}{2}-n)\vec{p}_n + t^2(1-\delta_{n0})\vec{p}_{n-1}\Big).
\end{aligned} \tag{16}$$

while $\vec{j}$ acts in the standard way.

Now, given some canonical transformation one can pose the question whether it comes from the point transformation on Lagrangian level. In most cases the answer is **no**, because the point transformations rewritten as a canonical ones have a special form:

- new coordinates are the functions of old ones and time only
- new momenta are the linear functions of old momenta with coordinate- and time-dependent coefficients (actually they can be more general affine functions if the Lagrangian transforms by a total derivative) .

However, the above statement is true only provided the time variable remains unchanged. If it changes the momenta can enter the expression for the variations of coordinate variables provided they appear only in the form of Poisson brackets of Hamiltonian with coordinates. Then the terms containing momenta can be removed from transformation formulas at the expense of admitting the time variation. The resulting modified point transformations coincide "on-shell" (i.e. provided the equations of motion hold) with the initial canonical ones.

To be more explicit, let us write out the form of infinitesimal transformations both on Lagrangian and Hamiltonian levels:

- Lagrangian level

$$\begin{aligned}
q'(t') &= q(t) + \delta q(t) \qquad t'(t) = t + \delta t(t) \\
\delta q(t) &= q'(t') - q(t) = q'(t) + \dot{q}'(t)\delta t - q(t) \\
&\simeq q'(t) - q(t) + \delta t\{q, H\}
\end{aligned} \tag{17}$$

- Hamiltonian level

$$\begin{aligned}
q'(t) &= q(t) + \delta_0 q(t), \qquad \delta_0 q(t) = \delta\lambda\{G, q(t)\} \\
p'(t) &= p(t) + \delta_0 q(t), \qquad \delta_0 p(t) = \delta\lambda\{G, p(t)\}
\end{aligned} \tag{18}$$

$$\delta_0 q(t) = \delta q(t) - \delta t\{q, H\}; \tag{19}$$

Note, that $q(t) \longmapsto q'(t')$ is the point transformation. Therefore, the form of $\delta q(t)$ can be found on Lagrangian level. Eq.(18) tells us then that the Hamiltonian form of the coordinate transformation may involve momentum variables provided they enter in specific form.

The careful inspection of formulas defining the canonical action of N-conformal Galilei group shows that they have the desired property described above and can be therefore rewritten as the point transformations. Below we quote the final results. Moreover, we write out the global form of the relevant transformations.

- transformations generated by $\vec{c}_k$'s:

$$\begin{aligned}
t' &= t, \\
\vec{q}_n'(t') &= \vec{q}_n(t) + \sum_{k=0}^{N-n}(-1)^{k+n-\frac{N+1}{2}}\frac{(k+n)!}{k!}t^k\vec{x}_{n+k}.
\end{aligned} \tag{20}$$

- transformations generated by h:

$$\begin{aligned} t' &= t + \tau, \\ \vec{q}_n'(t') &= \vec{q}_n(t). \end{aligned} \tag{21}$$

- transformations generated by dilatation d:

$$\begin{aligned} t' &= e^{-\lambda} t, \\ \vec{q}_n'(t') &= e^{\lambda(n-\frac{N}{2})} \vec{q}_n(t). \end{aligned} \tag{22}$$

- transformations generated by special conformal generator k:

$$\begin{aligned} t' &= \frac{t}{1-ct} \equiv t(c) \\ \vec{q}_n'(t') &= \sum_{k=0}^{n} \binom{n}{k} \frac{(N+k-1)!}{(N-1)!} \frac{c^k}{(1-ct)^{N+k}} \vec{q}_{n-k}(t) \end{aligned} \tag{23}$$

- finally, the action of rotation subgroup is standard.

As it is well-known [8] the coordinate space for higher-derivative theories consists of the original coordinate together with the supplementary ones which represent the consecutive time derivatives. It is easy to see that this structure is preserved by the above transformations. Namely, in all cases the following property holds:

$$\vec{q}_{n+1} = \dot{\vec{q}}_n \quad \text{implies} \quad \vec{q}\,'_{n+1}(t') = \frac{d\vec{q}_n'(t')}{dt'} \tag{24}$$

which allows us to go back from Ostrogradski coordinates to the original one $\vec{q} = \vec{q}_0$ entering the Lagrangian

$$L = \frac{m}{2} \left(\frac{d^{\frac{N+1}{2}} \vec{q}}{dt^{\frac{N+1}{2}}} \right)^2 . \tag{25}$$

In terms of this original coordinate $\vec{q}$ and t we find the following transformation rules:

- for $\vec{c}_k$'s:

$$t' = t, \quad \vec{q}\,'(t') = \vec{q}(t) + \sum_{k=0}^{N} (-1)^{k-\frac{N+1}{2}} t^k \vec{x}_k; \tag{26}$$

- for h:

$$t' = t + \tau, \quad \vec{q}\,'(t') = \vec{q}(t); \tag{27}$$

- for d:

$$t' = e^{-\lambda}t, \quad \vec{q}'(t') = e^{-\lambda\frac{N}{2}}\vec{q}(t); \tag{28}$$

- for k:

$$t' = \frac{t}{1-ct}, \quad \vec{q}'(t') = \frac{\vec{q}(t)}{(1-ct)^N}; \tag{29}$$

The above equations allow us to write out the differential realization of n-conformal Galilean algebra

$$\begin{gathered}\hat{H} = i\frac{\partial}{\partial t}, \quad \hat{D} = -i\left(\frac{N}{2}\vec{q}\frac{\partial}{\partial\vec{q}} + t\frac{\partial}{\partial t}\right), \\ \hat{K} = i\left(Nt\vec{q}\frac{\partial}{\partial q} + t^2\frac{\partial}{\partial t}\right), \; \hat{\vec{C}}_k = i(-1)^{k-\frac{N-1}{2}}t^k\frac{\partial}{\partial\vec{q}}.\end{gathered} \tag{30}$$

It is straightforward to check that the above operators obey the relevant algebra **without** central extension.
This is a general property - the geometrical realization of a given initial symmetry group in coordinate space leads, in general, to centrally extended group acting in phase space (with central charge acting trivially).

To complete the picture we can show that all integrals of motion (11) can be obtained via Noether theorem applied to higher derivative original Lagrangian

$$L = L(\vec{q}(t), \frac{d\vec{q}(t)}{dt}, \frac{d^2\vec{q}(t)}{dt^2}, ..., \frac{d^{\frac{N+1}{2}}\vec{q}(t)}{dt^{\frac{N+1}{2}}}). \tag{31}$$

Let us remind the form of Noether identity in the higher-derivative case. The symmetry condition reads

$$L(\vec{q}'(t'), \frac{d\vec{q}'}{dt'}, \ldots, \frac{d^{\frac{N+1}{2}}\vec{q}'}{dt'^{\frac{N+1}{2}}})\frac{dt'}{dt} = L(\vec{q}(t), \frac{d\vec{q}}{dt}, \ldots, \frac{d^{\frac{N+1}{2}}\vec{q}}{dt^{\frac{N+1}{2}}}) + \epsilon\frac{d}{dt}(\delta f), \tag{32}$$

where

$$\vec{q}' = \vec{q} + \epsilon\vec{\chi}(q,t), \quad t' = t + \epsilon g(t), \quad \text{and} \quad \delta f = \delta f(\vec{q}, \dot{\vec{q}}, \ddot{\vec{q}}, ..., \vec{q}^{(\frac{N-1}{2})}) \tag{33}$$

By a straightforward reasoning, as in the first-order case, one concludes that the corresponding integral of motion is of the form

$$\begin{gathered}C = Hg - \sum_{k=0}^{\frac{N-1}{2}}\vec{p}_k\vec{\chi}^{(k)} + \\ \sum_{n=2}^{\frac{N+1}{2}}\sum_{k=1}^{n-1}\sum_{l=0}^{k-1}\frac{d^{k-l-1}}{dt^{k-l-1}}(\dot{g}\vec{q}^{(n-k)})\left(-\frac{d^l}{dt^l}\right)\left(\frac{\partial L}{\partial\vec{q}^{(n)}}\right) + \delta f.\end{gathered} \tag{34}$$

where

$$\vec{p}_n = \sum_{j=0}^{\frac{N-1}{2}-n}\left(-\frac{d}{dt}\right)^j\left(\frac{\partial L}{\partial\vec{q}^{(n+j+1)}}\right), \quad n = 0, 1, \ldots, \frac{N-1}{2}; \tag{35}$$

are the Ostrogradski momenta, and

$$H = \sum_{l=0}^{\frac{N-1}{2}} \vec{p}_l \vec{q}^{(l+1)} - L, \tag{36}$$

is the Ostrogradski Hamiltonian.

In our case the generalized momenta and the Hamiltonian H, when written in terms of $\vec{q}$'s, read

$$\vec{p}_n = m(-1)^{\frac{N-1}{2}-n}\vec{q}^{(N-n)}, \quad n = 0, 1, \ldots, \frac{N-1}{2}, \tag{37}$$

$$H = \sum_{n=0}^{\frac{N-3}{2}} (-1)^{\frac{N-1}{2}-n}\vec{q}^{(N-n)}\vec{q}^{(n+1)} + \frac{m}{2}(\vec{q}^{(\frac{N+1}{2})})^2. \tag{38}$$

Now, one can identify δf for each symmetry transformation and check that the integrals of motion derived previously coincide with those given by (31). In detail, we obtain the following results:

- for the transformation generated by $\vec{c}_k$'s

$$g_k = 0, \quad \vec{\chi}_k = (-1)^{k-\frac{N+1}{2}} t^k, \qquad k = 0, \ldots, N$$

and

$$\delta \vec{f}_k = 0, \; k = 0, \ldots, \frac{N-1}{2};$$
$$\delta \vec{f}_k = m \sum_{n=\frac{N+1}{2}}^{k} (-t)^{k-n}\frac{k!}{(k-n)!}\vec{q}^{(N-n)}, \; k = \frac{N+1}{2}, \ldots, N. \tag{39}$$

The corresponding integrals of motion are of the form

$$\vec{C}_k = m \sum_{n=0}^{k} (-t)^{k-n}\frac{k!}{(k-n)!}\vec{q}^{(N-k)}, \; k = 0, \ldots, N. \tag{40}$$

- for time translations obviously $C = H$,
- for dilatation

$$\delta f = 0, \quad g = -t, \quad \vec{\chi} = -\frac{N}{2}\vec{q} \tag{41}$$

and the integral of motion reads

$$D = -tH + D(t) = -tH + m\sum_{k=0}^{\frac{N-1}{2}} (-1)^{\frac{N-1}{2}-k}(\frac{N}{2} - k)\vec{q}^{(N-k)}\vec{q}^{(k)}. \tag{42}$$

- for rotation, the conserved angular momentum reads

$$\vec{J} = m\sum_{k=0}^{\frac{N-1}{2}} (-1)^{\frac{N-1}{2}-k}\vec{q}^{(k)} \times \vec{q}^{(N-k)}. \tag{43}$$

- for special conformal transformation

$$\delta f = \frac{m}{2}\left(\frac{N+1}{2}\right)^2\left(\vec{q}^{(\frac{N-1}{2})}\right)^2, \quad g = t^2 \text{ and } \quad \vec{\chi} = tN\vec{q} \tag{44}$$

while the integral of motion takes the form

$$\begin{aligned} K = t^2H - 2tD(t) + K(t) = t^2H - 2tD(t) + \\ m\sum_{j=0}^{\frac{N-3}{2}}(j+1)(N-j)(-1)^{\frac{N-1}{2}-j}\vec{q}^{(j)}\vec{q}^{(N-j+1)} + \\ \frac{m}{2}\left(\frac{N+1}{2}\right)^2\left(\vec{q}^{(\frac{N-1}{2})}\right)^2. \end{aligned} \tag{45}$$

As in the case of first order theory, all integrals of motion can be obtained from the ones on the Hamiltonian level by expressing Ostrogradski momenta in terms of $\vec{q}$ and its time derivatives.

5. Conclusion

In the case of trivial $SL(2,\mathcal{R})$ and $SU(2)$ orbits the action of N-conformal Galilei group reduces to the one on the standard coordinate space with trivial topology. This is no longer the case if we select nontrivial orbit of $SL(2,\mathcal{R}) \times SU(2)$ (except the case of positive Casimir of $sl(2,\mathcal{R})$ and vanishing one of $SU(2)$).
Above we have considered the case when the symmetry algebra g_N admits central extension (i.e. N-odd). The case of N-even (and $d > 2$) is more complicated. The set of coadjoint orbits is more reach and their classification does not seem to be easy. Moreover, in most cases, the global Darboux coordinates do not exist. Therefore, the classification and detailed description of invariant dynamical systems remains in this case open.

[1] Andrzejewski K, Gonera J, Kijanka-Dec A 2013 *Phys. Rev.* **D 87** 065012;
[2] Kirillow A 1976 Elements of the Theory of Representation (Springer, New York);
Souriau J.M. 1997 *Structure of Dynamical Systems: A Symplectic View of Physics*(Birkhauser, Berlin);
Arnold V.I 1989 *Mathematical Methods of Classical Mechanics* (Springer, New York);
Giachetti R 1981 *Riv. Nuovo Cim.* **4**;
[3] Negro J, del Olmo M.A., Rodriguez-Marco A 1997 *J.Math.Phys.* **38** 3786;
Negro J, del Olmo M.A., Rodriguez-Marco A *J.Math.Phys.* **38** 3810;
[4] Gomis J, Kamimura K 2012 *Phys. Rev.* **D85** , 045023;
[5] Andrzejewski K, Gonera J, Maslanka P *Phys. Rev.* **D 86**, 065009;
[6] Andrzejewski K, Gonera J, arXiv: 1108.1299;
[7] de Alfaro V, Fubini S, Furlan G *Nuovo Cim.* **34A** 569;
[8] Ostrogradski M, 1850 *Mem. Acad. St. Petersburg* **4** 385;
Govaerts J, Rashid M S arXiv : 9403.009;
Nakamura T, Hamamoto S 1996 *Prog. Theor. Phys.* **95** 469;

Journal of Physics: Conference Series **474** (2013) 012004 doi:10.1088/1742-6596/474/1/012004

Representations of a_∞ and d_∞ with central charge 1 on the single neutral fermion Fock space $F^{\otimes \frac{1}{2}}$

Iana I Anguelova, Ben Cox, Elizabeth Jurisich

Department of Mathematics, College of Charleston, Charleston SC 29424, USA

E-mail: anguelovai@cofc.edu, coxbl@cofc.edu, jurische@cofc.edu

Abstract. We construct a new representation of the infinite rank Lie algebra a_∞ with central charge $c = 1$ on the Fock space $F^{\otimes \frac{1}{2}}$ of a single neutral fermion. We show that $F^{\otimes \frac{1}{2}}$ is a direct sum of irreducible integrable highest weight modules for a_∞ with central charge $c = 1$. We prove that as a_∞ modules $F^{\otimes \frac{1}{2}}$ is isomorphic to the Fock space $F^{\otimes 1}$ of the charged free fermions. As a corollary we obtain the decompositions of certain irreducible highest weight modules for d_∞ with central charge $c = \frac{1}{2}$ into irreducible highest weight modules for d_∞ with central charge $c = 1$.

1. Introduction

Our motivation for this paper was to better understand the various boson-fermion correspondences and their connection with the representation theory of certain infinite dimensional Lie algebras. The first to write on the topic of the relationship between a boson-fermion correspondence and representation theory were Date, Jimbo, Kashiwara, Miwa in [DKM81], [DJKM81a] and I. Frenkel in [Fre81]. Since then many attempts have been made to understand the boson-fermion correspondences as nothing else but an isomorphism of infinite dimensional Lie algebra modules. In his seminal paper I. Frenkel wrote: "The Boson-Fermion correspondence is nothing else but the **canonical** isomorphism between two realizations of the same representation of the affine Lie algebra $\hat{D}(2l)$ and in particular of its subalgebra $\widehat{gl}(l)$" (page 317 of [Fre81]). In the language that became commonly used later many had translated this to mean that the boson-fermion correspondence of type A is **just** an isomorphism between the vertex (the bosonic) and the spinor (the fermionic) realizations of the standard modules of $a_\infty = \widehat{gl}_\infty$ (the label "type A" derives from the $a_\infty = \widehat{gl}_\infty$, and is intended to distinguish this correspondence from the boson-fermion correspondence of type B for example, [DJKM81b]). This point of view of course had to be amended, as the charged free fermion Fock space $F^{\otimes 1}$ underlying the fermionic side of the boson-fermion correspondence of type A is actually an infinite direct sum of irreducible standard modules of $a_\infty = \widehat{gl}_\infty$ (for details on $F^{\otimes 1}$ see e.g. [Fre81], [KR87], [Kac98], [Wan99a], as well as Remark 3.2 in this paper). Starting with I. Frenkel's work in [Fre81], and later, the boson-fermion correspondence of type A (and the correspondence of type B) was related to different kinds of Howe-type dualities ([KWY98], [Wan99a], [Wan99b]). For instance, in [Wan99a] Wang wrote that "the $(GL_1, \hat{D})$ -duality in Theorem 5.3 is essentially the celebrated boson-fermion correspondence" ($\hat{D}$ denotes the universal central extension of the Lie algebra of differential operators on the circle, sometimes also labeled by $W_{1+\infty}$). But what we contend is that the boson-fermion correspondences are more than just isomorphisms between certain Lie algebra modules: a boson-fermion correspondence is first and foremost an isomorphism between two different chiral field theories, one fermionic (expressible in terms of free fermions and their descendants), the other bosonic (expressible in terms of exponentiated bosons). In fact, as I. Frenkel was careful to summarize in Theorem II.4.1 of his very influential paper [Fre81], "the canonical" isomorphism of the two $o(2l)$-current algebra modules in the bosonic and the fermionic Fock spaces **follows** from the boson-fermion correspondence (in fact that is what makes the isomorphism canonical), but not vice versa. But what we will show is that although the isomorphism of Lie algebra representations (and indeed various dualities) follow from a boson-fermion correspondence, the isomorphism as Lie algebra modules is **not equivalent** to a boson-fermion correspondence. To do that we consider a **single** neutral fermion Fock space $F^{\otimes\frac{1}{2}}$ and show that as modules for the Lie algebra a_∞, $F^{\otimes\frac{1}{2}} \cong F^{\otimes 1}$. Of course, it is known that even as super vertex algebras, and certainly as modules for the Lie algebras a_∞, d_∞ with central charge $c = 1$, as well as other affine Lie algebras, $F^{\otimes 1} \cong F^{\otimes\frac{1}{2}} \otimes F^{\otimes\frac{1}{2}}$. This fact is often and extensively used in many papers on vertex algebras, and it was once again I. Frenkel who used it first in [Fre81] in connection to representation theory. But, the representations of a_∞, d_∞ and other affine algebras on $F^{\otimes 1} \cong F^{\otimes\frac{1}{2}} \otimes F^{\otimes\frac{1}{2}}$ that are known in the literature do not reduce to representations on each of the $F^{\otimes\frac{1}{2}}$ factors. What was known is that $F^{\otimes\frac{1}{2}}$ is a representation of the Lie algebra d_∞ with central charge $c = \frac{1}{2}$ (this is one of the explanations for the label $\frac{1}{2}$ in $F^{\otimes\frac{1}{2}}$, the other being that $F^{\otimes\frac{1}{2}}$ is only a "half-infinite" Fock space, as opposed to $F^{\otimes 1}$). This is then what we do in this paper: First, we build a fermionic (spinor) representation of a_∞ with central charge $c = 1$ on $F^{\otimes\frac{1}{2}}$. Next we show how this representation decomposes into irreducible highest weight modules, which ultimately shows that as modules for the Lie algebra a_∞ with central charge $c = 1$, $F^{\otimes\frac{1}{2}} \cong F^{\otimes 1}$. This shows that it is not the a_∞-module structure that distinguishes these spaces—$F^{\otimes\frac{1}{2}}$ and $F^{\otimes 1}$ are identical as vector spaces, or even as a_∞ Lie algebra modules. The difference is in the vertex algebra structure (field theory) on $F^{\otimes\frac{1}{2}}$, versus the vertex algebra structure on $F^{\otimes 1}$. The field theory on $F^{\otimes 1}$ is local in the usual sense (at $z = w$, or as we can refer to it, 1-point local, see Definition 2.2); or more precisely $F^{\otimes 1}$ has a **super vertex algebra** structure (see e.g. [Kac98], [LL04], [FBZ04] for a precise definition of a super vertex algebra). On the other hand, even though $F^{\otimes\frac{1}{2}}$ has a super vertex algebra structure, this super vertex algebra structure is not enough to produce the new representations that we obtain below– to do that we at the minimum need to introduce 2-point locality (i.e., the fields we consider on $F^{\otimes\frac{1}{2}}$ are allowed to be multi-local, at both $z = w$ and $z = -w$). More precisely, there is a **twisted vertex algebra** structure on $F^{\otimes\frac{1}{2}}$ (see [Ang12], [ACJ13] for a precise definition of a twisted vertex algebra). This shows that the type of vertex algebra structure on $F^{\otimes 1}$ versus $F^{\otimes\frac{1}{2}}$ is of great importance, in particular the set of points of locality is a necessary part of the data describing any boson-fermion correspondence.

The outlay of the paper is as follows. First, we recall the necessary definitions and technical tools in Section 2. In Section 3 we introduce the infinite Lie algebras that we will work with, and the Fock space $F^{\otimes\frac{1}{2}}$ and its different gradings. Next we show that $F^{\otimes\frac{1}{2}}$ is a module for the Lie algebra a_∞ with central charge $c = 1$, and by restriction for the Lie algebra d_∞ with central charge $c = 1$. Next we show that each homogeneous component of $F^{\otimes\frac{1}{2}}$ is a highest weight module for a_∞ with central charge $c = 1$, which is moreover irreducible. That allows us to show that $F^{\otimes\frac{1}{2}}$ is completely reducible and to show its decomposition in terms of irreducible modules for a_∞ with central charge $c = 1$. Hence we can compare and conclude that as a_∞ modules with central charge $c = 1$ $F^{\otimes\frac{1}{2}} \cong F^{\otimes 1}$. Finally as a corollary we obtain the decomposition of certain $c = \frac{1}{2}$ modules for d_∞ in terms of irreducible highest weight d_∞ with central charge $c = 1$.

2. Notation and background

We work over the field of complex numbers $\mathbb{C}$.

The mathematical definitions of a field in a chiral quantum field theory and normal ordered products of fields are well known, they can be found for instance in [FLM88], [FHL93], [Kac98], [LL04] and others, we include them for completeness:

Definition 2.1 (Field) *A field $a(z)$ on a vector space V is a series of the form*

$$a(z) = \sum_{n\in\mathbf{Z}} a_{(n)} z^{-n-1}, \quad a_{(n)} \in \mathrm{End}(V), \quad \text{such that } a_{(n_v)}v = 0 \text{ for any } v \in V,\ n_v \gg 0. \tag{2.1}$$

Denote

$$a(z)_- := \sum_{n\geq 0} a_n z^{-n-1}, \quad a(z)_+ := \sum_{n<0} a_n z^{-n-1}. \tag{2.2}$$

Definition 2.2 *([ACJ13])* **(N-point local fields)** *Let ϵ be a primitive Nth root of unity. We say that a field $a(z)$ on a vector space V is* **even** *and N-point self-local at $1, \epsilon, \epsilon^2, \dots, \epsilon^{N-1}$, if there exist $n_0, n_1, \dots, n_{N-1} \in \mathbb{Z}_{\geq 0}$ such that*

$$(z-w)^{n_0}(z-\epsilon w)^{n_1}\cdots(z-\epsilon^{N-1}w)^{n_{N-1}}[a(z), a(w)] = 0. \tag{2.3}$$

In this case we set the **parity** *$p(a(z))$ of $a(z)$ to be 0.*
We set $\{a, b\} := ab + ba$. We say that a field $a(z)$ on V is N-point self-local at $1, \epsilon, \epsilon^2, \dots, \epsilon^{N-1}$ and **odd** *if there exist $n_0, n_1, \dots, n_{N-1} \in \mathbb{Z}_{\geq 0}$ such that*

$$(z-w)^{n_0}(z-\epsilon w)^{n_1}\cdots(z-\epsilon^{N-1}w)^{n_{N-1}}\{a(z), a(w)\} = 0. \tag{2.4}$$

In this case we set the **parity** *$p(a(z))$ to be 1. For brevity we will just write $p(a)$ instead of $p(a(z))$.*
Finally, if $a(z), b(z)$ are fields on V, we say that $a(z)$ and $b(z)$ are N-point mutually local at $1, \epsilon, \epsilon^2, \dots, \epsilon^{N-1}$ if there exist $n_0, n_1, \dots, n_{N-1} \in \mathbb{Z}_{\geq 0}$ such that

$$(z-w)^{n_0}(z-\epsilon w)^{n_1}\cdots(z-\epsilon^{N-1}w)^{n_{N-1}}\Big(a(z)b(w) - (-1)^{p(a)p(b)}b(w)a(z)\Big) = 0. \tag{2.5}$$

Definition 2.3 (Normal ordered product) *Let $a(z), b(z)$ be fields on a vector space V. Define*

$$:a(z)b(w): = a(z)_+b(w) + (-1)^{p(a)p(b)}b(w)a_-(z). \tag{2.6}$$

One calls this the "normal ordered product" of $a(z)$ and $b(w)$.

Remark 2.4 Let $a(z), b(z)$ be any fields on a vector space V. Then $:a(z)b(\lambda z):$ and $:a(\lambda z)b(z):$ are well defined fields on V for any $\lambda \in \mathbb{C}^*$.

For a rational function $f(z, w)$, with poles only at $z = 0$, $z = \epsilon^i w$, $0 \leq i \leq N-1$, we denote by $i_{z,w}f(z, w)$ the expansion of $f(z, w)$ in the region $|z| \gg |w|$ (the region in the complex z plane outside of all the points $z = \epsilon^i w$, $0 \leq i \leq N-1$), and correspondingly for $i_{w,z}f(z, w)$. The mathematical background of the well-known and often used (both in physics and in mathematics) notion of Operator Product Expansion (OPE) of product of two fields for the case of usual locality ($N = 1$) has been established for example in [Kac98], [LL04]. The following lemma extended the mathematical background to the case of N-point locality:

Lemma 2.5 *([ACJ13])* **(Operator Product Expansion (OPE) of N-point local fields)** *Let $a(z)$, $b(w)$ be N-point mutually local fields on a vector space V. Then exists fields $c_{jk}(w)$, $j=0,\dots,N-1; k=0,\dots,n_j-1$, such that we have*

$$a(z)b(w)=i_{z,w}\sum_{j=0}^{N-1}\sum_{k=0}^{n_j-1}\frac{c_{jk}(w)}{(z-\epsilon^j w)^{k+1}}+:a(z)b(w):. \tag{2.7}$$

We call the fields $c_{jk}(w)$, $j=0,\dots,N-1; k=0,\dots,n_j-1$ OPE coefficients. We will write the above OPE as

$$a(z)b(w)\sim\sum_{j=0}^{N-1}\sum_{k=0}^{n_j-1}\frac{c_{jk}(w)}{(z-\epsilon_j w)^{k+1}}. \tag{2.8}$$

The $\sim$ signifies that we have only written the singular part, and also we have omitted writing explicitly the expansion $i_{z,w}$, which we do acknowledge tacitly. Often also the following notation is used for short:

$$\lfloor ab\rfloor=a(z)b(w)-:a(z)b(w):=[a(z)_-,b(w)], \tag{2.9}$$

i.e., the *contraction* of any two fields $a(z)$ and $b(w)$ is in fact also the $i_{z,w}$ expansion of the singular part of the OPE of the two fields $a(z)$ and $b(w)$.

The OPE expansion of the product of two fields is very convenient, as it completely determines in a very compact manner the commutation relations between the modes of the two fields, and we will use it extensively in what follows. In particular, extending of the OPEs to the case of N-point local fields allows us to extend and use Wick's Theorem for N-point local fields:

Theorem 2.6 (Wick's Theorem, [BS83], [Hua98] or [Kac98]) *Let $a(z)$, $b(w)$ be N-point mutually local fields on a vector space V, satisfying*

(i) $[\lfloor a^i(z)b^j(w)\rfloor, c^k(x)_\pm]=[\lfloor a^ib^j\rfloor, c^k(x)_\pm]=0$, for all i,j,k and $c^k(x)=a^k(z)$ or $c^k(x)=b^k(w)$.

(ii) $[a^i(z)_\pm, b^j(w)_\pm]=0$ for all i and j.

Then

$$:a^1(z)\cdots a^M(z)::b^1(w)\cdots b^N(w):=$$
$$\sum_{s=0}^{\min(M,N)}\sum_{\substack{i_1<\cdots<i_s,\\ j_1\neq\cdots\neq j_s}}\pm\lfloor a^{i_1}b^{j_1}\rfloor\cdots\lfloor a^{i_s}b^{j_s}\rfloor:a^1(z)\cdots a^M(z)b^1(w)\cdots b^N(w):_{(i_1,\dots,i_s;j_1,\dots,j_s)}.$$

Here the subscript $(i_1,\dots,i_s;j_1,\dots,j_s)$ means that those factors $a^i(z)$, $b^j(w)$ with indices $i\in\{i_1,\dots,i_s\}$, $j\in\{j_1,\dots,j_s\}$ are to be omitted from the product $:a^1\cdots a^Mb^1\cdots b^N:$ and when $s=0$ we do not omit any factors. The plus or minus sign is determined as follows: each permutation of an adjacent odd field changes the sign.

3. The Fock space $F^{\otimes\frac{1}{2}}$ and representations of a_∞ and d_∞ with central charge 1

We recall the definitions and notation for the Fock space $F^{\otimes\frac{1}{2}}$ and the double-infinite rank Lie algebras a_∞ and d_∞ as in [Fre81], [DJKM81a], [Kac90], [Wan99b]; in particular we follow the notation of [Wan99b], [Wan99a].

Consider a single odd self-local field $\phi^D(z)$, which we index in the form $\phi^D(z)=\sum_{n\in\mathbb{Z}+\frac{1}{2}}\phi^D_n z^{-n-\frac{1}{2}}$. The OPE of $\phi^D(z)$ is given by

$$\phi^D(z)\phi^D(w)\sim\frac{1}{z-w}. \tag{3.1}$$

This OPE completely determines the commutation relations between the modes ϕ_n^D, $n \in \mathbb{Z} + \frac{1}{2}$:

$$\{\phi_m^D, \phi_n^D\} = \phi_m^D\phi_n^D + \phi_n^D\phi_m^D = \delta_{m,-n}1. \tag{3.2}$$

and so the modes generate a Clifford algebra Cl_D. The field $\phi^D(z)$ is usually called a "neutral fermion field". Now Cl_D has basis consisting of 1 and the products $\phi_{i_1}^D\phi_{i_2}^D\cdots\phi_{i_k}^D$ where $i_1 < i_2 < \cdots < i_k$, $i_j \in \mathbb{Z} + 1/2$. We introduce a $\mathbb{Z}$-grading dg on Cl_D by defining the following degree of a basis element:

$$dg(1) = 0,$$

$$dg(\phi^D_{n_k-\frac{1}{2}}\cdots\phi^D_{n_2-\frac{1}{2}}\phi^D_{n_1-\frac{1}{2}}) = \#\{i = 1, 2, \ldots, k|\ n_i = \text{odd}\} - \#\{i = 1, 2, \ldots, k|\ n_i = \text{even}\}.$$

Lemma 3.1 *The $\mathbb{Z}$-grading of Cl_D is an algebra grading. Furthermore, the operation left multiplication by an element $\phi_n^D \in CL_D$ for any $n \in \mathbb{Z} + 1/2$ is a homogenous operator on CL_D, of degree 1 if $n = 2k + 1/2$, and of degree -1 if $n = 2k - 1/2$ for some integer k.*

Proof: For any pair ϕ_m, ϕ_n $(m \neq n)$ we claim $dg(\phi_m\phi_n) = dg(\phi_n\phi_m)$. If $m \neq -n$ then $\phi_m\phi_n = -\phi_n\phi_m$, and one of these expressions appears in the given basis so determines the degree of both $\phi_m\phi_n$ and $\phi_n\phi_m$. If $m = -n$, $\phi_m\phi_n = 1 - \phi_n\phi_m$, and again, either the left or right hand side is a sum of basis vectors, since $dg(1) = 0$ the degree of $\phi_m\phi_n$ and $\phi_n\phi_m$ again agree. Thus the Clifford algebra relation (3.2) is compatible with the definition of dg. Now it is obvious from the definition of the grading that as an operator left multiplication by $\phi_n^D \in CL_D$ is a homogeneous operator of the given degree. □

The Fock space of the field $\phi^D(z)$ is the highest weight module of Cl_D with vacuum vector $|0\rangle$, so that $\phi_n^D|0\rangle = 0$ for $n > 0$. It is denoted by $F^{\otimes\frac{1}{2}}$ (see e.g. [DJKM81b], [FFR91], [KW94], [Wan99a], [Wan99b], [KWY98]). $F^{\otimes\frac{1}{2}}$ has basis

$$\{\phi^D_{-n_k-\frac{1}{2}}\cdots\phi^D_{-n_2-\frac{1}{2}}\phi^D_{-n_1-\frac{1}{2}}|0\rangle, |0\rangle|\ n_k > \cdots > n_2 > n_1 \geq 0, n_i \in \mathbb{Z}, \imath = 1, 2, \ldots, k\} \tag{3.3}$$

The space $F^{\otimes\frac{1}{2}}$ has a $\mathbb{Z}_2$ grading given by $k \ mod\ 2$,

$$F^{\otimes\frac{1}{2}} = F_{\bar{0}}^{\otimes\frac{1}{2}} \oplus F_{\bar{1}}^{\otimes\frac{1}{2}},$$

where $F_{\bar{0}}^{\otimes\frac{1}{2}}$ (resp. $F_{\bar{1}}^{\otimes\frac{1}{2}}$) denote the even (resp. odd) components of $F^{\otimes\frac{1}{2}}$. This $\mathbb{Z}_2$ grading can be extended to a $\mathbb{Z}_{\geq 0}$ grading $\tilde{L}$, called "length", by setting

$$\tilde{L}(\phi^D_{-n_k-\frac{1}{2}}\cdots\phi^D_{-n_2-\frac{1}{2}}\phi^D_{-n_1-\frac{1}{2}}|0\rangle) = k. \tag{3.4}$$

The space $F^{\otimes\frac{1}{2}}$ can be given a super vertex algebra structure, as is known from e.g. [FFR91], [KW94], [Kac98].

The $\mathbb{Z}$ grading dg on Cl_D induces a $\mathbb{Z}$ grading dg on $F^{\otimes\frac{1}{2}}$ by assigning $dg(|0\rangle) = 0$ and

$$dg(\phi^D_{-n_k-\frac{1}{2}}\cdots\phi^D_{-n_2-\frac{1}{2}}\phi^D_{-n_1-\frac{1}{2}}|0\rangle) = \#\{i = 1, 2, \ldots, k|\ n_i = \text{odd}\} - \#\{i = 1, 2, \ldots, k|\ n_i = \text{even}\}. \tag{3.5}$$

Denote the space of homogenous elements of degree $dg = n \in \mathbb{Z}$ by $F_{(n)}^{\otimes\frac{1}{2}}$, hence as vector spaces we have

$$F^{\otimes\frac{1}{2}} = \oplus_{n\in\mathbb{Z}} F_{(n)}^{\otimes\frac{1}{2}}. \tag{3.6}$$

Introduce also the special vectors $v_n \in F_{(n)}^{\otimes \frac{1}{2}}$ defined by

$$v_0 = |0\rangle \in F_{(0)}^{\otimes \frac{1}{2}}; \tag{3.7}$$

$$v_n = \phi^D_{-2n+1-\frac{1}{2}} \cdots \phi^D_{-3-\frac{1}{2}} \phi^D_{-1-\frac{1}{2}} |0\rangle \in F_{(n)}^{\otimes \frac{1}{2}}, \quad \text{for } n > 0; \tag{3.8}$$

$$v_{-n} = \phi^D_{-2n+2-\frac{1}{2}} \cdots \phi^D_{-2-\frac{1}{2}} \phi^D_{-\frac{1}{2}} |0\rangle \in F_{(-n)}^{\otimes \frac{1}{2}}, \quad \text{for } n > 0. \tag{3.9}$$

Note that the vectors $v_n \in F_{(n)}^{\otimes \frac{1}{2}}$ have minimal length $\tilde{L} = |n|$ among the vectors within $F_{(n)}^{\otimes \frac{1}{2}}$, and they are in fact the unique (up-to a scalar) vectors minimizing the length $\tilde{L}$, such that the index n_k is minimal too.

The Lie algebra $\bar{a}_\infty$ (sometimes denoted $\bar{gl}_\infty$ or just $\mathfrak{gl}$, see for instance [Wan99a], [Wan99b], [KWY98]) is the Lie algebra of infinite matrices

$$\bar{a}_\infty = \{(a_{ij})|\ i, j \in \mathbb{Z},\ a_{ij} = 0 \text{ for} |i - j| \gg 0\}. \tag{3.10}$$

As usual denote the elementary matrices by E_{ij}.

The algebra a_∞ (often denoted also by $\widehat{gl}_\infty$ or $\widehat{\mathfrak{gl}}$) is a central extension of $\bar{a}_\infty$ by a central element c, $a_\infty = \bar{a}_\infty \oplus \mathbb{C}c$, with cocycle given by

$$C(A, B) = Trace([J, A]B), \tag{3.11}$$

where the matrix $J = \sum_{i \leq 0} E_{ii}$. In particular

$$C(E_{ij}, E_{ji}) = -C(E_{ji}, E_{ij}) = 1, \quad \text{if } i \leq 0,\ j \geq 1$$
$$C(E_{ij}, E_{kl}) = 0 \quad \text{in all other cases.}$$

The commutation relations for the elementary matrices in a_∞ are

$$[E_{ij}, E_{kl}] = \delta_{jk} E_{il} - \delta_{li} E_{kj} + C(E_{ij}, E_{kl}) c.$$

The non-central generators have generating series

$$E^A(z, w) = \sum_{i,j \in \mathbb{Z}} E_{ij} z^{i-1} w^{-j}, \tag{3.12}$$

and relations

$$\begin{aligned}[E^A(z_1, w_1), E^A(z_2, w_2)] &= E^A(z_1, w_2)\delta(z_2 - w_1) - E^A(z_2, w_1)\delta(z_1 - w_2) \\ &+ \iota_{z_1,w_2} \frac{1}{z_1 - w_2} \iota_{w_1,z_2} \frac{1}{w_1 - z_2} c - \iota_{w_2,z_1} \frac{1}{z_1 - w_2} \iota_{z_2,w_1} \frac{1}{w_1 - z_2} c.\end{aligned} \tag{3.13}$$

Here we used the formal delta function notation $\delta(z - w) := \sum_{n \in \mathbb{Z}} z^n w^{-n-1} = \delta(w - z)$ (see e.g. [Kac98], [FBZ04], [ACJ13]).

Further, a_∞ has a triangular decomposition

$$a_\infty = a_\infty^- \oplus a_\infty^0 \oplus a_\infty^+. \tag{3.14}$$

Here $a_\infty^\pm$ consists of correspondingly the strictly upper (strictly lower) triangular infinite matrices; $a_\infty^0 = \mathfrak{gl}_0 \oplus \mathbb{C}c$ where $\mathfrak{gl}_0$ denotes the diagonal matrices.

The root system of a_∞ is $\Delta = \{\epsilon_i - \epsilon_j \mid i,j \in \mathbb{Z}, i \neq j\}$ where $\epsilon_i \in (\mathfrak{gl}_0)^*$ is defined by $\epsilon_i(E_{jj}) = \delta_{ij}$ $(i,j \in \mathbb{Z})$. There is a conjugate linear, involutive anti-automorphism $\omega \in \mathrm{End}(a_\infty)$ defined by $\omega(E_{ij}) = E_{ji}$ and this is called "the compact anti-involution".

For $¢\in \mathbb{C}$ and $\Lambda \in \bigoplus_{i\in\mathbb{Z}, i\neq 0}(\mathbb{C}E_{ii})^*$, set

$$\begin{aligned} {}^a\lambda_i &:= \Lambda(E_{ii}) \\ {}^aH_i &:= E_{ii} - E_{i+1,i+1} + \delta_{i,0}c \\ {}^ah_i &:= \Lambda({}^aH_i) = {}^a\lambda_i - {}^a\lambda_{i+1} + \delta_{i,0}¢ \end{aligned}$$

Define ${}^a\Lambda_j \in (a^0_\infty)^*$ by

$$\begin{aligned} {}^a\Lambda_j(E_{ii}) &= \begin{cases} 1, & \text{for } 0 < i \leq j, \\ -1, & \text{for } j < i \leq 0, \\ 0, & \text{otherwise,} \end{cases} \\ {}^a\Lambda_j(c) &= 0. \end{aligned}$$

Define also ${}^a\hat{\Lambda}_0 \in (a^0_\infty)^*$ by ${}^a\hat{\Lambda}_0(c) = 1$, ${}^a\hat{\Lambda}_0(E_{ii}) = 0$ for $i \in \mathbb{Z}$. Then the i-th fundamental weight is

$${}^a\hat{\Lambda}_j = {}^a\Lambda_j + {}^a\hat{\Lambda}_0, \quad i \in \mathbb{Z}.$$

Let $L(a_\infty; {}^a\Lambda, ¢) = L(\widehat{\mathfrak{gl}}_\infty; {}^a\Lambda, ¢)$ denote the highest weight a_∞-module with highest weight Λ and central charge $¢$.

The algebra $\bar{d}_\infty$ is defined as the **subalgebra** of $\bar{a}_\infty$, consisting of the infinite matrices preserving the bilinear form $D(v_i, v_j) = \delta_{i,1-j}$, i.e.,

$$\bar{d}_\infty = \{(a_{ij}) \in \bar{a}_\infty \mid a_{ij} = -a_{1-j,1-i}\}. \tag{3.15}$$

Denote by d_∞ the central extension of $\bar{d}_\infty$ by a central element c, $d_\infty = \bar{d}_\infty \oplus \mathbb{C}c$, with the same cocycle as for a_∞, (3.11). The commutation relations for the elementary matrices in d_∞ are obtained using the relations in a_∞: [1]

$$[E_{ij}, E_{kl}] = \delta_{jk}E_{il} - \delta_{li}E_{kj} + C(E_{ij}, E_{kl})c.$$

The generators for the algebra d_∞ can be written in terms of these elementary matrices as:

$$\{E_{i,j} - E_{1-j,1-i},\ i,j \in \mathbb{Z}; \text{and}\ \ c\}.$$

We can arrange the non-central generators in a generating series

$$E^D(z,w) = \sum_{i,j\in\mathbb{Z}} (E_{ij} - E_{1-j,1-i}) z^{i-1} w^{-j}. \tag{3.16}$$

The generating series $E^D(z,w)$ obeys the following relations:

$$E^D(z,w) = -E^D(w,z)$$

and

$$\begin{aligned} [E^D(z_1,w_1), E^D(z_2,w_2)] &= E^D(z_1,w_2)\delta(z_2 - w_1) - E^D(z_2,w_1)\delta(z_1 - w_2) \\ &\quad + E^D(w_2,w_1)\delta(z_1 - z_2) - E^D(z_1,z_2)\delta(w_1 - w_2) \\ &\quad + 2\iota_{z_1,w_2}\frac{1}{z_1 - w_2}\iota_{w_1,z_2}\frac{1}{w_1 - z_2}c - 2\iota_{z_2,w_1}\frac{1}{z_2 - w_1}\iota_{w_2,z_1}\frac{1}{w_2 - z_1}c \\ &\quad - 2\iota_{z_1,z_2}\frac{1}{z_1 - z_2}\iota_{w_1,w_2}\frac{1}{w_1 - w_2}c + 2\iota_{z_2,z_1}\frac{1}{z_1 - z_2}\iota_{w_2,w_1}\frac{1}{w_2 - w_1}c. \end{aligned}$$

[1] Note that in [Kac90] the commutation relation $[E_{ij}, E_{kl}] = \delta_{jk}E_{il} - \delta_{li}E_{kj} + \frac{1}{2}C(E_{ij}, E_{kl})c$ is used instead.

The assignment $E(z,w) \mapsto : \phi^D(z)\phi^D(w) :$, $c \mapsto \frac{1}{2}Id_{F^{\otimes\frac{1}{2}}}$ gives a representation of the Lie algebra d_∞ on $F^{\otimes\frac{1}{2}}$ (see e.g. [DJKM81b], [KW94], [KWY98]), which we denote by $r_{\frac{1}{2}}$. Further, it is known (see e.g. [KW94], [KWY98], [Wan99a]) that as d_∞ modules

$$F_{\bar{0}}^{\otimes\frac{1}{2}} \cong L(d_\infty; {}^d\hat{\Lambda}_0, \frac{1}{2}); \quad F_{\bar{1}}^{\otimes\frac{1}{2}} \cong L(d_\infty; {}^d\hat{\Lambda}_1, \frac{1}{2}). \tag{3.17}$$

where $L(d_\infty; {}^d\Lambda, ¢)$ denotes the highest weight d_∞-module with highest weight ${}^d\Lambda$ and central charge $¢$. The highest weights are defined by the following, using the symmetry in d_∞:

$$\begin{aligned}&{}^d\hat{\Lambda}_0(E_{i,i} - E_{1-i,1-i}) = 0 \quad \text{for any } \; i \in \mathbb{Z};\\ &{}^d\hat{\Lambda}_1(E_{i,i} - E_{1-i,1-i}) = 1; \quad \text{for } i = 1; \quad {}^d\hat{\Lambda}_1(E_{i,i} - E_{1-i,1-i}) = 0 \quad \text{for } i \neq 0, 1;\\ &{}^d\hat{\Lambda}_0(c) = {}^d\hat{\Lambda}_1(c) = \frac{1}{2}.\end{aligned}$$

As a d_∞-module with central charge $c = \frac{1}{2}$ $F^{\otimes\frac{1}{2}}$ then decomposes as

$$F^{\otimes\frac{1}{2}} = F_{\bar{0}}^{\otimes\frac{1}{2}} \oplus F_{\bar{1}}^{\otimes\frac{1}{2}} \cong L(d_\infty; {}^d\hat{\Lambda}_0, \frac{1}{2}) \oplus L(d_\infty; {}^d\hat{\Lambda}_1, \frac{1}{2}),$$

Next we will show that $F^{\otimes\frac{1}{2}}$ is also a module for a_∞ (and thus d_∞) with central charge $c = 1$.

Remark 3.2 It is well known (in the context of representation theory it was introduced by I. Frenkel in [Fre81] and extensively used afterwards) that

$$F^{\otimes\frac{1}{2}} \otimes F^{\otimes\frac{1}{2}} \cong F^{\otimes 1};$$

where $F^{\otimes 1}$ is the Fock space of **1 pair** of two charged fermions. The two charged fermions are the fields $\psi^+(z)$ and $\psi^-(z)$ with operator product expansions (OPEs):

$$\psi^+(z)\psi^-(w) \sim \frac{1}{z-w} \sim \psi^-(z)\psi^+(w), \quad \psi^+(z)\psi^+(w) \sim 0 \sim \psi^-(z)\psi^-(w)$$

where the 1 above denotes the identity map $Id_{F^{\otimes 1}}$. The modes ψ_n^+ and ψ_n^-, $n \in \mathbb{Z}$ of the fields $\psi^+(z)$ and $\psi^-(z)$, which we index as follows:

$$\psi^+(z) = \sum_{n\in\mathbf{Z}} \psi_n^+ z^{-n-1}, \quad \psi^-(z) = \sum_{n\in\mathbf{Z}} \psi_n^- z^{-n-1}, \tag{3.18}$$

form a Clifford algebra Cl_A with relations

$$\{\psi_m^+, \psi_n^-\} = \delta_{m+n,-1}1, \quad \{\psi_m^+, \psi_n^+\} = \{\psi_m^-, \psi_n^-\} = 0. \tag{3.19}$$

The Fock space $F^{\otimes 1}$ is the highest weight representation of Cl_A generated by the vacuum vector $|0\rangle$, so that $\psi_n^+|0\rangle = \psi_n^-|0\rangle = 0$ for $n \geq 0$ (see e.g. [Fre81], [KR87], [KW94], [Kac98] for more details on $F^{\otimes 1}$). It is well known (see e.g., [Kac98], [FBZ04], [LL04]) that $F^{\otimes 1}$ has a structure of a super vertex algebra (i.e., with a single point of locality at $z = w$ in the OPEs); this vertex algebra is often called "charged free fermion vertex algebra". It is also well known (introduced by I. Frenkel, [Fre81]; and [DJKM81a]) and extensively used (e.g., [KR87], [FFR91], [Kac90], [Wan99b], [Wan99a] among many others) that $F^{\otimes 1}$ is a module for the a_∞ algebra, moreover

$$F^{\otimes 1} \cong \oplus_{n\in\mathbb{Z}} L(a_\infty; {}^a\hat{\Lambda}_n, 1).$$

This isomorphism has often been referred to as equivalent to the well known charged free boson-fermion correspondence (an isomorphism between the super vertex algebra on $F^{\otimes 1}$ and the super vertex algebra of the rank one odd lattice). Here we are concerned with a more subtle point: a boson-fermion correspondence is not just an equivalence of Lie algebra representations, but an isomorphism of appropriate field theories (or vertex algebra structures). As shown and used by I. Frenkel ([Fre81]) one has

$$F^{\otimes 1} \cong F^{\otimes \frac{1}{2}} \otimes F^{\otimes \frac{1}{2}},$$

which is implemented by

$$\phi_1(z) = \frac{1}{\sqrt{2}}\left(\psi^+(z) + \psi^-(z)\right), \ \phi_2(z) = \frac{i}{\sqrt{2}}\left(\psi^+(z) - \psi^-(z)\right).$$

In today's language we would say that this fields map generates an isomorphism of super vertex algebras, a fact which is extensively used by many authors (e.g., [FFR91], [MTZ08], [Bar11] among many). But as we will show below, as a_∞ modules

$$F^{\otimes 1} \cong F^{\otimes \frac{1}{2}} \cong \oplus_{n\in\mathbb{Z}} L(a_\infty; {}^a\hat{\Lambda}_n, 1).$$

In other words, as vector spaces and as a_∞ modules $F^{\otimes \frac{1}{2}}$ and $F^{\otimes 1}$ are indistinguishable. On the other hand, in the case of $F^{\otimes 1}$ the boson-fermion correspondence (of type A) is an isomorphism of **super vertex algebras**, requiring locality only at $z = w$ (see e.g., [Kac98]). But, in the case of $F^{\otimes \frac{1}{2}}$ the boson-fermion correspondence (of type D-A) is an isomorphism of **twisted vertex algebras**, requiring locality at both $z = w$ and $z = -w$ (see [Ang12], [ACJ13]).

We obtain a representation of a_∞ on $F^{\otimes \frac{1}{2}}$ by introducing fields arising from a 2-point local twisted vertex algebra:

Proposition 3.3 *Let*

$$\phi^{+DA}(z) = \frac{\phi^D(z) - \phi^D(-z)}{2}, \quad \phi^{-DA}(z) = \frac{\phi^D(z) + \phi^D(-z)}{2}. \tag{3.20}$$

The assignment $zE(z^2, w^2) \mapsto : \phi^{+DA}(z)\phi^{-DA}(w) :$, $c \mapsto Id_{F^{\otimes \frac{1}{2}}}$ *gives a representation of the Lie algebra* a_∞ *on* $F^{\otimes \frac{1}{2}}$ *with central charge* $c = 1$.

Proof: We will use Wick's Theorem. We have $\phi^{+DA}(z) = \sum_{n\in\mathbb{Z}} \phi^D_{-2n+\frac{1}{2}} z^{2n-1}$ and $\phi^{-DA}(z) = \sum_{n\in\mathbb{Z}} \phi^D_{-2n-\frac{1}{2}} z^{2n}$, thus the modes $\phi^{+DA}(z)$ (resp. $\phi^{-DA}(z)$) are the operator coefficients of $\phi^D(z)$ in front of odd (resp. even) powers of the formal variable z. Hence, since $\phi^D(z)$ obeys the second condition of Wick's theorem, $\phi^{+DA}(z)$ and $\phi^{-DA}(z)$ obey it too. We also have

$$\phi^{+DA}(z)\phi^{+DA}(w) \sim 0, \quad \phi^{-DA}(z)\phi^{-DA}(w) \sim 0; \tag{3.21}$$

$$\phi^{+DA}(z)\phi^{-DA}(w) \sim \frac{1}{2}\left(\frac{1}{z-w} + \frac{1}{z+w}\right) \sim \frac{z}{z^2 - w^2}; \tag{3.22}$$

$$\phi^{-DA}(z)\phi^{+DA}(w) \sim \frac{1}{2}\left(\frac{1}{z-w} - \frac{1}{z+w}\right) \sim \frac{w}{z^2 - w^2}; \tag{3.23}$$

hence the first condition of Wick's theorem is also satisfied. Thus from Wick's theorem we have

$$\begin{aligned} &: \phi^{+DA}(z_1)\phi^{-DA}(w_1) :: \phi^{+DA}(z_2)\phi^{-DA}(w_2) : \\ &\quad \sim \frac{z_1}{z_1^2 - w_2^2} : \phi^{-DA}(w_1)\phi^{+DA}(z_2) : + \frac{z_2}{w_1^2 - z_2^2} : \phi^{+DA}(z_1)\phi^{-DA}(w_2) : \\ &\quad + \frac{z_1}{z_1^2 - w_2^2}\frac{z_2}{w_1^2 - z_2^2}. \end{aligned}$$

Hence

$$\begin{aligned}
&[: \phi^{+DA}(z_1)\phi^{-DA}(w_1) :, : \phi^{+DA}(z_2)\phi^{-DA}(w_2) :] \\
&= z_1\delta(z_1^2 - w_2^2) : \phi^{-DA}(w_1)\phi^{+DA}(z_2) : + z_2\delta(z_2^2 - w_1^2) : \phi^{+DA}(z_1)\phi^{-DA}(w_2) : \\
&\quad + \iota_{z_1,w_2}\frac{z_1}{z_1^2 - w_2^2}\iota_{w_1,z_2}\frac{z_2}{w_1^2 - z_2^2} - \iota_{w_2,z_1}\frac{z_1}{z_1^2 - w_2^2}\iota_{z_2,w_1}\frac{z_2}{w_1^2 - z_2^2};
\end{aligned}$$

and we use the fact that $: \phi^{-DA}(w)\phi^{+DA}(z) := - : \phi^{+DA}(z)\phi^{-DA}(w) :$. □

We will denote this new representation on $F^{\otimes\frac{1}{2}}$ by r_1. Since d_∞ is a subalgebra of a_∞, we have the following

Corollary 3.4 *$F^{\otimes\frac{1}{2}}$ is a module for d_∞ with central charge $c = 1$ via the restriction of the representation r_1.*

If we introduce a normal ordered product $: \phi^D_m\phi^D_n :$ on the modes ϕ^D_m of the field $\phi^D(z)$, compatible with the normal ordered product of fields (Definition 2.3), we have to have

$$: \phi^D(z)\phi^D(w) := \sum_{m,n\in\mathbf{Z}+\frac{1}{2}} : \phi^D_{-m-\frac{1}{2}}\phi^D_{-n-\frac{1}{2}} : z^m w^n,$$

and thus

$$: \phi^D_{-m-\frac{1}{2}}\phi^D_{-n-\frac{1}{2}} := \begin{cases} \phi^D_{-m-\frac{1}{2}}\phi^D_{-n-\frac{1}{2}} & \text{for } m+n \neq 1 \\ \phi^D_{-m-\frac{1}{2}}\phi^D_{-n-\frac{1}{2}} - 1 = -\phi^D_{-n-\frac{1}{2}}\phi^D_{-m-\frac{1}{2}} & \text{for } m+n = -1, n \geq 0, \\ \phi^D_{-m-\frac{1}{2}}\phi^D_{-n-\frac{1}{2}} & \text{for } m+n = -1, m \geq 0. \end{cases} \tag{3.24}$$

Hence the well known representation $r_{\frac{1}{2}}$ of d_∞ on $F^{\otimes\frac{1}{2}}$ with central charge $c = \frac{1}{2}$ is defined by

$$r_{\frac{1}{2}}(E_{m,n} - E_{1-n,1-m}) =: \phi^D_{-m+1/2}\phi^D_{n-\frac{1}{2}} :$$

for all $m, n \in \mathbb{Z}$. Now Proposition 3.3 gives us a new representation r_1 of a_∞ on $F^{\otimes\frac{1}{2}}$ given by

$$r_1(E_{m,n}) =: \phi^D_{-2m+\frac{1}{2}}\phi^D_{2n-\frac{1}{2}} : . \tag{3.25}$$

for all $m, n \in \mathbb{Z}$. Hence by restriction the representation of d_∞ on $F^{\otimes\frac{1}{2}}$ with central charge $c = 1$ is

$$r_1(E_{m,n} - E_{1-n,1-m}) :=: \phi^D_{-2m+\frac{1}{2}}\phi^D_{2n-\frac{1}{2}} : - : \phi^D_{2n-3/2}\phi^D_{-2m+3/2} : .$$

for all $m, n \in \mathbb{Z}$.

Proposition 3.5 *(The upper triangular elements annihilate v_n) For all $n \in \mathbb{Z}$, $a^+_\infty v_n = 0$.*

Proof: We need to prove that for any $k \geq 1$ and any $i, n \in \mathbb{Z}$, $r_1(E_{i,i+k})v_n = 0$. We have $r_1(E_{i,i+k}) =: \phi^D_{-2i+\frac{1}{2}}\phi^D_{2i+2k-\frac{1}{2}} :$. We start with $n = 0$, $v_0 = |0\rangle$. There are two cases: The case of $2i + 2k > 0$ is trivial. In the case $2i + 2k \leq 0$, then $: \phi^D_{-2i+\frac{1}{2}}\phi^D_{2i+2k-\frac{1}{2}} := -\phi^D_{2i+2k-\frac{1}{2}}\phi^D_{-2i+\frac{1}{2}}$. But $-2i \geq 2k \geq 2$, hence

$$r_1(E_{i,i+k})v_0 = r_1(E_{i,i+k})|0\rangle = -\phi^D_{2i+2k-\frac{1}{2}}\phi^D_{-2i+\frac{1}{2}}|0\rangle = 0.$$

We continue with $n>0$, where $v_n=\phi^D_{-2n+1-\frac{1}{2}}\cdots\phi^D_{-3-\frac{1}{2}}\phi^D_{-1-\frac{1}{2}}|0\rangle$. Again, there are two cases: first we will consider the case when $2i+2k\le 0$:

$$r_1(E_{i,i+k})\phi^D_{-2n+1-\frac{1}{2}}\cdots\phi^D_{-3-\frac{1}{2}}\phi^D_{-1-\frac{1}{2}}|0\rangle=-\phi^D_{2i+2k-\frac{1}{2}}\phi^D_{-2i+\frac{1}{2}}\phi^D_{-2n+1-\frac{1}{2}}\cdots\phi^D_{-3-\frac{1}{2}}\phi^D_{-1-\frac{1}{2}}|0\rangle.$$

Now since it is impossible to have $-2i+\frac{1}{2}=-(-2l+1-\frac{1}{2})$ for any $l\in\mathbb{Z}$, $\phi^D_{-2i+\frac{1}{2}}$ will anticommute with any $\phi^D_{-2l+1-\frac{1}{2}}$, and thus $\phi^D_{-2i+\frac{1}{2}}\phi^D_{-2n+1-\frac{1}{2}}\cdots\phi^D_{-3-\frac{1}{2}}\phi^D_{-1-\frac{1}{2}}|0\rangle=0$.

Next let $2i+2k>0$, then $:\phi^D_{-2i+\frac{1}{2}}\phi^D_{2i+2k-\frac{1}{2}}:=\phi^D_{-2i+\frac{1}{2}}\phi^D_{2i+2k-\frac{1}{2}}$, and

$$r_1(E_{i,i+k})\phi^D_{-2n+1-\frac{1}{2}}\cdots\phi^D_{-3-\frac{1}{2}}\phi^D_{-1-\frac{1}{2}}|0\rangle=\phi^D_{-2i+\frac{1}{2}}\phi^D_{2i+2k-\frac{1}{2}}\phi^D_{-2n+1-\frac{1}{2}}\cdots\phi^D_{-3-\frac{1}{2}}\phi^D_{-1-\frac{1}{2}}|0\rangle.$$

Now $\phi^D_{2i+2k-\frac{1}{2}}$ is an operator annihilating the vacuum, and unless we have $2i+2k-\frac{1}{2}=-(-2l+1-\frac{1}{2})$ for some $1\le l\le n$, $l\in\mathbb{Z}$, then it will anticommute with any of the $\phi^D_{-2l+1-\frac{1}{2}}$, $1\le l\le n$, and thus $\phi^D_{2i+2k-\frac{1}{2}}\phi^D_{-2n+1-\frac{1}{2}}\cdots\phi^D_{-3-\frac{1}{2}}\phi^D_{-1-\frac{1}{2}}|0\rangle=0$. If, on the other hand $2i+2k-\frac{1}{2}=-(-2l+1-\frac{1}{2})$ for some $1\le l\le n$, $l\in\mathbb{Z}$, then we have

$$\begin{aligned}r_1(E_{i,i+k})&\phi^D_{-2n+1-\frac{1}{2}}\cdots\phi^D_{-3-\frac{1}{2}}\phi^D_{-1-\frac{1}{2}}|0\rangle\\&=\phi^D_{-2i+\frac{1}{2}}\phi^D_{2i+2k-\frac{1}{2}}\phi^D_{-2n+1-\frac{1}{2}}\cdots\phi^D_{-2l+1-\frac{1}{2}}\cdots\phi^D_{-1-\frac{1}{2}}|0\rangle\\&=\pm\phi^D_{-2i+\frac{1}{2}}\phi^D_{-2n+1-\frac{1}{2}}\cdots\widehat{\phi^D_{-2l+1-\frac{1}{2}}}\phi^D_{-2l+3-\frac{1}{2}}\cdots\phi^D_{-1-\frac{1}{2}}|0\rangle;\end{aligned}$$

here $\widehat{\phi^D_{-2l+1-\frac{1}{2}}}$ denotes the fact that $\phi^D_{-2l+1-\frac{1}{2}}$ is absent. But then from $2i+2k-1/2=-(-2l+1-\frac{1}{2})$, we have $i+k=l$ and $2i+\frac{1}{2}=-2l+2k+1-\frac{1}{2}$, and we know from $k\ge 1$ that either $-2l+3-\frac{1}{2}\le-2i+\frac{1}{2}\le-1-\frac{1}{2}$, or $-2i+\frac{1}{2}\ge\frac{1}{2}$. If $-2i+\frac{1}{2}\ge\frac{1}{2}$ (i.e., $k\ge l$), then since $\phi^D_{-2i+\frac{1}{2}}$ anticommutes with all $\phi^D_{-2l+1-\frac{1}{2}}$, we have $\phi^D_{-2i+\frac{1}{2}}\phi^D_{-2n+1-\frac{1}{2}}\cdots\widehat{\phi^D_{-2l+1-\frac{1}{2}}}\phi^D_{-2l+3-\frac{1}{2}}\cdots\phi^D_{-1-\frac{1}{2}}|0\rangle$. If on the other hand $-2l+3-\frac{1}{2}\le-2i+\frac{1}{2}\le-1-\frac{1}{2}$, then that means $\phi^D_{-2i+\frac{1}{2}}=\phi^D_{-2l_1+1-\frac{1}{2}}$ for $1\le l_1\le l-1$ and

$$\begin{aligned}\phi^D_{-2i+\frac{1}{2}}&\phi^D_{-2n+1-\frac{1}{2}}\cdots\widehat{\phi^D_{-2l+1-\frac{1}{2}}}\phi^D_{-2l+3-\frac{1}{2}}\cdots\phi^D_{-1-\frac{1}{2}}|0\rangle\\&=\phi^D_{-2i+\frac{1}{2}}\phi^D_{-2n+1-\frac{1}{2}}\cdots\widehat{\phi^D_{-2l+1-\frac{1}{2}}}\phi^D_{-2l+3-\frac{1}{2}}\cdots\phi^D_{-2l_1+1-\frac{1}{2}}\cdots\phi^D_{-1-\frac{1}{2}}|0\rangle\\&=\phi^D_{-2l_1+1-\frac{1}{2}}\phi^D_{-2n+1-\frac{1}{2}}\cdots\widehat{\phi^D_{-2l+1-\frac{1}{2}}}\phi^D_{-2l+1-\frac{1}{2}}\cdots\phi^D_{-2l_1+1-\frac{1}{2}}\cdots\phi^D_{-1-\frac{1}{2}}|0\rangle\\&=\pm\phi^D_{-2n+1-\frac{1}{2}}\cdots\widehat{\phi^D_{-2l+1-\frac{1}{2}}}\phi^D_{-2l+1-\frac{1}{2}}\cdots\phi^D_{-2l_1+1-\frac{1}{2}}\phi^D_{-2l_1+1-\frac{1}{2}}\cdots\phi^D_{-1-\frac{1}{2}}|0\rangle=0;\end{aligned}$$

since $\phi^D_{-2l_1+1-\frac{1}{2}}\phi^D_{-2l_1+1-\frac{1}{2}}=0$.

Consider now $v_{-n}=\phi^D_{-2n+2-\frac{1}{2}}\cdots\phi^D_{-2-\frac{1}{2}}\phi^D_{-\frac{1}{2}}|0\rangle$, $n>0$. Again, first we consider the case $2i+2k>0$, when $:\phi^D_{-2i+\frac{1}{2}}\phi^D_{2i+2k-\frac{1}{2}}:=\phi^D_{-2i+\frac{1}{2}}\phi^D_{2i+2k-\frac{1}{2}}$.

$$r_1(E_{i,i+k})\phi^D_{-2n+2-\frac{1}{2}}\cdots\phi^D_{-2-\frac{1}{2}}\phi^D_{-\frac{1}{2}}|0\rangle=\phi^D_{-2i+\frac{1}{2}}\phi^D_{2i+2k-\frac{1}{2}}\phi^D_{-2n+2-\frac{1}{2}}\cdots\phi^D_{-2-\frac{1}{2}}\phi^D_{-\frac{1}{2}}|0\rangle.$$

Now since it is impossible to have $2i+2k-\frac{1}{2}=-(-2l+2-\frac{1}{2})$ for any $l\in\mathbb{Z}$, $\phi^D_{2i+2k-\frac{1}{2}}$ will anticommute with any $\phi^D_{-2l-\frac{1}{2}}$, and thus $\phi^D_{2i+2k-\frac{1}{2}}\phi^D_{-2n+2-\frac{1}{2}}\cdots\phi^D_{-2-\frac{1}{2}}\phi^D_{-\frac{1}{2}}|0\rangle=0$.

Next, let $2i+2k \le 0$, then : $\phi^D_{-2i+\frac{1}{2}}\phi^D_{2i+2k-\frac{1}{2}} := -\phi^D_{2i+2k-\frac{1}{2}}\phi^D_{-2i+\frac{1}{2}}$, and

$$r_1(E_{i,i+k})\phi^D_{-2n+2-\frac{1}{2}}\cdots\phi^D_{-2-\frac{1}{2}}\phi^D_{-\frac{1}{2}}|0\rangle = -\phi^D_{2i+2k-\frac{1}{2}}\phi^D_{-2i+\frac{1}{2}}\phi^D_{-2n+2-\frac{1}{2}}\cdots\phi^D_{-2-\frac{1}{2}}\phi^D_{-\frac{1}{2}}|0\rangle.$$

Now $\phi^D_{-2i+\frac{1}{2}}$ is an operator annihilating the vacuum, and unless we have $-2i+\frac{1}{2} = -(-2l-\frac{1}{2})$ for some $1 \le l \le n-1,\ l \in \mathbb{Z}$, then it will anticommute with any of the $\phi^D_{-2l-\frac{1}{2}},\ 1 \le l \le n-1$, and thus $\phi^D_{-2i+\frac{1}{2}}\phi^D_{-2n+2-\frac{1}{2}}\cdots\phi^D_{-2-\frac{1}{2}}\phi^D_{-\frac{1}{2}}|0\rangle = 0$. If, on the other hand $-2i+\frac{1}{2} = -(-2l-\frac{1}{2})$ for some $1 \le l \le n-1,\ l \in \mathbb{Z}$, then we have

$$\begin{aligned} r_1(E_{i,i+k})&\phi^D_{-2n+2-\frac{1}{2}}\cdots\phi^D_{-2-\frac{1}{2}}\phi^D_{-\frac{1}{2}}|0\rangle \\ &= -\phi^D_{2i+2k-\frac{1}{2}}\phi^D_{-2i+\frac{1}{2}}\phi^D_{-2n+2-\frac{1}{2}}\cdots\phi^D_{-2l+2-\frac{1}{2}}\cdots\phi^D_{-\frac{1}{2}}|0\rangle \\ &= \pm\phi^D_{2i+2k-\frac{1}{2}}\phi^D_{-2n+2-\frac{1}{2}}\cdots\widehat{\phi^D_{-2l-\frac{1}{2}}}\phi^D_{-2l+2-\frac{1}{2}}\cdots\phi^D_{-\frac{1}{2}}|0\rangle. \end{aligned}$$

But then from $-2i+1/2 = -(-2l-\frac{1}{2})$, we have $i=-l$ and $2i+2k-\frac{1}{2} = 2k-2l-\frac{1}{2}$, and we know from $i+k \le 0$ that $-2l+2-\frac{1}{2} \le 2i+2k-\frac{1}{2} \le -\frac{1}{2}$. That means $\phi^D_{2i+2k-\frac{1}{2}} = \phi^D_{-2l_1-\frac{1}{2}}$ for $0 \le l_1 \le l-1$ and

$$\begin{aligned} \phi^D_{2i+2k-\frac{1}{2}}&\phi^D_{-2n+2-\frac{1}{2}}\cdots\widehat{\phi^D_{-2l-\frac{1}{2}}}\phi^D_{-2l+2-\frac{1}{2}}\cdots\phi^D_{-\frac{1}{2}}|0\rangle \\ &= \phi^D_{2i+2k-\frac{1}{2}}\phi^D_{-2n+2-\frac{1}{2}}\cdots\widehat{\phi^D_{-2l-\frac{1}{2}}}\phi^D_{-2l+2-\frac{1}{2}}\cdots\phi^D_{-2l_1-\frac{1}{2}}\cdots\phi^D_{-\frac{1}{2}}|0\rangle \\ &= \phi^D_{-2l_1-\frac{1}{2}}\phi^D_{-2n+2-\frac{1}{2}}\cdots\widehat{\phi^D_{-2l-\frac{1}{2}}}\phi^D_{-2l+2-\frac{1}{2}}\cdots\phi^D_{-2l_1-\frac{1}{2}}\cdots\phi^D_{-\frac{1}{2}}|0\rangle \\ &= \pm\phi^D_{-2n+2-\frac{1}{2}}\cdots\widehat{\phi^D_{-2l-\frac{1}{2}}}\phi^D_{-2l+2-\frac{1}{2}}\cdots\phi^D_{-2l_1-\frac{1}{2}}\phi^D_{-2l_1-\frac{1}{2}}\cdots\phi^D_{-\frac{1}{2}}|0\rangle = 0; \end{aligned}$$

since $\phi^D_{-2l_1-\frac{1}{2}}\phi^D_{-2l_1-\frac{1}{2}} = 0$. □

Lemma 3.6 *(Calculating the weights) For all $i, n \in \mathbb{Z}$, $r_1(E_{ii})v_n = {}^a\hat{\Lambda}_n(E_{ii})v_n$.*

Proof: We want to prove

$r_1(E_{i,i})v_0 = 0 \cdot v_0$;
$r_1(E_{i,i})v_n = 1 \cdot v_n$ for $0 < i \le n$; $r_1(E_{i,i})v_n = 1 \cdot v_n$, for $i > n \ge 1$;
$r_1(E_{i,i})v_n = 0 \cdot v_n$ for $i \le 0,\ n > 0$;
$r_1(E_{i,i})v_{-n} = -1 \cdot v_{-n}$ for $-n+1 \le i \le 0$; $r_1(E_{i,i})v_{-n} = 0 \cdot v_{-n}$ for $-n+1 > i > 0$;
$r_1(E_{i,i})v_{-n} = 0 \cdot v_{-n}$ for $i > 0,\ n > 0$.

We have

$$r_1(E_{i,i}) =: \phi^D_{-2i+\frac{1}{2}}\phi^D_{2i-\frac{1}{2}} := \begin{cases} \phi^D_{-2i+\frac{1}{2}}\phi^D_{2i-\frac{1}{2}}, & \text{for } i > 0, \\ -\phi^D_{2i-\frac{1}{2}}\phi^D_{-2i+\frac{1}{2}}, & \text{for } i \le 0 \end{cases}$$

For $n=0$, $v_0 = |0\rangle$, and $r_1(E_{i,i}) =: \phi^D_{-2i+\frac{1}{2}}\phi^D_{2i-\frac{1}{2}} :$, and it is clear that $r_1(E_{i,i})|0\rangle = 0$.
Let $n > 0$ and $v_n = \phi^D_{-2n+1-\frac{1}{2}}\cdots\phi^D_{-3-\frac{1}{2}}\phi^D_{-1-\frac{1}{2}}|0\rangle$. If $i > 0$, $\phi^D_{2i-\frac{1}{2}}$ from $r_1(E_{i,i})$ will anticommute

with any $\phi^D_{-2l+1-\frac{1}{2}}$ in v_n, **except** for l such that $-2l+1-\frac{1}{2} = -(2i-\frac{1}{2})$, i.e., $l=i$. In the case there exist an l such that $l=i$, then

$$\begin{aligned} r_1(E_{i,i})v_n &= r_1(E_{i,i})\phi^D_{-2n+1-\frac{1}{2}}\cdots\phi^D_{-3-\frac{1}{2}}\phi^D_{-1-\frac{1}{2}}|0\rangle \\ &= \phi^D_{-2i+\frac{1}{2}}\phi^D_{2i-\frac{1}{2}}\phi^D_{-2n+1-\frac{1}{2}}\cdots\phi^D_{-2l+1-\frac{1}{2}}\cdots\phi^D_{-1-\frac{1}{2}}|0\rangle \\ &= (-1)^{n-l}\phi^D_{-2i+\frac{1}{2}}\phi^D_{-2n+1-\frac{1}{2}}\cdots\phi^D_{2i-\frac{1}{2}}\phi^D_{-2l+1-\frac{1}{2}}\cdots\phi^D_{-1-\frac{1}{2}}|0\rangle \\ &= (-1)^{n-l}\phi^D_{-2l+1-\frac{1}{2}}\phi^D_{-2n+1-\frac{1}{2}}\cdots\widehat{\phi^D_{-2l+1-\frac{1}{2}}}\cdots\phi^D_{-1-\frac{1}{2}}|0\rangle \\ &= \phi^D_{-2n+1-\frac{1}{2}}\cdots\phi^D_{-2l+1-\frac{1}{2}}\cdots\phi^D_{-1-\frac{1}{2}}|0\rangle = v_n \end{aligned}$$

Hence $r_1(E_{i,i})v_n = v_n$ for $i \le 0$ when $n \ge i$ (i.e., exist an l such that $l=i$), and $r_1(E_{i,i})v_n = 0$ for $i \le 0$ when $n < i$.

If $i \le 0$, then $r_1(E_{i,i}) = -\phi^D_{2i-\frac{1}{2}}\phi^D_{-2i+\frac{1}{2}}$ and $\phi^D_{-2i+\frac{1}{2}}$ will anticommute with **any** $\phi^D_{-2l+1-\frac{1}{2}}$ in v_n, as it is impossible to have $-2l+1-\frac{1}{2} = -(-2i+\frac{1}{2})$. Hence $r_1(E_{i,i})v_n = 0$ for $i \le 0$.

Let $n > 0$ and $v_{-n} = \phi^D_{-2n+2-\frac{1}{2}}\cdots\phi^D_{-2-\frac{1}{2}}\phi^D_{-\frac{1}{2}}|0\rangle$. If $i \le 0$, $\phi^D_{-2i+\frac{1}{2}}$ from $r_1(E_{i,i})$ will anticommute with any $\phi^D_{-2l+2-\frac{1}{2}}$ in v_{-n}, **except** for l such that $-2l+2-\frac{1}{2} = -(2i+\frac{1}{2})$, i.e., $l = -i-1$. In the case there exist an l such that $l=-i-1$, then

$$\begin{aligned} r_1(E_{i,i})v_n &= r_1(E_{i,i})\phi^D_{-2n+2-\frac{1}{2}}\cdots\phi^D_{-2-\frac{1}{2}}\phi^D_{-\frac{1}{2}}|0\rangle \\ &= -\phi^D_{2i-\frac{1}{2}}\phi^D_{-2i+\frac{1}{2}}\phi^D_{-2n+2-\frac{1}{2}}\cdots\phi^D_{-2l+2-\frac{1}{2}}\cdots\phi^D_{-\frac{1}{2}}|0\rangle \\ &= -(-1)^{n-l}\phi^D_{2i-\frac{1}{2}}\phi^D_{-2n+2-\frac{1}{2}}\cdots\phi^D_{-2i+\frac{1}{2}}\phi^D_{-2l+2-\frac{1}{2}}\cdots\phi^D_{-\frac{1}{2}}|0\rangle \\ &= -(-1)^{n-l}\phi^D_{-2l+2-\frac{1}{2}}\phi^D_{-2n+2-\frac{1}{2}}\cdots\widehat{\phi^D_{-2l+2-\frac{1}{2}}}\cdots\phi^D_{-\frac{1}{2}}|0\rangle \\ &= -\phi^D_{-2n+2-\frac{1}{2}}\cdots\phi^D_{-2l+2-\frac{1}{2}}\cdots\phi^D_{-\frac{1}{2}}|0\rangle = v_n \end{aligned}$$

Hence $r_1(E_{i,i})v_{-n} = -v_{-n}$ for $i \le 0$ when $n \ge -i-1$ (i.e., exist an l such that $l=-i-1$), and $r_1(E_{i,i})v_{-n} = 0$ for $i \le 0$ when $n < -i-1$.

If $i > 0$, then $r_1(E_{i,i}) = \phi^D_{-2i+\frac{1}{2}}\phi^D_{2i-\frac{1}{2}}$ and $\phi^D_{2i-\frac{1}{2}}$ will anticommute with **any** $\phi^D_{-2l+2-\frac{1}{2}}$, as it is impossible to have $-2l+2-\frac{1}{2} = -(2i-\frac{1}{2})$. Hence $r_1(E_{i,i})v_{-n} = 0$ for $i > 0$. □

Proposition 3.7 *For any $n \in \mathbb{Z}$, $F^{\otimes\frac{1}{2}}_{(n)}$ is an a_∞-submodule of $F^{\otimes\frac{1}{2}}$ and $F^{\otimes\frac{1}{2}}_{(n)} = U(a^-_\infty)v_n$.*

Proof: First, $F^{\otimes\frac{1}{2}}_{(n)}$ is an a_∞-submodule of $F^{\otimes\frac{1}{2}}$ as $r_1(E_{ij})$ is a homogenous operator acting on $F^{\otimes\frac{1}{2}}$ of degree 0 with respect to the grading given in Lemma 3.1 and in (3.5). Then certainly that gives us $U(a^-_\infty)v_n \subseteq F^{\otimes\frac{1}{2}}_{(n)}$. The proof that $F^{\otimes\frac{1}{2}}_{(n)} = U(a^-_\infty)v_n$ is similar for each $n \in \mathbb{Z}$, thus we will show it only for $n=0$. Let $v \in F^{\otimes\frac{1}{2}}_{(0)}$, without loss of generality here we can assume v is homogeneous, i.e. $v = \phi^D_{-n_k-\frac{1}{2}}\cdots\phi^D_{-n_2-\frac{1}{2}}\phi^D_{-n_1-\frac{1}{2}}|0\rangle$. Since $v \in F^{\otimes\frac{1}{2}}_{(0)}$, we have $k=2l$ and precisely half of the indexes $n_1, n_2, \ldots, n_k$ are even, the other half are odd. Thus we can write after eventual use of the anticommutation relations in Cl_D

$$v = \pm\phi^D_{-n^o_l-\frac{1}{2}}\phi^D_{-n^e_l-\frac{1}{2}}\cdots\phi^D_{-n^o_1-\frac{1}{2}}\phi^D_{-n^e_1-\frac{1}{2}}|0\rangle,$$

i.e., we have rearranged the factors in pairs, so that the indexes in the pairs are n_s^e and n_s^o, and we have $n_l^o > \cdots > n_1^o$, $n_l^e > \cdots > n_1^e$. Hence we have $n_s^e = 2q_s$ for some $q_s \in \mathbb{Z}$ and $n_s^o = 2p_s - 1$ for some $p_s \in \mathbb{Z}$. Thus

$$v = \pm\phi^D_{-n_l^o-\frac{1}{2}}\phi^D_{-n_l^e-\frac{1}{2}}\cdots\phi^D_{-n_1^o-\frac{1}{2}}\phi^D_{-n_1^e-\frac{1}{2}}|0\rangle = \pm r_1(E_{p_l,-q_l})\cdots r_1(E_{p_1,-q_1}|0\rangle;$$

which proves that $F^{\otimes\frac{1}{2}}_{(0)} = U(a_\infty^-)v_0$. The proof that $F^{\otimes\frac{1}{2}}_{(n)} = U(a_\infty^-)v_n$ is very similar for the other $n \in \mathbb{Z}$ and we will omit it. □

Proposition 3.8 *For any $n \in \mathbb{Z}$, $F^{\otimes\frac{1}{2}}_{(n)}$ is an irreducible submodule for the representation r_1 of a_∞ inside $F^{\otimes\frac{1}{2}}$. Moreover for any $n \in \mathbb{Z}$, $F^{\otimes\frac{1}{2}}_{(n)} \cong L(a_\infty; {}^a\hat{\Lambda}_n, 1)$.*

Proof: This proof uses the uniqueness property of the contragradient Hermitian symmetric form on the Verma module $V(a_\infty; {}^a\hat{\Lambda}_n, 1)$ (see [Jan79], [Kac90] or [MP95]). For a more direct, but calculational proof the reader can see the Appendix. It is well known that one can define ω the conjugate linear involutive anti-automorphism on Cl_D by $\omega(\phi^D_m) = \phi^D_{-m}$ for all $m \in \mathbb{Z} + \frac{1}{2}$ and $\omega(1) = 1$. Recall the module $F^{\otimes\frac{1}{2}}$ is defined to be the induced module

$$F^{\otimes\frac{1}{2}} = Cl_D \otimes_{\mathbb{C}1\otimes Cl_{D+}} \mathbb{C}|0\rangle$$

whereby $1|0\rangle = |0\rangle$ and $Cl_{D+}|0\rangle = 0$. The conjugate linear involutive anti-automorphism (or antilinear antiautomorphism) $\omega : Cl_D \to Cl_D$ defined by $\omega(\phi_n) = \phi_{-n}$ gives rise to a non-degenerate positive definite form $\langle \mid \rangle$ defined on $F^{\otimes\frac{1}{2}}$ whereby

$$\langle Xv \,|\, w\rangle = \langle v \,|\, \omega(X)w\rangle$$

for all $v, w \in F^{\otimes\frac{1}{2}}$ and $X \in Cl_D$. Observe that $F^{\otimes\frac{1}{2}}{}_{(n)} \perp F^{\otimes\frac{1}{2}}{}_{(m)}$ for $m \neq n$.

It is straightforward to check that

$$\omega(r_1(E_{m,n})) = r_1(E_{n,m}) = r_1(\omega(E_{m,n})). \tag{3.26}$$

Thus ω defined on Cl_D agrees with the compact anti-involution defined on a_∞ given earlier.

We have from Proposition 3.5, Lemma 3.6, and Proposition 3.7 that $F^{\otimes\frac{1}{2}}_{(n)}$ is a highest weight a_∞-module. By the universal mapping property of Verma modules there exists an a_∞-module homomorphism $\pi : V(a_\infty; {}^a\hat{\Lambda}_n, 1) \to F^{\otimes\frac{1}{2}}_{(n)}$ sending the highest weight vector $v_{{}^a\hat{\Lambda}_n}$ of the Verma module $V(a_\infty; {}^a\hat{\Lambda}_n, 1)$ to v_n in $F^{\otimes\frac{1}{2}}_{(n)}$. Moreover the Hermitian symmetric ω-contragradient form $\langle\ ,\ \rangle$ on $F^{\otimes\frac{1}{2}}_{(n)}$ pulls back to a Hermitian symmetric form $(\ ,\ )$ on the Verma module $V(a_\infty; {}^a\hat{\Lambda}_n, 1)$. In other words

$$(uv_{{}^a\hat{\Lambda}_n}, u'v_{{}^a\hat{\Lambda}_n}) := \langle\pi(uv_{{}^a\hat{\Lambda}_n}), \pi(u'v_{{}^a\hat{\Lambda}_n})\rangle = \langle r_1(u)v_n, r_1(u')v_n\rangle. \tag{3.27}$$

for all $u, u' \in a_\infty$. By (3.26), we have

$$\begin{aligned}(Xuv_{{}^a\hat{\Lambda}_n}, u'v_{{}^a\hat{\Lambda}_n}) &= \langle r_1(X)r_1(u)v_n, r_1(u')v_n\rangle \\ &= \langle r_1(u)v_n, \omega(r_1(X))r_1(u')v_n\rangle \\ &= \langle r_1(u)v_n, r_1(\omega(X)u')v_n\rangle \\ &= (uv_{{}^a\hat{\Lambda}_n}, \omega(X)u'v_{{}^a\hat{\Lambda}_n})\end{aligned}$$

Hence (,) is contragradient with respect to ω.

Now it is known that there is a unique contragradient Hermitian symmetric form on the Verma module $V(a_\infty; {}^a\hat{\Lambda}_n, 1)$ with $(v_{{}^a\hat{\Lambda}_n}, v_{{}^a\hat{\Lambda}_n}) = \langle v_n, v_n \rangle$ and its radical is precisely the unique maximal submodule $\overline{V(a_\infty; {}^a\hat{\Lambda}_n, 1)}$ of $V(a_\infty; {}^a\hat{\Lambda}_n, 1)$ (see [Jan79], [Kac90] or [MP95]). So

$$0 = (\overline{V(a_\infty; {}^a\hat{\Lambda}_n, 1)}, u' v_{{}^a\hat{\Lambda}_n}) := \langle \pi(\overline{V(a_\infty; {}^a\hat{\Lambda}_n, 1)}), \pi(u' v_{{}^a\hat{\Lambda}_n}) \rangle \tag{3.28}$$

for all $u' \in U(a_\infty)$. Since $\langle\ ,\ \rangle$ is nondegenerate and π is surjective one must have $\overline{V(a_\infty; {}^a\hat{\Lambda}_n, 1)} \subseteq \ker \pi$. Thus $L(a_\infty; {}^a\hat{\Lambda}_n, 1) = V(a_\infty; {}^a\hat{\Lambda}_n, 1)/\overline{V(a_\infty; {}^a\hat{\Lambda}_n, 1)} \cong F^{\otimes \frac{1}{2}}{}_{(n)}$. □

Remark 3.9 An alternative proof of the irreducibility of $F^{\otimes \frac{1}{2}}_{(n)}$ can be given as follows. In [Ang12] we considered the Heisenberg algebra $\mathcal{H}_\mathbb{Z}$, which is a subalgebra of a_∞ represented by

$$h_n \mapsto \sum_{i \in \mathbb{Z}} E_{i,i+n}$$

In [Ang12] we prove that each $F^{\otimes \frac{1}{2}}_{(n)}$ is irreducible under $\mathcal{H}_\mathbb{Z}$. Since $\mathcal{H}_\mathbb{Z}$ is a subalgebra of a_∞, then $F^{\otimes \frac{1}{2}}_{(n)}$ is irreducible under a_∞.

The previous three propositions can now be combined in the following

Theorem 3.10 *As a_∞ modules with central charge $c = 1$ $F^{\otimes \frac{1}{2}}_{(n)} \cong L(a_\infty; {}^a\hat{\Lambda}_n, 1)$ and $F^{\otimes \frac{1}{2}}$ decomposes into irreducible submodules as follows:*

$$F^{\otimes \frac{1}{2}} \cong \oplus_{n \in \mathbb{Z}} L(a_\infty; {}^a\hat{\Lambda}_n, 1).$$

Hence $F^{\otimes \frac{1}{2}} \cong F^{\otimes 1}$ as a_∞ modules with central charge $c = 1$.

Here $F^{\otimes 1}$ denotes the fermionic Fock space of the charged free fermions (see Remark 3.2 above, after [Fre81], [KR87], [KW94], [Kac98], [KWY98], [Wan99a]).

Since $F^{\otimes \frac{1}{2}} \cong F^{\otimes 1}$ as a_∞ modules with central charge $c = 1$, we can use Theorem 3.2 of [Wan99b] to get the decomposition of the irreducible central charge $c = \frac{1}{2}$ d_∞ modules $F^{\otimes \frac{1}{2}}_{\bar{0}}$ and $F^{\otimes \frac{1}{2}}_{\bar{1}}$ in terms of the new d_∞ central charge $c = 1$ action:

Corollary 3.11 *As d_∞ modules with central charge $c = 1$ we have*

$$F^{\otimes \frac{1}{2}}_{(n)} \cong L(d_\infty; {}^{\mathfrak{d}\mathfrak{d}}\hat{\Lambda}_n, 1) \quad \textit{for } n \neq 0 \qquad F^{\otimes \frac{1}{2}}_{(0)} \cong L(d_\infty; {}^{\mathfrak{d}\mathfrak{d}}\hat{\Lambda}_0, 1) \oplus L(d_\infty; {}^{\mathfrak{d}\mathfrak{d}}\hat{\Lambda}_{det}, 1); \tag{3.29}$$

where for $n \neq 0$ $F^{\otimes \frac{1}{2}}_{(n)}$ has highest weight vector v_n, and $F^{\otimes \frac{1}{2}}_{(0)}$ decomposes into two irreducible highest weight modules, with highest weight vectors v_0 and $\tilde{v}_0 = \phi^D_{-\frac{3}{2}} \phi^D_{-\frac{1}{2}} |0\rangle \in F^{\otimes \frac{1}{2}}_{(0)}$ (note $\tilde{v}_0$ is not a highest weight vector for the a_∞ action, only the d_∞ action). Thus as d_∞ modules with central charge $c = 1$

$$F^{\otimes \frac{1}{2}} = \left(\oplus_{\substack{n \in \mathbb{Z} \\ n \neq 0}} L(d_\infty; {}^{\mathfrak{d}\mathfrak{d}}\hat{\Lambda}_n, 1) \right) \bigoplus \left(L(d_\infty; {}^{\mathfrak{d}\mathfrak{d}}\hat{\Lambda}_0, 1) \oplus L(d_\infty; {}^{\mathfrak{d}\mathfrak{d}}\hat{\Lambda}_{det}, 1) \right),$$

and

$$F^{\otimes \frac{1}{2}}_{\bar{1}} = \oplus_{n \in \mathbb{Z}} L(d_\infty; {}^{\mathfrak{d}\mathfrak{d}}\hat{\Lambda}_{2n-1}, 1);$$

$$F^{\otimes \frac{1}{2}}_{\bar{0}} = \left(\oplus_{\substack{n \in \mathbb{Z} \\ n \neq 0}} L(d_\infty; {}^{\mathfrak{d}\mathfrak{d}}\hat{\Lambda}_{2n}, 1) \right) \bigoplus \left(L(d_\infty; {}^{\mathfrak{d}\mathfrak{d}}\hat{\Lambda}_0, 1) \oplus L(d_\infty; {}^{\mathfrak{d}\mathfrak{d}}\hat{\Lambda}_{det}, 1) \right).$$

Here, as in [Wan99b], the highest weights ${}^{\mathfrak{dd}}\hat{\Lambda}_n$ are obtained from the restrictions of ${}^{a}\hat{\Lambda}_n$, except for ${}^{\mathfrak{dd}}\hat{\Lambda}_{det}$ which is defined by

$${}^{\mathfrak{dd}}\hat{\Lambda}_{det}(E_{i,i} - E_{1-i,1-i}) = 2; \quad \text{for } i = 1, \quad {}^{\mathfrak{dd}}\hat{\Lambda}_{det}(E_{i,i} - E_{1-i,1-i}) = 0 \quad \text{for } i \neq 0, 1.$$

Proof: Follows directly from Theorem 3.2 of [Wan99b] and Theorem 3.10. □

4. Appendix

For readers who would like to see a more computational proof of Proposition 3.8 we present one below.

Alternate proof of Proposition 3.8: First we will prove that the action of a_∞ on $F^{\otimes \frac{1}{2}}$ will preserve the $\mathbb{Z}$ grading gd; that will show that each $F^{\otimes \frac{1}{2}}_{(n)}$ is a submodule for the r_1 action of a_∞ on $F^{\otimes \frac{1}{2}}$. Then we will show that the submodule $F^{\otimes \frac{1}{2}}_{(n)}$ is irreducible.

Let $v = \phi^D_{-n_k-\frac{1}{2}} \dots \phi^D_{-n_2-\frac{1}{2}}\phi^D_{-n_1-\frac{1}{2}}|0\rangle$ be any homogeneous vector in $F^{\otimes \frac{1}{2}}_{(n)}$. We have

$$r_1(E_{p,q}) =: \phi^D_{-2p+\frac{1}{2}}\phi^D_{2q-\frac{1}{2}} := \begin{cases} \phi^D_{-2p+\frac{1}{2}}\phi^D_{2q-\frac{1}{2}}, & \text{unless } q \leq 0 \text{ and } 2q - \frac{1}{2} = -(-2p + \frac{1}{2}) \\ -\phi^D_{2q-\frac{1}{2}}\phi^D_{-2p+\frac{1}{2}}, & \text{otherwise.} \end{cases}$$

Consider first the case when $-(2q - \frac{1}{2}) = -2p + \frac{1}{2}$ and $q \leq 0$. Then $-2p + \frac{1}{2} > 0$, and in

$$r_1(E_{p,q})v = -\phi^D_{2q-\frac{1}{2}}\phi^D_{-2p+\frac{1}{2}}\phi^D_{-n_k-\frac{1}{2}} \cdots \phi^D_{-n_2-\frac{1}{2}}\phi^D_{-n_1-\frac{1}{2}}|0\rangle$$

$\phi^D_{-2p+\frac{1}{2}}$ will either anticommute with all $\phi^D_{-n_s-\frac{1}{2}}$, $s = 1, \dots, k$, in which case $r_1(E_{p,q})v = 0$ as $\phi^D_{-2p+\frac{1}{2}}|0\rangle = 0$; or otherwise we will have $-2p + \frac{1}{2} = -(-n_s - \frac{1}{2})$, for some $s = 1, \dots, k$. In that case we also have $-n_s - \frac{1}{2} = 2q - \frac{1}{2}$ and

$$\begin{aligned} r_1(E_{p,q})v &= -\phi^D_{2q-\frac{1}{2}}\phi^D_{-2p+\frac{1}{2}}\phi^D_{-n_k-\frac{1}{2}} \cdots \phi^D_{-n_s-\frac{1}{2}} \cdots \phi^D_{-n_1-\frac{1}{2}}|0\rangle \\ &= \pm\phi^D_{2q-\frac{1}{2}}\phi^D_{-n_k-\frac{1}{2}} \cdots \widehat{\phi^D_{-n_s-\frac{1}{2}}} \cdots \phi^D_{-n_1-\frac{1}{2}}|0\rangle \\ &= \pm\phi^D_{-n_k-\frac{1}{2}} \cdots \phi^D_{-n_s-\frac{1}{2}} \cdots \phi^D_{-n_1-\frac{1}{2}}|0\rangle. \end{aligned}$$

This shows that in both cases when $-(2q - \frac{1}{2}) = -2p + \frac{1}{2}$ we have if $v \in F^{\otimes \frac{1}{2}}_{(n)}$ then $r_1(E_{p,q})v \in F^{\otimes \frac{1}{2}}_{(n)}$.

Next, consider the case when $-(2q - \frac{1}{2}) \neq -2p + \frac{1}{2}$, but still $q \leq 0$, which implies again that $r_1(E_{p,q}) = -\phi^D_{2q-\frac{1}{2}}\phi^D_{-2p+\frac{1}{2}}$ (anticommutation) and again we consider the possible cases for

$$r_1(E_{p,q})v = -\phi^D_{2q-\frac{1}{2}}\phi^D_{-2p+\frac{1}{2}}\phi^D_{-n_k-\frac{1}{2}} \cdots \phi^D_{-n_2-\frac{1}{2}}\phi^D_{-n_1-\frac{1}{2}}|0\rangle.$$

The first case is when $p < 0$ and $\phi^D_{-2p+\frac{1}{2}}$ anticommutes with all $\phi^D_{-n_s-\frac{1}{2}}$, $s = 1, \dots, k$, in which case $r_1(E_{p,q})v = 0$. The second case is again as above when $-2p + \frac{1}{2} = -(-n_s - \frac{1}{2})$, for some $s = 1, \dots, k$. We again have

$$\begin{aligned} r_1(E_{p,q})v &= -\phi^D_{2q-\frac{1}{2}}\phi^D_{-2p+\frac{1}{2}}\phi^D_{-n_k-\frac{1}{2}} \cdots \phi^D_{-n_s-\frac{1}{2}} \cdots \phi^D_{-n_1-\frac{1}{2}}|0\rangle \\ &= \pm\phi^D_{2q-\frac{1}{2}}\phi^D_{-n_k-\frac{1}{2}} \cdots \widehat{\phi^D_{-n_s-\frac{1}{2}}} \cdots \phi^D_{-n_1-\frac{1}{2}}|0\rangle. \end{aligned}$$

In other words, we have "removed" $\phi^D_{-n_s-\frac{1}{2}}$ with **even** index $n_s = 2p$. Now since $q < 0$, we either have $2q - \frac{1}{2} = -n_t - \frac{1}{2}$ for some $t = 1, \ldots, k, t \neq s$ in which case again we have $r_1(E_{p,q})v = 0$ as $\phi^D_{-2q-\frac{1}{2}}\phi^D_{-2q-\frac{1}{2}} = 0$. Or otherwise we have "added" $\phi^D_{-n_t-\frac{1}{2}}$ with **even** index $n_t = -2q$, which doesn't change the degree dg, as we have "removed" an even index $n_s = 2p$ and "added" an even index $n_t = -2q$. Thus in all cases when $q < 0$ we have if $v \in F^{\otimes\frac{1}{2}}_{(n)}$ then $r_1(E_{p,q})v \in F^{\otimes\frac{1}{2}}_{(n)}$.

Lastly, let $q \geq 1$. Similar argument as above show that either $r_1(E_{p,q})v = 0$, or

$$\begin{aligned} r_1(E_{p,q})v &= \phi^D_{-2p+\frac{1}{2}}\phi^D_{2q-\frac{1}{2}}\phi^D_{-n_k-\frac{1}{2}} \cdots \phi^D_{-n_t-\frac{1}{2}} \cdots \phi^D_{-n_1-\frac{1}{2}}|0\rangle \\ &= \pm\phi^D_{-2p+1-\frac{1}{2}}\phi^D_{-n_k-\frac{1}{2}} \cdots \widehat{\phi^D_{-n_t-\frac{1}{2}}} \cdots \phi^D_{-n_1-\frac{1}{2}}|0\rangle; \end{aligned}$$

where $2q - \frac{1}{2} = -(-n_t - \frac{1}{2})$ for some $t = 1, \ldots, k, t \neq s$. Thus $n_t = 2q - 1$ and we have "removed" an **odd** index n_t. Now if $-2p + 1 - \frac{1}{2} < 0$, we are "adding" back an **odd** index $-2p + 1$. If, on the other hand $-2p + 1 - \frac{1}{2} > 0$, then either we get 0, or we remove also an **even** index. Thus as a summary, in all cases we either remove an odd (even) index and add an odd (even) index back; we remove both an odd and an even index; or we get 0. Hence in all cases the action of $r_1(E_{p,q})$ will preserve the grading. Hence each $F^{\otimes\frac{1}{2}}_{(n)}$ is a submodule for the representation r_1 of a_∞ on $F^{\otimes\frac{1}{2}}$.

Further, from the observations above we can summarize the action r_1 on $F^{\otimes\frac{1}{2}}$ as follows. We have 3 nontrivial cases: in case 1 $r_1(E_{p,q})$ acting on a homogeneous vector v "**adds**" two factors $\phi^D_{-n_{s_1}-\frac{1}{2}}$ and $\phi^D_{-n_{s_2}-\frac{1}{2}}$, so that one of the indexes n_{s_1}, n_{s_2} is **even** , the other is **odd** (we will call it for short "adding an even and an odd index"). In case 2, we **replace** a factor $\phi^D_{-n_{s_1}-\frac{1}{2}}$ with another factor $\phi^D_{-n_{s_2}-\frac{1}{2}}$, where either both factors n_{s_1}, n_{s_2} are **even**, or both factors n_{s_1}, n_{s_2} are **odd** (we will call it for short "replacing even with even index" and "replacing odd with odd index"). And case 3 is when we "**remove**" two factors $\phi^D_{-n_{s_1}-\frac{1}{2}}$ and $\phi^D_{-n_{s_2}-\frac{1}{2}}$, so that one of the indexes n_{s_1}, n_{s_2} is **even** , the other is **odd** ("removing an even and an odd index"). Not that "adding an index" that is already present, or "removing an index" that was absent, will of course produce the 0 vector.

Now we want to prove that for each $n \in \mathbb{Z}$ $F^{\otimes\frac{1}{2}}_{(n)}$ is an irreducible module for a_∞. This will be done in two steps. The first step is to prove that each vector in $F^{\otimes\frac{1}{2}}_{(n)}$ can be generated from the "n-th vacuum vector" v_n, i.e., $F^{\otimes\frac{1}{2}}_{(n)} = U(a^-_\infty)v_n$. The second step is to prove that for any vector $v \in F^{\otimes\frac{1}{2}}_{(n)}$, we have $v_n \in U(a_\infty)v$.

The proof that $F^{\otimes\frac{1}{2}}_{(n)} = U(a^-_\infty)v_n$ is done in Proposition 3.7. Finally, we prove step 2, i.e., for any vector $v \in F^{\otimes\frac{1}{2}}_{(n)}$, we have $v_n \in U(a_\infty)v$. Let $v \in F^{\otimes\frac{1}{2}}_{(n)}$, not necessary homogeneous: $v = \sum_{hk} c_{hk}v^{hk}$, where v^{hk} are homogeneous vectors,

$$v^{hk} = \phi^D_{-n_k-\frac{1}{2}} \cdots \phi^D_{-n_2\ \frac{1}{2}}\phi^D_{-n_1-\frac{1}{2}}|0\rangle, \quad c_k \in \mathbb{C}.$$

It is clear that by the operation "adding an even and an odd index" we can reduce v only to a linear combination of homogeneous elements with minimal possible length $\tilde{L}$ among the v^{hk} by the following two steps: we would "add an even and an odd index" starting from the already existing indexes in the vector v^{hk} with the largest length $\tilde{L}$ (which will annihilate it), and then we would remove the same combination back (which will bring the elements with lower length back to their original length). This is always possible, as any two lengths within $F_{(n)}^{\otimes\frac{1}{2}}$ differ always by an even number; and we always have at least two differing elements $\phi^D_{-n_k-\frac{1}{2}}$ between the homogeneous elements with the two consecutive lengths, as well as two differing elements $\phi^D_{-n_k-\frac{1}{2}}$ between homogeneous elements with same lengths. Now among the remaining different vectors v^{hk} with minimal possible length $\tilde{L}$ we can use the operations of "replacing even index with even" and "replacing odd index with odd" until only a single v^{hk} with largest index n_k being the minimal possible remains. We will show how this algorithm on an example: consider a vector $v \in F_{(0)}^{\otimes\frac{1}{2}}$ that will illustrate these 3 steps,

$$\begin{aligned} v = {} & \phi^D_{-7-\frac{1}{2}}\phi^D_{-4-\frac{1}{2}}\phi^D_{-3-\frac{1}{2}}\phi^D_{-2-\frac{1}{2}}|0\rangle + \phi^D_{-5-\frac{1}{2}}\phi^D_{-3-\frac{1}{2}}\phi^D_{-2-\frac{1}{2}}\phi^D_{-0-\frac{1}{2}}|0\rangle \\ & + \phi^D_{-8-\frac{1}{2}}\phi^D_{-5-\frac{1}{2}}|0\rangle + \phi^D_{-6-\frac{1}{2}}\phi^D_{-1-\frac{1}{2}}|0\rangle + \phi^D_{-5-\frac{1}{2}}\phi^D_{-4-\frac{1}{2}}|0\rangle + \phi^D_{-9-\frac{1}{2}}\phi^D_{-8-\frac{1}{2}}|0\rangle. \end{aligned}$$

We have Step 1:

$$\begin{aligned} E_{4,-2}v = {} & \phi^D_{-7-\frac{1}{2}}\phi^D_{-4-\frac{1}{2}}v = \phi^D_{-7-\frac{1}{2}}\phi^D_{-4-\frac{1}{2}}\phi^D_{-7-\frac{1}{2}}\phi^D_{-4-\frac{1}{2}}\phi^D_{-3-\frac{1}{2}}\phi^D_{-2-\frac{1}{2}}|0\rangle \\ & + \phi^D_{-7-\frac{1}{2}}\phi^D_{-4-\frac{1}{2}}\phi^D_{-5-\frac{1}{2}}\phi^D_{-3-\frac{1}{2}}\phi^D_{-2-\frac{1}{2}}\phi^D_{-0-\frac{1}{2}}|0\rangle \\ & + \phi^D_{-7-\frac{1}{2}}\phi^D_{-4-\frac{1}{2}}\phi^D_{-8-\frac{1}{2}}\phi^D_{-5-\frac{1}{2}}|0\rangle + \phi^D_{-7-\frac{1}{2}}\phi^D_{-4-\frac{1}{2}}\phi^D_{-6-\frac{1}{2}}\phi^D_{-1-\frac{1}{2}}|0\rangle \\ & + \phi^D_{-7-\frac{1}{2}}\phi^D_{-4-\frac{1}{2}}\phi^D_{-5-\frac{1}{2}}\phi^D_{-4-\frac{1}{2}}|0\rangle + \phi^D_{-7-\frac{1}{2}}\phi^D_{-4-\frac{1}{2}}\phi^D_{-9-\frac{1}{2}}\phi^D_{-8-\frac{1}{2}}|0\rangle \\ = {} & \phi^D_{-7-\frac{1}{2}}\phi^D_{-4-\frac{1}{2}}\phi^D_{-5-\frac{1}{2}}\phi^D_{-3-\frac{1}{2}}\phi^D_{-2-\frac{1}{2}}\phi^D_{-0-\frac{1}{2}}|0\rangle \\ & + \phi^D_{-7-\frac{1}{2}}\phi^D_{-4-\frac{1}{2}}\phi^D_{-8-\frac{1}{2}}\phi^D_{-5-\frac{1}{2}}|0\rangle + \phi^D_{-7-\frac{1}{2}}\phi^D_{-4-\frac{1}{2}}\phi^D_{-6-\frac{1}{2}}\phi^D_{-1-\frac{1}{2}}|0\rangle \\ & + \phi^D_{-7-\frac{1}{2}}\phi^D_{-4-\frac{1}{2}}\phi^D_{-9-\frac{1}{2}}\phi^D_{-8-\frac{1}{2}}|0\rangle; \end{aligned}$$

Step 2:

$$\begin{aligned} E_{-2,4}E_{4,-2}v = {} & \phi^D_{4+\frac{1}{2}}\phi^D_{7+\frac{1}{2}}\phi^D_{-7-\frac{1}{2}}\phi^D_{-4-\frac{1}{2}}v \\ = {} & \phi^D_{4+\frac{1}{2}}\phi^D_{7+\frac{1}{2}}\phi^D_{-7-\frac{1}{2}}\phi^D_{-4-\frac{1}{2}}\phi^D_{-5-\frac{1}{2}}\phi^D_{-3-\frac{1}{2}}\phi^D_{-2-\frac{1}{2}}\phi^D_{-0-\frac{1}{2}}|0\rangle \\ & + \phi^D_{4+\frac{1}{2}}\phi^D_{7+\frac{1}{2}}\phi^D_{-7-\frac{1}{2}}\phi^D_{-4-\frac{1}{2}}\phi^D_{-8-\frac{1}{2}}\phi^D_{-5-\frac{1}{2}}|0\rangle \\ & + \phi^D_{4+\frac{1}{2}}\phi^D_{7+\frac{1}{2}}\phi^D_{-7-\frac{1}{2}}\phi^D_{-4-\frac{1}{2}}\phi^D_{-6-\frac{1}{2}}\phi^D_{-1-\frac{1}{2}}|0\rangle \\ & + \phi^D_{4+\frac{1}{2}}\phi^D_{7+\frac{1}{2}}\phi^D_{-7-\frac{1}{2}}\phi^D_{-4-\frac{1}{2}}\phi^D_{-9-\frac{1}{2}}\phi^D_{-8-\frac{1}{2}}|0\rangle \\ = {} & \phi^D_{-5-\frac{1}{2}}\phi^D_{-3-\frac{1}{2}}\phi^D_{-2-\frac{1}{2}}\phi^D_{-0-\frac{1}{2}}|0\rangle \\ & + \phi^D_{-8-\frac{1}{2}}\phi^D_{-5-\frac{1}{2}}|0\rangle + \phi^D_{-6-\frac{1}{2}}\phi^D_{-1-\frac{1}{2}}|0\rangle + \phi^D_{-9-\frac{1}{2}}\phi^D_{-8-\frac{1}{2}}|0\rangle; \end{aligned}$$

We repeat Step 1:

$$\begin{aligned} E_{3,-1}E_{-2,4}E_{4,-2}v = {} & \phi^D_{-5-\frac{1}{2}}\phi^D_{-2-\frac{1}{2}}E_{-2,4}E_{4,-2}v \\ = {} & \phi^D_{-5-\frac{1}{2}}\phi^D_{-2-\frac{1}{2}}\phi^D_{-5-\frac{1}{2}}\phi^D_{-3-\frac{1}{2}}\phi^D_{-2-\frac{1}{2}}\phi^D_{-0-\frac{1}{2}}|0\rangle \\ & + \phi^D_{-5-\frac{1}{2}}\phi^D_{-2-\frac{1}{2}}\phi^D_{-8-\frac{1}{2}}\phi^D_{-5-\frac{1}{2}}|0\rangle + \phi^D_{-5-\frac{1}{2}}\phi^D_{-2-\frac{1}{2}}\phi^D_{-6-\frac{1}{2}}\phi^D_{-1-\frac{1}{2}}|0\rangle \\ & + \phi^D_{-5-\frac{1}{2}}\phi^D_{-2-\frac{1}{2}}\phi^D_{-9-\frac{1}{2}}\phi^D_{-8-\frac{1}{2}}|0\rangle \\ = {} & \phi^D_{-5-\frac{1}{2}}\phi^D_{-2-\frac{1}{2}}\phi^D_{-6-\frac{1}{2}}\phi^D_{-1-\frac{1}{2}}|0\rangle + \phi^D_{-5-\frac{1}{2}}\phi^D_{-2-\frac{1}{2}}\phi^D_{-9-\frac{1}{2}}\phi^D_{-8-\frac{1}{2}}|0\rangle. \end{aligned}$$

We repeat Step 2:

$$E_{-1,3}E_{3,-1}E_{-2,4}E_{4,-2}v = \phi^D_{-6-\frac{1}{2}}\phi^D_{-1-\frac{1}{2}}|0\rangle + \phi^D_{-9-\frac{1}{2}}\phi^D_{-8-\frac{1}{2}}|0\rangle.$$

Finally Step 3:

$$\begin{aligned} E_{1,5}E_{-1,3}E_{3,-1}E_{-2,4}E_{4,-2}v &= \phi^D_{-1-\frac{1}{2}}\phi^D_{9+\frac{1}{2}}\phi^D_{-6-\frac{1}{2}}\phi^D_{-1-\frac{1}{2}}|0\rangle + \phi^D_{-1-\frac{1}{2}}\phi^D_{9+\frac{1}{2}}\phi^D_{-9-\frac{1}{2}}\phi^D_{-8-\frac{1}{2}}|0\rangle \\ &= \phi^D_{-8-\frac{1}{2}}\phi^D_{-1-\frac{1}{2}}|0\rangle. \end{aligned}$$

Hence similarly we can reduce any potentially nonhomogeneous vector v we started with to a homogeneous vector v^{hk} by successive action of $r_1(E_{pq})$. Thus we consider the homogeneous vector $v^{hk} = \phi^D_{-n_k-\frac{1}{2}}\dots\phi^D_{-n_2-\frac{1}{2}}\phi^D_{-n_1-\frac{1}{2}}|0\rangle$. If v^{hk} has length $\tilde{L}(v^{hk}) > |n|$, then $\tilde{L}(v^{hk}) = |n|+2l$ and we can use the the operation of "removing an even and an odd index" l times in succession until we get a vector of minimal length $\tilde{L} = |n|$. After that we just have to eventually "replace some even indexes with even" and "replace some odd indexes with odd" to produce the highest weight vector v_n. Hence, we have proved that $v_n \in U(a_\infty)v$ for any $v \in F^{\otimes\frac{1}{2}}_{(n)}$, which since $F^{\otimes\frac{1}{2}}_{(n)} = U(a^-_\infty)v_n$ proves that $F^{\otimes\frac{1}{2}}_{(n)}$ is an irreducible highest weight module for a_∞. The highest weights are calculated in Lemma 3.6, and that proves Proposition 3.8. □

References

[ACJ13] Iana I Anguelova, Ben Cox, and Elizabeth Jurisich. N-point locality for vertex operators: normal ordered products, operator product expansions, twisted vertex algebras. arXiv:1307.4830, submitted, 2013.

[Ang12] Iana I Anguelova. Twisted vertex algebras, bicharacter construction and boson-fermion correspondences. arXiv:1206.4026, submitted, 2012.

[Bar11] Katrina Barron. Twisted modules for N=2 supersymmetric vertex operator superalgebras arising from finite automorphisms of the N=2 Neveu-Schwarz algebra. arXiv:1110.0229, submitted, 2011.

[BS83] N N Bogoliubov and D V Shirkov. *Quantum fields.* Benjamin/Cummings Publishing Co. Inc. Advanced Book Program, Reading, MA, 1983. Translated from the Russian by D. B. Pontecorvo.

[DJKM81a] Etsurō Date, Michio Jimbo, Masaki Kashiwara, and Tetsuji Miwa. Transformation groups for soliton equations. III. Operator approach to the Kadomtsev-Petviashvili equation. *J. Phys. Soc. Japan*, 50(11):3806–3812, 1981.

[DJKM81b] Etsurō Date, Michio Jimbo, Masaki Kashiwara, and Tetsuji Miwa. Transformation groups for soliton equations. VI. KP hierarchies of orthogonal and symplectic type. *J. Phys. Soc. Japan*, 50(11):3813–3818, 1981.

[DKM81] Etsurō Date, Masaki Kashiwara, and Tetsuji Miwa. Transformation groups for soliton equations. II. Vertex operators and τ functions. *Proc. Japan Acad. Ser. A Math. Sci.*, 57(8):387–392, 1981.

[FBZ04] Edward Frenkel and David Ben-Zvi. *Vertex algebras and algebraic curves*, volume 88 of *Mathematical Surveys and Monographs.* American Mathematical Society, Providence, RI, second edition, 2004.

[FFR91] Alex J Feingold, Igor B Frenkel, and John F X Ries. *Spinor construction of vertex operator algebras, triality, and* $E_8^{(1)}$, volume 121 of *Contemporary Mathematics.* American Mathematical Society, Providence, RI, 1991.

[Fre81] Igor B Frenkel. Two constructions of affine Lie algebra representations and boson-fermion correspondence in quantum field theory. *J. Funct. Anal.*, 44(3):259–327, 1981.

[FHL93] Igor B Frenkel, Yi-Zhi Huang, and James Lepowsky. On axiomatic approaches to vertex operator algebras and modules. *Mem. Amer. Math. Soc.*, 104(494):viii+64, 1993.

[FLM88] Igor B Frenkel, James Lepowsky, and Arne Meurman. *Vertex operator algebras and the Monster*, volume 134 of *Pure and Applied Mathematics.* Academic Press Inc., Boston, MA, 1988.

[Hua98] Kerson Huang. *Quantum field theory.* John Wiley & Sons Inc., New York, 1998. From operators to path integrals.

[Jan79] Jens Carsten Jantzen. *Moduln mit einem höchsten Gewicht*, volume 750 of *Lecture Notes in Mathematics.* Springer, Berlin, 1979.

[Kac90] Victor Kac. *Infinite-dimensional Lie algebras.* Cambridge University Press, Cambridge, third edition, 1990.

[Kac98] Victor Kac. *Vertex algebras for beginners*, volume 10 of *University Lecture Series.* American Mathematical Society, Providence, RI, second edition, 1998.

[KR87] Victor Kac and A K Raina. *Bombay lectures on highest weight representations of infinite-dimensional Lie algebras*, volume 2 of *Advanced Series in Mathematical Physics.* World Scientific Publishing Co. Inc., Teaneck, NJ, 1987.

[KW94] Victor Kac and Weiqiang Wang. Vertex operator superalgebras and their representations. In *Mathematical aspects of conformal and topological field theories and quantum groups (South Hadley, MA, 1992)*, volume 175 of *Contemp. Math.*, pages 161–191. Amer. Math. Soc., Providence, RI, 1994.

[KWY98] Victor Kac, Weiqiang Wang, and Catherine H Yan. Quasifinite representations of classical Lie subalgebras of $W_{1+\infty}$. *Adv. Math.*, 139(1):56–140, 1998.

[LL04] James Lepowsky and Haisheng Li. *Introduction to vertex operator algebras and their representations*, volume 227 of *Progress in Mathematics.* Birkhäuser Boston Inc., Boston, MA, 2004.

[MP95] Robert V Moody and Arturo Pianzola. *Lie algebras with triangular decompositions.* Canadian Mathematical Society Series of Monographs and Advanced Texts. John Wiley & Sons Inc., New York, 1995. A Wiley-Interscience Publication.

[MTZ08] Geoffrey Mason, Michael P Tuite, and Alexander Zuevsky. Torus n-point functions for $\mathbb{R}$-graded vertex operator superalgebras and continuous fermion orbifolds. *Comm. Math. Phys.*, 283(2):305–342, 2008.

[Wan99a] Weiqiang Wang. Dual pairs and infinite dimensional Lie algebras. In *Recent developments in quantum affine algebras and related topics (Raleigh, NC, 1998)*, volume 248 of *Contemp. Math.*, pages 453–469. Amer. Math. Soc., Providence, RI, 1999.

[Wan99b] Weiqiang Wang. Duality in infinite-dimensional Fock representations. *Commun. Contemp. Math.*, 1(2):155–199, 1999.

Journal of Physics: Conference Series **474** (2013) 012005 doi:10.1088/1742-6596/474/1/012005

Operators versus functions: from quantum dynamical semigroups to tomographic semigroups

Paolo Aniello

Dipartimento di Fisica dell'Università di Napoli "Federico II" and INFN – Sezione di Napoli, Complesso Universitario di Monte S. Angelo, via Cintia, 80126 Napoli, Italy

E-mail: `paolo.aniello@na.infn.it`

Abstract. Quantum mechanics can be formulated in terms of phase-space functions, according to Wigner's approach. A generalization of this approach consists in replacing the density operators of the standard formulation with suitable functions, the so-called generalized Wigner functions or (group-covariant) tomograms, obtained by means of group-theoretical methods. A typical problem arising in this context is to express the evolution of a quantum system in terms of tomograms. In the case of a (suitable) open quantum system, the dynamics can be described by means of a quantum dynamical semigroup 'in disguise', namely, by a semigroup of operators acting on tomograms rather than on density operators. We focus on a special class of quantum dynamical semigroups, the twirling semigroups, that have interesting applications, e.g., in quantum information science. The 'disguised counterparts' of the twirling semigroups, i.e., the corresponding semigroups acting on tomograms, form a class of semigroups of operators that we call *tomographic semigroups*. We show that the twirling semigroups and the tomographic semigroups can be encompassed in a unique theoretical framework, a class of semigroups of operators including also the probability semigroups of classical probability theory, so achieving a deeper insight into both the mathematical and the physical aspects of the problem.

1. Introduction

As is well known, quantum mechanics admits a remarkable formulation in terms of functions living on phase space, as first recognized by Wigner in his seminal paper [1]. In this approach, further developed by Groenewold [2] and Moyal [3], every pure state $|\psi\rangle\langle\psi|$ — a rank-one projector, with ψ denoting a wave function (of a single spatial degree of freedom, for notational simplicity) — is replaced by a phase space function ϱ_ψ:

$$\hat{\rho}_\psi = |\psi\rangle\langle\psi| \;\;\mapsto\;\; \varrho_\psi(q,p) := \frac{1}{2\pi}\int_{\mathbb{R}} \mathrm{e}^{-\mathrm{i}px}\, \psi\Big(q-\frac{x}{2}\Big)^{*}\, \psi\Big(q+\frac{x}{2}\Big)\, \mathrm{d}x\,. \tag{1}$$

The (real) function ϱ_ψ is usually called the *Wigner function* — or distribution — associated with the pure state $\hat{\rho}_\psi$. More generally [4], with a state $\hat{\rho}$ (a density operator) and with an observable $\hat{A}$ one can suitably associate real functions ϱ and $\mathcal{A}$, respectively, in such a way that

$$\langle \hat{A}\rangle_{\hat{\rho}} = \mathrm{tr}(\hat{A}\,\hat{\rho}) = \int_{\mathbb{R}\times\mathbb{R}} \mathcal{A}(q,p)\, \varrho(q,p)\, \mathrm{d}q\, \mathrm{d}p\,; \tag{2}$$

namely, the expectation value $\langle \hat{A}\rangle_{\hat{\rho}}$ of the observable $\hat{A}$ in the state $\hat{\rho}$ can be represented by a 'formally classical' expression (the function ϱ may assume negative values so that, in general, it cannot be regarded as a genuine probability distribution).

This intriguing approach has several remarkable applications, ranging from the study of the classical limit of quantum mechanics [5] to quantum optics [6], and it can be generalized by replacing the usual Hilbert space operators, states and observables, of the standard formulation of quantum mechanics with suitable functions — not necessarily living on a standard phase space (a symplectic vector space, a symplectic manifold) — obtained by means of various techniques sometimes collected under the generic name of *dequantization*, a procedure which can be regarded as the 'reverse arrow' of quantization [7]. Specifically, Wigner's correspondence — operator $\mapsto$ function — turns out to be the reverse arrow of Weyl's quantization prescription [8]; see [9] for the technical details. Among all known (quantization-) dequantization schemes, a central role is played by those schemes based on group-theoretical methods [10–12], and quantization-dequantization *à la* Weyl-Wigner is no exception, as it will be briefly recalled later on. In this framework, the functions corresponding to quantum-mechanical states — the so-called generalized Wigner functions or (group-covariant quantum) *tomograms* — live on the symmetry group of a quantum system, or on an homogeneous space of that group, and the symmetry action of the group on tomograms admits a simple expression.

A natural problem that arises in this general approach is to express the temporal evolution — or the evolution equation — of a quantum system in terms of the dequantized objects, i.e., of tomograms. In the case of a (suitable) *open* quantum system, the temporal evolution can be described by means of a quantum dynamical semigroup 'in disguise', namely, by a semigroup of operators acting on quantum tomograms in such a way to mimic the action of a *standard* quantum dynamical semigroup — i.e., a completely positive trace-preserving semigroup of operators [13, 14] — on states.

In the present paper, we will focus, in particular, on a class of quantum dynamical semigroups — the so-called *twirling semigroups* — with interesting applications [15–17]. Each of these semigroups of operators is associated with a pair formed by a representation of a group and by a convolution semigroup of probability measures on that group. Quantum dynamical semigroups of this form were first considered by Kossakowski [18], in the early times of the theory of open quantum systems. The 'disguised counterparts' of the twirling semigroups — acting on tomograms — form a class of semigroups of operators the we call, for rather obvious reasons, *tomographic semigroups* [16, 17].

We will show that the twirling semigroups and the tomographic semigroups can be encompassed in a unique theoretical framework — a large class of semigroups of operators including also the probability semigroups of classical probability theory — the *randomly generated semigroups*. This unifying theoretical framework allows one to achieve a deeper insight into both the mathematical and the physical aspects of the topic.

As a remarkable example, we will study the case where the underlying symmetry group is the group of translations on phase space. In this case, the quantum tomograms are directly related to the Wigner functions, and the explicit form of tomographic semigroups reveals an interesting relation with classical Brownian motion.

The paper is organized as follows. In sect. 2, we introduce the general class of randomly generated semigroups. Next, in sect. 3, we describe the special class of dequantized objects (the group-covariant tomograms) that we consider. This class encompasses, in particular, the standard Wigner functions. We then derive the general form of tomographic semigroups — see sect. 4 — and we consider, in particular, the case of the temporal evolution of Wigner functions. Finally, in sect. 5, a few conclusions are drawn.

2. A unifying framework: the randomly generated semigroups

As anticipated, we will first consider a general class of semigroups of operators that contains, in particular, both the quantum dynamical semigroups we are interested in and their 'disguised counterparts'. For the technical details concerning probability theory on groups (convolution semigroups of measures, probability semigroups, Lévy-Kintchine formula), we send the reader to the classical references [19, 20].

Let G be a *locally compact group*; in particular, it may be a *Lie group* — e.g., the symmetry group related to some physical model — like the group of translations on phase space, the rotation group, a (finite-dimensional) unitary group, or the Galilei group. Suppose that G is endowed with a *convolution semigroup of measures*, i.e., a family $\{\mu_t\}_{t\in\mathbb{R}^+}$ of probability measures on G such that

$$\mu_t \circledcirc \mu_s = \mu_{t+s}, \quad t, s \geq 0, \tag{3}$$

with $\mu_t \circledcirc \mu_s$ denoting the convolution of the measure μ_t with the measure μ_s, and

$$\lim_{t\downarrow 0} \mu_t = \delta, \quad \delta \equiv \delta_e \ \text{(Dirac measure at the identity } e \text{ in } G). \tag{4}$$

Let, moreover, $\mathfrak{V}$ be a (weakly continuous) representation, or *anti*representation, of G in a real or complex Banach space $\mathcal{J}$ — in the case of an antirepresentation, of course, we will have that $\mathfrak{V}(g_1 g_2) = \mathfrak{V}(g_2)\,\mathfrak{V}(g_1)$, for all $g_1, g_2 \in G$. The set $\{\mu_t[\mathfrak{V}]\}_{t\in\mathbb{R}^+}$, with $\mu_t[\mathfrak{V}]\colon \mathcal{J} \to \mathcal{J}$ denoting the bounded linear operator defined by

$$\mu_t[\mathfrak{V}]\,\Psi := \int_G \mathfrak{V}(g)\,\Psi \,\mathrm{d}\mu_t(g), \quad \forall \Psi \in \mathcal{J}, \tag{5}$$

is a *semigroup of operators* (for a rigorous definition of the operator $\mu_t[\mathfrak{V}]$ see [16]); i.e.,

(i) $\mu_t[\mathfrak{V}]\,\mu_s[\mathfrak{V}] = \mu_{t+s}[\mathfrak{V}]$, $t, s \geq 0$ (one-parameter semigroup property);

(ii) $\mu_0[\mathfrak{V}] = I$ (I denoting the identity operator);

(iii) $\lim_{t\downarrow 0} \|\mu_t[\mathfrak{V}]\,\Psi - \Psi\| = 0$, $\forall \Psi \in \mathcal{J}$ (strong right continuity at $t = 0$).

We call a semigroup of operators of the form (5) a *randomly generated semigroup* [16, 17], associated with the pair $(\mathfrak{V}, \{\mu_t\}_{t\in\mathbb{R}^+})$.

Remark 1 Being a semigroup of operators, $\{\mu_t[\mathfrak{V}]\}_{t\in\mathbb{R}^+}$ is completely characterized by its *infinitesimal generator*, namely, by the closed linear operator $\mathfrak{A}$ in $\mathcal{J}$ defined by

$$\mathrm{Dom}(\mathfrak{A}) := \Big\{\Psi \in \mathcal{J}\colon \exists \lim_{t\downarrow 0} t^{-1}\big(\mu_t[\mathfrak{V}]\,\Psi - \Psi\big)\Big\},\ \mathfrak{A}\,\Psi := \lim_{t\downarrow 0} t^{-1}\big(\mu_t[\mathfrak{V}]\,\Psi - \Psi\big),\ \Psi \in \mathrm{Dom}(\mathfrak{A}). \tag{6}$$

Remark 2 It is worth noting that the set of the randomly generated semigroups associated with group representations coincides with the whole set of randomly generated semigroups; namely, with the randomly generated semigroups associated with either representations or antirepresentations. Indeed, given a representation (alternatively, an antirepresentation) $\mathfrak{V}$ of G, and denoting by $\check{\mathfrak{V}}$ the related *anti*representation (respectively, representation) defined by $\check{\mathfrak{V}}(g) := \mathfrak{V}(g^{-1})$, we have

$$\mu_t[\check{\mathfrak{V}}] = \check{\mu}_t[\mathfrak{V}], \tag{7}$$

where $\check{\mu}_t$ is the *adjoint* of the measure μ_t, namely, the probability measure determined by $\int_G f(g)\,\mathrm{d}\check{\mu}_t(g) = \int_G f(g^{-1})\,\mathrm{d}\mu_t(g)$, for every f belonging to the space $\mathsf{C}_{\mathrm{c}}(G;\mathbb{R})$ of continuous $\mathbb{R}$-valued functions on G with compact support (the set $\{\check{\mu}_t\}_{t\in\mathbb{R}^+}$ is a convolution semigroup of measures too).

The class of randomly generated semigroups contains, in particular, the following remarkable subclasses:

- The *probability semigroups*, that describe the statistical properties of Brownian motion [15, 21, 22]: in this case, $\mathfrak{V}(g) = \mathsf{R}_g$ or $\mathfrak{V}(g) = \mathsf{L}_g$, where R_g, L_g are the left and right translation operators, respectively, acting in the Banach space $\mathsf{C}_0(G;\mathbb{R})$ of continuous functions on G vanishing at infinity, i.e.,

$$\mathsf{R}_g f := f((\cdot)g), \ \mathsf{L}_g f := f(g^{-1}(\cdot)), \ f \in \mathsf{C}_0(G;\mathbb{R}). \tag{8}$$

  In the case where G is a Lie group, the infinitesimal generators of these semigroups of operators are given by the Lévy-Kintchine formula [15, 19, 20].
- The *twirling semigroups*, i.e., certain quantum dynamical semigroups introduced by Kossakowski [18] during the pioneering times of the theory of open quantum systems: in this case, given a unitary (or, in general, a projective) representation U of G, in a separable complex Hilbert space $\mathcal{H}$, and denoting by $\mathcal{B}_1(\mathcal{H})$ the Banach space of trace class operators in $\mathcal{H}$, we have:

$$\mathfrak{V}(g)\,\hat{\rho} = U \vee U(g)\,\hat{\rho} := U(g)\,\hat{\rho}\,U(g)^*, \quad \hat{\rho} \in \mathcal{B}_1(\mathcal{H}); \tag{9}$$

  namely, $\mathfrak{V}$ is nothing but the standard symmetry action of G on quantum states. Classical examples of twirling semigroups are the quantum dynamical semigroups that model a finite-dimensional quantum system either coupled to an infinite free boson bath with Gaussian time correlation functions [23], or in the limit of singular coupling to a reservoir at infinite temperature [24]. The infinitesimal generators of the twirling semigroups have been studied in detail recently [15].
- The *tomographic semigroups*: in this case, the representation $\mathfrak{V}$ is given by the symmetry action of the group G on the group-covariant tomograms associated with the representation U of relation (9); the explicit form of this action will be given in sect. 4.

As already mentioned, the statistical properties of Brownian motion are described in terms of convolution semigroups of measures. Therefore, definition (5) establishes a remarkable link between Brownian motion on groups and semigroups of operators.

3. Group-covariant quantum tomograms and Wigner functions

Although not immediately evident from its usual definition — see (1) — the Wigner function associated with a pure state has a group-theoretical content that allows one to extend this definition to a larger class of objects: the generalized Wigner functions or (group-covariant) tomograms.

Let us first briefly sketch the mathematical framework; for the details, see [10, 25–27], and references therein. Consider a *square integrable* (irreducible) projective representation U of a locally compact group G in a separable complex Hilbert space $\mathcal{H}$. Examples of groups admitting representations of this kind are the group of translations on phase space — see below — and the affine group (whose square integrable unitary representations are involved in wavelet theory [10, 28]), and *all* the irreducible unitary representations of compact groups are square integrable. By means of such a representation, one can construct a map

$$\mathcal{V}\colon \mathcal{B}_2(\mathcal{H}) \to \mathrm{L}^2(G) \tag{10}$$

— with $\mathcal{B}_2(\mathcal{H})$, $\mathrm{L}^2(G)$ denoting the Hilbert spaces, respectively, of Hilbert-Schmidt operators in $\mathcal{H}$, and of square integrable, with respect to the left Haar measure, $\mathbb{C}$-valued functions on G — called the *generalized Wigner map*, or *Wigner transform*, generated by the square integrable representation U. The map $\mathcal{V}$ is a (linear) *isometry*, and we will denote its range by $\mathrm{Ran}(\mathcal{V})$. It turns out that $\mathrm{Ran}(\mathcal{V})$ depends only on the unitary equivalence class of the representation U.

If the group G is *unimodular* (e.g., a compact group), then the isometry $\mathcal{V}$ maps a trace class operator $\hat{\rho} \in \mathcal{B}_1(\mathcal{H})$ to the function

$$(\mathcal{V}\hat{\rho})(g) = d_U^{-1}\ \mathrm{tr}(U(g)^*\hat{\rho}), \tag{11}$$

with $d_U > 0$ denoting a normalizing constant, depending on the representation U and, of course, on the normalization of the Haar measure. Since trace class operators are dense in $\mathcal{B}_2(\mathcal{H})$, formula (11) determines the Wigner transform completely, in the case of an unimodular group, while in the general case the definition of $\mathcal{V}$ is somewhat more complicated and involves a suitable positive selfadjoint operator associated with U, the so-called 'Duflo-Moore operator' (which, in the unimodular case, reduces to a multiple of the identity of the form $d_U I$, where d_U is the positive constant appearing in (11)).

The symmetry group that gives rise to the *standard* Wigner transform is the group of translations on phase space. For notational simplicity, we will consider the special case of the $(1+1)$-dimensional the additive group $\mathbb{R}\times\mathbb{R}$ (a single spatial degree of freedom and the associated momentum), but the extension to the $(n+n)$-dimensional case is straightforward. The irreducible unitary representations of this (abelian) group are obviously one-dimensional; hence, they are not relevant from the point of view of quantum mechanics. Therefore, one is lead to consider a suitable class of irreducible *projective* representations that are labeled by a nonzero real number whose (absolute) value can be interpreted as Planck's constant $\hbar$; see, e.g., [29,30]. Setting $\hbar = 1$, one has the projective representation

$$\mathbb{R}\times\mathbb{R} \ni (q,p) \mapsto U(q,p) := \exp(\mathrm{i}(p\hat{q} - q\hat{p})) \tag{12}$$

— where $\hat{q}$, $\hat{p}$ are the standard position and momentum operators in $\mathrm{L}^2(\mathbb{R})$ — the so-called *Weyl system* (in quantum optics, $U(q,p)$ is often called *displacement operator* [6,31]). The fact that we are dealing with a *projective* representation means, in this case, that

$$U(q+\tilde{q}, p+\tilde{p}) = \mathsf{m}(q,p;\tilde{q},\tilde{p})\, U(q,p)\, U(\tilde{q},\tilde{p}), \tag{13}$$

where the *multiplier* m is of the form

$$\mathsf{m}(q,p;\tilde{q},\tilde{p}) := \exp\Big(\frac{\mathrm{i}}{2}(q\tilde{p} - p\tilde{q})\Big). \tag{14}$$

However, the physically relevant symmetry action is given by the associated (non-projective) representation $U\vee U$ of $\mathbb{R}\times\mathbb{R}$ in the Banach space $\mathcal{B}_1(\mathrm{L}^2(\mathbb{R}))$ — see (9) — which extends in a natural way to a unitary representation $\underline{U\vee U}$ in the Hilbert space $\mathcal{B}_2(\mathrm{L}^2(\mathbb{R}))$; i.e.,

$$\underline{U\vee U}(q,p)\,\hat{A} := U(q,p)\,\hat{A}\,U(q,p)^*, \quad \hat{A} \in \mathcal{B}_2(\mathrm{L}^2(\mathbb{R})). \tag{15}$$

The Weyl system is a square integrable representation. Hence, one can define the associated map $\mathcal{V}$ by formula (11) (the additive group $\mathbb{R}\times\mathbb{R}$ is obviously unimodular). It turns out that, in this case, the isometry $\mathcal{V}$ that one obtains is not directly the standard Wigner transform but the so-called *Fourier-Wigner transform* [9]. The former map is related to the latter one by the *symplectic Fourier transform*, i.e., by the unitary operator $\mathcal{F}_{\mathrm{sp}} : \mathrm{L}^2(\mathbb{R}\times\mathbb{R}) \rightarrow \mathrm{L}^2(\mathbb{R}\times\mathbb{R})$ determined by

$$(\mathcal{F}_{\mathrm{sp}} f)(q,p) = \frac{1}{2\pi}\int_{\mathbb{R}\times\mathbb{R}} f(q',p')\, \mathrm{e}^{\mathrm{i}(qp'-pq')}\, \mathrm{d}q'\mathrm{d}p', \quad \forall f \in \mathrm{L}^1(\mathbb{R}\times\mathbb{R}) \cap \mathrm{L}^2(\mathbb{R}\times\mathbb{R}). \tag{16}$$

Recall, by the way, that $\mathcal{F}_{\mathrm{sp}}$ is both unitary and selfadjoint:

$$\mathcal{F}_{\mathrm{sp}} = \mathcal{F}_{\mathrm{sp}}^*, \quad \mathcal{F}_{\mathrm{sp}}^2 = I. \tag{17}$$

The symplectic Fourier transform, however, does not play any essential mathematical or conceptual role, here; it allows one to obtain the usual quantization-dequantization rules (*à la* Weyl-Wigner) for position and momentum. Therefore, one should not expect to have any analogous of this operator involved in the general group-theoretical quantization-dequantization framework (namely, for a generic locally compact group).

Going back to the definition of the map $\mathcal{V}$, we first note that $(2\pi)^{-1}\mathrm{d}q\mathrm{d}p$ is the Haar measure on $\mathbb{R}\times\mathbb{R}$ normalized in a such a way that $d_U = 1$ for the Weyl system U (this fact is a consequence of Moyal's identity, see [9]). Then, in this case, the generalized Wigner transform (the Fourier-Wigner transform) $\mathcal{V}$ is the isometry from $\mathcal{B}_2(\mathrm{L}^2(\mathbb{R}))$ into $\mathrm{L}^2(\mathbb{R}\times\mathbb{R}) \equiv \mathrm{L}^2\big(\mathbb{R}\times\mathbb{R}, (2\pi)^{-1}\mathrm{d}q\mathrm{d}p; \mathbb{C}\big)$ determined by

$$\big(\mathcal{V}\hat{\rho}\big)(q,p) = \mathrm{tr}(U(q,p)^*\hat{\rho}), \quad \forall \hat{\rho} \in \mathcal{B}_1(\mathrm{L}^2(\mathbb{R})). \tag{18}$$

As a consequence of a result of Pool [32], in this case $\mathrm{Ran}(\mathcal{V})$ coincides with the whole Hilbert space $\mathrm{L}^2(G) = \mathrm{L}^2(\mathbb{R}\times\mathbb{R})$; otherwise stated, the isometry $\mathcal{V}$ is actually a unitary operator.

The map $\mathcal{V}$ intertwines the unitary representation $\underline{U \vee U}$ with the representation $\mathcal{T}$ of $\mathbb{R}\times\mathbb{R}$ in $\mathrm{L}^2(\mathbb{R}\times\mathbb{R}))$ defined by

$$\big(\mathcal{T}(q,p)f\big)(\tilde{q},\tilde{p}) = \mathrm{e}^{-\mathrm{i}(q\tilde{p}-p\tilde{q})} f(\tilde{q},\tilde{p}), \quad \forall f \in \mathrm{L}^2(\mathbb{R}\times\mathbb{R}). \tag{19}$$

Moreover, $\mathcal{V}$ intertwines the standard involution $\mathfrak{J}$ in $\mathcal{B}_2(\mathcal{H})$ — namely, the adjoining map $\mathfrak{J}\colon \hat{A} \mapsto \hat{A}^*$ — with the complex conjugation $\mathsf{J}\colon \mathrm{L}^2(\mathbb{R}\times\mathbb{R}) \to \mathrm{L}^2(\mathbb{R}\times\mathbb{R})$ (the idempotent antiunitary operator) defined by

$$\big(\mathsf{J}f\big)(q,p) = f(-q,-p)^*, \quad \forall f \in \mathrm{L}^2(\mathbb{R}\times\mathbb{R}). \tag{20}$$

Therefore, one should not expect the image of a selfadjoint operator, via the map $\mathcal{V}$, to be a real function — as it happens for a Wigner distribution [4] — but rather to satisfy the relation $f(q,p) = f(-q,-p)^*$.

This fact is not surprising since, as anticipated, the *standard Wigner transform* — we will denote it by $\mathcal{W}$ — is the unitary operator obtained composing the map $\mathcal{V}$ with the symplectic Fourier transform:

$$\mathcal{W} := \mathcal{F}_{\mathrm{sp}}\,\mathcal{V}\colon \mathcal{B}_2(\mathrm{L}^2(\mathbb{R})) \to \mathrm{L}^2(\mathbb{R}\times\mathbb{R}). \tag{21}$$

In the case of a pure state $\hat{\rho}_\psi = |\psi\rangle\langle\psi|$, $\|\psi\| = 1$, by a simple calculation one checks that for the function $\varrho_\psi = \mathcal{W}\hat{\rho}_\psi$ the classical formula (1) holds.

Clearly, the unitary operator $\mathcal{W}$ intertwines the representation $\underline{U \vee U}$ with the unitary representation $\mathcal{S}$ of $\mathbb{R}\times\mathbb{R}$ in $\mathrm{L}^2(\mathbb{R}\times\mathbb{R})$ defined by

$$\mathcal{S}(q,p) := \mathcal{F}_{\mathrm{sp}}\,\mathcal{T}(q,p)\,\mathcal{F}_{\mathrm{sp}}, \quad \forall (q,p) \in \mathbb{R}\times\mathbb{R}. \tag{22}$$

It is easy to check that, explicitly, we have:

$$\big(\mathcal{S}(q,p)f\big)(\tilde{q},\tilde{p}) = f(\tilde{q}-q,\tilde{p}-p), \quad \forall f \in \mathrm{L}^2(\mathbb{R}\times\mathbb{R}). \tag{23}$$

Thus, the representation $\mathcal{S}$ is nothing but the regular representation of $\mathbb{R}\times\mathbb{R}$ in $\mathrm{L}^2(\mathbb{R}\times\mathbb{R})$; it acts by simply translating functions on phase space.

Remark 3 The Hilbert space $\mathrm{L}^2(\mathbb{R}\times\mathbb{R})$, regarded as the range of the Wigner map (the standard Wigner transform, or the map directly associated with the Weyl system, i.e., the Fourier-Wigner transform), carries a natural structure of an algebra, endowed with the operation induced by the composition of operators in $\mathcal{B}_2(\mathrm{L}^2(\mathbb{R}))$ (the domain of the unitary operators $\mathcal{V}$ and $\mathcal{W}$); precisely, it is a H*-algebra [12,33]). This algebra operation is usually called by physicists a 'star product' of functions, see [11, 12, 34] and references therein. In the case where $\mathrm{L}^2(\mathbb{R}\times\mathbb{R})$ is regarded as $\mathrm{Ran}(\mathcal{V})$, the star product is given by the expression

$$\big(f_1 \star f_2\big)(q,p) = \frac{1}{2\pi}\int_{\mathbb{R}\times\mathbb{R}} f_1(q',p')\, f_2(q-q',p-p')\, \exp\Big(\frac{\mathrm{i}}{2}(qp'-pq')\Big)\, \mathrm{d}q'\mathrm{d}p', \tag{24}$$

$\forall f_1, f_2 \in \mathrm{L}^2(\mathbb{R}\times\mathbb{R})$. Therefore, the star product associated with the Weyl system U is precisely the *twisted convolution* of functions [9], the space $\big(\mathrm{L}^2(\mathbb{R}\times\mathbb{R}),\star,\mathsf{J}\big)$ is a H*-algebra and $\mathcal{V}\colon \mathcal{B}_2(\mathcal{H})\to \mathrm{L}^2(\mathbb{R}\times\mathbb{R})$ is an isomorphism of H*-algebras. The unitary operators $\mathcal{W}$, $\mathcal{W}^*$ (i.e., respectively, the dequantization and the quantization map) induce a further star product

$$(\cdot)\circledast(\cdot)\colon \mathrm{L}^2(\mathbb{R}\times\mathbb{R})\times \mathrm{L}^2(\mathbb{R}\times\mathbb{R}) \ni (f_1,f_2)\mapsto \mathcal{W}\big((\mathcal{W}^*f_1)(\mathcal{W}^*f_2)\big)\in \mathrm{L}^2(\mathbb{R}\times\mathbb{R}), \tag{25}$$

known as the *twisted product* [35]. Since $\mathcal{W}=\mathcal{F}_{\mathrm{sp}}\mathcal{V}$ and $\mathcal{W}^* = \mathcal{V}^*\mathcal{F}_{\mathrm{sp}}$, we find that

$$f_1\circledast f_2 = \mathcal{F}_{\mathrm{sp}}\big((\mathcal{F}_{\mathrm{sp}}f_1)\star(\mathcal{F}_{\mathrm{sp}}f_2)\big). \tag{26}$$

Using this formula, one obtains that, for every pair of functions f_1, f_2 in $\mathrm{L}^1(\mathbb{R}\times\mathbb{R})\cap\mathrm{L}^2(\mathbb{R}\times\mathbb{R})$,

$$(f_1\circledast f_2)(q,p) = \frac{1}{\pi^2}\int_{\mathbb{R}\times\mathbb{R}}\mathrm{d}q'\mathrm{d}p'\int_{\mathbb{R}\times\mathbb{R}}\mathrm{d}q''\mathrm{d}p''\ \theta(q,p;q',p';q'',p'')\, f_1(q',p')\, f_2(q'',p''), \tag{27}$$

where

$$\theta(q,p;q',p';q'',p'') := \exp\big(\mathrm{i}2(qp'-pq'+q'p''-p'q''+q''p-p''q)\big). \tag{28}$$

The function $\theta\colon(\mathbb{R}\times\mathbb{R})\times(\mathbb{R}\times\mathbb{R})\times(\mathbb{R}\times\mathbb{R})\to\mathbb{T}$ is the so-called *Groenewold-Moyal integral kernel*. The symplectic Fourier transform intertwines the complex conjugation J defined by (20) with the standard complex conjugation in $\mathrm{L}^2(\mathbb{R}\times\mathbb{R})$, i.e., $\mathcal{F}_{\mathrm{sp}}\,\mathsf{J}\,\mathcal{F}_{\mathrm{sp}}f = f^*$. Then $\mathrm{L}^2(\mathbb{R}\times\mathbb{R})$, endowed with the twisted product and with the standard complex conjugation of $\mathbb{C}$-valued functions, is once again a H*-algebra [32].

4. From quantum dynamical semigroups to tomographic semigroups

In view of the importance, in several applications, of both the 'phase-space approach' to quantum mechanics, briefly outlined in sect. 3, and of the theory of open quantum systems [13, 14], it is a natural problem to translate the expressions of quantum dynamical semigroups (and of their infinitesimal generators) in the language of group-covariant tomograms.

Let G a locally compact group and let Δ_G be the *modular function* on G. Suppose that $\mathsf{m}: G\times G\to\mathbb{T}$ — with $\mathbb{T}$ denoting the circle group of complex numbers of modulus one — is a *multiplier* for G, i.e., that

$$\mathsf{m}(g,e)=\mathsf{m}(e,g)=1,\quad \forall g\in G, \tag{29}$$

and

$$\mathsf{m}(g_1,g_2g_3)\,\mathsf{m}(g_2,g_3)=\mathsf{m}(g_1g_2,g_3)\,\mathsf{m}(g_1,g_2),\quad \forall g_1,g_2,g_3\in G. \tag{30}$$

E.g., a multiplier for the group of translations on phase space is given by (14), the multiplier associated with the Weyl system. Consider now the map $\mathcal{T}_{\mathsf{m}}: G\to\mathcal{U}(\mathrm{L}^2(G))$ — with $\mathcal{U}(\mathrm{L}^2(G))$ denoting the unitary group of the Hilbert space $\mathrm{L}^2(G)$ — defined by

$$\big(\mathcal{T}_{\mathsf{m}}(g)f\big)(h) := \Delta_G(g)^{\frac{1}{2}}\ \check{\mathsf{m}}(g,h)\, f(g^{-1}hg),\quad f\in\mathrm{L}^2(G), \tag{31}$$

where the function $\check{\mathsf{m}}: G\times G\to\mathbb{T}$ is given by the expression

$$\check{\mathsf{m}}(g,h) := \mathsf{m}(g,g^{-1}h)^*\,\mathsf{m}(g^{-1}h,g),\quad \forall g,h\in G. \tag{32}$$

One can prove that the map $\mathcal{T}_{\mathsf{m}}$ is a (strongly continuous) unitary representation [12].

Therefore, given a convolution semigroup $\{\mu_t\}_{t\in\mathbb{R}^+}$ of measures on G, we can define the randomly generated semigroup associated with the pair $(\mathcal{T}_{\mathsf{m}}, \{\mu_t\}_{t\in\mathbb{R}^+})$, namely, the semigroup of operators

$$\mathfrak{T}_t^{\mathsf{m}} f := \mu_t[\mathcal{T}_{\mathsf{m}}]\, f = \int_G \mathcal{T}_{\mathsf{m}}(g)\, f\, \mathrm{d}\mu_t(g), \quad \forall f \in \mathrm{L}^2(G). \tag{33}$$

It is then natural to wonder whether the semigroup of operators $\{\mathfrak{T}_t^{\mathsf{m}}\}_{t\in\mathbb{R}^+}$ has a physical meaning. One can answer to this question at least in the case where there exists a square integrable projective representation U of the group G, with multiplier m.

Indeed, in this case it can be shown that the semigroup of operators $\{\mathfrak{T}_t^{\mathsf{m}}\}_{t\in\mathbb{R}^+}$ is essentially *the twirling semigroup associated with the pair* $(U, \{\mu_t\}_{t\in\mathbb{R}^+})$ [15], *but expressed in terms of the quantum tomograms associated with* U. Thus, it is quite natural to call $\{\mathfrak{T}_t^{\mathsf{m}}\}_{t\in\mathbb{R}^+}$ the *tomographic semigroup* associated with the multiplier m.

More precisely, it turns out that the tomograms form a closed subspace of $\mathrm{L}^2(G)$ — the range $\mathrm{Ran}(\mathcal{V})$ of the Wigner map $\mathcal{V}$ associated with U — *stable* under the action of both the unitary representation $\mathcal{T}_{\mathsf{m}}$ and the semigroup of operators $\{\mathfrak{T}_t^{\mathsf{m}}\}_{t\in\mathbb{R}^+}$. The range $\mathrm{Ran}(\mathcal{V})$ contains a distinct (dense) linear subspace $\mathfrak{S}(\mathcal{V})$ which is the image, via the isometry $\mathcal{V}$, of the Banach space $\mathcal{B}_1(\mathcal{H})$. Moreover, $\mathfrak{S}(\mathcal{V})$ contains a convex set $\mathfrak{S}_0(\mathcal{V})$ which is the image of the convex set of density operators in $\mathcal{H}$ (the physical states). The subspace $\mathfrak{S}(\mathcal{V})$ and the convex set $\mathfrak{S}_0(\mathcal{V})$ are again stable in the sense mentioned before. One can show that the representation $U \vee U$ is related to the representation $\mathcal{T}_{\mathsf{m}}$ by the intertwining property

$$\mathcal{T}_{\mathsf{m}}(g)\, \mathcal{V}\hat{\rho} = \mathcal{V}\, U \vee U(g)\, \hat{\rho}, \quad \forall \hat{\rho} \in \mathcal{B}_1(\mathcal{H}). \tag{34}$$

It follows that, for every $t \geq 0$, we have:

$$\mathfrak{T}_t^{\mathsf{m}}\, \mathcal{V}\hat{\rho} = \mathcal{V}\, \mu_t[U \vee U]\, \hat{\rho}, \quad \forall \hat{\rho} \in \mathcal{B}_1(\mathcal{H}), \tag{35}$$

where $\{\mu_t[U \vee U]\colon \mathcal{B}_1(\mathcal{H}) \to \mathcal{B}_1(\mathcal{H})\}_{t\in\mathbb{R}^+}$ is the twirling semigroup associated with the pair $(U, \{\mu_t\}_{t\in\mathbb{R}^+})$, i.e., the quantum dynamical semigroup determined by

$$\mu_t[U \vee U]\, \hat{\rho} = \int_G U \vee U(g)\, \hat{\rho}\, \mathrm{d}\mu_t(g). \tag{36}$$

Incidentally, we note that the twirling semigroup $\{\mu_t[U \vee U]\}_{t\in\mathbb{R}^+}$ extends in a natural way to the randomly generated semigroup $\{\mu_t[\underline{U \vee U}]\colon \mathcal{B}_2(\mathcal{H}) \to \mathcal{B}_2(\mathcal{H})\}_{t\in\mathbb{R}^+}$, for which an intertwining relation analogous to (35) holds.

As the subspace $\mathfrak{S}(\mathcal{V})$ is stable under the tomographic semigroup, we can further introduce a semigroup of operators $\{\mathfrak{T}_t^{\mathcal{V}}\}_{t\in\mathbb{R}^+}$ by setting

$$\mathfrak{T}_t^{\mathcal{V}} \varrho := \mathfrak{T}_t^{\mathsf{m}} \varrho, \quad \forall \varrho \in \mathfrak{S}(\mathcal{V}). \tag{37}$$

Note that we have denoted by ϱ a generic element of the space $\mathfrak{S}(\mathcal{V})$ in order to highlight the fact that it corresponds, via dequantization, to a trace class operator $\hat{\rho}$. We stress that, whereas the tomographic semigroup $\{\mathfrak{T}_t^{\mathsf{m}}\}_{t\in\mathbb{R}^+}$ can always be defined (for every multiplier m), by virtue of relation (35) the semigroup of operators $\{\mathfrak{T}_t^{\mathcal{V}}\}_{t\in\mathbb{R}^+}$ is the *bona fide* dequantized version of the twirling semigroup $\mu_t[U \vee U]$, whose definition relies on the fact that U is a square integrable representation; hence, it can be regarded as a quantum dynamical semigroup 'in disguise'. We will then call $\{\mathfrak{T}_t^{\mathcal{V}}\}_{t\in\mathbb{R}^+}$ a *proper tomographic semigroup*.

Let us consider the case where G is the group of translations on the $(1+1)$-dimensional phase space, i.e., the additive group $\mathbb{R} \times \mathbb{R}$. In this case, we can define *two* (proper) tomographic semigroups — one associated with the Fourier-Wigner transform ($\{\mathfrak{T}_t^{\mathcal{V}}\}_{t\in\mathbb{R}^+}$), the other with the standard Wigner transform ($\{\mathfrak{T}_t^{\mathcal{W}}\}_{t\in\mathbb{R}^+}$) — mutually related by the symplectic Fourier transform. We will now briefly describe these semigroups of operators.

To this aim, note that for the function $\breve{\mathsf{m}}\colon (\mathbb{R}\times\mathbb{R})\times(\mathbb{R}\times\mathbb{R})\to\mathbb{T}$, defined according to (32), we find the expression

$$\breve{\mathsf{m}}(q,p\,;\tilde{q},\tilde{p}) = \mathsf{m}(q,p\,;\tilde{q}-q,\tilde{p}-p)^*\,\mathsf{m}(\tilde{q}-q,\tilde{p}-p\,;q,p) = \exp\big(-\,\mathrm{i}(q\tilde{p}-p\tilde{q})\big), \tag{38}$$

and, therefore, as anticipated in sect. 3 the unitary representation $\mathcal{T}\equiv\mathcal{T}_{\mathsf{m}}\colon \mathbb{R}\times\mathbb{R}\to\mathcal{U}(\mathrm{L}^2(\mathbb{R}\times\mathbb{R}))$ has the form (19). Then, the proper tomographic semigroup $\{\mathfrak{T}_t^{\mathcal{V}}\}_{t\in\mathbb{R}^+}$ is given by

$$\big(\mathfrak{T}_t^{\mathcal{V}}\varrho\big)(\tilde{q},\tilde{p}) = \int_{\mathbb{R}\times\mathbb{R}} \big(\mathcal{T}(q,p)\,\varrho\big)(\tilde{q},\tilde{p})\;\mathrm{d}\mu_t(q,p) = \varrho(\tilde{q},\tilde{p})\int_{\mathbb{R}\times\mathbb{R}} \mathrm{e}^{\mathrm{i}(\tilde{q}p-\tilde{p}q)}\;\mathrm{d}\mu_t(q,p). \tag{39}$$

Hence, denoting by $\widetilde{\mu}_t$ the symplectic Fourier transform of the probability measure μ_t, we have:

$$\big(\mathfrak{T}_t^{\mathcal{V}}\varrho\big)(\tilde{q},\tilde{p}) = \widetilde{\mu}_t(\tilde{q},\tilde{p})\,\varrho(\tilde{q},\tilde{p}), \quad t\geq 0. \tag{40}$$

Remark 4 The expression (40) of the tomographic semigroup $\{\mathfrak{T}_t^{\mathcal{V}}\}_{t\in\mathbb{R}^+}$ has an interesting interpretation. In fact, according to Bochner's theorem [36], the (symplectic) Fourier transform of a probability measure on $\mathbb{R}\times\mathbb{R}$ is a *normalized* continuous positive definite function, i.e., a continuous positive definite function $f\colon\mathbb{R}\times\mathbb{R}\to\mathbb{C}$ such that $f(0,0)=1$. Hence, the tomographic semigroup $\{\mathfrak{T}_t^{\mathcal{V}}\}_{t\in\mathbb{R}^+}$ acts by (pointwise) multiplication by a positive definite function. Considering the fact that the space $\mathfrak{S}(\mathcal{V})$ where the semigroup of operators acts can be characterized as the set of continuous KLM-positive definite functions on $\mathbb{R}\times\mathbb{R}$ — where the acronym refers to the fundamental contributions of Kastler [37], and of Loupias and Miracle-Sole [38, 39] — here we have a nice interplay between the 'classical' positive definite functions and the 'quantum' KLM-positive definite functions.

Let us now express the tomographic semigroup in terms of (standard) Wigner functions — $\varrho=\mathcal{W}\hat{\rho}=\mathcal{F}_{\mathrm{sp}}\mathcal{V}\hat{\rho}$ — by setting $\mathfrak{T}_t^{\mathcal{W}} := \mathcal{F}_{\mathrm{sp}}\,\mathfrak{T}_t^{\mathcal{V}}\,\mathcal{F}_{\mathrm{sp}}$, i.e.,

$$\mathfrak{T}_t^{\mathcal{W}}\varrho = \mathcal{F}_{\mathrm{sp}}\int_{\mathbb{R}\times\mathbb{R}} \mathcal{T}(q,p)\,(\mathcal{F}_{\mathrm{sp}}\,\varrho)\;\mathrm{d}\mu_t(q,p). \tag{41}$$

By relations (22) and (23) we then find that

$$\big(\mathfrak{T}_t^{\mathcal{W}}\varrho\big)(\tilde{q},\tilde{p}) = \int_{\mathbb{R}\times\mathbb{R}} \big(\mathcal{S}(q,p)\,\varrho\big)(\tilde{q},\tilde{p})\;\mathrm{d}\mu_t(q,p) = \int_{\mathbb{R}\times\mathbb{R}} \varrho(\tilde{q}-q,\tilde{p}-p)\;\mathrm{d}\mu_t(q,p). \tag{42}$$

Thus, as in the case of (classical) Brownian motion in $\mathbb{R}^3$, we have a semigroup of operators acting on a function ϱ by taking the convolution of this function with a probability measure belonging to a convolution semigroup. It should be noted, however, that the function ϱ in (42) lives on phase space (rather than on configuration space) and is characterized as an element of the space $\mathfrak{S}(\mathcal{W})=\mathcal{F}_{\mathrm{sp}}\mathfrak{S}(\mathcal{V})$ that contains the quantum states.

5. Conclusions and perspectives

A state in classical statistical mechanics is a probability measure on phase space — typically, associated with a probability distribution (a function), the Liouville density — and, according to a celebrated result of Bochner, the (symplectic) Fourier transform of a classical state is a continuous positive definite function. In the quantum mechanical setting, states are usually realized as density operators, but an elegant description in terms of functions is still possible by a suitable dequantization scheme. In particular, the properties of square integrable representations allow one to map a density operator to a generalized Wigner function or group-covariant quantum tomogram. These tomograms are embedded in an algebra of functions, with the algebra operation — a star product — corresponding to the product of operators [12]. In the case of a Weyl system (a square integrable projective representation of the group of translations on phase space), a tomogram is given by a Fourier-Wigner distribution — a normalized continuous KLM-positive definite function — or by its symplectic Fourier transform, namely, by a standard Wigner distribution, a 'quasi-probability distribution' [6, 31].

We have explored the idea of re-expressing a remarkable class of quantum dynamical semigroups — the twirling semigroups — in terms of 'phase-space functions', i.e., of group-covariant quantum tomograms. The implementation of this idea leads to the notion of *tomographic semigroup*, a quantum dynamical semigroups 'in disguise'. Both the twirling semigroups and their disguised counterparts are contained in a larger class of semigroups of operators, the randomly generated semigroups. These semigroups of operators are associated with pairs of the type $(\mathfrak{V}, \{\mu_t\}_{t\in\mathbb{R}^+})$, where $\mathfrak{V}$ is a representation, or an antirepresentation, of a locally compact group G and $\{\mu_t\}_{t\in\mathbb{R}^+}$ is a convolution semigroup of measures on G. Interestingly, the class of randomly generated semigroups also contains the probability semigroups of classical probability theory, and a tomographic semigroup acting on Wigner distributions is formally similar to a probability semigroup associated with the group of translations on phase space, see (42).

We have chosen to keep the exposition at a mild level of mathematical sophistication in such a way to skip several technical details and to focus on the main ideas. E.g., we have omitted any explanation concerning the precise meaning of the integrals of vector-valued functions. For these details, the reader is referred to [15, 16].

The results outlined in the present paper may be extended in several directions. In particular, there is ongoing work devoted to the characterization of the infinitesimal generators of the tomographic semigroups. In this regard, it should be noted that, in spite of the formal similarity with the probability semigroups, the machinery and the techniques associated with the classical Lévy-Kintchine formula cannot be applied directly in the new setting, and one needs to solve certain nontrivial mathematical issues.

References

[1] Wigner E 1932 On the quantum correction for thermodinamic equilibrium *Phys. Rev.* **40** 749
[2] Groenewold H J 1946 On the principles of elementary quantum mechanics *Physica* **12** 405
[3] Moyal J E 1949 Quantum mechanics as a statistical theory *Proc. Camb. Phil. Soc.* **45** 99
[4] Hillery M, O'Connell R F, Scully M O and Wigner E P 1984 Distribution functions in physics: fundametals *Phys. Rep.* **106** 121
[5] Landsman N P 1998 *Mathematical Topics Between Classical and Quantum Mechanics* (New York: Springer)
[6] Schleich W P 2001 *Quantum Optics in Phase Space* (Berlin: Wiley-VCH)
[7] Ali S T and Englis M 2005 Quantization methods: a guide for physicists and analysts *Rev. Math. Phys.* **17** 391
[8] Weyl H 1927 Quantenmechanik und Gruppentheorie *Z. Phys.* **46** 1
[9] Folland G B 1989 *Harmonic Analysis in Phase Space* (Princeton, NJ: Princeton University Press)
[10] Ali S T, Antoine J P and Gazeau J P 2000 *Coherent States, Wavelets and Their Generalizations* (New York: Springer)
[11] Aniello P, Man'ko V I and Marmo G 2008 Frame transforms, star products and quantum mechanics on phase space *J. Phys. A: Math. Theor.* **41** 285304
[12] Aniello P 2009 Star products: a group-theoretical point of view *J. Phys. A: Math. Theor.* **42** 475210
[13] Holevo A S 2001 *Statistical Structure of Quantum Theory* (Berlin: Springer)
[14] Breuer H-P and Petruccione F 2002 *The Theory of Open Quantum Systems* (Oxford: Oxford University Press)
[15] Aniello P, Kossakowski A, Marmo G and Ventriglia F 2010 Brownian motion on Lie groups and open quantum systems *J. Phys. A: Math. Theor.* **43** 265301

[16] Aniello P 2011 On a certain class of semigroups of operators *Open Syst. & Inf. Dynamics* **18** 129
[17] Aniello P 2013 Quantum dynamical semigroups, group representations and convolution semigroups *Physica Scripta* **T153** 014003
[18] Kossakowski A 1972 On quantum statistical mechanics of non-Hamiltonian systems *Rep. Math. Phys.* **3** 247
[19] Grenander W 1963 *Probabilities on Algebraic Structures* (New York: Wiley)
[20] Heyer H 1977 *Probability Measures on Locally Compact Groups* (Berlin: Springer)
[21] Ito K 1950 Brownian motions in a Lie group *Proc. Jap. Acad.* **26** 4
[22] Nelson E 1967 *Dynamical Theories of Brownian Motion* (Princeton, NJ: Princeton University Press)
[23] Gorini V and Kossakowski A 1976 N-level system in contact with a singular reservoir *J. Math. Phys.* **17** 1298
[24] Frigerio A and Gorini V 1976 N-level system in contact with a singular reservoir. II *J. Math. Phys.* **17** 2123
[25] Aniello P, Cassinelli G, De Vito E and Levrero A 1998 Square-integrability of induced representations of semidirect products *Rev. Math. Phys.* **10** 301
[26] Aniello P 2006 Square integrable projective representations and square integrable representations modulo a relatively central subgroup *Int. J. Geom. Meth. Mod. Phys.* **3** 233
[27] Aniello P 2006 Extended wavelet transforms *Int. J. Geom. Meth. Mod. Phys.* **3** 341
[28] Daubechies I 1992 *Ten Lectures on Wavelets* (Philadelphia: SIAM)
[29] Emch G G 1984 *Mathematical and Conceptual Foundations of 20th-Century Physics* (Amsterdam: North-Holland)
[30] Aniello P 2010 On the notion of Weyl system *Journal of Russian Laser Research* **31** 102
[31] Aniello P, Man'ko V, Marmo G, Solimeno S and Zaccaria F 2000 On the coherent states, displacement operators and quasidistributions associated with deformed quantum oscillators *J. Opt. B: Quantum Semiclass. Opt.* **2** 718
[32] Pool J C T 1966 Mathematical aspects of Weyl correspondence *J. Math. Phys.* **7** 66
[33] Rickart C E 1960 *General Theory of Banach Algebras* (Princeton, NJ: Van Nostrand)
[34] Zachos C K, Fairlie D B and Curtright T L eds. 2005 *Quantum Mechanics in Phase Space* (Singapore: World Scientific)
[35] Grossmann A, Loupias G and Stein E M 1968 An algebra of pseudo-differential operators and quantum mechanics in phase space *Ann. Inst. Fourier* **18** 343
[36] Rudin W 1962 *Fourier Analysis on Gorups* (New York: Interscience Publishers)
[37] Kastler D 1965 The C^*-algebras of free boson field *Commun. Math. Phys.* **1** 14
[38] Loupias G and Miracle-Sole S 1966 C^*-algèbre des systèmes canoniques. I *Commun. Math. Phys* **2** 31
[39] Loupias G and Miracle-Sole S 1967 C^*-algèbre des systèmes canoniques. II *Ann. Inst. Henri Poincaré Sect. A* **6** 39

Journal of Physics: Conference Series **474** (2013) 012006 doi:10.1088/1742-6596/474/1/012006

Rational solutions from Padé approximants for the generalized Hunter-Saxton equation

H Aratyn [1], J F Gomes [2], D V Ruy [2] and A H Zimerman [2]

[1] Department of Physics, University of Illinois at Chicago, 845 W. Taylor St., Chicago, IL 60607-7059

[2]Instituto de Física Teórica-UNESP, Rua Dr Bento Teobaldo Ferraz 271, Bloco II, 01140-070, São Paulo, Brazil

E-mail: `aratyn@uic.edu`, `jfg@ift.unesp.br`, `daniloruy@ift.unesp.br`, `zimerman@ift.unesp.br`

Abstract. Exact rational solutions of the generalized Hunter-Saxton equation are obtained using Padé approximant approach for the traveling-wave and self-similarity reduction. A larger class of algebraic solutions are also obtained by extending a range of parameters within the solutions obtained by this approach.

1. Introduction

The generalized Hunter-Saxton (HS) equation was originally proposed in [1] as

$$u_{xt} = uu_{xx} + ku_x^2. \tag{1}$$

This model reproduces the original HS equation when $k = 1/2$, which is a model for wave propagation in a director field of nematic liquid crystals [2]. Besides, equation (1) appears in different applications, such as the study of the Einstein-Weyl space for $k = -1/3$ [5]. Solutions of (1) in a non-closed form were obtained in ref. [1] by a hodographic transformation, however the task of expressing these solutions in a closed form of the original variables may be complicated and indirect.

It is known that the Padé approximant usually gives a good representation of the solution near a regular point (see [3] for more details) and it has widely been used to construct approximative solutions of differential equations with already determined initial conditions [7, 8]. Moreover, an exact solution was found for the Boussinesq equation in [8] for a predetermined initial condition. In this note we employ the inverse path for study the generalized HS equation. We start with a rational expression given by the Padé approximant and seek for a set parameters or initial conditions that allows the ansatz to become an exact solution. Although there is no guarantee that a given ODE has rational or polynomial solutions, this approach provides a useful tool for searching for them.

The Padé approximant consists of a ratio of two polynomials that can be expressed as

$$[L/M](x) \equiv \frac{P_L(x)}{Q_M(x)} = \frac{\sum_{j=0}^{L} p_j x^j}{1 + {}_{j=1}^{M} q_j x^j}$$

where the polynomials $P_L(x)$ and $Q_M(x)$ are defined such that $[L/M](x)$ agrees with the Taylor expansion a function $f = f(x)$ up to degree $L + M$, i.e.

$$f(x) = \frac{P_L(x)}{Q_M(x)} + \mathcal{O}(x^{L+M+1}).$$

Thus, in order to determine the Padé approximant of a Taylor expansion at $x = 0$ given by $f = \sum_{j=0} a_j x^j$, we need to solve a system of $L + M$ equations, namely,

$$\sum_{m+n=j} a_m q_n - p_j = 0, \qquad j = 0, 1, ..., L + M.$$

Consider now an arbitrary ordinary differential equation (ODE):

$$E(z, v, v_z, ...; \mathcal{S}_0) = 0, \qquad \mathcal{S}_0 \equiv \text{Cartesian product of the sets of parameters}, \tag{2}$$

and let $v = v(z; \mathcal{S}_0 \times \mathcal{S}_1)$ be the general solution of this equation, where $\mathcal{S}_1$ is the Cartesian product of the sets of initial conditions. Let us assume the origin has a regular local representation, i.e. $v = \sum_{j=0}^{\infty} v_j z^j$. Substituting in the differential equation (2), we have

$$E(v; \mathcal{S}_0) = \sum_{j=0}^{\infty} E_j(v_0, ..., v_{j+n}) z^j = 0,$$

where n is the order of the equation. The above expression determines the coefficients v_j such that

$$v = \sum_{j=0}^{\infty} v_j(\mathcal{S}_0 \times \mathcal{S}_1) z^j .$$

Calculating the Padé approximant of this Taylor expansion, we have

$$v = \frac{P_L(z; \mathcal{S}_0 \times \mathcal{S}_1)}{Q_M(z; \mathcal{S}_0 \times \mathcal{S}_1)} + \mathcal{O}(z^{L+M+1}).$$

Let us assume now that there are subsets $\hat{\mathcal{S}}_0 \subset \mathcal{S}_0$ and $\hat{\mathcal{S}}_1 \subset \mathcal{S}_1$, such that the Padé approximant of $\hat{\mathcal{S}} = \hat{\mathcal{S}}_0 \times \hat{\mathcal{S}}_1$ leads to an exact solution, i.e.

$$v(z; \hat{\mathcal{S}}) = \frac{P_L(z; \hat{\mathcal{S}})}{Q_M(z; \hat{\mathcal{S}})}. \tag{3}$$

This ansatz allows us to seek solutions in the rational form without a need to consider further properties of the differential equation. Substituting (3) in (2), we have

$$E(z, v, v_z, ...; \hat{\mathcal{S}}_0) = \frac{\sum_{j=0}^{\Lambda_{L,M}} \hat{E}_j(\hat{\mathcal{S}}) z^j}{D(\hat{\mathcal{S}})} = 0, \tag{4}$$

where the degree $\Lambda_{L,M}$ depends of L, M and the particular form of the differential equation. In order to determine all elements $s_j \in \hat{\mathcal{S}}$ for which the ansatz (3) is true, we need to solve a system of algebraic equations,

$$\hat{E}_j(\hat{\mathcal{S}}) = 0, \qquad j = 0, 1, ..., L + M - n \tag{5}$$

$$\hat{E}_j(\hat{\mathcal{S}}) = 0, \qquad j = L + M - n + 1, ..., \Lambda_{L,M} \tag{6}$$

$$D(\hat{\mathcal{S}}) \neq 0 \tag{7}$$

Observe that the system (5) is identically satisfied by the definition of the Padé approximant. Thus, the exact solution (3) is determined by the choice of parameters $\hat{\mathcal{S}}_0$ and initial conditions $\hat{\mathcal{S}}_1$.

In the next section, we apply this approach to the traveling-wave and self-similarity reductions of the generalized Hunter-Saxton equation. Moreover, for these specific reductions, we are able to extend the rational solutions obtained by the Padé approximant method to a more general set of parameters.

2. Generalized Hunter-Saxton equation

2.1. Traveling-wave reduction of the generalized HS equation

Applying the traveling-wave reduction

$$z = x + \mu t, \qquad u(x,t) = u(z), \qquad \mu = \text{constant} \tag{8}$$

to the generalized Hunter-Saxton equation (1) yields:

$$-\mu u_{zz} + u u_{zz} + k u_z^2 = 0\,. \tag{9}$$

The Taylor expansion of a solution to the above equation near the origin ($z = 0$) can be written as

$$u = u_0 + u_1 z + \frac{k u_1^2}{2(\mu - u_0)} z^2 + \frac{k(2k+1)u_1^3}{6(\mu - u_0)^2} z^3 + \frac{k(6k^2+7k+2)u_1^4}{24(\mu - u_0)^3} z^4 + \ldots \qquad u_0 \neq \mu,$$

where u_0 and u_1 are arbitrary constants. In terms of the formalism from the previous section we have $\mathcal{S}_0 = (k, \mu)$ and $\mathcal{S}_1 = (u_0, u_1)$.

We now apply the Padé approximant approach to the two simplest cases [1/1] and [2/2]. Let us begin with the ansatz

$$u \equiv u(z; (k, \mu, u_0, u_1)) = \frac{P_1(z; \hat{\mathcal{S}})}{Q_1(z; \hat{\mathcal{S}})} \tag{10}$$

where

$$\begin{aligned} P_1(z; \hat{\mathcal{S}}) &= u_0 + \frac{(2\mu - (k+2)u_0)u_1}{2(\mu - u_0)} z \\ Q_1(z; \hat{\mathcal{S}}) &= 1 - \frac{k u_1}{2(\mu - u_0)} z \end{aligned}$$

For this ansatz, the system (6) is given by

$$\hat{E}_1(\hat{\mathcal{S}}) = 8k(k+2)u_1^3(\mu - u_0)^3 = 0, \tag{11}$$

with $\Lambda_{1,1} = 1$. The condition (7) reads here as

$$D(\hat{\mathcal{S}}) = k^4 u_1^4 z^4 - 8k^3 u_1^3(\mu - u_0)z^3 + 24k^2 u_1^2(\mu - u_0)^2 z^2 - 32k u_1(\mu - u_0)^3 z + 16(\mu - u_0)^4 \neq 0\,. \tag{12}$$

We see that $k = -2$ and $k = 0$ simultaneously solve conditions (11) and (12). Thus, by substituting these particular values of parameters on (10), we obtain exact solutions:

$$k = -2 \quad ; \quad u = \frac{(\mu - u_0)u_0 + z\mu u_1}{\mu - u_0 + u_1 z} \tag{13}$$

$$k = 0 \quad ; \quad u = u_0 + u_1 z\,. \tag{14}$$

Let us now take as an ansatz the Padé approximant of type [2/2] :

$$u \equiv u(z;(k,\mu,u_0,u_1)) = \frac{P_2(z;\hat{\mathcal{S}})}{Q_2(z;\hat{\mathcal{S}})}, \tag{15}$$

where

$$\begin{aligned} P_2(z;\hat{\mathcal{S}}) &= u_0 + \frac{(2\mu-(2k+3)u_0)u_1}{2(\mu-u_0)}z + \frac{(-6(k+1)\mu-(2k^2+7k+6)u_0)u_1^2}{12(\mu-u_0)^2}z^2 \\ Q_2(z;\hat{\mathcal{S}}) &= 1 - \frac{(2k+1)u_1}{2(\mu-u_0)}z + \frac{k(2k+1)u_1^2}{12(\mu-u_0)^2}z^2 . \end{aligned}$$

It follows that for this ansatz the system (6) is given by

$$\hat{E}_3(\hat{\mathcal{S}}) = 576k(k+2)(2k+1)(2k+3)u_1^5(\mu-u_0)^5 = 0 \tag{16}$$
$$\hat{E}_4(\hat{\mathcal{S}}) = -432k(k+1)(k+2)(2k+1)(2k+3)u_1^6(\mu-u_0)^4 = 0 \tag{17}$$
$$\hat{E}_5(\hat{\mathcal{S}}) = 24k(k+2)(2k+1)^2(2k+3)^2u_1^7(\mu-u_0)^3 = 0 \tag{18}$$

supplemented by condition (7), i. e.

$$\begin{aligned} D(\hat{\mathcal{S}}) &= k^4(2k+1)^4u_1^8z^8 - 24k^3(2k+1)^4u_1^7(\mu-u_0)z^7 \\ &+ 24k^2(2k+1)^3(20k+9)u_1^6(\mu-u_0)^2z^6 - 864k(2k+1)^3(3k+1)u_1^5(\mu-u_0)z^5 \\ &+ 432(2k+1)^2(38k^2+24k+3)u_1^4(\mu-u_0)^4z^4 \\ &- 10368(2k+1)^2(3k+1)u_1^3(\mu-u_0)z^3 + 3456(2k+1)(20k+9)u_1^2(\mu-u_0)z^2 \\ &- 41472(2k+1)u_1(\mu-u_0)^7z + 20736(\mu-u_0)^8 \neq 0 \end{aligned}$$

Substituting the solution of system (16-18) into the ansatz (15) yields exact solutions:

$$k=-2 \quad ; \quad u = \frac{(\mu-u_0)u_0 + z\mu u_1}{\mu-u_0+u_1z}$$

$$k=-3/2 \quad ; \quad u = \frac{4(u_0-\mu)u_0 + 4(\mu-u_0)\mu u_1z + \mu u_1^2z^2}{(2(\mu-u_0)+u_1z)^2}$$

$$k=-1/2 \quad ; \quad u = u_0 + u_1z - \frac{u_1^2z^2}{4(\mu-u_0)}$$

$$k=0 \quad ; \quad u = u_0 + u_1z$$

Notice that with the above choice of parameters the order of the Padé approximants is reduced accordingly.

Observe that solutions (13) and (14) obtained from the first ansatz are included in the above set of solutions. This is not a general case as it will be evident in the next section. These ansatzes can be worked out by an algebraic manipulation software such as Mathematica or Maple so we omit details for the system (5) and (6) in the remaining part of the text. Following similar procedure as above for the ansatz:

$$u \equiv u(z;(k,\mu,u_0,u_1)) = \frac{P_3(z;\hat{\mathcal{S}})}{Q_3(z;\hat{\mathcal{S}})}, \tag{19}$$

we obtain the solutions

$$k=-2 \quad ; \quad u=\mu+\frac{-\mu^2+2\mu u_0-u_0^2}{\mu+u_1 z-u_0}$$

$$k=-3/2 \quad ; \quad u=\mu-\frac{4(\mu^3-3\mu^2u_0+3\mu u_0^2-u_0^3)}{(2\mu+u_1z-2u_0)^2}$$

$$k=-4/3 \quad ; \quad u=\mu-\frac{27(\mu^4-4\mu^3u_0+6\mu^2u_0^2-4\mu u_0^3+u_0^4)}{(3\mu+u_1z-3u_0)^3}$$

$$k=-\frac{2}{3} \quad ; \quad u=\frac{u_1^3z^3}{27(\mu-u_0)^2}-\frac{u_1^2z^2}{3(\mu-u_0)}+u_0+u_1z$$

$$k=-\frac{1}{2} \quad ; \quad u=-\frac{u_1^2z^2}{4(\mu-u_0)}+u_0+u_1z$$

$$k=0 \quad ; \quad u=u_0+u_1z$$

We can see that all non-trivial solutions of the above system are expressed in terms of $\hat{\mathcal{S}}_1=\mathcal{S}_1$ and $\hat{\mathcal{S}}_0$ being independent of the initial conditions, therefore, they are general solutions.

A careful study of these solutions suggests a general expression that compactly expresses the whole set $\hat{\mathcal{S}}$. Indeed direct verification confirms that all solutions are reproduced by

$$u=\mu+\frac{(u_0-\mu)}{(1+\frac{(-1)^r zu_1}{r(-\mu+u_0)})^r}, \qquad r=-\frac{1}{(k+1)}. \tag{20}$$

Although this solution was obtained originally for integer values of r only it can be checked that it holds for any r different from 0. The closed general solution for $k=1/2$ was obtained recently in [6] and agrees with (20).

2.2. Self-similarity reduction of the generalized HS equation

In this section, we explore the self-similarity reduction of the generalized Hunter-Saxton equation. Observe that inserting the self-similarity reduction expression

$$u(x,t)=t^{-(1+\xi)}v(z), \qquad z=xt^{\xi}, \qquad \xi=\text{constant} \tag{21}$$

into (1) yields

$$v_z-\xi zv_{zz}+vv_{zz}+kv_z^2=0 \tag{22}$$

The local representation of the solution v around the origin is

$$v=v_0+v_1z-\frac{v_1(1+kv_1)}{2v_0}z^2+\frac{v_1(1+kv_1)(1-\xi+(1+2k)v_1)}{6v_0^2}z^3+\ldots, \tag{23}$$

where v_0 and v_1 are arbitrary constants, $\mathcal{S}_0=(k,\xi)$ and $\mathcal{S}_1=(v_0,v_1)$. Notice that for $v_1=-1/k$ the above series truncates leading to the solution

$$v=v_0-\frac{z}{k}.$$

We can apply the ansatz for the Padé approximants provided that $v_1\neq-\frac{1}{k}$. Therefore, considering the ansatz $v\equiv v(z;(k,\xi,v_0,v_1))=\frac{P_1(z;\hat{\mathcal{S}})}{Q_1(z;\hat{\mathcal{S}})}$, we have the set $\hat{\mathcal{S}}$ and the corresponding solution given by

$$(k,\xi,v_0,v_1)=(-2,-1/2,v_0,v_1) \quad ; \quad v=\frac{v_0(2v_0+z)}{2v_0+(1-2v_1)z}$$

Applying the same procedure to the ansatz $v \equiv v(z;(k,\xi,v_0,v_1)) = \frac{P_2(z;\hat{S})}{Q_2(z;\hat{S})}$, it give us

$$(k,\xi,v_0,v_1) = (-2,1,v_0,v_1) \quad ; \quad v = \frac{2v_0^2 - v_1 z^2}{2v_0 - 2v_1 z}$$

$$(k,\xi,v_0,v_1) = (-3/2,-1/3,v_0,v_1) \quad ; \quad v = \frac{2v_0[18u_0^2 + 12v_0 z + (2-3v_1)z^2]}{[6v_0 + (2-3v_1)z]^2}$$

$$(k,\xi,v_0,v_1) = (-1/2,1,v_0,v_1) \quad ; \quad v = v_0 + v_1 z + \frac{v_1(v_1-2)}{4v_0} z^2$$

In order to obtain more solutions which helps to look for a compact expression of all rational solutions, we also consider the ansatz $v \equiv v(z;(k,\xi,v_0,v_1)) = \frac{P_3(z;\hat{S})}{Q_3(z;\hat{S})}$. This yields the following solutions

$$(k,\xi,v_0,v_1) = (-3/2,1,v_0,v_1) \quad ; \quad v = \frac{2(6v_0^3 - 3v_0 v_1 z^2 + v_1^2 z^3)}{3(2v_0 - z v_1)^2}$$

$$(k,\xi,v_0,v_1) = (-4/3,-1/4,v_0,v_1) \quad ; \quad v = -\frac{3v_0[576u_0^3 + 432v_0^2 z - 36v_0(4v_1-3)z^2 + (3-4v_1)^2 z^3]}{[(4v_1-3)z - 12v_0]^3}$$

$$(k,\xi,v_0,v_1) = (-2/3,1/2,v_0,v_1) \quad ; \quad v = v_0 + v_1 z + \frac{v_1(2v_1-3)}{6v_0} z^2 + \frac{v_1(2v_1-3)^2}{108v_0^2} z^3$$

$$(k,\xi,v_0,v_1) = (-2/3,1,v_0,v_1) \quad ; \quad v = v_0 + v_1 z + \frac{v_1(2v_1-3)}{6v_0} z^2 + \frac{v_1^2(2v_1-3)}{54v_0^2} z^3.$$

We see that the parameters of the above general solutions can be organized as

$$k = -\frac{(r+1)}{r} \qquad \text{with} \qquad \xi = 1 \qquad \text{or} \qquad \xi = -\frac{1}{(r+1)}$$

In order to express the solutions in a compact form, let us define the constants u_0, z_0 as

$$v_0 = \frac{z_0 v_1}{r}, \qquad v_1 = \frac{r}{(r+1)} + \frac{(-1)^r r u_0}{z_0^{r+1}}$$

when $\xi = 1$ and

$$v_0 = \left(-\frac{1}{(r+1)} + \frac{v_1}{r}\right) z_0, \qquad v_1 = \frac{(-1)^r r u_0}{z_0^{r+1}}$$

when $\xi = -1/(r+1)$.

The solutions found by the Padé approximant approach for the self-similarity reduction can be summarized as

$$v = \frac{u_0}{(z-z_0)^r} + \frac{z_0 + rz}{r+1}, \qquad \text{when} \qquad \xi = 1, \tag{24}$$

$$v = \frac{u_0}{(z-z_0)^r} - \frac{z_0}{r+1}, \qquad \text{when} \qquad \xi = -\frac{1}{(r+1)} \tag{25}$$

where $r = -1/(k+1)$. This expression was derived for an integer r, however, one can verify that expressions (24) and (25) are still solutions for a non-integer r (different from 0 and -1).

Acknowledgments
The authors thank CNPq and Fapesp for support.

References
[1] Pavlov M V 2001 *Theor. and Math. Phys.* **128** 927
[2] Hunter J K and Saxton R 1991 , "Dynamics of director fields", *SIAM J. Appl. Math.*, **51**, No. 6, 1498-1521 (1991)
[3] Baker Jr G A and Graves-Moris P 1996 *Padé Approximants* (New York: Cambridge University Press)
[4] Hunter J K and Saxton R 1991 *SIAM J. Appl. Math.* **51** 1498
[5] Tod K P 2000 *J. Math. Phys.* **41** 5572
[6] Al-Ali E M 2013 *Int. Journal of Math. Analysis* **7** 1647
[7] Boyd J P 1997 *Computers in Physics* **11** 299
[8] Mousa M M and Kaltaye A 2009 *Int. J. of Eng. and Applied Sciences* **5** 237

Topology-preserving quantum deformation with non-numerical parameter

Marat Aukhadiev, Suren Grigoryan and Ekaterina Lipacheva
Kazan State Power Engineering University, Krasnoselskaya 51, Kazan, 420066, Russia
E-mail: m.aukhadiev@gmail.com, gsuren@inbox.ru, elipacheva@gmail.com

Abstract. We introduce a class of compact quantum semigroups, that we call semigroup deformations of compact Abelian qroups. These objects arise from reduced semigroup $\mathcal{C}^*$-algebras, the generalization of the Toeplitz algebra. We study quantum subgroups, quantum projective spaces and quantum quotient groups for such objects, and show that the group is contained as a compact quantum subgroup in the deformation of itself. The connection with the weak Hopf algebra notion is described. We give a grading on the $\mathcal{C}^*$-algebra of the compact quantum semigroups constructed.

1. Preliminaries

This work is devoted to constructing compact quantum semigroups using non-commutative $\mathcal{C}^*$-algebras, generated by isometries. The procedure starts with considering a compact Abelian group G and its Pontryagin dual group Γ. The method described in section 2 represents a sort of deformation of the group G, which preserves its topology and multiplication.

Unlike the standard quantization procedure, the deformation parameter here is not a number, but a set of elements. More precisely, any semigroup S generating group Γ can play a role of deformation parameter. It is shown that this deformation depends essentially on the choice of S, giving different results for different semigroups generating the same group G.

Another special feature of this method is that the object $\mathcal{QS}_r(S)$ obtained as deformation is not a quantum group, but a compact quantum semigroup with some unique properties. Notion of the compact quantum semigroup is given in section 1.1. We show that $\mathcal{QS}_r(S)$ can be regarded as an extension of G, since $G \subset \mathcal{QS}_r(S)$ as a compact quantum subgroup (see section 3). Similarly to the classical case this extension generates an action of G on $\mathcal{QS}_r(S)$. This action and quantum projective space are described in section 3.

1.1. Quantum semigroup

Let P be a compact semigroup, i.e. a Hausdorff compact space with continuous associative operation $(x, y) \to xy$. Denote by $C(P)$ the algebra of continuous functions on P. Then $C(P)$ is a commutative unital $\mathcal{C}^*$-algebra, which encodes all the topological information about the space P. We identify $C(P \times P)$ with $C(P) \otimes C(P)$. In what follows symbol $\otimes$ denotes minimal $\mathcal{C}^*$-tensor product. Define the map $\Delta\colon C(P) \to C(P) \otimes C(P)$:

$$\Delta(f)(x, y) = f(xy). \qquad (1.1)$$

Obviously, Δ is a continuous unital *-homomorphism. The associativity of multiplication in P is encoded in the following property of Δ, called *coassociativity*:

$$(\Delta \otimes \mathrm{id})\Delta = (\mathrm{id} \otimes \Delta)\Delta$$

Thus, all the information about the compact group P is encoded in the pair $(C(P), \Delta)$.

Now, let $\mathcal{A}$ be a commutative unital C^*-algebra. Then, by Gelfand-Naimark theorem, $\mathcal{A}$ is *-isomorphic to the algebra of all continuous functions $C(P)$ on some compact Hausdorff space P. If $\mathcal{A}$ is endowed with a *-homomorphism $\Delta\colon \mathcal{A} \to \mathcal{A} \otimes \mathcal{A}$, satisfying the coassociativity condition, then equation $f(xy) = \Delta(f)(x, y)$ endows P with a compact semigroup structure.

By the term "quantization" we mean a passage from the commutative algebra $C(P)$ to a noncommutative unital C^*-algebra $\mathcal{A}$. Due to noncommutativity this algebra does not corresppond to any semigroup. Nevertheless, algebra $\mathcal{A}$ can be regarded as an algebra of continuous functions on some virtual compact geometrical object. This object is called a *quantum space.* In other words, quantum space is an object of a category, dual to the category of C^*-algebras (see [13] for details). Usually quantum space correspoding to C^*-algebra $\mathcal{A}$ is denoted by $\mathcal{QS}(\mathcal{A})$ [9].

A unital *-homomorphism $\Delta\colon \mathcal{A} \to \mathcal{A} \otimes \mathcal{A}$ satisfying coassociativity condition is called *comultiplication.* Analogously to the classical case, Δ turns quantum space $\mathcal{QS}(\mathcal{A})$ into a quantum semigroup. Then the algebra $\mathcal{A}$ with its comultiplication represents *an algebra of functions on a quantum semigroup* [8]. The pair $(\mathcal{A}, \Delta)$ is called *a compact quantum semigroup* [5].

Compact quantum semigroup $(\mathcal{A}, \Delta)$ is called *cocommutative,* if $\sigma\Delta = \Delta$, where $\sigma\colon \mathcal{A} \otimes \mathcal{A} \to \mathcal{A} \otimes \mathcal{A}$ is a flip: $\sigma(a \otimes b) = b \otimes a$.

By definition of S.L. Woronowicz [12], the compact quantum semigroup $(\mathcal{A}, \Delta)$ is *a compact quantum group* if each of the sets

$$(1 \otimes \mathcal{A})\Delta(\mathcal{A}), \ (\mathcal{A} \otimes 1)\Delta(\mathcal{A})$$

is linear dense in $\mathcal{A} \otimes \mathcal{A}$. In the classical case (when $\mathcal{A}$ is commutative) $(\mathcal{A}, \Delta)$ corresponds to an ordinary group, the density conditions of the above-mentioned sets corresponds to existence of inverse elements in P.

1.2. Generalized groups and weak Hopf algebras

A semigroup is called *generalized group* (or *inverse semigroup*), if it contains with each element x a unique x^*, called *generalized inverse* to x, such that

$$xx^*x = x, \ x^*xx^* = x^*.$$

This definition, and theory of generalized groups was established in 1952 by V.V. Wagner [11].

In 1998 F.Li [4] defined the notion, naturally generalizing Hopf algebra. This generalization is based on weakening the antipode condition, and called *a weak Hopf algebra.* Let A be an algebra with a comultiplication Δ. Denote by m the multiplication map $A \otimes A \otimes A \to A$, $m(a \otimes b \otimes c) = abc$.

Suppose there exists a linear map $T\colon A \to A$ such that

$$m(\mathrm{id} \otimes T \otimes \mathrm{id})(\Delta \otimes \mathrm{id})\Delta = \mathrm{id}, \tag{1.2}$$

$$m(T \otimes \mathrm{id} \otimes T)(\Delta \otimes \mathrm{id})\Delta = T \tag{1.3}$$

Then (A, Δ, T) is called *a weak Hopf algebra*, and T is *a weak antipode.* Note that weak Hopf algebra is not bialgebra, since a counit is not required. Obviously, $T(1) = 1$. Every Hopf algebra admits a unique antipode. However, this is not true for weak Hopf algebra. Other weak Hopf algebra properties are shown in [6].

The analogous connection between Hopf algebra and group takes place here as well. Weak Hopf algebras are closely related to generalized groups.

Consider a generalized group M and the semigroup algebra $\mathbb{C}[M]$. Algebra $\mathbb{C}[M]$ admits the following comultiplication and weak antipode:

$$\Delta(x) = x \otimes x, \tag{1.4}$$

$$\Delta(\sum_i \lambda_i x_i) = \sum_i \lambda_i x_i \otimes x_i,$$

$$T(x) = x^*.$$

Clearly, equations (1.2, 1.3) hold and $\mathbb{C}[M]$ is a weak Hopf algebra, called *a generalized group bialgebra.*

Let A be a weak Hopf algebra with a weak antipode T. Then the set of group-like elements in A, i.e. elements satisfying (1.4), forms a generalized group, where inverse of any x is $T(x)$. Thus, every generalized group is associated with a classical weak Hopf algebra, and every weak Hopf algebra contains a generalized group.

A.M. Vershik proved a result which shows the connection between finite generalized group bialgebras and finite-dimensional bialgebras, it can be found in [10].

2. Quantum deformation

2.1. Step 1. Algebra C(G)

Let G be a compact Abelian group. We denote by Γ a group isomorphic to the group of characters of G. For the group operation in Γ we use additive notation. For each $a \in \Gamma$ let χ^a denote the corresponding character of G. Consider $\mathcal{C}^*$-algebra $C(G)$ of continuous complex-valued functions on G with pointwise multiplication. The linear span of characters $\chi^a, a \in \Gamma$ is dense in $C(G)$. As mentioned in section 1.1, $C(G)$ is endowed with the natural comultiplication Δ_G, which encodes multiplication of G by the formula (1.1).

Obviously, this algebra is commutative. In order to deform this algebra and make it non-commutative, we first immerse it into non-commutative surroundings by means of representation in the algebra of bounded linear operators on a Hilbert space. To this end consider Hilbert space $H = L^2(G, d\mu)$ of square-integrable functions on G with respect to μ, where μ is a shift-invariant normed measure. For any $f, g \in H$ the scalar product is given by

$$(f, g) = \int_G f(\alpha)\overline{g(\alpha)}d\mu.$$

Denote by $\{e_a\}_{a\in\Gamma}$ an orthonormal basis of H, corresponding to the characters $\chi^a, a \in \Gamma$ of the group G.

Let $B(H)$ denote an algebra of all bounded linear operators on H. For $f \in C(G)$ define an operator $L_f \colon H \to H$ by

$$L_f g = f \cdot g.$$

Then a map $f \to L_f$ is a representation of the algebra $C(G)$ in $B(H)$. Obviously, the image of $C(G)$ under this representation is a commutative $\mathcal{C}^*$-algebra:

$$L_f L_g = L_{fg} = L_g L_f.$$

2.2. *Step 2. Reduced semigroup C^*-algebra*

Before we deform algebra $C(G)$ in $B(H)$, we choose a deformation parameter. Unlike the standard quantization procedure, a parameter here is not a number, but a semigroup. We take a semigroup $S \subset \Gamma$ with zero, which generates Γ, i.e. $\Gamma = \{a - b|\ a, b \in S\}$. The motivation to use such semigroup lies in a widely developed theory of reduced semigroup C^*-algebras $C^*_r(S)$, established by G.J. Murphy [7]. $C^*_r(S)$ is a C^*-algebra generated by the regular isometric representation of the semigroup S. One can ask if other representations of S could be used for our purposes. The answer is positive, and for example we can use isometric representations of S. But the situation with such representations becomes more complicated (see [2]). In fact, all these representations generate in some sense not equal C^*-algebras. So, we would get a different deformation in each case.

Example 1 *Suppose Γ is the group of integers $\mathbb{Z}$. As an example of such semigroup S we could choose the semigroup of non-negative integers $\mathbb{Z}_+ \subset \Gamma$. The Pontryagin dual of $\mathbb{Z}$, its group of characters, would be the unit circle $G = S^1$. Note that for the same group Γ we could choose other semigroups, e.g. $S = \mathbb{Z}_+ \setminus \{1\} = \{0, 2, 3, 4, ...\}$ or $S = \{0, 3, 6, 7, 8, ...\}$. Indeed, all such semigroups generate $\mathbb{Z}$, since they contain elements with difference between them equal to unit. Note, in all these cases we have $S \cap (-S) = \{0\}$.*

Example 2 *Consider $S = \mathbb{Z} \times \mathbb{Z}_+$ with standard additive operation. The generated group is $\mathbb{Z} \times \mathbb{Z}$, with dual $S^1 \times S^1$. Note that we can produce another group $\Gamma_S = S \cap (-S)$, as for any other semigroup S. This group is isomorphic to $\mathbb{Z}$.*

Let $H_S \subset L^2$ be a Hilbert space with basis $\{e_a\}_{a \in S}$, and $P_S \colon L^2 \to H_S$ – a projection. Define for $f \in C(G)$:

$$T_f = P_S L_f P_S. \tag{2.1}$$

Reduced semigroup C^*-algebra is the C^*-algebra generated by all T_f for $f \in C(G)$, denoted by $C^*_r(S)$. We regard this algebra as an S-deformation of the algebra $C(G)$.

Consider the semigroup S in Example 1. In this case one can show that H_S is the standard Hardi space, and $C^*_r(S)$ is *-isomorphic to the Toeplitz algebra $\mathcal{T}$ [3], i.e. the C^*-algebra generated by one isometric operator on a Hilbert space.

Isometric operators $T_a = P_S L_{\chi^a} P_S$ for $a \in S$ and T^*_a are generators in $C^*_r(S)$. These operators act on the basis in the following way:

$$T_a e_b = e_{a+b},\ T^*_a e_b = \begin{cases} e_c, & \text{if } b = c + a, \\ 0 & \text{otherwise.} \end{cases} \tag{2.2}$$

A finite product of T_a, T^*_b, $a, b \in S$ is called *a monomial*. Monomials form a generalized group denoted by S^*_r. Linear span of S^*_r is dense *-subalgebra in $C^*_r(S)$, denoted by $\mathcal{P}(S)$. By virtue of (2.2), for any monomial V there exists $a \in \Gamma$ such that for any e_b, $b \in S$ we have either $Ve_b = 0$ or $Ve_b = e_{b+a}$. So, we define the index of monomial V, $\mathrm{ind}V = a$. Obviously, $\mathrm{ind}(VW) = \mathrm{ind}V + \mathrm{ind}W$ and $\mathrm{ind}V^* = -\mathrm{ind}V$.

The C^*-algebra $C^*_r(S)$ is unital and non-commutative, so by reasoning of section 1.1 it can be regarded as an algebra of functions on a compact quantum space. Let us denote this quantum space by $\mathcal{QS}_r(S)$.

2.3. *Step 3. Quantum semigroup*

At this stage we have deformed G, as a compact space, and obtained the compact quantum space $\mathcal{QS}_r(S)$. The algebra $C^*_r(S)$ is regarded as algebra of functions on this quantum space. Semigroup S plays a role of deformation parameter.

Besides that G is a compact space, it is endowed with multiplication. So, it is natural to ask whether $\mathcal{QS}_r(S)$ is endowed with an analogue of operation in G. In order to answer to this question, one must give a structure of compact quantum semigroup on $\mathcal{QS}_r(S)$, following definition of Section 1.1.

Define comultiplication $\Delta\colon S_r^* \to S_r^* \otimes S_r^*$ by

$$\Delta(V) = V \otimes V. \tag{2.3}$$

Theorem 1 *The map Δ extends to a unital *-homomorphism $\Delta\colon \mathcal{C}_r^*(S) \to \mathcal{C}_r^*(S) \otimes \mathcal{C}_r^*(S)$. The pair $(\mathcal{C}_r^*(S), \Delta)$ is a compact quantum semigroup.*

Proof *Firstly, note that $Mon(S^2) = S_r^* \times S_r^*$, where $Mon(S^2)$ is a generalized group generated by monomials in $\mathcal{C}_r^*(S^2)$, and $P(S^2) = \mathcal{P}(S) \odot \mathcal{P}(S)$. One can show that $\mathcal{C}_r^*(S^2)$ and $\mathcal{C}_r^*(S) \otimes \mathcal{C}_r^*(S)$ are *-isomorphic.*

Define $\Delta\colon \mathcal{P}(S) \to \mathcal{P}(S) \otimes \mathcal{P}(S)$ by relation

$$\Delta(\sum_{i=1}^{n} \lambda_i V_i) = \sum_{i=1}^{n} \lambda_i V_i \otimes V_i,$$

where V_i are monomials.

In order to show that Δ extends to the whole $\mathcal{C}^*$-algebra, divide the basis $\{e_c \otimes e_d\}_{c,d \in S}$ of the Hilbert space $l^2(S) \otimes l^2(S)$ to equivalence classes in the following way.

We say that elements $e_c \otimes e_d$ and $e_l \otimes e_k$ of the basis in $l^2(S) \otimes l^2(S)$ are equivalent *($e_c \otimes e_d \sim e_l \otimes e_k$), if $e_{c+a} \otimes e_{d+a} = e_{l+b} \otimes e_{k+b}$ for some $a, b \in S$. One can easily verify that $e_c \otimes e_d \sim e_l \otimes e_k$ and $e_l \otimes e_k \sim e_m \otimes e_n$ imply $e_c \otimes e_d \sim e_m \otimes e_n$. Note that $e_c \otimes e_d \sim e_l \otimes e_k$ if and only if $d - c = k - l$. Denote by H_c Hilbert subspace in $l^2(S) \otimes l^2(S)$, generated by family $\{e_l \otimes e_k | k - l = c,\ k, l \in S\}$. Obviously, $l^2(S) \otimes l^2(S) = \bigoplus_{c \in \Gamma} H_c$. Each of the spaces H_c is invariant under the action of *-algebra $\Delta(\mathcal{P}(S))$. Indeed, let V be a monomial, $e_l \otimes e_k \in H_c$ and $(V \otimes V) e_l \otimes e_k \neq 0$. Consequently,*

$$(V \otimes V)(e_l \otimes e_k) = e_{l+c} \otimes e_{k+c},$$

where $c = \mathrm{ind} V$.

Fixing $a \in S, Q \in \mathcal{P}(S)$, we show the commutativity of the following diagram:

$$\begin{array}{ccc} H_a & \xrightarrow{\rho_a} & l^2(S) \\ {\scriptstyle \Delta_a(Q) = \Delta(Q)|_{H_a}} \downarrow & & \downarrow {\scriptstyle P_a Q} \\ H_a & \xrightarrow{\rho_a} & l^2(S) \end{array} \tag{2.4}$$

where ρ_a is a projection onto the first component:

$$\rho_a(\sum_{i=1}^{n} \lambda_i e_{a_i} \otimes e_{b_i}) = \sum_{i=1}^{n} \lambda_i e_{a_i},$$

$P_a\colon l^2(S) \to \rho_a(H_a) \subset l^2(S)$ is an orthogonal projection. Obviously, ρ_a is an isometric embedding of H_a into $l^2(S)$. It is sufficient to check 2.4 for basic elements $e_c \otimes e_d$ in H_a. Suppose $Q = \sum_{i=1}^{n} \lambda_i V_i$. Then

$$\Delta(Q)(e_c \otimes e_d) = \sum_{i=1}^{n} \lambda_i' e_{c+a_i} \otimes e_{d+a_i},$$

where $a_i = \mathrm{ind}V_i$ *and*

$$\lambda'_i = \begin{cases} \lambda_i, & \text{if } c + a_i, d + a_i \in S \\ 0, & \text{if either } c + a_i \notin S, \text{ or } d + a_i \notin S. \end{cases}$$

Consequently,

$$\rho_a(\Delta)(Q)(e_c \otimes e_d) = \sum_{i=1}^{n} \lambda'_i e_{c+a_i}.$$

On the other hand,

$$((P_a Q)\rho_a)(e_c \otimes e_d) = P_a(Q(e_c)) = P_a(\sum_{i=1}^{n} \lambda_i e_{c+a_i}) = \sum_{i=1}^{n} \lambda'_i e_{c+a_i}.$$

So, the commutativity of diagram (2.4) is proved. Further, $\|\rho_a \Delta_a(Q)\| = \|P_a Q \rho_a\|$. *Due to the fact that* ρ_a *is embedding and* P_a *is projection, we have* $\|\Delta_a(Q)\| \leq \|Q\|$ *for all* $a \in \Gamma$.

Note that the space H_0 *is generated by* $e_a \otimes e_a$ *for all* $a \in S$. *Operator* $\rho_0 \colon H_0 \to l^2(S)$ *is a bijection and* P_0 *is an identity. Therefore,* $\|\Delta_0(Q)\| = \|Q\|$ *for all* $Q \in \mathcal{P}(S)$. *Thus, operator* $\Delta \colon \mathcal{P}(S) \to \mathcal{P}(S) \otimes \mathcal{P}(S)$ *extends to a continuous embedding* $\Delta \colon C^*_{red}(S) \to C^*_{red}(S) \otimes C^*_{red}(S)$.

One can show that $(\mathcal{C}^*_r(S), \Delta)$ is a compact quantum group if and only if $S = \Gamma$. In this case the $\mathcal{C}^*$-algebra is commutative and one can identify $(\mathcal{C}^*_r(\Gamma), \Delta) \cong G$.

Thus, to each semigroup S we associate quantum semigroup $\mathcal{QS}_r(S)$, and in the limit case to each group Γ we associate its group of characters G, i.e. the Pontryagin dual.

Theorem 2 *The* **-algebra* $\mathcal{P}(S)$ *is a weak Hopf algebra with comultiplication* Δ.

Proof *Indeed, the* **-algebra* $\mathcal{P}(S)$ *is a semigroup algebra* $\mathbb{C}[M]$ *for* $M = S^*_r$. *Since* S^*_r *is a generalized group, we use the weak Hopf algebra structure on* $\mathbb{C}[M]$ *introduced in Section 1.2.*

This is an analogue of the well-known fact, proved by S.L. Woronowicz [12], which states that the $\mathcal{C}^*$-algebra of the compact quantum group contains a dense Hopf *-algebra.

3. Quantum subgroups, quantum quotient groups and quantum projective spaces

3.1. Quantum subgroup and quantum projective space

We say that π^* is *a morphism* taking a compact quantum semigroup $(\mathcal{A}_2, \Delta_2)$ to $(\mathcal{A}_1, \Delta_1)$, if there exists a *-homomorphism $\pi \colon \mathcal{A}_1 \to \mathcal{A}_2$, verifying the commutativity of the following diagram:

$$\begin{array}{ccc} \mathcal{A}_1 & \xrightarrow{\Delta_1} & \mathcal{A}_1 \otimes \mathcal{A}_1 \\ \pi\downarrow & & \downarrow\pi\otimes\pi \\ \mathcal{A}_2 & \xrightarrow{\Delta_2} & \mathcal{A}_2 \otimes \mathcal{A}_2 \end{array}$$

π^* is called a dual morphism to π.

If π is a surjection, then $(\mathcal{A}_2, \Delta_2)$ is called *a compact quantum subsemigroup* in $(\mathcal{A}_1, \Delta_1)$, and π^* is *an embedding.*

An action of a compact quantum semigroup $(\mathcal{A}, \Delta)$ on a unital C^*-algebra B is a unital *-homomorphism $\delta \colon \mathcal{B} \to \mathcal{B} \otimes \mathcal{A}$ such that

$$(\delta \otimes \mathrm{id})\delta = (\mathrm{id} \otimes \Delta)\delta.$$

Action δ is called *ergodic*, if the fixed point algebra

$$\mathcal{B}^\delta = \{b \in \mathcal{B} | \delta(b) = b \otimes I\}$$

coincides with $\mathbb{C}$, where I denotes the unit in $\mathcal{A}$.

Obviously, any compact quantum subsemigroup $(\mathcal{A}', \Delta')$ in a compact quantum semigroup $(\mathcal{A}, \Delta)$ naturally acts on the algebra $\mathcal{A}$:

$$\delta = (\mathrm{id} \otimes \pi)\Delta.$$

Then the fixed point algebra $\mathcal{A}^\delta$ is called *a quantum left coset space.* Space $\mathcal{A}^\delta$ with $\Delta|_{\mathcal{A}^\delta} \colon \mathcal{A}^\delta \to \mathcal{A} \otimes \mathcal{A}^\delta$ is *a quantum left quotient space* $(\mathcal{A}, \Delta)/(\mathcal{A}', \Delta')$.

Consider compact Abelian group G. In the classical compact quantum group $(C(G), \Delta_G)$ the comultiplication is given by (1.1).

Let us show that G is a compact (quantum) subgroup in $(\mathcal{C}_r^*(S), \Delta)$. Further $\mathcal{K}$ denotes the commutator *-ideal in the algebra $\mathcal{C}_r^*(S)$.

Theorem 3 *$(C(G), \Delta_G)$ is a compact quantum subgroup in the compact quantum semigroup $(\mathcal{C}_r^*(S), \Delta)$.*

Proof *One can verify that the quotient map $\pi\colon \mathcal{C}_r^*(S) \to \mathcal{C}_r^*(S)/\mathcal{K}$ is a *-epimorphism on algebra $C(G)$. Here we have $\pi(T_a) = \chi^a$, where χ^a is a character of group G. For any $a \in S$ we have*

$$(\pi \otimes \pi)\Delta(T_a) = (\pi \otimes \pi)(T_a \otimes T_a) = \chi^a \otimes \chi^a.$$

On the other and, the classical comultiplication on $C(G)$ acts on χ^a in the following way:

$$\Delta_G(\chi^a)(\alpha, \beta) = \chi^a(\alpha\beta) = \chi^a(\alpha)\chi^a(\beta) = (\chi^a \otimes \chi^a)(\alpha \otimes \beta).$$

Since characters generate the algebra $C(G)$, we obtain $(\pi \otimes \pi)\Delta = \Delta_G \pi$. It is sufficient to show that $\mathcal{K}$ is a coideal. To this end, consider monomials V, W. Then we have

$$\Delta(VW - WV) = VW \otimes VW - WV \otimes WV =$$

$$= (VW - WV) \otimes VW + WV \otimes (VW - WV) \in \mathcal{K} \otimes \mathcal{C}_r^*(S) + \mathcal{C}_r^*(S) \otimes \mathcal{K}.$$

*Since $\mathcal{K}$ is a closed ideal and Δ is a *-homomorphism, the required injection holds for all elements.*

Thus, $(C(G), \Delta_G)$ is a compact quantum subgroup in $(\mathcal{C}_r^*(S), \Delta)$. Formally, $G \subset \mathcal{QS}_r(S)$.

In the example 1, the algebra $\mathcal{C}_r^*(\mathbb{Z}_+)$ is isomorphic to the Toeplitz algebra. The corresponding compact quantum semigroup is described in details in [1]. Here we obtain that the underlying compact quantum semigroup of the Toeplitz algebra contains the unit circle with the natural group structure as a compact quantum subgroup.

We will see now that $\mathcal{QS}_r(S)$ is endowed with an action of G. Take $a \in S$ and define

$$\tau = (\mathrm{id} \otimes \pi)\Delta. \tag{3.1}$$

This map extends to *-homomorphism $\tau\colon \mathcal{C}_r^*(S) \to \mathcal{C}_r^*(S) \otimes C(G)$. Indeed, τ is a coaction of G on $\mathcal{QS}_r(S)$:

$$(\mathrm{id} \otimes \Delta_G)\tau(T_a) = (\mathrm{id} \otimes \Delta_G)(\mathrm{id} \otimes \pi)(T_a \otimes T_a) =$$
$$= (\mathrm{id} \otimes \pi \otimes \pi)(\mathrm{id} \otimes \Delta)\Delta(T_a) = (\mathrm{id} \otimes \tau)\tau(T_a).$$

Futher one can search for the quantum projective space, i.e. the left coset space with respect to the action τ. The corresponding $\mathcal{C}^*$-algebra is given by the set $\{A \in \mathcal{C}_r^*(S) | \pi(A) = 1\}$. By virtue of (3.1), this set is a closed linear span of idempotents in S_r^*, the semigroup of projections. Denote this $\mathcal{C}^*$-algebra by $\mathcal{A}_0$. By virtue of (2.3), the comultiplication Δ induces the comultiplication on $\mathcal{A}_0$ and obviously the compact quantum semigroup structure. Denote $(\mathcal{A}_0, \Delta)$ by P. The results in [3] imply that $\mathcal{A}_0$ is a commutative unital $\mathcal{C}^*$-algebra and we obtain the following statement.

Proposition 1 *The quantum projective space P produced by the action of the compact group G on the compact quantum semigroup $\mathcal{QS}_r(S)$ is a compact semigroup.*

Using the semigroup S_r^* and the C^*-algebra $\mathcal{A}_0$ we reconstruct the whole C^*-algebra $C_r^*(S)$. Consider monomial V with $\text{ind}V = a - b,\ a, b \in S$. Then $T_b V T_a^* \in \mathcal{A}_0$ and the relation

$$V = T_b^*(T_b V T_a^*)T_a$$

implies the following statement.

Proposition 2 *Any monomial V can be represented as $V = T_b^* V_0 T_a$, where $\text{ind}V = a - b,\ a, b \in S$ and $V_0 \in \mathcal{A}_0$.*

Denote by $\mathcal{A}_c$, $c \in \Gamma$ the *-algebra $T_b^* \mathcal{A}_0 T_a$, where $c = a - b,\ a, b \in S$. By virtue of Proposition 2, $\mathcal{A}_c$ is generated by all monomials with index equal to c. So, we obtain a Γ-grading on the C^*-algebra $C_r^*(S)$, given by the Fell bundle $\{\mathcal{A}_c | c \in \Gamma\}$:

$$C_r^*(S) = \overline{\bigoplus_{c \in \Gamma} \mathcal{A}_c},\quad \mathcal{A}_c \cdot \mathcal{A}_b \subseteq \mathcal{A}_{c+b},\quad c, b \in \Gamma.$$

3.2. Quantum quotient group

The quantum projective space given by the C^*-algebra $\mathcal{A}_0$ is in fact a quantum quotient semigroup, which is produced by $\mathcal{QS}_r(S)/G$ and the action of G on $C_r^*(S)$. But in some cases we can obtain a quantum quotient group of $\mathcal{QS}_r(S)$, given by the following procedure.

For the semigroup S denote $\Gamma_S = S \cap (-S)$. One can easily verify that Γ_S is a subgroup in S. By virtue of [3], the center $Z(C_r^*(S))$ of the C^*-algebra $C_r^*(S)$ is *-isomorphic to $C(G_S)$, where G_S is the compact group of characters of Γ_S.

One can show that the restriction of the comultiplication Δ on $C(G_S)$ gives a compact quantum group structure, which coincides with the compact group structure on G_S. This means that G_S is a compact quantum quotient group of the compact quantum semigroup $\mathcal{QS}_r(S)$ by some quantum space.

Consider the C^*-algebra $C_r^*(S)$ generated by semigroup S in example 2. As mentioned above, Γ_S is isomorphic to $\mathbb{Z}$, with the dual G_S being isomorphic to the unit circle S^1. So, in this case the quotient of the compact quantum semigroup $(C_r^*(S), \Delta)$ is the unit circle with the natural group structure.

Acknowledgments

This research is partially supported by RFBR grant 12-01-97016.

References

[1] Aukhadiev M A, Grigoryan S A and Lipacheva E V 2011 Infinite-dimensional compact quantum semigroup *Lobachevskii J. Math.* **32** 304–16

[2] Aukhadiev M A and Tepoyan V H 2012 Isometric representations of totally ordered semigroups *Lobachevskii J. Math.* **33** 239–43

[3] Grigoryan S A and Salakhutdinov A F 2009 C^*-algebras generated by semigroups *Izv. Vyssh. Uchebn. Zaved. Mat.* **10** 68–71

[4] Duplij S and Li F 2001 Regular solutions of quantum Yang-Baxter equation from weak Hopf algebras *PCzech. J. Phys.* **51** 1306–11

[5] A Maes and A Van Daele 1998 Notes on compact quantum groups *Nieuw Arch. Wisk.* **4** 73–112

[6] Li F and Duplij S Weak Hopf algebras and singular solutions of quantum Yang-Baxter equation *Commun. Math. Phys.* **225** 191–217

[7] Murphy G J 1987 Ordered groups and Toeplitz algebras *J. Operator Theory* **18** 303–26

[8] Murphy G J and Tuset L 2004 Aspects of compact quantum group theory *Proc. Amer. Math. Soc.* **132** 3055–67

[9] Sołtan P 2009 On quantum semigroup actions on finite quantum spaces *Infin. Dimens. Anal. Quantum Probab. Relat. Top.* **12** 503–9

[10] Vershik A M 2007 Krein duality, positive 2-algebras, and the dilation of comultiplications *Funct. Anal. Appl.* **41** 99–114

[11] Wagner V V 1952 Generalized groups *Doklady Akademii Nauk SSSR* **84** 1119–22

[12] Woronowicz S L 1998 Compact quantum groups *Symetries quantiques* (Amsterdam: North-Holland) pp 845–84

[13] Woronowicz S L 1991 Unbounded elements affiliated with C^*-algebras and non-compact quantum groups *Commun. Math. Phys.* **136** 399–432

A maximally superintegrable deformation of the N-dimensional quantum Kepler–Coulomb system

A Ballesteros[1], A Enciso[2], F J Herranz[1], O Ragnisco[3] and D Riglioni[4]

[1] Departamento de Física, Universidad de Burgos, E-09001 Burgos, Spain
[2] Instituto de Ciencias Matemáticas, CSIC, E-28049 Madrid, Spain
[3] Dipartimento di Fisica, Università di Roma Tre and Istituto Nazionale di Fisica Nucleare sezione di Roma Tre, Via Vasca Navale 84, I-00146 Roma, Italy
[4] Centre de Recherches Mathématiques, Université de Montréal, H3T 1J4 2920 Chemin de la tour, Montreal, Canada

E-mail: angelb@ubu.es, aenciso@icmat.es, fjherranz@ubu.es, ragnisco@fis.uniroma3.it, riglioni@crm.umontreal.ca

Abstract. The N-dimensional quantum Hamiltonian

$$\hat{H} = -\frac{\hbar^2|\mathbf{q}|}{2(\eta+|\mathbf{q}|)}\nabla^2 - \frac{k}{\eta+|\mathbf{q}|}$$

is shown to be exactly solvable for any real positive value of the parameter η. Algebraically, this Hamiltonian system can be regarded as a new maximally superintegrable η-deformation of the N-dimensional Kepler–Coulomb Hamiltonian while, from a geometric viewpoint, this superintegrable Hamiltonian can be interpreted as a system on an N-dimensional Riemannian space with nonconstant curvature. The eigenvalues and eigenfunctions of the model are explicitly obtained, and the spectrum presents a hydrogen-like shape for positive values of the deformation parameter η and of the coupling constant k.

1. Introduction

Let us consider the N-dimensional (ND) classical Hamiltonian given by

$$\mathcal{H}_\eta(\mathbf{q},\mathbf{p}) = \mathcal{T}_\eta(\mathbf{q},\mathbf{p}) + \mathcal{U}_\eta(\mathbf{q}) = \frac{|\mathbf{q}|\mathbf{p}^2}{2(\eta+|\mathbf{q}|)} - \frac{k}{\eta+|\mathbf{q}|}, \tag{1}$$

where k and η are real parameters, $\mathbf{q} = (q_1,\ldots,q_N)$, $\mathbf{p} = (p_1,\ldots,p_N) \in \mathbb{R}^N$ are conjugate coordinates and momenta, and $\mathbf{q}^2 \equiv |\mathbf{q}|^2 = \sum_{i=1}^N q_i^2$. We recall that $\mathcal{H}_\eta$ has been proven to be a maximally superintegrable Hamiltonian by making use of symmetry techniques [1]. This means that $\mathcal{H}_\eta$ is endowed with the maximum possible number of $(2N-1)$ functionally independent constants of motion (including $\mathcal{H}_\eta$ itself).

Explicitly, $(2N-3)$ of such integrals are provided by the radial symmetry of the system, namely,

$$\mathcal{C}^{(m)} = \sum_{1\le i<j\le m}(q_ip_j-q_jp_i)^2, \qquad \mathcal{C}_{(m)} = \sum_{N-m<i<j\le N}(q_ip_j-q_jp_i)^2, \qquad m=2,\ldots,N; \tag{2}$$

such that $\mathcal{C}^{(N)} = \mathcal{C}_{(N)} \equiv \mathbf{L}^2$ is the square of the total angular momentum. Furthermore, $\mathcal{H}_\eta$ is endowed with an ND Laplace–Runge–Lenz vector $\mathbf{R}$. This means there exist N additional constants of motion coming from the components of $\mathbf{R}$, which are given by

$$\mathcal{R}_i = \sum_{j=1}^{N} p_j(q_j p_i - q_i p_j) + \frac{q_i}{|\mathbf{q}|}(\eta\mathcal{H}_\eta + k), \qquad i = 1, \dots, N. \tag{3}$$

The squared modulus of $\mathbf{R}$ is radially symmetric, and turns out to be expressible in terms of $\mathcal{H}_\eta$ and $\mathbf{L}^2$:

$$\mathbf{R}^2 = \sum_{i=1}^{N} \mathcal{R}_i^2 = 2\mathbf{L}^2\mathcal{H}_\eta + (\eta\mathcal{H}_\eta + k)^2.$$

We remark that each of the three sets $\{\mathcal{H}_\eta, \mathcal{C}^{(m)}\}$, $\{\mathcal{H}_\eta, \mathcal{C}_{(m)}\}$ $(m = 2, \dots, N)$ and $\{\mathcal{R}_i\}$ $(i = 1, \dots, N)$ is formed by N functionally independent functions in involution, and the set $\{\mathcal{H}_\eta, \mathcal{C}^{(m)}, \mathcal{C}_{(m)}, \mathcal{R}_i\}$ for $m = 2, \dots, N$ with a fixed i provides the set of $(2N-1)$ functionally independent functions. As it was shown in [1], the set of constants of the motion $\mathcal{R}_i$ can be obtained explicitly by appying a Stäckel transform [2, 3] to the nondeformed Kepler–Coulomb (KC) Hamiltonian.

This maximally superintegrable Hamiltonian is obviously endowed with an $\mathfrak{so}(N)$ Lie–Poisson symmetry, since it can be constructed on an ND spherically symmetric space. In particular, we can consider the $N(N-1)/2$ generators of rotations $J_{ij} = q_i p_j - q_j p_i$ with $i < j$ and $i, j = 1, \dots, N$ which span the $\mathfrak{so}(N)$ Lie–Poisson algebra with Poisson brackets given by

$$\{J_{ij}, J_{ik}\} = J_{jk}, \qquad \{J_{ij}, J_{jk}\} = -J_{ik}, \qquad \{J_{ik}, J_{jk}\} = J_{ij}, \qquad i < j < k.$$

Hence the $(2N-3)$ angular momentum integrals $\mathcal{C}^{(m)}$ and $\mathcal{C}_{(m)}$ (2) turn out to be the quadratic Casimirs of some rotation subalgebras $\mathfrak{so}(m) \subset \mathfrak{so}(N)$. Moreover we observe that the 'η-deformation' of the Laplace–Runge–Lenz vector $\mathcal{R}_i$ (3) closes the same Poisson algebra as its nondeformed counterpart, namely:

$$\{J_{ij}, \mathcal{R}_k\} = \delta_{ik}\mathcal{R}_j - \delta_{jk}\mathcal{R}_i, \qquad \{\mathcal{R}_i, \mathcal{R}_j\} = -2\mathcal{H}_\eta J_{ij}.$$

Therefore, since

$$\{J_{ij}, \mathcal{H}_\eta\} = \{\mathcal{R}_i, \mathcal{H}_\eta\} = 0, \qquad \forall i, j,$$

the Hamiltonian $\mathcal{H}_\eta$ behaves as a 'constant' with respect to the $N(N+1)/2$ 'generators' $\{J_{ij}, \mathcal{R}_i\}$. This fact makes possible to identify the classical integrals of the motion with the generators of an $\mathfrak{so}(N+1)$ algebra similarly to what happens with the usual ND Euclidean KC system (see [4] and references therein).

We stress that maximally superintegrable Hamiltonians in N dimensions are quite scarce, even on the Euclidean space. The two representative examples of this class of systems are the KC system and the isotropic harmonic oscillator, for which all bounded trajectories are periodic (Bertrand's Theorem). In this respect, we recall that a maximally superintegrable 'deformation/generalization' of the ND isotropic oscillator was firstly presented in [5] and its quantum counterpart was constructed and solved in [6, 7]. In fact, on the same footing of such a maximally superintegrable oscillator system, the system $\mathcal{H}_\eta$ (1) can be regarded as a genuine (maximally superintegrable) η-deformation of the ND usual KC system, since the limit $\eta \to 0$ of $\mathcal{H}_\eta$ (1) yields

$$\mathcal{H}_0 = \frac{1}{2}\mathbf{p}^2 - \frac{k}{|\mathbf{q}|}.$$

Moreover, from a geometric perspective the kinetic energy term $\mathcal{T}_\eta$ can be interpreted as the one generating the geodesic motion of a particle with unit mass on a conformally flat space $\mathcal{M}^N = (\mathbb{R}^N, g)$, which is the complete Riemannian manifold with metric

$$\mathrm{d}s^2 = \left(1 + \frac{\eta}{|\mathbf{q}|}\right) \mathrm{d}\mathbf{q}^2 \tag{4}$$

and nonconstant scalar curvature given by

$$R = \eta(N-1)\frac{4(N-3)r + 3(N-2)\eta}{4r(\eta + r)^3},$$

where we have introduced the radial coordinate $r = |\mathbf{q}|$. Here it is straightforward to check that the limit $\eta \to 0$ of (4) returns the flat Euclidean metric $\mathrm{d}s^2 = \mathrm{d}\mathbf{q}^2$ with $R = 0$. In fact, we stress that $\mathcal{H}_\eta$ can be naturally related to the Taub-NUT system [8, 9, 10, 11, 12, 13, 14, 15, 16, 17, 18, 19] since $\mathcal{M}^N$ can be regarded as the (Riemannian) ND Taub-NUT space [20].

The aim of this contribution is to anticipate the main results concerning a maximally superintegrable quantization of the classical Hamiltonian (1), since a complete study of this new exactly solvable quantum system will be given elsewhere [21]. In the next Section, the properties of the classical system, including its effective potential, are presented. In Section 3 the corresponding quantum Hamiltonian is constructed by imposing the existence of the quantum analog of the full set of $(2N-1)$ classical integrals of the motion (2) and (3). Finally, the explicit solution of the spectral problem is sketched in Section 4.

2. The classical Hamiltonian and its effective potential

Firstly, it should be remarked that, quite surprisingly, the potential $\mathcal{U}_\eta$ (1) with $\eta \neq 0$ can be considered as the ND generalization of an 'intrinsic' oscillator on the curved space $\mathcal{M}^N$. This statement comes from the approach introduced in [20], that generalizes the Bertrand's Theorem [22] to 3D conformally flat Riemannian spaces, thus providing a more general notion of KC and harmonic oscillator potentials [23, 24, 25]. To be self-contained, let us briefly recall these ideas by considering a 3D spherically symmetric space $\mathcal{M}^3$ with coordinates (q_1, q_2, q_3) and equipped with a metric

$$g_{ij} = f(|\mathbf{q}|)^2 \delta_{ij},$$

such that $f(|\mathbf{q}|) = f(r)$ is the conformal factor. Then the corresponding Laplace–Beltrami operator on $\mathcal{M}^3$ is given by

$$\Delta_{\mathcal{M}^3} = \sum_{i,j=1}^{3} \frac{1}{\sqrt{g}} \frac{\partial}{\partial q_i} \sqrt{g}\, g^{ij} \frac{\partial}{\partial q_j},$$

where g^{ij} is the inverse of the metric tensor g_{ij}, and g is the determinant of g_{ij}. The radial symmetric Green function $U(|\mathbf{q}|) = U(r)$ on $\mathcal{M}^3$ (up to multiplicative and additive constants) is defined as the positive nonconstant solution to the equation

$$\Delta_{\mathcal{M}^3} U(r) = 0 \quad \text{on} \quad \mathcal{M}^3 \backslash \{\mathbf{0}\},$$

namely,

$$U(r) = \int^r \frac{\mathrm{d}r'}{r'^2 f(r')}. \tag{5}$$

The prescription for the ND case [5] is to keep the very same definitions given in [24, 25, 26, 27, 28, 29] for the KC and oscillator potentials on $\mathcal{M}^3$. In particular, the *intrinsic KC potential* on the ND space $\mathcal{M}^N$ will be defined by

$$\mathcal{U}_{\mathrm{KC}}(r) := A\, U(r) + B, \tag{6}$$

while the *intrinsic oscillator potential* is defined to be proportional to the inverse square of the KC potential

$$\mathcal{U}_{\rm O}(r) := \frac{C}{U^2(r)} + D, \tag{7}$$

where A, B, C and D are real constants.

Hence according to the above definitions and by considering the Taub-NUT metric (4), with conformal factor

$$f(r) = \sqrt{1 + \frac{\eta}{r}},$$

it is straightforward to obtain that the corresponding intrinsic potentials read

$$\mathcal{U}_{\rm KC}(r) = A\sqrt{1 + \frac{\eta}{r}} + B, \qquad \mathcal{U}_{\rm O}(r) = C\frac{r}{r+\eta} + D.$$

Therefore, whenever $\eta \neq 0$, we find that the potential (1) corresponds to $\mathcal{U}_{\rm O}$ provided that $C = k/\eta$ and $D = -C$:

$$\mathcal{U}_\eta(r) = \frac{k}{\eta}\left(\frac{r}{r+\eta} - 1\right) = -\frac{k}{\eta + r},$$

which shows that $\mathcal{U}_\eta$ can be interpreted as an intrinsic oscillator on $\mathcal{M}^N$ with metric (4).

In order to understand the dynamical properties of the system (such as the existence of bounded states and any other critical features) we shall make its radial symmetry manifest by introducing the hyperspherical coordinates r, θ_j $(j = 1, \ldots, N-1)$:

$$q_j = r\cos\theta_j \prod_{k=1}^{j-1} \sin\theta_j, \quad 1 \le j < N, \qquad q_N = r\prod_{k=1}^{N-1} \sin\theta_k. \tag{8}$$

Their corresponding canonical momenta p_r, p_{θ_j} can be straightforwardly computed, and in these hyperspherical variables the ND Hamiltonian (1) takes the following 1D radial form

$$\mathcal{H}_\eta(r, p_r) = \mathcal{T}_\eta(r, p_r) + \mathcal{U}_\eta(r) = \frac{r}{2(\eta + r)}\left(p_r^2 + \frac{\mathbf{L}^2}{r^2}\right) - \frac{k}{\eta + r}, \tag{9}$$

where the total angular momentum is given by

$$\mathbf{L}^2 = \sum_{j=1}^{N-1} p_{\theta_j}^2 \prod_{k=1}^{j-1} \frac{1}{\sin^2\theta_k}.$$

The contribution to the dynamics of both of the non-flat metric and the potential can be better understood if we consider a new set of canonical variables Q, P given by

$$Q(r) = \sqrt{r(\eta + r)} + \eta\log(\sqrt{r} + \sqrt{r+\eta}),$$
$$P(r, p_r) = \sqrt{\frac{r}{\eta + r}}\, p_r.$$

In terms of these new variables the Hamiltonian (9) is written as

$$\mathcal{H}_\eta(Q, P) = \frac{1}{2}P^2 + \frac{\mathbf{L}^2}{2r(Q)(\eta + r(Q))} + \mathcal{U}(r(Q)) \equiv \frac{1}{2}P^2 + \mathcal{U}_{\rm eff}(Q),$$

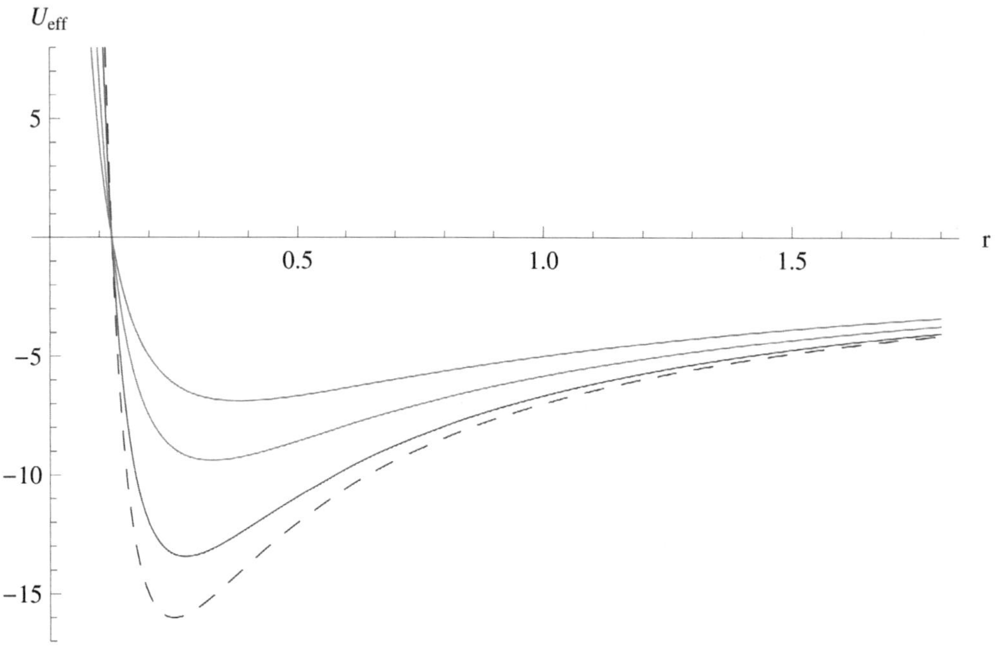

Figure 1. The effective potential $\mathcal{U}_{\text{eff}}$ (10) with $k = 8$ and $\mathbf{L}^2 = 2$, for $\eta = \{0, 0.05, 0.2, 0.4\}$. The dashed line corresponds to the KC potential ($\eta = 0$) and increasing values of η lead to higher values of the potential.

where the effective potential is thus given by

$$\mathcal{U}_{\text{eff}}(Q(r)) = \frac{\mathbf{L}^2}{2r(\eta + r)} - \frac{k}{\eta + r}. \tag{10}$$

Consequently, the radial motion can be described as the 1D dynamics of a particle under the effective potential $\mathcal{U}_{\text{eff}}(Q(r))$. As it can be appreciated from figure 1, the radial equation admits a hydrogen-like potential for $\eta > 0$ and $k > 0$, which can be interpreted as a genuine η-deformation of the effective potential for the KC system.

3. A maximally superintegrable quantization

In order to obtain the quantum analog of the kinetic energy term $\mathcal{T}_\eta$ (1) we have to deal with the unavoidable ordering problems in the canonical quantization process that come from the nonzero curvature of the underlying space (see, e.g [7] and references therein). A detailed analysis of the different possible quantization prescriptions together with a proof of their equivalence through gauge transformations will be presented in a forthcoming paper [21]. One of this prescriptions consists in the so called 'direct' or Schrödinger quantization [15], under which the quantum Hamiltonian $\mathcal{H}_\eta$ keeps the maximal superintegrability property and is therefore endowed with $(2N-1)$ algebraically independent operators that commute with $\mathcal{H}_\eta$. This prescription has been already succesfully used in the case of a curved (Darboux III) oscillator system [6, 7] and makes use of all the algebraic machinery coming from the symmetries of the classical Hamiltonian. This result can be stated as follows.

Theorem 1. *Let $\hat{\mathcal{H}}_\eta$ be the quantum Hamiltonian given by*

$$\hat{\mathcal{H}}_\eta = \frac{|\hat{\mathbf{q}}|}{2(\eta + |\hat{\mathbf{q}}|)}\,\hat{\mathbf{p}}^2 - \frac{k}{\eta + |\hat{\mathbf{q}}|} = \frac{|\mathbf{q}|}{2(\eta + |\mathbf{q}|)}\left(-\hbar^2\nabla^2 - \frac{2k}{|\mathbf{q}|}\right), \tag{11}$$

where $\hat{\mathbf{q}} = \mathbf{q}$, $\hat{\mathbf{p}} = -i\hbar\nabla$ *and* $\nabla = (\partial_1, \ldots, \partial_N)$ *such that* $[\hat{q}_i, \hat{p}_j] = i\hbar\delta_{ij}$. *For any value of* η *and* k *it is verified that:*
(i) $\hat{\mathcal{H}}_\eta$ *commutes with the following operators* $(m = 2, \ldots, N; i = 1, \ldots, N)$

$$\hat{\mathcal{C}}^{(m)} = \sum_{1\leq i<j\leq m} (\hat{q}_i\hat{p}_j - \hat{q}_j\hat{p}_i)^2, \qquad \hat{\mathcal{C}}_{(m)} = \sum_{N-m\leq i<j\leq N} (\hat{q}_i\hat{p}_j - \hat{q}_j\hat{p}_i)^2, \tag{12}$$

$$\hat{\mathcal{R}}_i = \frac{1}{2}\sum_{j=1}^{N} \hat{p}_j(\hat{q}_j\hat{p}_i - \hat{q}_i\hat{p}_j) + \frac{1}{2}\sum_{j=1}^{N}(\hat{q}_j\hat{p}_i - \hat{q}_i\hat{p}_j)\hat{p}_j + \frac{\hat{q}_i}{\hat{\mathbf{q}}}\left(\eta\hat{\mathcal{H}}_\eta + k\right),$$

where $\hat{\mathcal{C}}^{(N)} = \hat{\mathcal{C}}_{(N)} \equiv \hat{\mathbf{L}}^2$ *is the total quantum angular momentum and*

$$\hat{\mathbf{R}}^2 = \sum_{i=1}^{N} \hat{\mathcal{R}}_i^2 = 2\hat{\mathcal{H}}_\eta\left(\hat{\mathbf{L}}^2 + \hbar^2\frac{(N-1)^2}{4}\right) + \left(\eta\hat{\mathcal{H}}_\eta + k\right)^2.$$

(ii) Each of the three sets $\{\hat{\mathcal{H}}_\eta, \hat{\mathcal{C}}^{(m)}\}$, $\{\hat{\mathcal{H}}_\eta, \hat{\mathcal{C}}_{(m)}\}$ $(m = 2, \ldots, N)$ *and* $\{\hat{\mathcal{R}}_i\}$ $(i = 1, \ldots, N)$ *is formed by* N *algebraically independent commuting operators.*
(iii) The set $\{\hat{\mathcal{H}}_\eta, \hat{\mathcal{C}}^{(m)}, \hat{\mathcal{C}}_{(m)}, \hat{\mathcal{R}}_i\}$ *for* $m = 2, \ldots, N$ *with a fixed index* i *is formed by* $2N - 1$ *algebraically independent operators*
(iv) $\hat{\mathcal{H}}_\eta$ *is formally self-adjoint on the Hilbert space* $L^2(\mathcal{M}^N)$, *endowed with the scalar product*

$$\langle\Psi|\Phi\rangle = \int_{\mathbb{R}^N} \overline{\Psi(\mathbf{q})}\Phi(\mathbf{q})\left(1 + \frac{\eta}{|\mathbf{q}|}\right) d\mathbf{q}.$$

The proof of this result can be obtained through direct computation. Recall that in [7] we have considered a similar problem on the conformally flat Darboux III space.

Therefore, the Hamiltonian (11) leads to the following Schrödinger equation

$$\left(\frac{-\hbar^2|\mathbf{q}|}{2(\eta + |\mathbf{q}|)}\nabla^2 - \frac{k}{\eta + |\mathbf{q}|}\right)\Psi(\mathbf{q}) = E\Psi(\mathbf{q}),$$

which in hyperspherical variables (8) turns into

$$\frac{r}{2(\eta + r)}\left(-\hbar^2\partial_r^2 - \frac{\hbar^2(N-1)}{r}\partial_r + \frac{\hat{\mathbf{L}}^2}{r^2} - \frac{2k}{r}\right)\Psi(r, \boldsymbol{\theta}) = E\Psi(r, \boldsymbol{\theta}), \tag{13}$$

where $\boldsymbol{\theta} = (\theta_1, \ldots, \theta_{N-1})$.

We remark that all the above results are well defined for any value of the parameters η and k. Nevertheless, the explicit solution of the quantum Hamiltonian depends on the sign of both of them. In particular, hereafter we shall restrict ourselves to consider $\eta > 0$, which implies that the variable $r \in (0, +\infty)$, and $k > 0$ which corresponds to the proper 'curved' hydrogen atom potential.

4. Spectrum and eigenfunctions

In view of the effective potential introduced in (10), one should expect that the quantum Hamiltonian (11) should have both a discrete and a continuous spectrum, and this is indeed the case. When the Schrödinger equation (13) is expressed in hyperspherical variables, it can be solved by factorizing the wave function into radial and angular components

$$\Psi(r, \boldsymbol{\theta}) = \psi(r)Y(\boldsymbol{\theta}),$$

and by considering the separability provided by the first integrals $\hat{C}_{(m)}$ (12) with eigenvalue equations given by

$$\hat{C}_{(m)}\Psi = c_m\Psi, \qquad m = 2, \ldots, N.$$

From it, we obtain that $Y(\boldsymbol{\theta})$ solves completely the angular part and this corresponds, as expected, to the hyperspherical harmonics satisfying that

$$\hat{C}_{(N)}Y(\boldsymbol{\theta}) \equiv \hat{\mathbf{L}}^2 Y(\boldsymbol{\theta}) = \hbar^2 l(l+N-2)Y(\boldsymbol{\theta}), \quad l = 0, 1, 2\ldots$$

where l is the angular momentum quantum number. It can be proven [7] that the eigenvalues c_m are related with the $N-1$ quantum numbers of the angular observables through

$$c_k \leftrightarrow l_{k-1}, \qquad k = 2, \ldots, N-1, \qquad c_N \leftrightarrow l,$$

which means that

$$Y(\boldsymbol{\theta}) \equiv Y^{c_N}_{c_{N-1},\ldots,c_2}(\theta_1, \theta_2, \ldots, \theta_{N-1}) \equiv Y^{l}_{l_{N-2},\ldots,l_1}(\theta_1, \theta_2, \ldots, \theta_{N-1}).$$

As a consequence, the radial Schrödinger equation (13) is given by

$$\frac{r}{2(\eta+r)}\left(-\hbar^2\frac{\mathrm{d}^2}{\mathrm{d}r^2} - \frac{\hbar^2(N-1)}{r}\frac{\mathrm{d}}{\mathrm{d}r} + \frac{\hbar^2 l(l+N-2)}{r^2} - \frac{2k}{r}\right)\psi(r) = E\psi(r),$$

which can be written in the form

$$\left(-\hbar^2\frac{\mathrm{d}^2}{\mathrm{d}r^2} - \frac{\hbar^2(N-1)}{r}\frac{\mathrm{d}}{\mathrm{d}r} + \frac{\hbar^2 l(l+N-2)}{r^2} - \frac{2K}{r}\right)\psi(r) = 2E\psi(r), \tag{14}$$

where the new 'coupling constant' K turns out to be energy–dependent

$$K = k + \eta E. \tag{15}$$

In this way we find that the equation (14) is formally equivalent to the Schrödinger equation of the radial hydrogen atom, whose bounded eigenfunctions are given, in terms of the generalised Laguerre polynomials L_n^α, by:

$$\psi_{n,l}(r) = r^l \exp\left(-\frac{Kr}{\hbar^2\left(n+l+\frac{N-1}{2}\right)}\right) L_n^{2l+N-2}\left(\frac{2Kr}{\hbar^2\left(n+l+\frac{N-1}{2}\right)}\right). \tag{16}$$

Notice that the eigenfunctions do not only depend on the usual quantum numbers n, l but also on the eigenvalue E through K (15).

If we substitute these functions within the equation (14), the following algebraic equation is obtained:

$$E = -\frac{K^2}{2\hbar^2\left(n+l+\frac{N-1}{2}\right)^2} = -\frac{(k+\eta E)^2}{2\hbar^2\left(n+l+\frac{N-1}{2}\right)^2}.$$

By solving such a quadratic equation in terms of E we obtain the bounded discrete spectrum of the system, whose eigenvalues depend both on the 'deformation' parameter η and on the quantum numbers n, l. Namely,

$$E^\eta_{n,l} = \frac{-\hbar^2\left(n+l+\frac{N-1}{2}\right)^2 - \eta k + \sqrt{\hbar^4\left(n+l+\frac{N-1}{2}\right)^4 + 2\eta k\hbar^2\left(n+l+\frac{N-1}{2}\right)^2}}{\eta^2}. \tag{17}$$

In this way, the eigenfunctions, $\psi^{\eta}_{n,l}(r)$, can be explicitly obtained by introducing (17) into K (15) and, next, by substituting the latter in (16). Note that in the limit $\eta \to 0$ the following well–known expression for the energies is recovered:

$$E^0_{n,l} = -\frac{k^2}{2\hbar^2\left(n+l+\frac{N-1}{2}\right)^2}.$$

And the first-order effect of the deformation on the spectrum can be appreciated through a power series expansion in η:

$$E^{\eta}_{n,l} = E^0_{n,l} + \eta\,\frac{k^3}{2\hbar^4\left(n+l+\frac{N-1}{2}\right)^4} + o(\eta^2). \tag{18}$$

Let us remark that since the spectrum of $\hat{\mathcal{H}}_\eta$ is bounded from below (as the classical effective potential indicates), for a sufficiently large k we can safely assume that $K = k + \eta E > 0$. Moreover the condition $\Psi \in L^2(\mathcal{M}^N)$ translates to

$$\int_{\mathbb{R}^N} |\Psi(\mathbf{q})|^2 \left(1 + \frac{\eta}{|\mathbf{q}|}\right) \mathrm{d}\mathbf{q} < \infty.$$

Note also that the degeneracy of this spectrum is exactly the same as in the ND hydrogen atom, which is a strong signature of the maximal superintegrability of this quantum system. The explicit expressions for the wave functions corresponding to the continous spectrum will be given in the forthcoming paper [21].

Finally let us stress that if we consider the following reparametrization

$$\eta \to \frac{1}{\sqrt{\lambda}}, \qquad k \to \frac{\omega^2}{\sqrt{\lambda}}.$$

the spectrum (17) turns out to have the same dependence on the principal quantum number (which in this case corresponds to $\mathcal{N} = n+l$) as the bounded spectrum for the intrinsic oscillator on a Darboux III space that was given in [6, 7]. Indeed, this fact can be understood in terms of the classification of intrinsic KC and oscillator systems on non-Euclidean spaces provided in [30]. According to such classification, the Taub-NUT and the Darboux III oscillator are superintegrable systems of type II with the same parameters λ, δ (Taub-NUT: $\gamma = \frac{1}{2}, \lambda = 0, \delta$; Darboux III: $\gamma = 1, \lambda = 0, \delta$). However as showed in [31] for the intrinsic KC systems, the parameter γ only affects the form of the principal quantum number $\mathcal{N}$ (Taub-NUT $\mathcal{N} = n + l$; Darboux III: $\mathcal{N} = 2n + l$) but not the overall dependence on $\mathcal{N}$ which indeed turns out to be the same.

Acknowledgments
This work was partially supported by the Spanish MINECO through the Ramón y Cajal program (A.E.) and under grants MTM2010-18556 (A.B and F.J.H.), AIC-D-2011-0711 (MINECO-INFN) (A.B, F.J.H. and O.R.) and FIS201-22566 (A.E.), by the ICMAT Severo Ochoa under grant SEV-2011-0087 (A.E.), by Banco Santander-UCM under grant GR35/10-A-910556 (A.E.), and by a postdoctoral fellowship by the Laboratory of Mathematical Physics of the CRM, Université de Montréal (D.R.).

Journal of Physics: Conference Series **474** (2013) 012008 doi:10.1088/1742-6596/474/1/012008

References

[1] Ballesteros A, Enciso A, Herranz F J, Ragnisco O and Riglioni D 2011 *SIGMA* **7** 048
[2] Kalnins E G, Kress J M and Miller W Jr 2005 *J. Math. Phys.* **46** 053510
[3] Kalnins E G, Kress J M and Miller W Jr 2006 *J. Math. Phys.* **47** 043514
[4] Ballesteros A, and Herranz F J 2009 *J. Phys. A: Math. Theor.* **42** 245203
[5] Ballesteros A, Enciso A, Herranz F J and Ragnisco O 2008 *Physica D* **237** 505
[6] Ballesteros A, Enciso A, Herranz F J, Ragnisco O and Riglioni D 2011 *Phys. Lett. A* **375** 1431
[7] Ballesteros A, Enciso A, Herranz F J, Ragnisco O and Riglioni D 2011 *Ann. Phys.* **326** 2053
[8] Manton N S 1982 *Phys. Lett. B* **110** 54
[9] Atiyah M F and N.J. Hitchin N J 1985 *Phys. Lett. A* **107** 21
[10] Gibbons G W and Manton N S 1986 *Nucl. Phys. B* **274** 183
[11] Fehér L G and Horváthy P A 1987 *Phys. Lett. B* **183** 182
[12] Gibbons G W and Ruback P J 1988 *Comm. Math. Phys.* **115** 267
[13] Iwai T and Katayama N 1994 *J.. Phys. A: Math. Gen.* **27** 3179
[14] Iwai T and Katayama N 1995 *J. Math. Phys.* **36** 1790
[15] Iwai T, Uwano Y and Katayama N 1996 *J. Math. Phys.* **37** 608
[16] Bini D, Cherubini C and Jantzen R T 2002 *Class. Quantum Grav.* **19** 5481
[17] Bini D, Cherubini C, Jantzen R T and Mashhoon B 2003 *Class. Quantum Grav.* **20** 457
[18] Gibbons G W and Warnick C M 2007 *J. Geom. Phys.* **57** 2286
[19] Jezierski J and Lukasik M 2007 *Class. Quantum Grav.* **24** 1331
[20] Ballesteros A, Enciso A, Herranz F J and Ragnisco O 2009 *Ann. Phys.* **324** 1219
[21] Ballesteros A, Enciso A, Herranz F J, Ragnisco O and Riglioni D In preparation
[22] Bertrand J 1873 *C.R. Acad. Sci. Paris* **77** 849
[23] Perlick V 1992 *Class. Quantum Grav.* **9** 1009
[24] Ballesteros A, Enciso A, Herranz F J and Ragnisco O 2008 *Class. Quantum Grav.* **25** 165005
[25] Ballesteros A, Enciso A, Herranz F J and Ragnisco O 2009 *Commun. Math. Phys.* **290** 1033
[26] Enciso A and Peralta-Salas D 2007 *J. Geom. Phys.* **57** 1679
[27] Enciso A and Peralta-Salas D 2009 *Indiana Univ. Math. J.* **58** 1947
[28] Li P and Tam L F 1987 *Amer. J. Math.* **109** 1129
[29] Li P and Tam L F 1995 *J. Differential Geom.* **41** 277
[30] Ballesteros A, Enciso A, Herranz F J, Ragnisco O and Riglioni D 2013 *Nankai Series in Pure, Appl. Math. Theor. Phys.* **11** ed C Bai, J P Gazeau and Mo-Lin Ge (Singapore: World Scientific) p 211, arXiv:1304.4544
[31] Riglioni D, *J. Phys. A: Math. Theor.* **46** 265207

On permutation-twisted free fermions and two conjectures

Katrina Barron
Department of Mathematics, University of Notre Dame, Notre Dame, IN 46556, USA
E-mail: kbarron@nd.edu

Nathan Vander Werf
Department of Mathematics, University of Notre Dame, Notre Dame, IN 46556, USA
E-mail: nvanderw@nd.edu

Abstract. We conjecture that the category of permutation-twisted modules for a multi-fold tensor product vertex operator superalgebra and a cyclic permutation of even order is isomorphic to the category of parity-twisted modules for the underlying vertex operator superalgebra. This conjecture is based on our observations of the cyclic permutation-twisted modules for free fermions as we discuss in this work, as well as previous work of the first author constructing and classifying permutation-twisted modules for tensor product vertex operator superalgebras and a permutation of odd order. In addition, we observe that the transposition isomorphism for two free fermions corresponds to a lift of the -1 isometry of the integral lattice vertex operator superalgebra corresponding to two free fermions under boson-fermion correspondence. We conjecture that all even order cyclic permutation automorphisms of free fermions can be realized as lifts of lattice isometries under boson-fermion correspondence. We discuss the role of parity stability in the construction of these twisted modules and prove that in general, parity-unstable weak twisted modules for a vertex operator superalgebras come in pairs that form orthogonal invariant subspaces of parity-stable weak twisted modules, clarifying their role in many other settings.

1. Introduction and preliminaries

Let V be a vertex operator (super)algebra, and for a fixed positive integer k, consider the tensor product vertex operator (super)algebra $V^{\otimes k}$ (see [FLM3], [FHL]). Any element g of the symmetric group S_k acts in a natural way on $V^{\otimes k}$ as a vertex operator (super)algebra automorphism, and thus it is appropriate to consider g-twisted $V^{\otimes k}$-modules. This is the setting for permutation orbifold conformal field theory, and for permutation orbifold superconformal field theory if the vertex operator superalgebra is not just super, but is also supersymmetric, i.e. is a representation of a Neveu-Schwarz super-extension of the Virasoro algebra.

In [BDM], the first author along with Dong and Mason constructed and classified the g-twisted $V^{\otimes k}$-modules for V a vertex operator algebra and $g \in S_k$. In particular, it was proved that the category of weak $(1\ 2\ \cdots\ k)$-twisted $V^{\otimes k}$-modules is isomorphic to the category of weak V-modules. In [Bar12], the first author extended this result to a construction and classification of $(1\ 2\ \cdots\ k)$-twisted $V^{\otimes k}$-modules for V a vertex operator superalgebra and k odd. However, as was shown by Barron in [Bar12], the results of [BDM] for permutation-twisted tensor product vertex operator algebras and the results of Barron for the odd order case in the super setting, do not extend in a natural way to the super setting for permutation automorphisms of even order. Rather, the construction of the twisted modules for even order permutations is fundamentally different whenever V has nontrivial odd subspace.

In this paper, following first the construction of [DZ2], we study the case of $V_{fer}^{\otimes k}$ a k-fold tensor product of the free fermion vertex operator superalgebra for k even and $(1\ 2\ \cdots\ k)$ acting as a vertex operator superalgebra automorphism to gain further insight into the problem. These models, along with observations from [Bar12], provide the basis for a conjecture we make on the nature of the classification of permutation-twisted modules in general. Our main conjecture is that for k even, the category of weak $(1\ 2\ \cdots\ k)$-twisted $V^{\otimes k}$-modules is isomorphic to the category of weak parity-twisted V-modules. This contrasts to the case when k is odd where, as shown in [Bar12], the category of $(1\ 2\ \cdots\ k)$-twisted $V^{\otimes k}$-modules is isomorphic to the category of weak untwisted V-modules.

Permutation-twisted modules for free fermions, as well as parity-twisted modules for free fermions, give examples of how "parity-unstable" twisted modules arise. Motivated by these examples, we prove that parity-unstable weak twisted modules arise as orthogonal pairs of invariant subspaces of a parity-stable weak twisted module. Furthermore, if these two parity-unstable weak twisted modules are ordinary then they will always have the same graded dimension but are not isomorphic, meaning they can not be detected via techniques which only produce the graded dimensions of the twisted modules. This simplifies much of the work in [DZ1], [DZ3], [DH], and shows how from a categorical standpoint, all modules should be defined so as to be parity-stable, and then what is referred to as "parity-unstable modules" in, for instance [DZ1], [DZ3], [DH], should be referred to as "parity-unstable invariant subspaces" of the parity-stable modules.

We use "boson-fermion correspondence" to formulate another conjecture regarding whether one can realize the permutation-twisted modules for free fermions in the even cyclic case via two different constructions. Boson-fermion correspondence refers to the fact that the two free fermion vertex operator superalgebra is isomorphic to the rank one lattice vertex operator superalgebra with length one generator, i.e. one fermion propagating on a circle. Thus one can use the work of [DZ2] to construct the permutation-twisted modules for permutation-twisted free fermions, or one can try to use boson-fermion correspondence and the theory of twisted modules for a lattice vertex operator superalgebra and a lift of a lattice isometry as developed in [DL2] and [X]. If the automorphism on the lattice is the lift of a lattice isometry, then one has an overlap of construction techniques which can potentially give insight into the general theory of the construction and classification of twisted modules.

However, then the question arises: When does the permutation correspond to a lift of a lattice isometry? In this paper we provide an example of when it does, and make a conjecture that any cyclic permutation of even order acting on free fermions is conjugate to a lift of a lattice isometry under boson-fermion correspondence. This gives an alternative construction to even order cyclic permutation-twisted modules for free fermion vertex operator superalgebras based on twisted lattice constructions, i.e. space-time orbifold constructions, versus worldsheet orbifold constructions that has potentially for being extended in general to further explore the connection between the space-time geometry of the lattice versus the worldsheet geometry of propagating strings, following, for instance [BHL], but in a new and different setting which explicitly involves the supergeometry underlying vertex operator superalgebras. This is particularly interesting and relevant for cases involving supersymmetric vertex operator superalgebras, cf. [Bar1] – [Bar13].

It is important, at the same time to note the following: We show (see Remark 4.1 below) that the construction of twisted modules for lattice vertex operator superalgebras and a lift of a lattice isometry can not be used in general to construct the permutation-twisted modules for lattice vertex operator superalgebras directly for the case of permutations of even order. This is because if ν is a permutation isometry on the lattice of even order k, due to certain properties of ν, which we note in Remark 4.1, the construction of [DL2], [X], only holds for ν lifted to an automorphism $\hat{\nu}$ of the vertex operator superalgebra such that $\hat{\nu}$ is of order $2k$. Thus it is impossible for $\hat{\nu}$ to correspond to the permutation automorphism of order k. Details are given in Section 4 and Section 5, and in particular in Remark 4.1. In particular, we give general criteria for a lattice isometry of order k to lift to a vertex operator superalgebra automorphism of order k. In the process of this, we clarify aspects of the construction of twisted modules for a lift of a lattice isometry and a positive definite integral lattice following [DL2] and [X].

1.1. Background

Twisted vertex operators were discovered and used in [LW]. Twisted modules for vertex operator algebras arose in the work of I. Frenkel, J. Lepowsky and A. Meurman [FLM1], [FLM2], [FLM3] for the case of a lattice vertex operator algebra and the lattice isometry -1, in the course of the construction of the moonshine module vertex operator algebra (see also [Bo]). This structure came to be understood as an "orbifold model" in the sense of conformal field theory and string theory. Twisted modules are the mathematical counterpart of "twisted sectors", which are the basic building blocks of orbifold models in conformal field theory and string theory (see [DHVW1], [DHVW2], [DFMS], [DVVV], [DGM], as well as [KS], [FKS], [Ban1], [Ban2], [BHS], [dBHO], [HO], [GHHO], [Ban3] and [HH]). Orbifold theory plays an important role in conformal field theory and in superextensions, and is also a way of constructing a new vertex operator (super)algebra from a given one.

Formal calculus arising from twisted vertex operators associated to a an even lattice was systematically developed in [Le1], [FLM2], [FLM3] and [Le2], and the twisted Jacobi identity was formulated and shown to hold for these operators (see also [DL2]). These results led to the introduction of the notion of g-twisted V-module [FFR], [D], for V a vertex operator algebra and g an automorphism of V. This notion records the properties of twisted operators obtained in [Le1], [FLM1], [FLM2], [FLM3] and [Le2], and provides an axiomatic definition of the notion of twisted sectors for conformal field theory. In general, given a vertex operator algebra V and an automorphism g of V, it is an open problem as to how to construct a g-twisted V-module.

The focus of this paper is the study of permutation-twisted sectors for free fermion vertex operator superalgebras. A theory of twisted operators for integral lattice vertex operator superalgebras and finite automorphisms that are lifts of a lattice isometry were studied in [DL2] and [X], and the general theory of twisted modules for vertex operator superalgebras was developed by Li in [Li2]. Certain specific examples of permutation-twisted sectors in superconformal field theory have been studied from a physical point of view in, for instance, [FKS], [BHS], [MS1], [MS2].

The case of V a vertex operator superalgebra and $V \otimes V$ being permuted by the (1 2) transposition is the mirror map if $V \otimes V$, in addition to being a vertex operator superalgebra is also N=2 supersymmetric (see for example, [Bar11]). This is one of the motivations for studying this construction in detail in the case of free fermions. Although the free fermion vertex operator superalgebras are not N=2 supersymmetric, they can be used to achieve supersymmetry via tensoring with an appropriate bosonic theory as, for example, in [Bar11]. In particular, a mirror-twisted module for an N=2 supersymmetric vertex operator superalgebra is naturally a representation of the "mirror-twisted N=2 superconformal Lie superalgebra", cf. [Bar10] and [Bar11]. In [Bar13], further classifications and constructions involving this transposition mirror map as well as other mirror maps for N=2 supersymmetric vertex operator superalgebras arising from free fermions are given.

1.2. The notion of vertex operator superalgebra

In this section, we recall the notion of vertex operator superalgebra, following the notation and terminology of, for instance [LL], [Bar3]. Let x, x_0, x_1, x_2, etc., denote commuting independent formal variables. Let $\delta(x) = \sum_{n\in\mathbb{Z}} x^n$. We will use the binomial expansion convention, namely, that any expression such as $(x_1 - x_2)^n$ for $n \in \mathbb{C}$ is to be expanded as a formal power series in nonnegative integral powers of the second variable, in this case x_2.

A *vertex operator superalgebra* is a $\frac{1}{2}\mathbb{Z}$-graded vector space $V = \coprod_{n\in\frac{1}{2}\mathbb{Z}} V_n$, satisfying $\dim V < \infty$ and $V_n = 0$ for n sufficiently negative, that is also $\mathbb{Z}_2$-graded by *sign*

$$V = V^{(0)} \oplus V^{(1)}, \quad \text{with} \quad V^{(j)} = \coprod_{n\in\mathbb{Z}+\frac{j}{2}} V_n,$$

and equipped with a linear map

$$V \longrightarrow (\operatorname{End} V)[[x, x^{-1}]], \qquad v \mapsto Y(v,x) = \sum_{n\in\mathbb{Z}} v_n x^{-n-1}, \tag{1}$$

and with two distinguished vectors $\mathbf{1} \in V_0$, (the *vacuum vector*) and $\omega \in V_2$ (the *conformal element*) satisfying the following conditions for $u, v \in V$:

$$u_n v = 0 \quad \text{for } n \text{ sufficiently large;} \tag{2}$$

$$Y(\mathbf{1}, x) = 1; \tag{3}$$

$$Y(v,x)\mathbf{1} \in V[[x]] \quad \text{and} \quad \lim_{x\to 0} Y(v,x)\mathbf{1} = v; \tag{4}$$

$$x_0^{-1}\delta\left(\frac{x_1 - x_2}{x_0}\right) Y(u,x_1)Y(v,x_2) - (-1)^{|u||v|} x_0^{-1}\delta\left(\frac{x_2 - x_1}{-x_0}\right) Y(v,x_2)Y(u,x_1) \tag{5}$$

$$= x_2^{-1}\delta\left(\frac{x_1 - x_0}{x_2}\right) Y(Y(u,x_0)v, x_2)$$

(the *Jacobi identity*), where $|v| = j$ if $v \in V^{(j)}$ for $j \in \mathbb{Z}_2$;

$$[L(m), L(n)] = (m-n)L(m+n) + \frac{1}{12}(m^3 - m)\delta_{m+n,0}c \tag{6}$$

for $m, n \in \mathbb{Z}$, where $L(n) = \omega_{n+1}$, for $n \in \mathbb{Z}$, i.e., $Y(\omega, x) = \sum_{n\in\mathbb{Z}} L(n)x^{-n-2}$, and $c \in \mathbb{C}$ (the *central charge* of V);

$$L(0)v = nv = (\operatorname{wt} v)v \quad \text{for } n \in \tfrac{1}{2}\mathbb{Z} \text{ and } v \in V_n; \tag{7}$$

$$\frac{d}{dx} Y(v,x) = Y(L(-1)v, x). \tag{8}$$

This completes the definition. We denote the vertex operator superalgebra just defined by $(V, Y, \mathbf{1}, \omega)$, or briefly, by V.

Given two vertex operator superalgebras $(V_1, Y_1, \mathbf{1}^{(1)}, \omega^{(1)})$ and $(V_2, Y_2, \mathbf{1}^{(2)}, \omega^{(2)})$, we have that $(V_1 \otimes V_2,\ Y,\ \mathbf{1}^{(1)} \otimes \mathbf{1}^{(2)},\ \omega^{(1)} \otimes \mathbf{1}^{(2)} + \mathbf{1}^{(1)} \otimes \omega^{(2)})$ is a vertex operator superalgebra, where Y is given by

$$Y(u_1 \otimes u_2, x)(v_1 \otimes v_2) = (-1)^{|u_2||v_1|} Y_1(u_1,x)v_1 \otimes Y_2(u_2,x)v_2, \tag{9}$$

for $u_1 \otimes u_2,\ v_1 \otimes v_2 \in V_1 \otimes V_2$.

Remark 1.1 As a consequence of the definition of vertex operator superalgebra, we have that $\operatorname{wt}(v_n u) = \operatorname{wt} u + \operatorname{wt} v - n - 1$, for $u, v \in V$ and $n \in \mathbb{Z}$. This implies that $v_n \in (\operatorname{End} V)^{(j)}$ if and only if $v \in V^{(j)}$ for $j \in \mathbb{Z}_2$, i.e. that $v_n V^{(j)} \subseteq V^{(j+|v|) \bmod 2}$.

1.3. Automorphisms of vertex operator superalgebras and the notion of twisted module

An *automorphism* of a vertex operator superalgebra V is a linear automorphism g of V preserving $\mathbf{1}$ and ω such that the actions of g and $Y(v,x)$ on V are compatible in the sense that

$$gY(v,x)g^{-1} = Y(gv,x) \tag{10}$$

for $v \in V$. Then $gV_n \subset V_n$ for $n \in \frac{1}{2}\mathbb{Z}$.

If g has finite order, V is a direct sum of the eigenspaces V^j of g,

$$V = \coprod_{j\in\mathbb{Z}/k\mathbb{Z}} V^j, \tag{11}$$

where $k \in \mathbb{Z}_+$ is a period of g (i.e., $g^k = 1$ but k is not necessarily the order of g) and $V^j = \{v \in V \mid gv = \eta^j v\}$, for η a fixed primitive kth root of unity.

Note that we have the following δ-function identity

$$x_2^{-1}\delta\left(\frac{x_1-x_0}{x_2}\right)\left(\frac{x_1-x_0}{x_2}\right)^k = x_1^{-1}\delta\left(\frac{x_2+x_0}{x_1}\right)\left(\frac{x_2+x_0}{x_1}\right)^{-k} \tag{12}$$

for any $k \in \mathbb{C}$.

Next we review the notions of weak, weak admissible and ordinary g-twisted V-module for a vertex operator superalgebra V and an automorphism g of V of finite order k, as well as the notion of "parity stability" for these various kinds of g-twisted V-modules. These are the "standard" definitions, following, for instance [DZ1], [DZ3], [DH]. However, below we argue (see Remark 2.2), that the more natural notion of "weak g-twisted V-module" should be that of a "weak parity-stable g-twisted V-module", and similarly for the notions of weak admissible or ordinary g-twisted V-module.

Let $(V, Y, \mathbf{1}, \omega)$ be a vertex operator superalgebra and let g be an automorphism of V of period $k \in \mathbb{Z}_+$. A *weak g-twisted V-module* is a vector space M equipped with a linear map

$$V \longrightarrow (\mathrm{End}\, M)[[x^{1/k}, x^{-1/k}]], \qquad v \mapsto Y^g(v,x) = \sum_{n\in\frac{1}{k}\mathbb{Z}} v_n^g x^{-n-1}, \tag{13}$$

satisfying the following conditions for $u, v \in V$ of homogeneous sign and $w \in M$:

$$v_n^g w = 0 \quad \text{for } n \text{ sufficiently large}; \tag{14}$$

$$Y^g(\mathbf{1}, x) = 1; \tag{15}$$

$$\begin{aligned} x_0^{-1}\delta\left(\frac{x_1-x_2}{x_0}\right)Y^g(u,x_1)Y^g(v,x_2) - (-1)^{|u||v|}x_0^{-1}\delta\left(\frac{x_2-x_1}{-x_0}\right)Y^g(v,x_2)Y^g(u,x_1) \\ = x_2^{-1}\frac{1}{k}\sum_{j\in\mathbb{Z}/k\mathbb{Z}}\delta\left(\eta^j\frac{(x_1-x_0)^{1/k}}{x_2^{1/k}}\right)Y^g(Y(g^j u,x_0)v,x_2) \end{aligned} \tag{16}$$

(the *twisted Jacobi identity*) where η is a fixed primitive kth root of unity.

We denote a weak g-twisted V-module by (M, Y^g), or briefly, by M.

If we take $g = 1$, then we obtain the notion of weak V-module. Note that the notion of weak g-twisted V-module for a vertex operator superalgebra is equivalent to the notion of g-twisted V-module for V as a vertex superalgebra, cf. [Li2]. In particular, the term "weak" simply implies that we are making no assumptions about a grading on M.

Journal of Physics: Conference Series **474** (2013) 012009 doi:10.1088/1742-6596/474/1/012009

It follows from the twisted Jacobi identity that

$$Y^g(v,x) = \sum_{n\in\mathbb{Z}+\frac{j}{k}} v_n^g x^{-n-1} \tag{17}$$

for $j \in \mathbb{Z}/k\mathbb{Z}$ and $v \in V^j$, and thus we have

$$Y^g(gv,x) = \lim_{x^{1/k}\to\eta^{-1}x^{1/k}} Y^g(v,x), \tag{18}$$

where the limit stands for formal substitution.

Let (M_1, Y_1^g) and (M_2, Y_2^g) be two weak g-twisted V-modules. A *g-twisted V-module homomorphism* from M_1 to M_2, is a linear map $f : M_1 \longrightarrow M_2$ such that

$$f(Y_1^g(v,x)w) = Y_2^g(v,x)f(w) \tag{19}$$

for $v \in V$ and $w \in M_1$.

A weak g-twisted V-module may or may not have additional grading structures. These possible grading structures fall into two different types: 1. Those involving the $\mathbb{Z}_2$ grading structure, i.e. by sign or parity. 2. Those involving the weight grading structure. The first type leads to the notion of parity stability for a weak g-twisted V-module which detects whether the module has a $\mathbb{Z}_2$-grading that is compatible with the $\mathbb{Z}_2$-grading of V. The second type leads to the notion of weak admissible g-twisted V-module which detects whether the module has a $\frac{1}{2k}\mathbb{Z}$-grading compatible with the $\frac{1}{2}\mathbb{Z}$-grading of V where k is the order of g. This second type also leads to the notion of ordinary g-twisted V-modules which detects whether the g-twisted V-module is graded by eigenvectors of the twisted $L^g(0)$ operator.

We now give the details for these different module definitions.

A *weak admissible* g-twisted V-module is a weak g-twisted V-module M which carries a $\frac{1}{2k}\mathbb{Z}$-grading

$$M = \coprod_{n\in\frac{1}{2k}\mathbb{Z}} M(n) \tag{20}$$

such that $v_m^g M(n) \subseteq M(n + \mathrm{wt}\ v - m - 1)$ for homogeneous $v \in V$, and $M(n) = 0$ for n sufficiently small. If $g = 1$, we have the notion of weak admissible V-module.

An *ordinary g-twisted V-module* is a weak g-twisted V-module M which is $\mathbb{C}$-graded

$$M = \coprod_{\lambda\in\mathbb{C}} M_\lambda \tag{21}$$

such that for each λ, $\dim M_\lambda < \infty$ and $M_{n/k+\lambda} = 0$ for all sufficiently negative integers n. In addition,

$$L^g(0)w = \lambda w \qquad \text{for } w \in M_\lambda, \tag{22}$$

where $L^g(n) = \omega_{n+1}^g$ are the modes for the twisted vertex operator corresponding to the Virasoro element. We will usually refer to an ordinary g-twisted V-module, as just a g-twisted V-module. We call a g-twisted V-module M *simple* or *irreducible* if the only submodules are 0 and M.

For an ordinary g-twisted V-module, M, we have the notion of *graded dimension* or *q-dimension*, denoted $\dim_q M$, and defined to be

$$\dim_q M = tr_M q^{L^g(0)-c/24} = q^{-c/24} \sum_{\lambda\in\mathbb{C}} (\dim M_\lambda) q^\lambda. \tag{23}$$

A weak, weak admissible or ordinary g-twisted V-module M is said to be *parity stable* if there exists a $\mathbb{Z}_2$-grading on M that is compatible with the $\mathbb{Z}_2$-grading of V in the following sense:

$$v_m^g M^{(j)} \subseteq M^{(j+|v|)\bmod 2}. \tag{24}$$

In this case, setting $|w| = j$ for $w \in M^{(j)}$, defining the parity map on M by

$$\sigma_M : M \longrightarrow M, \qquad w \mapsto (-1)^{|w|}w, \tag{25}$$

and defining $Y^g \circ \sigma_V$ by

$$Y^g \circ \sigma_V(v,x) = Y^g(\sigma_V(v),x) = (-1)^{|v|}Y^g(v,x), \tag{26}$$

we have that (M, Y^g) is isomorphic to $(\sigma_M(M), Y^g \circ \sigma_V)$ as weak (or weak admissible or ordinary) g-twisted V-modules. Note that a vertex operator superalgebra V is always a parity-stable V-module by Remark 1.1.

2. Parity-unstable modules arise as pairs of invariant subspaces of parity-stable modules

The notion of parity stability features prominently in, for instance, [DZ1]–[DZ3], [DH]. However, in this section, we show that all parity-unstable weak twisted modules appear as invariant subspaces of parity-stable weak twisted modules. Thus it is enough to study the parity-stable weak twisted modules and then restrict to the invariant subspaces of such modules to study the parity-stable ones. This theorem was motivated by constructions involving free fermions such as those given below in Sections 3.2 and 3.3.

Theorem 2.1 *Let V be a vertex operator superalgebra and g an automorphism. Suppose (M, Y_M) is a parity-unstable weak g-twisted V-module. Then $(M, Y_M \circ \sigma_V)$ is a parity-unstable weak g-twisted V-module which is not isomorphic to (M, Y_M). Moreover $(M, Y_M) \oplus (M, Y_M \circ \sigma_V)$ is a parity-stable weak g-twisted V-module. In addition, if (M, Y_M) is weak admissible or ordinary, then $(M, Y_M \circ \sigma_V)$ and hence $(M, Y_M) \oplus (M, Y_M \circ \sigma_V)$ are weak admissible or ordinary. In the case that (M, Y_M) is ordinary, then (M, Y_M) and $(M, Y_M \circ \sigma_V)$ have the same graded dimension.*

Proof: Suppose (M, Y_M) is a parity-unstable weak g-twisted V-module. Then it follows immediately that $(M, Y_M \circ \sigma_V)$ is a weak g-twisted V-module. If $(M, Y_M \circ \sigma_V)$ were parity stable, that would imply that there exists σ_M as in (25) such that $(\sigma_M(M), Y_M \circ \sigma_V \circ \sigma_V) = (M, Y_M)$ is isomorphic to $(M, Y_M \circ \sigma_V)$, implying (M, Y_M) is parity stable. Thus $(M, Y_M \circ \sigma_V)$ is parity unstable.

Now consider $(M, Y_M) \oplus (M, Y_M \circ \sigma_V)$, and let

$$\sigma_{M\oplus M} : M \oplus M \longrightarrow M \oplus M, \quad \sigma_{M\oplus M} : (w_1, w_2) \mapsto (w_2, w_1) \tag{27}$$

so that $M \oplus M$ has a $\mathbb{Z}_2$ grading with respect to $\sigma_{M\oplus M}$ given by

$$(M \oplus M)^{(0)} = \{(w,w) \mid w \in M\} \quad \text{and} \quad (M \oplus M)^{(1)} = \{(w,-w) \mid w \in M\}. \tag{28}$$

Then $(\sigma_{M\oplus M}(M\oplus M), (Y_M \oplus (Y_M \circ \sigma_V)) \circ \sigma_V)$ is obviously isomorphic to $(M \oplus M, Y_M \oplus (Y_M \circ \sigma_V))$.

The rest of the theorem follows in a straightforward way from the definitions. □

Remark 2.2 Requiring weak twisted modules to be parity stable as part of the definition gives the more canonical notion of weak twisted module from a categorical point of view, for instance to allow for the tensor product of modules for two vertex operator superalgebras be a module for the tensor product vertex operator superalgebra. (See e.g. (9)). In particular, the notion of a weak V-module corresponding to a *representation* of V as a vertex superalgebra only holds for parity-stable weak g-twisted V-modules, in that the vertex operators acting on a weak g-twisted V-module have coefficients in End M such that, the operators v_m^g have a $\mathbb{Z}_2$-graded structure compatible with that of V. For instance the operators v_0^g, for $v \in V$, give a representation of the Lie superalgebra generated by v_0 in End V if and only if M is parity stable. This corresponds to V acting as endomorphisms in the category of vector spaces (i.e., via even or odd endomorphisms) rather than in the category of $\mathbb{Z}_2$-graded vectors spaces (i.e., as grade-preserving and thus strictly even endomorphisms). However, it is interesting to note that, as is shown in Section 7, for a lift of a lattice isometry, the twisted modules for a lattice vertex operator superalgebra naturally sometimes give rise to parity-unstable modules. Thus the notion of "parity-unstable module" does naturally arise in certain constructions.

3. Permutation-twisted free fermion vertex operator superalgebras and a conjecture

We first recall the notion of free fermion vertex operator superalgebras following the notation of [Bar11], but also in the spirit of [DZ2]. Then we recall the construction of parity-twisted modules and construct the permutation-twisted modules following [DZ2]. Finally we make a conjecture based on this example on the nature of the construction of $(1\ 2\ \cdots k)$-twisted $V^{\otimes k}$-modules for k even and V any vertex operator superalgebra based on the example of free fermions.

3.1. Free Fermion vertex operator superalgebras

Let $\mathfrak{h}$ be finite-dimensional vector space over $\mathbb{C}$ equipped with a nondegenerate symmetric bilinear form $\langle\cdot,\cdot\rangle$. Let d denote the dimension of $\mathfrak{h}$, let t and x denote formal commuting variables, and let $U(\cdot)$ denote the universal enveloping algebra for a Lie superalgebra $(\cdot)$.

Form the affine Lie superalgebra

$$\hat{\mathfrak{h}}^f = \mathfrak{h} \otimes t^{1/2}\mathbb{C}[t,t^{-1}] \oplus \mathbb{C}\mathbf{k},$$

with $\mathbb{Z}_2$-grading given by $\mathrm{sgn}(\alpha \otimes t^n) = 1$ for $n \in \mathbb{Z} + \frac{1}{2}$, and $\mathrm{sgn}(\mathbf{k}) = 0$, and Lie super-bracket relations

$$[\mathbf{k}, \hat{\mathfrak{h}}^f] = 0, \quad \text{and} \quad [\alpha \otimes t^m, \beta \otimes t^n] = \langle \alpha, \beta\rangle \delta_{m+n,0}\mathbf{k} \tag{29}$$

for $\alpha, \beta \in \mathfrak{h}$ and $m, n \in \mathbb{Z} + \frac{1}{2}$. Then $\hat{\mathfrak{h}}^f$ is a $((\mathbb{Z}+\frac{1}{2}) \cup \{0\})$-graded Lie superalgebra

$$\hat{\mathfrak{h}}^f = \coprod_{n \in (\mathbb{Z}+\frac{1}{2})\cup\{0\}} \hat{\mathfrak{h}}^f_n$$

where $\hat{\mathfrak{h}}^f_n = \mathfrak{h} \otimes t^{-n}$, for $n \in \mathbb{Z} + \frac{1}{2}$, and $\hat{\mathfrak{h}}^f_0 = \mathbb{C}\mathbf{k}$. It has graded subalgebras

$$\hat{\mathfrak{h}}^f_+ = \mathfrak{h} \otimes t^{-1/2}\mathbb{C}[t^{-1}] \quad \text{and} \quad \hat{\mathfrak{h}}^f_- = \mathfrak{h} \otimes t^{1/2}\mathbb{C}[t].$$

Note that $\hat{\mathfrak{h}}^f = \hat{\mathfrak{h}}^f_- \oplus \hat{\mathfrak{h}}^f_+ \oplus \mathbb{C}\mathbf{k}$, and note that $\hat{\mathfrak{h}}^f$ is a Heisenberg superalgebra.

Let $\mathbb{C}$ be the $(\hat{\mathfrak{h}}^f_- \oplus \mathbb{C}\mathbf{k})$-module such that $\hat{\mathfrak{h}}^f_-$ acts trivially and $\mathbf{k}$ acts as 1. Let

$$V_{fer}^{\otimes d} = U(\hat{\mathfrak{h}}^f) \otimes_{U(\hat{\mathfrak{h}}^f_- \oplus \mathbb{C}\mathbf{k})} \mathbb{C} \simeq \bigwedge(\hat{\mathfrak{h}}^f_+),$$

so that $V_{fer}^{\otimes d}$ is naturally isomorphic as a $\hat{\mathfrak{h}}^f$-module to the algebra of polynomials in the anticommuting elements of $\hat{\mathfrak{h}}_+^f$.

Let $\alpha \in \mathfrak{h}$ and $n \in \mathbb{Z} + \frac{1}{2}$. We will use the notation

$$\alpha(n) = \alpha \otimes t^n.$$

Then $V_{fer}^{\otimes d}$ is a $\hat{\mathfrak{h}}^f$-module with action induced from the supercommutation relations (29) given by

$$\mathbf{k}\beta(-m)\mathbf{1} = \beta(-m)\mathbf{1} \quad (30)$$
$$\alpha(n)\beta(-m)\mathbf{1} = \langle\alpha, \beta\rangle\delta_{m,n}\mathbf{1} \quad (31)$$
$$\alpha(-n)\beta(-m)\mathbf{1} = -\beta(-m)\alpha(-n)\mathbf{1} \quad (32)$$

for $\alpha, \beta \in \mathfrak{h}$ and $m, n \in \mathbb{N} + \frac{1}{2}$. That is letting $\{\alpha^{(1)}, \alpha^{(2)}, \ldots, \alpha^{(d)}\}$ be an orthonormal basis for $\mathfrak{h}$, we have

$$V_{fer}^{\otimes d} = \bigwedge \left[\alpha^{(j)}(-n) \;\middle|\; j = 1, \ldots d,\ n \in \mathbb{N} + \tfrac{1}{2}\right] \quad (33)$$

where $\mathbf{k}$ acts as 1, and for $j = 1, \ldots, d$ and $n \in \mathbb{N} + \frac{1}{2}$, the operator $\alpha^{(j)}(n)$ acts as the partial derivative with respect to $a^{(j)}(-n)$, and the operator $\alpha^{(j)}(-n)$ acts as multiplication.

For $\alpha \in \mathfrak{h}$, set

$$\alpha(x) = \sum_{n \in \frac{1}{2} + \mathbb{Z}} \alpha(n) x^{-n-\frac{1}{2}}, \quad (34)$$

Define the *normal ordering* operator ${}^{\circ}_{\circ} \cdot {}^{\circ}_{\circ}$ on products of the operators $\alpha(n)$ by

$${}^{\circ}_{\circ}\alpha(m)\beta(n){}^{\circ}_{\circ} = \begin{cases} \alpha(m)\beta(n) & \text{if } m \leq n \\ -\beta(n)\alpha(m) & \text{if } m > n \end{cases} \quad (35)$$

for $m, n \in \mathbb{Z} + \frac{1}{2}$.

For $v = \alpha_1(-n_1)\alpha_2(-n_2)\cdots\alpha_m(-n_m)\mathbf{1} \in V_{fer}$, for $\alpha_j \in \mathfrak{h}$, $n_j \in \mathbb{N} + \frac{1}{2}$, and $j = 1, \ldots, m$ and $m \in \mathbb{N}$, define the vertex operator corresponding to v to be

$$Y(v, x) = {}^{\circ}_{\circ}\left(\partial_{n_1 - \frac{1}{2}}\alpha_1(x)\right)\left(\partial_{n_2 - \frac{1}{2}}\alpha_2(x)\right)\cdots\left(\partial_{n_m - \frac{1}{2}}\alpha_m(x)\right){}^{\circ}_{\circ}, \quad (36)$$

where for $n \in \mathbb{N}$, we use the notation $\partial_n = \frac{1}{n!}\left(\frac{d}{dx}\right)^n$.

Note that

$$[\alpha^{(j)}(x_1), \alpha^{(k)}(x_2)] = \delta_{j,k}\left(\frac{1}{(x_1 - x_2)} - \frac{1}{(-x_2 + x_1)}\right) \quad (37)$$

implying that the $\alpha^{(j)}(x) = Y(\alpha^{(j)}(-1/2)\mathbf{1}, x)$, for $j = 1, \ldots, d$, are mutually local. And, in fact, setting

$$\omega = \frac{1}{2}\sum_{j=1}^{d} \alpha^{(j)}(-3/2)\alpha^{(j)}(-1/2)\mathbf{1}, \quad (38)$$

we have that $(V_{fer}^{\otimes d}, Y, \mathbf{1}, \omega)$ is a vertex operator superalgebra with central charge $d/2$. $V_{fer}^{\otimes d}$ is called the *d free fermion vertex operator superalgebra.*

When d is even, $V_{fer}^{\otimes d}$ is precisely the vertex operator superalgebra studied in [FFR] denoted $CM(\mathbb{Z} + \frac{1}{2})$, although in [FFR] a polarized basis for $\mathfrak{h}$ is used as we will do also below, as in Equation (43).

The graded dimension of $V_{fer}^{\otimes d}$ using the $\frac{1}{2}\mathbb{Z}$-grading of $V_{fer}^{\otimes d}$ by eigenvalues of $L(0)$ is

$$\dim_q V_{fer}^{\otimes d} = q^{-c/24} \sum_{n\in\frac{1}{2}\mathbb{Z}} \dim(V_{fer}^{\otimes d})_n q^n = q^{-d/48} \prod_{n\in\mathbb{Z}_+} (1+q^{n-1/2})^d = \mathfrak{f}(q)^d, \tag{39}$$

where $\mathfrak{f}(q)$ is a classical Weber function [YZ]. A simple calculation shows that in fact $\mathfrak{f}(q) = \frac{\eta(q)^2}{\eta(q^2)\eta(q^{1/2})}$, where $\eta(q) = q^{1/24}\prod_{n\in\mathbb{Z}_+}(1-q^n)$ is the Dedekind η-function.

In addition, the *superdimension* of a vertex operator superalgebra $V = V^0 \oplus V^1$ is sometimes of interest. It is defined to be $\mathrm{sdim}_q V = \dim_q V^{(0)} - \dim_q V^{(1)}$. Thus the superdimension of V_{fer} is

$$\mathrm{sdim}_q V_{fer} = q^{-d/48} \prod_{n\in\mathbb{Z}_+} (1-q^{n-1/2})^d = \mathfrak{f}_1(q)^d \tag{40}$$

where $\mathfrak{f}_1(q)$ is also a classical Weber function. Observe that $\mathfrak{f}_1(q) = \frac{\eta(q^{1/2})}{\eta(q)}$.

Remark 3.1 In addition to the two classical Weber functions, $\mathfrak{f}$ and $\mathfrak{f}_1$, there is a third classical Weber function, denoted $\mathfrak{f}_2$ and given by

$$\mathfrak{f}_2(q) = \sqrt{2}q^{1/24} \prod_{n\in\mathbb{Z}_+} (1+q^n) = \sqrt{2}\frac{\eta(q^2)}{\eta(q)}. \tag{41}$$

This third classical Weber function, $\mathfrak{f}_2$, will appear in Section 3.2. These three Weber functions, $\mathfrak{f}, \mathfrak{f}_1$, and $\mathfrak{f}_2$, form a set that is $SL_2(\mathbb{Z})$-invariant up to permutation and multiplication by 48th roots of unity [YZ].

Finally, we note that V_{fer}, and thus $V_{fer}^{\otimes d}$, is not only rational, but is self-dual as a vertex operator superalgebra (cf. [FFR], [KW], [Li1]), i.e., the only irreducible V_{fer}-module is V_{fer} itself.

3.2. Parity-twisted free fermions

Form the affine Lie superalgebra

$$\hat{\mathfrak{h}}^f[\sigma] = \mathfrak{h} \otimes \mathbb{C}[t, t^{-1}] \oplus \mathbb{C}\mathbf{k},$$

with $\mathbb{Z}_2$-grading given by $\mathrm{sgn}(\alpha \otimes t^n) = 1$ for $n \in \mathbb{Z}$, and $\mathrm{sgn}(\mathbf{k}) = 0$, and Lie super-bracket relations

$$[\mathbf{k}, \hat{\mathfrak{h}}^f[\sigma]] = 0, \quad \text{and} \quad [\alpha\otimes t^m, \beta \otimes t^n] = \langle \alpha, \beta\rangle \delta_{m+n,0}\mathbf{k} \tag{42}$$

for $\alpha, \beta \in \mathfrak{h}$ and $m, n \in \mathbb{Z}$. Then $\hat{\mathfrak{h}}^f[\sigma]$ is a $\mathbb{Z}$-graded Lie superalgebra

$$\hat{\mathfrak{h}}^f[\sigma] = \coprod_{n\in\mathbb{Z}} \hat{\mathfrak{h}}^f[\sigma]_n$$

where $\hat{\mathfrak{h}}^f[\sigma]_0 = \mathfrak{h} \oplus \mathbb{C}\mathbf{k}$, and $\hat{\mathfrak{h}}^f[\sigma]_n = \mathfrak{h} \otimes t^{-n}$ for $n \neq 0$. And $\hat{\mathfrak{h}}^f[\sigma]$ is a Heisenberg superalgebra.

If $\dim \mathfrak{h} = d$ is even, i.e. $d = 2l$, then we can choose a polarization of $\mathfrak{h}$ into maximal isotropic subspaces $\mathfrak{a}^\pm$. That is $\mathfrak{a}^\pm$ both have dimension l, and satisfy $\langle \mathfrak{a}^+, \mathfrak{a}^+\rangle = \langle \mathfrak{a}^-, \mathfrak{a}^-\rangle = 0$, and we can choose a basis of $\mathfrak{a}^-$, given by $\{\beta_-^{(1)}, \beta_-^{(2)}, \ldots, \beta_-^{(l)}\}$, and a dual basis for $\mathfrak{a}^+$, given by $\{\beta_+^{(1)}, \beta_+^{(2)}, \ldots, \beta_+^{(l)}\}$ such that $\langle \beta_-^{(j)}, \beta_+^{(n)}\rangle = \delta_{j,n}$.

If $\dim \mathfrak{h} = d$ is odd, i.e. $d = 2l+1$, then we can choose a polarization of $\mathfrak{h}$ into maximal isotropic subspaces $\mathfrak{a}^\pm$, each of dimension l, and a one-dimensional space $\mathfrak{e}$, so that $\mathfrak{h} = \mathfrak{a}^- \oplus \mathfrak{a}^+ \oplus \mathfrak{e}$, and such that $\langle \mathfrak{a}^\pm, \mathfrak{e}\rangle = 0$, and $\mathfrak{e} = \mathbb{C}\epsilon$ with $\langle \epsilon, \epsilon\rangle = 2$.

Remark 3.2 If $\{\alpha^{(1)}, \alpha^{(2)}, \ldots, \alpha^{(d)}\}$ is an orthonormal basis for $\mathfrak{h}$ with respect to the symmetric bilinear form, then a polarization for $\mathfrak{h}$ can be given as follows: For d either $2l$ or $2l+1$, set

$$\beta_{\pm}^{(j)} = \frac{1}{\sqrt{2}}\left(\alpha^{(j)} \pm i\alpha^{(j+l)}\right) \tag{43}$$

for $j = 1, 2, \ldots l$. Then $\mathfrak{a}^{\pm} = \mathrm{span}_{\mathbb{C}}\{\beta_{\pm}^{(1)}, \beta_{\pm}^{(2)}, \ldots, \beta_{\pm}^{(l)}\}$ gives a decomposition into maximal polarized spaces. If $d = 2l+1$, then set $\epsilon = \sqrt{2}\alpha^{(d)}$. Note that (43) is equivalent to $\alpha^{(j)} = \frac{1}{\sqrt{2}}\left(\beta_{+}^{(j)} + \beta_{-}^{(j)}\right)$ and $\alpha^{(j+l)} = \frac{-i}{\sqrt{2}}\left(\beta_{+}^{(j)} - \beta_{-}^{(j)}\right)$ for $j = 1, \ldots, l$.

Then $\hat{\mathfrak{h}}^f[\sigma]$ has the following graded subalgebras

$$\hat{\mathfrak{h}}^f[\sigma]_+ = \mathfrak{h} \otimes t^{-1}\mathbb{C}[t^{-1}] \qquad \text{and} \qquad \hat{\mathfrak{h}}^f[\sigma]_- = \mathfrak{h} \otimes t\mathbb{C}[t],$$

and we have $\hat{\mathfrak{h}}^f[\sigma] = \hat{\mathfrak{h}}^f[\sigma]_- \oplus \mathfrak{h} \oplus \hat{\mathfrak{h}}^f[\sigma]_+ \oplus \mathbb{C}\mathbf{k}$. In addition, $\hat{\mathfrak{h}}^f[\sigma]$ has the subalgebras

$$\hat{\mathfrak{h}}^f[\sigma]_+ \oplus \mathfrak{a}^+ \qquad \text{and} \qquad \hat{\mathfrak{h}}^f[\sigma]_- \oplus \mathfrak{a}^-$$

for d even and

$$\hat{\mathfrak{h}}^f[\sigma]_+ \oplus \mathfrak{a}^+ \oplus \mathfrak{e} \qquad \text{and} \qquad \hat{\mathfrak{h}}^f[\sigma]_- \oplus \mathfrak{a}^-$$

for d odd.

Let $\mathbb{C}$ be the $(\hat{\mathfrak{h}}^f[\sigma]_- \oplus \mathfrak{a}^- \oplus \mathbb{C}\mathbf{k})$-module such that $\hat{\mathfrak{h}}^f[\sigma]_- \oplus \mathfrak{a}^-$ acts trivially and $\mathbf{k}$ acts as 1. Set

$$M_\sigma = U(\hat{\mathfrak{h}}^f[\sigma]) \otimes_{U(\hat{\mathfrak{h}}^f[\sigma]_- \oplus \mathfrak{a}^- \oplus \mathbb{C}\mathbf{k})} \mathbb{C}. \tag{44}$$

Then as a vector space, we have

$$M_\sigma \overset{\text{vec.sp.}}{\simeq} \begin{cases} \bigwedge(\hat{\mathfrak{h}}^f[\sigma]_+ \oplus \mathfrak{a}^+) & \text{if } d \text{ is even} \\ \bigwedge(\hat{\mathfrak{h}}^f[\sigma]_+ \oplus \mathfrak{a}^+ \oplus \mathfrak{e}) & \text{if } d \text{ is odd} \end{cases}, \tag{45}$$

where if d is even, this is also an associative algebra isomorphism, but if d is odd it is not; rather, if d is odd, M_σ is a Clifford algebra but not an exterior algebra.

Let $\alpha \in \mathfrak{h}$ and $n \in \mathbb{Z}$. We use the notation

$$\overline{\alpha(n)} = \alpha \otimes t^n \in \hat{\mathfrak{h}}^f[\sigma]$$

where the overline is meant to distinguish elements of $\hat{\mathfrak{h}}^f[\sigma]$ from elements of $\hat{\mathfrak{h}}$, used to construct the free bosonic theory.

Then M_σ is a $\hat{\mathfrak{h}}^f[\sigma]$-module. For d even, the action induced from the supercommutation relations (42) is given by

$$\mathbf{k}\overline{\beta(-m)}\mathbf{1} = \overline{\beta(-m)}\mathbf{1} \tag{46}$$
$$\overline{\alpha(n)}\,\overline{\beta(-m)}\mathbf{1} = \langle\alpha, \beta\rangle\delta_{m,n}\mathbf{1} \tag{47}$$
$$\overline{\alpha(-n)}\,\overline{\beta(-m)}\mathbf{1} = -\overline{\beta(-m)}\,\overline{\alpha(-n)}\mathbf{1} \tag{48}$$

for either (i) $\alpha, \beta \in \mathfrak{h}$ and $m, n \in \mathbb{Z}_+$; (ii) $\alpha \in \mathfrak{h}$, $\beta \in \mathfrak{a}^+$, $m = 0$, and $n \in \mathbb{Z}_+$; or (iii) $\alpha \in \mathfrak{a}^-$, $\beta \in \mathfrak{h}$, $n = 0$, and $m \in \mathbb{Z}_+$; and

$$\overline{\alpha(0)}\,\overline{\beta(0)}\mathbf{1} = \langle\alpha, \beta\rangle\mathbf{1} \tag{49}$$

if $\alpha \in \mathfrak{a}^-$ and $\beta \in \mathfrak{a}^+$, and where here $\mathbf{1} = \mathbf{1}_{M_\sigma} = 1$.

For d odd, the action induced from the supercommutation relations are given by (46)–(49) as well as

$$\mathbf{k}\overline{\epsilon(0)}\mathbf{1} = \overline{\epsilon(0)}\mathbf{1}, \qquad \overline{\alpha(0)}\,\overline{\epsilon(0)}\mathbf{1} = 0, \tag{50}$$

$$\overline{\beta(0)}\,\overline{\epsilon(0)}\mathbf{1} = -\overline{\epsilon(0)}\,\overline{\beta(0)}\mathbf{1}, \qquad \overline{\epsilon(0)}\,\overline{\epsilon(0)} = \frac{1}{2}\langle\epsilon,\epsilon\rangle\mathbf{1}, \tag{51}$$

for $\alpha \in \mathfrak{a}^-$, $\beta \in \mathfrak{a}^+$, and $\epsilon \in \mathfrak{e}$.

In particular, letting $\{\beta_\pm^{(1)}, \beta_\pm^{(2)}, \ldots, \beta_\pm^{(l)}\}$ be the bases for the polarized spaces $\mathfrak{a}^\pm$ as defined in Remark 3.2, and if d is odd, letting $\mathfrak{e} = \mathbb{C}\epsilon$ with $\langle\epsilon,\epsilon\rangle = 2$, then we have

$$M_\sigma = \bigwedge\left[\overline{\beta_-^{(j)}(-m)}\mathbf{1},\, \overline{\beta_+^{(j)}(-n)}\mathbf{1} \;\middle|\; m \in \mathbb{Z}_+,\, n \in \mathbb{N}, \text{ and } j = 1,\ldots,l\right]$$

for d even, and in this case, the identification is as an associative algebra. For d odd, we have

$$M_\sigma = \bigwedge\left[\overline{\beta_-^{(j)}(-m)}\mathbf{1},\, \overline{\beta_+^{(j)}(-n)}\mathbf{1},\, \overline{\epsilon(-n)}\mathbf{1} \;\middle|\; m \in \mathbb{Z}_+,\, n \in \mathbb{N}, \text{ and } j = 1,\ldots,l\right]$$

where in this case, the identification is as a vector space but not as an associative algebra. As an associative algebra with identity, M_σ for d odd is the Clifford algebra generated by $\hat{\mathfrak{h}}^f[\sigma]_+ \oplus \mathfrak{a}^+ \oplus \mathfrak{e}$ with the corresponding symmetric bilinear form.

That is, for d even, k acts as 1, and for $j = 1,\ldots,l$, and $n \in \mathbb{Z}_+$, the operator $\overline{\beta_\pm^{(j)}(n)}$ acts as the partial derivative with respect to $\overline{\beta_\mp^{(j)}(-n)}$, the operator $\overline{\beta_\pm^{(j)}(-n)}$ acts as multiplication by $\overline{\beta_\pm^{(j)}(-n)}$, the operator $\overline{\beta_-^{(j)}(0)}$ acts as the partial derivative with respect to $\overline{\beta_+^{(j)}(0)}$, and the operator $\overline{\beta_+^{(j)}(0)}$ acts via multiplication. If d is odd, then we have the operators as above in addition to the operators $\overline{\epsilon(n)}$ for $n \in \mathbb{Z}_+$, which act as two times the partial derivative with respect to $\overline{\epsilon(-n)}$, and the operators $\overline{\epsilon(-n)}$ for $n \in \mathbb{N}$, which act as multiplication with the condition that $\overline{\epsilon(0)\epsilon(0)} = 1$.

For $\alpha \in \mathfrak{h}$, set

$$\alpha(x)^\sigma = \sum_{n\in\mathbb{Z}} \overline{\alpha(n)}x^{-n-\frac{1}{2}}. \tag{52}$$

Then for the orthonormal basis of $\mathfrak{h}$, $\alpha^{(j)}$, for $j = 1,\ldots,d$, we have

$$[\alpha^{(j)}(x_1)^\sigma, \alpha^{(k)}(x_2)^\sigma] = \delta_{j,k}\, x_1^{1/2}x_2^{-1/2}\left(\frac{1}{(x_1 - x_2)} - \frac{1}{(-x_2 + x_1)}\right) \tag{53}$$

for $j,k = 1,\ldots,d$, implying that the $\alpha^{(j)}(x)^\sigma$, for $j = 1,\ldots,d$, are mutually local.

For $v \in V_{fer}^{\otimes d}$, define $Y^\sigma(v,x) : M_\sigma \longrightarrow M_\sigma[[x^{1/2}, x^{-1/2}]]$ as follows: For $\alpha \in \mathfrak{h}$, $n \in \mathbb{N} + 1/2$, and $u \in V_{fer}^{\otimes d}$, let

$$\begin{aligned} Y^\sigma(\alpha(-n)u, x) &= Y^\sigma(\alpha_{-n-1/2}u, x) = \mathrm{Res}_{x_1}\mathrm{Res}_{x_0}\left(\frac{x_1 - x_0}{x}\right)^{1/2} x_0^{-n-1/2} \\ &\cdot \left(x_0^{-1}\delta\left(\frac{x_1 - x}{x_0}\right)\alpha(x_1)^\sigma Y^\sigma(u,x) - (-1)^{|u|}x_0^{-1}\delta\left(\frac{x - x_1}{-x_0}\right)Y^\sigma(u,x)\alpha(x_1)^\sigma\right). \end{aligned} \tag{54}$$

Then since $V_{fer}^{\otimes d} = \langle\alpha^{(j)}(-1/2)\mathbf{1} \mid j = 1,\ldots,d\rangle$, equation (54) defines $Y^\sigma(v,x)$ iteratively for any $v \in V_{fer}^{\otimes d}$.

Recalling that the Virasoro element, ω_{fer}, for the free fermionic vertex operator superalgebra $V_{fer}^{\otimes d}$ is given by (38), we have

$$Y^{\sigma}(\omega_{fer},x) = \frac{1}{2}\sum_{j=1}^{d} Y^{\sigma}(\alpha^{(j)}(-1/2)_{-2}\alpha^{(j)}(-1/2)\mathbf{1},x) = \sum_{n\in\mathbb{Z}} L^{\sigma}(n)x^{-n-2}, \tag{55}$$

and thus

$$L^{\sigma}(m) = \sum_{j=1}^{d}\sum_{\substack{n\in\mathbb{Z}\\ n>-\frac{m}{2}}} \left(n+\frac{m}{2}\right)\overline{\alpha^{(j)}(-n)}\,\overline{\alpha^{(j)}(n+m)} + \frac{d}{16}\delta_{m,0}. \tag{56}$$

From this it follows that $\left[L^{\sigma}(-1), Y^{\sigma}(\alpha^{(j)}(-1/2)\mathbf{1},x)\right] = \frac{d}{dx}Y^{\sigma}(\alpha^{(j)}(-1/2)\mathbf{1},x)$. Thus from [Li2], we have that M_σ is a weak σ-twisted module for $V_{fer}^{\otimes d}$. It is also admissible. In [FFR], if d is even, M_σ is denoted by $CM(\mathbb{Z})$.

By [Li2] as well as [DZ2], in the case that $d = \dim\mathfrak{h}$ is even, M_σ is irreducible and is the only irreducible admissible σ-twisted module for $V_{fer}^{\otimes d}$, up to isomorphism. It is parity stable and is also an ordinary σ-twisted $V_{fer}^{\otimes d}$-module, as we will see below when we discuss the $L^{\sigma}(0)$-grading and the $\mathbb{Z}_2$-grading.

In the case that d is odd, M_σ is irreducible as a parity-stable module but reduces to the direct sum of two irreducible parity-unstable subspaces, and these two are the only irreducible admissible parity-unstable σ-twisted modules for $V_{fer}^{\otimes d}$, up to isomorphism. In this case, setting

$$W = \bigwedge\left[\overline{\beta_-^{(j)}(-m)\mathbf{1}},\ \overline{\beta_+^{(j)}(-n)\mathbf{1}},\ \overline{\epsilon(-m)\mathbf{1}}\ \middle|\ m\in\mathbb{Z}_+,\ n\in\mathbb{N},\ \text{and } j=1,\dots,l\right]$$

and letting $W = W^0 \oplus W^1$ be the decomposition of W into even and odd subspaces, these two irreducibles are given by

$$M_\sigma^{\pm} = \left(1 \pm \overline{\epsilon(0)1}\right)W^0 \oplus \left(1 \mp \overline{\epsilon(0)1}\right)W^1, \tag{57}$$

and we have $M_\sigma = M_\sigma^- \oplus M_\sigma^+$. That $M_\sigma^{\pm}$ are in fact ordinary σ-twisted modules for $V_{fer}^{\otimes d}$ and parity-unstable, we shall see now by discussing the $L^{\sigma}(0)$-grading and the $\mathbb{Z}_2$-grading.

In terms of the polarization of $\mathfrak{h}$ with respect to the basis $\alpha^{(j)}$, we have from equation (56)

$$L^{\sigma}(0) = \sum_{j=1}^{l}\sum_{m\in\mathbb{Z}_+}\left(m\overline{\beta_+^{(j)}(-m)}\,\overline{\beta_-^{(j)}(m)} + m\overline{\beta_-^{(j)}(-m)}\,\overline{\beta_+^{(j)}(m)}\right) + L' + \frac{d}{16} \tag{58}$$

where if d is even, $L' = 0$, and if d is odd, $L' = \frac{1}{2}\sum_{m\in\mathbb{Z}_+} m\overline{\epsilon(-m)}\,\overline{\epsilon(m)}$. Thus for $j = 1,\dots,l$, and $m\in\mathbb{Z}_+$, the $L^{\sigma}(0)$ grading is given by

$$\mathrm{wt}\,\mathbf{1} = \mathrm{wt}\,\overline{\beta_+^{(j)}(0)\mathbf{1}} = \frac{d}{16}, \quad \text{and} \quad \mathrm{wt}\,\overline{\beta_\pm^{(j)}(-m)\mathbf{1}} = m + \frac{d}{16}, \tag{59}$$

for $d = 2l$, and if d is odd, we also have

$$\mathrm{wt}\,\overline{\epsilon(0)\mathbf{1}} = \frac{d}{16}, \quad \text{and} \quad \mathrm{wt}\,\overline{\epsilon(-m)\mathbf{1}} = m + \frac{d}{16}. \tag{60}$$

Therefore, for d even, the graded dimension of M_σ is

$$\dim_q M_\sigma = q^{-c/24}\sum_{\lambda\in\mathbb{C}}(M_\sigma)_\lambda q^\lambda \ = \ q^{-d/48}q^{d/16}2^{d/2}\prod_{n\in\mathbb{Z}_+}(1+q^n)^d = \mathfrak{f}_2(q)^d \tag{61}$$

where $\mathfrak{f}_2$ is a classical Weber function as discussed in Remark 3.1. For d odd, the graded dimension of M_σ is

$$\dim_q M_\sigma = q^{-d/48} q^{d/16} 2^{(d+1)/2} \prod_{n\in\mathbb{Z}_+} (1+q^n)^d = \sqrt{2}\mathfrak{f}_2(q)^d, \tag{62}$$

and the grading of each of the two submodules $M_\sigma^{\pm}$ is exactly half that of the graded dimension of M_σ.

Lemma 3.3 *If d is even, then the unique up to equivalence irreducible parity-twisted module for d free fermions M_σ is a parity-stable twisted module. If d is odd, then the two unique up to equivalence irreducible parity-twisted modules $M_\sigma^{\pm}$ for d free fermions are parity-unstable invariant subspace of M_σ. In addition, $M_\sigma = M_\sigma^+ \oplus M_\sigma^-$ is a parity-stable twisted module and is irreducible as a parity-stable twisted module.*

Proof: We first show that M_σ is parity stable for d even or odd. Define a $\mathbb{Z}_2$-grading on M_σ via the natural $\mathbb{Z}_2$-grading on $\bigwedge(\hat{\mathfrak{h}}[\sigma]_+ \oplus \mathfrak{a}^+)$ for d even and via the natural $\mathbb{Z}_2$-grading on $\bigwedge(\hat{\mathfrak{h}}[\sigma]_+ \oplus \mathfrak{a}^+ \oplus \mathfrak{e})$ for d odd. That is $w = \beta_1(-n_1)\beta_2(-n_2)\cdot\beta_m(-n_m)1 \in M^\sigma$ for $n_j \in \mathbb{Z}_+$ if $\beta_j \in \hat{\mathfrak{h}}[\sigma]_+$ and $n_j \in \mathbb{N}$ if $\beta_j \in \mathfrak{a}^+ \oplus \mathfrak{e}$ has odd parity if m is odd and even parity if m is even.

Then $Y^\sigma(\alpha^{(j)}(-1/2)\mathbf{1}, x)\cdot w = \alpha^{(j)}(x)^\sigma \cdot w = \sum_{n\in\mathbb{Z}} \overline{\alpha(n)} \cdot w x^{-n-1/2}$ is contained in $M_\sigma^{(m+1) \bmod 2}[[x^{1/2}, x^{-1/2}]]$, implying $v_n^\sigma \cdot M_\sigma^{(j)} \subset M_\sigma^{(j+|v|) \bmod 2}$ for all $v \in V_{fer}^{\otimes d}$ and $w \in M_\sigma$.

However, for d odd, considering the irreducible modules $M_\sigma^{\pm}$, we have, for instance

$$Y^\sigma(\epsilon(-1/2)\mathbf{1}, x)\cdot(1 \pm \overline{\epsilon(0)}1) = \pm(1 \pm \overline{\epsilon(0)}1)x^{-1/2} + (1 \mp \overline{\epsilon(0)}1) \sum_{-n\in\mathbb{Z}_+} \overline{\epsilon(n)} 1 x^{-n-1/2}. \tag{63}$$

Thus there exists no $\mathbb{Z}_2$-grading on $M_\sigma^{\pm}$ such that this lowest weight vector $(1 \pm \overline{\epsilon(0)}1)$ has a parity compatible with the $\mathbb{Z}_2$-grading of $V_{fer}^{\otimes d}$. □

From Theorem 2.1, we have that $(M_\sigma^-, Y^\sigma \circ \sigma)$ must be a parity-unstable parity-twisted module that is isomorphic to (M_σ^+, Y^σ). This isomorphism is given explicitly by

$$\begin{aligned} f : M_\sigma^+ &\longrightarrow M_\sigma^- \\ \left((1+\overline{\epsilon(0)}1)w_0, (1-\overline{\epsilon(0)}1)w_1\right) &\mapsto \left((1-\overline{\epsilon(0)}1)w_0, -(1+\overline{\epsilon(0)}1)w_1\right), \end{aligned} \tag{64}$$

for $w_j \in W^{(j)}$ for $j = 0, 1$.

3.3. Permutation-twisted modules for free fermions

Now we turn our attention to tensor product vertex operator superalgebras. Let $V = (V, Y, \mathbf{1}, \omega)$ be a vertex operator superalgebra, and let k be a fixed positive integer. Then $V^{\otimes k}$ is also a vertex operator superalgebra, and the permutation group S_k acts naturally on $V^{\otimes k}$ as signed automorphisms. That is $(j\ j+1)\cdot(v_1 \otimes v_2 \otimes \cdots \otimes v_k) = (-1)^{|v_j||v_{j+1}|}(v_1 \otimes v_2 \otimes \cdots v_{j-1} \otimes v_{j+1} \otimes v_j \otimes v_{j+2} \otimes \cdots \otimes v_k)$, and we take this to be a left action so that, for instance

$$\begin{aligned} (1\,2\cdots k) : V \otimes V \otimes \cdots \otimes V &\longrightarrow V \otimes V \otimes \cdots \otimes V \\ v_1 \otimes v_2 \otimes \cdots \otimes v_k &\mapsto (-1)^{|v_1|(|v_2|+\cdots+|v_k|)} v_2 \otimes v_3 \otimes \cdots \otimes v_k \otimes v_1. \end{aligned} \tag{65}$$

Letting $V = V_{fer}$, we have that $g = (1\,2\cdots k)$ acting as a signed permutation on $V_{fer}^{\otimes k}$ is a lift of the following permutation on $\mathfrak{h}$, the k-dimensional Heisenberg Lie superalgebra used to construct $V_{fer}^{\otimes k}$: Let $\alpha^{(j)}$, for $j = 1, \ldots, d = k$ be an orthonormal basis for $\mathfrak{h}$ as before. Then

$$\begin{aligned} (1\,2\cdots k) : \mathbb{C}\alpha^{(1)} \oplus \mathbb{C}\alpha^{(2)} \oplus \cdots \oplus \mathbb{C}\alpha^{(k)} &\longrightarrow \mathbb{C}\alpha^{(1)} \oplus \mathbb{C}\alpha^{(2)} \oplus \cdots \oplus \mathbb{C}\alpha^{(k)} \\ (c_1, c_2, \ldots, c_k) &\mapsto (c_2, c_3, \ldots, c_k, c_1), \end{aligned} \tag{66}$$

that is $(1\,2\cdots k)\alpha^{(j)} = \alpha^{(j-1)}$ for $j = 1, \ldots, k$ where $\alpha^{(-1)}$ is understood to be $\alpha^{(k)}$.

Defining $\mathfrak{h}^0 = \{h \in \mathfrak{h} \mid gh = h\}$, we have that $\mathfrak{h}^0 = \mathbb{C}\beta$ with $\beta = \sum_{j=1}^k \alpha^{(j)}$.

Thinking of $\mathfrak{h}$ as a purely odd super vector space, we also have a parity map on $\mathfrak{h}$ denoted by $\sigma_\mathfrak{h}$ which of course just acts as multiplication by -1. Then defining $\mathfrak{h}^{0*} = \{h \in \mathfrak{h} \mid g\sigma_\mathfrak{h} h = h\}$, we have that $\mathfrak{h}^{0*} = \{(c_1, c_2, \ldots, c_k) \in \mathfrak{h} \mid c_j = -c_{j+1}$ for $1 \leq j \leq k-1$ and $c_k = -c_1\}$. So that $\mathfrak{h}^{0*}$ is of dimension 1 if k is even and is of dimension 0 if k is odd.

Since from the first author's work in [Bar12], we already have a unified construction and classification of all $(1\ 2\ \cdots\ k)$-twisted $V^{\otimes k}$ modules for k odd and V any vertex operator superalgebra, we turn our attention here to the case when k is even, following [DZ2]. In this case, according to [DZ2], we should obtain two equivalence classes of irreducible parity-unstable $(1\ 2\ \cdots\ k)$-twisted $V_{fer}^{\otimes k}$ modules for k even.

Letting $g = (1\ 2\ \cdots\ k)$ for k odd, we consider the $g\sigma_\mathfrak{h}$-eigenspaces

$$\mathfrak{h}^f_{(n)} = \{h \in \mathfrak{h} \mid g\sigma_\mathfrak{h} h = \eta^n h\} \subset \mathfrak{h}, \tag{67}$$

for η a fixed primitive kth root of unity. And so in terms of our discussion above, $\mathfrak{h}^{0*} = \mathfrak{h}^f_{(0)}$. (In the notation of [DZ2], we have $\mathfrak{h}^f_{(n)} = H^{n*}$.) We use the f superscript to denote this fermionic setting as opposed to the bosonic setting of the lattice we will encounter latter in Section 5.

We have $\mathfrak{h} = \coprod_{n\in\mathbb{Z}/k\mathbb{Z}} \mathfrak{h}^f_{(n)}$, where we identify $\mathfrak{h}^f_{(n \bmod k)}$ with $\mathfrak{h}^f_{(n)}$ for $n \in \mathbb{Z}$. For $n \in \mathbb{Z}/k\mathbb{Z}$, denote by $P_n : \mathfrak{h} \longrightarrow \mathfrak{h}^f_{(n)}$, the projection onto $\mathfrak{h}^f_{(n)}$, and for $h \in \mathfrak{h}$ and $n \in \mathbb{Z}$, set $h_{(n)} = P_{(n \bmod k)} h$. In general, we have that for $h \in \mathfrak{h}$ and $n \in \mathbb{Z}$,

$$h_{(n)} = \frac{1}{k}\sum_{j=0}^{k-1} \eta^{-nj}(g\sigma_\mathfrak{h})^j h. \tag{68}$$

Then it is clear that dim $\mathfrak{h}^f_{(n)} = 1$, for $0 \leq n \leq k-1$. In fact, $\alpha^{(1)}_{(n)}$ can be taken as a basis for each $\mathfrak{h}^f_{(n)}$.

Viewing $\mathfrak{h}$ as an abelian Lie superalgebra concentrated in the odd component, let

$$\hat{\mathfrak{h}}^f[g] = \coprod_{n\in\frac{1}{k}\mathbb{Z}} \mathfrak{h}^f_{(kn)} \otimes t^n \oplus \mathbb{C}\mathbf{k} \tag{69}$$

with $\mathbb{Z}_2$-grading given by $\mathrm{sgn}(\alpha \otimes t^n) = 1$ for $n \in \frac{1}{k}\mathbb{Z}$, and $\mathrm{sgn}(\mathbf{k}) = 0$, and Lie super-bracket relations

$$[\mathbf{k}, \hat{\mathfrak{h}}^f[g]] = 0, \quad \text{and} \quad [\alpha \otimes t^m, \beta \otimes t^n] = \langle \alpha, \beta\rangle \delta_{m+n,0}\mathbf{k} \tag{70}$$

for $\alpha \in \mathfrak{h}^f{}_{(km)}$, $\beta \in \mathfrak{h}^f{}_{(kn)}$, and $m, n \in \frac{1}{k}\mathbb{Z}$. Then $\hat{\mathfrak{h}}^f[g]$ is a $\frac{1}{k}\mathbb{Z}$-graded Lie superalgebra

$$\hat{\mathfrak{h}}^f[g] = \coprod_{n\in\frac{1}{k}\mathbb{Z}} \hat{\mathfrak{h}}^f[g]_n$$

where $\hat{\mathfrak{h}}^f[g]_0 = \mathfrak{h}^f_{(0)} \oplus \mathbb{C}\mathbf{k}$, and $\hat{\mathfrak{h}}^f[g]_n = \mathfrak{h}_{(kn)} \otimes t^{-n}$ for $n \neq 0$. And $\hat{\mathfrak{h}}^f[g]$ is a Heisenberg superalgebra.

Then $\hat{\mathfrak{h}}^f[g]$ has the following graded subalgebras

$$\hat{\mathfrak{h}}^f[g]_+ = \coprod_{n<0} \mathfrak{h}^f{}_{(kn)} \otimes t^n, \quad \text{and} \quad \hat{\mathfrak{h}}^f[g]_- = \coprod_{n>0} \mathfrak{h}^f{}_{(kn)} \otimes t^n, \tag{71}$$

and we have $\hat{\mathfrak{h}}^f[g] = \hat{\mathfrak{h}}^f[g]_- \oplus \mathfrak{h}^f_{(0)} \oplus \hat{\mathfrak{h}}^f[g]_+$.

Let $\mathbb{C}$ be the $\hat{\mathfrak{h}}^f[g]_- \oplus \mathbb{C}\mathbf{k}$-module such that $\hat{\mathfrak{h}}^f[g]_-$ acts trivially and $\mathbf{k}$ acts as 1. Set

$$M_g = U(\hat{\mathfrak{h}}^f[g]) \otimes_{U(\hat{\mathfrak{h}}^f[g]_- \oplus \mathbb{C}\mathbf{k})} \mathbb{C}. \tag{72}$$

Then as a vector space, we have

$$M_g \overset{\text{vec.sp.}}{\simeq} \bigwedge(\hat{\mathfrak{h}}^f[g]_+ \oplus \mathfrak{h}^f_{(0)}). \tag{73}$$

Let $\alpha \in \mathfrak{h}$ and $n \in \frac{1}{k}\mathbb{Z}$. We use the notation

$$\alpha(n)^g = \alpha \otimes t^n \in \hat{\mathfrak{h}}^f[g].$$

Then M_g is a $\hat{\mathfrak{h}}^f[g]$-module. The action induced from the supercommutation relations (70) is given by

$$\begin{aligned}
\mathbf{k}\beta(-l)^g\mathbf{1} &= \beta(-l)^g\mathbf{1} && (74)\\
\alpha(n)^g\beta(-m)^g\mathbf{1} &= \langle \alpha_{(kn)}, \beta_{(km)}\rangle \delta_{m,n}\mathbf{1} && (75)\\
\alpha(-n)^g\beta(-m)^g\mathbf{1} &= -\beta(-m)^g\alpha(-n)^g\mathbf{1} && (76)\\
\alpha(0)^g\beta(0)^g\mathbf{1} &= \frac{1}{2}\langle \alpha_{(0)}, \beta_{(0)}\rangle \mathbf{1} && (77)
\end{aligned}$$

for $\alpha, \beta \in \mathfrak{h}$, $m, n \in \frac{1}{k}\mathbb{Z}_+$ and $l \in \frac{1}{k}\mathbb{N}$. (Here $\mathbf{1} = 1 \in M_g$.) As an associative algebra with identity, M_g is the Clifford algebra generated by $\hat{\mathfrak{h}}^f[g]_+ \oplus \mathfrak{h}^f_{(0)}$ with the corresponding symmetric bilinear form.

For $\alpha \in \mathfrak{h}$, set

$$\alpha(x)^g = \sum_{n \in \frac{1}{k}\mathbb{Z}} \alpha(n)^g x^{-n-\frac{1}{2}}. \tag{78}$$

Then for the orthonormal basis of $\mathfrak{h}$, $\alpha^{(j)}$, for $j = 1, \ldots, k$, we have

$$[\alpha^{(j)}_{(km)}(x_1)^g, \alpha^{(l)}_{(kn)}(x_2)^g] = \frac{1}{k}\delta_{j,l}\delta_{m,-n}x_1^{m+1/2}x_2^{-m-1/2}\left(\frac{1}{(x_1 - x_2)} - \frac{1}{(-x_2 + x_1)}\right) \tag{79}$$

for $j, l = 1, \ldots, k$ and $m, n \in \frac{1}{k}\mathbb{Z}$ implying that the $\alpha^{(j)}(x)^g$, for $j = 1, \ldots, k$, are mutually local.

For $v \in V_{fer}^{\otimes k}$, define $Y^g(v, x) : M_g \longrightarrow M_g[[x^{1/k}, x^{-1/k}]]$ as follows: For $\alpha \in \mathfrak{h}^f_{(r)}$, $n \in \mathbb{N} + \frac{1}{2}$, and $u \in V_{fer}^{\otimes k}$, let

$$\begin{aligned}
Y^g(\alpha(-n)u, x) = Y^g(\alpha_{-n-1/2}u, x) = \mathrm{Res}_{x_1}\mathrm{Res}_{x_0}\left(\frac{x_1 - x_0}{x}\right)^{r/k} x_0^{-n-1/2} \qquad (80)\\
\cdot \left(x_0^{-1}\delta\left(\frac{x_1 - x}{x_0}\right)\alpha(x_1)^g Y^g(u, x) - (-1)^{|u|}x_0^{-1}\delta\left(\frac{x - x_1}{-x_0}\right)Y^g(u, x)\alpha(x_1)^g\right).
\end{aligned}$$

Then since $V_{fer}^{\otimes k} = \langle \alpha^{(j)}(-1/2)\mathbf{1} \mid j = 1, \ldots, k\rangle$, equation (80) defines $Y^g(v, x)$ iteratively for any $v \in V_{fer}^{\otimes k}$.

Recalling that the Virasoro element, ω_{fer}, for the free fermionic vertex operator superalgebra $V_{fer}^{\otimes d}$ is given by (38), a nontrivial computation shows that

$$L^g(m) = k\sum_{r=0}^{k-1} \sum_{\substack{n \in \mathbb{Z}\\ n > -\frac{m}{2}}} \left(n + \frac{m}{2} - \frac{r}{k}\right) \alpha^{(1)}_{(r)}(-n + r/k)\alpha^{(1)}_{(-r)}(n + m - r/k) + \frac{k^2 + 2}{48k}\delta_{m,0}. \tag{81}$$

One should compare this with (56).

From (81) it follows that $\left[L^g(-1), Y^g(\alpha^{(j)}(-1/2)\mathbf{1}, x)\right] = \frac{d}{dx} Y^g(\alpha^{(j)}(-1/2)\mathbf{1}, x)$, and thus, following [Li2], M_g is a weak g-twisted $V_{fer}^{\otimes k}$-module. It is also admissible.

Similarly to the situation in the parity-twisted case, the admissible g-twisted module, M_g, reduces as the direct sum of two irreducible parity unstable admissible g-twisted modules, and according to[DZ2], these two irreducibles are the only irreducible admissible g-twisted modules for $V_{fer}^{\otimes k}$, up to isomorphism.

Note that $M_g = \bigwedge\left[\alpha^{(1)}_{(r)}(-n)^g 1 \mid r = 0, \dots, k-1,\ n \in \mathbb{N} + \frac{r}{k}\right]$. Thus setting $\alpha = k\alpha^{(1)}_{(0)} = \alpha^{(1)} - \alpha^{(2)} + \alpha^{(3)} + \cdots + \alpha^{(k-1)} - \alpha^{(k)}$, then $\mathfrak{h}^f_{(0)} = \mathfrak{h}^{0*} = \mathbb{C}\alpha$. Let $\epsilon = \frac{1}{\sqrt{2k}}\alpha$ so that $\mathfrak{h}^f_{(0)} = \mathbb{C}\epsilon$ and $\langle \epsilon, \epsilon \rangle = 1$. Set

$$W = \bigwedge\left[\epsilon(-m)^g 1 \,\middle|\, \text{for } m \in \mathbb{Z}_+, \right] \tag{82}$$

and let $W = W^0 \oplus W^1$ be the decomposition of W into even and odd subspaces. Then these parity-unstable subspaces of the irreducible parity-stable module M_g are given by

$$\begin{aligned} M_g^{\pm} = {} & \left((1 \pm \epsilon(0)^g 1)\, W^0 \oplus (1 \mp \epsilon(0)^g 1)\, W^1\right) \\ & \otimes \bigwedge\left[\alpha^{(1)}_{(r)}(-n)^g 1 \,\middle|\, r = 1, \dots, k-1,\ n \in \mathbb{N} + \tfrac{r}{k}, \right]. \end{aligned} \tag{83}$$

Then we have $M_g = M_g^- \oplus M_g^+$ is an ordinary parity-stable irreducible g-twisted $V_{fer}^{\otimes k}$-module and $M_g^{\pm}$ are parity unstable invariant subspaces of M_g, i.e. parity unstable irreducible g-twisted $V_{fer}^{\otimes k}$-modules.

From (81), we have that the $L^g(0)$ grading on M_g is given by

$$\mathrm{wt}\,\mathbf{1} = \frac{k^2+2}{48k}, \quad \text{and} \quad \mathrm{wt}\,\alpha^{(1)}_{(r)}(-n)^g\mathbf{1} = n + \frac{k^2+2}{48k}, \tag{84}$$

for $n \in \mathbb{N} + \frac{r}{k}$ and $r = 0, \dots, k-1$. Thus the graded dimension of M_g is

$$\dim_q M_g \;=\; 2q^{-k/48} q^{(k^2+2)/(48k)} \prod_{n \in \frac{1}{k}\mathbb{Z}_+} (1 + q^n) \;=\; \sqrt{2}\mathfrak{f}_2(q^{1/k}). \tag{85}$$

3.4. A conjecture for $(1\ 2\ \cdots\ k)$*-twisted* $V^{\otimes k}$*-modules for* k *even and* V *any vertex operator superalgebra*

We make the following two observations, Remarks 3.4 and 3.5, to motivate the conjecture we are about to make.

Remark 3.4 Comparing the graded dimension of the σ twisted V_{fer}-module, M_σ, to the graded dimension of the $(1\ 2\ \cdots\ k)$-twisted $V_{fer}^{\otimes k}$ module, M_g, we have that

$$\dim_q M_g = \sqrt{2}\mathfrak{f}_2(q^{1/k}) = \dim_{q^{1/k}} M_\sigma. \tag{86}$$

This relationship of $q \rightarrow q^{1/k}$ between graded dimensions was in the past observed in the vertex operator algebra setting between untwisted V-modules and $(1\ 2\ \cdots\ k)$-twisted $V^{\otimes k}$-modules for k even or odd, and was one of the original motivations to the proof that these two categories of modules are in fact isomorphic given in [BDM]. That is, it had been observe that the graded dimension of a $(1\ 2\ \cdots\ k)$-twisted $V^{\otimes k}$-module was the same as the graded dimension of a V-module but with q replaced by $q^{1/k}$. In [Bar12], the first author showed that in the case when V is a vertex operator superalgebra, the extension of the construction in [BDM] to an

Journal of Physics: Conference Series **474** (2013) 012009 doi:10.1088/1742-6596/474/1/012009

isomorphism of categories between untwisted V-modules and $(1\ 2\ \cdots\ k)$-twisted $V^{\otimes k}$-modules exists in general for a vertex operator superalgebra V if and only if k is an odd positive integer. Motivated by the relationship between M_g and M_σ observed here, we make the conjecture below, Conjecture 3.6, that for k even, the categorical correspondence is between $(1\ 2\ \cdots\ k)$-twisted $V^{\otimes k}$-modules and parity-twisted V-modules.

Remark 3.5 In addition, we have that both the modules M_σ and M_g split into parity-unstable invariant subspaces. This was another one of the motivating examples for Theorem 2.1 as well as further evidence to bolster our conjecture given below, Conjecture 3.6.

These two observations given above, as well as recent constructions given by the second author, and certain observations given by the first author in [Bar12], leads us to the following conjecture.

Conjecture 3.6 *If V is a vertex operator superalgebra, and k is an even positive integer, then the category of weak (parity-stable) $(1\ 2\ \cdots\ k)$-twisted $V^{\otimes k}$-modules for k even is isomorphic to the category of weak (parity-stable) parity-twisted V-modules. In addition, the subcategories of weak admissible and ordinary $(1\ 2\ \cdots\ k)$-twisted $V^{\otimes k}$-modules are isomorphic to the subcategories of weak admissible and ordinary parity-twisted V-modules, respectively. Furthermore, all the various subcategories of parity-unstable invariant subspaces, i.e. parity-unstable submodules, coincide.*

Note that this is in contrast to the results of the first author in [Bar12] where we prove the following:

Theorem 3.7 *([Bar12]) If V is a vertex operator superalgebra, and k is an odd positive integer, then the category of weak (parity-stable) $(1\ 2\ \cdots\ k)$-twisted $V^{\otimes k}$-modules is isomorphic to the category of weak (parity-stable) V-modules. In addition, the subcategories of weak admissible and ordinary $(1\ 2\ \cdots\ k)$-twisted $V^{\otimes k}$-modules are isomorphic to the subcategories of weak admissible and ordinary V-modules, respectively. Furthermore, all the various subcategories of parity-unstable invariant subspaces, i.e. parity-unstable submodules, coincide.*

In addition, in [Bar12], an explicit construction of the weak, weak admissible, and ordinary $(1\ 2\ \cdots\ k)$-twisted $V^{\otimes k}$-modules for k odd is given in terms of the weak, weak admissible and ordinary V-modules.

4. Lattice vertex operator superalgebras

We recall the notion of a lattice vertex operator superalgebra following the notation and terminology of [FLM3] and using the setting and results of, e.g. [Le1], [FLM2], [DL1], [X], and [DL2].

Let L be a positive-definite integral lattice, with nondegenerate symmetric $\mathbb{Z}$-bilinear form $\langle\cdot,\cdot\rangle$. We introduce a lattice L with an isometry ν, and two central extensions, $\hat{L}$ and $\hat{L}_\nu$. (There should be no confusion between this use of the symbol L and the operators $L(n)$ for the Virasoro Algebra). The lattice L together with the central extension $\hat{L}$ will be used to construct a vertex operator super algebra V_L. The central extension $\hat{L}_\nu$ will be used in Section 5 to construct a space V_L^T on which V_L acts via twisted vertex operators. In Section 7.1, ν will be specified to the -1 isometry and a certain lift $\hat{\nu}$ on $\hat{L}_\nu$ to construct the twisted modules we are interested in.

Let k be a fixed positive integer. The following initial assumptions and conditions are assumed.

1. Let L be a positive definite integral lattice with nondegenerate symmetric $\mathbb{Z}$-valued bilinear form $\langle\cdot,\cdot\rangle$, i.e. L is a finitely generated abelian group with positive definite symmetric $\mathbb{Z}$-bilinear

form $\langle \cdot, \cdot \rangle : L \times L \to \mathbb{Z}$.
2. Let ν be an isometry of L with period k (k need not be the order of ν, and in fact will be a period that is not the order in the particular case in which we will be interested).
3. We fix a primitive kth root of unity η. Set $\eta_0 = (-1)^k \eta$, so that η_0 is a primitive $2k$th root of unity if k is odd, and $\eta_0 = \eta$ remains a primitive kth root of unity if k is even.

Since L is integral, we can give it a natural $\mathbb{Z}_2$-grading

$$L = L^0 \cup L^1, \quad L^j = \{\alpha \in L \mid \langle \alpha, \alpha \rangle \in 2\mathbb{Z} + j\}, \tag{87}$$

and L^0 is an even sublattice of L. We will use the notation $|\alpha| = j$ for $\alpha \in L^j$.

Note that

$$\sum_{j=0}^{k-1} \langle \nu^j \alpha, \alpha \rangle \in \begin{cases} |\alpha| + 2\mathbb{Z} & \text{if } k \text{ is odd} \\ |\alpha| + \langle \nu^{k/2}\alpha, \alpha \rangle + 2\mathbb{Z} & \text{if } k \text{ is even} \end{cases} . \tag{88}$$

In addition,

$$\sum_{j=0}^{k-1} \langle j\nu^j \alpha, \alpha \rangle \in \begin{cases} k\mathbb{Z} & \text{if } k \text{ is odd} \\ \frac{k}{2}\langle \nu^{k/2}\alpha, \alpha \rangle + k\mathbb{Z} & \text{if } k \text{ is even} \end{cases} . \tag{89}$$

Remark 4.1 If k is even and $\langle \nu^{k/2}\alpha, \alpha \rangle \in 2\mathbb{Z} + |\alpha|$, which can always be arranged by doubling k if necessary, then the expressions in 88 and 89 are in $|\alpha| + 2\mathbb{Z}$ and $k\mathbb{Z}$, respectively. For the purposes of this paper, we always assume that if k is even, then $\langle \nu^{k/2}\alpha, \alpha \rangle \in 2\mathbb{Z} + |\alpha|$. That is we do indeed double k if necessary. However in the setting of permutation-twisted modules for lattice vertex operator superalgebras, this can not be done. That is, following but extending [BHL], taking L to be the orthogonal sum of k copies of K for k even and considering $\nu = (1\ 2\ \cdots\ k)$ acting on L in the natural way, then we have $\langle \nu^{k/2}\alpha, \alpha \rangle \in 2\mathbb{Z}$. But doubling k results in a lift that is of order $2k$, i.e. that is not the permutation automorphism on the tensor product lattice vertex operator superalgebra. This is another illustration of the fundamental difference between the nonsuper case or the super case for k odd versus the super case for k even in the permutation twisted setting.

Let $q = k$ if k is even, and $q = 2k$ if k is odd. We define the ν-invariant functions

$$C_0 : L \times L \longrightarrow \mathbb{C}^\times, \quad (\alpha, \beta) \mapsto (-1)^{\langle \alpha,\alpha \rangle \langle \beta,\beta \rangle + \langle \alpha,\beta \rangle}, \tag{90}$$

$$\begin{aligned} C : L \times L \longrightarrow \mathbb{C}^\times, \quad (\alpha, \beta) &\mapsto (-1)^{\langle \alpha,\alpha \rangle \langle \beta,\beta \rangle + \sum_{j=0}^{k-1} \langle \nu^j \alpha,\beta \rangle} \eta^{\sum_{j=0}^{k-1} \langle j\nu^j \alpha,\beta \rangle} \\ &= (-1)^{\langle \alpha,\alpha \rangle \langle \beta,\beta \rangle} \prod_{j=0}^{k-1} (-\eta^j)^{\langle \nu^j \alpha,\beta \rangle}. \end{aligned} \tag{91}$$

Note that C_0 and C are bilinear into the abelian group $\mathbb{C}^\times$; i.e., $C(\alpha+\beta, \gamma) = C(\alpha, \gamma)C(\beta, \gamma)$ and $C(\alpha, \beta + \gamma) = C(\alpha, \beta)C(\alpha, \gamma)$, for $\alpha, \beta, \gamma \in L$, and similarly for C_0. In addition, we have $C_0(\alpha, \alpha) = 1$, and by (88) and (89), we have $C(\alpha, \alpha) = 1$. Moreover, $C(\beta, \alpha) = C(\alpha, \beta)^{-1}$.

The maps C_0 and C determine uniquely (up to equivalence) two central extensions of L by the cyclic group $\langle \eta_0 \rangle$,

$$1 \to \langle \eta_0 \rangle \to \hat{L} \xrightarrow{-} L \to 1, \tag{92}$$

$$1 \to \langle \eta_0 \rangle \to \hat{L}_\nu \xrightarrow{-} L \to 1, \tag{93}$$

with commutator maps c_0 and c_0^ν, respectively, i.e., such that

$$aba^{-1}b^{-1} = C_0(\bar{a}, \bar{b}) \quad \text{for} \quad a, b \in \hat{L}, \tag{94}$$

$$aba^{-1}b^{-1} = C(\bar{a}, \bar{b}) \quad \text{for} \quad a, b \in \hat{L}_\nu. \tag{95}$$

There is a natural set-theoretic identification (which is not an isomorphism of groups unless $k = 1$ or $k = 2$) between the groups $\hat{L}$ and $\hat{L}_\nu$ such that the respective group multiplications $\times$ and $\times_\nu$ are related by

$$a \times b = \prod_{0<j<k/2} (-\eta^j)^{\langle \nu^{-j}\bar{a}, \bar{b}\rangle} a \times_\nu b \qquad \text{for} \quad a, b \in \hat{L}. \tag{96}$$

Note that this is the exact same relationship as in the even lattice case treated in [FLM2], [Le1], and [BHL]. Observe further that since C_0 is ν-invariant, if we replace the map $^-$ in (92) by $\nu \circ {}^-$, we obtain another central extension of L by $\langle \eta_0 \rangle$ with commutator map C_0. By uniqueness of the central extension of L, there is an automorphism $\hat{\nu}$ of $\hat{L}$ (fixing η_0) such that $\hat{\nu}$ is a lifting of ν, i.e., such that

$$\overline{(\hat{\nu} a)} = \nu \bar{a} \quad \text{for} \quad a \in \hat{L}. \tag{97}$$

The map $\hat{\nu}$ is also an automorphism of $\hat{L}_\nu$ satisfying

$$\overline{(\hat{\nu} a)} = \nu \bar{a} \quad \text{for} \quad a \in \hat{L}_\nu. \tag{98}$$

Moreover, we may choose the lifting $\hat{\nu}$ of ν so that

$$\hat{\nu} a = a \quad \text{if} \quad \nu \bar{a} = \bar{a} \tag{99}$$

(see (134) below).

We now use the central extension $\hat{L}$ to construct a vertex operator superalgebra V_L equipped with an automorphism $\hat{\nu}$ of period k, induced from the automorphism $\hat{\nu}$ of $\hat{L}$. This is essentially a specialized case of the "unrelativised operators" in Section 2 of [DL1], [DL2] and of [X].

Embed L canonically in the $\mathbb{C}$-vector space $\mathfrak{h} = \mathbb{C} \otimes_{\mathbb{Z}} L$, and extend the $\mathbb{Z}$-bilinear form on L to a $\mathbb{C}$-bilinear form $\langle \cdot, \cdot \rangle$ on $\mathfrak{h}$. The corresponding affine Lie algebra is

$$\hat{\mathfrak{h}} = \mathfrak{h} \otimes \mathbb{C}[t, t^{-1}] \oplus \mathbb{C}\mathbf{k}, \tag{100}$$

with brackets determined by

$$[\mathbf{k}, \hat{\mathfrak{h}}] = 0 \qquad \text{and} \qquad [\alpha \otimes t^m, \beta \otimes t^n] = \langle \alpha, \beta \rangle m \delta_{m+n,0} \mathbf{k} \tag{101}$$

for $\alpha, \beta \in \mathfrak{h}$, and $m, n \in \mathbb{Z}$. Then $\hat{\mathfrak{h}}$ has a $\mathbb{Z}$-gradation, the *weight gradation*, given by $\mathrm{wt}\,(\alpha \otimes t^n) = -n$ and $\mathrm{wt}\,\mathbf{k} = 0$, for $\alpha \in \mathfrak{h}$ and $n \in \mathbb{Z}$.

Set

$$\hat{\mathfrak{h}}^+ = \mathfrak{h} \otimes t\mathbb{C}[t] \;\text{ and }\; \hat{\mathfrak{h}}^- = \mathfrak{h} \otimes t^{-1}\mathbb{C}[t^{-1}]. \tag{102}$$

The subalgebra $\hat{\mathfrak{h}}_{\mathbb{Z}} = \hat{\mathfrak{h}}^+ \oplus \hat{\mathfrak{h}}^- \oplus \mathbb{C}\mathbf{k}$ of $\hat{\mathfrak{h}}$ is a Heisenberg algebra, in the sense that its commutator subalgebra equals its center, which is one-dimensional. Consider the induced $\hat{\mathfrak{h}}$-module, irreducible even under $\hat{\mathfrak{h}}_{\mathbb{Z}}$, given by

$$M(1) = U(\hat{\mathfrak{h}}) \otimes_{U(\mathfrak{h} \otimes \mathbb{C}[t] \oplus \mathbb{C}\mathbf{k})} \mathbb{C} \simeq S(\hat{\mathfrak{h}}^-) \quad \text{(linearly)}, \tag{103}$$

where $\mathfrak{h} \otimes \mathbb{C}[t]$ acts trivially on $\mathbb{C}$ and $\mathbf{k}$ acts as 1, $U(\cdot)$ denotes universal enveloping algebra and $S(\cdot)$ denotes symmetric algebra. The $\hat{\mathfrak{h}}$-module $M(1)$ is $\mathbb{Z}$-graded so that $\mathrm{wt}\,1 = 0$ (where we write 1 for $1 \otimes 1$)

$$M(1) = \coprod_{n \in \mathbb{N}} M(1)_n, \tag{104}$$

where $M(1)_n$ denotes the homogeneous subspace of weight n.

Form the induced $\hat{L}$-module and $\mathbb{C}$-algebra

$$\mathbb{C}\{L\} \;=\; \mathbb{C}[\hat{L}] \otimes_{\mathbb{C}[\langle \eta_0 \rangle]} \mathbb{C} \;\simeq\; \mathbb{C}[L] \qquad \text{(linearly)}, \tag{105}$$

where $\mathbb{C}[\cdot]$ denotes group algebra. For $a \in \hat{L}$, write $\iota(a)$ for the image of a in $\mathbb{C}\{L\}$. Then the action of $\hat{L}$ on $\mathbb{C}\{L\}$ is given by

$$a \cdot \iota(b) = \iota(a)\iota(b) = \iota(ab) \tag{106}$$

for $a, b \in \hat{L}$. We give $\mathbb{C}\{L\}$ the $\mathbb{C}$-gradation determined by

$$\mathrm{wt}\, \iota(a) = \frac{1}{2}\langle \bar{a}, \bar{a} \rangle \quad \text{ for } \ a \in \hat{L}. \tag{107}$$

Also define a grading-preserving action of $\mathfrak{h}$ on $\mathbb{C}\{L\}$ by

$$h \cdot \iota(a) = \langle h, \bar{a} \rangle \iota(a) \tag{108}$$

for $h \in \mathfrak{h}$, and define

$$x^h \cdot \iota(a) = x^{\langle h, \bar{a} \rangle} \iota(a) \tag{109}$$

for $h \in \mathfrak{h}$. Set

$$V_L \;=\; M(1) \otimes_{\mathbb{C}} \mathbb{C}\{L\} \;\simeq\; S(\hat{\mathfrak{h}}^-) \otimes \mathbb{C}[L] \qquad \text{(linearly)} \tag{110}$$

and give V_L the tensor product $\mathbb{C}$-gradation

$$V_L = \coprod_{n \in \mathbb{C}} (V_L)_n. \tag{111}$$

We have $\mathrm{wt}\, \iota(1) = 0$, where we identify $\mathbb{C}\{L\}$ with $1 \otimes \mathbb{C}\{L\}$. Then $\hat{L}$, $\hat{\mathfrak{h}}_{\mathbb{Z}}$, h, x^h ($h \in \mathfrak{h}$) act naturally on V_L by acting on either $M(1)$ or $\mathbb{C}\{L\}$ as indicated above. In particular, $\mathbf{k}$ acts as 1.

For $\alpha \in \mathfrak{h}$, $n \in \mathbb{Z}$, we write $\alpha(n)$ for the operator on V_L determined by $\alpha \otimes t^n$. For $\alpha \in \mathfrak{h}$, set

$$\alpha(x) = \sum_{n \in \mathbb{Z}} \alpha(n) x^{-n-1}. \tag{112}$$

We use a normal ordering procedure, indicated by open colons, which signify that the enclosed expression is to be reordered if necessary so that all the operators $\alpha(n)$, for $\alpha \in \mathfrak{h}$, with $n < 0$, as well as the operator a for $a \in \hat{L}$, are to be placed to the left of all the operators $\alpha(n)$ and x^α, for $\alpha \in \mathfrak{h}$ and $n \geq 0$, before the expression is evaluated. For $a \in \hat{L}$, set

$$Y(a, x) = {}^\circ_\circ\, e^{\int (\bar{a}(x) - \bar{a}(0)x^{-1})} a x^{\bar{a}} \,{}^\circ_\circ, \tag{113}$$

using an obvious formal integration notation. Let $a \in \hat{L}$, $\alpha_1, \ldots, \alpha_m \in \mathfrak{h}$, $n_1, \ldots, n_m \in \mathbb{Z}_+$ and set

$$v \;=\; \alpha_1(-n_1) \cdots \alpha_m(-n_m) \otimes \iota(a) \;=\; \alpha_1(-n_1) \cdots \alpha_m(-n_m) \cdot \iota(a) \;\in\; V_L. \tag{114}$$

Define

$$Y(v, x) = {}^\circ_\circ\, (\partial_{n_1 - 1} \alpha_1(x)) \cdots (\partial_{n_m - 1} \alpha_m(x))\, Y(a, x) \,{}^\circ_\circ, \tag{115}$$

where again, for $n \in \mathbb{N}$, we use the notation $\partial_n = \frac{1}{n!}\left(\frac{d}{dx}\right)^n$. This gives us a well-defined linear map

$$V_L \to (\mathrm{End}\, V_L)[[x, x^{-1}]], \quad v \mapsto Y(v, x) = \sum_{n \in \mathbb{Z}} v_n x^{-n-1}. \tag{116}$$

Set $\mathbf{1} = 1 = 1 \otimes 1 \in V_L$ and $\omega = \frac{1}{2}\sum_{i=1}^{\dim \mathfrak{h}} h_i(-1)h_i(-1)\mathbf{1}$, where $\{h_j \,|\, j = 1, \ldots, \dim \mathfrak{h}\}$ is an orthonormal basis of $\mathfrak{h}$. Then $V_L = (V_L, Y, \mathbf{1}, \omega)$ is a vertex operator superalgebra of central charge $c = \dim \mathfrak{h} = \mathrm{rank}\, L$. For a proof that this is a vertex operator superalgebra, see for instance Chapter 6.1 of [X].

Remark 4.2 The construction of the vertex operator superalgebra V_L depends on the central extension (92) subject to (94), and hence on the choices of $k \in \mathbb{Z}_+$ and the primitive root of unity η. But it is a standard fact that V_L is independent of these choices, up to isomorphism of vertex operator superalgebras preserving the $\hat{\mathfrak{h}}$-module structure; see for instance Proposition 6.5.5, and also Remarks 6.5.4 and 6.5.6, of [LL]. In particular, V_L as constructed above is essentially the same as V_L constructed from a central extension of the type (92) subject to (94) but with the kernel of the central extension replaced by the group $\langle \pm 1 \rangle$. For the purpose of constructing twisted modules, it is valuable to have this flexibility, and we will use this property of lattice vertex superalgebras below in Section 5.

5. Twisted modules for a lattice vertex operator superalgebra and a lift of a lattice isometry

Following [Le1], [FLM2], [FLM3], [DL2], [X], we recall the construction and classification of $\hat{\nu}$-twisted V_L-modules for a general lattice isometry ν and a lift $\hat{\nu}$.

Following [Le1], but extended to integral lattices, we note that the automorphism ν of L acts in a natural way on $\mathfrak{h}$, on $\hat{\mathfrak{h}}$ (fixing $\mathbf{k}$) and on $M(1)$, preserving the gradations, and for $u \in \hat{\mathfrak{h}}$ and $m \in M(1)$,

$$\nu(u \cdot m) = \nu(u) \cdot \nu(m). \tag{117}$$

The automorphism ν of L lifted to the automorphism $\hat{\nu}$ of $\hat{L}$ satisfies

$$\hat{\nu}(h \cdot \iota(a)) = \nu(h) \cdot \hat{\nu}\iota(a), \tag{118}$$

for $h \in \mathfrak{h}$ and $a \in \hat{L}$, and for $b \in \hat{L}$ we have

$$\hat{\nu}(\iota(a)\iota(b)) = \hat{\nu}(a \cdot \iota(b)) = \hat{\nu}(a) \cdot \hat{\nu}\iota(b) = \hat{\nu}\iota(a)\hat{\nu}\iota(b), \tag{119}$$

$$\hat{\nu}(x^h \cdot \iota(a)) = x^{\nu(h)} \cdot \hat{\nu}\iota(a). \tag{120}$$

Thus we have a natural grading-preserving automorphism of V_L, which we also call $\hat{\nu}$, which acts via $\nu \otimes \hat{\nu}$, and this action is compatible with the other actions

$$\hat{\nu}(a \cdot v) = \hat{\nu}(a) \cdot \hat{\nu}(v) \tag{121}$$
$$\hat{\nu}(u \cdot v) = \nu(u) \cdot \hat{\nu}(v) \tag{122}$$
$$\hat{\nu}(x^h \cdot v) = x^{\nu(h)} \cdot \hat{\nu}(v) \tag{123}$$

for $a \in \hat{L}$, $u \in \hat{\mathfrak{h}}$, $h \in \mathfrak{h}$, and $v \in V_L$, so that $\hat{\nu}$ is an automorphism of the vertex operator superalgebra V_L.

Recalling our fixed primitive kth root of unity η from Section 4, for $n \in \mathbb{Z}$ set

$$\mathfrak{h}_{(n)} = \{h \in \mathfrak{h} \mid \nu h = \eta^n h\} \subset \mathfrak{h}, \tag{124}$$

so that $\mathfrak{h} = \coprod_{n \in \mathbb{Z}/k\mathbb{Z}} \mathfrak{h}_{(n)}$, where we identify $\mathfrak{h}_{(n \bmod k)}$ with $\mathfrak{h}_{(n)}$, for $n \in \mathbb{Z}$. Then in general,

$$\mathfrak{h}_{(n)} = \{h + \eta^{-n}\nu h + \eta^{-2n}\nu^2 h + \cdots + \eta^{-(k-1)n}\nu^{k-1}h \mid h \in \mathfrak{h}\}. \tag{125}$$

For $n \in \mathbb{Z}/k\mathbb{Z}$, denote by

$$P_n : \mathfrak{h} \longrightarrow \mathfrak{h}_{(n)}, \tag{126}$$

the projection onto $\mathfrak{h}_{(n)}$, and for $h \in \mathfrak{h}$ and $n \in \mathbb{Z}$, set $h_{(n)} = P_{(n \bmod k)}h$. In general, we have that for $h \in \mathfrak{h}$ and $n \in \mathbb{Z}$,

$$h_{(n)} = \frac{1}{k}\sum_{j=0}^{k-1} \eta^{-nj}\nu^j h. \tag{127}$$

Viewing $\mathfrak{h}$ as an abelian Lie algebra, consider the ν-twisted affine Lie algebra

$$\hat{\mathfrak{h}}[\nu] = \coprod_{n \in \frac{1}{k}\mathbb{Z}} \mathfrak{h}_{(kn)} \otimes t^n \oplus \mathbb{C}\mathbf{k} \tag{128}$$

with brackets determined by

$$[\mathbf{k}, \hat{\mathfrak{h}}[\nu]] = 0 \quad \text{and} \quad [\alpha \otimes t^m, \beta \otimes t^n] = \langle \alpha, \beta \rangle m \delta_{m+n,0} \mathbf{k} \tag{129}$$

for $\alpha \in \mathfrak{h}_{(km)}$, $\beta \in \mathfrak{h}_{(kn)}$, and $m, n \in \frac{1}{k}\mathbb{Z}$.

Define the *weight gradation* on $\hat{\mathfrak{h}}[\nu]$ by $\mathrm{wt}\,(\alpha \otimes t^n) = -n$, $\mathrm{wt}\,\mathbf{k} = 0$, for $n \in \frac{1}{k}\mathbb{Z}$, $\alpha \in \mathfrak{h}_{(kn)}$. Set

$$\hat{\mathfrak{h}}[\nu]^+ = \coprod_{n>0} \mathfrak{h}_{(kn)} \otimes t^n, \qquad \hat{\mathfrak{h}}[\nu]^- = \coprod_{n<0} \mathfrak{h}_{(kn)} \otimes t^n. \tag{130}$$

Now the subalgebra

$$\hat{\mathfrak{h}}[\nu]_{\frac{1}{k}\mathbb{Z}} = \hat{\mathfrak{h}}[\nu]^+ \oplus \hat{\mathfrak{h}}[\nu]^- \oplus \mathbb{C}\mathbf{k} \tag{131}$$

of $\hat{\mathfrak{h}}[\nu]$ is a Heisenberg algebra. Form the induced $\hat{\mathfrak{h}}[\nu]$-module

$$S[\nu] = U(\hat{\mathfrak{h}}[\nu]) \otimes_{U(\coprod_{n \geq 0} \mathfrak{h}_{(kn)} \otimes t^n \oplus \mathbb{C}\mathbf{k})} \mathbb{C} \simeq S(\hat{\mathfrak{h}}[\nu]^-) \qquad \text{(linearly)}, \tag{132}$$

where $\coprod_{n \geq 0} \mathfrak{h}_{(kn)} \otimes t^n$ acts trivially on $\mathbb{C}$ and $\mathbf{k}$ acts as 1. Then $S[\nu]$ is irreducible under $\hat{\mathfrak{h}}[\nu]_{\frac{1}{k}\mathbb{Z}}$.

Following [DL2], Section 6, we give the module $S[\nu]$ the natural $\mathbb{Q}$-grading (by weights) compatible with the action of $\hat{\mathfrak{h}}[\nu]$ and such that

$$\mathrm{wt}\, 1 = \frac{1}{4k^2} \sum_{j=1}^{k-1} j(k-j) \mathrm{dim}\,(\mathfrak{h}_{(j)}). \tag{133}$$

Following Sections 5 and 6 of [Le1] extended to this setting, we have that the automorphisms of $\hat{L}_\nu$ covering the identity automorphism of L are precisely the maps $\rho^* : a \to a\rho(\bar{a})$ for a homomorphism $\rho : L \to \langle \eta_0 \rangle$. Similarly, there is a homomorphism $\rho_0 : L \cap \mathfrak{h}_{(0)} \to \langle \eta_0 \rangle$ such that $\hat{\nu} a = a\rho_0(\bar{a})$ if $\nu\bar{a} = \bar{a}$. Now ρ_0 can be extended to a homomorphism $\rho : L \to \langle \eta_0 \rangle$ since the map $1 - P_0$ induces an isomorphism from $L/(L \cap \mathfrak{h}_{(0)})$ to the free abelian group $(1 - P_0)L$. Multiplying $\hat{\nu}$ by the inverse of ρ_0^* gives us an automorphism $\hat{\nu}$ of $\hat{L}_\nu$ satisfying (98) and

$$\hat{\nu} a = a \quad \text{if} \quad \nu\bar{a} = \bar{a}, \tag{134}$$

as in (99).

Next, we wish to construct a space U_T for $\hat{L}_\nu$ and $\mathfrak{h}_{(0)}$ to act upon which will be a subspace of our twisted module. Set

$$N = \{\alpha \in L \mid \langle \alpha, \mathfrak{h}_{(0)} \rangle = 0\} = (1 - P_0)\mathfrak{h} \cap L, \tag{135}$$

$$M = (1 - \nu)L \subset N, \tag{136}$$

$$R = \{\alpha \in N \mid C_N(\alpha, N) = 1\}, \tag{137}$$

where C_N denotes the map C restricted to $N \times N$. Note that $M \subset R = Z(N)$ are all subgroups of L, where $Z(N)$ denotes the center of N. Also, it is clear that $M \subset L^0$, the even sublattice of

L. For $\alpha \in \mathfrak{h}$, we have $\sum_{j=0}^{k-1} \nu^j \alpha \in \mathfrak{h}_{(0)}$, and $N \subset \sum_{j=1}^{k-1} \mathfrak{h}_{(j)}$ and thus for $\alpha, \beta \in N$, the commutator map C, defined by (91) on N, simplifies to

$$C_N(\alpha, \beta) = (-1)^{\langle \alpha,\alpha\rangle\langle \beta,\beta\rangle} \eta^{\sum_{j=0}^{k-1}\langle j\nu^j \alpha,\beta\rangle}. \tag{138}$$

Denote by $\hat{Q}$ the subgroup of $\hat{L}_\nu$ obtained by pulling back any subgroup Q of L. Then $\{a\hat{\nu}a^{-1} \,|\, a \in \hat{L}_\nu\} \subset \hat{M} \subset (\hat{L}^0)_\nu$. Note that by (134), we have that $\{a\hat{\nu}a^{-1} \,|\, a \in \hat{L}_\nu\} \cap \langle \eta_0 \rangle = 1$.

For $a \in \hat{L}_\nu$ define

$$\tau(a\hat{\nu}a^{-1}) = \eta^{k\langle \bar{a},\bar{a}\rangle/2 - \sum_{j=0}^{k-1}\langle \nu^j \bar{a},\bar{a}\rangle/2} = \eta^{k\langle \bar{a},\bar{a}\rangle/2 - k\langle \bar{a}_{(0)},\bar{a}_{(0)}\rangle/2}. \tag{139}$$

In addition, for $b \in \hat{M}$, let

$$\tau(\eta_0^j b) = \tau(\eta_0^j)\tau(b) = \eta_0^j \tau(b) \quad \text{for } j = 1, \dots, q. \tag{140}$$

Then we have the following proposition:

Proposition 5.1 *The map $\tau : \hat{M} \to \mathbb{C}^\times$ given by (139) and satisfying (140) is a well-defined group homomorphism. Moreover, τ is the unique group homomorphism from $\hat{M}$ to $\mathbb{C}^\times$ satisfying (139) and (140). In addition, if $\langle \nu^{k/2}\alpha, \alpha\rangle \in 2\mathbb{Z} + |\alpha|$ for all $\alpha \in L$, then the image of τ lies in $\langle \eta \rangle$.*

Proof: We first show that τ is well-defined. Suppose $a\hat{\nu}a^{-1} = b\hat{\nu}b^{-1}$. Then $(1-\nu)\bar{a} = (1-\nu)\bar{b}$, which implies $(1-\nu^j)\bar{a} = (1-\nu^j)\bar{b}$, for $j = 1, \dots, k-1$. Thus

$$\begin{aligned} 2\langle \bar{a},\bar{a}\rangle - \langle \nu^j\bar{a},\bar{a}\rangle - \langle \nu^{k-j}\bar{a},\bar{a}\rangle &= \langle (1-\nu^j)\bar{a}, (1-\nu^j)\bar{a}\rangle = \langle (1-\nu^j)\bar{b}, (1-\nu^j)\bar{b}\rangle \\ &= 2\langle \bar{b},\bar{b}\rangle - \langle \nu^j\bar{b},\bar{b}\rangle - \langle \nu^{k-j}\bar{b},\bar{b}\rangle \end{aligned}$$

which implies that

$$k\langle \bar{a},\bar{a}\rangle/2 - \sum_{j=0}^{k-1}\langle \nu^j\bar{a},\bar{a}\rangle/2 = k\langle \bar{b},\bar{b}\rangle/2 - \sum_{j=0}^{k-1}\langle \nu^j\bar{b},\bar{b}\rangle/2.$$

Therefore $\tau(a\hat{\nu}a^{-1}) = \tau(b\hat{\nu}b^{-1})$, proving that τ is well defined.

For $a, b \in \hat{L}_\nu$, we have

$$\begin{aligned} \tau(a\hat{\nu}a^{-1})\tau(b\hat{\nu}b^{-1}) &= \eta^{k\langle \bar{a},\bar{a}\rangle/2 - \sum_{j=0}^{k-1}\langle \nu^j\bar{a},\bar{a}\rangle/2 + k\langle \bar{b},\bar{b}\rangle/2 - \sum_{j=0}^{k-1}\langle \nu^j\bar{b},\bar{b}\rangle/2} \\ &= \eta^{k\langle \bar{a}+\bar{b},\bar{a}+\bar{b}\rangle/2 - \sum_{j=0}^{k-1}\langle \nu^j(\bar{a}+\bar{b}),\bar{a}+\bar{b}\rangle/2 + \sum_{j=0}^{k-1}\langle \nu^j\bar{a},\bar{b}\rangle} \\ &= C(\bar{a}-\nu\bar{a},\bar{b})\tau((ba)\hat{\nu}(ba)^{-1}) \\ &= C(\bar{a}-\nu\bar{a},\bar{b})\tau(C(\bar{a}-\nu\bar{a},\bar{b})^{-1}(a\hat{\nu}a^{-1})(b\hat{\nu}b^{-1})) \\ &= \tau((a\hat{\nu}a^{-1})(b\hat{\nu}b^{-1})). \end{aligned}$$

This proves τ is a group homomorphism. Since $\hat{M}$ is the subgroup of $\hat{L}_\nu$ which is a lift of M, the uniqueness follows immediately from (139) and (140).

The last statement follows from (88). □

Next we extend τ to $\hat{R}$, and then to a maximal abelian subgroup $\hat{J}$ of $\hat{N}$. We first observe that if $\alpha \in N$, then there exists $h \in \mathfrak{h}$, such that

$$k\alpha = kh - kh_{(0)} = kh - \sum_{j=0}^{k-1}\nu^j h = \sum_{j=1}^{k-1}(h - \nu^j h). \tag{141}$$

Furthermore for $j = 1, \ldots, k$, we have $h - \nu^j h = (h - \nu h) + (\nu h - \nu^2 h) + \cdots + (\nu^{j-1} h - \nu^j h) \in (1-\nu)\mathfrak{h}$. Therefore $k\alpha \in (1-\nu)\mathfrak{h}$. Writing $h = c\beta$ for $c \in \mathbb{C}$ and $\beta \in L$, we have that $k\alpha \in L$ and $k\alpha = c(k\beta - \sum_{j=0}^{k-1} \nu^j \beta)$. It follows that $c \in \mathbb{Z}$ and thus $k\alpha \in (1-\nu)L = M$. That is

$$kN \subset (1-\nu)L = M. \tag{142}$$

Therefore N/M is a finitely generated torsion group, i.e. it is finite. Thus R/M is a finite group. Also N/M finite implies that $\hat{N}/\hat{M}$ and $\hat{N}/\ker\tau$ are finite as well. (The last statement follows from the fact that $\tau(a^k \hat{\nu} a^{-k}) = 1$ for all $a\hat{\nu}a^{-1} \in \hat{M}$.)

We wish to construct an irreducible $\hat{N}$-module, T, on which $\hat{M}$ acts as multiplication by the character τ.

The following is just a restatement of Proposition 6.2 of [Le1], but extended to our setting, and follows directly from Theorem 5.5.1 of [FLM3].

Proposition 5.2 *There are exactly $|R/M|$ extensions of τ to a homomorphism $\chi : \hat{R} \to \mathbb{C}^\times$. For each such χ, there is a unique (up to equivalence) irreducible $\hat{N}$-module on which $\hat{R}$ acts according to χ, and every irreducible $\hat{N}$-module on which $\hat{M}$ acts according to τ is equivalent to one of these. Every such module has dimension $|N/R|^{1/2}$. Supposing that T is an irreducible module for $\hat{M}$ such that $\hat{M}$ acts as τ, to construct the $\hat{N}$-module structure for T corresponding to χ, let J be any subgroup of N (necessarily containing R) that is maximal such that C_N is trivial on J. Then $\hat{J}$ is a maximal abelian subgroup of $\hat{N}$. Let $\psi : \hat{J} \to \mathbb{C}^\times$ be any homomorphism extending χ and denote by $\mathbb{C}_\psi$ the $\hat{J}$-module $\mathbb{C}$ with character ψ. Then T is isomorphic to the induced $\hat{N}$-module*

$$T = \mathbb{C}[\hat{N}] \otimes_{\mathbb{C}[\hat{J}]} \mathbb{C}_\psi \simeq \mathbb{C}[N/J] \qquad \textit{(linearly)}. \tag{143}$$

Let T be any $\hat{N}$-module on which $\hat{M}$ acts as multiplication by the character τ as given by Proposition 5.2. Form the induced $\hat{L}_\nu$-module

$$U_T = \mathbb{C}[\hat{L}_\nu] \otimes_{\mathbb{C}[\hat{N}]} T. \tag{144}$$

Since T can be viewed as a module for the finite group $\hat{N}/\ker\tau$, we have that T is completely reducible. Then the structure of T follows from Proposition 5.2, and in the irreducible case,

$$U_T = \mathbb{C}[\hat{L}_\nu] \otimes_{\mathbb{C}[\hat{N}]} T = \mathbb{C}[\hat{L}_\nu] \otimes_{\mathbb{C}[\hat{J}]} \mathbb{C}_\psi \simeq \mathbb{C}[L/J] \qquad \text{(linearly)}. \tag{145}$$

The action of $\hat{L}_\nu$ on U_T is given by

$$a \cdot b \otimes r = ab \otimes r, \tag{146}$$

for $a, b \in \hat{L}_\nu$, and $r \in T$, and of course

$$(a\hat{\nu}a^{-1}) \cdot b \otimes r = C(\bar{a} - \nu\bar{a}, \bar{b})(b(a\hat{\nu}a^{-1})) \otimes r = C(\bar{a} - \nu\bar{a}, \bar{b}) b \otimes \tau(a\hat{\nu}a^{-1}) r. \tag{147}$$

Let $\hat{\lambda} \in \mathfrak{h}_{(0)}$ be any fixed element such that

$$\langle \alpha, \hat{\lambda} \rangle \in \frac{1}{k}\mathbb{Z} \tag{148}$$

for all $\alpha \in L$.

Define the following action of $\mathfrak{h}_{(0)}$ on U_T by

$$h \cdot b \otimes r = \langle h, \bar{b} + \hat{\lambda} \rangle b \otimes r \tag{149}$$

for $a, b \in \hat{L}_\nu$, $r \in T$, $h \in \mathfrak{h}_{(0)}$. Then as operators on U_T,

$$ha = a(\langle h, \bar{a} \rangle + h) \tag{150}$$

for $a \in \hat{L}_\nu$ and $h \in \mathfrak{h}_{(0)}$.

Remark 5.3 For instance, we could take $\hat{\lambda} = 0$, but in general it can be nonzero, and this gives us other $\mathfrak{h}_{(0)}$-module structures, and will result in other $\hat{\nu}$-twisted V_L-module structures. However, for the specialized case we are interested in, since $\mathfrak{h}_{(0)} = 0$ (see (177)), this $\hat{\lambda}$ will be zero.

Note that the projection map P_0 (recall (126)) induces an isomorphism from L/N to $P_0 L$, and thus we have a natural isomorphism

$$U_T = \mathbb{C}[P_0 L] \otimes_{\mathbb{C}} T, \tag{151}$$

of $\mathfrak{h}_{(0)} \cup \hat{L}_\nu$-modules. We extend U_T to a $\hat{\mathfrak{h}}[\nu]$-module by letting $\hat{\mathfrak{h}}[\nu]_{\frac{1}{k}\mathbb{Z}}$ (recalling (131)) act trivially.

Remark 5.4 In the case that $R = N$, we have a linear isomorphism $U_T \simeq \mathbb{C}[P_0 L]$. Also $P_0 L = \frac{1}{k}\left(L \cap \mathfrak{h}_{(0)}\right)$, and so in the case when $R = N$ we have $U_T \simeq \mathbb{C}\left[\frac{1}{k}\left(L \cap \mathfrak{h}_{(0)}\right)\right]$. This is the case in, for instance, the important setting of permutation-twisted modules for lattice vertex operator superalgebras [BDM], [BHL].

Now note that we can write

$$U_T = \coprod_{\alpha \in P_0 L} U_\alpha, \tag{152}$$

where

$$U_\alpha = \{u \in U_T \mid h \cdot u = \langle h, \alpha + \hat{\lambda} \rangle u \text{ for } h \in \mathfrak{h}_{(0)}\}, \tag{153}$$

and the actions of $\hat{L}_\nu$ and $\mathfrak{h}_{(0)}$ are compatible in the sense that

$$a \cdot U_\alpha \subset U_{\alpha + \bar{a}_{(0)}} \tag{154}$$

for $a \in \hat{L}_\nu$ and $\alpha \in P_0 L$.

We define an End U_T-valued formal Laurent series x^h for $h \in \mathfrak{h}_{(0)}$ as follows

$$x^h \cdot u = x^{\langle h, \alpha \rangle} u \quad \text{for} \quad \alpha \in \mathfrak{h}_{(0)} \quad \text{and} \quad u \in U_\alpha. \tag{155}$$

Then from (150),

$$x^h a = a x^{\langle h, \bar{a} \rangle + h} \quad \text{for} \quad a \in \hat{L}_\nu \tag{156}$$

as operators on U_T. Also, for $h \in h_{(0)}$, if $\langle h, \bar{a}_{(0)} \rangle \in \mathbb{Z}$ for all $a \in L$, define the operator η^h on U_T by

$$\eta^h \cdot u = \eta^{\langle h, \alpha \rangle} u \tag{157}$$

for $u \in U_\alpha$ with $\alpha \in P_0 L$.

Then for $a \in \hat{L}_\nu$, and using (147), we have

$$\hat{\nu} a = a \eta^{-\sum_{j=0}^{k-1} \nu^j \bar{a} + k\langle \bar{a}, \bar{a} \rangle/2 - \sum_{j=0}^{k-1} \langle \nu^j \bar{a}, \bar{a} \rangle/2} = a \eta^{-k\bar{a}_{(0)} + k\langle \bar{a}, \bar{a} \rangle/2 - k\langle \bar{a}_{(0)}, \bar{a}_{(0)} \rangle/2} \tag{158}$$

as operators on U_T.

Then we have

$$\hat{\nu}^j a = a\eta^{-jk\bar{a}_{(0)}+jk\langle\bar{a},\bar{a}\rangle/2-jk\langle\bar{a}_{(0)},\bar{a}_{(0)}\rangle/2} \tag{159}$$

and thus

$$\hat{\nu}^k a = a, \tag{160}$$

for all $a \in \hat{L}_\nu$ acting as operators in $\operatorname{End} U_T$, where we recall that we had from the beginning doubled k if necessary (see Remark 4.1). And thus $\hat{\nu}^k = 1$ on $\hat{L}_\nu$ as well.

It is shown in, for instance, [X] in Chapter 6.2, that U_T is an irreducible $\hat{L}_\nu \cup \mathfrak{h}_{(0)}$-module when T is irreducible.

Define a $\mathbb{C}$-gradation on U_T by

$$\operatorname{wt} u = \frac{1}{2}\langle\alpha,\alpha\rangle \quad \text{for } \alpha \in P_0 L \text{ and } u \in U_\alpha. \tag{161}$$

Then $\hat{\nu}$ preserves this gradation of U_T since $\nu(\alpha) = \alpha$ for $\alpha \in P_0 L \subset \mathfrak{h}_{(0)}$.

Form the space

$$\begin{aligned} V_L^T &= S[\nu] \otimes U_T = \left(U(\hat{\mathfrak{h}}[\nu]) \otimes_{U(\coprod_{n\geq 0}\mathfrak{h}_{(kn)}\otimes t^n \oplus \mathbb{C}\mathbf{c})} \mathbb{C}\right) \otimes \left(\mathbb{C}[\hat{L}_\nu] \otimes_{\mathbb{C}[\hat{N}]} \mathbb{C}_\tau\right) \\ &\simeq S(\hat{\mathfrak{h}}[\nu]^-) \otimes_{\mathbb{C}} (\mathbb{C}[P_0 L] \otimes_{\mathbb{C}} T), \end{aligned} \tag{162}$$

which is naturally graded (by weights), using the weight gradations of $S[\nu]$ and U_T. We let $\hat{L}_\nu$, $\hat{\mathfrak{h}}[\nu]_{\frac{1}{k}\mathbb{Z}}$, $\mathfrak{h}_{(0)}$ and x^h, for $h \in \mathfrak{h}_{(0)}$, act on V_L^T by acting on either $S[\nu]$ or U_T, as described above.

For $\alpha \in \mathfrak{h}$ and $n \in \frac{1}{k}\mathbb{Z}$, write $\alpha^T(n)$ or $\alpha_{(kn)}(n)$ for the operator on V_L^T associated with $\alpha_{(kn)} \otimes t^n$, and set

$$\alpha^T(x) = \sum_{n\in\frac{1}{k}\mathbb{Z}} \alpha^T(n)x^{-n-1} = \sum_{n\in\frac{1}{k}\mathbb{Z}} \alpha_{(kn)}(n)x^{-n-1}. \tag{163}$$

Following [Le1] and [FLM2], for $\alpha \in L$, define

$$\rho(\alpha) = \begin{cases} 2^{\langle\nu^{k/2}\alpha,\alpha\rangle/2} \prod_{0<j<k/2} (1-\eta^{-j})^{\langle\nu^j\alpha,\alpha\rangle} & \text{if } k \in 2\mathbb{Z} \\ \prod_{0<j<k/2} (1-\eta^{-j})^{\langle\nu^j\alpha,\alpha\rangle} & \text{if } k \in 2\mathbb{Z}+1 \end{cases}. \tag{164}$$

Then $\rho(\nu\alpha) = \rho(\alpha)$.

Using the normal-ordering procedure described above, define the *$\hat{\nu}$-twisted vertex operator* $Y^{\hat{\nu}}(a,x)$ for $a \in \hat{L}$ acting on V_L^T as follows

$$Y^{\hat{\nu}}(a,x) = k^{-\langle\bar{a},\bar{a}\rangle/2}\rho(\bar{a}) \,{}^{\circ}_{\circ}\, e^{\int(\bar{a}^T(x)-\bar{a}^T(0)x^{-1})} a x^{\bar{a}_{(0)}+\langle\bar{a}_{(0)},\bar{a}_{(0)}\rangle/2-\langle\bar{a},\bar{a}\rangle/2} \,{}^{\circ}_{\circ}. \tag{165}$$

Note that on the right-hand side of (165), we view a as an element of $\hat{L}_\nu$ using our set-theoretic identification between $\hat{L}$ and $\hat{L}_\nu$ given by (96).

For $\alpha_1,\ldots,\alpha_m \in \mathfrak{h}$, $n_1,\ldots,n_m \in \mathbb{Z}_+$ and $v = \alpha_1(-n_1)\cdots\alpha_m(-n_m)\cdot\iota(a) \in V_L$, set

$$W(v,x) = {}^{\circ}_{\circ} \left(\partial_{n_1-1}\alpha_1^T(x)\right)\cdots\left(\partial_{n_m-1}\alpha_m^T(x)\right) Y^{\hat{\nu}}(a,x) \,{}^{\circ}_{\circ}, \tag{166}$$

where the right-hand side is an operator on V_L^T. Extend to all $v \in V_L$ by linearity.

Define constants $c_{mnr} \in \mathbb{C}$ for $m, n \in \mathbb{N}$ and $r = 0, \ldots, k-1$ by the formulas

$$\sum_{m,n\geq 0} c_{mn0} x^m y^n = -\frac{1}{2}\sum_{j=1}^{k-1} \log\left(\frac{(1+x)^{1/k} - \eta^{-j}(1+y)^{1/k}}{1-\eta^{-j}}\right), \tag{167}$$

$$\sum_{m,n\geq 0} c_{mnr} x^m y^n = \frac{1}{2}\log\left(\frac{(1+x)^{1/k} - \eta^{-r}(1+y)^{1/k}}{1-\eta^{-r}}\right) \quad \text{for } r \neq 0. \tag{168}$$

Let $\{\beta_1, \ldots, \beta_{\dim \mathfrak{h}}\}$ be an orthonormal basis of $\mathfrak{h}$, and set

$$\Delta_x = \sum_{m,n\geq 0}\sum_{r=0}^{k-1}\sum_{j=1}^{\dim\mathfrak{h}} c_{mnr}(\nu^{-r}\beta_j)(m)\beta_j(n)x^{-m-n}. \tag{169}$$

Then e^{Δ_x} is well defined on V_L since $c_{00r} = 0$ for all r, and for $v \in V_L$, $e^{\Delta_x}v \in V_L[x^{-1}]$. Note that Δ_x is independent of the choice of orthonormal basis. Then $\hat{\nu}\Delta_x = \Delta_x\hat{\nu}$ and hence $\hat{\nu}e^{\Delta_x} = e^{\Delta_x}\hat{\nu}$ on V_L.

For $v \in V_L$, the *$\hat{\nu}$-twisted vertex operator* $Y^{\hat{\nu}}(v, x)$ is defined by

$$Y^{\hat{\nu}}(v,x) = W(e^{\Delta_x}v, x). \tag{170}$$

Then this yields a well-defined linear map

$$V_L \longrightarrow (\mathrm{End}\, V_L^T)[[x^{1/k}, x^{-1/k}]], \quad v \mapsto Y^{\hat{\nu}}(v,x) = \sum_{n\in\frac{1}{k}\mathbb{Z}} v_n^{\hat{\nu}} x^{-n-1} \tag{171}$$

where $v_n^{\hat{\nu}} \in \mathrm{End}\, V_L^T$.

From [DL2], [X] we have that $(V_L^T, Y^{\hat{\nu}})$ is an irreducible $\hat{\nu}$-twisted V_L-module.

6. An isomorphism between $V_{fer} \otimes V_{fer}$ and $V_{\mathbb{Z}\alpha}$

In this section we present an isomorphism between the two free fermion vertex operator superalgebra $V_{fer} \otimes V_{fer}$ and the lattice vertex operator superalgebra $V_{\mathbb{Z}\alpha}$ with $\langle \alpha, \alpha\rangle = 1$. The fact that these two vertex operator superalgebras are isomorphic is commonly referred to as "boson-fermion correspondence" [F], [FFR]. More specifically this isomorphism is a correspondence between two free fermions and a fermion constrained to the circle $\mathbb{R}/\mathbb{Z}\alpha$.

To express this isomorphism, we polarize our two free fermions using the transformation

$$\alpha^{\pm} = \frac{1}{\sqrt{2}}(\alpha_{(1)} \mp i\alpha_{(2)}) \tag{172}$$

or equivalently $\alpha_{(1)} = \frac{1}{\sqrt{2}}(\alpha^+ + \alpha^-)$ and $\alpha_{(2)} = \frac{i}{\sqrt{2}}(\alpha^+ - \alpha^-)$. This polarization puts us in the setting of [FFR]. In keeping with [Bar6], [Bar7], [Bar11], [Bar10], we call $\alpha^{\pm}$ the "homogeneous" basis for $\mathfrak{h} = \mathrm{span}_{\mathbb{C}}\{\alpha_1, \alpha_2\}$.

Consider the vertex operator subalgebra of $V_{fer} \otimes V_{fer}$ generated by the vector $\alpha^+(-1/2)\alpha^-(-1/2)$. Denote this vertex operator algebra by $\langle \alpha^+(-1/2)\alpha^-(-1/2)\rangle$. In addition, consider the free, rank one bosonic vertex operator algebra $V_{bos} = S(\hat{\mathfrak{h}}_+) = \langle \alpha(-1)\rangle$. Then V_{bos} is isomorphic to $\langle \alpha^+(-1/2)\alpha^-(-1/2)\rangle$ as vertex operator algebras with isomorphism given by $\alpha(-1) \mapsto \alpha^+(-1/2)\alpha^-(-1/2)$.

Then, for $n \in \mathbb{Z}$, the spaces $V_{bos} \otimes e^{n\alpha}$ are irreducible modules for V_{bos}, and $V_{\mathbb{Z}\alpha} = \coprod_{n\in\mathbb{Z}} V_{bos} \otimes e^{n\alpha}$. An isomorphism $\varphi : V_L \longrightarrow V_{fer} \otimes V_{fer}$ is given by

$$\varphi: \begin{array}{lcl} 1\otimes e^{n\alpha} & \mapsto & \alpha^+(-n+1/2)\alpha^+(-n+3/2)\cdots\alpha^+(-1/2)\cdot 1 \\ 1\otimes e^{-n\alpha} & \mapsto & \alpha^-(-n+1/2)\alpha^-(-n+3/2)\cdots\alpha^-(-1/2)\cdot 1 \end{array} \tag{173}$$

for $n \in \mathbb{Z}_+$.

Here $e^{n\alpha}$ is chosen as a section of $\hat{L}$ for convenience of notation. That is letting $e : L = \mathbb{Z}\alpha \longrightarrow \hat{L}$, $e : n\alpha \mapsto e_{n\alpha}$ be a section of $\hat{L}$, this choice of section allows us to identify $\mathbb{C}\{L\}$ with the group algebra $\mathbb{C}[L]$ by the linear isomorphism

$$\mathbb{C}[L] \longrightarrow \mathbb{C}\{L\}, \quad e^{n\alpha} \mapsto \iota(e_{n\alpha}). \tag{174}$$

But in a slight abuse of notation, we write $e^{n\alpha}$ for $\iota(e_{n\alpha})$.

The vertex operator subalgebra $\langle \alpha^+(-1/2)\alpha^-(-1/2)\rangle$ consists of those vectors in $V_{fer} \otimes V_{fer}$ which have an equal number of positive and negative $\alpha^{\pm}(-m)$ terms for $m \in \mathbb{Z}_+ - \frac{1}{2}$. Then the modules corresponding to $V_{bos} \otimes e^n$ consist of those vectors in $V_{fer} \otimes V_{fer}$ that have n more positive terms than negative terms if $n > 0$ and that have n more negative terms than positive terms for $n < 0$.

The isomorphism $V_{fer} \otimes V_{fer} \cong V_{\mathbb{Z}\alpha}$ implies that $V_{fer}^{\otimes 2d} \cong V_{\mathbb{Z}\alpha}^{\otimes d}$ for $d \in \mathbb{Z}_+$. That is, the lattice vertex operator superalgebra corresponding to the orthogonal rank d lattice $\bigoplus_{j=1}^d \mathbb{Z}\alpha^{(j)}$ with $\langle \alpha^{(j)}, \alpha^{(k)}\rangle = \delta_{j,k}$ is isomorphic to the $2d$ free boson vertex operator superalgebra.

7. Construction and classification of the (1 2)-twisted $V_{fer} \otimes V_{fer}$-modules through boson-fermion correspondence and a conjecture

In this section, we use the isomorphism of $V_{fer} \otimes V_{fer} \cong V_{\mathbb{Z}\alpha}$ to construct the (1 2)-twisted $V_{fer} \otimes V_{fer}$-modules by first transferring the signed permutation automorphism (1 2) on $V_{fer} \otimes V_{fer}$ to the corresponding automorphism $\varphi \circ (1\ 2) \circ \varphi^{-1}$ on $V_{\mathbb{Z}\alpha}$, observing that this automorphism is a lift of the -1 lattice isometry, and then using the construction of such twisted modules recalled in Section 5.

The transposition (1 2) acting as a signed permutation on $V_{fer} \otimes V_{fer}$ is given by $(1\ 2) : u \otimes v \mapsto (-1)^{|u||v|} v \otimes u$ for $u, v \in V_{fer}$. In terms of the polarization (172), this automorphism is given by

$$(1\ 2) : \alpha^{\pm}(-1/2) \mapsto \alpha^{\mp}(-1/2).$$

7.1. *The automorphism $\hat{\nu} = \varphi \circ (1\ 2) \circ \varphi^{-1}$ of $V_{\mathbb{Z}\alpha}$ corresponding to (1 2) on V_{fer}*

Let $\varphi \circ (1\ 2) \circ \varphi^{-1}$ be the automorphism of $V_{\mathbb{Z}\alpha}$ corresponding to (1 2) on $V_{fer} \otimes V_{fer}$. Then this automorphism is uniquely determined by

$$\begin{aligned} \varphi \circ (1\ 2) \circ \varphi^{-1} : \qquad V_{\mathbb{Z}\alpha} &\longrightarrow V_{\mathbb{Z}\alpha} \\ \alpha(-1)1 \otimes 1 &\mapsto -\alpha(-1)1 \otimes 1 \\ 1 \otimes e^{n\alpha} &\mapsto (-i)^n (1 \otimes e^{-n\alpha}). \end{aligned} \tag{175}$$

In particular, letting ν be the lattice isometry

$$\nu : \mathbb{Z}\alpha \longrightarrow \mathbb{Z}\alpha, \quad n\alpha \mapsto -n\alpha, \tag{176}$$

then $\hat{\nu} = \varphi \circ (1\ 2) \circ \varphi^{-1}$ is a lift of the lattice isometry ν to a central extension $\hat{L}_\nu$ of $L = \mathbb{Z}\alpha$ by the cycle group $\langle i \rangle$ of order 4.

7.2. *Constructing the $\hat{\nu}$-twisted $V_{\mathbb{Z}\alpha}$-modules*

We now specialize the construction of twisted modules for a lattice vertex operator superalgebra and a lift of a lattice isometry given in Section 5 to the following setting:

- Let $L = \mathbb{Z}\alpha$ with $\langle \alpha, \alpha \rangle = 1$.
- Let $k = 4$ and let $\eta = \eta_0 = i$.

Journal of Physics: Conference Series **474** (2013) 012009 doi:10.1088/1742-6596/474/1/012009

• Let $\nu = -1$ on $L = \mathbb{Z}\alpha$.
• Let $\hat{\nu} = \varphi \circ (1\ 2) \circ \varphi^{-1}$.

We follow Section 5 to construct and classify the $\hat{\nu}$-twisted $V_{\mathbb{Z}\alpha}$-modules.

We have

$$\mathfrak{h}_{(0)} = \mathfrak{h}_{(1)} = \mathfrak{h}_{(3)} = 0 \quad \text{and} \quad \mathfrak{h}_{(2)} = \mathfrak{h}, \tag{177}$$

and $C(\alpha, \beta) = 1$, for all $\alpha, \beta \in L$. Furthermore, we have $N = R = L$ and $M = 2L$.

For $a \in \hat{L}_\nu$ we have

$$\tau(a\hat{\nu}a^{-1}) = i^{2\langle \bar{a}, \bar{a}\rangle} = (-1)^{\langle \bar{a}, \bar{a}\rangle}. \tag{178}$$

But in addition, from (175), we have, for $a = e^{n\alpha}$,

$$\tau(a\hat{\nu}a^{-1}) = \tau(e^{n\alpha}\hat{\nu}e^{-n\alpha}) = \tau(e^{n\alpha}i^n e^{n\alpha}) = i^n \tau(e^{2n\alpha}). \tag{179}$$

Therefore,

$$\tau(e^{2n\alpha}) = (-1)^{\langle n\alpha, n\alpha\rangle} i^{-n} = (-1)^{n^2}(-i)^n = i^n. \tag{180}$$

Next we extend τ to $\hat{R} = \hat{L}_\nu$, thereby constructing an irreducible $\hat{L}_\nu$-module, T, on which $\hat{M} = 2\hat{L}_\nu$ acts as multiplication by the character τ. From Proposition 5.2, there are exactly $|R/M| = |L/2L| = 2$ extensions of τ to a homomorphism $\chi : \hat{L}_\nu \to \mathbb{C}^\times$, and every irreducible $\hat{L}_\nu$-module on which $\hat{M}$ acts as τ is equivalent to one of these. Furthermore, since $N = R = L$, we have $\chi = \psi$ in Proposition 5.2, and thus the modules $T = \mathbb{C}_\psi = \mathbb{C}_\chi$ will be precisely these two modules.

It is clear from (180), that the two choices for χ are

$$\chi_\pm : \hat{L}_\nu \longrightarrow \mathbb{C}^\times, \quad e^{n\alpha} \mapsto \pm(e^{\pi i/4})^n \tag{181}$$

for the primitive eighth root of unity $e^{\pi i/4}$ and $n \in \mathbb{Z}$.

Denote these two inequivalent irreducible $\hat{L}_\nu$-modules on which $\hat{L}_\nu$ act as χ_+ and χ_-, respectively, by $\mathbb{C}_+$ and $\mathbb{C}_-$, respectively. Then we have two choices for U_T up to isomorphism, namely $U_T = \mathbb{C}_\pm$.

Note that (158) does reduce to $\hat{\nu}a = ai^{2\langle \bar{a}, \bar{a}\rangle} = a(-1)^{\langle \bar{a}, \bar{a}\rangle} = (-1)^{|a|}a$, as operators on either $\mathbb{C}_\pm$, and $\hat{\nu}^2 = 1$ on $\hat{L}_\nu$.

Form the two $\hat{\nu}$-twisted $V_{\mathbb{Z}\alpha}$ modules

$$M_\pm = S[\nu] \otimes \mathbb{C}_\pm \simeq S[\nu]. \tag{182}$$

Note that in this setting, we have

$$\alpha^T(x) = \sum_{n \in \mathbb{Z}+\frac{1}{2}} \alpha^T(n) x^{-n-1}. \tag{183}$$

Then the $\hat{\nu}$-twisted vertex operators are given by (165) and (170). We denote these two different $\hat{\nu}$-twisted vertex operators by $Y_\pm^{\hat{\nu}}$ on $M\pm$, respectively.

And note that of course we have

$$V_{\mathbb{Z}\alpha} \longrightarrow (\mathrm{End}\, M_\pm)[[x^{1/2}, x^{-1/2}]], \quad v \mapsto Y_\pm^{\hat{\nu}}(v, x) = \sum_{n \in \frac{1}{2}\mathbb{Z}} v_n^{\hat{\nu},\pm} x^{-n-1} \tag{184}$$

where $v_n^{\hat{\nu},\pm} \in \mathrm{End}\, M_\pm$.

From (133), we have

$$\mathrm{wt}\,1 \;=\; \frac{1}{64}\sum_{j=1}^{3} j(4-j)\dim(\mathfrak{h}_{(j)}) = \frac{1}{16}. \tag{185}$$

Following (161), we have that the $\mathbb{C}$-gradation on $\mathbb{C}_\pm$ is zero, and thus the weight grading of $\mathbb{C}_\pm$ is $\frac{1}{16}$. In fact, we have that

$$L_\pm^{\hat{\nu}}(0) = \sum_{n\in\mathbb{N}} \alpha^T(-n-1/2)\alpha^T(n+1/2) + \frac{1}{16}, \tag{186}$$

and thus

$$\dim_q M_\pm = q^{-1/24}q^{1/16}\prod_{n\in\mathbb{Z}_+}(1+q^{n/2}) = \frac{\eta(q)}{\eta(q^{1/2})} = \sqrt{2}^{-1}\mathfrak{f}_2(q^{1/2}). \tag{187}$$

In particular, M_+ and M_- have the same graded dimension. Note also that this graded dimension is the same as the graded dimension for the g-twisted free boson vertex operator algebra module where g is uniquely determined by -1 on the generator; see for instance [Bar11] Section 5.2. It is also the graded dimension with q replaced by $q^{1/2}$ for either of the two unique up to equivalence irreducible parity-twisted V_{fer}-modules as constructed in Section 3.2; see Remark 3.4.

From [DL2], [X] we have that $(M_\pm, Y_\pm^{\hat{\nu}})$ are each irreducible $\hat{\nu}$-twisted $V_{\mathbb{Z}\alpha}$-modules, and they are the only irreducible $\hat{\nu}$-twisted $V_{\mathbb{Z}\alpha}$-modules. In addition, although these two modules, M_+ and M_- have the same graded dimension, they are not isomorphic to each other as $\hat{\nu}$-twisted $V_{\mathbb{Z}\alpha}$- modules. That is, if $f : (M_+, Y_+^{\hat{\nu}}) \longrightarrow (M_-, Y_-^{\hat{\nu}})$ is a twisted module isomorphism, then $f\circ Y_+(v,x)\circ f^{-1} = Y_-^{\hat{\nu}}(v,x) = (-1)^{|v|}Y_+^{\hat{\nu}}(v,x)$ for all $v\in V$. This would imply that there exists a $\mathbb{Z}_2$-grading on M_+, given by $M_\pm = M_\pm^{(0)}\oplus M_+^{(1)}$, such that $|Y_+(v,x)w| = (|v|+|w|)\mathrm{mod}\,2$ for all $w\in M_+$, where $|w| = j$ for $w\in M_+^{(j)}$, for $j=0,1$. But if $w\in M_+$ is a nonzero vector which is homogeneous with respect to the $\mathbb{Z}_2$-grading, then writing $w = u\otimes t$ with $u\in S[\nu]$ and $t\in\mathbb{C}_+$, we have

$$\begin{aligned} Y_+^{\hat{\nu}}(e^\alpha,x)w &= \frac{1}{2}\rho(\alpha)\left({}^\circ_\circ e^{\int(\alpha^T(x)-\alpha^T(0)x^{-1})}{}^\circ_\circ x^{-1/2}u\right)\otimes\chi_+(e^\alpha)\cdot t \\ &= r\left({}^\circ_\circ e^{\int(\alpha^T(x)-\alpha^T(0)x^{-1})}{}^\circ_\circ x^{-1/2}u\right)\otimes t \end{aligned} \tag{188}$$

for a constant $r\in\mathbb{C}^\times$. Taking $r^{-1}\mathrm{Res}_x x^{-1/2}$ of both sides we obtain $u\otimes t = w$. Since $|e^\alpha| = 1$, this implies that $|w| = (1+|w|)\mathrm{mod}\,2$, a contradiction. Thus there exists no such $\mathbb{Z}_2$-grading that would give an isomorphism between M_+ and M_-.

It follows that

Lemma 7.1 *The modules $M_\pm$ are isomorphic as ordinary parity-unstable* $(1\ 2)$*-twisted* $V_{fer}^{\otimes 2}$*-modules to the modules* $M_{(1\ 2)}^\pm$ *presented in Section 3.3 following [DZ2].*

Remark 7.2 We observe that using the construction of [DZ2], one first constructs a parity-stable irreducible $(1\ 2)$-twisted $V_{fer}^{\otimes 2}$-module and then identifies two invariant subspaces which are parity-unstable irreducible $(1\ 2)$-twisted $V_{fer}^{\otimes 2}$-modules. However using boson-fermion correspondence and the theory of lattice vertex operator superalgebras, one first directly constructs a pair of parity-unstable irreducible $(1\ 2)$-twisted $V_{fer}^{\otimes 2}$-modules. In addition, from the lattice construction it is immediately obvious that $(M_+, Y_+^{\hat{\nu}})$ is isomorphic to $(M_-, Y_-^{\hat{\nu}}\circ\sigma_V)$ as parity-unstable $\hat{\nu}$-twisted modules, wheres the isomorphism is less straightforward in the [DZ2] construction.

7.3. A conjecture

For $k > 2$ even, if one tries to directly lift $(1\ 2\ \cdots\ k)$ to the lattice vertex operator superalgebra $V_{\mathbb{Z}\alpha}^{\otimes k} = V_{\mathbb{Z}\alpha\oplus\cdots\oplus\mathbb{Z}\alpha}$ in the obvious way extending what we did in the case $k = 2$ in the last section, one does not get a lift of a lattice isometry.

For example if one tries to directly lift, say $(1\ 2\ 3\ 4)$ to the lattice vertex operator superalgebra $V_{\mathbb{Z}\alpha}^{\otimes 2} = V_{\mathbb{Z}\alpha\oplus\mathbb{Z}\alpha}$ in the obvious way, we have that for instance under the map $(\varphi \otimes \varphi) \circ (1\,2\,3\,4) \circ (\varphi \otimes \varphi)^{-1}$, we have

$$(\alpha(-1)1 \otimes 1) \otimes (1 \otimes 1) \mapsto \frac{1}{2}\left(\left(1 \otimes (e^{\alpha} + e^{-\alpha})\right) \otimes \left(1 \otimes (e^{\alpha} - e^{-\alpha})\right)\right). \tag{189}$$

However, we note that all the $(1\ 2\ \cdots\ k)$-twisted $V_{fer}^{\otimes k}$-modules, $M^{\pm}_{(1\,2\,\cdots\,k)}$, of Section 3.3 for k even, have a structure that would imply that they could be realized as g-twisted $V_{\mathbb{Z}\alpha}^{\otimes k/2}$-modules for g a lift of some lattice isometry. In particular, they look like $S[\nu] \otimes \mathbb{C}_{\pm}$ for S a symmetric algebra and $\mathbb{C}_{\pm}$ a one dimensional space on which the odd generating operator acts as $\pm c$ for a constant c.

This, as well as recent constructions of other permutation-twisted modules for free fermions given by the second author, lead us to the following conjecture:

Conjecture 7.3 *The $(1\ 2\ \cdots\ k)$ permutation automorphism of $V_{fer}^{\otimes k}$ for k even is conjugate to a lift of a lattice isometry on $V_{\mathbb{Z}\alpha}^{\otimes k/2}$ via boson-fermion correspondence.*

Note that from (189), this conjecture is nontrivial. In addition, we stress that although the permutation automorphisms for free fermions can be realized as lifts of isometries on $\mathfrak{h}$ as in [DZ2] and Section 3.3, this conjecture goes further to state that they can be be realized as lifts of isometries on the lattice underlying the purely bosonic part of the vertex operator superalgebra. This is a much stronger statement, and allows for the full theory of twisted modules for lattice vertex operator superalgebras to come to bear.

8. Construction and classification of the $\sigma \circ (1\ 2)$-twisted $V_{fer} \otimes V_{fer}$-modules

Let $\hat{\nu} = \varphi \circ (1\ 2) \circ \varphi^{-1}$ be the automorphism of $V_{\mathbb{Z}\alpha}$ corresponding to $(1\ 2)$ on $V_{fer} \otimes V_{fer}$ given explicitly by (175), where φ is the isomorphism between $V_{fer} \otimes V_{fer}$ and $V_{\mathbb{Z}\alpha}$ given by (173). Then the automorphism $\sigma \circ (1\ 2)$ of $V_{fer} \otimes V_{fer}$ corresponds to the automorphism $\sigma \circ \hat{\nu} = \varphi \circ \sigma \circ (1\ 2) \circ \varphi^{-1}$ given by

$$\begin{aligned} \varphi \circ \sigma \circ (1\,2) \circ \varphi^{-1}: \qquad V_{\mathbb{Z}\alpha} &\longrightarrow V_{\mathbb{Z}\alpha} \\ \alpha(-1) \otimes 1 &\mapsto -\alpha(-1) \\ 1 \otimes e^{n\alpha} &\mapsto i^n(1 \otimes e^{-n\alpha}). \end{aligned} \tag{190}$$

Then $\sigma \circ \hat{\nu} = \varphi \circ \sigma \circ (1\,2) \circ \varphi^{-1}$ is also a lift of the lattice isometry $\nu = -1$ to a central extension of $L = \mathbb{Z}\alpha$ by the cycle group $\langle i \rangle$ of order 4. Repeating the construction of Section 7.2, we have that there are exactly two inequivalent irreducible $\sigma \circ \hat{\nu}$-twisted $V_{\mathbb{Z}\alpha}$ given by

$$M^{\sigma}_{\pm} = S[\nu] \otimes \mathbb{C}^{\sigma}_{\pm} \simeq S[\nu]. \tag{191}$$

where $\mathbb{C}^{\sigma}_{\pm}$ are the irreducible $\hat{L}_{\nu}$-modules constructed as follows: Define the characters

$$\chi^{\sigma}_{\pm} : \hat{L}_{\nu} \longrightarrow \mathbb{C}^{\times}, \quad e^{n\alpha} \mapsto \pm(e^{3\pi i/4})^n \tag{192}$$

for the primitive eighth root of unity $e^{3\pi i/4}$, cf. (181). Denote the two inequivalent irreducible $\hat{L}_{\nu}$-modules on which $\hat{L}_{\nu}$ act as χ^{σ}_{+} and χ^{σ}_{-}, respectively, by $\mathbb{C}^{\sigma}_{+}$ and $\mathbb{C}^{\sigma}_{-}$, respectively.

Proposition 8.1 *The modules $M_{\pm}$ are isomorphic as ordinary parity-unstable $\sigma(1\ 2)$-twisted $V_{fer}^{\otimes 2}$-modules to the modules $M_{\sigma(1\ 2)}^{\pm}$ following [DZ2], and are the only irreducible parity-unstable $\sigma(1\ 2)$-twisted $V_{fer}^{\otimes 2}$-modules up to equivalence.*

We further conjecture that in general, for k even, (not just two) that $\sigma \circ (1\ 2\ \cdots\ k)$ can be realized as a conjugate of a lift of a lattice isometry under boson-fermion correspondence.

References

[Ban1] Bantay P, Algebraic aspects of orbifold models, *Int. J. Mod. Phys.* **A9** (1994), 1443–1456.

[Ban2] Bantay P, Characters and modular properties of permutation orbifolds, *Phys. Lett.* **B419** (1998), 175–178.

[Ban3] Bantay P, Permutation orbifolds and their applications, in: *Vertex Operator Algebras in Mathematics and Physics, Proc. workshop, Fields Institute for Research in Mathematical Sciences, 2000*, ed. by S. Berman, Y. Billig, Y.-Z. Huang and J. Lepowsky, Fields Institute Communications, Vol. 39, Amer. Math. Soc., 2003, 13–23.

[Bar1] Barron K, A supergeometric interpretation of vertex operator superalgebras, *Int. Math. Res. Notices*, 1996 No. 9, Duke University Press, 409–430.

[Bar2] Barron K, *The supergeometric interpretation of vertex operator superalgebras*, Ph.D. thesis, Rutgers University, 1996.

[Bar3] Barron K, "$N = 1$ Neveu-Schwarz vertex operator superalgebras over Grassmann algebras and with odd formal variables" in *Representations and Quantizations: Proceedings of the International Conference on Representation Theory, July 1998*, Shanghai China, Springer-Verlag, 9–39.

[Bar4] Barron K, The moduli space of $N = 1$ superspheres with tubes and the sewing operation, *Memoirs Amer. Math. Soc.*, **162**, No. 772, 2003.

[Bar5] Barron K, The notion of $N = 1$ supergeometric vertex operator superalgebra and the isomorphism theorem, *Commun. in Contemp. Math.*, Vol. 5, No. 4, (2003), 481–567.

[Bar6] Barron K, The moduli space of N=2 super-Riemann spheres with tubes, *Commun. Contemp. Math.* **9** (2007), 857–940.

[Bar7] Barron K, Axiomatic aspects of N=2 vertex superalgebras with odd formal variables, *Commun. in Alg.* **38** (2010), 1199–1268.

[Bar8] Barron K, On uniformization of N=2 superconformal and N=1 superanalytic DeWitt super-Riemann surfaces, arXiv:0807.2826v3.

[Bar9] Barron K, Automorphism groups of N=2 superconformal super-Riemann spheres, *J. Pure Appl. Algebra* **214** (2010), 1973–1987.

[Bar10] Barron K, On twisted modules for N=2 supersymmetric vertex operator superalgebras, in: "Lie Theory and Its Applications in Physics", Proceedings of the IX-th International Workshop; ed. V. Dobrev, Springer 2013, 411–420.

[Bar11] Barron K, Twisted modules for N=2 supersymmetric vertex operator superalgebras arising from finite automorphisms of the N=2 Neveu-Schwarz algebra, arXiv:1110.0229v3.

[Bar12] Barron K, Twisted modules for tensor product vertex operator superalgebras and permutation automorphisms of odd order, in preparation.

[Bar13] Barron K, Mirror maps and mirror-twisted modules for certain N=2 supersymmetric vertex operator superalgebras and triality, in preparation.

[BDM] Barron K, Dong C, and Mason G, Twisted sectors for tensor products vertex operator algebras associated to permutation groups, *Comm. Math. Phys.* **227** (2002), 349–384.

[BHL] Barron K, Huang Y, and Lepowsky J, An equivalence of two constructions of permutation-twisted modules for lattice vertex operator algebras, *J. Pure Appl. Algebra* **210** (2007), 797–826.

[Bo] Borcherds R, Vertex algebras, Kac-Moody algebras, and the Monster, *Proc. Natl. Acad. Sci. USA* **83** (1986), 3068–3071.

[BHS] Borisov L, Halpern M, and Schweigert C, Systematic approach to cyclic orbifolds, *Internat. J. Modern Phys.* **A 13** (1998), 125–168.

[dBHO] de Boer J, Halpern M, and Obers N, The operator algebra and twisted KZ equations of WZW orbifolds, *J. High Energy Phys.* **10** (2001), no. 11.

[DVVV] Dijkgraaf R, Vafa C, Verlinde E, and Verlinde H, The operator algebra of orbifold models, *Comm. Math. Phys.* **123** (1989), 485–526.

[DFMS] Dixon L, Friedan D, Martinec E, and Shenker S, The conformal field theory of orbifolds, *Nucl. Phys.* **B282** (1987), 13–73.

[DHVW1] Dixon L, Harvey J, Vafa C, and Witten E, Strings on orbifolds, *Nucl. Phys.* **B 261** (1985) 678–686.

[DHVW2] Dixon L, Harvey J, Vafa C, and Witten E, Strings on orbifolds, II, *Nucl. Phys.* **B 274** (1986) 285–314.

[DGM] Dolan D, Goddard P, and Montague P, Conformal field theory of twisted vertex operators, *Nucl. Phys.* **B338** (1990), 529–601.

[D] Dong C, Twisted modules for vertex algebras associated with even lattices, *J. Algebra* **165** (1994), 90–112.

[DH] Dong C and Han J, Some finite properties for vertex operator superalgebras, *Pacific J. Math.* **258** (2012), 269–290.

[DL1] Dong C and Lepowsky J, *Generalized Vertex Algebras and Relative Vertex Operators,* Progress in Math. Vol. 112, Birkhäuser, Boston, 1993.

[DL2] Dong C and Lepowsky J, The algebraic structure of relative twisted vertex operators, *J. Pure Appl. Algebra* **110** (1996), 259–295.

[DZ1] Dong C and Zhao Z, Modularity in orbifold theory for vertex operator superalgebras, Comm. Math. Phys. **260** (2005), no. 1, 227–256.

[DZ2] Dong C and Zhao Z, Twisted representations of vertex operator superalgebras, *Commun. in Contemp. Math.* **8** (2006), 101–121.

[DZ3] Dong C and Zhao Z, Modularity of trace functions in orbifold theory for $\mathbb{Z}$-graded vertex operator superalgebras, in: " Moonshine: the first quarter century and beyond", London Math. Soc. Lecture Note Ser., 372, Cambridge Univ. Press, Cambridge, 2010, 128–143.

[FFR] Feingold A, Frenkel I, and Ries J, Spinor Construction of Vertex Operator Algebras, Triality, and $E_8^{(1)}$, *Contemp. Math.* **121** (1991).

[F] Frenkel I, Two constructions of affine Lie algebra representations and boson-fermion correspondence in quantum field theory, *J. Funct. Anal.* **44** (1981), 259–327.

[FHL] Frenkel I, Huang Y, and Lepowsky J, On axiomatic approaches to vertex operator algebras and modules, *Mem. Amer. Math. Soc.* **104**, No. 494, 1993.

[FLM1] Frenkel I, Lepowsky J, and Meurman A, A natural representation of the Fischer-Griess Monster with the modular function J as character, *Proc. Natl. Acad. Sci. USA* **81** (1984), 3256–3260.

[FLM2] Frenkel I, Lepowsky J, and Meurman A, Vertex operator calculus, in: *Mathematical Aspects of String Theory, Proc. 1986 Conference, San Diego,* ed. by S.-T. Yau, World Scientific, Singapore, 1987, 150–188.

[FLM3] Frenkel I, Lepowsky J, and Meurman A, Vertex Operator Algebras and the Monster, *Pure and Applied Math.,* Vol. 134, Academic Press, 1988.

[FKS] Fuchs J, Klemm A, and Schmidt M, Orbifolds by cyclic permutations in Gepner type superstrings and in the corresponding Calabi-Yau manifolds, *Ann. Phys.* **214** (1992), 221–257.

[GHHO] Ganor O, Halpern M, Helfgott C, and Obers N, The outer-automorphic WZW orbifolds on $\mathfrak{so}(2n)$, including five triality orbifolds on $\mathfrak{so}(8)$, *J. High Energy Phys.* **12** (2002), no. 19.

[HH] Halpern M and Helfgott C, The general twisted open WZW string, *Internat. J. Modern Phys.* **A20** (2005), 923–992.

[HO] Halpern M and Obers N, Two large examples in orbifold theory: abelian orbifolds and the charge conjugation orbifold on $\mathfrak{su}(n)$, *Internat. J. Modern Phys.* **A17** (2002), 3897–3961.

[KW] Kac V and Wang W, Vertex operator superalgebras and representations, *Contemp. Math.* **175** (1994), 161–191.

[KS] Klemm A and Schmidt M, Orbifolds by cyclic permutations of tensor product conformal field theories, *Phys. Lett.* **B245** (1990), 53– 58.

[Le1] Lepowsky J, Calculus of twisted vertex operators, *Proc. Nat. Acad. Sci. USA* **82** (1985), 8295–8299.

[Le2] Lepowsky J, Perspectives on vertex operators and the Monster, in: *Proc. 1987 Symposium on the Mathematical Heritage of Hermann Weyl, Duke Univ., Proc. Symp. Pure Math., American Math. Soc.,* **48** (1988).

[LL] Lepowsky J and Li H, "Introduction to vertex operator algebras and their representations", *Progress in Math.* **227**, Birkhäuser Boston, Inc., Boston, MA, 2004.

[LW] Lepowsky J and Wilson R, Construction of the affine Lie algebra $A_1^{(1)}$, *Comm. Math. Phys.* **62** (1978), 43–53.

[Li1] Li H, Local systems of vertex operators, vertex superalgebras and modules, *J. Pure Appl. Algebra* **109** (1996), 143–195.

[Li2] Li H, Local systems of twisted vertex operators, vertex operator superalgebras and twisted modules, in: "Moonshine, the Monster, and related topics (South Hadley, MA, 1994), *Contemp. Math* **193** (1996), 203–236.

[MS1] Maio M and Schellekens A, Permutation orbifolds of N=2 supersymmetric minimal models, *Nuclear Phys* **B 845 (2011)**, 212–245.

[MS2] Maio M and Schellekens A, Permutation orbifolds of heterotic Gepner models, *Nuclear Phys.* **B 848**

(2011), no. 3, 594–628.

[X] Xu X, *Introduction to vertex operator superalgebras and their modules*, Math. and its Applications, Vol. 456, Kluwer Acad. Pub., Dordrecht, 1998.

[YZ] Yui N and Zagier D, On the singular values of Weber modular functions, *Math. Comp.* **66** (1997), 1645–1662.

Infinite-dimensional symmetries of a general class of variable coefficient evolution equations in 2+1 dimensions

P. Basarab-Horwath[1], **F. Güngör**[2], **C. Özemir**[3]

[1] Department of Mathematics, Linköping University, S-581 83 Linköping, Sweden
[2] Department of Mathematics, Faculty of Science and Sciences, Istanbul Technical University, 34469 Istanbul, Turkey
[3] Department of Mathematics, Faculty of Science and Letters, Istanbul Technical University, 34469 Istanbul, Turkey

E-mail: pehor@mai.liu.se, gungorf@itu.edu.tr, ozemir@itu.edu.tr

Abstract. We consider generalized KP-Burgers equations and attempt to identify subclasses admitting Virasoro or Kac-Moody type algebras as their symmetries. We give reductions to ODEs constructed from invariance requirement under these infinite-dimensional Lie symmetry algebras and integrate them in cases where it is possible. We also look at the conditions under which the equation passes the Painlevé test and construct some exact solutions by truncation.

1. Introduction

In this paper we give a classification of infinite-dimensional symmetries of the family of equations

$$\begin{aligned}&(u_t + p(y,t)uu_x + q(y,t)u_{xx} + r(y,t)u_{xxx})_x + \sigma(y,t)u_{yy}\\&+ a(y,t)u_y + b(y,t)u_{xy} + c(y,t)u_{xx} + e(y,t)u_x + f(y,t)u + h(y,t) = 0.\end{aligned} \tag{1.1}$$

with the assumption $p(y,t) \neq 0$, $q(y,t) \neq 0$, $r(y,t) \neq 0$, $\sigma(y,t) \neq 0$. Eq. (1.1) includes generalizations of the 2+1-dimensional Burgers and KP (Kadomtsev-Petviashvili) equations for the choice of $r = 0$ and $q = 0$, respectively. For derivations of constant coefficient KP-Burgers equations in different physical applications we refer, for instance, to Ref. [1, 2]

From group theoretical point of view, the case $q = 0$, $r = r(t)$ was studied in [3], whereas the case $q = q(t)$, $r = 0$ was recently done in [4]. In the former, the equation has a Virasoro algebra for some specific choice of the coefficients. In the latter, this remarkable algebra does not arise. However, both cases contain infinite-dimensional symmetry algebras of Kac-Moody type [5]. In the present work we wish to analyze the situation where both q and r can survive with an additional y-dependence.

We note that for $q(y,t) = 0$ this family of equations is analyzed in [6] in the context of its integrability properties and exact solutions.

2. Canonical Forms and Determining Equations

We shall convert Eq. (1.1) to some canonical form using point transformations preserving the differential structure of the equation. These type of transformations are known as equivalence, allowed or form-invariant transformations. It is a straightforward calculation to show that they are given by

$$\begin{aligned} &u(x,y,t) = R(y,t)\tilde{u}(\tilde{x},\tilde{y},\tilde{t}) - \frac{\dot{\alpha}}{\alpha p}x + S_0(y,t), \\ &\tilde{x} = \alpha(t)x + \beta(y,t), \quad \tilde{y} = Y(y,t), \quad \tilde{t} = T(t), \\ &\alpha \neq 0, \quad R \neq 0 \quad Y_y \neq 0, \quad \dot{T} \neq 0 \end{aligned} \tag{2.1}$$

together with some restrictions on $\alpha(t)$ and coefficients of the equation:

$$\Big(\frac{ap_y}{p} + \frac{\sigma p_{yy}}{p} - \frac{2\sigma p_y^2}{p^2} - f\Big)\dot{\alpha} = 0. \tag{2.2}$$

With this transformation coefficient functions are mapped to

$$\begin{aligned} &\tilde{p}(\tilde{y},\tilde{t}) = p(y,t)\frac{R\alpha}{\dot{T}}, \quad \tilde{q}(\tilde{y},\tilde{t}) = q(y,t)\frac{\alpha^2}{\dot{T}}, \quad \tilde{r}(\tilde{y},\tilde{t}) = r(y,t)\frac{\alpha^3}{\dot{T}}, \\ &\tilde{\sigma}(y,t) = \sigma(y,t)\frac{Y_y^2}{\alpha\dot{T}}, \\ &\tilde{a}(\tilde{y},\tilde{t}) = \frac{1}{\alpha\dot{T}}\{aY_y + \sigma Y_{yy} + 2\sigma Y_y\frac{R_y}{R}\}, \\ &\tilde{b}(\tilde{y},\tilde{t}) = \frac{1}{\alpha\dot{T}}\{(b\alpha + 2\sigma\beta_y)Y_y + \alpha Y_t\}, \\ &\tilde{c}(\tilde{y},\tilde{t}) = \frac{1}{\alpha\dot{T}}\{c\alpha^2 + \beta_t\alpha + pS_0\alpha^2 + \sigma\beta_y^2 + b\alpha\beta_y\}, \\ &\tilde{e}(\tilde{y},\tilde{t}) = \frac{1}{\alpha R\dot{T}}\{R\alpha e - R\dot{\alpha} + R_t\alpha + aR\beta_y + \sigma R\beta_{yy} + b\alpha R_y + 2\sigma R_y\beta_y\}, \\ &\tilde{f}(\tilde{y},\tilde{t}) = \frac{1}{\alpha R\dot{T}}(fR + aR_y + \sigma R_{yy}), \\ &\tilde{h}(\tilde{y},\tilde{t}) = \frac{1}{\alpha R\dot{T}}\{h - \frac{\partial}{\partial t}\left(\frac{\dot{\alpha}}{\alpha p}\right) + \frac{1}{p}\left(\frac{\dot{\alpha}}{\alpha}\right)^2 + \sigma S_{0,yy} + aS_{0,y} + fS_0 - e\frac{\dot{\alpha}}{\alpha p} + b\frac{\dot{\alpha}p_y}{\alpha p^2}\}. \end{aligned} \tag{2.3}$$

We can choose $R(y,t) = \dot{T}/(p\alpha)$ to normalize $\tilde{p}(\tilde{y},\tilde{t}) = 1$. With this simplification the condition (2.2) reduces to $\dot{\alpha}f = 0$. If we set

$$Y_y^2 = |\frac{\alpha\dot{T}}{\sigma}|$$

we can normalize $\tilde{\sigma}(\tilde{y},\tilde{t}) = \varepsilon = \mp 1$. Furthermore, choosing $\beta(y,t)$ and $S_0(y,t)$ suitably we can have

$$\tilde{e}(\tilde{y},\tilde{t}) = \tilde{h}(\tilde{y},\tilde{t}) = 0.$$

Therefore, instead of Eq. (1.1) we can equivalently work on

$$\begin{aligned} &(u_t + uu_x + q(y,t)u_{xx} + r(y,t)u_{xxx})_x + \varepsilon u_{yy} \\ &+ a(y,t)u_y + b(y,t)u_{xy} + c(y,t)u_{xx} + f(y,t)u = 0, \quad \varepsilon = \mp 1 \end{aligned} \tag{2.4}$$

which we call the canonical form of (1.1). Allowed transformations for this class are found as

$$\begin{aligned} &\tilde{x} = \alpha(t)x + \beta(y,t), \qquad \tilde{y} = \epsilon\sqrt{\alpha\dot{T}}y + \gamma(t), \quad \epsilon = \mp 1, \\ &\tilde{t} = T(t), \quad u = \frac{\dot{T}}{\alpha}\tilde{u} - \frac{\dot{\alpha}}{\alpha}x + S_0(y,t) \end{aligned} \tag{2.5}$$

with the constraints

$$\dot{\alpha} f = 0, \quad \varepsilon\beta_{yy} + a\beta_y + \alpha\frac{\ddot{T}}{\dot{T}} - 2\dot{\alpha} = 0, \quad \varepsilon S_{0,yy} + aS_{0,y} + fS_0 = \frac{d}{dt}(\frac{\dot{\alpha}}{\alpha}) - (\frac{\dot{\alpha}}{\alpha})^2. \tag{2.6}$$

We shall restrict ourselves to the more manageable case $r_y = 0$ in the rest of the article.

2.1. *Canonical class when $r = r(t)$*

In this case we can immediately normalize $\tilde{r}(\tilde{t}) = 1$ by choosing $\dot{T} = \alpha^3(t)\, r(t)$ in (2.3). We note that this normalization will lead to an enormous simplicity in the determining equations. We thus consider the canonical family

$$\begin{aligned}&(u_t + uu_x + q(y,t)u_{xx} + u_{xxx})_x + \varepsilon u_{yy}\\ &+ a(y,t)u_y + b(y,t)u_{xy} + c(y,t)u_{xx} + f(y,t)u = 0, \quad \varepsilon = \mp 1\end{aligned} \tag{2.7}$$

with its allowed transformations

$$\begin{aligned}&\tilde{x} = \alpha(t)x + \beta(y,t), \qquad \tilde{y} = \epsilon\alpha^2 y + \gamma(t), \quad \epsilon = \mp 1,\\ &\tilde{t} = \int \alpha^3(t)dt, \quad u = \alpha^2\tilde{u} - \frac{\dot{\alpha}}{\alpha}x + S_0(y,t),\end{aligned} \tag{2.8}$$

obeying the constraints

$$\dot{\alpha} f = 0, \quad \varepsilon\beta_{yy} + a\beta_y + \dot{\alpha} = 0, \quad \varepsilon S_{0,yy} + aS_{0,y} + fS_0 = \frac{d}{dt}(\frac{\dot{\alpha}}{\alpha}) - (\frac{\dot{\alpha}}{\alpha})^2. \tag{2.9}$$

2.2. *Canonical class when $q = q(t)$, $r = r(t)$, $a = f = 0$*

For later convenience we note a further simplification of (2.7) when $q = q(t)$ and $a = f = 0$. If f vanishes identically, we have no condition on $\alpha(t)$ that we can choose $\alpha(t) := q(t)$ to simplify $\tilde{q} = \frac{q\alpha^2}{\dot{T}} = \frac{q}{\alpha} = 1$ in (2.3). Assuming further $a = 0$, we have the canonical family

$$(u_t + uu_x + u_{xx} + u_{xxx})_x + \varepsilon u_{yy} + b(y,t)u_{xy} + c(y,t)u_{xx} = 0, \quad \varepsilon = \pm 1 \tag{2.10}$$

with the equivalence transformations

$$\begin{aligned}&\tilde{x} = x + \beta_1(t)y + \beta_2(t), \qquad \tilde{y} = \epsilon y + \gamma(t),\\ &\tilde{t} = t + T_0, \qquad u = \tilde{u} + S_1(t)y + S_2(t).\end{aligned} \tag{2.11}$$

2.3. *The infinitesimal and determining equations*

Now we proceed to obtain the symmetry algebra of (2.7). Let us denote the infinitesimal as

$$V = \tau\partial_t + \xi\partial_x + \eta\partial_y + \phi\partial_u. \tag{2.12}$$

Here the coefficients τ, ξ, η, ϕ depend on the variables t, x, y, u. We see that V has the form

$$V = \tau(t)\partial_t + \left(\frac{\dot{\tau}}{3}x + \xi_0(y,t)\right)\partial_x + \left(\frac{2}{3}\dot{\tau}y + \eta_0(t)\right)\partial_y + \left(-\frac{2}{3}\dot{\tau}u + \frac{\ddot{\tau}}{3}x + S(y,t)\right)\partial_u, \tag{2.13}$$

where

$$S(y,t) = -\tau c_t - (\frac{2}{3}\dot{\tau}y + \eta_0)c_y + \xi_{0,t} + b\xi_{0,y} - \frac{2}{3}c\dot{\tau} \tag{2.14}$$

subject to the determining equations

$$\dot{\tau}(q + 2yq_y) + 3\tau q_t + 3\eta_0 q_y = 0, \tag{2.15}$$

$$3\tau a_t + (2\dot{\tau} y + 3\eta_0)a_y + 2a\dot{\tau} = 0, \tag{2.16}$$

$$-3\dot{\eta}_0 - 2y\ddot{\tau} + 3\tau b_t + (2\dot{\tau} y + 3\eta_0)b_y + b\dot{\tau} - 6\varepsilon\xi_{0,y} = 0, \tag{2.17}$$

$$\ddot{\tau} + 3a\xi_{0,y} + 3\varepsilon\xi_{0,yy} = 0, \tag{2.18}$$

$$f\ddot{\tau} = 0, \tag{2.19}$$

$$4f\dot{\tau} + 3f_t\tau + f_y(2\dot{\tau} y + 3\eta_0) = 0, \tag{2.20}$$

$$\dddot{\tau} + 3fS + 3aS_y + 3\varepsilon S_{yy} = 0. \tag{2.21}$$

3. Symmetry Algebras

3.1. Search for the Virasoro symmetries

The set of determining equations (2.16)-(2.21) are almost the same with those obtained in [3]. Equation (2.15) stems from the extra term $q(y,t)u_{xx}$. We are going to try to satisfy Eqs. (2.15)-(2.21) without imposing any condition on $\tau(t)$. This requirement leading to a similar analysis carried out in [3] implies that we must have

$$\xi_0(y,t) = -\frac{\varepsilon}{6}\ddot{\tau} y^2 + \mu(t)y + \xi(t), \quad \mu(t) = \frac{\varepsilon}{6}\left(b\dot{\tau} + 3\tau\dot{b} - 3\dot{\eta}_0\right), \tag{3.1}$$

where $\tau(t), \xi(t)$ and $\eta_0(t)$ are arbitrary. Using these results in (2.21) we find

$$2\dot{\tau}(3c_{yy} + yc_{yyy}) + 3\eta_0 c_{yyy} + 3\tau c_{tyy} = 0. \tag{3.2}$$

We have the following splitting:
(i) $c_{tyy} = 0$. This means we have

$$c(y,t) = k_0 y^2 + c_1(t)y + c_0(t) \tag{3.3}$$

with k_0 a constant and (3.2) is satisfied for arbitrary τ if $k_0 = 0$. This is the form of $c(t)$ obtained in [3].
(ii) $3c_{yy} + yc_{yyy} = 0$. We find by integration

$$c(y,t) = \frac{c_2(t)}{y} + c_1(t)y + c_0(t). \tag{3.4}$$

The substitution of $c(y,t)$ into (3.2) results in

$$3c_2\eta_0 - y\tau\dot{c}_2 = 0. \tag{3.5}$$

$c_2 = 0$ corresponds to the previous case. If $c_2 \neq 0$, we need to have $\eta_0(t) = 0$ and $c_2(t) = l_0$, a constant. We therefore have

$$c(y,t) = \frac{l_0}{y} + c_1(t)y + c_0(t), \tag{3.6}$$

where

$$\eta_0(t) = \begin{cases} \text{free}, & l_0 = 0, \\ 0, & l_0 \neq 0. \end{cases} \tag{3.7}$$

Now Eq. (2.15) has the form

$$\dot{\tau}(q + 2yq_y) + 3\tau q_t + 3\eta_0 q_y = 0, \tag{3.8}$$

which implies that $q + 2yq_y = 0$ and $q_t = 0$ from which we have $q(y,t) = \frac{q_0}{\sqrt{y}}$ with $q_0 = $ const. and the condition $q_0\eta_0(t) = 0$. From the assumption $q(y,t) \neq 0$ we should have $\eta_0(t) = 0$.

We state the results of this part as theorems.

Theorem 3.1 *The canonical generalized KP-Burgers equation* (2.7) *allows the Virasoro algebra as a symmetry algebra if and only if the coefficients satisfy*

$$a = f = 0, \quad q(y,t) = \frac{q_0}{\sqrt{y}}, \quad b = b(t), \quad c = c_0(t) + c_1(t)y + \frac{l_0}{y}. \tag{3.9}$$

Theorem 3.2 *The canonical generalized KP-Burgers equation*

$$(u_t + uu_x + \frac{q_0}{\sqrt{y}}u_{xx} + u_{xxx})_x + \varepsilon u_{yy} + b(t)u_{xy} + (c_0(t) + c_1(t)y + \frac{l_0}{y})u_{xx} = 0 \tag{3.10}$$

with $\varepsilon = \pm 1$, $l_0 \in \mathbb{R}$ *and* $b(t)$, $c_0(t)$ *and* $c_1(t)$ *arbitrary smooth functions is invariant under an infinite dimensional Lie point symmetry group. Its Lie algebra is realized by vector fields of the form*

$$\hat{V} = T(\tau) + X(\xi), \tag{3.11}$$

where $\tau(t)$ *and* $\xi(t)$ *are arbitrary smooth functions of time and we have*

$$\begin{aligned} T(\tau) = {} & \tau(t)\partial_t + \frac{1}{6}[3\varepsilon\dot{b}y\tau + (2x + \varepsilon by)\dot{\tau} - \varepsilon\ddot{\tau}y^2]\partial_x \\ & + \frac{2}{3}\dot{\tau}y\partial_y + \frac{1}{6}\{[-6\dot{c}_0 + 3\varepsilon b\dot{b} + (-6\dot{c}_1 + 3\varepsilon\ddot{b})y]\tau \\ & + [-4u + \varepsilon b^2 - 4c_0 + 4(\varepsilon\dot{b} - 2c_1)y]\dot{\tau} + (2x - \varepsilon by)\ddot{\tau} - \varepsilon y^2\dddot{\tau}\}\partial_u, \end{aligned} \tag{3.12}$$

$$X(\xi) = \xi(t)\partial_x + \dot{\xi}(t)\partial_u. \tag{3.13}$$

Remark 3.1 *Theorems 1 and 2 are also valid for* $q_0 = 0$*, therefore they should be noted as different canonical classes in [3]. For the case* $q_0 = 0$ *studied in [3], there are three arbitrary functions in the vector field. Two of these functions exactly give rise to the generators* $T(\tau)$ *and* $X(\xi)$*. The last arbitrary function is nothing but* $\eta_0(t)$*, and the above analysis shows that restricting* $\eta_0(t) = 0$ *allows the terms* $q_0u_{xxx}/\sqrt{y}$ *and* l_0u_{xx}/y *in* (3.10).

The canonical class (3.10) can further be simplified with the allowed transformations (2.8). Since $a = f = 0$, conditions (2.9) can be formulated as

$$\beta(y,t) = -\frac{\varepsilon\dot{\alpha}}{2}y^2 + \beta_1(t)y + \beta_2(t), \quad S_0(y,t) = \frac{\varepsilon}{2}\chi(t)y^2 + S_1(t)y + S_2(t) \tag{3.14}$$

with $\chi(t) = \frac{d}{dt}(\frac{\dot{\alpha}}{\alpha}) - (\frac{\dot{\alpha}}{\alpha})^2$. Using (2.3) we get

$$\tilde{b}(\tilde{t}) = \frac{1}{\alpha^3}(\epsilon b\alpha^2 + 2\epsilon\varepsilon\alpha\beta_1 + \dot{\gamma}), \tag{3.15}$$

$$\begin{aligned} \tilde{c}(\tilde{y},\tilde{t}) = {} & \frac{l_0\alpha^{-2}}{y} + \alpha^{-4}(c_0\alpha^2 + \alpha\dot{\beta}_2 + S_2\alpha^2 + \varepsilon\beta_1^2 + b\alpha\beta_1) \\ & + \alpha^{-4}(c_1\alpha^2 + \alpha\dot{\beta}_1 + S_1\alpha^2 - 2\dot{\alpha}\beta_1 - \varepsilon b\alpha\dot{\alpha})y \\ & + \alpha^{-4}\varepsilon(-\frac{1}{2}\alpha\ddot{\alpha} + \frac{1}{2}\alpha^2\chi + \dot{\alpha}^2)y^2. \end{aligned} \tag{3.16}$$

Choosing γ appropriately, we can have $\tilde{b}(\tilde{t}) = 0$. If $l_0 \neq 0$, we choose $\alpha^2(t) = |l_0|$ to make the normalization $\tilde{l}_0 = \mp 1$. Coefficient of y^2 vanishes, and there exist functions $S_1(t)$ and $S_2(t)$ such that the coefficients of y^0 and y also vanishes. Similar arguments apply if $l_0 = 0$ in which $\alpha(t)$ is chosen as an arbitrary constant. In summary, we can consider

$$b(t) = c_0(t) = c_1(t) = 0. \tag{3.17}$$

3.2. The function $\eta_0(t)$ free

From (2.15) we see that $\eta_0(t)$ is free only if we have $q_y = 0$, that is, $q = q(t)$. Then (2.15) is easy to solve. We momentarily leave it aside and focus on the other equations. Since (2.16)-(2.21) are the same with those given in [3], we skip the details of the calculations and state that (2.16)-(2.20) can be solved for $\eta_0(t)$ free if

$$\xi_0(y,t) = -\frac{\varepsilon}{6}\ddot{\tau}\,y^2 + \nu(t)y + \xi(t),\ \nu(t) = \frac{\varepsilon}{6}\big(- 3\dot{\eta}_0 + 3\tau\dot{b}_0 + 3\eta_0 b_1 + b_0\dot{\tau}\big), \tag{3.18}$$

$$b(y,t) = b_1(t)y + b_0(t), \quad (\tau b_1)^{\cdot} = 0, \quad a(y,t) = f(y,t) \equiv 0. \tag{3.19}$$

Since we have come to the situation that $a = f = 0$, this means the canonical equation (2.10) can be studied instead of (2.7). Furthermore, equivalence transformations (2.11) are valid. If we use $q(t) = 1$ in (2.15), we find that $\tau(t) = \tau_0 = \text{const.}$ (2.21) simplifies to

$$\eta_0 c_{yyy} + \tau_0 c_{tyy} = 0. \tag{3.20}$$

Obviously we need to have $c(y,t) = c_0(t) + c_1(t)y + c_2(t)y^2$. Then the only relations to be satisfied are

$$\tau_0\,\dot{b}_1 = 0, \qquad \tau_0\,\dot{c}_2 = 0. \tag{3.21}$$

(i) If $\tau_0 = 0$, we have the symmetry algebra in [3] and state it as a theorem.

Theorem 3.3 *The equation*

$$\begin{aligned}(u_t + uu_x + u_{xx} + u_{xxx})_x + \varepsilon u_{yy} + (b_1(t)y + b_0(t))u_{xy}\\ + (c_2(t)y^2 + c_1(t)y + c_0(t))u_{xx} = 0,\end{aligned} \tag{3.22}$$

where $\varepsilon = \pm 1$ and b_0, b_1, c_0, c_1, c_2 are arbitrary functions of time, is invariant under an infinite-dimensional Lie point symmetry group depending on two arbitrary functions. Its Lie algebra has a Kac-Moody structure and is realized by vector fields of the form

$$\hat{V} = X(\xi) + Y(\eta), \tag{3.23}$$

where $\xi(t)$ and $\eta(t)$ are arbitrary smooth functions of time and

$$X(\xi) = \xi\partial_x + \dot{\xi}\partial_u, \tag{3.24}$$

$$\begin{aligned}Y(\eta) = \frac{\varepsilon}{2}y(-\dot{\eta} + b_1\eta)\partial_x + \eta\partial_y + \{[-2c_2\eta\\ + \frac{\varepsilon}{2}(-\ddot{\eta} + \dot{b}_1\eta + b_1^2\eta)]y - c_1\eta + \frac{\varepsilon}{2}b_0(-\dot{\eta} + b_1\eta)\}\partial_u.\end{aligned} \tag{3.25}$$

It is possible to make

$$b_0(t) = c_0(t) = c_1(t) = 0 \tag{3.26}$$

using transformations (2.11).

Remark 3.2 *This result shows that the Kac-Moody algebra (3.24)-(3.25) of the generalized KP equation [3] and generalized Burgers equation [4] is actually admitted by the KP-Burgers equation.*

(ii) For the case $\tau_0 \neq 0$ we proceed with the case (3.26) and find $b_1(t) = k_1$, $c_2(t) = k_2$ as constants. So, the equation

$$(u_t + uu_x + u_{xx} + u_{xxx})_x + \varepsilon u_{yy} + k_1 y\, u_{xy} + k_2 y^2 u_{xx} = 0 \tag{3.27}$$

has the obvious symmetry $T = \partial_t$ in addition to (3.24) and (3.25).

3.3. One function in symmetry algebra

If we let $\tau(t)$ and $\eta_0(t)$ be constrained by the determining equations, but let some freedom remain in $\xi_0(y,t)$, then we have the following:

Theorem 3.4 *Equation* (2.4) *is invariant under an infinite dimensional Abelian group generated by the vector field*

$$X(\xi) = \xi(t)\partial_x + \dot{\xi}(t)\partial_u \tag{3.28}$$

for $f(y,t) = 0$ *and* a, b, c, q, r *arbitrary.*

4. Some Applications

In this Section we apply method of reduction and truncating Painlevé expansion to find exact solutions.

4.1. Reduction for the Virasoro case

The solution invariant under the group $\exp(T(\tau) + X(\xi))$ for $\tau \neq 0$ is given by

$$\begin{aligned} \rho &= x\tau^{-1/3} + \frac{\epsilon}{6}\dot{\tau}\tau^{-4/3}y^2 - \int_0^t \xi(s)\tau(s)^{-4/3}ds, \quad \theta = y^{-3/2}\tau, \\ u &= \tau^{-2/3}w(\rho,\theta) - \frac{\varepsilon}{18}\tau^{-2}y^2(3\tau\ddot{\tau} - 2\dot{\tau}^2) + \frac{x}{3}\frac{\dot{\tau}}{\tau} + \frac{\xi}{\tau}, \end{aligned} \tag{4.1}$$

where w satisfies

$$[w_{\rho\rho\rho} + q_0\theta^{1/3}w_{\rho\rho} + l_0\theta^{2/3}w_\rho + ww_\rho]_\rho + \frac{15\varepsilon}{4}\theta^{7/3}w_\theta + \frac{9\varepsilon}{4}\theta^{10/3}w_{\theta\theta} = 0. \tag{4.2}$$

This equation admits a two-dimensional Lie algebra generated by $X_1 = \partial_\rho$, $X_2 = \rho\partial_\rho - 3\theta\partial_\theta - 2w\partial_w$. The solutions invariant under the subalgebra X_1 are straightforward to obtain. They are given by

$$w(\theta) = w_1 + w_0\theta^{-2/3}$$

which produces solutions

$$u(x,y,t) = w_0\tau^{-4/3}y^{-2/3} + w_1\tau^{-2/3} - \frac{\varepsilon}{18}\tau^{-2}y^2(3\tau\ddot{\tau} - 2\dot{\tau}^2) + \frac{x}{3}\frac{\dot{\tau}}{\tau} + \frac{\xi}{\tau}$$

depending on two arbitrary functions of time $\tau(t)$, $\xi(t)$ and two constants w_0, w_1.

Invariance under X_2 implies that the solution will have the form

$$w(\rho,\theta) = \theta^{2/3}F(z), \quad z = \rho\theta^{1/3}, \tag{4.3}$$

where $F(z)$ satisfies the fourth order ODE

$$F'''' + q_0F''' + (l_0 + \frac{\varepsilon}{4}z^2)F'' + \frac{7\varepsilon}{4}zF' + (FF')' + 2\varepsilon F = 0. \tag{4.4}$$

Unfortunately, we have not been able to integrate it further, neither found a first integral.

We mention that the projective action on the solution u under the subgroup of the full symmetry group

$$C = T(t^2) = t^2\partial_t + \frac{1}{6}(4xt - 2\epsilon y^2)\partial_x + \frac{4}{3}ty\partial_y + \frac{1}{6}(4x - ut)\partial_u \tag{4.5}$$

is given by

$$\begin{aligned}
\tilde{t} &= t(1-\lambda t)^{-1}, \quad \tilde{y} = (1-\lambda t)^{-4/3} y, \\
\tilde{x} &= (1-\lambda t)^{-2/3}(x + \frac{\epsilon}{3}\frac{y^2}{t}) - \frac{\epsilon}{3}\frac{y^2}{t}(1-\lambda t)^{-5/3}, \\
\tilde{u} &= (1-\lambda t)^{4/3}\{u + \frac{2}{3}(x + \frac{\epsilon}{3}\frac{y^2}{t})\frac{\lambda}{1-\lambda t} + \frac{\epsilon}{6}\frac{\lambda y^2 t}{(1-\lambda t)^2}(\lambda t - 2)\},
\end{aligned} \tag{4.6}$$

where λ is the group parameter. These type of solution formulas have proved useful in constructing blow-up profiles for evolution equations.

4.2. Reduction for the Kac-Moody case

A solution of (3.22) invariant under $\exp(X(\xi) + Y(\eta))$ will have the form

$$\begin{aligned}
u &= \left[\frac{\varepsilon}{4}\left(\dot{b} + b^2 - \frac{\ddot{\eta}}{\eta}\right) - c\right] y^2 + \frac{\dot{\xi}}{\eta} y + F(z,t), \\
z &= x + \frac{\varepsilon}{4}(-b + \frac{\dot{\eta}}{\eta}) y^2 - \frac{\xi}{\eta} y.
\end{aligned} \tag{4.7}$$

We assumed the simplification (3.26) and re-labeled $b_1(t) = b(t)$, $c_2(t) = c(t)$. The reduced equation is found to be

$$(F_t + FF_z + F_{zz} + F_{zzz})_z + \varepsilon\frac{\xi^2}{\eta^2} F_{zz} + \frac{1}{2}\left(\frac{\dot{\eta}}{\eta} - b\right) F_z - 2\varepsilon c + \frac{1}{2}\left(\dot{b} + b^2 - \frac{\ddot{\eta}}{\eta}\right) = 0. \tag{4.8}$$

It is possible to eliminate the term F_{zz} by putting

$$F(z,t) = \tilde{F}(\tilde{z},\tilde{t}), \quad \tilde{z} = z + \beta(t), \quad \tilde{t} = t, \quad \dot{\beta}(t) = -\varepsilon\frac{\xi^2}{\eta^2}. \tag{4.9}$$

Further choosing $\dot{\eta}/\eta = b$ gives rise to the one-dimensional nonhomogeneous KdV-Burgers equation

$$(F_t + FF_z + F_{zz} + F_{zzz})_z = 2\varepsilon c(t). \tag{4.10}$$

4.3. Painlevé property and an exact solution

We checked whether (2.4) has the Painlevé property, using the package PainleveTestV2.m [7] with a Kruskal anzats in x, and we saw that the equation passes the test when

$$\begin{aligned}
&q(y,t) = a(y,t) = f(y,t) = 0, \quad b(y,t) = b_0(t) + b_1(t) y, \\
&r(y,t) = r_0 \exp\left(\frac{2}{3}\int b_1(t) dt\right), \quad c(y,t) = \frac{\varepsilon}{3}\left(\frac{2}{3} b_1^2 + \dot{b}_1\right) y^2 + c_1(t) y + c_0(t).
\end{aligned} \tag{4.11}$$

Since r is a function of t, it can be normalized to 1, which means that we can make $b_1(t) \to 0$. They are exactly the coefficients of the class which are invariant under the Kac-Moody-Virasoro algebra [3].

According to these results, our canonical equation (3.10) with the Virasoro algebra cannot have the Painlevé property ($q_0 \neq 0$).

We also found that

$$(u_t + uu_x + u_{xx} + u_{xxx})_x + \varepsilon u_{yy} + b(t) y u_{xy} + c(t) y^2 u_{xx} = 0 \tag{4.12}$$

fails to have the Painlevé property. In search for any possible exact solutions we truncate the Painlevé series for this equation at the first term and propose a solution of the form

$$u(x,y,t) = \frac{u_0(x,y,t)}{\Phi^2(x,y,t)}, \tag{4.13}$$

where Φ defines the singularity manifold. We substitute (4.13) in (4.12) and require that the coefficients of Φ^j, $j=-6,-5,..,-1$ vanish identically. For $c(t)=0$ it has been possible to solve these system of equations in the form

$$\begin{aligned} u_0 &= -\frac{12k_1^2}{25}\exp\Big[-\frac{2x}{5}-\frac{12t}{125}+2k_2ye^{-\int b(t)dt}+10\varepsilon k_2^2\int e^{-2\int b(t)dt}dt\Big], \\ \Phi &= k_1\exp\Big[-\frac{x}{5}-\frac{6t}{125}+k_2ye^{-\int b(t)dt}+5\varepsilon k_2^2\int e^{-2\int b(t)dt}dt\Big]+k_3, \end{aligned} \tag{4.14}$$

where k_1, k_2, k_3 are arbitrary constants. We checked whether the same approach works for (3.10) with (3.17). It turned out that a solution of the type (4.13) is not possible in this case.

4.4. The special case $a=b=c=e=f=h=0$
We finally consider the equation

$$(u_t + p(t)uu_x + q(t)u_{xx} + r(t)u_{xxx})_x + \sigma(t)u_{yy} = 0, \tag{4.15}$$

where all the coefficients are assumed to be nonzero. Allowed transformations are again of the form (2.1), while the coefficients map to

$$\tilde{p}(\tilde{t}) = p(t)\frac{R\alpha}{\dot{T}}, \quad \tilde{q}(\tilde{t}) = q(t)\frac{\alpha^2}{\dot{T}}, \quad \tilde{r}(\tilde{t}) = r(t)\frac{\alpha^3}{\dot{T}}, \quad \tilde{\sigma}(t) = \sigma(t)\frac{Y_y^2}{\alpha\dot{T}}. \tag{4.16}$$

In order that no additional terms appear in the equation, we must have

$$\sigma Y_{yy} = 0, \tag{4.17}$$
$$2\sigma\beta_y Y_y + \alpha Y_t = 0, \tag{4.18}$$
$$\beta_t\alpha + pS\alpha^2 + \sigma\beta_y^2 = 0, \tag{4.19}$$
$$-R\dot{\alpha} + \dot{R}\alpha + \sigma R\beta_{yy} = 0, \tag{4.20}$$
$$-\frac{d}{dt}\left(\frac{\dot{\alpha}}{\alpha p}\right) + \frac{1}{p}\left(\frac{\dot{\alpha}}{\alpha}\right)^2 + \sigma S_{yy} = 0. \tag{4.21}$$

Normalization $\tilde{p}(\tilde{t}) = \tilde{r}(\tilde{t}) = 1$ is again possible if we choose $\dot{T} = \alpha^3 r$ and $R = \alpha^2\frac{r}{p}$. Furthermore, we see that $Y(y,t) = \mu(t)y + \nu(t)$ so that $\tilde{\sigma}$ has no dependence on y. We solve (4.18) for $\beta(y,t)$ and (4.19) for $S(y,t)$ to obtain

$$\beta(y,t) = -\frac{\alpha\dot{\mu}}{4\sigma\mu}y^2 - \frac{\alpha\dot{\nu}}{2\sigma\mu}y + \beta_1(t), \tag{4.22}$$

$$S(y,t) = -\frac{1}{p\alpha^2}(\alpha\beta_t + \sigma\beta_y^2). \tag{4.23}$$

(4.20) is satisfied if $\mu(t) = \mu_0\left(\frac{\alpha r}{p}\right)^2$. There remains Eq. (4.21), which takes the form

$$\frac{\dot{\alpha}}{\alpha}\left(4\frac{\dot{p}}{p} - 3\frac{\dot{r}}{r} - \frac{\dot{\sigma}}{\sigma}\right) = \frac{\ddot{p}}{p} - \frac{\ddot{r}}{r} + \frac{\dot{p}^2}{p^2} + 3\frac{\dot{r}^2}{r^2} - 4\frac{\dot{p}\dot{r}}{pr} - \frac{\dot{p}\dot{\sigma}}{p\sigma} + \frac{\dot{r}\dot{\sigma}}{r\sigma}. \tag{4.24}$$

(i) Suppose $\sigma \neq \sigma_0 \frac{p^4}{r^3}$, where σ_0 is an arbitrary constant. Then we can solve $\alpha(t)$ from (4.24) and the normalization process is completed, which means we have obtained the canonical form

$$(u_t + uu_x + q(t)u_{xx} + u_{xxx})_x + \sigma(t)u_{yy} = 0. \tag{4.25}$$

(ii) If $\sigma = \sigma_0 \frac{p^4}{r^3}$, then the left-hand side of (4.24) vanishes and from the right-hand side we have the condition

$$\frac{\ddot{r}}{r} - \frac{\ddot{p}}{p} + 3\frac{\dot{p}^2}{p^2} - 3\frac{\dot{p}\dot{r}}{pr} = 0. \tag{4.26}$$

When we consider the canonical version (4.25), we see that the infinitesimal generator takes the form

$$V = (c_1 t + c_2)\partial_t + (\frac{c_1}{3}x - \frac{\dot{\eta}_0}{2\sigma}y + \xi_0)\partial_x + (\frac{2c_1}{3}y + \eta_0)\partial_y - \left(\frac{2c_1}{3}u + y\frac{d}{dt}\left(\frac{\dot{\eta}_0}{2\sigma}\right) - \dot{\xi}_0\right)\partial_u. \tag{4.27}$$

Here $\xi_0(t)$ and $\eta_0(t)$ are arbitrary functions, and we have the following two restricting equations

$$(c_1 t + c_2)\sigma_t = 0, \qquad (c_1 t + c_2)q^3 = q_0, \tag{4.28}$$

where q_0 is an arbitrary constant. Let us consider the case $c_1 = c_2 = 0$, then there is no condition on $q(t)$ and $\sigma(t)$. The symmetry algebra is an infinite-dimensional one generated by the vector field

$$V = X(\xi) + Y(\eta), \tag{4.29}$$

where $X(\xi)$ and $Y(\eta)$ are given by

$$X(\xi) = \xi\partial_x + \dot{\xi}\partial_u, \tag{4.30}$$

$$Y(\eta) = -\frac{\dot{\eta}}{2\sigma}y\partial_x + \eta\partial_y - y\frac{d}{dt}\left(\frac{\dot{\eta}}{2\sigma}\right)\partial_u. \tag{4.31}$$

Using invariance under $Y(\eta)$, we perform a reduction of (4.25) seeking solutions in the form

$$u = -\frac{y^2}{2\eta}\frac{d}{dt}\left(\frac{\dot{\eta}}{2\sigma}\right) + F(z,t), \quad z = x + \frac{\dot{\eta}}{4\sigma\eta}y^2, \tag{4.32}$$

and the reduced equation reads

$$(F_t + FF_z + qF_{zz} + F_{zzz})_z + \frac{\dot{\eta}}{2\eta}F_z + \frac{\dot{\eta}\dot{\sigma}}{2\eta\sigma} - \frac{\ddot{\eta}}{2\eta} = 0. \tag{4.33}$$

Choosing $\dot{\eta}(t) = \sigma(t)$ the inhomogeneity is disposed of and we end up with the equation

$$(F_t + FF_z + qF_{zz} + F_{zzz})_z + \tilde{\sigma}(t)F_z = 0, \tag{4.34}$$

where $\tilde{\sigma}(t) = \frac{\sigma(t)}{2\int \sigma(t)\,dt}$. An integration gives a one-dimensional variable coefficient KP-Burgers equation

$$F_t + FF_z + qF_{zz} + F_{zzz} + \tilde{\sigma}(t)F = f(t).$$

References

[1] J.-K. Xue. Kadomtsev-Petviashvili (KP) Burgers equation in a dusty plasmas with non-adiabatic dust charge fluctuation. *The European Physical Journal D - Atomic, Molecular, Optical and Plasma Physics*, 26:211–214, 2003.

[2] H. Leblond. KP lumps in ferromagnets: a three-dimensional KdV-Burgers model. *J. of Phys. A: Math. and Gen.*, 35(47):10149, 2002.

[3] F. Güngör and P. Winternitz. Generalized Kadomtsev-Petviashvili equation with an infinite dimensional symmetry algebra. *J. Math. Anal. and Appl.*, 276:314–328, 2002.

[4] F. Güngör. Infinite-dimensional Symmetries of a Two-dimensional Generalized Burgers Equation. *J. Math. Phys.*, 51:083504:1–12, 2010.

[5] F. Güngör. On the Virasoro structure of symmetry algebras of nonlinear partial differential equations. *SIGMA*, 2:14–, 2006.

[6] Shou-Fu Tian and Hong-Qing Zhang. On the integrability of a generalized variable-coefficient Kadomtsev-Petviashvili equation. *J. Phys. A: Math. and Gen.*, 45:055203, 2012.

[7] D. Baldwin and W. Hereman. *PainleveTest.m: A Mathematica package for the Painlevé test of systems of nonlinear ordinary and partial differential equations.* http://inside.mines.edu/ whereman/, 2001, updated: 2009.

Journal of Physics: Conference Series **474** (2013) 012011 doi:10.1088/1742-6596/474/1/012011

Interpolation via symmetric exponential functions

Agata Bezubik[1], **Severin Pošta**[2]

[1] Insitute of Mathematics, University of Białystok, Akademicka 2, 15-267 Białystok, Poland
[2] Department of mathematics Faculty of nuclear sciences and physical engineering, Czech Technical University in Prague, Trojanova 13, 120 00 Prague, Czech Republic

E-mail: agatab@math.uwb.edu.pl, severin.posta@fjfi.cvut.cz

Abstract. Complex valued functions on the Euclidean space $\mathbb{R}^n$, symmetric or antisymmetric with respect to the permutation group S_n, are often dealt with in various branches of physics, such as quantum theory or theory of integrable systems. One often needs to approximate such functions with series consisting of various special functions which satisfy nice properties. Questions of uniform convergence of such approximations are crucial for applications. In this article a family of special functions called the symmetric exponential functions are used for such approximation and the uniform convergence of their sums is considered.

1. Introduction. Symmetric exponential functions

The symmetric or antisymmetric functions with respect to the permutation group S_n appear very often, namely in quantum theory or theory of integrable systems. We often need to approximate such functions. In this article, we study approximations with so called symmetric exponential functions, which are defined as permanents of matrices whose entries are exponential functions of one variable. They are introduced in general for any integer n in [4]. Such approximations can be done in three different ways: expansions into the corresponding Fourier series, integral Fourier transforms and multivariate finite Fourier transforms. We deal with finite Fourier transforms and study the uniform convergence of the sum to the original function.

The symmetric and antisymmetric exponential functions [4, 3] are closely related to the symmetric [5], antisymmetric [6] and $E-$orbit functions [7] and to the corresponding orthogonal polynomials [1]. Through the inclusion of the alternaning subgroup $Alt_n \subset S_n$, they are also related to alternating exponential functions [2, 8].

2. Two dimensional symmetric exponential functions

The two dimensional symmetric exponential functions $E^+_{(k,l)} : \mathbb{R}^2 \to \mathbb{C}$, with $k, l \in \mathbb{Z}$ and $k \geq l$, are defined by the formula

$$E^+_{(k,l)}(x,y) = \begin{vmatrix} e^{2\pi ikx} & e^{2\pi iky} \\ e^{2\pi ilx} & e^{2\pi ily} \end{vmatrix}^+ = e^{2\pi i(kx+ly)} + e^{2\pi i(ky+lx)}.$$

Similar antisymmetric functions can be defined with the help of determinant instead of permanent. Similar construction is possible in each dimension. We consider dimension two for simplicity here.

From the explicit formula we see the symmetry of $E^+_{(k,l)}$ with respect to permutation of $x \leftrightarrow y$,

$$E^+_{(k,l)}(x,y) = E^+_{(k,l)}(y,x),$$

and also with respect to $k \leftrightarrow l$,

$$E^+_{(k,l)}(x,y) = E^+_{(l,k)}(x,y).$$

Therefore we can consider only functions with $k \geq l$.

One has also

$$E^+_{(k,l)}(x+r, y+s) = E^+_{(k,l)}(x,y) \text{ for all } r,s \in \mathbb{Z}.$$

Due to this periodicity and symmetry, the functions $E^+_{(k,l)}$ can be considered on the closure of the so called fundamental domain $F(S_2^{\mathrm{aff}})$ only, where

$$F(S_2^{\mathrm{aff}}) = \{(x,y) \in (0,1) \times (0,1) | x > y\}.$$

The functions $E^+_{(k,l)}$ are mutually orthogonal on $F(S_2^{\mathrm{aff}})$, i. e.

$$\int_{F(S_2^{\mathrm{aff}})} E^+_{(k,l)}(x,y)\overline{E^+_{(m,n)}(x,y)}\mathrm{d}x\mathrm{d}y = G_{kl}\delta_{km}\delta_{ln},$$
$$k,l,m,n \in \mathbb{Z},\ k \geq l,\ m \geq n,$$

where G_{kl} is defined by

$$G_{kl} = \begin{cases} 2 & \text{if } k = l, \\ 1 & \text{otherwise.} \end{cases}$$

Every function $f : \mathbb{R}^2 \to \mathbb{C}$ which is symmetric and periodic, i. e.

$$f(x,y) = f(y,x), \qquad f(x+r, y+s) = f(x,y), \quad x,y \in \mathbb{R},\ r,s \in \mathbb{Z},$$

can be under certain conditions expanded in fourier-like series

$$f(x,y) = \sum_{\substack{k,l \in \mathbb{Z} \\ k \geq l}} c_{kl} E^+_{(k,l)}(x,y),$$

where

$$c_{kl} = G_{kl}^{-1} \int_{F(S_2^{\mathrm{aff}})} f(x,y)\overline{E^+_{(k,l)}(x,y)}\mathrm{d}x\mathrm{d}y.$$

Besides the continuous orthogonality, one has also discrete orthogonality of symmetric exponential functions over the grid of the form

$$L_N = \left\{ \left(\frac{m}{N}, \frac{n}{N}\right) \,\middle|\, m,n = 0,...,N-1; m \geq n \right\}$$

which is given by

$$\sum_{\substack{m,n=0 \\ m \geq n}}^{N-1} G_{mn}^{-1} E^+_{(k,l)}\left(\frac{m}{N}, \frac{n}{N}\right)\overline{E^+_{(k',l')}\left(\frac{m}{N}, \frac{n}{N}\right)} = G_{kl}N^2\delta_{kk'}\delta_{ll'},$$

where $k, l, k', l' \in \{0, ..., N-1\}$, $k \geq l$, $k' \geq l'$.

Note that for $N = 2M + 1$ odd, discrete orthogonality formulas also hold for the set of functions where $k, l \in \{-M, ..., M\}$, $k \geq l$.

Now when we have any function $f : L_N \to \mathbb{C}$ we can compute discrete-like-Fourier transform coefficients

$$\beta^+_{kl} = \frac{1}{G_{kl}N^2} \sum_{\substack{m,n=0\\ m\geq n}}^{N-1} G^{-1}_{mn} f\Big(\frac{m}{N}, \frac{n}{N}\Big) \overline{E^+_{k,l}\Big(\frac{m}{N}, \frac{n}{N}\Big)},$$

where $k, l = -M, ..., M$, $k \geq l$. Then we can define the function

$$\psi_N(x, y) = \sum_{\substack{k,l=-M\\ k\geq l}}^{M} \beta^+_{kl} E^+_{(k,l)}(x, y).$$

Immediately from orthogonality relation we obtain that this Nth interpolation function coincides with f on the grid L_N:

$$\psi_N\Big(\frac{m}{N}, \frac{n}{N}\Big) = f\Big(\frac{m}{N}, \frac{n}{N}\Big).$$

Example of grid L_N is depicted on the following figure 1 for $N = 10$.

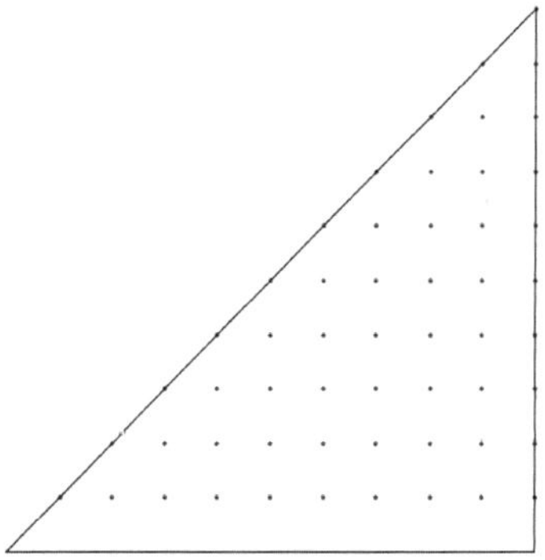

Figure 1. Grid L_{10}

We give a simple example of discrete interpolation. Let us define

$$f(x, y) = \begin{cases} 0 & \text{if } \left(x - \frac{2}{3}\right)^2 + \left(y - \frac{1}{3}\right)^2 \geq \frac{1}{25}, \\ \exp\left(\dfrac{1}{100\left(\sqrt{\left(x-\frac{2}{3}\right)^2 + \left(y-\frac{1}{3}\right)^2} - \frac{1}{10}\right)^2 - 1} + 1\right) & \text{otherwise.} \end{cases}$$

The function f is shown on figure 2.

For $N = 3$ using described procedure we construct

$$\psi_3(x, y) = -\frac{1}{18}(-1)^{5/6}\left(\sqrt{3} + i\right)\left(-1 + e^{2i\pi x}\right)$$
$$\left(-1 + e^{2i\pi y}\right) e^{-2i\pi(x+y)} \left(2e^{2i\pi(x+y)} + e^{2i\pi x} + e^{2i\pi y} + 2\right)$$

which is shown on figure 3. To see how the sum converges, we show another two finer approximations. For $N = 23$ it is drawn on figure 4, for $N = 55$ we get even better approximation, which is shown on figure 5. Approximation errors of the form $\int_{F(S_2^{\mathrm{aff}})} |f - \psi_N|^2$ of computed functions fall rapidly with increasing N.

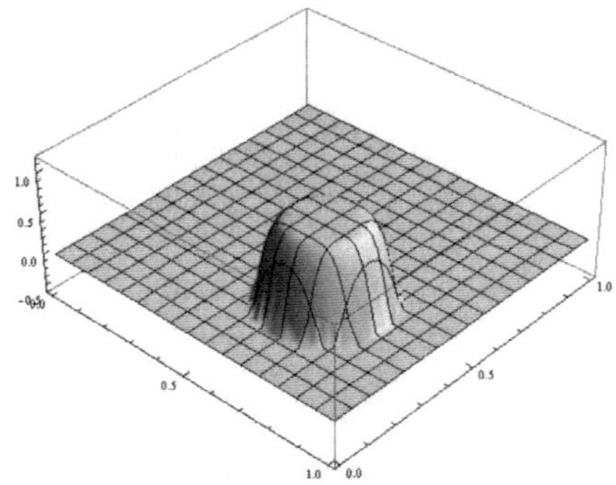

Figure 2. Function f

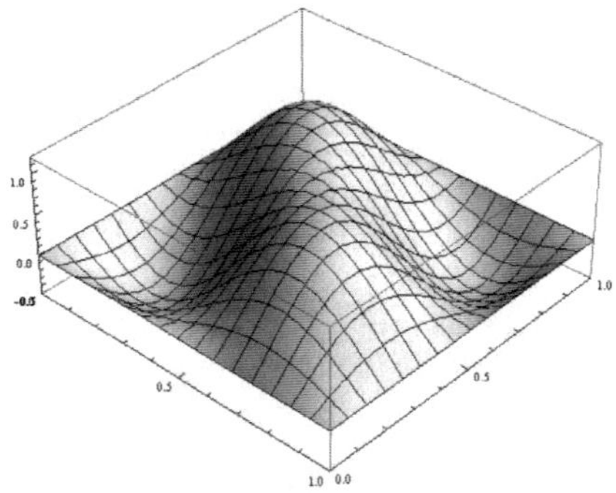

Figure 3. Function ψ_7

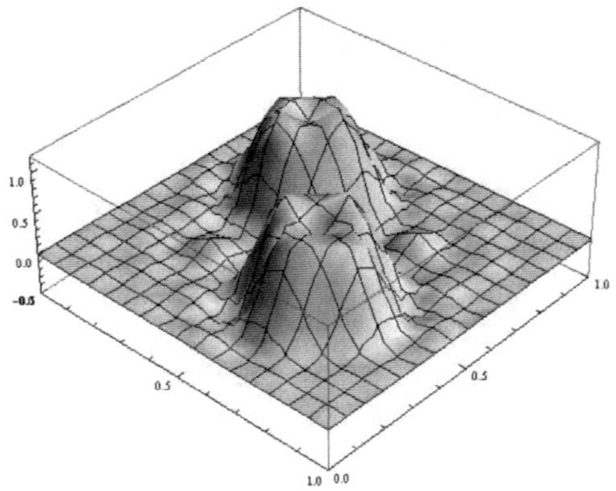

Figure 4. Function ψ_{23}

N	Approximation error
3	0.01690682
23	0.00272393
55	0.00016751

Table 1. Approx. errors

3. Uniform convergence

In one dimension, if f is m-times continuously differentiable, $m \geq 2$, and

$$f(x) = \sum_{k\in\mathbb{Z}} c_k e^{2\pi i k x},$$

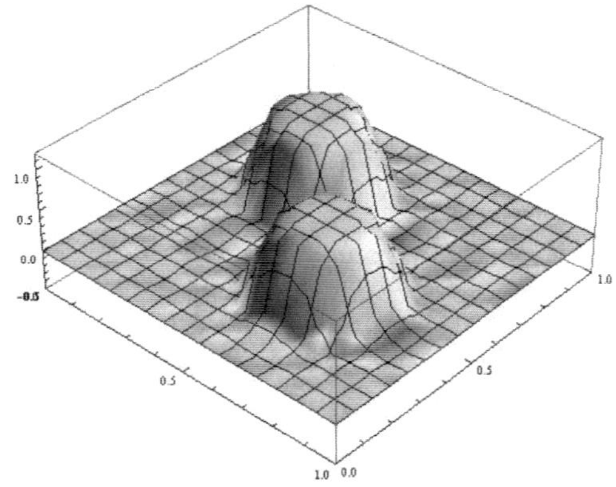

Figure 5. Function ψ_{55}

decay of fourier coefficient is $|c_k| \leq ck^{-m}$, where c is some constant. Therefore, if we define for $N = 2M + 1$

$$\psi_N(x) = \sum_{|k| \leq M} d_k e^{2\pi ikx},$$

where

$$d_k = (f, e^{2\pi ikx})_N,$$

and

$$(f,g)_N = \frac{1}{N} \sum_{j=0}^{N-1} f\left(\frac{j}{N}\right) \overline{g\left(\frac{j}{N}\right)},$$

we obtain

$$|f(x) - \psi_N(x)| \leq \sum_{|k| \leq M} |c_k - d_k| + \sum_{|k| > M} |c_k|,$$

where

$$|c_k - d_k| = |c_k - \sum_{l \in \mathbb{Z}} c_l (e^{2\pi ilx}, e^{2\pi ikx})_N|,$$

which can be simplified due to $(e^{2\pi ilx}, e^{2\pi ikx})_N = 1$ if $k \equiv l \,(\bmod N)$ and 0 otherwise:

$$|c_k - d_k| \leq \sum_{l \neq 0} |c_{k+lN}| \leq \frac{\text{const}}{N^m}.$$

Because

$$\sum_{|k| > M} |c_k| \leq \frac{\text{const}}{N^{m-1}},$$

we have

$$|f(x) - \psi_N(x)| \leq \frac{\text{const}}{N^{m-1}} \to 0 \text{ as } N \to +\infty \text{ if } m \geq 2.$$

Similarly, in n dimensions, if f is m-times continuously differentiable and we have

$$f(x) = \sum_{k \in \mathbb{Z}^n} c_k e^{2\pi i(k,x)},$$

since

$$D^l e^{2\pi i(k,x)} = \prod_{j=1}^{n} (2\pi i k_j)^{l_j} e^{2\pi i(k,x)}$$

we see that $|\prod_{j=1}^{n} k_j^{l_j} c_k|$ is bounded for every multi-index l for which $\sum_{j=1}^{n} l_j \leq m$.

If we now have $2m$-times continuously differentiable function on $\mathbb{R}^2$, symmetric and periodic, expanded in pointwise convergent Fourier series as

$$f(x,y) = \sum_{k \geq l} c_{kl} E^+_{(k,l)}(x,y),$$

we can apply the same procedure. For $N = 2M + 1$ we take

$$\psi_N(x,y) = \sum_{\substack{k,l=-M \\ k \geq l}}^{M} \beta^+_{kl} E^+_{(k,l)}(x,y),$$

and study the difference

$$|f(x) - \psi_N(x)|.$$

If we analyze this difference similarly as above, one can see that when m is equal to 2, it converges to zero as $N \to \infty$.

4. Conclusion

The complex valued functions on the Euclidean space $\mathbb{R}^n$, symmetric or antisymmetric with respect to the permutation group S_n, are very well suited for approximate of functions with the same symmetry properties, which appear in applications. Uniform convergence happens whenever sufficient smoothness of original approximated function is assured. Our approach was demonstrated on the simplest example of symmetric two dimensional exponential functions, however it is clear that similar analysis can be performed within any families of special functions such as symmetric, antisymmetric, alternating exponential functions, symmetric, antisymmetric, alternating multivariate trigonometric functions, as well as for families of C-, S-, and E-functions of any compact semisimple Lie group (families introduced and studied by Patera, Moody, Klimyk, Hrivnák and others).

Acknowledgments

SP acknowledges support of CTU Project SGS12/198/OHK4/3T/14 and P201/10/1509, project of the Grant Agency of the Czech Republic. AB acknowledges support of 7AMB13PL035, project of the Czech ministry of education, youth and sports.

References

[1] Bezubik, Agata; Hrivnák, Jiří; Pošta, Severin: Two dimensional symmetric and antisymmetric exponential functions and orthogonal polynomials. J. Phys.: Conf. Ser. 411 (2013), 012007.

[2] Hrivnák Jiří; Patera, Jiří, Pošta, Severin: Three-variable exponential functions of the alternating group. J. Phys. A 45 (2012), no. 4, 045201, 10 pp.

[3] Hrivnák, Jiří; Patera, Jiří: Two-dimensional symmetric and antisymmetric generalizations of exponential and cosine functions. J. Math. Phys. 51 (2010), 023515.

[4] Klimyk, Anatoliy; Patera, Jiří: (Anti)symmetric multivariate exponential functions and corresponding Fourier transforms. J. Phys. A 40 (2007), no. 34, 10473–10489.

[5] Klimyk Anatoliy; Patera, Jiří: Orbit functions, Symmetry Integrability Geom. Methods Appl. 2 (2006), Paper 006.

[6] Klimyk Anatoliy; Patera, Jiří: Antisymmetric orbit functions, Symmetry Integrability Geom. Methods Appl. 3 (2007), Paper 023.

[7] Klimyk Anatoliy; Patera, Jiří: E-orbit functions, Symmetry Integrability Geom. Methods Appl. 4 (2008), Paper 002.

[8] Klimyk Anatoliy; Patera, Jiří: Alternating group and multivariate exponential functions, in Groups and Symmetries; From Neolithic Scots to John McKay, AMS-CRM Proceedings and Lectures Notes Series, 47, 233-246, eds. J. Harnad and P. Winternitz; arXiv:0907.0601.

Journal of Physics: Conference Series **474** (2013) 012012 doi:10.1088/1742-6596/474/1/012012

On low-energy effective action in three-dimensional $\mathcal{N} = 2$ and $\mathcal{N} = 4$ supersymmetric electrodynamics

I. L. Buchbinder[1], **B. S. Merzlikin**[1,2], **I. B. Samsonov**[2,3]

[1] Department of Theoretical Physics, Tomsk State Pedagogical University, Tomsk 634061, Russia

[2] Department of Higher Mathematics and Mathematical Physics, Tomsk Polytechnic University, 634050 Tomsk, Russia

[3] INFN, Sezione di Padova, 35131 Padova, Italy

E-mail: joseph@tspu.edu.ru, merzlikin@tspu.edu.ru, samsonov@mph.phtd.tpu.ru

Abstract. We discuss general structure of low-energy effective actions in $\mathcal{N} = 2$ and $\mathcal{N} = 4$ three-dimensional supersymmetric electrodynamics (SQED) in gauge superfield sector. There are specific terms in the effective action having no four-dimensional analogs. Some of these terms are responsible for the moduli space metric in the Coulomb branch of the theory. We find two-loop quantum corrections to the moduli space metric in the $\mathcal{N} = 2$ SQED and show that in the $\mathcal{N} = 4$ SQED the moduli space does not receive two-loop quantum corrections.

1. Introduction and conclusions

Three-dimensional supersymmetric gauge theories with extended supersymmetry have attracted considerable attention recently [1, 2]. On the one hand, they exhibit reach duality properties [3, 4, 5, 6, 7, 8, 9, 10, 11], which restrict their low-energy dynamics severely. On the other hand, they have deep relations to the dynamics of multiple M2 branes described in terms of the BLG [12, 13, 14, 15, 16, 17, 18] and ABJM [19, 20] models.

In the present paper we study low-energy dynamics in $\mathcal{N} = 2$ and $\mathcal{N} = 4$ SQEDs without Chern-Simons term. When the Chern-Simons therm vanishes, the Coulomb branch exists and the low-energy effective action is given by supersymmetric Euler-Heisenberg effective action for the gauge superfield. In four-dimensional $\mathcal{N} = 1$ and $\mathcal{N} = 2$ SQEDs the Euler-Heisenberg effective action was studied in [21, 22, 23, 24, 25, 26]. It is interesting to compare the forms of Euler-Heisenberg effective actions among three- and four-dimensional SQEDs.

Recall that in $\mathcal{N} = 1$, $d = 4$ supersymmetric gauge theories the gauge superfield V has Grassmann-odd superfield strengths W_α and $\bar{W}_{\dot{\alpha}}$ which carry spinorial indices. In three-dimensional case these two Grassmann-odd superfield strengths remain, but there is also a Grassmann-even superfield G which is a scalar with respect to the Lorentz group. To study the Euler-Heisenberg effective action it is sufficient to consider constant on-shell superfield strengths $\partial_m G = \partial_m W_\alpha = \partial_m \bar{W}_\alpha$ constrained by $D^2 W_\alpha = \bar{D}^2 \bar{W}_\alpha = 0$. [1] Hence, the most general action with the superfields W_α, $\bar{W}_\alpha$ and G subject to these constraints reads

$$\int d^3x d^4\theta \left[A(G, D_\gamma W_\delta) + W^\alpha \bar{W}^\beta B_{\alpha\beta}(G, D_\gamma W_\delta) + W^2 \bar{W}^2 C(G, D_\gamma W_\delta) \right], \tag{1.1}$$

[1] In components, these constraints correspond, in particular, to constant Maxwell field strength, $\partial_m F_{np} = 0$.

where A, B and C are some functions. It is important to note that the first two terms in (1.1) have no four-dimensional analogs and are specific for three-dimensional gauge theories in $\mathcal{N}=2$, $d=3$ superspace.

In the present paper we will discuss only the first term in (1.1) under additional constraints, $W_\alpha = \bar{W}_\alpha = 0$, $G = const$. In this case (1.1) reduces to a simple expression

$$\int d^3x d^4\theta \, f(G) \,, \tag{1.2}$$

where $f(G)$ is a real function of one real argument which we refer to as the effective potential. This function has good geometrical meaning [7]: it is responsible for the moduli space metric for the $\mathcal{N}=2$, $d=3$ gauge multiplet. In particular, the classical action for the gauge superfield V implies $f(G) = \frac{1}{e^2}G^2$, where e is the gauge coupling constant. This quadratic function corresponds to flat moduli space metric while loop quantum corrections deform the flat moduli space. For a long time only one-loop quantum corrections to $f(G)$ were known [3, 4]. In our recent publication [27] we found two-loop effective potential in $\mathcal{N}=2$ SQED:

$$f(G) = \frac{1}{g^2}G^2 + \frac{1}{2\pi}(G\ln(G+\sqrt{G^2+m^2}) - \sqrt{G^2+m^2}) - \frac{e^2}{8\pi^2}\ln(G^2+m^2) \,. \tag{1.3}$$

Here m is the mass of the chiral matter superfield. The corresponding moduli space metric reads

$$ds^2 = \frac{1}{2}g(r)dr^2 + \frac{1}{2}\frac{1}{g(r)}d\sigma^2 \,, \quad g(r) = \frac{1}{e^2} + \frac{1}{4\pi}\frac{1}{\sqrt{r^2+m^2}} + \frac{e^2}{8\pi^2}\frac{r^2-m^2}{(r^2+m^2)^2} \,. \tag{1.4}$$

Equations (1.3) and (1.4) represent one of the main results of [27] which will be reviewed in the present contribution.

The $\mathcal{N}=4$, $d=3$ electrodynamics, as compared with $\mathcal{N}=2$ SQED, has extra propagating chiral superfield Φ which, together with V, forms the $\mathcal{N}=4$ gauge multiplet. Therefore, an extension of (1.2) to the $\mathcal{N}=4$ supersymmetric case reads

$$\int d^3x d^4\theta \, f(G, \Phi\bar{\Phi}) \,. \tag{1.5}$$

This effective potential in known to be one-loop exact in $\mathcal{N}=4$ supersymmetric gauge theories [3, 7]. For $\mathcal{N}=4$ SQED, one-loop quantum contributions to (1.5) were computed in [28],

$$f(G, \Phi\bar{\Phi}) = \frac{1}{e^2}(G^2 - \frac{1}{2}\bar{\Phi}\Phi) + \frac{1}{2\pi}\Big[G\ln(G+\sqrt{G^2+\bar{\Phi}\Phi}) - \sqrt{G^2+\Phi\bar{\Phi}}\Big] \,. \tag{1.6}$$

In the present paper, by analyzing the structure of Feynman graphs in $\mathcal{N}=2$, $d=3$ superspace, we show that two-loop quantum contributions to (1.5) vanish, confirming the non-renormalizability of the effective potential beyond one loop. The moduli space metric corresponding to the effective potential (1.6) is the well-known Taub-NUT metric [29].

It is natural to generalize the present results to the non-Abelian case and to find two-loop quantum correction to the moduli space metric (1.4) in $\mathcal{N}=2$ supersymmetric Yang-Mills models with different matter superfields. Another interesting extension could be to find all-loop quantum contributions to (1.5) and derive the effective potential $f(G)$ in a closed form. Such an effective potential would give non-perturbative expression for the moduli space metric in $\mathcal{N}=2$ SQED. Another tempting problem is to find effective Kähler potential in three-dimensional gauge theories in $\mathcal{N}=2$, $d=3$ superspace.

The present contribution is based essentially on the results of our recent work [27]. Here we will employ superspace notations adopted in this paper.

2. $\mathcal{N}=2$ supersymmetric electrodynamics

2.1. Classical action

The classical action of the $\mathcal{N}=2$, $d=3$ supersymmetric electrodynamics reads

$$S_{\mathcal{N}=2}=\frac{1}{e^2}\int d^7z\,G^2-\int d^7z\,(\bar{Q}_+e^{2V}Q_++\bar{Q}_-e^{-2V}Q_-)-\left(m\int d^5z\,Q_+Q_-+c.c.\right), \quad (2.1)$$

where $Q_\pm$ are chiral superfields with opposite charges with respect to the gauge superfield V. G is the superfield strength for the gauge superfield V,

$$G=\frac{i}{2}\bar{D}^\alpha D_\alpha V\,. \quad (2.2)$$

This superfield is real $G^*=G$ and linear,

$$D^2G=\bar{D}^2G=0\,. \quad (2.3)$$

The action (2.1) appears by virtue of the dimensional reduction from the $\mathcal{N}=1$, $d=4$ electrodynamics [30, 31].

We are interested in the part of low-energy effective action which depends on the gauge superfield only, $\Gamma=\Gamma[V]$, while the chiral superfields $Q_\pm$ are integrated out. For this problem the background field method in the $\mathcal{N}=2$, $d=3$ superspace [32] appears to be useful. We split the gauge superfield V into the background V and quantum v parts

$$V\to V+e\,v\,. \quad (2.4)$$

Upon this splitting the Maxwell term in (2.1) changes as

$$\frac{1}{e^2}\int d^7z\,G^2\to\frac{1}{e^2}\int d^7z\,G^2+\frac{i}{e}\int d^7z\,vD^\alpha W_\alpha+\frac{1}{8}\int d^7z\,vD^\alpha\bar{D}^2D_\alpha v\,. \quad (2.5)$$

The operator $D^\alpha\bar{D}^2D_\alpha$ in the last term is degenerate and requires gauge fixing. In particular, the Fermi-Feynman gauge is implemented by the following gauge-fixing term

$$S_{\rm gf}=-\frac{1}{16}\int d^7z\,v\{D^2,\bar{D}^2\}v\,. \quad (2.6)$$

Adding this term to (2.1) we get

$$S_{\rm quantum} = S_2+S_{\rm int}\,, \quad (2.7)$$

$$S_2 = -\int d^7z\,(v\Box v+\bar{\mathcal{Q}}_+\mathcal{Q}_++\bar{\mathcal{Q}}_-\mathcal{Q}_-)-\left(m\int d^5z\,\mathcal{Q}_+\mathcal{Q}_-+c.c.\right), \quad (2.8)$$

$$S_{\rm int} = -2\int d^7z\left[e\left(\bar{\mathcal{Q}}_+\mathcal{Q}_+-\bar{\mathcal{Q}}_-\mathcal{Q}_-\right)v+e^2\left(\bar{\mathcal{Q}}_+\mathcal{Q}_++\bar{\mathcal{Q}}_-\mathcal{Q}_-\right)v^2\right]+O(e^3)\,, \quad (2.9)$$

where $\mathcal{Q}_\pm$ and $\bar{\mathcal{Q}}_\pm$ are covariantly (anti)chiral superfields with respect to the background gauge superfield,

$$\bar{\mathcal{Q}}_+=\bar{Q}_+e^{2V}\,,\quad \mathcal{Q}_+=Q_+\,,\quad \bar{\mathcal{Q}}_-=\bar{Q}_-e^{-2V}\,,\quad \mathcal{Q}_-=Q_-\,. \quad (2.10)$$

2.2. Propagators and heat kernels

The action S_2 (2.8) is responsible for the propagators,

$$\begin{aligned} i\langle \mathcal{Q}_+(z)\mathcal{Q}_-(z')\rangle &= -mG_+(z,z')\,, \\ i\langle \bar{\mathcal{Q}}_+(z)\bar{\mathcal{Q}}_-(z')\rangle &= mG_-(z',z)\,, \\ i\langle \mathcal{Q}_+(z)\bar{\mathcal{Q}}_+(z')\rangle &= G_{+-}(z,z') = G_{-+}(z',z)\,, \\ i\langle \bar{\mathcal{Q}}_-(z)\mathcal{Q}_-(z')\rangle &= G_{-+}(z,z')\,, \end{aligned} \tag{2.11}$$

where the Green's functions G_+ and G_{+-} obey the equations

$$(\Box_\pm + m^2)G_\pm(z,z') = -\delta_\pm(z,z')\,, \tag{2.12}$$

$$\frac{1}{4}\nabla^2 G_{+-}(z,z') + m^2 G_-(z,z') = -\delta_-(z,z')\,, \tag{2.13}$$

$$\frac{1}{4}\bar{\nabla}^2 G_{-+}(z,z') + m^2 G_+(z,z') = -\delta_+(z,z')\,. \tag{2.14}$$

Here $\delta_\pm(z,z')$ are (anti)chiral delta-functions and $\Box_\pm$ are d'Alembertians acting in the space of (anti)chiral superfields,

$$\Box_+ = \nabla^m\nabla_m + G^2 + \frac{i}{2}(\nabla^\alpha W_\alpha) + iW^\alpha\nabla_\alpha\,, \tag{2.15}$$

$$\Box_- = \nabla^m\nabla_m + G^2 - \frac{i}{2}(\bar{\nabla}^\alpha \bar{W}_\alpha) - i\bar{W}^\alpha\bar{\nabla}_\alpha\,. \tag{2.16}$$

The Green's functions (2.12)–(2.14) can be expressed in terms of corresponding heat kernels,

$$G_\pm(z,z') = i\int_0^\infty ds\, K_\pm(z,z'|s)e^{is(m^2+i\epsilon)}\,, \tag{2.17}$$

$$G_{+-}(z,z') = i\int_0^\infty ds\, K_{+-}(z,z'|s)e^{is(m^2+i\epsilon)}\,, \tag{2.18}$$

where the standard $\epsilon \to +0$ prescription is assumed. These heat kernels were computed exactly for the on-shell gauge superfield background $D^\alpha W_\alpha = 0$ subject to $\partial_m W_\alpha = 0$, [27]

$$K_+(z,z'|s) = \frac{1}{8(i\pi s)^{3/2}}\frac{sB}{\sinh(sB)}e^{isG^2}\mathcal{O}(s)e^{\frac{i}{4}(F\coth(sF))_{mn}\rho^m\rho^n - \frac{1}{2}\bar{\zeta}^\beta\rho_{\beta\gamma}W^\gamma}\zeta^2 I(z,z') \tag{2.19}$$

$$K_{+-}(z,z'|s) = -\frac{1}{8(i\pi s)^{3/2}}\frac{sB}{\sinh(sB)}e^{isG^2}\mathcal{O}(s)e^{\frac{i}{4}(F\coth(sF))_{mn}\tilde{\rho}^m\tilde{\rho}^n + R(z,z')}I(z,z')\,, \tag{2.20}$$

where

$$\zeta^\alpha = (\theta-\theta')^\alpha\,, \quad \bar{\zeta}^\alpha = (\bar{\theta}-\bar{\theta}')^\alpha\,, \quad \rho^m = (x-x')^m - i\gamma^m_{\alpha\beta}\zeta^\alpha\bar{\theta}'^\beta + i\gamma^m_{\alpha\beta}\theta'^\alpha\bar{\zeta}^\beta \tag{2.21}$$

are components of the supersymmetric interval and $I(z,z')$ is the parallel transport propagator [33] in $\mathcal{N}=2$, $d=3$ superspace [27]. Among its properties there are

$$I(z,z) = 1\,, \qquad I(z,z')I(z',z) = 1\,, \tag{2.22}$$

which will be useful in quantum loop computations. The heat kernel (2.20) contains also a version of supersymmetric interval $\tilde{\rho}^m$ which is chiral with respect to the first argument and is antichiral with respect to the second one,

$$\tilde{\rho}^m = \rho^m + i\zeta^\alpha\gamma^m_{\alpha\beta}\bar{\zeta}^\beta\,, \qquad D'_\alpha\tilde{\rho}^m = \bar{D}_\alpha\tilde{\rho}^m = 0\,. \tag{2.23}$$

The two-point function $R(z,z')$ in (2.20) was found in the form

$$\begin{aligned} R(z,z') &= -i\zeta\bar\zeta G + \frac{7i}{12}\bar\zeta^2\zeta W + \frac{i}{12}\zeta^2\bar\zeta\bar W - \frac{1}{2}\bar\zeta^\alpha\tilde\rho_{\alpha\beta}W^\beta - \frac{1}{2}\zeta^\alpha\tilde\rho_{\alpha\beta}\bar W^\beta \\ &+\frac{1}{12}\zeta^\alpha\bar\zeta^\beta[\tilde\rho_\beta^\gamma D_\alpha W_\gamma - 7\tilde\rho_\alpha^\gamma D_\gamma W_\beta]\,. \end{aligned} \tag{2.24}$$

The heat kernels (2.19) and (2.20) contain also the operator

$$\mathcal{O}(s) = e^{s(\bar W^\alpha\bar\nabla_\alpha - W^\alpha\nabla_\alpha)}\,, \tag{2.25}$$

and we use the following notations for derivatives of superfield strengths:

$$N_{\alpha\beta} = D_{(\alpha}W_{\beta)}\,,\quad \bar N_{\alpha\beta} = \bar D_{(\alpha}\bar W_{\beta)}\,,\quad B^2 = \frac{1}{2}N_\alpha^\beta N_\beta^\alpha\,. \tag{2.26}$$

The propagator $G_0(z,z')$ for the gauge superfield v obeys the equation

$$2i\langle v(z)\,v(z')\rangle = G_0(z,z')\,,\qquad \Box G_0(z,z') = -\delta^7(z-z')\,. \tag{2.27}$$

The corresponding heat kernel reads

$$G_0(z,z') = i\int_0^\infty ds\,K_0(z,z'|s)\,e^{-s\epsilon}\,,\qquad K_0(z,z'|s) = \frac{1}{(4i\pi s)^{3/2}}e^{\frac{i\rho^m\rho_m}{4s}}\zeta^2\bar\zeta^2\,. \tag{2.28}$$

Here ρ^m, ζ^α and $\bar\zeta^\alpha$ are the components of supersymmetric interval (2.21).

2.3. Low-energy effective action

Consider loop expansion of the effective action in $\mathcal{N}=2$, $d=3$ SQED,

$$\Gamma_{\mathcal{N}=2} = \Gamma^{(1)}_{\mathcal{N}=2} + \Gamma^{(2)}_{\mathcal{N}=2} + \ldots\,, \tag{2.29}$$

where $\Gamma^{(1)}_{\mathcal{N}=2}$ and $\Gamma^{(2)}_{\mathcal{N}=2}$ are one- and two-loop quantum correction and dots correspond to higher-order terms. It is clear that the one-loop effective action is given by the loop of chiral superfields in the gauge superfield background,

$$\Gamma^{(1)}_{\mathcal{N}=2} = i\,\mathrm{Tr}\ln(\Box_+ + m^2) = -i\int_0^\infty\frac{ds}{s}\int d^3x d^2\theta d^2\bar\theta\,K_+(z,z|s)\,, \tag{2.30}$$

where $K_+(z,z')$ is the chiral heat kernel (2.19). At coincident superspace points it reads

$$K_+(z,z|s) = \frac{1}{8(i\pi s)^{3/2}}s^2W^2e^{isG^2}\frac{\tanh(sB/2)}{sB/2}\,. \tag{2.31}$$

In fact, for computing the effective potential it is sufficient to consider the expression (2.31) for $B=0$. Taking into account this simplification we calculate the integral over ds in (2.30),

$$\Gamma^{(1)}_{\mathcal{N}=2} = \frac{1}{2\pi}\int d^7z\Big[G\ln(G+\sqrt{G^2+m^2}) - \sqrt{G^2+m^2}\Big]\,. \tag{2.32}$$

Consider now the two-loop effective action

$$\Gamma^{(2)}_{\mathcal{N}=2} = -2e^2\int d^7z\,d^7z'[G_{+-}(z,z')G_{+-}(z',z) + m^2G_+(z,z')G_-(z,z')]G_0(z,z')\,. \tag{2.33}$$

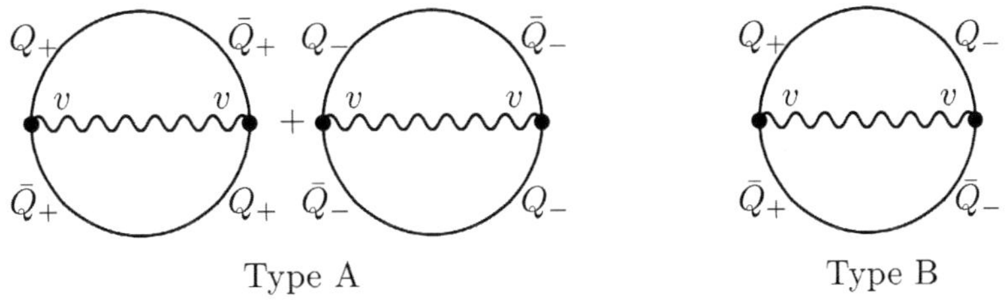

Figure 1. Two-loop supergraphs in $\mathcal{N}=2$ supersymmetric electrodynamics.

The two terms in r.h.s. correspond to the Feynman graphs of Type A and B in fig. 1, respectively. For obtaining two-loop quantum corrections to the effective potential it is sufficient to consider the propagators (2.19) and (2.20) with $W_\alpha = \bar{W}_\alpha = 0$, $F_{mn} = 0$, $N_{\alpha\beta} = 0$ and $B = 0$, leaving only the dependence on the superfield strengths G,

$$K_+(z,z'|s) = \frac{1}{8(i\pi s)^{3/2}} e^{isG^2} e^{\frac{i}{4s}\rho^m\rho_m} \zeta^2 I(z,z'), \tag{2.34}$$

$$K_{+-}(z,z'|s) = -\frac{1}{8(i\pi s)^{3/2}} e^{isG^2} e^{\frac{i}{4s}\tilde{\rho}^m\tilde{\rho}_m - i\zeta\bar{\zeta}G} I(z,z'). \tag{2.35}$$

Note that only the first term in (2.24) contributes to (2.35) in the case of such a simple background.

The heat kernel (2.28) for the photon propagator contains the delta-function for Grassmann variables $\zeta^2\bar{\zeta}^2$. Therefore, it is sufficient to consider the heat kernels (2.34) and (2.35) at coincident Grassmann points,

$$K_+(z,z'|s)|_{\zeta=\bar{\zeta}=0} = 0, \tag{2.36}$$

$$K_{+-}(z,z'|s)|_{\zeta=\bar{\zeta}=0} = -\frac{1}{8(i\pi s)^{3/2}} e^{isG^2} e^{\frac{i}{4s}\rho^m\rho_m} I(z,z')|_{\zeta=\bar{\zeta}=0}. \tag{2.37}$$

Hence, only the diagrams of Type A in fig. 1 contribute to the effective potential. Substituting now (2.37) and (2.28) into (2.33) we find

$$\begin{aligned}\Gamma^{(2)}_{\mathcal{N}=2} &= \frac{2ie^2}{(4i\pi)^{9/2}} \int d^7z d^3\rho \int_0^\infty \frac{ds\,dt\,du}{(stu)^{3/2}} e^{i(s+t)(G^2+m^2)} e^{\frac{i}{4}(\frac{1}{s}+\frac{1}{t}+\frac{1}{u})\rho^2} \\ &= \frac{2e^2}{(4\pi)^3} \int d^7z \int_0^\infty \frac{ds\,dt\,du}{(st+su+tu)^{3/2}} e^{i(s+t)(G^2+m^2)}.\end{aligned} \tag{2.38}$$

Here we did the Gaussian integration over $d^3\rho$,

$$\int d^3\rho\, e^{\frac{i}{4}(\frac{1}{s}+\frac{1}{t}+\frac{1}{u})\rho^2} = -\left(\frac{4i\pi}{\frac{1}{s}+\frac{1}{t}+\frac{1}{u}}\right)^{3/2}. \tag{2.39}$$

Hence, after computing the integrals over the parameters s, t, u we get

$$\Gamma^{(2)}_{\mathcal{N}=2} = -\frac{e^2}{8\pi^2} \int d^7z \ln(G^2+m^2). \tag{2.40}$$

Journal of Physics: Conference Series **474** (2013) 012012 doi:10.1088/1742-6596/474/1/012012

Finally, putting together (2.32) and (2.40) we get the two-loop effective potential,

$$\begin{aligned}\Gamma_{\mathcal{N}=2} &= \int d^7z\, f(G)\,, \\ f(G) &= \frac{1}{g^2}G^2 + \frac{1}{2\pi}(G\ln(G+\sqrt{G^2+m^2}) - \sqrt{G^2+m^2}) - \frac{e^2}{8\pi^2}\ln(G^2+m^2)\,. \end{aligned} \quad (2.41)$$

Here the term $\frac{1}{e^2}G^2$ is a part of the classical action (2.1) in the gauge superfield sector. We stress that the effective potential (2.41) was known before only in the one-loop approximation [3, 4] while the two-loop correction to this result was found in our work [27].

2.4. Two-loop moduli space metric

The moduli space is an important object in gauge theories which allows one to study various issues of dualities [3, 4, 5, 6, 7, 8, 9, 10, 11] (see also [1, 2] for recent discussions of these problems). The perturbative quantum corrections to the moduli space metric in the $\mathcal{N}=2$, $d=3$ gauge theories are known only up the one-loop order [3, 4, 7]. The two-loop effective action (2.41) allows us to study two-loop moduli space in the $\mathcal{N}=2$, $d=3$ electrodynamics (2.1).

The moduli space in $\mathcal{N}=2$, $d=3$ SQED is a two-dimensional Kähler manifold. It can be parametrized by two real coordinates r and σ. The coordinate r is naturally identified with the vev of the scalar field ϕ which is a part of the $\mathcal{N}=2$, $d=3$ gauge multiplet, $r=\langle\phi\rangle$. This scalar is the lowest component of the superfield strength G,

$$G|_{\theta\to 0} = \phi\,. \quad (2.42)$$

Another scalar field a appears upon dualizing the Abelian vector A_m,

$$\partial_m a \propto \varepsilon_{mnp}F^{np}\,, \quad (2.43)$$

where F_{mn} is the Maxwell field strength corresponding to the Abelian vector field A_m. The coordinate σ corresponds to the vev of this scalar, $\sigma=\langle a\rangle$. The moduli space metric is parametrized by r and σ,

$$ds^2 = g_{rr}(r,\sigma)dr^2 + g_{\sigma\sigma}(r,\sigma)d\sigma^2\,. \quad (2.44)$$

Our aim is to find the explicit form of the functions $g_{rr}(r,\sigma)$ and $g_{\sigma\sigma}(r,\sigma)$.

The procedure of deriving the metric (2.44) from the low-energy effective action is described in [7]: Given the function $f(G)$ one dualizes the linear superfield G into a chiral superfield Φ following standard procedure of [29]. The chiral superfield serves as the Lagrange multiplier for the linearity constraint (2.3),

$$S_{\rm low-energy} = \int d^7z\,[f(G) - G(\Phi+\bar\Phi)]\,. \quad (2.45)$$

The superfield G is treated now as unconstrained. Varying (2.45) with respect to G we get

$$\Phi+\bar\Phi = f'(G) = \frac{2}{e^2}G + \frac{1}{2\pi}\ln(G+\sqrt{G^2+m^2}) - \frac{e^2}{4\pi^2}\frac{G}{G^2+m^2}\,. \quad (2.46)$$

From this equation the superfield G should be expressed in terms of $\Phi+\bar\Phi$ and substituted back to (2.45). This yields a sigma-model action,

$$S_{\rm low-energy} = \int d^7z\, K(\Phi+\bar\Phi)\,, \quad (2.47)$$

with some function $K(\Phi+\bar{\Phi})$. We do not need the manifest expression for K since the Kähler metric is defined rather by its second derivative,

$$ds^2 = K''\, d\Phi\, d\bar{\Phi}\,. \tag{2.48}$$

This metric should be expressed in terms of r and σ where $r = \langle G\rangle$ and σ can be identified with the imaginary part of Φ, $\sigma = \mathrm{Im}\Phi$. Using the fact that the inverse Legendre transform is a Legendre transform, we have

$$K'(\Phi+\bar{\Phi}) = r\,. \tag{2.49}$$

From this equation and from (2.46) we conclude

$$K''(\Phi+\bar{\Phi}) = \left(\frac{\partial(\Phi+\bar{\Phi})}{\partial r}\right)^{-1} = \frac{1}{2}\frac{1}{g(r)}\,, \tag{2.50}$$

where

$$g(r) = \frac{1}{e^2} + \frac{1}{4\pi}\frac{1}{\sqrt{r^2+m^2}} + \frac{e^2}{8\pi^2}\frac{r^2-m^2}{(r^2+m^2)^2}\,. \tag{2.51}$$

Finally, we note that (2.46) implies that

$$d\Phi = g(r)dr + id\sigma\,, \qquad d\bar{\Phi} = g(r)dr - id\sigma\,. \tag{2.52}$$

Substituting now (2.50) and the latter identities into (2.48) we find the moduli space metric in the form

$$ds^2 = \frac{1}{2}g(r)dr^2 + \frac{1}{2}\frac{1}{g(r)}d\sigma^2\,. \tag{2.53}$$

In the massless limit the function $g(r)$ in (2.51) simplifies such that

$$ds^2|_{m=0} = \frac{1}{2}\left(\frac{1}{e^2}+\frac{1}{4\pi r}+\frac{e^2}{8\pi^2 r^2}\right)dr^2 + \frac{1}{2}\left(\frac{1}{e^2}+\frac{1}{4\pi r}+\frac{e^2}{8\pi^2 r^2}\right)^{-1} d\sigma^2\,. \tag{2.54}$$

Equation (2.54) shows that the one-loop metric is corrected by the two-loop contribution $\frac{e^2}{8\pi^2 r^2}$. It is naturally to expect that the n-loop correction could be of the form $c_n\frac{1}{e^2}(\frac{e^2}{r})^n$, with some coefficient c_n. It is very tempting to compute such higher-loop coefficients c_n and to find a closed expression for all-loop moduli space metric both for the Abelian and non-Abelian $\mathcal{N}=2$, $d=3$ gauge theories.

3. $\mathcal{N}=4$ supersymmetric electrodynamics

3.1. Classical action

$\mathcal{N}=4$ supersymmetric gauge multiplet consists of the $\mathcal{N}=2$ gauge superfield V and a chiral superfield Φ. The classical action of the $\mathcal{N}=4$ supersymmetric electrodynamics reads

$$\begin{aligned} S_{\mathcal{N}=4} &= \frac{1}{e^2}\int d^7z(G^2 - \frac{1}{2}\bar{\Phi}\Phi) + S_Q\,, \\ S_Q &= -\int d^7z(\bar{\mathcal{Q}}_+\mathcal{Q}_+ + \bar{\mathcal{Q}}_-\mathcal{Q}_-) - \int d^5z\, \mathcal{Q}_+\Phi\mathcal{Q}_- + \int d^5\bar{z}\, \bar{\mathcal{Q}}_+\bar{\Phi}\bar{\mathcal{Q}}_-\,, \end{aligned} \tag{3.1}$$

We are interested in the part of low-energy effective action in this model which depends both on V and Φ. Therefore, we make the background-quantum splitting for both these superfields,

$$V \to V + e\,v\,, \qquad \Phi \to \Phi + e\,\phi\,, \tag{3.2}$$

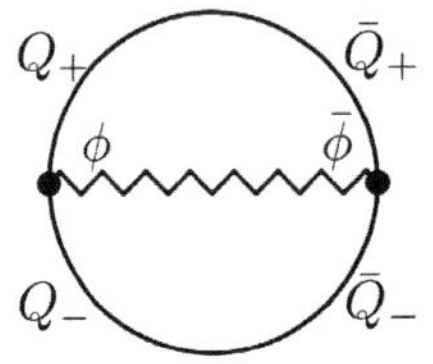

Type C

Figure 2. Two-loop supergraph in $\mathcal{N}=4$ supersymmetric electrodynamics which involves chiral superfield propagator $\langle\phi\bar{\phi}\rangle$.

while the hypermultiplet $(\mathcal{Q}_+, \mathcal{Q}_-)$ is considered as the 'quantum' superfield which should be integrated out in the path integral. We impose the following constraints on the background superfield:

$$\partial_m G = 0\,,\quad W_\alpha = \bar{W}_\alpha = 0\,,\quad D_\alpha \Phi = 0\,. \tag{3.3}$$

These constraints are sufficient for obtaining the effective potential of the form (1.5).

Upon quantization in the Fermi-Feynman gauge (2.6), we end up with the following action for 'quantum' superfields,

$$\begin{aligned} S_{\rm quantum} &= S_2 + S_{\rm int}\,, & (3.4)\\ S_2 &= -\int d^7z(v\Box v + \frac{1}{2}\bar{\phi}\phi + \bar{\mathcal{Q}}_+\mathcal{Q}_+ + \bar{\mathcal{Q}}_-\mathcal{Q}_-) - \left(\int d^5z\, \mathcal{Q}_+\Phi\mathcal{Q}_- + c.c.\right), & (3.5)\\ S_{\rm int} &= -2\int d^7z\left[e\left(\bar{\mathcal{Q}}_+\mathcal{Q}_+ - \bar{\mathcal{Q}}_-\mathcal{Q}_-\right)v + e^2\left(\bar{\mathcal{Q}}_+\mathcal{Q}_+ + \bar{\mathcal{Q}}_-\mathcal{Q}_-\right)v^2\right]\\ &\quad -e\int d^5z\, \mathcal{Q}_+\phi\mathcal{Q}_- + e\int d^5\bar{z}\, \bar{\mathcal{Q}}_+\bar{\phi}\bar{\mathcal{Q}}_+ + O(e^3)\,. & (3.6)\end{aligned}$$

The propagators for the chiral matter superfields and for the gauge superfield V are the same as in the $\mathcal{N}=2$ electrodynamics (2.11) and (2.27). However, the chiral superfield ϕ propagates now,

$$\langle\phi(z)\bar{\phi}(z')\rangle = -\frac{i}{8}\bar{D}^2D^2G_0(z,z')\,. \tag{3.7}$$

3.2. Low-energy effective action

The one-loop effective action $\Gamma^{(1)}_{\mathcal{N}=4}$ receives contributions only from the chiral matter superfields which are the same as in the $\mathcal{N}=2$ electrodynamics (2.1). The only difference is that the chiral superfield is dynamical now while in the $\mathcal{N}=2$ case it was 'frozen' to be equal to the mass parameter m. Therefore we can easily generalize the one-loop effective action (2.32) to the $\mathcal{N}=4$ case,

$$\Gamma^{(1)}_{\mathcal{N}=4} = \frac{1}{2\pi}\int d^7z\left[G\ln(G+\sqrt{G^2+\bar{\Phi}\Phi}) - \sqrt{G^2+\Phi\bar{\Phi}}\right]. \tag{3.8}$$

The two-loop effective action in the $\mathcal{N}=4$ electrodynamics differs from the $\mathcal{N}=2$ case because the chiral superfield Φ propagates now. Indeed, apart from the diagrams in fig. 1, there is an additional two-loop diagram with quantum $\langle\phi\bar{\phi}\rangle$ line, see fig. 2. This extra graph yields additional contributions to the two-loop effective action which we denote by $\Gamma_{\rm C}$,

$$\Gamma^{(2)}_{\mathcal{N}=4} = \Gamma_{\rm A} + \Gamma_{\rm B} + \Gamma_{\rm C}\,, \tag{3.9}$$

$$\Gamma_{\rm A} = -2e^2 \int d^7z\, d^7z'\, G_{+-}(z,z')G_{+-}(z',z)G_0(z,z')\,, \tag{3.10}$$

$$\Gamma_{\rm B} = -2e^2 \int d^7z\, d^7z'\, \bar\Phi\Phi\, G_{+}(z,z')G_{-}(z,z')G_0(z,z')\,, \tag{3.11}$$

$$\Gamma_{\rm C} = 2e^2 \int d^7z\, d^7z'\, G_{+-}(z,z')G_{+-}(z,z')G_0(z,z')\,. \tag{3.12}$$

We are interested in the part of the low-energy effective action depending on constant background superfields G and Φ which we refer to as the effective potential. It is easy to see that the part of the low-energy effective action (3.11) does not contribute to the effective potential because of (2.36). Hence, it is necessary to consider only

$$\Gamma_{\rm A} + \Gamma_{\rm B} = 2e^2 \int d^7zd^7z'[G_{+-}(z,z')G_{+-}(z,z') - G_{+-}(z,z')G_{+-}(z',z)]G_0(z,z')\,. \tag{3.13}$$

The heat kernel $K_{+-}(z,z'|s)$ given by (2.37) is symmetric with respect to the superspace points. Hence, the two terms in the r.h.s. of (3.13) exactly cancel each other. Thus, we conclude that there are no two-loop quantum corrections to the effective potential (3.8). This result is a direct check of the general statement that the moduli space metric in the $\mathcal{N}=4$, $d=3$ gauge theories is one-loop exact [3, 7]. This metric is known as the Taub-NUT metric which was originally derived in [29] from geometrical principles.

Acknowledgments

The authors are grateful to the RFBR grant Nr. 12-02-00121 and LRSS grant Nr. 224.2012.2. The study was partially supported by the Ministry of education and science of Russian Federation, project 14.B.21.0774. The work of I.L.B. and I.B.S. was also supported by the RFBR grants Nrs. 13-02-90430 and 13-02-91330 and DFG grant LE 838/12/1. The work of I.B.S. was supported by Marie Curie research fellowship Nr. 909231, "QuantumSupersymmetry" and by the Padova University Project CPDA119349.

References

[1] K. Intriligator and N. Seiberg, *Aspects of 3d N=2 Chern-Simons-matter theories*, `arXiv:1305.1633 [hep-th]`.
[2] O. Aharony, S. S. Razamat, N. Seiberg and B. Willett, *3d dualities from 4d dualities*, `arXiv:1305.3924 [hep-th]`.
[3] K. A. Intriligator and N. Seiberg, *Mirror symmetry in three-dimensional gauge theories*, Phys. Lett. B **387** (1996) 513, `hep-th/9607207`.
[4] J. de Boer, K. Hori, H. Ooguri and Y. Oz, *Mirror symmetry in three-dimensional gauge theories, quivers and D-branes*, Nucl. Phys. B **493** (1997) 101, `hep-th/9611063`.
[5] J. de Boer, K. Hori, H. Ooguri, Y. Oz and Z. Yin, *Mirror symmetry in three-dimensional theories, SL(2,Z) and D-brane moduli spaces*, Nucl. Phys. B **493** (1997) 148, `hep-th/9612131`.
[6] J. de Boer, K. Hori, Y. Oz and Z. Yin, *Branes and mirror symmetry in N=2 supersymmetric gauge theories in three-dimensions*, Nucl. Phys. B **502** (1997) 107, `hep-th/9702154`.
[7] J. de Boer, K. Hori and Y. Oz, *Dynamics of N=2 supersymmetric gauge theories in three-dimensions*, Nucl. Phys. B **500** (1997) 163, `hep-th/9703100`.
[8] O. Aharony, A. Hanany, K. A. Intriligator, N. Seiberg and M. J. Strassler, *Aspects of N=2 supersymmetric gauge theories in three-dimensions*, Nucl. Phys. B **499** (1997) 67, `hep-th/9703110`.
[9] O. Aharony, *IR duality in d=3 N=2 supersymmetric USp(2N(c)) and U(N(c)) gauge theories*, Phys. Lett. B **404** (1997) 71, `hep-th/9703215`.
[10] A. Karch, *Seiberg duality in three-dimensions*, Phys. Lett. B **405** (1997) 79, `hep-th/9703172`.

[11] A. Giveon and D. Kutasov, *Seiberg duality in Chern-Simons theory*, Nucl. Phys. B **812** (2009) 1, arXiv:0808.0360 [hep-th].
[12] J. Bagger, N. Lambert, *Modeling multiple M2's*, Phys. Rev. D75 (2007) 045020, hep-th/0611108.
[13] J. Bagger, N. Lambert, *Gauge symmetry and supersymmetry of multiple M2-branes*, Phys. Rev. D77 (2008) 065008, arXiv:0711.0955.
[14] J. Bagger, N. Lambert, *Comments on multiple M2-branes*, JHEP 02 (2008) 105, arXiv:0712.3738.
[15] J. Bagger, N. Lambert, *Three-algebras and N=6 Chern-Simons gauge theories*, Phys. Rev. D79 (2009) 025002, arXiv:0807.0163.
[16] A. Gustavsson, *Algebraic structures on parallel M2-branes*, Nucl. Phys. B811 (2009) 66, arXiv:0709.1260.
[17] A. Gustavsson, *Selfdual strings and loop space Nahm equations*, JHEP 04 (2008) 083, arXiv:0802.3456.
[18] A. Gustavsson, *One-loop corrections to Bagger-Lambert theory*, Nucl. Phys. B807 (2009) 315, arXiv:0805.4443.
[19] O. Aharony, O. Bergman, D.L. Jafferis, J. Maldacena, *N=6 superconformal Chern-Simons-matter theories, M2-branes and their gravity duals*, JHEP 0810 (2008) 091, arXiv:0806.1218.
[20] M. Benna, I. Klebanov, T. Klose, M. Smedback, *Superconformal Chern-Simons theories and AdS_4/CFT_3 correspondence*, JHEP 0809 (2008) 072, arXiv:0806.1519.
[21] T. Ohrndorf, *The effective lagrangian of supersymmetric Yang-Mills theory*, Phys. Lett. B **176** (1986) 421.
[22] K. -i. Shizuya and Y. Yasui, *Construction of effective actions in superspace*, Phys. Rev. D **29** (1984) 1160.
[23] I. N. McArthur and T. D. Gargett, *A 'Gaussian' approach to computing supersymmetric effective actions*, Nucl. Phys. B **497** (1997) 525, hep-th/9705200.
[24] N. G. Pletnev and A. T. Banin, *Covariant technique of derivative expansion of one loop effective action*, Phys. Rev. D **60** (1999) 105017, hep-th/9811031.
[25] S. M. Kuzenko and I. N. McArthur, *Low-energy dynamics in N=2 super QED: Two loop approximation*, JHEP **0310** (2003) 029, hep-th/0308136.
[26] S. M. Kuzenko and S. J. Tyler, *Supersymmetric Euler-Heisenberg effective action: Two-loop results*, JHEP **0705** (2007) 081, hep-th/0703269.
[27] I. L. Buchbinder, B. S. Merzlikin and I. B. Samsonov, *Two-loop low-energy effective actions in N=2 and N=4 three-dimensional SQED*, JHEP **1307** (2013) 012, arXiv:1305.4815 [hep-th].
[28] I. L. Buchbinder, N. G. Pletnev and I. B. Samsonov, *Effective action of three-dimensional extended supersymmetric matter on gauge superfield background*, JHEP **1004** (2010) 124, arXiv:1003.4806 [hep-th].
[29] N. J. Hitchin, A. Karlhede, U. Lindström and M. Roček, *Hyperkahler metrics and supersymmetry*, Commun. Math. Phys. **108** (1987) 535.
[30] I. L. Buchbinder and S. M. Kuzenko, *Ideas and methods of supersymmetry and supergravity: Or a walk through superspace*, Bristol, UK: IOP (1998) 656 p.
[31] S. J. Gates, M. T. Grisaru, M. Roček and W. Siegel, *Superspace or One Thousand and One Lessons in Supersymmetry*, Benjamin/Cummings, Reading, MA, 1983, 548 p.
[32] I. L. Buchbinder, N. G. Pletnev and I. B. Samsonov, *Background field formalism and construction of effective action for $\mathcal{N} = 2, d = 3$ supersymmetric gauge theories*, Phys. Part. Nucl. **44** (2013) 234, arXiv:1206.5711 [hep-th].
[33] S. M. Kuzenko and I. N. McArthur, *On the background field method beyond one loop: A manifestly covariant derivative expansion in super Yang-Mills theories*, JHEP **0305** (2003) 015, hep-th/0302205.

Journal of Physics: Conference Series **474** (2013) 012013 doi:10.1088/1742-6596/474/1/012013

Root Structures of Infinite Gauge Groups and Supersymmetric Field Theories

Sultan Catto[1,2,†], Yasemin Gürcan[3], Amish Khalfan[4] and Levent Kurt[5]

[1]Physics Department, The Graduate School, City University of New York, New York, NY 10016-4309
[2]Theoretical Physics Group, Rockefeller University, 1230 York Avenue, New York, NY 10021-6399
[3] Chairperson, Physics Department, Brownson 129, Manhattanville College, Purchase, NY 10577
[4] Physics Department, Yeshiva University, New York, NY 10033
5 Department of Science, Borough of Manhattan Community College,
The City University of New York, 199 Chambers Street, New York, NY 10007

† Work supported in part by the PSC-CUNY Research Awards, and DOE contracts No. DE-AC-0276 ER 03074 and 03075; and NSF Grant No. DMS-8917754.

Abstract.

We show the relationship between critical dimensions of supersymmetric fundamental theories and dimensions of certain Jordan algebras. In our approach position vectors in spacetime or in superspace are endowed with algebraic properties that are present only in those critical dimensions. A uniform construction of super Poincaré groups in these dimensions will be shown. Some applications of these algebraic methods to hidden symmetries present in the covariant and interacting string Lagrangians and to superparticle will be discussed. Algebraic methods we develop will be shown to generate the root structure of some infinite groups that play the role of gauge groups in a second quantized theory of strings.

1. Introduction

In the following we will put together a few remarks pertaining to some mathematical aspects of physical theories. They will revolve around a theme relating space-time structure to internal symmetries. Chiral symmetry, analyticity properties related to causality and conformal invariance, supersymmetry between elementary excitations belong to those set of ideas. Properties like analyticity and color are generalized in terms of division algebras. In turn these lead to mathematical structures that unify Minkowski space and internal charge space only within certain geometries with critical dimensions that occur in supergravity theories as well as in string and superstring models.

As an introduction to this kind of analysis we shall exploit the relation of generalized integers to some Lie groups. This will be followed by the representations of the Lorentz and Poincarè groups by means of division algebras in higher dimensions. Then we will take a closer look at their extensions to conformal groups and super Poincarè groups. Jordan algebras will be shown to arise naturally. Finally we will give examples of applications to particle dynamics and two-dimensional field theories that can be related to string theories.

The interest in exceptional structures arose in physics from a simple observation that the color group $SU(3)$ which is exact, and most probably not accidental, is connected with the octonion algebra [1].

Indeed if one tries to form three anti commuting Grassmann numbers by using three pairs of octonionic imaginary units, the algebra of these split units admits $SU(3)$ as its automorphism group. One is therefore led to construct quark states with such units and reformulate QCD in terms of these octonion valued fields [2].

A phenomenological manifestation of octonionic structure (based on Cayley numbers) has been found in connection with quark dynamics inside hadrons leading to an effective dynamical supersymmetry that was shown in our earlier papers [3],[4] (for further reviews and applications, see also [5], [6] and [7]). It is based on $SU(m/n)$ supergroups yielding combined classification of mesons and baryons. Also, further applications of split octonions to supersymmetric potential models and skyrmions have recently been discussed in literature [8],[9],[10]. Octonionic structures also appear in constructions of extended Hilbert spaces and associated projective geometries [11], [12], [13]. For a brief history of dynamical supersymmetries we refer to an article by Francesco Iachello [14], and for an extensive paper on octonions and their applications to a paper by John Baez [15].

The fitting of color into larger groups involving also flavor degrees of freedom in grand unified theories can then be done by embedding octonions into larger algebras that would have grand unified theory groups as their automorphism groups. For instance the automorphism group of the algebra of 2×2 octonionic hermitian matrices is $SO(10)$ while E_6 leaves invariant the algebra of 3×3 hermitian octonionic matrices, the so called exceptional Jordan algebra.

Cancellation of anomalies in 10 dimensional supergravity or superstring theories [16] also requires the restriction of internal symmetries of the Yang-Mills matter to very special group like $E_8 \times E_8$ associated with a remarkable 16-dimensional lattice. All those occurrences suggest that unified theory of gravity and gauge interactions might best be formulated by means of an extension of such algebras that would also incorporate supersymmetry from the start.

Such theories can only be formulated in very special dimensions like $d = 3, 4, 6$ and 10 for supersymmetric theories [17] and $d = 5, 8, 14$ and 26 for their bosonic counterparts. It now seems that $d = 10$ and $d = 26$ associated with the octonionic structures are the only ones that can be quantified consistently although other possibilities might exist [18]. There is another kind of invariance that involves auxiliary variables. In relativistic theories, there are unphysical degrees of freedom, like the $s = 0$ component of a 4-vector field that is necessary for a covariant description of $s = 1$ particle. They contribute to the energy with the wrong sign and cause the occurrence of negative probabilities. They are either eliminated by subsidiary conditions that will in general destroy the covariance or the gauge invariance of the theory or both. Or, alternatively they are kept and further unphysical (ghost) degrees of freedom are added to cancel their contributions. The ghost enlarged action may now behave like the BRST symmetry [19],[20] that transforms ghosts, unphysical particles and physical particles into each other. That generalized action may now exhibit an exact supersymmetry which has nothing to do with the physical spectrum or the physical invariance of the Lagrangian. In this case the symmetry is used for the covariant and gauge invariant quantization of the theory. It is useful and elegant rather than fundamental.

Ingredients of the theory, exact conformal invariance in the 2-dimensional submanifold, presence of new space-time symmetries that mix states with different spin, exact duality symmetry between exchange and resonance channels, a non-locality of the basic objects (open or closed charged strings) that interact locally, were all suggested as approximate or phenomenological feature of a theory based on the QCD Lagrangian. It turns out that such theories are linked with unique exceptional mathematical structures that enjoy unique algebraic properties, like surprising relations to the octonion algebra.

In short, inspired by the intricate symmetries appearing in the rich world of hadronic physics, we were led to invent a theory that has all these new symmetries as basic ingredients, and hope that such a maximally symmetric theory would be capable of unification of all fundamental forces, including gravity. The constraints on such a theory are so great that if some very restricted solutions exist they have a good chance of being related to some exceptional structures in mathematics that unify various aspects of algebra, geometry, analysis and number theory.

2. Relation of some remarkable groups to division algebras

Hurwitz' Theorem [21] assures us that there are four division algebras with quadratic norm, namely $\mathcal{R}$ (reals), $\mathcal{C}$ (complex numbers), $\mathcal{H}$ (quaternions) and $\mathcal{O}$ (octonions).

The first two are commutative and associative. The third is associative but not commutative while the last is neither commutative nor associative. Each algebra has a subring of integers. In the case of complex numbers we have the Gaussian integers of the form $m + in$ with m and n real integers. The theory of quaternionic and octonionic integers with norm and twice the scalar part being integers was developed by Hurwitz, Wedderburn, Dickson and Coxeter.

We can first consider Lie groups that leave the norm and the trace form (twice the scalar part) of the elements of the 4 division algebras invariant. Then we look at their automorphism group and finally we associate generalized integers of unit norm with the root system of Lie groups. Results are in the following table:

	$\mathcal{R}$	$\mathcal{C}$	$\mathcal{H}$	$\mathcal{O}$
Norm Group	-	$O(2) \sim U(1)$	$O(4) \sim SU(2) \times SU(2)$	$O(8)$
Trace Group	-	-	$O(3) \sim SU(2)$	$O(7)$
Automorphism Group	-	-	$SU(2)$	G_2
Root System	$O(3)$	$O(4)$	$O(8)$	E_8

Table I - Lie Groups associated with division algebras

The first three lines are well known. The last one requires some explanation. Consider a semisimple Lie group G and its Cartan subalgebra $H = (H_1, \ldots, H_r)$ where r is the rank of G. In the Cartan basis we consider the non hermitian generators E_k and their conjugates E_{-k}. E_k is associated with the root vector r^k_m such that

$$[H_m, E_{\pm k}] = \pm r^{(k)}_m E_{\pm k}. \tag{1}$$

When the rank r is $1, 2, 4$ or 8 we can combine the operators H_m and its "eigenvalues" $r^{(k)}_m$ into elements of a division algebra with imaginary units e_i:

$$H = H_0 + e_i H_i, \qquad r^{(k)} = r^k_0 + e_i r^{(k)}_i \tag{2}$$

so that

$$[H, E_{\pm k}] = \pm r^{(k)} E_{\pm k}. \tag{3}$$

In the case of $G = SU(2),\ O(4),\ O(8)$ and E_8 all roots have the same length that can be normalized to unity. It turns out that in all four cases the roots are integer elements of the Hurwitz algebras.

In the case of $SU(2)$ there are two E generators

$$E_{\pm 1} = J_1 \pm iJ_2 \tag{4}$$

and one Cartan subalgebra operator $H = J_3$ with eigenvalues ± 1 which are real integers with unit norm.

$O(4)$ has roots $(\pm 1, 0)$ and $(0, \pm 1)$ that combine into the four Gaussian integers ± 1 and $\pm i$ that are the rational points on the circle S^1.

Similarly the 24 roots of $O(8)$ correspond to the integer quaternions $\pm e_1$, $\pm e_2$, $\pm e_3$, (8 such roots) and the remaining ones of the form

$$\frac{1}{2}(\pm 1 \pm e_1 \pm e_2 \pm e_3) \tag{5}$$

(16 such roots). They are the rational points on the unit sphere S^4. The principal positive roots can be chosen to be

$$e_1 = e^{\frac{\pi}{2} e_1}, \qquad e_2 = e^{\frac{\pi}{2} e_2}, \qquad e_3 = e^{\frac{\pi}{2} e_3} \tag{6}$$

and

$$\ell = exp(\frac{\pi}{3} \frac{e_1 + e_2 + e_3}{\sqrt{3}}) = \frac{1}{2}(1 + e_1 + e_2 + e_3) \tag{7}$$

corresponding to the four points of the Dynkin diagram

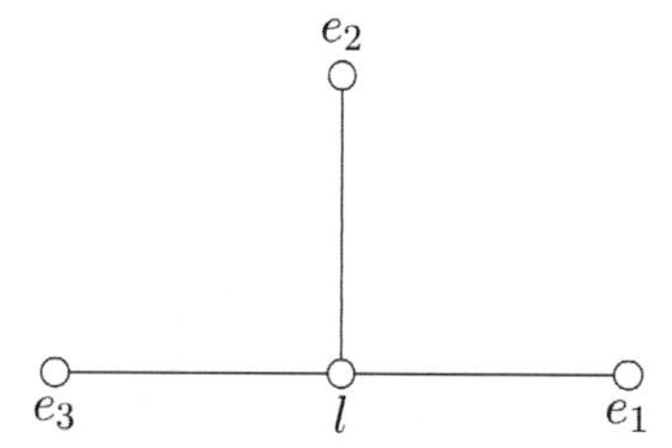

Fig. 1. Dynkin diagram of O(8)

The triality of the $O(8)$ group reflects the symmetry of the three quaternionic units $e_i = -i\,\sigma_i$, where σ_i are the Pauli matrices.

The Cartan subalgebra is more conveniently represented by the quaternionic operator

$$H = H_0' \frac{1 + e_1 + e_2 + e_3}{2} + e_1 H_1' + e_2 H_2' + e_3 H_3'. \tag{8}$$

All the roots can be obtained from the principal positive roots by Weyl reflections. If α and β are roots represented by quaternions $\beta' = -\alpha\bar{\beta}\bar{\alpha}$ is the quaternion form of β reflected with respect to α. Obviously β' has unit norm and its trace form is ± 1 or 0 like that of β.

The case of E_8 was worked out by Coxeter [22] in connection with 8-dimensional regular solids. There are 240 rational points on the unit sphere S^7 represented by integer octonions that correspond to the 240 roots of E_8. We first introduce octonionic imaginary units e_α ($\alpha = 1, \ldots, 7$) with multiplication rule

$$e_\alpha e_\beta = -\delta_{\alpha\beta} + \psi_{\alpha\beta\gamma} e_\gamma \tag{9}$$

where the octonion structure constants form an antisymmetrical tensor of third rank with

$$\psi_{123} = \psi_{246} = \psi_{435} = \psi_{367} = \psi_{651} = \psi_{572} = \psi_{714} = 1 \tag{10}$$

and other elements zero. Now define the octonions

Journal of Physics: Conference Series **474** (2013) 012013 doi:10.1088/1742-6596/474/1/012013

$$\begin{aligned} l_1 &= \tfrac{1}{2}(e_1 - e_4), \quad l_2 = \tfrac{1}{2}(e_2 - e_5), \quad l_3 = \tfrac{1}{2}(e_3 - e_6) \\ l_4 &= \tfrac{1}{2}(e_1 + e_4), \quad l_5 = \tfrac{1}{2}(e_2 + e_5), \quad l_6 = \tfrac{1}{2}(e_3 + e_6) \\ l_7 &= \tfrac{1}{2}(1 + e_7), \quad l_8 = \overline{l_7} = \tfrac{1}{2}(1 - e_7). \end{aligned} \tag{11}$$

We have

$$\begin{aligned} l_1 &= e_1 l_7 = l_8 e_1, \quad l_2 = e_2 l_7 = l_8 e_2, \\ l_3 &= e_3 l_7 = l_8 e_3, \quad l_4 = e_1 l_8 = l_7 e_1, \\ l_5 &= e_2 l_8 = l_7 e_2, \quad l_6 = e_3 l_8 = l_7 e_3. \end{aligned} \tag{12}$$

Now consider

$$\rho_{rs} = \pm l_r \pm l_s \qquad (r \neq s) \tag{13}$$

and

$$\sigma = \frac{1}{2}(\pm l_1 \pm l_2 \pm l_3 \pm l_4 \pm l_5 \pm l_6 \pm l_7 \pm l_8) \tag{14}$$

with an odd number of minus signs in σ.

A special choice of σ gives

$$\begin{aligned} h &= \frac{1}{2}(l_1 + l_2 + l_3 + l_4 + l_5 + l_6 + l_7 - l_8) \\ &= \frac{1}{2}(e_1 + e_2 + e_3 + e_7). \end{aligned} \tag{15}$$

Special ρ's are $1, e_7, e_2, e_6$, which, together with h, $e_1 h$, $e_2 h$ and $e_7 h$ correspond to one possible set of principal positive roots. These are shown in the figure below:

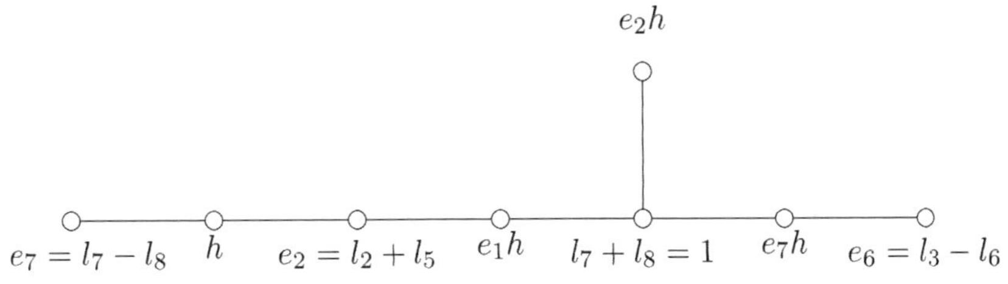

Fig. 2. The E_8 Dynkin Diagram

It is shown easily that the 112 octonions ρ_{rs} and the 128 octonions σ are integer octonions with norm 1. It is possible to choose the Cartan subalgebra H_a $(a = 1, \ldots, 8)$ such that these 240 octonions are the roots of E_8 expressed as eigenvalues of the operator

$$H = H_0 + e_\alpha H_\alpha \quad (\alpha = 1, \ldots, 7) \tag{16}$$

for the adjoint representation. The 112 roots ρ_{rs} are those of the subgroup $Spin(16)$ while σ are the weights associated with the 120 dimensional spinor representation of $Spin(16)$.

Now consider integer octonions of length $2, 3$ and 4. Their corresponding numbers 2160, 6720 and 17280 have been calculated by Coxeter. They correspond exactly to the number or weights with maximum length 4, 6 and 8 in the fundamental representations (00000010), (01000000) and (00000001) with respective dimensions 3875, 30380 and 147250, provided that the length of the root is normalized to 2 in the adjoint representation. For example in the 3875 dimensional representation the multiplicity of the weight length 2 is 7 and that of weight length zero associated with generators in the Cartan subalgebra is 35. These numbers can be found in the table by Brenner, Moody and Patera [23]. Hence the weight of that representation consists of 2160 integer octonions of norm 2, $240 \times 7 = 1680$ integer octonions of norm 1 so that we have

$$3875 = 35 \times 1 + 240 \times 7 + 2160. \tag{17}$$

Thus the representations of E_8 are directly related to the number theory of octonions.

If we consider a octonionic spinor with two octonionic components, then integer octonionic spinors are associated with representations of $E_8 \times E_8$. We shall come back to this point later.

It is also possible to select the roots corresponding to the subgroup E_6 of E_8 which is of some importance in the compactification of superstrings. To this end we introduce a new root l_0 associated with the extended Dynkin diagram

$$\begin{aligned} l_0 &= \frac{1}{2}(l_1 + l_2 + l_3 + l_4 + l_5 + l_6 + l_7 + l_8) \\ &= \frac{1}{2}(1 + e_1 + e_2 + e_3). \end{aligned} \tag{18}$$

Now consider the $2 \times 36 = 72$ octonionic integers of the form

$$\tau_{ab} = -\tau_{ba} = \pm(l_a - l_b)\,, \qquad (a, b = 0, 1, \ldots, 8) \tag{19}$$

These are the subset of the E_8 roots that form the root system of E_6. The remaining 168 roots of E_8 can be put in the form of a symmetrical tensor

$$\zeta_{abc} = \pm(l_a + l_b + l_c - \frac{1}{3}\sum_{r=0}^{8} l_r) \tag{20}$$

with a, b, c all different.

3. Lorentz Vectors as 2×2 Hermitian matrices

Now we consider 2×2 hermitian matrices of the form

$$V = \begin{pmatrix} \alpha & \bar{v} \\ v & \beta \end{pmatrix} = V^\dagger \tag{21}$$

and

$$\bar{V} = \begin{pmatrix} \beta & -\bar{v} \\ -v & \alpha \end{pmatrix} \tag{22}$$

with α, β real

$$\alpha = v_0 + v_{d-1}, \qquad \beta = v_0 - v_{d-1} \tag{23}$$

and v belonging to a Hurwitz division algebra such that

$$N^2(V) = V\bar{V} = DetV = v_0^2 - v_1^2 - \cdots - v_{d-1}^2. \tag{24}$$

Then V represents a vector in $d = 3, 4, 6$ and 10 dimensions which are exactly the dimensions in which classical superstrings can exist [24]. The linear transformations that leave $N(V)$ invariant are the Lorentz transformations in the same dimensions and for $\mathcal{R}$, $\mathcal{C}$ and $\mathcal{H}$ they are isomorphic to [25],[26]

$$\begin{array}{lcl} d=3 & : & O(2,1) \sim SL(2,\mathcal{R}) \\ d=4 & : & O(3,1) \sim SL(2,\mathcal{C}) \\ d=6 & : & O(5,1) \sim SL(2,\mathcal{H}) \quad \text{or} \quad SL(2,\mathcal{Q}) \end{array} \tag{25}$$

These Lorentz transformations can be represented by

$$T_L\, V = V' = L\, V\, L^\dagger = V'^\dagger \tag{26}$$

where

$$L = \begin{pmatrix} l_{11} & l_{12} \\ l_{21} & l_{22} \end{pmatrix} \tag{27}$$

$$\text{Det}L = 1, \qquad l_{\alpha\beta} \in \mathcal{R}, \mathcal{C}, \mathcal{H} \tag{28}$$

In the octonionic case the form (26) is not the most general linear transformation owing to the non associativity of octonions. It must be complemented by their automorphism group G_2 with 14 parameters which requires associators in its expression. In this case of $d = 10$, L has $4 \times 8 - 1 = 31$ parameters that add up to the 45 parameters of $O(9,1)$ with the inclusion of G_2.

This point is not clear in the literature. With this modification we can represent $Spin(9,1)$ by $SL(2,\mathcal{Q})$.

In all four cases the lowest spinor representations of the Lorentz groups are 2×1 matrices

$$\psi = \begin{pmatrix} \psi_1 \\ \psi_2 \end{pmatrix} \tag{29}$$

where ψ_α $(\alpha = 1, 2)$ belong to the 4 division algebras and transforms as

$$T_L\psi = \psi' = L\psi. \tag{30}$$

In the octonionic case this must be supplemented by the G_2 transformations of the imaginary units. It follows that spinors have 2, 4, 8 and 16 real components respectively in $d = 3, 4, 6$ and 10. Such spinors are real in $d = 3$, Weyl or Majorana in $d = 4$, Weyl in $d = 6$ and both Majorana and Weyl in $d = 10$.

We also note that vectors in these dimensions are elements of 2×2 Jordan algebras with the commutative product defined by

$$W = U \cdot V = \frac{1}{2}\{UV\} = \frac{1}{2}(UV + VU) \tag{31}$$

However, this product is only invariant under the rotation subgroup of the Lorentz group. The parity conjugate vector

$$\bar{V} = V^{-1}\, DetV \tag{32}$$

obeys the transformation law

Journal of Physics: Conference Series **474** (2013) 012013 doi:10.1088/1742-6596/474/1/012013

$$\bar{V}' = L^{\dagger -1}\, \bar{V}\, L^{-1} \tag{33}$$

In the associative cases we can construct 2×2 matrices with the transformation law

$$F = V\, \bar{U} \to L\, F\, L^{-1} \tag{34}$$

under Lorentz transformations and the triple product

$$X = \frac{1}{2}(V\, \bar{U}\, W + W\, \bar{U}\, V) \to L\, X\, L^{\dagger} \tag{35}$$

is again a vector. It can be rewritten by means of Jordan products as

$$X = (V \cdot \bar{U}) \cdot W + V \cdot (\bar{U} \cdot W) - (V \cdot W) \cdot \bar{U} \tag{36}$$

This is nothing but the isotopic Jordan product of V and W with respect to $\bar{U}$ introduced by Jacobson [27]. The equation (36) is also valid in the octonionic case for $O(9, 1)$ vectors. It becomes a quadratic Jordan algebra in the case $V = W$.

Now we can ask a question about the nature of the discrete subgroups of the Lorentz groups in these critical dimensions. The case of $d = 3$ is well known. Real integer 2×2 matrices with unit determinant give the modular group Γ and automorphic forms (or theta functions) are representations of the cosets $SL(2, R)/\Gamma$. In the case of $SL(2, C)$ the discrete subgroup has 2×2 matrices of Gaussian integers as elements forming the Klein group. For the quaternonic case $d = 6$, the discrete group involves integer quaternions. Those in turn are associated with towers of $O(8)$ representations which could be related to the infinite roots of an affine Lie algebra associated with $O(8)$. Similarly we may speculate that unimodular 2×2 integer octonionic hermitian matrices might be related to the affine group E_{10}.

4. Representation of Poincarè groups by triangular 4×4 matrices

Let X be a position vector represented by a hermitian 2×2 matrix. The Poincarè group is represented by the 4×4 matrix

$$\begin{aligned} &\begin{pmatrix} I & X' \\ 0 & I \end{pmatrix} = \\ &= \begin{pmatrix} L & \frac{1}{2}AL^{\dagger -1} \\ 0 & L^{\dagger -1} \end{pmatrix} \begin{pmatrix} I & X \\ 0 & I \end{pmatrix} \begin{pmatrix} L^{-1} & \frac{1}{2}L^{-1}A \\ 0 & L^{\dagger} \end{pmatrix} \\ &= \begin{pmatrix} I & LXL^{\dagger} + A \\ 0 & I \end{pmatrix} \end{aligned} \tag{37}$$

where X and A are hermitian and I is the 2×2 unit matrix. Then $dX$and the momentum P transform like vectors.

By a Lorentz transformation P can be diagonalized to take the standard form

$$\Pi = \begin{pmatrix} \alpha & 0 \\ 0 & \epsilon\alpha \end{pmatrix}, \tag{38}$$

where $\epsilon = 1, 0, -1$ respectively for a time-like, light-like and space-like momentum vector. The subgroup of the Lorentz group that leaves Π invariant is the little group given by the following table:

	$SL(2,\mathcal{R})$	$SL(2,\mathcal{C})$	$SL(2,\mathcal{Q})$	$SL(2,\mathcal{O})$
massive:$\epsilon=1$	$U(1)$	$SU(2)$	$Spin(5)=Sp(2,\mathcal{Q})$	$Spin(9)$
light-like: $\epsilon=0$	T_1	$E(2)=U(1)\odot T_2$	$E(4)=Spin(4)\odot T_4$	$E(8)=Spin(8)\odot T_8$
tachyon: $\epsilon=-1$	$SO(1,1)$	$SO(2,1)$	$SO(4,1)$	$SO(8,1)$
	$d=3$	$d=4$	$d=6$	$d=10$

Table 2.

where $E(d-2)$ is the euclidean group of motion in $d-2$ dimensions, T_{d-2} is the translation group and $\odot$ denotes semi-direct product. The $\epsilon=1$ groups are the spin groups for massive states. The compact parts of the little groups for $\epsilon=0$ are the helicity groups for light-like states and the $\epsilon=-1$ groups are little groups for tachyons.

The position vector X can be expressed in terms of the 2×2 components Ψ and Σ of a "twistor" in the following way (Ψ is non singular)

$$\begin{pmatrix} I & X \\ 0 & I \end{pmatrix} = \begin{pmatrix} \Psi & \Sigma \\ 0 & \Psi^{\dagger -1} \end{pmatrix} \begin{pmatrix} I & I \\ 0 & I \end{pmatrix} \begin{pmatrix} \Psi^{-1} & \Sigma^{\dagger} \\ 0 & \Psi^{\dagger} \end{pmatrix} \tag{39}$$

where the triangular matrix made of Ψ and Σ has the Poincarè transformation law

$$\begin{pmatrix} \Psi' & \Sigma' \\ 0 & \Psi'^{\dagger -1} \end{pmatrix} = \begin{pmatrix} L & \frac{1}{2}AL^{\dagger -1} \\ 0 & L^{\dagger -1} \end{pmatrix} \begin{pmatrix} \Psi & \Sigma \\ 0 & \Psi^{\dagger -1} \end{pmatrix} \tag{40}$$

so that

$$\Psi' = L\,\Psi, \qquad \Sigma' = L\,\Sigma + \frac{1}{2}A\,L^{\dagger -1}\,\Psi^{\dagger -1} \tag{41}$$

We have

$$X = \Psi\Psi^{\dagger} + \Psi\Sigma^{\dagger} + \Sigma\Psi^{\dagger} \tag{42}$$

giving

$$X' = \Psi'\Psi'^{\dagger} + \Psi'\Sigma'^{\dagger} + \Sigma'\Psi'^{\dagger} = L\,X\,L^{\dagger} + A \tag{43}$$

If we put

$$S = \begin{pmatrix} \Psi & \Sigma \\ 0 & \Psi^{\dagger -1} \end{pmatrix}, \qquad \Gamma = \begin{pmatrix} 0 & I \\ I & 0 \end{pmatrix} \tag{44}$$

$$\bar{S} = \Gamma S^{\dagger}\Gamma = \begin{pmatrix} \Psi^{-1} & \Sigma^{\dagger} \\ 0 & \Psi^{\dagger} \end{pmatrix} \tag{45}$$

and

$$\Xi = \bar{\Xi} = \begin{pmatrix} I & X \\ 0 & I \end{pmatrix} \tag{46}$$

$$H = \bar{H} = \begin{pmatrix} I & I \\ 0 & I \end{pmatrix}, \qquad H^n = \begin{pmatrix} I & nI \\ 0 & I \end{pmatrix} \tag{47}$$

we can write

$$\Xi = SH\bar{S} = Z\,\bar{Z} \quad \text{with} \quad Z = SH^{\frac{1}{2}}. \tag{48}$$

Journal of Physics: Conference Series **474** (2013) 012013 doi:10.1088/1742-6596/474/1/012013

5. The Super Poincarè Group and its imbedding

We now replace the position X by the superspace coordinates $Z(X, \theta, \theta^\dagger)$ where θ is a spinor. Because in $d = 3, 4, 6, 10,$ X can be regarded as a 2×2 hermitian matrix and θ as a 2×1 matrix over the Hurwitz algebras, the super-position vector can be represented by a 3×3 hermitian matrix

$$J = \begin{pmatrix} X & \theta \\ \theta^\dagger & s \end{pmatrix} = J^\dagger \tag{49}$$

with the addition of a scalar s. The superspace variable is now upgraded to the 6×6 matrix

$$Z = \begin{pmatrix} I & J \\ 0 & I \end{pmatrix} \qquad (I = 3 \times 3 \text{ unit matrix}) \tag{50}$$

where J is an element of a graded Jordan algebra. In a super-Lorentz transformation we have

$$\begin{aligned} T_\Lambda Z &= Z' = \begin{pmatrix} \Lambda & 0 \\ 0 & \Lambda^{\dagger -1} \end{pmatrix} \begin{pmatrix} I & J \\ 0 & I \end{pmatrix} \begin{pmatrix} \Lambda^{-1} & 0 \\ 0 & \Lambda^\dagger \end{pmatrix} \\ &= \begin{pmatrix} I & \Lambda J \Lambda^\dagger \\ 0 & I \end{pmatrix} \end{aligned} \tag{51}$$

and in a generalized translation

$$\begin{aligned} T_B Z &= \begin{pmatrix} I & \frac{1}{2}B \\ 0 & I \end{pmatrix} \begin{pmatrix} I & J \\ 0 & I \end{pmatrix} \begin{pmatrix} I & \frac{1}{2}B \\ 0 & I \end{pmatrix} \\ &= \begin{pmatrix} I & J + B \\ 0 & I \end{pmatrix} \end{aligned} \tag{52}$$

Det$Z$ is invariant under this transformation. Since dJ is invariant under translations in superspace, Det(dJ) is super Poincarè invariant.

$$\begin{array}{ccccc} & d = 3 & d = 4 & d = 6 & d = 10 \\ (dimJ) - 1: & 5 & 8 & 14 & 26 \\ & (3 \times 1 + 2) & (3 \times 2 + 2) & (3 \times 4 + 2) & (3 \times 8 + 2) \end{array} \tag{53}$$

Superspace vectors will transform like dJ. Let

$$W = \begin{pmatrix} V & \psi_L \\ \psi_L^\dagger & g \end{pmatrix} \tag{54}$$

Under translations only the argument of W will transform if W depends on space-time variables. Under the Lorentz group

$$\Lambda(L) = \begin{pmatrix} L & 0 \\ 0 & 1 \end{pmatrix}, \qquad \Lambda(L)^\dagger = \begin{pmatrix} L^\dagger & 0 \\ 0 & 1 \end{pmatrix} \tag{55}$$

so that

$$W' = \Lambda(L)\, W\, \Lambda(L)^\dagger = \begin{pmatrix} LVL^\dagger & L\psi_L \\ \psi_L^\dagger L^\dagger & g \end{pmatrix} \tag{56}$$

which shows V, ψ_L and g to transform respectively like a vector, a left handed spinor and a scalar. Under left super-transformations

$$\Lambda(\xi_L) = exp\begin{pmatrix} I & \xi_L \\ 0 & 0 \end{pmatrix} = \begin{pmatrix} I & \xi_L \\ 0 & 1 \end{pmatrix} \tag{57}$$

and

$$\Lambda(\xi_L)^\dagger = \begin{pmatrix} I & 0 \\ \xi_L^\dagger & 1 \end{pmatrix} \tag{58}$$

we have

$$\begin{aligned} W'' &= \Lambda(\xi_L)\, W\, \Lambda(\xi_L)^\dagger = \\ &\begin{pmatrix} V + \psi_L \xi_L^\dagger + \xi_L \psi_L^\dagger + g\xi_L\xi_L^\dagger & \psi_L + g\xi_L \\ \psi_L^\dagger + g\xi_L^\dagger & g \end{pmatrix} \end{aligned} \tag{59}$$

If we take $g = 1$, this coincides with the super Poincarè transformation.

We can also consider the right supertransformations

$$\Lambda(\eta_R) = exp\begin{pmatrix} 0 & 0 \\ \eta_R^\dagger & 0 \end{pmatrix} = \begin{pmatrix} I & 0 \\ \eta_r^\dagger & 1 \end{pmatrix} \tag{60}$$

and

$$\Lambda(\eta_R)^\dagger = \begin{pmatrix} I & \eta_R \\ 0 & 1 \end{pmatrix} \tag{61}$$

so that

$$\begin{aligned} W''' &= \Lambda(\eta_R) W \Lambda(\eta_R)^\dagger = \\ &\begin{pmatrix} V & \psi_L + V\eta_R \\ \psi_L^\dagger + \eta_R^\dagger V & g + \psi_L^\dagger \eta_R + \eta_R^\dagger \psi_L + \eta_R^\dagger V \eta_R \end{pmatrix} \end{aligned} \tag{62}$$

Under this transformation the vector field is invariant while the scalar field and the spinor field transform.

Thus, the super-Lorentz transformations include the super Poincarè transformations as a sub-supergroup.

We can also consider the transformation of W under dilatation

$$W'''' = \Lambda(\sigma) W \Lambda(\sigma)^\dagger \tag{63}$$

with

$$\Lambda(\sigma) = exp\begin{pmatrix} -\frac{1}{2}I\sigma & 0 \\ 0 & \sigma \end{pmatrix} = \begin{pmatrix} e^{-\frac{\sigma}{2}} & & \\ & e^{-\frac{\sigma}{2}} & \\ & & e^{-\sigma} \end{pmatrix} \tag{64}$$

such that the super determinant of Λ is one. Letting $\lambda = e^\sigma$ we have

$$W'''' = \begin{pmatrix} \lambda^{-1} V & \lambda^{-\frac{3}{2}} \psi_L \\ \lambda^{-\frac{3}{2}} \psi_L^\dagger & \lambda^{-2} g \end{pmatrix} \tag{65}$$

These are correct transformation laws for vector and spinor fields. Note that g transforms like a metric field under dilatations and is invariant under super-Poincarè transformations.

Journal of Physics: Conference Series **474** (2013) 012013 doi:10.1088/1742-6596/474/1/012013

Under the Super-Poincarè group, dilatations and left-supertransformations the super determinant of W is invariant.

Let us diagonalize W

$$W = \begin{pmatrix} I & g^{-1}\psi_L \\ 0 & 1 \end{pmatrix} \begin{pmatrix} V - g^{-1}\psi_L\psi_L^\dagger & 0 \\ 0 & g \end{pmatrix} \begin{pmatrix} I & 0 \\ g^{-1}\psi_L^\dagger & 1 \end{pmatrix} \tag{66}$$

We have

$$\begin{aligned} \text{SDet}W &= \text{SDet}\begin{pmatrix} V - g^{-1}\psi_L\psi_L^\dagger & 0 \\ 0 & g \end{pmatrix} \\ &= g^{-1}\text{Det}(V - g^{-1}\psi_L\psi_L^\dagger). \end{aligned} \tag{67}$$

On the other hand, using $\hat{\psi}_L = -i\sigma_2\psi_L^*$,

$$DetV = V\bar{V} \tag{68}$$

$$\begin{aligned} &\text{Det}(V - g^{-1}\psi_L\psi_L^\dagger) = \\ &= \text{Det}V - \frac{1}{2}g^{-2}\hat{\psi}_L^\dagger\psi_L\psi_L^\dagger\hat{\psi}_L + g^{-1}\psi_L^\dagger\bar{V}\psi_L \end{aligned} \tag{69}$$

and

$$\text{SDet}W = g^{-1}\bar{V}V + g^{-2}\psi_L^\dagger\bar{V}\psi_L - \frac{1}{2}g^{-3}\hat{\psi}_L^\dagger\psi_L\psi_L^\dagger\hat{\psi}_L \tag{70}$$

which is obviously invariant under dilatations. Note that $g^2\text{SDet}W$ is still super Poincarè invariant and transforms like a scalar density of weight 4 under dilatations. We have

$$g^2\ \text{SDet}W = g\bar{V}V + \psi_L^\dagger\bar{V}\psi_L - \frac{1}{2}g^{-1}\hat{\psi}_L^\dagger\psi_L\psi_L^\dagger\hat{\psi}_L. \tag{71}$$

Let us note that if the components of ψ_L are not treated like Grassmann variables, in the case one assumes that $\psi_\alpha, \psi_\alpha^*$ $(\alpha = 1, 2)$ all commute, but nevertheless transform like a spinor under Lorentz transformations, then, the invariant under the 3×3 Lorentz transformation matrix would be

$$\text{Det}W = gV\bar{V} - \psi_L^\dagger\bar{V}\psi_L \tag{72}$$

with no quartic spinor term.

The transformation matrix is now a 3×3 unimodular matrix that belongs to $SL(3, \mathcal{R})$ and $SL(3, \mathcal{C})$ respectively in the real and complex cases. In the quaternionic case we have $SL(3, \mathcal{H})$ which is isomorphic to $SU^*(6)$. Finally in the octonionic case $SL(3, \mathcal{O})$ does not give all the linear transformations on $\mathcal{O}$ since we can also act on the octonions in J by associators that come in the expression of the G_2 automorphism group of octonions. In all cases the dimension of the group is given by

$$D = 8d + \dim\,(\text{aut}) \tag{73}$$

where d is the dimension of the division algebra. The automorphism groups are non trivial only for quaternions ($SU(2)$ with dimension 3) and octonions (G_2 with dimension 14). This gives for D

$$D(\mathcal{R}) = 8, \quad D(\mathcal{C}) = 16, \quad D(\mathcal{H}) = 35, \quad D(\mathcal{O}) = 78 \tag{74}$$

which are the dimensions of the Lie algebras of $SL(3,\mathcal{R})$, $SL(3,\mathcal{C})$, $SU^*(6)$ and $E_{6,-26}$ respectively. The non compact group $E_{6,-26}$ has $O(9,1)$ as a subgroup as well as the antomorphism group F_4 of the Jordan algebra of 3×3 traceless octonionic hermitian matrices. $E_{6,-26}$ means that the group has rank 6 and the number of non compact generators minus the number of compact generators is -26.

6. Case of E_{10} and future prospects

If instead of $D = 11$ we start from $D = 10$, then the whole $9 + 1$ dimensional space can be regarded as the Cartan torus of E_{10}, the hyperbolic extension of E_8. This corresponds to a space-time with some periodic boundary conditions, the period being associated with some quantum theoretical cutoff. The actual group of this $9 + 1$ dimensional space-time would be $O(9,1)$ regarded as a torus it would lie in the Cartan disk of the invisible E_{10}.

If we start from $D = 26 = (25 + 1)$ space-time (dimension of the string) then torus compactification to $D = 10$ would give the root lattices of $E_8 \times E_8$ or $O(32)$ as the internal group of the 16 compactified small dimensions. We can compactify the remaining 10 dimensions to a torus with large spatial periods and large or (∞) time period and identify it with part of E_{10} i.e. with the Lorentz group $O(9,1)$.

Hence the groups of the compactification $O(9,1) \times O(8) \times O(8)$, or $O(9,1) \times O(16)$ are lifted to $E_{10} \times E_8 \times E_8$ or $E_{10} \times O(32)$. Here E_{10} is the external group of space-time group $O(9,1)$, and in a similar way we identify with $E_8 \times E_8$ and $O(32)$ the internal groups $O(8) \times O(8)$ and $O(16)$, respectively.

External and internal groups are supported respectively by their 10 dimensional Lorentzian and 16 dimensional Euclidean lattices. They are imbedded into a $25 + 1$ Lorentzian lattice $L(25,1)$. This is the Conway-Sloane lattice. It is regarded as corresponding to a new Lorentzian extension of the hyperbolic group E_{10}. Cartan generators can be introduced to give the structure of a new infinite Lie algebra with a rank 26 generalization (called E_∞ by Conway and Sloane). It has sublattices $E_{10} \times E_8 \times E_8$, $E_{10} \times O(32)$, $E_{10} \times O(16) \times O(16)$, i.e. the groups of known heterotic string theories. Hence a new string theory could be associated with E_∞ it would fuse external and internal symmetries together. Note that E_∞ also contains the Leech lattice (not a group!) and the Monster.

At this point it may be useful to summarize some known results (mostly due to Kac) about hyperbolic groups:

Characterized by Dynkin diagrams such that if a point is erased the remaining diagram is either an extended Dynkin diagram (associated with affine algebra $\hat{G}$) or a Dynkin diagram of a Lie group.

They stop at rank 10 corresponding to hyperbolic extension of E_8. Among those the strictly hyperbolic groups are those which are not $\hat{G}$, but give Lie group G when one point is erased. They stop at rank 4.

Extensions which correspond to even Lorentzian lattices and are not hyperbolic groups stop at rank 26 (the Conway-Sloane lattice) [$8k + 2$, families with $k = 1, 2, 3$]. Again we find the critical dimensions 4, 10, 26 for hyperbolic and Lorentzian algebras.

Construction of the root lattices of E_8, $E_9 = \hat{E}_8$ and E_{10}, as well as association of the Conway-Sloane lattice with some of the lattices described above and with discrete Jordan algebras over octonions is deferred to another publication. Remarkable associations between symmetries of superstrings and the lattices generated by discrete Jordan algebras suggest that all known superstring theories are related and originate from a more general theory related to the Conway-Sloane transhyperbolic group.

7. References

[1] M. Günaydın and F. Gürsey, J. Math. Phys. **14** (1973) 1651.
[2] S. Catto, H.Y. Cheung and F. Gürsey, Mod. Phys. Lett **A6** (1991) 3485.
[3] S. Catto and F. Gürsey, Nuovo Cimento **A86** (1985) 201.
[4] S. Catto and F. Gürsey, Nuovo Cimento **A99** (1988) 685.
[5] M. Anselmino, et al. Rev. Mod. Phys. **65** (1993) 1199.
[6] E.Klempt, *Baryon Resonances and Strong QCD.* arXiv:nucl-ex/0203002.
[7] C-H. Tze and F. Gürsey, *On the Role of Division, Jordan and Related Algebras in Particle Physics*, World Scientific, 1996.

[8] S. Catto, Acta Polytechnica **51** (2011) 77. See also: *Effective Lagrangians based on QCD: Potential Models and Skyrmions.* arXiv:hep-th/0302101.
[9] S. Catto, *Workshop on Scalar Mesons and Related Topics*, Eds. G. Rupp, E. van Beveren, P. Bicudo, B. Hiller and F. Kleefeld *AIP Proc. 1030,* (2008) 184. See also:*Symmetries and Mass Predictions of the Supersymmetric Quark Model.* arXiv:hep-ph/0302153.
[10] S. Catto and F. Gürsey, Lett. Nuovo Cim. **35** (1982) 241.
[11] S. Catto and Y.S. Choun, *Effective Supersymmetry Based on* $SU(3)^C \times S$ *Superalgebra.* Lie Theory and Its Applications, Ed. V. Dobrev, AIP Conf. Proc. 1243 (2010) 233.
[12] S. Catto, Y.S. Choun and L. Kurt, *Invariance Properties of the Exceptional Quantum Mechanics* (F_4) *and its Generalization to Complex Jordan Algebra* (E_8). Springer Proc. in Mathematics and Statistics, 36 (2013) 469.
[13] S. Catto, *Exceptional Projective Geometries and Internal Symmetries.* arXiv:hep-th/0302079.
[14] F. Iachello, Proceedings of ICGTMP, Eds. M.A. del Olmo, M. Santander and J. Mateos Guilarte (1993) 13, CIEMAT, Anales de Fisica Monografias, Madrid.
[15] J. Baez, Bull. Am. Math. Soc. **39(2)** (2002) 145-205.
[16] For a review see M.B. Green, J.H. Schwarz and E. Witten,"Superstring Theory I and II," (Cambridge U. Press, 1987).
[17] M.B. Green and J.H. Schwarz, *Phys. Lett.***136B**,367(1984. For super Yang-Mills theories see L. Brink, J.H. Schwarz and J. Scherk, *Nucl. Phys.***B121**, 253(1977), and F. Gliozzi, J. Scherk and D. Olive, *Nucl. Phys.***B122**, 253(1977).
[18] F. Ardalan and F. Mansouri, *Phys. Rev.***D9**, 3341(1974); *Phys. Rev. Lett.***56**, 2456(1986).
[19] C. Becchi, A. Rouet and R. Stora, *Phys. Lett.***52B**, 344(1974); I. Tyutin, Lebedev preprint *FIAN No. 39(1975).*
[20] S. Catto, *Constraint superalgebras and Their Application to Gauge Field Theories and String Theories.* arXiv:hep-th/0302088.
[21] A. Hurwitz, "Über die Composition der quadratischen Formen von beliebig vielen Variabeln," *Nachrichten von der königlichen Gesellshaft der wissenshaften in Göttingen* (1898); *Über die Zahlentheorie der Quaternionen*, Mathematische Werke, vol. 2(1933).
[22] H. S. M. Coxeter, *Duke Math. J.* **13**(1946)561.
[23] M.R. Brenner, R.V. Moody and J. Patera, *Tables of Dominant Weight Multiplicities of Simple Lie Algebras.* M. Dekker, New York, 1985.
[24] For a review see P. Goddard and D. Olive, *Int. J. Mod. Phys.* **A1** (1986) 303.
[25] F. Gürsey, *Mod. Phys. Lett.* **A3** (1988) 1155.
[26] F. Gürsey, *Mod. Phys. Lett.* **A2** (1987) 967.
[27] N. Jacobson, "Structure and Representations of Jordan Algebras" *Am. Math. Soc.,* Providence 1968; "Composition algebras and their automorphisms," *Rendiconti del Circolo Matematico di Palermo,* Vol. 7 (1958), Ser. 11; "Exceptional Lie Algebras," (M. Dekker, New York, 1971).

8. Acknowledgments

One of us (SC) would like to thank Professor Čestmir Burdik for invitation to Prague and for his kind hospitality. He is also indebted to Prof. Vladimir Dobrev for useful discussions.

Journal of Physics: Conference Series **474** (2013) 012014 doi:10.1088/1742-6596/474/1/012014

Monodromy operators and symmetric correlators

D Chicherin[1] **and R Kirschner**[2]

[1] St. Petersburg Department of Steklov Mathematical Institute of Russian Academy of Sciences, Fontanka 27, 191023 St. Petersburg, Russia and
Chebyshev Laboratory, St. Petersburg State University, 14th Line, 29b, St. Petersburg, 199178 Russia

[2] Institut für Theoretische Physik, Universität Leipzig, PF 100 920, D-04009 Leipzig, Germany.

E-mail: `Roland.Kirschner@itp.uni-leipzig.de`

Abstract. Yangian symmetric correlators are defined as eigenfunctions of monodromy operators. Examples and relations are given and an evaluation method of the symmetry condition is described.

1. Yangian symmetry

We consider N-point correlators in n dimensions, i.e. functions of nN variables interpreted as N points $\mathbf{x}_i$ with coordinates $(x_{a,i})_{a=1}^n = (x_{1,i}, x_{2,i}, \cdots, x_{n,i})$. We shall study a symmetry condition imposed on the correlators in terms of the monodromy matrix of a N-site spin chain with Jordan-Schwinger-type [1, 2, 3, 4, 5, 6, 7, 8, 9, 10] representations of $g\ell_n$. Correlators with this Yangian symmetry are known to appear as kernels of Yang-Baxter operators, as evolution kernels descibing high-energy asymptotics of QCD and composite operator renormalisation and as scattering amplitudes [11, 12, 13, 14].

The notion of Yangian algebra was introduced by Drinfeld [18] in general form. The case related to $g\ell_n$ was worked out earlier by L. Faddeev and collaborators [19, 20, 21] in the QISM formulation. The relation of the latter formulation to the one in algebra generators by Drinfeld is well explained in [22].

In this contribution we recall the concept and shortly review some results of our paper [23]. Beyond this, we describe in detail a method of evaluating the symmetry condition and work out non-trivial examples.

Associated with the coordinates we shall work with nN Heisenberg canonical pairs $\mathbf{p}_i = (p_{a,i})_{a=1}^n = (p_{1,i}, p_{2,i}, \cdots, p_{n,i})$ and $\mathbf{x}_i = (x_{a,i})_{a=1}^n = (x_{1,i}, x_{2,i}, \cdots, x_{n,i})$, $[p_{a,i}, x_{b,j}] = \delta_{a,b}\delta_{i,j}$. We consider the partition of the index set $N = \{1, 2, \cdots, N\}$ labeling these points in two non-overlapping subsets I and J, $I \cup J = N$, $I \cap J = \varnothing$. The partition of the set of N sites into the subsets I, J can also be denoted as signature, i.e. by a sequence of symbols $''+''$, $''-''$ where the symbol $''+''$ is put for a site in I and $''-''$ for a site in J.

We define the action of $g\ell_n$ on the points in dependence of their signature as

$$\delta x_{c,i} = \left[\left[\mathrm{L}_i^+\right]_{a,b}, x_{c,i}\right], \; i \in I, \quad \delta x_{c,j} = \left[\left[\mathrm{L}_j^-\right]_{a,b}, x_{c,j}\right], \; j \in J$$

$$\left[\mathrm{L}_i^+\right]_{a,b} = p_{a,i}x_{b,i}\,, \quad \left[\mathrm{L}_j^-\right]_{a,b} = -x_{a,j}p_{b,j}\,. \tag{1}$$

We denote the inner product by $(\cdot)$,

$$(ji) \equiv (ij) \equiv (\mathbf{x}_i \cdot \mathbf{x}_j) \equiv x_{a,i}x_{a,j} \,. \tag{2}$$

and notice that for $i \in I, j \in J$ it is $g\ell_n$ invariant in the sense $[\mathrm{L}_i^+ + \mathrm{L}_j^-, (\mathbf{x}_i \cdot \mathbf{x}_j)] = 0$. Monomials of the form

$$\Phi_{I,J} = \prod_{i\in I, j\in J} (\mathbf{x}_i \cdot \mathbf{x}_j)^{\lambda_{ij}} \tag{3}$$

are $g\ell_n$ invariant and general $g\ell_n$ invariant correlators are superpositions of such monomials with varying exponents which can take complex values. In general the coordinates are complex valued.

The L matrices are defined in terms these generator matrices by adding a spectral parameter being a complex number,

$$\mathrm{L}_k^+(u_k) = u_k + \mathbf{p}_k\mathbf{x}_k \,, \quad \mathrm{L}_k^-(u_k) = u_k - \mathbf{x}_k\mathbf{p}_k \,, \tag{4}$$

or in component form

$$\left[\mathrm{L}_k^+(u_k)\right]_{a,b} = u_k\delta_{a,b} + p_{a,k}x_{b,k} \,, \quad \left[\mathrm{L}_k^-(u_k)\right]_{a,b} = u_k\delta_{a,b} - x_{a,k}p_{b,k} \,. \tag{5}$$

The L matrices are well known as the basic tool of treating the spin chain by the Quantum Inverse Scattering Method (QISM) [19, 20, 24, 25].

We are going to study symmetry conditions on correlators going beyond the mentioned $g\ell_n$ invariance to be formulated in terms of the monodromy matrix

$$\mathrm{T}_{1\cdots N}^{\alpha_1\cdots\alpha_N}(u_1,\cdots,u_N) = \mathrm{L}_1^{\alpha_1}(u_1)\mathrm{L}_2^{\alpha_2}(u_2)\cdots\mathrm{L}_N^{\alpha_N}(u_N)\,, \tag{6}$$

where $\alpha_1\alpha_2\cdots\alpha_N$ denotes the signature, i.e. $\alpha_k = \pm$ at $k = 1,\cdots,N$. The lower indices refer to the chain site with the representation (the quantum space) where the operators act nontrivially.

A *Yangian symmetric correlator* is defined as a $g\ell_n$ invariant correlator of definite dilatation weights being a solution of the monodromy eigenvalue relation

$$\mathrm{T}(u_1,\cdots,u_N)\cdot\Phi(u_1,\cdots,u_N) = E(u_1,\cdots,u_N)\,\Phi(u_1,\cdots,u_N)\,. \tag{7}$$

Here the matrix T with operator elements acts on the correlator function resulting in the r.h.s. proportional to the unit matrix.

2. Similarity transformations and eigenfunctions

The study of similarity transformations of the monodromy matrix by monomial expressions (3) allows to obtain simple symmetric correlators. The action of the similarity transformation on a single L-operator depends on whether its label falls into the sets $i \in I$ or $j \in J$,

$$\begin{aligned}(\mathbf{x}_i\cdot\mathbf{x}_j)^{-\lambda_{ij}}\mathrm{L}_i^+(u_i)(\mathbf{x}_i\cdot\mathbf{x}_j)^{\lambda_{ij}} &= \mathrm{L}_i^+(u_i) + \lambda_{ij}\mathrm{l}_{j\,i}\\ (\mathbf{x}_i\cdot\mathbf{x}_j)^{-\lambda_{ij}}\mathrm{L}_j^-(u_j)(\mathbf{x}_i\cdot\mathbf{x}_j)^{\lambda_{ij}} &= \mathrm{L}_j^-(u_j) - \lambda_{ij}\mathrm{l}_{j\,i}\end{aligned} \tag{8}$$

where we use the abbreviation $\mathrm{l}_{ij} \equiv \frac{\mathbf{x}_i\mathbf{x}_j}{(\mathbf{x}_i\cdot\mathbf{x}_j)}$. In the following we also adopt the notation $\Phi \circ T = \Phi T \Phi^{-1}$.

It is easy to perform the similarity transformation of the monodromy matrix of the two-site chain, $\mathrm{T}^{+-}(u_1,u_2)$ (6), using (8)

$$(\mathbf{x}_1\cdot\mathbf{x}_2)^{\lambda_{12}} \circ \mathrm{L}_1^+(u_1)\,\mathrm{L}_2^-(u_2) = \left(\mathrm{L}_1^+(u_1+\lambda_{12}) + \lambda_{12}(\mathrm{l}_{21}-1) \quad \mathrm{L}_2^-(u_2-\lambda_{12}) + \lambda_{12}(1-\mathrm{l}_{21})\right).$$

Then taking into account the relations $\mathbf{x}_1 \mathrm{l}_{21} = \mathbf{x}_1$, $\mathrm{l}_{21}\mathbf{x}_2 = \mathbf{x}_2$, $\mathrm{l}_{21}(\mathrm{l}_{21} - 1) = 0$ we obtain

$$(\mathbf{x}_1 \cdot \mathbf{x}_2)^{\lambda_{12}} \circ \mathrm{T}^{+-}(u_1, u_2) = \mathrm{T}^{+-}(u_1 + \lambda_{12}, u_2 - \lambda_{12}) + \lambda_{12}\,(u_1 - u_2 + \lambda_{12})\,(1 - \mathrm{l}_{21})\,. \quad (9)$$

At the special value $\lambda_{12} = u_2 - u_1$ the remainder in the previous formula vanishes

$$(\mathbf{x}_1 \cdot \mathbf{x}_2)^{u_2 - u_1} \circ \mathrm{T}^{+-}(u_1, u_2) = \mathrm{T}^{+-}(u_2, u_1)\,. \quad (10)$$

Such similarity transformation leads to the permutation of the two spectral parameters $u_1 \leftrightarrow u_2$. Then applying both sides of (9) to the basic state 1 one obtains the eigenvalue relation for the monodromy matrix (see (7))

$$\mathrm{T}^{+-}(u_1, u_2) \cdot \Phi^{+-} = u_1(u_2 + 1)\,\Phi^{+-}\,, \quad \Phi^{+-} = (\mathbf{x}_1 \cdot \mathbf{x}_2)^{u_2 - u_1}\,. \quad (11)$$

Next we turn to the configuration $-+$ considering the similarity transformation of the monodromy matrix $\mathrm{T}^{-+}(u_1, u_2)$. Similar to the previous calculation one obtains easily

$$(\mathbf{x}_1 \cdot \mathbf{x}_2)^{\lambda_{21}} \circ \mathrm{T}^{-+}(u_1, u_2) = \mathrm{T}^{-+}(u_1, u_2) + \lambda_{21}\,\mathrm{l}_{12}\,(u_1 - u_2 - \lambda_{21} - (\mathbf{p}_1 \cdot \mathbf{x}_1) - (\mathbf{x}_2 \cdot \mathbf{p}_2))\,. \quad (12)$$

Contrary to (10) the spectral parameters stay untouched and the operator remainder term does not turn to zero for any nonzero λ_{21}. Taking $\lambda_{21} = u_1 - u_2 - n$ and acting with both sides of the latter equation on the basic state 1 we find the eigenvalue relation

$$\mathrm{T}^{-+}(u_1, u_2) \cdot \Phi^{-+} = u_1(u_2 + 1)\,\Phi^{-+}\,, \quad \Phi^{-+} = (\mathbf{x}_1 \cdot \mathbf{x}_2)^{u_1 - u_2 - n}\,. \quad (13)$$

We shall see further that the relations (10) and (12) are crucial in our discussion. The first one implies parameter permutation while in the second the parameters are untouched.

Let us consider the three point correlator with signature $++-$. The similarity transformation of the monodromy matrix $\mathrm{T}^{++-}(u_1, u_2, u_3)$ by $\Phi^{++-} = (\mathbf{x}_1 \cdot \mathbf{x}_3)^{\lambda_{13}}(\mathbf{x}_2 \cdot \mathbf{x}_3)^{\lambda_{23}}$ is performed in two steps. At first we permute $u_2 \leftrightarrow u_3$ (9),

$$(\mathbf{x}_2 \cdot \mathbf{x}_3)^{\lambda_{23}} \circ \mathrm{T}^{++-}(u_1, u_2, u_3) = \mathrm{T}^{++-}(u_1, u_3, u_2)\,, \quad \lambda_{23} = u_3 - u_2\,.$$

Then we act on a constant function in second space 1_2,

$$\mathrm{T}^{++-}(u_1, u_3, u_2) \cdot 1_2 = (u_3 + 1)\,\mathrm{T}^{+-}_{13}(u_1, u_2)$$

reducing the number of chain sites by one and apply the second similarity transformation by means of (9) obtaining

$$E^{++-} = u_1(u_2 + 1)(u_3 + 1)\,, \quad \Phi^{++-} = (13)^{u_2 - u_1}(23)^{u_3 - u_2}. \quad (14)$$

Now we consider the 4-point correlator with signature $+-+-$. The pattern to implement the similarity transformation of T^{+-+-} by

$$\Phi^{+-+-} = (\mathbf{x}_1 \cdot \mathbf{x}_2)^{\lambda_{12}}(\mathbf{x}_1 \cdot \mathbf{x}_4)^{\lambda_{14}}(\mathbf{x}_2 \cdot \mathbf{x}_3)^{\lambda_{32}}(\mathbf{x}_3 \cdot \mathbf{x}_4)^{\lambda_{34}} \quad (15)$$

is analogous to the above calculations of the three-point correlation functions. At first the transpositions $u_1 \leftrightarrow u_2$ and $u_3 \leftrightarrow u_4$ are performed due to (9)

$$(\mathbf{x}_1 \cdot \mathbf{x}_2)^{\lambda_{12}}(\mathbf{x}_3 \cdot \mathbf{x}_4)^{\lambda_{34}} \circ \mathrm{T}^{+-+-}(u_1, u_2, u_3, u_4) = \mathrm{T}^{+-+-}(u_2, u_1, u_4, u_3)$$

at $\lambda_{12} = u_2 - u_1$, $\lambda_{34} = u_4 - u_3$. Then we apply (12)

$$(\mathbf{x}_2 \cdot \mathbf{x}_3)^{\lambda_{32}} \circ \mathrm{T}^{+-+-}(u_2, u_1, u_4, u_3) = \mathrm{T}^{+-+-}(u_2, u_1, u_4, u_3) + \hat{r}$$

and notice that at $\lambda_{32} = u_1 - u_4 - n$ the remainder $\hat{r}$ vanishes after acting on a basic state in the second and the third spaces $1_2 1_3$

$$\mathrm{T}^{+-+-}(u_2, u_1, u_4, u_3) \cdot 1_2 1_3 = u_1(u_4 + 1)\, \mathrm{T}^{+-}_{14}(u_2, u_3)\,.$$

Finally we permute $u_2 \leftrightarrow u_3$ (9),

$$(\mathbf{x}_1 \cdot \mathbf{x}_4)^{\lambda_{14}} \circ \mathrm{T}^{+-}_{14}(u_2, u_3) = \mathrm{T}^{+-}_{14}(u_3, u_2)$$

at $\lambda_{14} = u_3 - u_2$. Thus we have shown that (7) holds with

$$E^{+-+-} = u_1 u_2 (u_3 + 1)(u_4 + 1)\,, \quad \Phi^{+-+-} = (12)^{u_2 - u_1}(14)^{u_3 - u_2}(23)^{u_1 - u_4 - n}(34)^{u_4 - u_3}\,. \quad (16)$$

3. A general method

We consider the symmetry condition (7) for N-point correlators with the signature $+ + - ... -$

$$\mathrm{L}^+_1(u_1)\mathrm{L}^+_2(u_2)\mathrm{L}^-_3(u_3)...\mathrm{L}^-_N(u_N)\Phi = E\Phi.$$

We use the inversion relations

$$\begin{aligned} \mathrm{L}^+(u)\,\mathrm{L}^+(-u - 1 - (\mathbf{x}\cdot\mathbf{p})) &= u(-u - 1 - (\mathbf{x}\cdot\mathbf{p})), \\ \mathrm{L}^-(u)\,\mathrm{L}^-(-u - 1 + (\mathbf{p}\cdot\mathbf{x})) &= u(-u - 1 + (\mathbf{p}\cdot\mathbf{x})) \end{aligned} \quad (17)$$

to rewrite the condition in the form

$$\mathrm{L}^-_3(u_3)...\mathrm{L}^-_N(u_N)\;\Phi =$$

$$\frac{E}{u_1 u_2 (u_1 + 1 + (\mathbf{x}_1\cdot\mathbf{p}_1))(u_2 + 1 + (\mathbf{x}_2\cdot\mathbf{p}_2))}\mathrm{L}^+_2(-u_2 - 1 - (\mathbf{x}_2\cdot\mathbf{p}_2))\mathrm{L}^+_1(-u_1 - 1 - (\mathbf{x}_1\cdot\mathbf{p}_1))\;\Phi$$

We look for the solution in the link-integral form

$$\Phi_{I,J} = \int \phi(c) \exp\left(-\sum c_{ij}\,(\mathbf{x}_i\cdot\mathbf{x}_j)\right) \prod \mathrm{d}c_{ij}\,. \quad (18)$$

In particular, using the integral formula for the Gamma function

$$y^\lambda = \frac{\Gamma(1+\lambda)}{2\pi i}\int_{\mathcal{C}} \mathrm{d}c\,(-c)^{-\lambda-1} e^{-cy} \quad (19)$$

where the contour $\mathcal{C}$ encircles clockwise the positive real semi-axis starting at $+\infty - i\epsilon$, surrounding 0, and ending at $+\infty + i\epsilon$, the monomial ansatz (3) acquires the link form with $\phi(c) = \prod c_{ij}^{-1-\lambda_{ij}}$. We need the more general ansatz

$$\phi(c_{ij}) = \prod c_{ij}^{-1-\lambda_{ij}}\; f(z^{i_1 i_2}_{j_1 j_2}) \quad (20)$$

involving a function of the cross ratios

$$z^{i_1 i_2}_{j_1 j_2} = \frac{c_{i_1 j_1} c_{i_2 j_2}}{c_{i_1 j_2} c_{i_2 j_1}}, \quad i_1, i_2 \in I, j_1, j_2 \in J$$

In the considered case we have $i_1 = 1, i_2 = 2$ and the upper labels can be omitted. We choose as a complete set of independent cross ratios $z_j = z_{jj+1}, j = 3, ..., N-1$.

Applying the similarity relations (8) and

$$\mathrm{L}^+(u)\cdot 1 = (u+1)\mathrm{I}, \quad \mathrm{L}^-(u)\cdot 1 = u\mathrm{I}$$

the symmetry condition with the ansatz (20) can be written as

$$\int \exp\left(-\sum c_{ij}\,(\mathbf{x}_i\cdot\mathbf{x}_j)\right)\ \psi(c,\mathbf{x})\prod \mathrm{d}c_{ij} = 0,$$

$$\psi(c,\mathbf{x}) = \left(\prod_3^N\left(1+\frac{1}{u_j}D_{1j}\mathrm{l}_{j1}+\frac{1}{u_j}D_{2j}\mathrm{l}_{j2}\right) - \tilde{E}\left(1+\frac{1}{\tilde{u}_2}\sum_3^N D_{2j}\mathrm{l}_{j2}\right)\left(1+\frac{1}{\tilde{u}_1}\sum_3^N D_{1j}\mathrm{l}_{j1}\right)\right)\phi,$$

where

$$\tilde{u}_i = u_i + \sum_{j=3}^N \lambda_{ij},\ i=1,2,\ \ \tilde{E} = \frac{E\tilde{u}_1\tilde{u}_2}{\prod_{s=1}^N u_s\ (\tilde{u}_1+1)(\tilde{u}_2+1)}, \qquad D_{ij} = \frac{\partial}{\partial c_{ij}}\ c_{ij}.$$

The generalisation to other split signature configurations is evident. The dependence on $\mathbf{x}$ of ψ is in the matrices l_{ji}. The products of the latter are expressed again in those matrices multiplied by cross ratios now in $(\mathbf{x}_i\cdot\mathbf{x}_j)$,

$$X_{j_1j_2} = \frac{(\mathbf{x}_1\cdot\mathbf{x}_{j_1})(\mathbf{x}_2\cdot\mathbf{x}_{j_2})}{(\mathbf{x}_1\cdot\mathbf{x}_{j_2})(\mathbf{x}_2\cdot\mathbf{x}_{j_1})}.$$

For example, we have

$$\mathrm{l}_{j_2 2}\mathrm{l}_{j_1 1} = X_{j_2j_1}\mathrm{l}_{j_2 1}.$$

In the integrand of the link integral a factor of the coordinate cross ratio $X_{j_2j_1}$ can be transformed by partial integration to a factor of a cross ratio in the link variables resulting in the rule

$$D_{1j_1}D_{2j_2}X_{j_2j_1} \to D_{1j_2}D_{2j_1}z_{j_1j_2}. \tag{21}$$

We transform also the operators D_{1j}, D_{2j} by expressing the differention with respect to the link variables c_{1j}, c_{2j} in terms of differentiation with respect to the cross ratios,

$$D_{1j}\phi(c) = \prod c_{ij}^{-1-\lambda_{ij}}\ (-\lambda_{1j} + z_j\partial_{z_j} - z_{j-1}\partial_{z_{j-1}})\ f(z).$$

Note that $f(z)$ involves just $z_3, ..., z_{N-1}$ and the differtiations with index values $j=2, j=N$ appearing formally here have to be substituted by zero. For the condition to hold it is sufficient that the transformed integrand of the link integral vanishes, i.e.

$$\left(1+\sum_3^N \frac{1}{u_j}(\tilde{\mathrm{l}}_{j1}+\tilde{\mathrm{l}}_{j2}) + \sum_{k=2}^{N-2}\ \sum_{3\le j_1<j_2...j_k\le N}\ \prod_{s=2}^k \frac{1}{u_{j_s}}\cdot\right. \tag{22}$$

$$\left.\cdot[-\lambda_{1j_s} - \lambda_{2j_s}z_{j_s,j_{s-1}} + (d_{j_s}-d_{j_{s-1}})(1-z_{j_s,j_{s-1}})]\ \ [\tilde{\mathrm{l}}_{j_11}+\tilde{\mathrm{l}}_{j_12}z_{j_1j_k}]\right)f(z)$$

$$= \tilde{E}\left(1+\frac{1}{\tilde{u}_1}\sum_3^N\tilde{\mathrm{l}}_{j1} + \frac{1}{\tilde{u}_2}\sum_3^N\tilde{\mathrm{l}}_{j2} + \frac{1}{\tilde{u}_1\tilde{u}_2}[\sum_3^N(-\lambda_{2j}+d_{j-1}-d_j)\frac{1}{z_{3j}}][\sum_3^N\tilde{\mathrm{l}}_{j1}z_{3j}]\right)f(z).$$

We use the abbreviations

$$\tilde{\mathrm{l}}_{ji} = (-\lambda_{ij}-1+D_{ij})\mathrm{l}_{ji},\ \ d_j = z_j\partial_{z_j},\ \ d_2=0, d_N=0.$$

Since the coordinate dependent matrices l_{ji} and the unit matrix are linearly independent the matrix condition implies a sequence of conditions. In the non-degenerate case, where none of the expressions $(-1-\lambda_{ij}+D_{ij})f(z), i=1,2, j=3,...N$ vanishes identically, also $\tilde{\mathrm{l}}_{ji}$ are linearly independent. The inherent Yangian structure amounts in a hierarchy of the conditions owing to the maximal inverse power of spectral parameters involved (not regarding the u-factors in $\tilde{E}$), and the less involved lower order conditions fix the solution essentially. The higher order conditions then appear as consistency relations, which can be fulfilled by adjusting the involved parameters, the exponents λ_{ij} and the spectral parameters u_s.

Let us show some explicit calculation steps in the case $N=5$. The condition of order zero is fulfilled if $\tilde{E}=1$. The condition arising as coefficient at $\tilde{\mathrm{l}}_{52}$ is of first order and implies just $u_5=\tilde{u}_2$. The conditions coming along with $\mathrm{l}_{52}, \mathrm{l}_{41}, \mathrm{l}_{42}$ are of second order and read in this ordering

$$(\tilde{u}_1-\tilde{u}_2+\lambda_{25}+\lambda_{24}z_4+\lambda_{23}z_3z_4-d_3(1-z_3)z_4-d_4(1-z_4))\,f(z_3,z_4)=0,$$

$$(u_5-u_4-\lambda_{25}-\lambda_{15}z_4+d_4(1-z_4))\,f(z_3,z_4)=0,$$

$$((\tilde{u}_1-u_4)\lambda_{25}+(u_4u_5-\tilde{u}_1u_5+\tilde{u}_1\lambda_{15}u_4\lambda_{24})z_4-u_4\lambda_{23}z_3z_4-$$
$$-(\tilde{u}_1-u_4)d_4(1-z_4)+u_4d_3(1-z_3)z_4)\,f(z_3,z_4)=0.$$

In the last condition the derivative terms can be substituted by using the first ones resulting in the conditions on the parameters $(u_4-\tilde{u}_1)u_5=0, \quad (u_4-\tilde{u}_1)(u_4-u_5)=0$. Let us continue with the case $\tilde{u}_1=u_4$. Then the condition coming with $\mathrm{l}_{52}, \mathrm{l}_{41}$ decouple and result in the solution

$$f(z_3,z_4)=z_3^{\lambda_{24}-\lambda_{15}}(1-z_3)^{-1-\lambda_{23}-\lambda_{24}+\lambda_{15}}\;z_4^{u_4-u_5+\lambda_{25}}(1-z_4)^{-1-\lambda_{15}-\lambda_{25}+u_5-u_4}. \tag{23}$$

We quote the results for the 3 and 4 point correlators with signature of the considered form. In the case $N=4$ we have the above form (20) with

$$f(z)=z^{u_3-u_4+\lambda_{24}}(1-z)^{-1-\lambda_{23}-\lambda_{24}-u_3+u_4},\quad \lambda_{14}=\lambda_{23}, \tilde{u}_1=u_3, \tilde{u}_2=u_4,\quad z=z_{34}.$$

In the case $N=3$

$$f=1, u_3=\tilde{u}_2, \lambda_{23}=\tilde{u}_2-\tilde{u}_1.$$

The latter reproduces the above result (14) for the symmetric 3-point correlator. The 4-point correlator obtained here and the one obtained in the previous section (16) are related by the signature transposition to be described below.

4. Relations between symmetric correlators

There are several ways to generate symmetric correlators from given ones.

One source of such relations is the structure of the L operators. In particular they obey the inversion relations (17) and the relation of matrix transposition (see (5))

$$\left[\mathrm{L}^{\pm}(u)\right]_{ba}=-\left[\mathrm{L}^{\mp}(-u-1)\right]_{ab}\,. \tag{24}$$

These relations give rise to relations for the monodromy matrices and to transformations of the symmetry condition. This implies in particular that a symmetric correlator $\Phi^{\alpha_1\cdots\alpha_N}$ is proportional to the one with opposite signature $-\alpha_1\cdots-\alpha_N$ and transformed spectral parameters and also to the one with the reversed order of points. Further interesting relations are the ones involving cyclic shifts of the points reminicent to crossing relations [15, 16, 17]. For example, if the first point is attributed sign + the relation for the cyclic shift moving this point to the last position reads

$$\mathrm{T}_{2\cdots N1}^{\alpha_2\cdots\alpha_N+}(u_2,\cdots,u_N,u_1-n)\cdot\Phi=E_0\cdot\Phi\,,\quad E_0=\frac{(u_1+1-n)(u_1'+1)}{u_1u_1'}E(u_1,\cdots,u_N)\,,\tag{25}$$

where $u_1'=-u_1-1-\sum_J\lambda_{1j}$.

The operation of elementary canonical transformation $\mathtt{C}_i$

$$\mathtt{C}_i^{-1}\,\mathbf{x}_i\,\mathtt{C}_i=-\mathbf{p}_i\,,\quad \mathtt{C}_i^{-1}\,\mathbf{p}_i\,\mathtt{C}_i=\mathbf{x}_i.\tag{26}$$

acting at site i interchanges $\mathrm{L}_i^{\pm}(u)$ with $\mathrm{L}_i^{\mp}(u)$ and $\Phi(\cdots,\mathbf{x}_i,\cdots)\cdot 1$ by $\Phi(\cdots,\mathbf{p}_i,\cdots)\cdot\delta(\mathbf{x}_i)$. In the discussion of this operation we restrict the related coordinate variables to real values. Applying this operation at one site i leads from a regular to a singular correlator. Let us apply the operation to the sites i,j of different signature. The symmetric correlator with + at i and − at j is transformed into a symmetric correlator with the transposed signature, − at i and + at j. The result is a regular correlator again, a singular one appears only intermediately.

To perform the transposition we write the original correlator in link integral form and $\delta(\mathbf{x}_i)\delta(\mathbf{x}_j)$ in Fourier integrals.

$$\Phi\to\mathtt{C}_{i,j}\,\Phi=\int \mathrm{d}c\,\mathrm{d}\mathbf{p}_i\mathrm{d}\mathbf{p}_j\,\phi(c)e^{-Q}=\int\mathrm{d}c\,\phi(c)e^{-Q'}\,,$$

$$Q\equiv c_{i_1j_1}(\mathbf{x}_{i_1}\cdot\mathbf{x}_{j_1})-c_{ij}(\mathbf{p}_i\cdot\mathbf{p}_j)+ic_{ij_1}(\mathbf{p}_i\cdot\mathbf{x}_{j_1})+ic_{i_1j}(\mathbf{x}_{i_1}\cdot\mathbf{p}_j)-i(\mathbf{p}_i\cdot\mathbf{x}_i)-i(\mathbf{p}_j\cdot\mathbf{x}_j)\,.$$

Here i_1,j_1 is summed over the ranges I,J respectively avoiding the fixed values i,j. Performing the integrals over $\mathbf{p}_i,\mathbf{p}_j$ we obtain a link integral of the above form with the quadratic form in the exponential replaced by

$$Q'=c'_{i'j'}(\mathbf{x}_{i'}\cdot\mathbf{x}_{j'})\,,\quad c'_{i_1j_1}=c_{i_1j_1}-\frac{c_{ij_1}c_{i_1j}}{c_{ij}}\,,\quad c'_{i_1i}=\frac{c_{i_1j}}{c_{ij}}\,,\quad c'_{jj_1}=\frac{c_{ij_1}}{c_{ij}}\,,\quad c'_{ji}=-\frac{1}{c_{ij}}\,.$$

The last step of the transposition operation is the change of integration variables $c_{i'j'}\to c'_{i'j'}$, where i' run over I and j' run over J. In particular one can check that the 4-point correlators with the signatures $+-+-$ and $++--$ are related in this way.

Correlation functions can be regarded as kernels of integral operators provided the integration can be defined.

$$\left[\hat{\mathrm{R}}\cdot\psi\right](\mathbf{x}_{K+1},\cdots,\mathbf{x}_N)=\int\mathrm{d}\mathbf{x}_1\cdots\mathrm{d}\mathbf{x}_K\,\mathrm{R}(\mathbf{x}_1,\cdots,\mathbf{x}_K,\mathbf{x}_{K+1},\cdots,\mathbf{x}_N)\,\psi(\mathbf{x}_1,\cdots,\mathbf{x}_K)\,.\tag{27}$$

The subsequent action of these operators is then also defined. This means that the convolution of kernels by the given integration prescription results in further correlation functions. We assume that the integration allows for a simple integration by parts defining the transposition rule,

$$\int\mathrm{d}\mathbf{x}\,\phi(\mathbf{x})\mathrm{L}^{\pm}(u)\psi(\mathbf{x})=\int\mathrm{d}\mathbf{x}\,\psi(\mathbf{x})\mathrm{L}^{\pm T}(u)\phi(\mathbf{x}),\tag{28}$$

$$\left[\mathrm{L}^{\pm T}(u)\right]_{ab}=-\left[\mathrm{L}^{\pm}(-u-1)\right]_{ab}\,.\tag{29}$$

We impose the condition that the dilatation weight of an integrated point plus the weight n of the measure adds up to zero. By this restriction on the dilatation weights the integration becomes compatible with the symmetry, i.e. the reduction to the projective space carrying the $s\ell_n$ irreducible representations should be always defined. With the transposition relation (29) this results in the compatibility of the symmetry conditions with the convolution; the result of this convolution of symmetric correlators is a symmetric correlator.

This procedure of connecting solutions has reminiscence to the BCFW [26] prescription for super YM amplitudes in the formulation by integral [12].

Yangian symmetric correlators are the solutions of the eigenvalue relations of monodromy operators restricted to definite dilatation weights at the points. To each of the n-dimensional points a dilatation weight, a spectral parameter and a signature are associated. The individual spectral parameters are useful ingredients; their role is not restricted to the one of expansion parameter in the monodromy operators as being generating functions of the Yangian algebra generators.

The Jordan-Schwinger type representations of $g\ell_n$ is a simple basic case of higher rank (A_{n-1}) symmetry. Generic irreducible representations of this type are characterized by just one parameter (related to the dilatation weight). Such representations are the building units by which general representations can be constructed iteratively. Extensions of the describied construction of symmetric correlators to general types of $g\ell_n$ representations or to quantum deformation are straightforward.

The simplicity of the Jordan-Schwinger case is manifest in the simple form of the L-matrices, leading by elementary steps to relations for similarity transformations, inversion, matrix transposition and operator conjugation. These transformations result in relations for monodromy operators and consequently for symmetric correlators.

The n degrees of freedom associated with any chain site can be related to a n dimensional position space. In this way the applications of the corresponding integrable dynamical system are not necessarily restricted to one or two (discrete) dimensions. The one-dimensional structure of the associated spin chain is reflected in the cyclicity property of the correlators.

Acknowledgments

The authors are grateful to Sergey Derkachov for useful discussions.

The work of D.C. is supported by the Chebyshev Laboratory (Department of Mathematics and Mechanics, St.-Petersburg State University) under RF government grant 11.G34.31.0026, by JSC "Gazprom Neft" and by Dmitry Zimin's "Dynasty" Foundation. He thanks Leipzig University for hospitality and DAAD for support.

References

[1] Jordan P 1935 Der Zusammenhang der symmetrischen und linearen Gruppen und das Mehrkörperproblem *Zeitschr. f. Physik*, **94** 331

[2] Schwinger J 1952 *notes reprinted in: Quantum Mechanics of Angular Momentum* Biedenharn L C and van Dam H (eds.) (London: Academic Press 1965)

[3] Holstein T and Primakoff H 1940 Field dependence of the intrinsic domain magnetization of a ferromagnet *Phys. Rev.* **58** 1098

[4] Gelfand I M and Naimark M A 1950 *Unitary representations of the classical groups, Trudy Math. Inst. Steklov, Vol.* **36**, (Moscow-Leningrad); (German translation: 1957 Berlin:Akademie Verlag)

[5] Borel A and Weil A 1954 *Sem. Bourbaki, Representations lineaires et espaces homogenes Kählerians des groupes de Lie compactes*

[6] Hecht K T 1987 *The vector coherent state method and its application to problems of higher symmetry. Lecture Notes in Physics* **290** (Berlin:Springer)

[7] Biedenharn L C and Lohe M A 1992 An extension of the Borel-Weil construction to the quantum group $U_q(n)$ *Commun. Math. Phys.* **146** 483; 1995 *Quantum group symmetry and q-tensor algebras* (Singapore:World Scientific)

[8] Dobrev V K, Truini P and Biedenharn L C 1994 Representation theory approach to the polynomial solutions of q difference equations:U-q(sl(3)) and beyond *J. Math. Phys.* **35** 6058 [arXiv:q-alg/9502001]

[9] Karakhanyan D and Kirschner R 2010 Jordan-Schwinger representations and factorised Yang-Baxter operators *SIGMA* **6** 029 [arXiv:0910.5144 [hep-th]]

[10] Kirschner R 2013 Integrable chains with Jordan-Schwinger representations *J. Phys. Conf. Ser.* **411** 012018

[11] Drummond J M Henn J Korchemsky G P and Sokatchev E 2010 Dual superconformal symmetry of scattering amplitudes in N=4 super-Yang-Mills theory *Nucl. Phys. B* **828** 317 [arXiv:0807.1095 [hep-th]]
Drummond J M Henn J M and Plefka J 2009 Yangian symmetry of scattering amplitudes in N=4 super Yang-Mills theory *JHEP* **0905** 046 [arXiv:0902.2987 [hep-th]]
Drummond J M and Ferro L 2010 Yangians, Grassmannians and T-duality *JHEP* **1007** 027 [arXiv:1001.3348 [hep-th]]
[12] Arkani-Hamed N Cachazo F Cheung C and Kaplan J 2010 The S-Matrix in Twistor Space *JHEP* **1003** 110 [arXiv:0903.2110 [hep-th]
[13] Arkani-Hamed N, Cachazo F, Cheung C and Kaplan J 2010 A Duality For The S Matrix *JHEP* **1003** 020 [arXiv:0907.5418 [hep-th]]
[14] Ferro L Lukowski T Meneghelli C Plefka J and Staudacher M 2012 Harmonic R-matrices for Scattering Amplitudes and Spectral Regularization arXiv:1212.0850 [hep-th].
[15] Berg B Karowski M and Weisz P 1979 Construction of Green Functions from an Exact S Matrix *Phys. Rev. D* **19** 2477
[16] Zamolodchikov A B and Zamolodchikov A B 1979 Factorized s Matrices in Two-Dimensions as the Exact Solutions of Certain Relativistic Quantum Field Models *Annals Phys.* **120** 253
[17] Ghoshal S and Zamolodchikov A B 1994 Boundary S matrix and boundary state in two-dimensional integrable quantum field theory *Int. J. Mod. Phys. A* **9** 3841 [Erratum-ibid. A **9** 4353] [hep-th/9306002]
[18] Drinfeld V G 1985 Hopf algebras and the quantum Yang-Baxter equation *Sov. Math. Dokl.* **32** 254 [*Dokl. Akad. Nauk Ser. Fiz.* **283** 1060]
Drinfeld V G 1988 A New realization of Yangians and quantized affine algebras *Sov. Math. Dokl.* **36** 212
[19] Takhtajan L A and Faddeev L D 1979 The Quantum method of the inverse problem and the Heisenberg XYZ model *Russ. Math. Surveys* **34** 11 [*Usp. Mat. Nauk* **34** 13]
[20] Kulish P P and Sklyanin E K 1982 Quantum Spectral Transform Method. Recent Developments *Lect. Notes Phys.* **151** 61
[21] Tarasov V O 1985 Irreducible Monodromy Matrices For The R Matrix Of The Xxz Model And Local Lattice Quantum Hamiltonians *Theor. Math. Phys.* **63** 440 [*Teor. Mat. Fiz.* **63** 175]
Tarasov V O 1993 Cyclic monodromy matrices for sl(n) trigonometric R matrices *Commun. Math. Phys.* **158** 459 [hep-th/9211105].
[22] Molev A Nazarov M and Olshansky G 1996 Yangians and classical Lie algebras *Russ. Math. Surveys* **51** 205 [hep-th/9409025]
[23] Chicherin D and Kirschner R 2013 Yangian symmetric correlators arXiv:1306.0711 [math-ph]
[24] Kulish P P Reshetikhin N Y and Sklyanin E K 1981 Yang-Baxter Equation and Representation Theory 1 *Lett. Math. Phys.* **5** 393
[25] Faddeev L D 1996 *How algebraic Bethe ansatz works for integrable model* hep-th/9605187
[26] Britto B Cachazo F and Feng B 2005 New recursion relations for tree amplitudes of gluons *Nucl. Phys. B* **715** 499 [hep-th/0412308]
Britto R Cachazo F Feng B and Witten E 2005 Direct proof of tree-level recursion relation in Yang-Mills theory *Phys. Rev. Lett.* **94** 181602 [hep-th/0501052].

Integrable Hamiltonian systems generated by antisymmetric matrices

Alina Dobrogowska
Institute of Mathematics, University of Białystok
Akademicka 2, 15-267 Białystok, Poland
E-mail: alaryzko@alpha.uwb.edu.pl

Abstract. We construct a family of integrable systems generated by the Casimir functions of Lie algebra of skew-symmetric matrices, where the Lie bracket is deformed by a symmetric matrix.

1. Introduction

Let $\mathcal{A}(n)$ be the vector space of antisymmetric $n \times n$ matrices and $Sym(n)$ be the vector space of symmetric $n \times n$ matrices. The $(\mathcal{A}(n), [\cdot,\cdot]_S)$ is a Lie algebra with the S–bracket defined by a deformeded commutator

$$[X, Y]_S = XSY - YSX \tag{1}$$

for fixed $S \in Sym(n)$ and $X, Y \in \mathcal{A}(n)$, (see [4, 5]).

In this paper we construct the family of integrable systems — a hierarchy generated by the Casimir functions on the dual of Lie algebra $\mathcal{A}(n)$. We prove that the integrals of this family of Hamiltonian systems are in involution.

The idea of considering these systems comes from [1]. In this paper we present more general case, which reduces to the case considering in [1] if we put that the matrix $S = \mathbf{1}$. Also in [3] the authors studied similar systems in the complex setting and for matrices with a different internal structure.

2. Hierarchy generated by Casimir functions

We identify $\mathcal{A}(n)$ with its dual $\mathcal{A}^*(n) \cong \mathcal{A}(n)$ using natural non-degenerate pairing by trace of the product

$$\langle X, \rho \rangle = \mathrm{Tr}(\rho X), \quad \rho \in \mathcal{A}^*(n), \quad X \in \mathcal{A}(n). \tag{2}$$

We shall write a general element $\mathcal{A}(n)$ as

$$X = \begin{pmatrix} A & B \\ -B^\top & C \end{pmatrix}, \tag{3}$$

where $A \in \mathcal{A}(2)$, $C \in \mathcal{A}(n-2)$ and $B \in \mathrm{Mat}_{2\times(n-2)}(\mathbb{R})$. Having Lie algebra $(\mathcal{A}(n), [\cdot,\cdot]_S)$ one defines the Lie-Poisson bracket on $C^\infty(\mathcal{A}(n))$ by

$$\{f, g\}_S = \mathrm{Tr}\left(X \left[\frac{\partial f}{\partial X}, \frac{\partial g}{\partial X}\right]_S\right), \quad f, g \in C^\infty(\mathcal{A}(n)), \tag{4}$$

where

$$\frac{\partial f}{\partial X} = \begin{pmatrix} \frac{\partial f}{\partial A} & \frac{\partial f}{\partial B} \\ -\frac{\partial f}{\partial B^\top} & \frac{\partial f}{\partial C} \end{pmatrix}, \tag{5}$$

In order to obtain the second Poisson bracket ("frozen" Poisson bracket) we fix the element $X_0 \in \mathcal{A}(n)$ and put

$$\{f, g\}_{FS} = \mathrm{Tr}\left(X_0 \left[\frac{\partial f}{\partial X}, \frac{\partial g}{\partial X} \right]_S \right), \quad f, g \in C^\infty(\mathcal{A}(n)). \tag{6}$$

It is a general fact that the Lie-Poisson bracket and frozen bracket are compatible in the sense that their linear combination

$$\alpha\{\cdot,\cdot\}_S + \beta\{\cdot,\cdot\}_{FS} \tag{7}$$

is also a Poisson bracket.

We shall choose

$$X_0 = \frac{1}{2}\begin{pmatrix} A_0 & 0 \\ 0 & 0 \end{pmatrix}, \quad S = \begin{pmatrix} S_1 & S_3 \\ S_3^\top & S_2 \end{pmatrix}, \tag{8}$$

where A_0 is 2×2 matrix defined by

$$A_0 := \begin{pmatrix} 0 & 1 \\ -1 & 0 \end{pmatrix}. \tag{9}$$

After simple calculation we show that the Poisson bracket (6) can be written in the form

$$\begin{aligned} \{f, g\}_{FS} = \mathrm{Tr}\left(\left(\frac{\partial g}{\partial A} A_0 \frac{\partial f}{\partial B} - \frac{\partial f}{\partial A} A_0 \frac{\partial g}{\partial B} \right) S_3^\top \right) + \\ + \mathrm{Tr}\left(\frac{\partial f}{\partial B^\top} A_0 \frac{\partial g}{\partial B} S_2 \right). \end{aligned} \tag{10}$$

Basic assumption. From now we put the block S_3 equal to zero ($S_3 \equiv 0$). After reducing to this case we obtain

$$\{f, g\}_{FS} = \mathrm{Tr}\left(\frac{\partial f}{\partial B^\top} A_0 \frac{\partial g}{\partial B} S_2 \right). \tag{11}$$

Thus, we can think of this bracket as being defined on $C^\infty(\mathrm{Mat}_{2\times(n-2)}(\mathbb{R}))$ (thus $\mathcal{A}(n) \to \mathrm{Mat}_{2\times(n-2)}(\mathbb{R})$ is injective smooth Poisson map).

In the case when $\det S \neq 0$ the Casimir functions for the Lie–Poisson bracket (4) are given by

$$C_k(X) = \frac{1}{2k} \mathrm{Tr}(XS^{-1})^{2k}, \quad k = 1, 2, \ldots \tag{12}$$

see [5]. For the degenerate case when $S_1 \equiv 0$ we know only some Casimir functions of the following form

$$C_k(X) = \frac{1}{k} \mathrm{Tr}\left(B^\top B S_2^{-1} \right)^k, \quad k = 1, 2, \ldots, \tag{13}$$

(see [2] for the case $S_2 = 1$). In this case the Lie-Poisson bracket (4) can be rewritten in the form

$$\begin{aligned} \{f, g\}_S = & 2\,\mathrm{Tr}\left(\frac{\partial f}{\partial B^\top} A \frac{\partial g}{\partial B} S_2 + \frac{\partial g}{\partial C} C \frac{\partial f}{\partial C} S_2 \right) + \\ & + 2\,\mathrm{Tr}\left(\left(\frac{\partial f}{\partial B^\top} B \frac{\partial g}{\partial C} - \frac{\partial g}{\partial B^\top} B \frac{\partial f}{\partial C} \right) S_2 \right). \end{aligned} \tag{14}$$

Now, we show that the functions given by (13) are Casimir functions for the bracket (14). Since the derivative of C_k is

$$\frac{\partial C_k}{\partial B} = 2BS_2^{-1}\left(B^\top BS_2^{-1}\right)^{k-1}, \tag{15}$$

$$\frac{\partial C_k}{\partial B^\top} = 2S_2^{-1}\left(B^\top BS_2^{-1}\right)^{k-1}B^\top, \tag{16}$$

$$\frac{\partial C_k}{\partial C} = 0, \tag{17}$$

we have

$$\begin{aligned}\{C_k, C_l\}_S &= \mathrm{Tr}\left(\frac{\partial C_k}{\partial B^\top}A\frac{\partial C_l}{\partial B}S_2\right) = \\ &= 4\,\mathrm{Tr}\left(S_2^{-1}\left(B^\top BS_2^{-1}\right)^{k-1}B^\top ABS_2^{-1}\left(B^\top BS_2^{-1}\right)^{l-1}S_2\right) = \\ &= 4\,\mathrm{Tr}\left(B^\top ABS_2^{-1}\left(B^\top BS_2^{-1}\right)^{k+l-2}\right) = 0 = \{C_k, C_l\}_{FS},\end{aligned} \tag{18}$$

because the matrix $B^\top AB$ is antisymmetric and $S_2^{-1}\left(B^\top BS_2^{-1}\right)^{k+l-2}$ is symmetric. Moreover we have the following proposition.

Proposition 1 *The Casimir functions C_k defined by (12) or (13) for the Lie-Poisson bracket (4) considered as functions of B are in involution with respect to the frozen bracket (11).*

Proof 1 *Since the derivative of C_k given by (12) is*

$$\frac{\partial C_k}{\partial B} = -2P_+\left(XS^{-1}\right)^{2k-1}P_-, \tag{19}$$

$$\frac{\partial C_k}{\partial B^\top} = 2P_-\left(S^{-1}X\right)^{2k-1}P_+, \tag{20}$$

where P_+, P_- are the orthogonal projectors given, in block matrix notation, by

$$P_+ = \begin{pmatrix} \mathbf{1} & 0 \\ 0 & 0 \end{pmatrix}, \quad P_- = \begin{pmatrix} 0 & 0 \\ 0 & \mathbf{1} \end{pmatrix}. \tag{21}$$

After a direct calculation we obtain

$$\begin{aligned}\{C_k, C_l\}_{FS} &= \mathrm{Tr}\left(\frac{\partial C_k}{\partial B^\top}A\frac{\partial C_l}{\partial B}S_2\right) = \\ &= -4\,\mathrm{Tr}\left(P_-\left(S^{-1}X\right)^{2k-1}P_+A_0P_+\left(XS^{-1}\right)^{2l-1}P_-S_2\right) = \\ &= -4\,\mathrm{Tr}\left(P_-\left(XS^{-1}\right)^{2k-2}XP_+A_0P_+\left(XS^{-1}\right)^{2l-1}P_-\right) = \\ &= -4\,\mathrm{Tr}\left(\left(XS^{-1}\right)^{2k-2}XP_+A_0P_+\left(XS^{-1}\right)^{2l-1}\right) + \\ &+ 4\,\mathrm{Tr}\left(P_+\left(XS^{-1}\right)^{2k-2}XP_+A_0P_+\left(XS^{-1}\right)^{2l-1}P_+\right) = \\ &= -4\,\mathrm{Tr}\left(\left(XS^{-1}\right)^{k+l-2}XP_+A_0P_+X\left(S^{-1}X\right)^{k+l-2}S^{-1}\right) = 0.\end{aligned} \tag{22}$$

Above vanishes because in the first term we have a product of three antisymmetric 2×2 matrices which is also antisymmetric and in the second term we have a product of an antisymmetric matrix $\left(XS^{-1}\right)^{k+l-2}XP_+A_0P_+X\left(S^{-1}X\right)^{k+l-2}$ and symmetric matrix S^{-1}.

The proof of the involution of the functions (13) with respect to the frozen Poisson bracket (11), was given before this proposition.

Proposition 2 *The smooth functions* $\delta_k : \mathrm{Mat}_{2\times(n-2)}(\mathbb{R}) \to \mathbb{R}$ *defined by*

$$\delta_k(B) = \mathrm{Tr}\left(BS_2^{-1}\left(CS_2^{-1}\right)^{2k-1}B^\top A_0\right) \tag{23}$$

are in involution with respect to the frozen Poisson bracket (11)

$$\{\delta_k, \delta_l\}_{FS} = 0. \tag{24}$$

Proof 2 *Since the derivative of* δ_k *is*

$$\frac{\partial \delta_k}{\partial B} = 2A_0BS_2^{-1}\left(CS_2^{-1}\right)^{2k-1}, \tag{25}$$

$$\frac{\partial \delta_k}{\partial B^\top} = 2S_2^{-1}\left(CS_2^{-1}\right)^{2k-1}B^\top A_0, \tag{26}$$

we have

$$\begin{aligned}\{\delta_k, \delta_l\}_{FS} &= \mathrm{Tr}\left(\frac{\partial \delta_k}{\partial B^\top}A_0\frac{\partial \delta_l}{\partial B}S_2\right) = \\ &= 4\,\mathrm{Tr}\left(S_2^{-1}\left(CS_2^{-1}\right)^{2k-1}B^\top A_0A_0A_0BS_2^{-1}\left(CS_2^{-1}\right)^{2l-1}S_2\right) = \\ &= -4\,\mathrm{Tr}\left(B^\top A_0BS_2^{-1}\left(CS_2^{-1}\right)^{2(k+l)-2}\right) = 0,\end{aligned} \tag{27}$$

because the matrix $B^\top A_0 B$ *is antisymmetric and* $S_2^{-1}\left(CS_2^{-1}\right)^{2(k+l)-2}$ *is symmetric.*

Proposition 3 *Assume that* $S_1 = 1$. *Then the functions* δ_k *and* C_l *(given by (12)),* $k, l = 1, 2, \ldots,$ *are in involution with respect to the frozen Poisson bracket (11)*

$$\{\delta_k, C_l\}_{FS} = 0. \tag{28}$$

Proof 3 *First, we show that* δ_1 *commutes with* C_k *given by (12)*

$$\begin{aligned}\{\delta_1, C_k\}_{FS} &= \mathrm{Tr}\left(\frac{\partial \delta_1}{\partial B^\top}A_0\frac{\partial C_k}{\partial B}\right) = \\ &= -4\,\mathrm{Tr}\left(S_2^{-1}CS_2^{-1}B^\top A_0A_0P_+\left(XS^{-1}\right)^{2k-1}P_-S_2\right) = \\ &= 4\,\mathrm{Tr}\left(CS_2^{-1}B^\top P_+\left(XS^{-1}\right)^{2k-1}P_-\right) = \\ &= -4\,\mathrm{Tr}\left(P_-\left(XS^{-1}\right)P_-XP_+\left(XS^{-1}\right)^{2k-1}P_-\right) = \\ &= -4\,\mathrm{Tr}\left(\left(XS^{-1}\right)P_-XP_+\left(XS^{-1}\right)^{2k-1}\right) + \\ &+ 4\,\mathrm{Tr}\left(P_+\left(XS^{-1}\right)P_-XP_+\left(XS^{-1}\right)^{2k-1}P_+\right) = \\ &= -4\,\mathrm{Tr}\left(P_-XP_+\left(XS^{-1}\right)^{2k}\right) + \\ &- 4\,\mathrm{Tr}\left(P_-XP_+\left(XS^{-1}\right)^{2k-1}\left(P_-XP_+\right)^\top S^{-1}\right) = \\ &= -4\,\mathrm{Tr}\left(XP_+\left(XS^{-1}\right)^{2k}\right) + \\ &+ 4\,\mathrm{Tr}\left(P_+XP_+\left(XS^{-1}\right)^{2k}P_+\right) = \\ &= -4\,\mathrm{Tr}\left(P_+\left(XS^{-1}\right)^{2k}XP_+\right) = 0,\end{aligned} \tag{29}$$

because $P_+\left(XS^{-1}\right)^{2k}XP_+$ *is antisymmetric,* P_+XP_+ *is antisymmetric and* $P_+\left(XS^{-1}\right)^{2k}P_+$ *is symmetric. Second, the functions* δ_k *and* C_l *satisfy the following recursion formula*

$$\{\delta_k, C_l\}_{FS} = \{\delta_{k-1}, C_{l+1}\}_{FS}. \tag{30}$$

Thus the relation (28) is valid for any k.

Proposition 4 *The functions* δ_k *and* C_l *(given by (13)),* $k, l = 1, 2, \dots,$ *are in involution with respect to the frozen Poisson bracket (11)*

$$\{\delta_k, C_l\}_{FS} = 0. \tag{31}$$

Proof 4 *For the functions* C_k *given by (13) we have*

$$\{\delta_k, C_l\}_{FS} = \mathrm{Tr}\left(\frac{\partial \delta_k}{\partial B^\top} A_0 \frac{\partial C_l}{\partial B}\right) = \tag{32}$$

$$= 4\,\mathrm{Tr}\left(S_2^{-1}\left(CS_2^{-1}\right)^{2k-1} B^\top A_0 A_0 B S_2^{-1}\left(B^\top B S_2^{-1}\right)^{l-1} S_2\right) =$$

$$= -4\,\mathrm{Tr}\left(\left(CS_2^{-1}\right)^{2k} CS_2^{-1}\left(B^\top B S_2^{-1}\right)^{l}\right) = 0, \tag{33}$$

because the matrix $\left(CS_2^{-1}\right)^{2k}C$ *is antisymmetric and* $S_2^{-1}\left(B^\top B S_2^{-1}\right)^l$ *is symmetric.*

We obtain a hierarchy of Hamilton's equations generated by Hamiltonians C_k given by (12) or (13) with respect the frozen Poisson bracket (11)

$$\frac{\partial B}{\partial t_k} = A_0 \frac{\partial C_k}{\partial B} S_2, \quad k = 1, 2, \dots \tag{34}$$

Example 1 *In this example we consider the case when* X *is* 5×5*-matrix which we denote*

$$X = \begin{pmatrix} 0 & a & p_1 & p_2 & p_3 \\ -a & 0 & q_1 & q_2 & q_3 \\ -p_1 & -q_1 & 0 & -c_3 & c_2 \\ -p_2 & -q_2 & c_3 & 0 & -c_1 \\ -p_3 & -q_3 & -c_2 & c_1 & 0 \end{pmatrix} \tag{35}$$

and matrix S *is degenerate, that mean* $S_1 = 0$ *and*

$$S_2 = \begin{pmatrix} e_1 & 0 & 0 \\ 0 & e_2 & 0 \\ 0 & 0 & e_3 \end{pmatrix}. \tag{36}$$

The frozen Poisson bracket in this case is

$$\{f,g\}_{FS}(p_i,q_i) = e_1\left(\frac{\partial f}{\partial p_1}\frac{\partial g}{\partial q_1} - \frac{\partial f}{\partial q_1}\frac{\partial g}{\partial p_1}\right) + e_2\left(\frac{\partial f}{\partial p_2}\frac{\partial g}{\partial q_2} - \frac{\partial f}{\partial q_2}\frac{\partial g}{\partial p_2}\right) +$$

$$+ e_3\left(\frac{\partial f}{\partial p_3}\frac{\partial g}{\partial q_3} - \frac{\partial f}{\partial q_2}\frac{\partial g}{\partial p_3}\right). \tag{37}$$

The integrals in involution are

$$C_1 = \left(S_2^{-1}\vec{p}\right) \cdot \vec{p} + \left(S_2^{-1}\vec{q}\right) \cdot \vec{q}, \tag{38}$$

$$C_2 = \frac{1}{2}C_1^2 - \left(\left(S_2^{-1}\vec{q}\right) \times \left(S_2^{-1}\vec{p}\right)\right) \cdot \left(\vec{q} \times \vec{p}\right), \tag{39}$$

$$\delta_1 = -2\vec{C} \cdot \left(\left(S_2^{-1}\vec{q}\right) \times \left(S_2^{-1}\vec{p}\right)\right). \tag{40}$$

Hamilton's equations for the Hamiltonian C_1 are

$$\frac{\partial \vec{p}}{\partial t} = 2\vec{q}, \tag{41}$$

$$\frac{\partial \vec{q}}{\partial t} = -2\vec{p}. \tag{42}$$

Hamilton's equations for the Hamiltonian C_2 are

$$\frac{\partial \vec{p}}{\partial t} = 2\left(C_1\vec{q} + \left(S_2^{-1}\vec{p}\right) \times \left(\vec{p} \times \vec{q}\right)\right), \tag{43}$$

$$\frac{\partial \vec{q}}{\partial t} = 2\left(-C_1\vec{p} + \left(S_2^{-1}\vec{q}\right) \times \left(\vec{p} \times \vec{q}\right)\right). \tag{44}$$

References

[1] A. Dobrogowska, T.S. Ratiu, *Integrable systems of Neumann type*, J. Dyn. Diff. Equat., DOI : 10.1007/s10884-013-9314-5, (2013).

[2] A. Odzijewicz, A. Dobrogowska, *Integrable Hamiltonian systems related to the Hilbert–Schmidt ideal*, J. Geom. Phys., **61**, 1426-1445 (2011).

[3] A. Odzijewicz, T. Goliński, *Hierarchy of integrable Hamiltonians describing the nonlinear n-wave interaction*, J. Phys. A Math. Theor., **45**, no. 4 045204 (2012).

[4] A.B. Yanovski, *Linear bundles of lie algebras and their applications*, J. Math. Phys., **41**, No. 12, (2000).

[5] V.V. Trofimov, A.T. Fomenko, *Algebra and geometry of integrable Hamiltonian differential equations*, Faktorial, Moscow, Russian (1995).

Maxwell group and HS field theory

Sergey Fedoruk[1,2], **Jerzy Lukierski**[3]

[1] On leave of absence from V. N. Karazin Kharkov National University, Ukraine
[2] Bogoliubov Laboratory of Theoretical Physics, JINR, Joliot-Curie 6, 141980 Dubna, Moscow region, Russia
[3] Institute for Theoretical Physics, University of Wrocław, pl. Maxa Borna 9, 50-204 Wrocław, Poland

E-mail: fedoruk@theor.jinr.ru, lukier@ift.uni.wroc.pl

Abstract. We consider the master fields for HS multiplets defined on 10-dimensional tensorial extension $\tilde{\mathcal{M}}$ of $D = 4$ space-time described as a coset $\tilde{\mathcal{M}} = \mathcal{M}/Sl(2;\mathbb{C})$ of 16-parameter Maxwell group $\mathcal{M}$. The tensorial coordinates provide a geometrization of the coupling to constant uniform EM fields. We describe the spinorial model in extended space-time $\tilde{\mathcal{M}}$ and by its first quantization we obtain new infinite HS-Maxwell multiplets with their massless components coupled to each other through constant EM background. We conclude our report by observing that three-dimensional spinorial model with a pair of spinors should provide after quantization $D = 3$ massive HS-Maxwell multiplets.

1. Introduction

In order to introduce the field-theoretic description of infinite number of relativistic quantum fields with all spins it is convenient to consider the enlargement of space-time. There were proposed various extensions of Minkowski space-time with vectorial [1, 2] and tensorial [3, 4, 5, 6, 7] auxiliary coordinates, with master fields describing the infinite-dimensional spin multiplets by Taylor expansion in additional variables. In such a way one connects the master HS fields with enlarged $D = 4$ Poincare algebra, with new generators describing the shifts in auxiliary variables.

In this talk we shall follow old derivation of HS field multiplets by quantization of spinorial superparticle model, invariant under SUSY with six tensorial central charges[1] [4, 5, 7]. Analogous tensorial charges occur e.g. in $D = 11$ in M-algebra [8] which as it is postulated describes the algebraic structure of eleven-dimensional M-theory. The master fields in $D = 4$ are described in such approach by fields on extended space-time $(\mathrm{x}_\mu, \mathrm{z}_{\mu\nu})$, where $\mathrm{z}_{\mu\nu} = \mathrm{z}_{[\mu\nu]}$ are six auxiliary translations generated by tensorial central charges. Further in order to describe integer and half-integer spins it is convenient to express the tensorial central charges as bilinears in $D = 4$

[1] We point out that tensorial central charges are commuting with all generators except the Lorentz generators.

Weyl spinorial variables λ_α, $\bar{\lambda}_{\dot{\alpha}} = (\overline{\lambda_\alpha})$, $\alpha = 1, 2$[2]

$$\mathrm{Z}_{\mu\nu} = \lambda^\alpha (\sigma_{\mu\nu})_\alpha{}^\beta \lambda_\beta + \bar{\lambda}_{\dot{\alpha}} (\tilde{\sigma}_{\mu\nu})^{\dot{\alpha}}{}_{\dot{\beta}} \bar{\lambda}^{\dot{\beta}} \tag{1}$$

in analogy to the Penrose formula for massless fourmomenta

$$\mathrm{P}_\mu = \lambda^\alpha (\sigma_\mu)_{\alpha\dot{\beta}} \bar{\lambda}^{\dot{\beta}} . \tag{2}$$

Finally we will arrive at master fields depending on spinorially extended space-time $(x_\mu, \lambda_\alpha, \bar{\lambda}_{\dot{\alpha}})$, with their Taylor expansions describing HS fields with arbitrary spin.

The description of HS with tensorial coordinates has been further generalized by M. Vasiliev [6, 9, 10] who followed Fronsdal observation [3] that the infinite HS free multiplets can be described by single irrep of $D = 4$ generalized conformal algebra $Sp(8)$, containing as its subalgebra the generalized Poincare algebra with tensorial central charges. Subsequently the extended space-time method was applied to the description in AdS space-time and there were derived the multiplets of free HS fields on AdS space [11, 12]. At present only two free HS field multiplets – in Minkowski and AdS space-time – are explicitly described, and both were derived by the method of quantization of spinorial model of Shirafuji type.[3]

In our presentation we shall describe the spinorial model defined on other version of ten-dimensional extended space-time $(x_\mu, z_{\mu\nu})$, with coordinates $z_{\mu\nu}$ generated by the tensorial central charges $Z_{\mu\nu}$ introduced in alternative way in the enlargement of Poincare algebra – by replacing commuting four-momenta P_μ by noncommutative ones P_μ

$$[\mathrm{P}_\mu, \mathrm{P}_\nu] = 0 \qquad \Rightarrow \qquad [P_\mu, P_\nu] = Z_{\mu\nu} \tag{3}$$

where

$$[Z_{\mu\nu}, Z_{\lambda\rho}] = 0 , \qquad [P_\mu, Z_{\lambda\rho}] = 0 . \tag{4}$$

Formulae (3), (4) describe the enlargement of $D = 4$ Poincare algebra to $D = 4$ Maxwell algebra firstly proposed more than 40 years ago [15, 16, 17]. Using Maurer-Cartan (MC) forms for corresponding Maxwell group one can introduce [7] the Maxwell-invariant spinorial particle model and perform its first quantization.[4] By analysis of the constraints in extended phase space we shall obtain in Schrödinger realization the master fields for Maxwell-HS free fields which are coupled to each other by constant uniform EM field.

The plan of our talk is the following. In Sect. 2 we will recall the notion of Maxwell group, Maxwell algebra and present the corresponding MC one-forms. Using this geometric framework we present in Sect. 3 the spinorial model [7] and its first quantization with complete discussion of occurring first and second class constraints. We will use the conversion method [18] which interprets canonical pair of second class constraints as describing the system with gauge-fixed local gauge transformations (one constraint from the canonical pair is considered as first class constraint and generating the gauge transformations, second as introducing the gauge-fixing condition). In such a way we get gauge-equivalent formulation of our model with nine first

[2] We shall use $D = 4$ two-spinor notation, i.e. $P_{\alpha\dot{\beta}} = \sigma^\mu_{\alpha\dot{\beta}} P_\mu$, $Z_{\alpha\beta} = \sigma^{\mu\nu}_{\alpha\beta} Z_{\mu\nu}$, $\bar{Z}_{\dot{\alpha}\dot{\beta}} = \tilde{\sigma}^{\mu\nu}_{\dot{\alpha}\dot{\beta}} Z_{\mu\nu}$, where $(\sigma^\mu)_{\alpha\dot{\alpha}} = (1_2, \vec{\sigma})_{\alpha\dot{\alpha}}$, $(\tilde{\sigma}^\mu)^{\dot{\alpha}\alpha} = \epsilon^{\alpha\beta}\epsilon^{\dot{\alpha}\dot{\beta}}(\sigma^\mu)_{\beta\dot{\beta}} = (1_2, -\vec{\sigma})^{\dot{\alpha}\alpha}$, $\sigma^{\mu\nu} = i\,\sigma^{[\mu}\tilde{\sigma}^{\nu]}$, $\tilde{\sigma}^{\mu\nu} = i\,\tilde{\sigma}^{[\mu}\sigma^{\nu]}$ $\sigma^{\mu\nu}_{\alpha\beta} = \epsilon_{\beta\gamma}(\sigma^{\mu\nu})_\alpha{}^\gamma$, $\tilde{\sigma}^{\mu\nu}_{\dot{\alpha}\dot{\beta}} = \epsilon_{\dot{\beta}\dot{\gamma}}(\tilde{\sigma}^{\mu\nu})^{\dot{\gamma}}{}_{\dot{\alpha}}$. We use weight coefficient in (anti)symmetrization, i.e. $A_{(\alpha}B_{\beta)} = \frac{1}{2}(A_\alpha B_\beta + A_\beta B_\alpha)$, $A_{[\alpha}B_{\beta]} = \frac{1}{2}(A_\alpha B_\beta - A_\beta B_\alpha)$.

[3] The formula (2) was firstly inserted in massless spinorial Shirafuji model [13] which was the first one describing the link between the model describing standard relativistic massless superparticle and the one describing free twistorial dynamics in $D = 4$ supertwistor space. The formula (1) has been firstly employed in the superparticle model in [4].

[4] Different Maxwell-invariant particle model without using spinorial variables was made in [14].

class constraints. In Sect. 4 we consider in detail the wave function of such a model satisfying on enlarged space-time nine wave equations. Finally we express the solution of quantum-mechanical model in terms of local $D = 4$ HS fields, which will be named free Maxwell-HS fields. In such a way we obtain three infinite-dimensional coupled sets of Maxwell-HS fields: one describing all HS bosonic fields ($s = 0, 1, 2, \ldots$) and two infinite set of chiral and antichiral fermionic HS fields with half-integer spins ($s = 1/2, 3/2, 5/2, \ldots$). In Sect. 5 we will discuss the considered model in dual representation as the one describing more explicitly Minkowski HS fields interacting with constant EM field. Sect. 6 is devoted to the outlook. In particular we conjecture that suitable reduction $D = 4 \to D = 3$ of considered in [7] $D = 4$ spinorial model can provide the infinite-dimensional $D = 3$ Maxwell HS massive multiplets.

2. Maxwell algebra and covariant Maurer-Cartan one-forms

The Maxwell algebra [15, 16] is the semi-direct sum of Lorentz algebra with the generators $M_{\alpha\beta} = M_{\beta\alpha}$, $\bar{M}_{\dot{\alpha}\dot{\beta}} = (M_{\alpha\beta})^+$ satisfying relations [2]

$$\begin{aligned} [M_{\alpha\beta}, M_{\gamma\delta}] &= i\left(\epsilon_{\alpha\gamma} M_{\beta\delta} + \epsilon_{\beta\delta} M_{\alpha\gamma}\right), \\ [\bar{M}_{\dot{\alpha}\dot{\beta}}, \bar{M}_{\dot{\gamma}\dot{\delta}}] &= i\left(\epsilon_{\dot{\alpha}\dot{\gamma}} \bar{M}_{\dot{\beta}\dot{\delta}} + \epsilon_{\dot{\beta}\dot{\delta}} \bar{M}_{\dot{\alpha}\dot{\gamma}}\right), \\ [M_{\alpha\beta}, \bar{M}_{\dot{\gamma}\dot{\delta}}] &= 0, \end{aligned} \tag{5}$$

and the ten-dimension sector with generators of the Poincare translations $P_{\alpha\dot{\beta}} = (P_{\beta\dot{\alpha}})^+$ and six tensorial generators $Z_{\alpha\beta} = Z_{\beta\alpha}$, $\bar{Z}_{\dot{\alpha}\dot{\beta}} = (Z_{\alpha\beta})^+$. The last ten generators transform under Lorentz algebra as it is indicated in the algebraic relations

$$\begin{aligned} [M_{\alpha\beta}, P_{\gamma\dot{\gamma}}] = -i\epsilon_{\gamma(\alpha} P_{\beta)\dot{\gamma}}\,, \qquad & [\bar{M}_{\dot{\alpha}\dot{\beta}}, P_{\gamma\dot{\gamma}}] = -i\epsilon_{\dot{\gamma}(\dot{\alpha}} P_{\gamma\dot{\beta})}\,, \\ [M_{\alpha\beta}, Z_{\gamma\delta}] &= i\left(\epsilon_{\alpha\gamma} Z_{\beta\delta} + \epsilon_{\beta\delta} Z_{\alpha\gamma}\right), \\ [\bar{M}_{\dot{\alpha}\dot{\beta}}, \bar{Z}_{\dot{\gamma}\dot{\delta}}] &= i\left(\epsilon_{\dot{\alpha}\dot{\gamma}} \bar{Z}_{\dot{\beta}\dot{\delta}} + \epsilon_{\dot{\beta}\dot{\delta}} \bar{Z}_{\dot{\alpha}\dot{\gamma}}\right), \\ [M_{\alpha\beta}, \bar{Z}_{\dot{\gamma}\dot{\delta}}] = 0, \quad & [\bar{M}_{\dot{\alpha}\dot{\beta}}, Z_{\gamma\delta}] = 0, \end{aligned} \tag{6}$$

Differently than in the Poincare algebra with commuting translation operators, the commutators of the quantities $P_{\alpha\dot{\beta}}$ yield new tensorial generators (see also (3))

$$[P_{\alpha\dot{\alpha}}, P_{\beta\dot{\beta}}] = 2i\, e\left(\epsilon_{\dot{\alpha}\dot{\beta}} Z_{\alpha\beta} + \epsilon_{\alpha\beta} \bar{Z}_{\dot{\alpha}\dot{\beta}}\right), \tag{7}$$

where e is a constant interpreted further as describing EM coupling. Six tensorial charges $Z_{\alpha\beta}$, $\bar{Z}_{\dot{\alpha}\dot{\beta}}$ commute with each other and with the Poincare translation generators, i.e. the following commutators (see also (4))

$$[Z_{\alpha\beta}, Z_{\gamma\delta}] = [Z_{\alpha\beta}, \bar{Z}_{\dot{\alpha}\dot{\beta}}] = [Z_{\alpha\beta}, P_{\gamma\dot{\gamma}}] = 0 \tag{8}$$

are valid.

The $D = 4$ Maxwell algebra define the Maxwell group $\mathcal{M}$, in standard way by exponential representation. We define the $D = 4$ proper Maxwell group $\mathcal{M}_0$, determining Maxwell tensorial space, as ten-dimensional coset

$$\mathcal{M}_0 = \frac{\mathcal{M}}{O(3,1)} = e^{i(z^{\alpha\beta} Z_{\alpha\beta} + \bar{z}^{\dot{\alpha}\dot{\beta}} \bar{Z}_{\dot{\alpha}\dot{\beta}})} e^{i x^{\alpha\dot{\beta}} P_{\alpha\dot{\beta}}} \tag{9}$$

with generators $P_{\alpha\dot\beta}$, $Z_{\alpha\beta}$, $\bar Z_{\dot\alpha\dot\beta}$. The coset coordinates have the following transformations under the space-time translations (parameters $a^{\alpha\dot\beta}$) and tensorial Maxwell shifts (parameters $b^{\alpha\beta}$, $\bar b^{\dot\alpha\dot\beta}$)

$$\begin{gathered}\delta x^{\alpha\dot\beta} = a^{\alpha\dot\beta}\,,\\ \delta z^{\alpha\beta} = b^{\alpha\beta} + x^{(\alpha\dot\gamma}a^{\beta)}_{\dot\gamma}\,, \qquad \delta \bar z^{\dot\alpha\dot\beta} = \bar b^{\dot\alpha\dot\beta} + x^{\gamma(\dot\alpha}a^{\dot\beta)}_{\gamma}\,.\end{gathered} \tag{10}$$

The Lorentz transformations with the parameters $\ell^{\alpha\beta}$, $\bar\ell^{\dot\alpha\dot\beta}$ look as follows

$$\delta x^{\alpha\dot\beta} = \ell^{\alpha\gamma}x^{\dot\beta}_{\gamma} + \bar\ell^{\dot\alpha\dot\gamma}x^{\beta}_{\dot\gamma}\,, \qquad \delta z^{\alpha\beta} = 2\ell^{\alpha\gamma}z^{\beta}_{\gamma}\,, \qquad \delta\bar z^{\dot\alpha\dot\beta} = 2\bar\ell^{\dot\alpha\dot\beta}\bar z^{\beta}_{\dot\gamma}\,. \tag{11}$$

Using the parametrization (9) and the algebraic relations (7), (8) we can define the Maurer-Cartan (MC) one-forms

$$\mathcal{M}_0^{-1}d\mathcal{M}_0 = i\Big(\omega^{\alpha\dot\beta}P_{\alpha\dot\beta} + \omega^{\alpha\beta}Z_{\alpha\beta} + \bar\omega^{\dot\alpha\dot\beta}\bar Z_{\dot\alpha\dot\beta}\Big)\,. \tag{12}$$

Explicit formulae for MC one-forms defined by (12) are

$$\begin{gathered}\omega^{\alpha\dot\beta} = dx^{\alpha\dot\beta}\,,\\ \omega^{\alpha\beta} = dz^{\alpha\beta} + e\,x^{(\alpha\dot\gamma}dx^{\beta)}_{\dot\gamma}\,, \qquad \bar\omega^{\dot\alpha\dot\beta} = d\bar z^{\dot\alpha\dot\beta} + e\,x^{\gamma(\dot\alpha}dx^{\dot\beta)}_{\gamma}\,.\end{gathered} \tag{13}$$

Corresponding covariant derivatives have the form

$$\begin{gathered}D_{\alpha\dot\beta} = -i\left(\frac{\partial}{\partial x^{\alpha\dot\beta}} + e\,x^{\gamma}_{\dot\beta}\frac{\partial}{\partial z^{\alpha\gamma}} + e\,x^{\dot\gamma}_{\alpha}\frac{\partial}{\partial \bar z^{\dot\beta\dot\gamma}}\right),\\ D_{\alpha\beta} = -i\,\frac{\partial}{\partial z^{\alpha\beta}}\,, \qquad \bar D_{\dot\alpha\dot\beta} = -i\,\frac{\partial}{\partial \bar z^{\dot\alpha\dot\beta}}\,.\end{gathered} \tag{14}$$

Since the space-time and tensorial translations are the shifts of the group parameters in Maxwell tensorial space, they do not change the MC one-forms (13) and the form of operators (14). We add that the Lorentz symmetry acts on MC one-forms and covariant derivatives in standard way, by the linear transformations.

3. Particle action, constraints and the Casimirs

We will consider the model of massless HS particle which propagates in the Maxwell tensorial space $X = (x^{\alpha\dot\alpha}, z^{\alpha\beta}, \bar z^{\dot\alpha\dot\beta})$ enlarged by the pair of spinorial variables. Our model is described by the following Maxwell-invariant particle action

$$S = S_X + S_\lambda\,. \tag{15}$$

First term in (15)

$$S_X = \int \left(\lambda_\alpha\bar\lambda_{\dot\beta}\,\omega^{\alpha\dot\beta} + a\lambda_\alpha\lambda_\beta\,\omega^{\alpha\beta} + \bar a\bar\lambda_{\dot\alpha}\bar\lambda_{\dot\beta}\,\bar\omega^{\dot\alpha\dot\beta}\right)\,. \tag{16}$$

is the generalization of the $D = 4$ spinorial particle model defined on flat tensorial space-time in [4, 5, 6] (it is a tensorial generalization of Shirafuji model [13]). The components of the *commuting* Weyl spinor λ_α, $\bar\lambda_{\dot\alpha} = (\overline{\lambda_\alpha})$ can be further considered as parts of the $D = 4$ twistor, and the model can be rewritten also as free $D = 4$ twistor particle model [4, 5].

In the action (16) the parameter a is complex. Because e in (16) is dimensionless, we should choose the tensorial coordinates $(z^{\alpha\beta}, \bar{z}^{\dot{\alpha}\dot{\beta}})$ having mass dimensionality equal to -2, $[z^{\alpha\beta}] = [\bar{z}^{\dot{\alpha}\dot{\beta}}] = -2$. Further from (16) follows that $[\lambda^{\alpha}] = [\bar{\lambda}^{\dot{\alpha}}] = \frac{1}{2}$, and one can deduce that the complex parameter a is mass-like, $[a] = 1$. This mass-like parameter can be chosen real, $a = \bar{a} = m$, if we take into account $U(1)$ phase transformations of the spinors $\lambda_{\alpha} = e^{i\varphi}\lambda_{\alpha}$, $\bar{\lambda}_{\dot{\alpha}} = e^{-i\varphi}\bar{\lambda}_{\dot{\alpha}}$.

Second term in (15) has the form

$$S_{\lambda} = \int \left(\lambda_{\alpha} dy^{\alpha} + \bar{\lambda}_{\dot{\alpha}} d\bar{y}^{\dot{\alpha}}\right). \tag{17}$$

Such term defines the variables $(y^{\alpha}, \bar{y}^{\dot{\alpha}})$, $\bar{y}^{\dot{\alpha}} = (\overline{y^{\alpha}})$, and $(\lambda_{\alpha}, \bar{\lambda}_{\dot{\alpha}})$ as canonically conjugated pairs. One can show that without second term (17), the action (16) possesses a complicated structure of the constrains: in such a case follow the second class constraints $y^{\alpha} \approx 0$, $\bar{y}^{\dot{\alpha}} \approx 0$ which make the quantization quite difficult [4, 5]. Adding the term (17) removes these complications.

Thus, the full Lagrangian of the model considered here is defined by [7]

$$\begin{aligned} L &= \lambda_{\alpha}\bar{\lambda}_{\dot{\beta}}\, \dot{x}^{\alpha\dot{\beta}} + m\lambda_{\alpha}\lambda_{\beta}\left(\dot{z}^{\alpha\beta} + e\, x^{\alpha\dot{\gamma}}\dot{x}^{\beta}_{\dot{\gamma}}\right) + m\bar{\lambda}_{\dot{\alpha}}\bar{\lambda}_{\dot{\beta}}\left(\dot{\bar{z}}^{\dot{\alpha}\dot{\beta}} + e\, x^{\gamma\dot{\alpha}}\dot{x}^{\dot{\beta}}_{\gamma}\right) \\ &\quad + \lambda_{\alpha}\dot{y}^{\alpha} + \bar{\lambda}_{\dot{\alpha}}\dot{\bar{y}}^{\dot{\alpha}} \end{aligned} \tag{18}$$

and describes the particle motion in generalized coordinate space $(x^{\alpha\dot{\beta}}, z^{\alpha\beta}, \bar{z}^{\dot{\alpha}\dot{\beta}}, y^{\alpha}, \bar{y}^{\dot{\alpha}})$, with generalized momenta $(p_{\alpha\dot{\beta}}, f_{\alpha\beta}, \bar{f}_{\dot{\alpha}\dot{\beta}}, \lambda_{\alpha}, \bar{\lambda}_{\dot{\alpha}})$ having the following nonvanishing Poisson brackets (PB)

$$\{x^{\alpha\dot{\alpha}}, p_{\beta\dot{\beta}}\}_P = \delta^{\alpha}_{\beta}\delta^{\dot{\alpha}}_{\dot{\beta}}, \quad \{z^{\alpha\beta}, f_{\gamma\delta}\}_P = \delta^{(\alpha}_{\gamma}\delta^{\beta)}_{\delta}, \quad \{\bar{z}^{\dot{\alpha}\dot{\beta}}, \bar{f}_{\dot{\gamma}\dot{\delta}}\}_P = \delta^{(\dot{\alpha}}_{\dot{\gamma}}\delta^{\dot{\beta})}_{\dot{\delta}}, \tag{19}$$

$$\{y^{\alpha}, \lambda_{\beta}\}_P = \delta^{\alpha}_{\beta}, \qquad \{\bar{y}^{\dot{\alpha}}, \bar{\lambda}_{\dot{\beta}}\}_P = \delta^{\dot{\alpha}}_{\dot{\beta}}. \tag{20}$$

The canonical definitions of vectorial and tensorial momenta $(p_{\alpha\dot{\beta}}, f_{\alpha\beta}, \bar{f}_{\dot{\alpha}\dot{\beta}})$ provide from the action density (18) the constraints

$$T_{\alpha\dot{\beta}} \equiv p_{\alpha\dot{\beta}} + em\,\lambda_{\alpha}\lambda_{\gamma}x^{\gamma}_{\dot{\beta}} + em\,\bar{\lambda}_{\dot{\beta}}\bar{\lambda}_{\dot{\gamma}}x^{\dot{\gamma}}_{\alpha} - \lambda_{\alpha}\bar{\lambda}_{\dot{\beta}} \approx 0\,, \tag{21}$$

$$T_{\alpha\beta} \equiv f_{\alpha\beta} - m\lambda_{\alpha}\lambda_{\beta} \approx 0\,, \tag{22}$$

$$\bar{T}_{\dot{\alpha}\dot{\beta}} \equiv \bar{f}_{\dot{\alpha}\dot{\beta}} - m\bar{\lambda}_{\dot{\alpha}}\bar{\lambda}_{\dot{\beta}} \approx 0\,. \tag{23}$$

In first order formulation we can introduce in spinorial sector of the model (18) "momenta for momenta" and consider the constraints $p^{\alpha}_{\lambda} - y^{\alpha} \approx 0$, $\bar{p}^{\dot{\alpha}}_{\lambda} - \bar{y}^{\dot{\alpha}} \approx 0$ and $p_{y\,\alpha} \approx 0$, $\bar{p}_{y\,\dot{\alpha}} \approx 0$. Subsequent introduction of Dirac brackets eliminates however the variables p_{λ}, $\bar{p}_{\lambda}$, p_y, $\bar{p}_y$ from the phase space and we get only PB (20).

Inserting (22), (23) in (21) we obtain the following representation of the constraints (21)-(23)

$$\begin{aligned} T_{\alpha\dot{\beta}} &= D_{\alpha\dot{\beta}} - \lambda_{\alpha}\bar{\lambda}_{\dot{\beta}} \approx 0\,, \\ T_{\alpha\beta} &= D_{\alpha\beta} - \lambda_{\alpha}\lambda_{\beta} \approx 0\,, \\ T_{\dot{\alpha}\dot{\beta}} &= D_{\dot{\alpha}\dot{\beta}} - \bar{\lambda}_{\dot{\alpha}}\bar{\lambda}_{\dot{\beta}} \approx 0\,, \end{aligned} \tag{24}$$

where

$$D_{\alpha\dot{\beta}} = p_{\alpha\dot{\beta}} + e\, x^{\gamma}_{\dot{\beta}} f_{\alpha\gamma} + e\, x^{\dot{\gamma}}_{\alpha}\bar{f}_{\dot{\beta}\dot{\gamma}}\,, \qquad D_{\alpha\beta} = f_{\alpha\beta}\,, \qquad D_{\dot{\alpha}\dot{\beta}} = \bar{f}_{\dot{\alpha}\dot{\beta}} \tag{25}$$

are the classical counterparts of the Maxwell-covariant derivatives (14). We see that the constraints (24) are the Maxwell-covariant generalization of the constraints leading to so-called unfolded equations for HS fields [5, 6].

We stress that present model has important difference in comparison with the one describing the HS particle of [5, 6], because the vectorial constraints do not commute (do have nonvanishing PB). There are the following nonvanishing Poisson brackets of the vectorial constraints (21)

$$\left\{T_{\alpha\dot\alpha}, T_{\beta\dot\beta}\right\}_P = 2e\left(\epsilon_{\dot\alpha\dot\beta}f_{\alpha\beta} + \epsilon_{\alpha\beta}\bar f_{\dot\alpha\dot\beta}\right) \approx 2e\,m\left(\epsilon_{\dot\alpha\dot\beta}\lambda_\alpha\lambda_\beta + \epsilon_{\alpha\beta}\bar\lambda_{\dot\alpha}\bar\lambda_{\dot\beta}\right). \tag{26}$$

Other tensorial constraints (22), (23) are the same as in [5, 6] and commute with all the constraints (21)-(23). Thus, the tensorial constraints (22), (23) are first class whereas the vectorial constraints (21) are the superposition of two first class and two second class constraints. In limit $e \to 0$ the Lagrangian (18) becomes the HS particle Lagrangian from [5, 6] where the covariant derivatives (25) form Abelian algebra and all constraints (21)-(23) are first class.

To perform quantization of our model it is important to project out the first and second class constraints present in (21). If we wish to preserve the Lorentz covariance this separation requires the use of additional spinorial variables. In order to have a basis in two-dimensional spinor space, we introduce second spinor, u_α, as firstly proposed in [5]. This auxiliary spinor satisfies the normalization condition

$$\lambda^\alpha u_\alpha = 1\,. \tag{27}$$

The nonvanishing PBs of u_α

$$\{u_\alpha, y^\beta\}_P = u_\alpha u^\beta \tag{28}$$

preserve the normalization (27).

Using this spinorial basis $(\lambda_\alpha, u_\alpha)$ we can introduce in Lorentz-covariant way the following projections

$$\begin{aligned} T_{\lambda\bar\lambda} &\equiv \lambda^\alpha T_{\alpha\dot\alpha}\bar\lambda^{\dot\alpha}\,, \qquad & T_{u\bar u} &\equiv u^\alpha T_{\alpha\dot\alpha}\bar u^{\dot\alpha}\,, \\ T_{\lambda\bar u} &\equiv \lambda^\alpha T_{\alpha\dot\alpha}\bar u^{\dot\alpha}\,, \qquad & T_{u\bar\lambda} &\equiv u^\alpha T_{\alpha\dot\alpha}\bar\lambda^{\dot\alpha}\,. \end{aligned} \tag{29}$$

Their unique nonvanishing PB is the following

$$\{T_{\lambda\bar u} + T_{u\bar\lambda}, T_{u\bar u}\}_P = 4e\,m \tag{30}$$

and one can conclude that the constraints

$$T_{\lambda\bar u} - T_{u\bar\lambda}\,, \qquad T_{\lambda\bar\lambda} \tag{31}$$

are first class, whereas

$$T_{\lambda\bar u} + T_{u\bar\lambda}\,, \qquad T_{u\bar u} \tag{32}$$

are second class.

Introduction of the Dirac brackets for the second class constraints leads to complicated structure of the quantum-mechanical algebra. As a way out we use the conversion method [18] in which the canonical pair of second class constraints can be considered as system where one second class constraint is interpreted as gauge-fixing condition for the gauge transformations which are generated by the other constraint. In the following we will consider the constraint $T_{u\bar u} \approx 0$ as gauge fixing condition, and the constraint $T_{\lambda\bar u} + T_{u\bar\lambda} \approx 0$ as generating new gauge degree of freedom. Finally we will consider the classical gauge-equivalent system which is described by the constraints

$$T_{\lambda\bar u} \approx 0\,, \qquad T_{u\bar\lambda} \approx 0\,, \qquad T_{\lambda\bar\lambda} \approx 0 \tag{33}$$

replacing the vectorial constraints (21).

The constraints (33) are in fact linear combinations of the projections $T_{\alpha\dot\beta}\bar\lambda^{\dot\beta}$, $\lambda^{\beta}T_{\beta\dot\alpha}$ of the constraints (21) on the Weyl spinor components λ^{α}, $\bar\lambda^{\dot\alpha}$. So, we can avoid using the auxiliary spinor u_α in definition of the constraints and as the result, we describe our model by the following set of first class constraints

$$S_\alpha \equiv T_{\alpha\dot\beta}\bar\lambda^{\dot\beta} = \left(p_{\alpha\dot\beta} + e\, f_{\alpha\gamma}x^{\gamma}_{\dot\beta}\right)\bar\lambda^{\dot\beta} \approx D_{\alpha\dot\beta}\bar\lambda^{\dot\beta} \approx 0\,, \tag{34}$$

$$\bar S_{\dot\alpha} \equiv \lambda^{\beta}T_{\beta\dot\alpha} = \lambda^{\beta}\left(p_{\beta\dot\alpha} + e\,\bar f_{\dot\alpha\dot\gamma}x^{\dot\gamma}_{\beta}\right) \approx \lambda^{\beta}D_{\beta\dot\alpha} \approx 0\,, \tag{35}$$

$$T_{\alpha\beta} \equiv f_{\alpha\beta} - m\lambda_\alpha\lambda_\beta \approx 0\,, \tag{36}$$

$$\bar T_{\dot\alpha\dot\beta} \equiv \bar f_{\dot\alpha\dot\beta} - m\bar\lambda_{\dot\alpha}\bar\lambda_{\dot\beta} \approx 0\,. \tag{37}$$

We observe that the four constraints (34), (35) are not independent, because they satisfy the relation

$$\lambda^{\alpha}S_\alpha = \bar\lambda^{\dot\alpha}\bar S_{\dot\alpha}\,, \tag{38}$$

i.e. we get only three independent first class constraints. It appears that the condition (38) does not enter into the derivation of the spectrum of our model.

In the transition to this new system of the constraints, we should be careful not to loose any of the constraints. In particular, performing the projections (34), (35) we are omitting the contribution to the vectorial constraint (21) which does not depend on spinor variables. Such contribution is described by the following new constraint

$$T \equiv T_{\alpha\dot\beta}T^{\alpha\dot\beta} \approx 0\,. \tag{39}$$

This quadratic constraint is of first class. Indeed, the constraint (39) can be represented by the formula $T = T_{\lambda\bar\lambda}T_{u\bar u} - T_{\lambda\bar u}T_{u\bar\lambda} \approx 0$ and therefore due to (33) it is first class. Thus, we should add the constraint (39) to the first class constraints (34)-(37). After quantization this constraint will provide the Maxwell extension of the Klein-Gordon (KG) equation.

One can find a physical interpretation of the system of the first class constraints (34)-(37), (39). For that purpose we will look for the values of the Casimir operators for the symmetry algebra in our model, i.e. the Casimirs of the the Maxwell algebra [16, 19, 20]

$$C_1^{Max} = P_{\alpha\dot\beta}P^{\alpha\dot\beta} + 4e\Big(M_{\alpha\beta}Z^{\alpha\beta} + \bar M_{\dot\alpha\dot\beta}\bar Z^{\dot\alpha\dot\beta}\Big)\,, \tag{40}$$

$$C_2^{Max} = Z_{\alpha\beta}Z^{\alpha\beta}\,, \qquad C_3^{Max} = \bar Z_{\dot\alpha\dot\beta}\bar Z^{\dot\alpha\dot\beta}\,, \tag{41}$$

$$\begin{aligned} C_4^{Max} = {} & 2Z^{\alpha\beta}\bar Z^{\dot\alpha\dot\beta}P_{\alpha\dot\alpha}P_{\beta\dot\beta} - \tfrac{1}{2}\left(Z^{\gamma\delta}Z_{\gamma\delta} + \bar Z^{\dot\gamma\dot\delta}\bar Z_{\dot\gamma\dot\delta}\right)P^{\alpha\dot\alpha}P_{\alpha\dot\alpha} \\ & + 2e\left(Z^{\gamma\delta}Z_{\gamma\delta} - \bar Z^{\dot\gamma\dot\delta}\bar Z_{\dot\gamma\dot\delta}\right)\left(Z^{\alpha\beta}M_{\alpha\beta} - \bar Z^{\dot\alpha\dot\beta}\bar M_{\dot\alpha\dot\beta}\right). \end{aligned} \tag{42}$$

Using the transformations (10), (11) and the following transformations of spinors

$$\delta y^{\alpha} = \ell^{\alpha\beta}y_\beta\,, \quad \delta\bar y^{\dot\alpha} = \bar\ell^{\dot\alpha\dot\beta}\bar y_{\dot\beta}\,, \qquad \delta\lambda_\alpha = -\ell_{\alpha\beta}\lambda^{\beta}\,, \quad \delta\bar\lambda_{\dot\alpha} = -\bar\ell_{\dot\alpha\dot\beta}\bar\lambda^{\dot\beta} \tag{43}$$

we obtain from the action (18) the Noether charges in our model

$$\begin{aligned} P_{\alpha\dot\beta} &= -p_{\alpha\dot\beta} + e\,x^{\dot\gamma}_{\alpha}\bar f_{\dot\gamma\dot\beta} + e\,f_{\alpha\gamma}x^{\gamma}_{\dot\beta}\,, \\ Z_{\alpha\beta} &= -f_{\alpha\beta}\,, \qquad\qquad \bar Z_{\dot\alpha\dot\beta} = -\bar f_{\dot\alpha\dot\beta}\,, \\ M_{\alpha\beta} &= x^{\dot\gamma}_{(\alpha}p_{\beta)\dot\gamma} + 2z^{\gamma}_{(\alpha}f_{\beta)\gamma} + y_{(\alpha}\lambda_{\beta)}\,, \\ \bar M_{\dot\alpha\dot\beta} &= x^{\gamma}_{(\dot\alpha}p_{\gamma\dot\beta)} + 2\bar z^{\dot\gamma}_{(\dot\alpha}\bar f_{\dot\beta)\dot\gamma} + \bar y_{(\dot\alpha}\bar\lambda_{\dot\beta)}\,, \end{aligned} \tag{44}$$

which represent the dynamical phase space realization of Maxwell algebra generators. If we insert the expressions (44) in the Casimirs (40), (41), (42) and take into account the first class constraints (34)-(36), (39) we find that

$$C_1^{Max} \approx 0\,, \qquad C_2^{Max} \approx 0\,, \qquad C_3^{Max} \approx 0\,, \qquad C_4^{Max} \approx 0\,. \tag{45}$$

Thus, our Maxwell-Shirafuji model, defined by the first class constrains (34)-(36), (39), describes the Maxwell-HS particle multiplet with all vanishing eigenvalues of Casimirs (40)-(42), similarly as massless scalar relativistic particle model is characterized by vanishing Casimirs of Poincare algebra.

4. First quantization of the particle model and interacting HS fields

Phase space coordinates after quantization become the operators, and for simplicity we will denote them further by the same letter (without hats). The Poisson brackets algebra (19), (20) generates the following quantum-mechanical algebra

$$[x^{\alpha\dot\alpha}, p_{\beta\dot\beta}] = i\delta^\alpha_\beta\delta^{\dot\alpha}_{\dot\beta}, \quad [z^{\alpha\beta}, f_{\gamma\delta}] = i\delta^{(\alpha}_\gamma\delta^{\beta)}_\delta, \quad [\bar z^{\dot\alpha\dot\beta}, \bar f_{\dot\gamma\dot\delta}] = i\delta^{(\dot\alpha}_{\dot\gamma}\delta^{\dot\beta)}_{\dot\delta}, \tag{46}$$

$$[y^\alpha, \lambda_\beta] = i\delta^\alpha_\beta, \quad [\bar y^{\dot\alpha}, \bar\lambda_{\dot\beta}] = i\delta^{\dot\alpha}_{\dot\beta}\,. \tag{47}$$

Further we consider Schrödinger-type representation in which the operators $x^{\alpha\dot\beta}$, $z^{\alpha\beta}$, $\bar z^{\dot\alpha\dot\beta}$, $y^\alpha, \bar y^{\dot\alpha}$ are realized as multiplications by c-numbers whereas the operators of quantized momenta are represented as partial derivatives

$$\begin{gathered} p_{\alpha\dot\beta} = -i\frac{\partial}{\partial x^{\alpha\dot\beta}} \equiv -i\partial_{\alpha\dot\beta}\,, \\ \lambda_\alpha = -i\frac{\partial}{\partial y^\alpha} \equiv -i\partial_\alpha\,, \qquad \bar\lambda_{\dot\alpha} = -i\frac{\partial}{\partial \bar y^{\dot\alpha}} \equiv -i\bar\partial_{\dot\alpha}\,, \end{gathered} \tag{48}$$

$$f_{\alpha\beta} = -i\frac{\partial}{\partial z^{\alpha\beta}} \equiv -i\partial_{\alpha\beta}\,, \qquad \bar f_{\dot\alpha\dot\beta} = -i\frac{\partial}{\partial \bar z^{\dot\alpha\dot\beta}} \equiv -i\bar\partial_{\dot\alpha\dot\beta}\,. \tag{49}$$

The physical spectrum of the wave function

$$\Phi = \Phi(x^{\alpha\dot\beta}, z^{\alpha\beta}, \bar z^{\dot\alpha\dot\beta}, y^\alpha, \bar y^{\dot\alpha}) \tag{50}$$

is defined by the quantum counterpart of first class constraints (34)-(36):

$$iD_{\alpha\dot\beta}\bar\partial^{\dot\beta}\,\Phi \;=\; \left(\partial_{\alpha\dot\beta} + e\,\partial_{\alpha\gamma}\,x^{\gamma}_{\dot\beta}\right)\bar\partial^{\dot\beta}\,\Phi = 0\,, \tag{51}$$

$$iD_{\beta\dot\alpha}\partial^{\beta}\,\Phi \;=\; \left(\partial_{\beta\dot\alpha} + e\,\bar\partial_{\dot\alpha\dot\gamma}\,x^{\dot\gamma}_{\beta}\right)\partial^{\beta}\,\Phi = 0\,, \tag{52}$$

$$\left(\partial_{\alpha\beta} + im\partial_\alpha\partial_\beta\right)\Phi = 0\,, \qquad \left(\bar\partial_{\dot\alpha\dot\beta} + im\bar\partial_{\dot\alpha}\bar\partial_{\dot\beta}\right)\Phi = 0\,, \tag{53}$$

where $D_{\alpha\dot\beta}$ is the Maxwell-covariant derivative (see (25), (14)).

The solutions of eqs. (53) are described by compact formula

$$\Phi(x,z,\bar z,y,\bar y) = e^{-im\left(z^{\alpha\beta}\partial_\alpha\partial_\beta + \bar z^{\dot\alpha\dot\beta}\bar\partial_{\dot\alpha}\bar\partial_{\dot\beta}\right)}\,\Phi_0(x,y,\bar y)\,. \tag{54}$$

From expression (54) follows that the tensorial coordinates $z^{\mu\nu} = (z^{\alpha\beta}, \bar z^{\dot\alpha\dot\beta})$ are the auxiliary gauge degrees of freedom and the gauge-independent degrees of freedom are present in the HS master field $\Phi_0(x,y,\bar y)$. Residual equations (51), (52) for unconstrained field $\Phi_0(x,y,\bar y)$ take the form

$$\left[\partial_{\alpha\dot\beta}\bar\partial^{\dot\beta} + iem\,(\partial^\beta x_{\beta\dot\beta}\bar\partial^{\dot\beta})\partial_\alpha\right]\Phi_0 \;=\; 0\,, \tag{55}$$

$$\left[\partial^\beta\partial_{\beta\dot\alpha} + iem\,(\partial^\beta x_{\beta\dot\beta}\bar\partial^{\dot\beta})\bar\partial_{\dot\alpha}\right]\Phi_0 \;=\; 0\,. \tag{56}$$

Spinor variables y^α, $\bar y^{\dot\alpha}$ are besides space-time coordinates, the additional spinorial variables. We consider the following Taylor expansion of HS master field with respect to the additional spinor variables

$$\Phi_0(x,y,\bar y) = \sum_{k,n=0}^{\infty}\frac{1}{k!n!}\,y^{\alpha_1}...y^{\alpha_k}\,\bar y^{\dot\beta_1}...\bar y^{\dot\beta_n}\,\phi^{(k,n)}_{\alpha_1...\alpha_k\,\dot\beta_1...\dot\beta_n}(x)\,, \tag{57}$$

where Maxwell-HS component fields $\phi^{(k,n)}_{\alpha_1...\alpha_k\,\dot\beta_1...\dot\beta_n}(x)$ are the usual $D=4$ space-time spin-tensor fields.

Let us find now the solution of the equations (55), (56).

In the beginning we present two simple equations, which are a direct consequence of the equations (55), (56).

1) Contraction (55) (or (56)) with ∂^α (or $\bar\partial^{\dot\alpha}$) and use of two-spinor identities $\partial^\alpha\partial_\alpha=\bar\partial^{\dot\alpha}\bar\partial_{\dot\alpha}=0$ leads to the relation

$$\partial^{\alpha\dot\alpha}\partial_\alpha\bar\partial_{\dot\alpha}\,\Phi_0 = 0\,. \tag{58}$$

The equation (58) provides the generalized Lorentz conditions for the component fields defined by (57)

$$\partial^{\alpha_1\dot\beta_1}\,\phi^{(k,n)}_{\alpha_1...\alpha_k\,\dot\beta_1...\dot\beta_n}(x) = 0\,, \qquad k,n\geq 1\,. \tag{59}$$

For example, if $n=k=1$ the equation (59) is the standard Lorentz condition for the four-vector field.

2) By considering the difference of the equations (55) multiplied by $\bar\partial_{\dot\alpha}$ and the equations (56) multiplied by ∂_α we obtain

$$\partial^\beta_{\dot\alpha}\,\partial_\alpha\partial_\beta\,\Phi_0 = \partial^{\dot\beta}_\alpha\,\bar\partial_{\dot\alpha}\bar\partial_{\dot\beta}\,\Phi_0\,. \tag{60}$$

From (60) we get the following equations for the component fields

$$\partial^{\alpha_k}_{\dot\beta_{n-1}}\,\phi^{(k,n-2)}_{\alpha_1...\alpha_{k-1}\alpha_k\,\dot\beta_1...\dot\beta_{n-2}} = \partial^{\dot\beta_n}_{\alpha_{k-1}}\,\phi^{(k-2,n)}_{\alpha_1...\alpha_{k-2}\,\dot\beta_1...\dot\beta_{n-1}\dot\beta_n}\,, \qquad k,n\geq 2\,. \tag{61}$$

The equations for lowest component fields described by Lorentz spins $(2,0)+(0,2)$ are as follows

$$\partial^\beta_{\dot\alpha}\,\phi^{(2,0)}_{\alpha\beta} = \partial^{\dot\beta}_\alpha\,\phi^{(0,2)}_{\dot\alpha\dot\beta}\,. \tag{62}$$

They represent half of the Maxwell equations (self-dual part) for the free electromagnetic field strength.

Full set of the equations for component fields of the HS master field (57), generated by the conditions (51), (52), are

$$\partial^{\alpha_k}_{\dot\beta_{n+1}}\,\phi^{(k,n)}_{\alpha_1...\alpha_k\,\dot\beta_1...\dot\beta_n} = iem\,x^{\alpha_k\dot\beta_{n+2}}\,\phi^{(k,n+2)}_{\alpha_1...\alpha_k\,\dot\beta_1...\dot\beta_{n+1}\dot\beta_{n+2}}\,, \qquad k\geq 1\,; \tag{63}$$

$$\partial^{\dot\beta_n}_{\alpha_{k+1}}\,\phi^{(k,n)}_{\alpha_1...\alpha_k\,\dot\beta_1...\dot\beta_n} = iem\,x^{\alpha_{k+2}\dot\beta_n}\,\phi^{(k+2,n)}_{\alpha_1...\alpha_{k+1}\alpha_{k+2}\,\dot\beta_1...\dot\beta_n}\,, \qquad n\geq 1\,. \tag{64}$$

They represent the Maxwell-invariant generalizations of well-known Dirac-Pauli-Fierz equations [21, 22].

For completing the analysis of constraints it is necessary to impose on the wave function the scalar constraint (39). This additional condition leads to additional equation for first scalar component of the wave function (57). The constraint (39) implies the following equation (see details in [7])

$$\left[\partial_{\alpha\dot{\alpha}}\partial^{\alpha\dot{\alpha}} - 2e^2m^2\left(\partial_\alpha x^{\alpha\dot{\alpha}}\bar{\partial}_{\dot{\alpha}}\right)^2\right]\Phi_0 = 0 \tag{65}$$

for the HS master field (57). The equation (65) provides the following infinite set of field equations for the HS component fields

$$\partial^{\gamma\dot{\gamma}}\partial_{\gamma\dot{\gamma}}\,\phi^{(k,n)}_{\alpha_1\ldots\alpha_k\,\dot{\beta}_1\ldots\dot{\beta}_n} = 2e^2m^2\,x^{\alpha_{k+1}\dot{\beta}_{n+1}}\,x^{\alpha_{k+2}\dot{\beta}_{n+2}}\,\phi^{(k+2,n+2)}_{\alpha_1\ldots\alpha_k\alpha_{k+1}\alpha_{k+2}\,\dot{\beta}_1\ldots\dot{\beta}_n\dot{\beta}_{n+1}\dot{\beta}_{n+2}}\,, \tag{66}$$

which are the Maxwell-invariant generalization of massless Klein-Gordon equation. One can show that the equations (66) for all component fields $\phi^{(k,n)}(x)$, except $\phi^{(0,0)}(x)$, are the consequences of the Maxwell-Dirac equations (63), (64). The only equation in (66) which is not following from (63), (64), describe the Maxwell-Klein-Gordon equation

$$\partial^{\gamma\dot{\gamma}}\partial_{\gamma\dot{\gamma}}\,\phi^{(0,0)} = 2e^2m^2\,x^{\alpha_1\dot{\beta}_1}\,x^{\alpha_2\dot{\beta}_2}\,\phi^{(2,2)}_{\alpha_1\alpha_2\,\dot{\beta}_1\dot{\beta}_2} \tag{67}$$

for the scalar field $\phi^{(0,0)}(x)$.

The Maxwell-Weyl equations (63), (64) and the Maxwell-Klein-Gordon equations (67 link the fields $\phi^{(k+2p,n+2r)}(x)$ with fixed k, n and $p = 0,1,2,...$, $r = 0,1,2,....$ We obtain infinite-dimensional Maxwell-HS multiplets with minimal Lorentz spin $(\frac{k}{2},\frac{n}{2})$ described by the field $\phi^{(k,n)}(x)$ (see the Maxwell symmetry transformations of the components of Maxwell-HS multiplets in [7]). In particular, we get the maximal (with largest number of components) bosonic Maxwell HS multiplet with scalar field $\phi^{(0,0)}(x)$ and the component fields with even Lorents spins $(j_1, j_2) = (p, r)$. Further, there are two maximal fermionic multiplets: chiral, with Maxwell-Weyl field $\phi^{(1,0)}_\alpha(x)$, and the antichiral one, with Maxwell-Weyl field $\phi^{(0,1)}_{\dot{\alpha}}(x)$.

In limit $e \to 0$ the equations (51), (52) coincide with massless Dirac-Pauli-Fierz equations for standard massless conformal HS fields

$$e = 0\ : \qquad \partial^{\alpha_1\dot{\gamma}}\,\phi^{(k,n)}_{\alpha_1\ldots\alpha_k\,\dot{\beta}_1\ldots\dot{\beta}_n}(x) = 0\,, \qquad \partial^{\gamma\dot{\beta}_1}\,\phi^{(k,n)}_{\alpha_1\ldots\alpha_k\,\dot{\beta}_1\ldots\dot{\beta}_n}(x) = 0\,. \tag{68}$$

Moreover, in the limit $e \to 0$ all the component fields, including 'mixed' fields $\phi^{(k>0,n>0)}_{\alpha_1\ldots\alpha_k\,\dot{\beta}_1\ldots\dot{\beta}_n}(x)$, are independent. This property differs from the case of standard HS particle [4, 5, 6] where the spectrum of independent fields contains only the scalar field $\phi^{(0,0)}(x)$, the spin-tensor fields with undotted indices $\phi^{(k,0)}_{\alpha_1\ldots\alpha_k}(x)$ $(k>0)$ and spin-tensor field with dotted indices $\phi^{(0,n)}_{\dot{\alpha}_1\ldots\dot{\alpha}_n}(x)$ $(n>0)$. In standard HS particle model due to the presence of one more first class constraint the 'mixed' fields $\phi^{(k>0,n>0)}_{\alpha_1\ldots\alpha_k\,\dot{\beta}_1\ldots\dot{\beta}_n}(x)$ are expressed by space-time derivatives of the fields $\phi^{(0,0)}(x)$, $\phi^{(k>0,0)}_{\alpha_1\ldots\alpha_k}(x)$, $\phi^{(0,n>0)}_{\dot{\alpha}_1\ldots\dot{\alpha}_n}(x)$ which follow from four components of unfolded equation $\left(\partial_{\alpha\dot{\beta}} + i\partial_\alpha\bar{\partial}_{\dot{\beta}}\right)\Phi = 0$. In our case the "mixed" fields stay independent because there are only three linearly independent components of unfolded equation relating Maxwell-HS fields with different Lorentz spins (k,n).

5. HS field equations in dual representation

In this section we consider the representation with the generalized coordinates $x^{\alpha\dot{\beta}}$, $f_{\alpha\beta}$, $\bar{f}_{\dot{\alpha}\dot{\beta}}$, $y^\alpha, \bar{y}^{\dot{\alpha}}$, the momenta (48) and

$$z^{\alpha\beta} = i\frac{\partial}{\partial f_{\alpha\beta}}\,, \qquad \bar{z}^{\dot{\alpha}\dot{\beta}} = i\frac{\partial}{\partial \bar{f}_{\dot{\alpha}\dot{\beta}}}\,. \tag{69}$$

The wave function has the following functional form

$$\tilde{\Psi} = \tilde{\Psi}(x^{\alpha\dot{\beta}}, f_{\alpha\beta}, \bar{f}_{\dot{\alpha}\dot{\beta}}, y^{\alpha}, \bar{y}^{\dot{\alpha}}) \tag{70}$$

and is related with the formula (50) by Fourier transform in the tensorial variables $(f_{\alpha\beta}, \bar{f}_{\dot{\alpha}\dot{\beta}})$.

First class constraints (34)-(36) yield in this representation the following equations

$$iD_{\alpha\dot{\beta}}\bar{\partial}^{\dot{\beta}}\,\tilde{\Psi} = \left(\partial_{\alpha\dot{\beta}} + i\,e\,f_{\alpha\gamma}\,x^{\gamma}_{\dot{\beta}}\right)\bar{\partial}^{\dot{\beta}}\,\tilde{\Psi} = 0\,, \tag{71}$$

$$iD_{\beta\dot{\alpha}}\partial^{\beta}\,\tilde{\Psi} = \left(\partial_{\beta\dot{\alpha}} + i\,e\,\bar{f}_{\dot{\alpha}\dot{\gamma}}\,x^{\dot{\gamma}}_{\beta}\right)\partial^{\beta}\,\tilde{\Psi} = 0\,, \tag{72}$$

$$\left(f_{\alpha\beta} + m\partial_{\alpha}\partial_{\beta}\right)\tilde{\Psi} = 0\,, \qquad \left(\bar{f}_{\dot{\alpha}\dot{\beta}} + m\bar{\partial}_{\dot{\alpha}}\bar{\partial}_{\dot{\beta}}\right)\tilde{\Psi} = 0\,. \tag{73}$$

For the analysis of equations (71)-(73) we introduce Taylor expansion of the wave function (70) with respect to spinor variables y^{α} and $\bar{y}^{\dot{\alpha}}$

$$\tilde{\Psi}(x,f,\bar{f},y,\bar{y}) = \sum_{k,n=0}^{\infty} \frac{1}{k!\,n!}\, y^{\alpha_1}...y^{\alpha_k}\,\bar{y}^{\dot{\beta}_1}...\bar{y}^{\dot{\beta}_n}\,\tilde{\Psi}^{(k,n)}_{\alpha_1...\alpha_k\,\dot{\beta}_1...\dot{\beta}_n}(x,f,\bar{f})\,. \tag{74}$$

Then the constraints (73) lead to expression of all components $\tilde{\Psi}^{(k,n)}$, $k>1$, $n>1$ in terms of four HS master fields $\tilde{\Psi}^{(0,0)}$, $\tilde{\Psi}^{(1,0)}$, $\tilde{\Psi}^{(0,1)}$ and $\tilde{\Psi}^{(1,1)}$. We obtain the equations

$$\tilde{\Psi}^{(2,0)}_{\alpha\beta} = -m^{-1}f_{\alpha\beta}\tilde{\Psi}^{(0,0)}\,, \qquad \tilde{\Psi}^{(0,2)}_{\dot{\alpha}\dot{\beta}} = -m^{-1}\bar{f}_{\dot{\alpha}\dot{\beta}}\tilde{\Psi}^{(0,0)}\,,$$

$$\tilde{\Psi}^{(2,1)}_{\alpha\beta\dot{\gamma}} = -m^{-1}f_{\alpha\beta}\tilde{\Phi}^{(0,1)}_{\dot{\gamma}}\,, \qquad \tilde{\Psi}^{(1,2)}_{\gamma\dot{\alpha}\dot{\beta}} = -m^{-1}\bar{f}_{\dot{\alpha}\dot{\beta}}\tilde{\Psi}^{(1,0)}_{\gamma}\,,$$

$$\tilde{\Psi}^{(3,1)}_{\alpha\beta\gamma\dot{\gamma}} = -m^{-1}f_{\alpha\beta}\tilde{\Psi}^{(1,1)}_{\gamma\dot{\gamma}}\,, \qquad \tilde{\Psi}^{(1,3)}_{\gamma\dot{\alpha}\dot{\beta}\dot{\gamma}} = -m^{-1}\bar{f}_{\dot{\alpha}\dot{\beta}}\tilde{\Psi}^{(1,1)}_{\gamma\dot{\gamma}}\,, \qquad \text{etc.}$$

In the expansion (74) the independent HS master fields are

i) one "generalized spin zero" field

$$\tilde{\Psi}^{(0,0)}(x,f,\bar{f})\,, \tag{75}$$

ii) two "generalized spin half" fields

$$\tilde{\Psi}^{(1,0)}_{\alpha}(x,f,\bar{f})\,, \qquad \tilde{\Psi}^{(0,1)}_{\dot{\beta}}(x,f,\bar{f}) \tag{76}$$

iii) one "generalized spin one" field

$$\tilde{\Psi}^{(1,1)}_{\alpha\dot{\beta}}(x,f,\bar{f})\,. \tag{77}$$

The equations (73) provide for the fields (75), (76), (77) the kinematic constraints:

$$f^{\alpha\beta}f_{\alpha\beta}\tilde{\Psi}^{(0,0)}(x,f,\bar{f}) = 0\,, \qquad \bar{f}^{\dot{\alpha}\dot{\beta}}\bar{f}_{\dot{\alpha}\dot{\beta}}\tilde{\Psi}^{(0,0)}(x,f,\bar{f}) = 0\,, \tag{78}$$

$$f_{\alpha}{}^{\beta}\tilde{\Psi}^{(1,0)}_{\beta}(x,f,\bar{f}) = 0\,, \qquad \bar{f}_{\dot{\alpha}}{}^{\dot{\beta}}\tilde{\Psi}^{(0,1)}_{\dot{\beta}}(x,f,\bar{f}) = 0\,, \tag{79}$$

$$f_{\alpha}{}^{\beta}\tilde{\Psi}^{(1,1)}_{\beta\dot{\beta}}(x,f,\bar{f}) = 0\,, \qquad \bar{f}_{\dot{\alpha}}{}^{\dot{\beta}}\tilde{\Psi}^{(1,1)}_{\beta\dot{\beta}}(x,f,\bar{f}) = 0\,. \tag{80}$$

From the equations (71), (72) one obtains for the unconstrained component (76), (77) the relations

$$\begin{aligned} &\left(\partial^{\alpha\dot\beta} - i\,e\,f^{\alpha\gamma}\,x_{\gamma}^{\dot\beta}\right)\tilde\Psi^{(0,1)}_{\dot\beta}(x,f,\bar f) = 0\,, \\ &\left(\partial^{\beta\dot\alpha} - i\,e\,\bar f^{\dot\alpha\dot\gamma}\,x_{\dot\gamma}^{\beta}\right)\tilde\Psi^{(1,0)}_{\beta}(x,f,\bar f) = 0\,, \end{aligned} \tag{81}$$

$$\begin{aligned} &\left(\partial^{\alpha\dot\beta} - i\,e\,f^{\alpha\gamma}\,x_{\gamma}^{\dot\beta}\right)\tilde\Psi^{(1,1)}_{\beta\dot\beta}(x,f,\bar f) = 0\,, \\ &\left(\partial^{\beta\dot\alpha} - i\,e\,\bar f^{\dot\alpha\dot\gamma}\,x_{\dot\gamma}^{\beta}\right)\tilde\Psi^{(1,1)}_{\beta\dot\beta}(x,f,\bar f) = 0\,, \end{aligned} \tag{82}$$

which have the form of the Dirac equations in a constant electromagnetic field, with electromagnetic potential $A_\mu = f_{\mu\nu}x^\nu$. As generalization of the standard approach for Dirac spin-half field, the wave functions in (81), (82) depend also on continuous electromagnetic field strength coordinates $f_{\alpha\beta}$, $\bar f_{\dot\alpha\dot\beta}$. We do not see yet the relation of our description of HS fields interacting with constant EM field to the approaches proposed in recent papers on EM coupling of HS fields [23, 24], however to find such a link would be desirable.

Generalized spin-zero field $\tilde\Psi^{(0,0)}(x,f,\bar f)$ is described by generalized Klein-Gordon equation, which follows from the constraint

$$\hat T^{\alpha\dot\alpha}\hat T_{\alpha\dot\alpha}\,\tilde\Psi \approx 0\,. \tag{83}$$

Taking into account that

$$\begin{aligned} \hat T^{\alpha\dot\alpha}\hat T_{\alpha\dot\alpha} \;=\; & -\partial^{\alpha\dot\alpha}\partial_{\alpha\dot\alpha} + 2ie\left(f^{\alpha\beta}x_{\beta}^{\dot\alpha} + \bar f^{\dot\alpha\dot\beta}x_{\dot\beta}^{\alpha}\right)\partial_{\alpha\dot\alpha} \\ & +\tfrac{1}{2}e^2\left(f^{\alpha\beta}f_{\alpha\beta} + \bar f^{\dot\alpha\dot\beta}\bar f_{\dot\alpha\dot\beta}\right)x^{\gamma\dot\gamma}x_{\gamma\dot\gamma} - 2e^2 f_{\alpha\beta}\bar f_{\dot\alpha\dot\beta}x^{\beta\dot\alpha}x^{\alpha\dot\beta} \\ & -2\left(i\partial^{\alpha\dot\alpha} + ef^{\alpha\beta}x_{\beta}^{\dot\alpha} + e\bar f^{\dot\alpha\dot\beta}x_{\dot\beta}^{\alpha}\right)\partial_\alpha\bar\partial_{\dot\alpha}\,, \end{aligned} \tag{84}$$

we obtain the following generalized Klein-Gordon equation for "generalized spin zero" field

$$\begin{aligned} \Big[-\Box + 2ie\left(f^{\alpha\beta}x_{\beta}^{\dot\alpha} + \bar f^{\dot\alpha\dot\beta}x_{\dot\beta}^{\alpha}\right)\partial_{\alpha\dot\alpha} & \\ +\tfrac{1}{2}e^2\left(f^2+\bar f^2\right)x^2 - 2e^2 f_{\alpha\beta}\bar f_{\dot\alpha\dot\beta}x^{\beta\dot\alpha}x^{\alpha\dot\beta}\Big]\,\tilde\Psi^{(0,0)} = 0\,, & \end{aligned} \tag{85}$$

where $\Box := \partial^{\alpha\dot\alpha}\partial_{\alpha\dot\alpha}$, $x^2 := x^{\alpha\dot\alpha}x_{\alpha\dot\alpha}$, $f^2 := f^{\alpha\beta}f_{\alpha\beta}$, $\bar f^2 := \bar f^{\dot\alpha\dot\beta}\bar f_{\dot\alpha\dot\beta}$. It should be emphasized that due to the equations (82) and the constraints (80) the last term in the operator (84) does not contribute to the equation (85) and finally we obtain standard Klein-Gordon equation with coupling to constant EM field.

From the equations (81), (82) and (85) follows that the link of different spins is due to EM coupling proportional to electric charge e. Further one can show that if the torsion in six tensorial dimensions of Maxwell space-time depends only on the $D=4$ space-time coordinates it can be interpreted as a coupling to $D=4$ Abelian gauge potential. Let us introduce the following "block-diagonal" 10-bein $E_{AB}{}^{CD} = \left(\delta_\alpha^\gamma\delta_{\dot\beta}^{\dot\delta}, E_{\alpha\dot\beta}{}^{\gamma\delta}, E_{\alpha\dot\beta}{}^{\dot\gamma\dot\delta}\right)$ in the tensorial space $\left(x^{\alpha\dot\beta}, z^{\alpha\beta}, \bar z^{\dot\alpha\dot\beta}\right)$ and corresponding covariant derivatives

$$\nabla_{\alpha\dot\beta} = \partial_{\alpha\dot\beta} + E_{\alpha\dot\beta}{}^{\gamma\delta}(x)\nabla_{\gamma\delta} + E_{\alpha\dot\beta}{}^{\dot\gamma\dot\delta}(x)\nabla_{\dot\gamma\dot\delta}\,, \tag{86}$$

$$\nabla_{\alpha\beta} = \partial_{\alpha\beta}\,, \qquad \nabla_{\dot\alpha\dot\beta} = \partial_{\dot\alpha\dot\beta}\,. \tag{87}$$

If we consider the plane wave solutions in additional dimensions, one can replace (see (48)) the derivatives (87) by constant tensors $f_{\alpha\beta}$, $\bar f_{\dot\alpha\dot\beta}$ which represent additional tensorial momenta. In such a case the derivative (86) can be written as the Abelian gauge-covariant derivative

$$\nabla_{\alpha\dot\beta} = \partial_{\alpha\dot\beta} + eA_{\alpha\dot\beta}(x)\,, \tag{88}$$

where $e\mathrm{A}_{\alpha\dot\beta}(x) = E_{\alpha\dot\beta}{}^{\gamma\delta}(x) f_{\gamma\delta} + E_{\alpha\dot\beta}{}^{\dot\gamma\dot\delta}(x)\bar f_{\dot\gamma\dot\delta}$. In Maxwell tensorial space additional tensorial coordinates are twisted by a constant torsion proportional to e, the functions $E_{\alpha\dot\beta}{}^{\gamma\delta}(x)$ and $E_{\alpha\dot\beta}{}^{\dot\gamma\dot\delta}(x)$ are linear in x, and we obtain in (88) the Abelian gauge field four-potential $\mathrm{A}_{\alpha\dot\beta}$ describing constant electromagnetic field strength ($\bar f_{\dot\alpha\dot\beta} = (f_{\alpha\beta})^\dagger$)

$$\mathrm{A}_{\alpha\dot\beta} = f_\alpha^{\ \gamma} x_{\gamma\dot\beta} + \bar f_{\dot\beta}^{\ \dot\gamma} x_{\alpha\dot\gamma}\,. \tag{89}$$

It can be added that the space-time translations $x_{\alpha\dot\beta} \to x_{\alpha\dot\beta} + a_{\alpha\dot\beta}$ modify (89) by a constant term, which can be however compensated by the standard Abelian gauge transformation of $\mathrm{A}_{\alpha\dot\beta}$, what leads finally to the translational invariance of the covariant derivative (88).

The solutions of the equations (78)-(80) can be represented as the Fourier transforms

$$\tilde\Psi^{(0,0)}(x,f,\bar f) = \int d^6 z\, e^{-i(fz+\bar f\bar z)}\, \Psi^{(0,0)}(x,z,\bar z)\,, \tag{90}$$

$$\tilde\Psi^{(1,0)}_{\gamma}(x,f,\bar f) = \int d^6 z\, e^{-i(fz+\bar f\bar z)}\, \Psi^{(1,0)}_{\gamma}(x,z,\bar z)\,, \tag{91}$$

$$\tilde\Psi^{(0,1)}_{\dot\gamma}(x,f,\bar f) = \int d^6 z\, e^{-i(fz+\bar f\bar z)}\, \Psi^{(0,1)}_{\dot\gamma}(x,z,\bar z)\,, \tag{92}$$

$$\tilde\Psi^{(1,1)}_{\gamma\dot\gamma}(x,f,\bar f) = \int d^6 z\, e^{-i(fz+\bar f\bar z)}\, \Psi^{(1,1)}_{\gamma\dot\gamma}(x,z,\bar z)\,, \tag{93}$$

where

$$\Psi^{(0,0)} = \sum_{k,n=0}^{\infty} \frac{1}{k!n!} z^{\alpha_1\beta_1}...z^{\alpha_k\beta_k}\, \bar z^{\dot\alpha_1\dot\beta_1}...\bar z^{\dot\alpha_n\dot\beta_n}\, \phi^{(2k,2n)}_{(\alpha_1\beta_1...\alpha_k\beta_k)\,(\dot\alpha_1\dot\beta_1...\dot\alpha_n\dot\beta_n)}(x)\,,$$

$$\Psi^{(1,0)}_{\gamma} = \sum_{k,n=0}^{\infty} \frac{1}{k!n!} z^{\alpha_1\beta_1}...z^{\alpha_k\beta_k}\, \bar z^{\dot\alpha_1\dot\beta_1}...\bar z^{\dot\alpha_n\dot\beta_n}\, \phi^{(2k+1,2n)}_{(\gamma\alpha_1\beta_1...\alpha_k\beta_k)\,(\dot\alpha_1\dot\beta_1...\dot\alpha_n\dot\beta_n)}(x)\,,$$

$$\Psi^{(0,1)}_{\dot\gamma} = \sum_{k,n=0}^{\infty} \frac{1}{k!n!} z^{\alpha_1\beta_1}...z^{\alpha_k\beta_k}\, \bar z^{\dot\alpha_1\dot\beta_1}...\bar z^{\dot\alpha_n\dot\beta_n}\, \phi^{(2k,2n+1)}_{(\alpha_1\beta_1...\alpha_k\beta_k)\,(\dot\gamma\dot\alpha_1\dot\beta_1...\dot\alpha_n\dot\beta_n)}(x)\,,$$

$$\Psi^{(1,1)}_{\gamma\dot\gamma} = \sum_{k,n=0}^{\infty} \frac{1}{k!n!} z^{\alpha_1\beta_1}...z^{\alpha_k\beta_k}\, \bar z^{\dot\alpha_1\dot\beta_1}...\bar z^{\dot\alpha_n\dot\beta_n}\, \phi^{(2k+1,2n+1)}_{(\gamma\alpha_1\beta_1...\alpha_k\beta_k)\,(\dot\gamma\dot\alpha_1\dot\beta_1...\dot\alpha_n\dot\beta_n)}(x)$$

have polynomial dependence on z, $\bar z$ and the component HS fields $\phi^{(k,n)}_{(\alpha...)\,(\dot\alpha...)}$ are symmetric with respect to all undotted and dotted spinor indices (compare with HS fields in (57) considered in previous subsection). The fields (90)-(93) have complicated, nonpolynomial dependence on the variables f, $\bar f$, with generalized Dirac equations (see (81)-(82)) and Klein-Gordon equation (85) providing the explicit functional dependence of the Maxwell-HS component fields on the constant values of EM field.

6. Outlook

In this paper we considered the spinorial particle model in ten-dimensional tensorial space-time with torsion described by the tensorial coset space $X = (x^{\alpha\dot\alpha}, z^{\alpha\beta}, \bar z^{\dot\alpha\dot\beta})$ of $D = 4$ Maxwell group and additional spinorial variables λ^α, $\bar\lambda^{\dot\alpha}$. We performed the canonical quantization of the model with supplemented kinetic term for λ^α, $\bar\lambda^{\dot\alpha}$. By using the phase space formulation we specified the set of first and second class constraints. It appears that in first quantized theory the first class constraints will describe the set of field equations for new higher spin multiplets in the tensorial space X defining new HS Maxwell dynamics. Such equations describe the generalization of the known "unfolded equations" [5, 6, 9] for massless HS free fields with flat space-time derivatives $\partial_{\alpha\dot{\beta}}$ replaced by the Maxwell-covariant derivatives $D_{\alpha\dot{\beta}}$ (see (14)). Note that the Maxwell-covariant description of $D=4$ Maxwell-HS fields requires the presence of particular space-time-dependent coupling terms between different spin fields which can be also interpreted as following from the electromagnetic covariantization of space-time derivatives in the presence of constant EM background field strength.

As an interesting question which we plan to study is the derivation of massive HS fields from twistorial model of Shirafuji type, with two spinors which are necessary in $D=4$ in order to define in spinorial framework the time-like four-momentum. The idea of describing massive particles by multispinors is well-known from the consideration of Penrose and his collaborators [25, 26, 27, 28, 29] and further was considered as well as in the supersymmetric case in [30, 31, 32, 33, 34, 35, 36]. In order to illustrate the derivation of massive spinorial model we shall argue following [30] that $D=3$ massive Shirafuji model can be obtained by dimensional reduction of $D=4$ spinorial model considered in [7].

For that purpose we should perform the decompositions of $D=4$ tensors in term of corresponding $D=3$ objects [30]. $D=4$ four-vector $X_{\alpha\dot{\beta}}$ has the following representation

$$X_{\alpha\dot{\beta}} = x_{\alpha\beta} - ix_2\epsilon_{\alpha\beta}\,, \qquad X^{\dot{\alpha}\beta} = x^{\alpha\beta} - ix_2\epsilon^{\alpha\beta}\,, \tag{94}$$

where the tensor $x^{\alpha\beta} = x^{\beta\alpha} = (\overline{x^{\alpha\beta}})$ describe $D=3$ three-vector and $\epsilon_{\alpha\beta} = -\epsilon_{\beta\alpha}$, $\epsilon^{\alpha\beta} = -\epsilon^{\beta\alpha}$, $\epsilon_{12} = \epsilon^{21} = 1$ represent $D=3$ invariant tensors raising or lowering spinor indices: $x^{\gamma}_{\alpha} = \epsilon_{\alpha\beta}x^{\beta\gamma}$. The $D=3$ decomposition of $z_{\mu\nu}$ after using the spinorial notation looks as follows:

$$z^{\alpha\beta} = z_1^{\alpha\beta} + iz_2^{\alpha\beta}\,, \qquad \bar{z}^{\dot{\alpha}\dot{\beta}} = z_1^{\alpha\beta} - iz_2^{\alpha\beta}\,, \tag{95}$$

where $(\overline{z_1^{\alpha\beta}}) = z_1^{\alpha\beta}$, $(\overline{z_2^{\alpha\beta}}) = z_2^{\alpha\beta}$ are real. Further $D=4$ Weyl spinor

$$\lambda^{\alpha} = u^{\alpha} + iv^{\alpha}\,, \qquad \bar{\lambda}^{\dot{\alpha}} = u^{\alpha} - iv^{\alpha} \tag{96}$$

provides two $D=3$ Majorana (real) spinors $(\overline{u^{\alpha}}) = u^{\alpha}$, $(\overline{v^{\alpha}}) = v^{\alpha}$. Similarly we have

$$y^{\alpha} = \tfrac{1}{2}\left(\rho^{\alpha} - i\varrho^{\alpha}\right), \qquad \bar{y}^{\dot{\alpha}} = \tfrac{1}{2}\left(\rho^{\alpha} + i\varrho^{\alpha}\right), \tag{97}$$

where $(\overline{\rho^{\alpha}}) = \rho^{\alpha}$, $(\overline{\varrho^{\alpha}}) = \varrho^{\alpha}$ are Majorana spinors.

Let us consider the $D=4$ Shirajuji action [13] for spinless massless particle

$$\begin{aligned} S^{D=4}_{m=0} &= \int d\tau \lambda_{\alpha}\bar{\lambda}_{\dot{\beta}}\dot{X}^{\dot{\beta}\alpha} \\ &= \int d\tau \Big[\left(u_{\alpha}u_{\beta} + v_{\alpha}v_{\beta}\right)\dot{x}^{\alpha\beta} + 2(u_{\alpha}v^{\alpha})\,\dot{\varphi} \Big]\,, \end{aligned} \tag{98}$$

where we denote $\varphi \equiv x_2$ (see (94)). From (98) follows by varying of φ that $\frac{d}{d\tau}(u_{\alpha}v^{\alpha}) = 0$. We will further fix the constant on-shell value of $(u_{\alpha}v^{\alpha})$ by the replacement in (98)

$$2(u_{\alpha}v^{\alpha})\,\dot{\varphi} \qquad \rightarrow \qquad -\lambda(u_{\alpha}v^{\alpha} - m)\,. \tag{99}$$

Because the three-momentum in (98) is given by the formula

$$p_{\alpha\beta} = u_{\alpha}u_{\beta} + v_{\alpha}v_{\beta} \tag{100}$$

we get from (99) the mass-shell condition

$$p^2 \equiv p^\mu p_\mu = \tfrac{1}{2}\, p_{\alpha\beta} p^{\alpha\beta} = (u_\alpha v^\alpha)^2 = m^2 \,. \tag{101}$$

In such a way we obtain the spinorial description of $D=3$ massive particle model with linear mass-shell condition

$$S^{D=3}_{m\neq 0} = \int d\tau \Big[p_{\alpha\beta}\dot{x}^{\alpha\beta} - \lambda \left(\sqrt{p^2} - m \right) \Big] \,. \tag{102}$$

The model (102) is equivalent to the standard massive particle model. Indeed, the action (102) after substituting the algebraic equations obtained by consecutive varying of p_μ and λ

$$p_\mu = 2m\dot{x}_\mu/\lambda \,, \qquad \lambda = 2\sqrt{\dot{x}^2} \tag{103}$$

provides (up to scale renormalization of mass parameter) the standard action for massive spinless particle

$$S^{D=3}_{m\neq 0} = m \int d\tau \, \sqrt{\dot{x}^2} \,. \tag{104}$$

Following [32] we write $D=3$ massive Shirafuji model as follows

$$S^{D=3}_{m\neq 0} = \int d\tau \Big[\left(u_\alpha u_\beta + v_\alpha v_\beta \right) \dot{x}^{\alpha\beta} - \lambda (u_\alpha v^\alpha - m) \Big] \tag{105}$$

with canonical three-momentum $p_{\alpha\beta} = \dfrac{\partial L}{\partial \dot{x}^{\alpha\beta}}$ given by the formula (100). One gets the field equations

$$\dot{p}_{\alpha\beta} = 0 \,, \tag{106}$$

$$\dot{x}^{\alpha\beta} u_\beta - \lambda v^\alpha = 0 \,, \qquad \dot{x}^{\alpha\beta} v_\beta + \lambda u^\alpha = 0 \,, \tag{107}$$

$$u_\alpha v^\alpha = m \,. \tag{108}$$

Assuming general formula

$$\dot{x}^{\alpha\beta} = a u^\alpha u^\beta + b(u^\alpha v^\beta + v^\alpha u^\beta) + c v^\alpha v^\beta \tag{109}$$

and substituting in (107) one gets after using (108) that $b=0$, $a=c=\lambda/m$ and we obtain (see also [29])

$$\dot{x}^{\alpha\beta} = \lambda p^{\alpha\beta}/m \,. \tag{110}$$

If we introduce $D=3$ twistors describing the following pair of $O(3,2) \cong Sp(4)$ real spinors

$$t^1_A = \begin{pmatrix} u_\alpha \\ \omega^\beta \end{pmatrix} , \qquad t^2_A = \begin{pmatrix} v_\alpha \\ \mu^\beta \end{pmatrix} , \tag{111}$$

where

$$\omega^\alpha = x^{\alpha\beta} u_\alpha \,, \qquad \mu^\alpha = x^{\alpha\beta} v_\alpha \,, \tag{112}$$

the action (105) can be rewritten modulo total derivative as a free bitwistorial particle model

$$S^{D=3}_{m\neq 0} = \int d\tau \Big[\left(u_\alpha \dot{\omega}^\alpha - \dot{u}_\alpha \omega^\alpha \right) + \left(v_\alpha \dot{\mu}^\alpha - \dot{v}_\alpha \mu^\alpha \right) - \lambda \left(u_\alpha v^\alpha - m \right) \Big] \tag{113}$$

where mass-shell term break $O(3,2)$ symmetry representing $D=3$ conformal invariance by the so-called "infinity twistor" I^{AB} [25, 26]

$$u_\alpha v^\alpha = t^1_A I^{AB} t^2_B \,. \tag{114}$$

In order to obtain massive $D=3$ Maxwell HS multiplets we should generalize the model (105) in three steps:

i) following [7] to enlarge $D = 3$ space-time to the generalized space-time with one or two additional vector variables (see (95)) describing $D = 3$ tensorial central charge coordinates;

ii) after the replacement (3) by using Maxwell-covariant CM one-forms add the terms depending on constant EM field in order to obtain $D = 3$ Maxwell-invariant Lagrangian;

iii) add the free action for additional spinorial variables ρ^α, ϱ^α (see (97)) in order to simplify the quantization procedure.

We plan to consider in our next publication the massive extension of the model considered in [7]. The construction of massive Shirafuji model one can obtain for $D = 3$ (real Majorana spinors), $D = 4$ (complex Weyl spinors) or $D = 6$ (quaternionic Weyl spinors), by using respective complex and quaternionic generalization of the formulae (94)-(97) (see also [37, 38]).

Acknowledgements

We acknowledge a support from the grant of the Bogoliubov-Infeld Programme and RFBR grants 12-02-00517, 13-02-90430 (S.F.), as well as Polish National Center of Science (NCN) research project No. 2011/01/ST2/03354 (J.L.). One of the author (J.L.) would like to thank Cestmir Burdik for warm hospitality at the conference in Prague and CERN Theory Division, where present paper has been completed.

References

[1] C. Fronsdal, *Massless Fields with Integer Spin*, Phys. Rev. **D18** (1978) 3624.

[2] A. Jevicki, K. Jin, Q. Ye, *Collective Dipole Model of AdS/CFT and Higher Spin Gravity*, J. Phys. **A44** (2011) 465402, `arXiv:1106.3983 [hep-th]`.

[3] C. Fronsdal, *Massless Particles, Orthosymplectic Symmetry And Another Type Of Kaluza-klein Theory*, in "Essays On Supersymmetry", Ed. C.Fronsdal, 163-265, Calif. Univ. Los Angeles, 1985.

[4] I. Bandos, J. Lukierski, *Tensorial Central Charges and New Superparticle Models with Fundamental Spinor Coordinates*, Mod. Phys. Lett. **A14** (1999) 1257, `arXiv:hep-th/9811022`.

[5] I. Bandos, J. Lukierski, D. Sorokin, *Superparticle Models with Tensorial Central Charges*, Phys. Rev. **D61** (2000) 045002, `arXiv:hep-th/9904109`.

[6] M.A. Vasiliev, *Conformal Higher Spin Symmetries of 4d Massless Supermultiplets and osp(L, 2M) Invariant Equations in Generalized (Super)Space*, Phys. Rev. **D66** (2002) 066006, `arXiv:hep-th/0106149`.

[7] S. Fedoruk, J. Lukierski, *New spinorial particle model in tensorial space-time and interacting higher spin fields*, JHEP **1302** (2013) 128, `arXiv:1210.1506 [hep-th]`.

[8] E. Sezgin, *The M algebra*, Phys. Lett. **B392** (1997) 323, `arXiv:hep-th/9609086`.

[9] M.A. Vasiliev, *Higher-Spin Theories and Sp(2M) Invariant Space–Time*, the talk presented at the third Sakharov Conference in Physics, Moscow, 26-29 June 2002, `arXiv:hep-th/0301235`.

[10] M.A. Vasiliev, *On Conformal, SL(4,R) and Sp(8,R) Symmetries of 4d Massless Fields*, Nucl. Phys. **B793** (2008) 469, `arXiv:0707.1085 [hep-th]`.

[11] M. Plyushchay, D. Sorokin, M. Tsulaia, *Higher Spins from Tensorial Charges and $OSp(N|2n)$ Symmetry*, JHEP **0304** (2003) 013, `arXiv:hep-th/0301067`.

[12] M. Plyushchay, D. Sorokin, M. Tsulaia, *GL Flatness of $OSp(1|2n)$ and Higher Spin Field Theory from Dynamics in Tensorial Spaces*, the Proceedings of the International Workshop "Supersymmetries and Quantum Symmetries" (SQS'03, Dubna, 24-29 July, 2003), `arXiv:hep-th/0310297`.

[13] T. Shirafuji, *Lagrangian Mechanics Of Massless Particles With Spin*, Prog. Theor. Phys. **70** (1983) 18.

[14] S. Fedoruk, J. Lukierski, *New particle model in extended space-time and covariantization of planar Landau dynamics*, Phys. Lett. **B718** (2012) 646, `arXiv:1207.5683 [hep-th]`.

[15] H. Bacry, P. Combe, J.L. Richard, *Group-theoretical analysis of elementary particles in an external electromagnetic field. 1. The relativistic particle in a constant and uniform field* , Nuovo Cim. **A67** (1970) 267.

[16] R. Schrader, *The Maxwell group and the quantum theory of particles in classical homogeneous electromagnetic fields*, Fortsch. Phys. **20** (1972) 701.

[17] J. Beckers, V. Hussin, *Minimal Electromagnetic Coupling Schemes. II. Relativistic And Nonrelativistic Maxwell Groups*, J. Math. Phys. **24** (1983) 1295.

[18] L.D. Faddeev, S.L. Shatashvili, *Realization of the Schwinger Term in the Gauss Law and the Possibility of Correct Quantization of a Theory with Anomalies*, Phys. Lett. **B167** (1986) 225.

[19] D.V. Soroka, V.A. Soroka, *Tensor extension of the Poincare algebra*, Phys. Lett. **B607** (2005) 302, arXiv:hep-th/0410012.
[20] J. Gomis, K. Kamimura, J. Lukierski, *Deformations of Maxwell algebra and their Dynamical Realizations*, JHEP **0908** (2009) 039, arXiv:0906.4464 [hep-th].
[21] P.A.M. Dirac, *Relativistic wave equations*, Proc. Roy. Soc. Lond. **A155** (1936) 447.
[22] M. Fierz, W. Pauli, *On relativistic wave equations for particles of arbitrary spin in an electromagnetic field*, Proc. Roy. Soc. Lond. **A173** (1939) 211.
[23] M. Porrati, R. Rahman, A. Sagnotti, *String Theory and The Velo-Zwanziger Problem*, Nucl. Phys. **B846** (2011) 250, arXiv:1011.6411 [hep-th].
[24] I.L. Buchbinder, T.V. Snegirev, Yu.M. Zinoviev, *Cubic interaction vertex of higher-spin fields with external electromagnetic field*, Nucl. Phys. **B864** (2012) 694, arXiv:1204.2341 [hep-th].
[25] R. Penrose, *Twistor algebra*, J. Math. Phys. **8** (1967) 345.
[26] R. Penrose, M.A.H. MacCallum, *Twistor theory: An Approach to the quantization of fields and space-time*, Phys. Rept. **6** (1972) 241.
[27] R. Penrose, *The Twistor Program*, Rept. Math. Phys. **12** (1977) 65.
[28] L.P. Hughston, *Twistors And Particles*, Lecture Notes In Physics, Vol. **97**, Springer-Verlag, Berlin, 1979.
[29] A. Bette, *On a point-like relativistic and spinning particle*, J. Math. Phys. **25** (1984) 2456; *Directly interacting massless particles: A Twistor approach*, J. Math. Phys. **37** (1996) 1724, arXiv:hep-th/9601017.
[30] D.P. Sorokin, D.V. Volkov, *(Anti)commuting spinors and supersymmetric dynamics of semions*, Nucl. Phys. **B409** (1993) 547.
[31] S. Fedoruk, V.G. Zima, *Massive Superparticle with Tensorial Central Charges*, Mod. Phys. Lett. **A15** (2000) 2281, arXiv:hep-th/0009166.
[32] S. Fedoruk, V.G. Zima, *Uniform Twistor-Like Formulation of Massive and Massless Superparticles with Tensorial Central Charges*, Nucl. Phys. Proc. Suppl. **102** (2001) 233, arXiv:hep-th/0104178.
[33] S. Fedoruk, V.G. Zima, *Bitwistor formulation of massive spinning particle*, Journal of Kharkiv University **585** (2003) 39, arXiv:hep-th/0308154.
[34] A. Bette, J.A. de Azcarraga, J. Lukierski, C. Miquel-Espanya, *Massive relativistic particle model with spin and electric charge from two twistor dynamics*, Phys. Lett. **B595** (2004) 491, arXiv:hep-th/0405166.
[35] J.A. de Azcarraga, A. Frydryszak, J. Lukierski, C. Miquel-Espanya, *Massive relativistic particle model with spin from free two-twistor dynamics and its quantization*, Phys. Rev. **D73** (2006) 105011, arXiv:hep-th/0510161.
[36] S. Fedoruk, A. Frydryszak, J. Lukierski, C. Miquel-Espanya, *Extension of the Shirafuji model for Massive Particles with Spin*, Int. J. Mod. Phys. **A21** (2006) 4137, arXiv:hep-th/0510266.
[37] T. Kugo, P.K. Townsend, *Supersymmetry and the Division Algebras*, Nucl. Phys. **B221** (1983) 357.
[38] I. Bengtsson, M. Cederwall, *Particles, Twistors and the Division Algebras*, Nucl. Phys. **B302** (1988) 81.

Cohomological obstructions in locally variational field theories

Mauro Francaviglia[1], **Marcella Palese and Ekkehart Winterroth**
Department of Mathematics, University of Torino, via C. A.lberto 10, I-10124 Torino , Italy
E-mail: marcella.palese@unito.it, ekkehart.winterroth@unito.it

Abstract. We study cohomological obstructions to the existence of global conserved quantities. In particular, we show that, if a given local variational problem is supposed to admit global solutions, certain cohomology classes cannot appear as obstructions. Vice versa, we obtain a new type of cohomological obstruction to the existence of global solutions for a variational problem.

2000 MSC: 58A20,58E30,46M18.
keywords: jet, gauge-natural bundle, conserved quantities, cohomology.

1. Introduction

We are interested in studying corservation laws associated with invariance properties of *global* field equations which are *locally variational*, this means that locally they can be derived variationally from a Lagrangian, which is then only defined on an open set of the relevant field space; this is what we call a *local presentation* of a given variational problem. The requirement of *globality* for field equations, of course, determine the relationship between a Local Lagrangian defined on one open set and another local Lagrangian defined on another open set providing the same global field equations: we require two local presentations to differ by a variationally closed form, *i.e.* locally by a variationally trivial Lagrangian (a local divergence). Noether currents for such local presentations and corresponding *conserved currents associated with each local presentation* have been characterized [6]; although *they are associated with invariance of global field equations*, such conserved currents are, in principle, *local currents*.

In this paper we show that there exists cohomological obstructions for such local currents be globalized and we shall see how such obstructions are also related with the *existence of global solutions* for a given global field equation. To investigate when local variational objects can be globalized, we use a geometric formulation of the calculus of variations as a subsequence of the de Rham sequence of differential forms on finite order prologations of fibered manifolds.

We assume the r-th order prolongation of a fibered manifold $\pi : \mathbf{Y} \to \mathbf{X}$, with $\dim \mathbf{X} = n$ and $\dim \mathbf{Y} = n + m$, to be the configuration space; this means that *fields are assumed to be (local) sections* of $\pi^r : J_r\mathbf{Y} \to \mathbf{X}$. Due to the affine bundle structure of $\pi^{r+1}_r : J_{r+1}\mathbf{Y} \to J_r\mathbf{Y}$, we have a natural splitting $J_r\mathbf{Y} \times_{J_{r-1}\mathbf{Y}} T^*J_{r-1}\mathbf{Y} = J_r\mathbf{Y} \times_{J_{r-1}\mathbf{Y}} (T^*\mathbf{X} \oplus V^*J_{r-1}\mathbf{Y})$, which induces natural *spittings* in horizontal and vertical parts of vector fields, forms and of the exterior differential

[1] Mauro Francaviglia suddenly passed away in June 2013; this paper is the outcome of our recent joint research work and we dedicate it to his memory.

on $J_r\mathbf{Y}$. Starting from this splitting one can define sheaves of contact forms Θ^*_r defined by the kernel of the horizontalization; the sheaves Θ^*_r form an exact subsequence of the de Rham sequence on $J_r\mathbf{Y}$ and one can define the quotient sequence

$$0 \to I\!R_{\mathbf{Y}} \to \ldots \to^{\mathcal{E}_{n-1}} \Lambda^n_r/\Theta^n_r \to^{\mathcal{E}_n} \Lambda^{n+1}_r/\Theta^{n+1}_r \to^{\mathcal{E}_{n+1}} \Lambda^{n+2}_r/\Theta^{n+2}_r \to^{\mathcal{E}_{n+2}} \ldots \to 0$$

the r–th order *variational sequence* on $\mathbf{Y} \to \mathbf{X}$ which is an acyclic resolution of the constant sheaf $I\!R_{\mathbf{Y}}$; see [13].

The quotient sheaves in the variational sequence can be represented as sheaves $\mathcal{V}^k_r$ of k-forms on jet spaces of higher order, see *e.g.* [12, 13, 16]. Lagrangians are sheaf sections $\lambda \in (\mathcal{V}^n_r)_{\mathbf{Y}}$, while $\mathcal{E}_n$ is called the Euler-Lagrange morphism. The Euler-Lagrange equations are then $\mathcal{E}_n(\lambda) \circ j_{2r+1}\sigma = 0$ for (local) sections $\sigma : \mathbf{X} \to \mathbf{Y}$. Sections $\eta \in (\mathcal{V}^{n+1}_r)_{\mathbf{Y}}$ are called *source forms* or also *dynamical forms*, $\mathcal{E}_{n+1}$ is called the Helmholtz morphism. The equations $\eta \circ j_{2r+1}\sigma = 0$ are locally variational if and only if Helmholtz conditions $\mathcal{E}_{n+1}(\eta) = 0$ hold true.

1.1. Inverse and local variational problems

Since the variational sequence is an exact resolution, it is possible to determine cohomological obstructions to global inverse problems: *i.e.* given a closed section of a quotient sheaf of the variational sequence, when this section is also globally exact. In fact, the cohomology of the complex of global sections $H^*_{VS}(\mathbf{Y})$ is naturally isomorphic to both the Čech cohomology and the de Rham cohomology $H^*_{dR}(\mathbf{Y})$ [13]. Thus if the cohomology of $\mathbf{Y}$ is trivial, each local inverse problem is also global; we are here interested in the case of a nontrivial cohomology of $\mathbf{Y}$. Let $\mathbf{K}^p_r \doteq Ker\,\mathcal{E}_p$; we have a natural short exact sequence of sheaves which gives rise in a standard way [3] to a long exact sequence in Čech cohomology, where the *connecting homomorphism* $\delta_p = i^{-1} \circ \mathfrak{d} \circ \mathcal{E}_p^{-1}$ is the mapping of cohomologies in the corresponding diagram of cochain complexes (here $\mathfrak{d}$ is the usual *coboundary operator*), see *e.g.* [2, 6, 27]. Every $\eta \in (\mathcal{E}_p(\mathcal{V}^p_r))_{\mathbf{Y}}$ (*i.e.* locally variational) defines a cohomology class $\delta_p(\eta) \in H^1(\mathbf{Y}, \mathbf{K}^p_r) \simeq H^{p+1}_{VS}(\mathbf{Y}) \simeq H^{p+1}_{dR}(\mathbf{Y})$ (in particular, $\mathcal{E}_n(\mathcal{V}^n_r)$ is the sheave of Euler–Lagrange morphisms and $\eta \in (\mathcal{E}_n(\mathcal{V}^n_r))_{\mathbf{Y}}$ if and only if $\mathcal{E}_{n+1}(\eta) = 0$, which are Helmholtz conditions). In the following we shall use the notation $\delta(\eta) \equiv \delta_n(\eta)$; we furthermore point out that every $\mu \in (d_H(\mathcal{V}^{n-1}_r))_{\mathbf{Y}}$ (*i.e.* locally variationally trivial) defines a cohomology class $\delta'(\mu) \equiv \delta_{n-1}(\mu) \in H^1(\mathbf{Y}, \mathbf{K}^{n-1}_r) \simeq H^n_{VS}(\mathbf{Y}) \simeq H^n_{dR}(\mathbf{Y})$.

It is clear that η is globally variational if and only if $\delta_p(\eta) = 0$. Every nonvanishing cohomology class in $H^p_{dR}(\mathbf{Y})$ gives rise to local variational problems: two local variational problems of degree p are *equivalent* if and only if they give rise to the same variational class of forms as the image of the corresponding morphism $\mathcal{E}_p$ in the variational sequence.

2. Currents variationally associated with locally variational field equations

We shall explicate now how cohomology enters in globality problems concerned with conserved quantities. We shall examine this aspect within Noether formalism [20] (see [11] for an epistemological discussion). Noether's Theorems already pointed out the cohomological content of the invariance of the Euler-Lagrange expressions and dealt with the study of the relationship between currents associated with symmetries of *global* Lagrangians and of corresponding Euler-Lagrange expressions; see, in particular, [1].

At the classical (infinitesimal) level, Noether Theorems can be formulated by means of the Lie derivative of a Lagrangian density and in fact consist of deriving (by an integration by parts procedure) a suitable version of the Cartan formula for the Lie derivative of horizontal n-forms on prolongations of fibered manifolds. We notice that the underling concept is that of finding a Lepage equivalent [18] for which a suitable Cartan formula can be written, then factorize modulo contact structures [14, 15, 16, 17, 19] (this procedure is involved with the Legendre transformation and thus with the appearing of a *momentum* at any degree of forms). This is the basic idea underling the definition of a *variational Lie derivative* operator $\mathcal{L}_{j_r\Xi}$ and of a *quotient Cartan formula* (a *variation formula*) defined on the sheaves of the variational sequence. This enable us to define symmetries of classes of forms of any degree in the variational sequence and corresponding conservation theorems; see [7] and [21] for details.

Having a suitable version of the Cartan formula and using naturality properties, for $\eta \in \mathcal{V}^p_r$ a locally variational p-form such that $\delta_p(\eta_\lambda) \neq 0$, we have that $\delta_p(\mathcal{L}_\Xi \eta_\lambda) = 0$, *i.e.* the variational Lie derivative 'trivializes' cohomology classes [6, 25, 27]. This suggests us that, if we know that the first variational derivative of a local presentation is closed, then the second variational derivative define a trivial cohomology class. In fact, the first variational derivative – with respect to symmetries of Euler-Lagrange expressions – of a local presentation is closed; therefore a trivial cohomology class is defined by the variational derivative of currents [9, 10].

Let then η_λ be a global Euler–Lagrange morphism for a local variational problem λ_i; within the previously stated geometric framework, Noether's First Theorem reads

$$\mathcal{L}_\Xi \lambda_i = \Xi_V \rfloor \eta_\lambda + d_H \epsilon_i \,.$$

Analogously, a suitable representation of the Lie derivative of classes of dynamical forms can be obtained by a quotient Cartan formula; for details, see [21]. It is perhaps worthy of mention that this is related with an integration by parts formula for $\Xi_V \rfloor \eta_\lambda$; such relationship has been worked out for gauge-natural theories, see *e.g.* [4, 5, 8, 23, 22, 24, 26].

Since we assume η_λ to be closed, the quotient Cartan formula reduces to $\mathcal{L}_\Xi \eta_\lambda = \mathcal{E}_n(\Xi_V \rfloor \eta_\lambda)$, and if Ξ is such that $\mathcal{L}_\Xi \eta_\lambda = 0$, then $\mathcal{E}_n(\Xi_V \rfloor \eta_\lambda) = 0$; therefore *locally*

$$\Xi_V \rfloor \eta_\lambda = d_H \nu_i \,.$$

Notice that $\mathfrak{d}(\Xi_V \rfloor \eta_\lambda) = 0$, but in general we have the obstruction

$$\delta^{'}(\Xi_V \rfloor \eta_\lambda) \equiv \delta_{n-1}(\Xi_V \rfloor \eta_\lambda) \neq 0 \,,$$

so that the current ν_i is a *local* object and it is conserved along the solutions of Euler–Lagrange equations (critical sections) [6].

On the other hand, *and independently* (see [20]), we get *locally*

$$\mathcal{L}_\Xi \lambda_i = d_H \beta_i \,,$$

thus we can write $\Xi_V \rfloor \eta_\lambda + d_H(\epsilon_i - \beta_i) = 0$, where ϵ_i is the usual *canonical* Noether current.

Definition 1 We call the (local) current $\epsilon_i - \beta_i$ a *Noether–Bessel-Hagen current.*

Note that when Ξ is only a symmetry of a dynamical form but not a symmetry of the Lagrangian, the so-called *strong Noether current* $\nu_i + \epsilon_i$ *is not a conserved current* and it is such that $d_H(\nu_i + \epsilon_i)$ is locally equal to $d_H \beta_i$ (*i.e.* they are variationally equivalent).

At this point it deserves attention that we are studing conserved quantities associated with symmetries of global dynamical forms being global Euler-Lagrange forms of local presentations. We identified an obstruction to globality of such currents and it is a natural question whether and when we could instead be able to associate global conserved currents with globally invariant dynamics. It is natural to expect that somehow some restrictions have to be made. We can, for example, proceed in two different manner:

(i) look for an additional principle inspired by the fact that $\delta^{'}(\mathcal{L}_\Xi(\Xi_V \rfloor \eta_\lambda)) = 0$ and therefore restrict the space of "admissible" generalized symmetries;

(ii) look for conditions on the cohomology of the configuration space and on the existence of global critical sections and therefore restrict the "admissible" fields.

The first approach provided us with the following results [9, 10].

Definition 2 A conserved current *variationally associated with locally variational invariant field equations* is a current associated with *a symmetry of the field equation which is also required to be a symmetry for the variational derivative of the local problem generating such a field equation*, *i.e.* a current associated to a vector field Ξ satisfying

$$\mathcal{L}_{\Xi}\eta_{\lambda} = 0 \quad and \quad \mathcal{L}_{\Xi}\mathcal{L}_{\Xi}\lambda_i = 0 \quad but \quad \mathcal{L}_{\Xi}\lambda_i \neq 0\,.$$

We see that the condition $\mathcal{L}_{\Xi}\mathcal{L}_{\Xi}\lambda_i = 0$ implies only $\mathfrak{d}d_H\nu_i = 0$, *i.e.* $d_H\nu_i$ is global.

If we require the stronger condition $\mathcal{L}_{\Xi}\mathfrak{d}\lambda_i = 0$ (*i.e.* the Lie derivative drags the local problems in such a way that their variational derivative coincide on the intersections of two open sets), we get $\mathfrak{d}d_H\beta_i = 0$, *i.e.* the divergence of the strong Noether current is *global*. This is also equivalent to say that the coboundary of the strong Noether currents is locally exact. Under this condition we have the conservation law $d_H\mathcal{L}_{\Xi}(\nu_i + \epsilon_i) = 0$, where $\mathcal{L}_{\Xi}(\nu_i + \epsilon_i)$, the *variation of the strong Noether currents*, is a *local representative* of a global conserved current.

3. Cohomological obstructions and existence of global critical sections

In this section we now investigate what is possible to state following the second approach. We have seen that what we called the Noether–Bessel-Hagen current $\epsilon_i - \beta_i$ is a *local* object and it is a current associated with a generalized symmetry (conserved along critical sections, since we locally have $\nu_i = \epsilon_i - \beta_i + d_H\psi_{ij}$). In the following we shall denote by $[\cdot,\cdot]$ a class in de Rham cohomology. In the previous section, we saw that the current ν_i (as well as the Noether-Bessel-Hagen current) is variationally equivalent to a global (conserved) current if and only if $0 = [\Xi_V \rfloor \mathcal{E}_n(\lambda_i)] \in H^n_{dR}(\mathbf{Y})$.

We shall investigate the relationship between the existence of global critical sections and the existence of global Noether–Bessel-Hagen currents. We show that for a large class of theories, if a given local variational problem is supposed to admit global solutions, the above cohomology class is trivial. Vice versa, we obtain a new type of cohomological obstruction to the existence of global critical sections for a variational problem.

In the following we shall need to see certain cohomology classes in $H^n_{dR}(\mathbf{Y})$ as the pull-back of cohomology classes in $H^n_{dR}(\mathbf{X})$, *i.e.* defined by *closed differential forms on the base manifold*. For this reason we shall assume some restrictive conditions, which in fact can be relaxed; for details see *e.g.* [28]. Note that $J_r\mathbf{Y}$ and $\mathbf{Y}$ have the same de Rham cohomology; accordingly, to simplify the notation, we shall omit the jet prolongation order. We also stress that our reasoning works well in cohomology in degree n, but not greater.

Let us then assume σ be a *global section* of $\pi : \mathbf{Y} \to \mathbf{X}$; for any global section we have of course $\sigma^* \circ \pi^* = id_{H^n_{dR}(\mathbf{X})}$. Let furthermore $H^n_{dR}(\mathbf{Y}) \sim \pi^*(H^n_{dR}(\mathbf{X}))$, *i.e.* the pull-back $\pi^* : H^n_{dR}(\mathbf{X}) \to H^n_{dR}(\mathbf{Y})$ is an isomorphism of cohomology groups.

Proposition 1 *Under the assumption of the isomorphism above, if a given (local) variational problem admits global critical sections (solutions), all conservation laws derived from symmetries of global field equations admit global Noether–Bessel-Hagen conserved currents.*

PROOF. Let then σ be a global *critical* section; being a critical section the locally variational source form η vanishes along its corresponding jet prolongation $j\sigma$ and we have that also $\Xi_V \rfloor \eta$ vanishes along $j\sigma$. Since it is a global section we get that its pull-back to $\mathbf{X}$ is identically zero, so that $[j\sigma^*\Xi_V \rfloor \eta] = 0$ and, hence, since π^* is an isomorphism, we have $[\Xi_V \rfloor \eta] = 0$. The vanishing of $\Xi_V \rfloor \eta$ along $j\sigma$ implies the vanishing of the pull-back of a differential form representing the cohomology class $[\Xi_V \rfloor \eta]$ and hence of the image of the cohomology class in $H^n_{dR}(\mathbf{X})$. This works well in spite of the fact that η, and thus also $\Xi_V \rfloor \eta$, is not a differential form, but a representative of an equivalence class of differential forms in the quotient Λ^*_r/Θ^*_r.

This is apparently a little problem, because we need to compare the cohomology of the base manifold with the cohomology of the total space, and while $H^n_{dR}(\mathbf{Y}) \simeq H^n_{VS}(\mathbf{Y})$, in principle we do not have a similar isomorphism for the cohomology groups $H^n_{dR}(\mathbf{X})$. This is overcome by the peculiar structure of quotient space in the variational sequence: for all sections $j\sigma^*(\Theta^*_r) = 0$, we can correctly define $j\sigma^*[\eta] \doteq [j\sigma^*\eta]$; in fact, as explained before since Θ^n_r consists of contact n-forms the pull-back σ^* factorizes over Λ^n_r/Θ^n_r. It is furthermore a well known fact that we can always find a closed form $\alpha \in \Lambda^n_r$ which represents the cohomology class $[\Xi_V \rfloor \eta]$ and which projects onto $\Xi_V \rfloor \eta \in \Lambda^n_r/\Theta^n_r$. Thus, $j\sigma^*(\alpha) = 0$ and the corresponding class vanishes in $H^n_{dR}(\mathbf{X})$; therefore $[\Xi \rfloor \eta_\lambda] = 0$, then there exist global conserved currents for all generalized symmetry Ξ. Note that the obstruction does not depend on the section because of the isomorphism.

We notice that without the assumption of the isomorphism of cohomology groups, we can anyway say that $[\Xi \rfloor \eta_\lambda] \notin \pi^* H^n_{dR}(\mathbf{X})$. Of course, in this case we cannot infer in the same way the existence of global currents.

Example 1 In the case of the geometric formulation of Chern-Simons theories $\mathbf{Y}$ takes the form $J^1\mathbf{P}/\mathbf{G}$ for a principal bundles $\mathbf{P}$ over an arbitrary odd dimensional $\mathbf{X}$. Since $J^1\mathbf{P}/\mathbf{G}$ is an affine bundle over $\mathbf{X}$, $J^1\mathbf{P}/\mathbf{G}$ and $\mathbf{X}$ are homotopy equivalent and, in particular, $H^n_{dR}(J^1\mathbf{P}/\mathbf{G}) \sim H^n_{dR}(\mathbf{X})$. We can therefore state that Chern Simons theories, which admit global critical sections, always admit global conserved quantities (Noether–Bessel-Hagen currents).

We can use the result above also conversely to define a cohomological obstruction to the existence of global solutions for a given problem.

Corollary 1 *Let σ be a section of $\mathbf{Y}$ over $\mathbf{X}$. Let α be a closed differential form which represents $[\Xi_V \rfloor \eta]$, the obstruction to the existence of global conserved quantities for conservation laws associated with the problem η and its symmetry Ξ, in $H^n_{dR}(\mathbf{Y})$. If the class $j\sigma^*[\alpha] \in H^n_{dR}(\mathbf{X})$ does not vanish, then there are no global solutions in the homotopy class of σ.*

Now, as discussed before, if $\pi : \mathbf{Y} \to \mathbf{X}$ admits a global section σ, then this section (as well as any other section) defines a projection $\sigma^* : H^n_{dR}(\mathbf{Y}) \mapsto H^n_{dR}(\mathbf{X})$; the non vanishing of $j\sigma^*([\Xi_V \rfloor \eta])$ is an obstruction to the existence of global solutions in the homotopy class of σ. Note that, in this case, the obstruction *does depend* on the section; $j\sigma^*[\alpha]$ may vanish for sections from one homotopy class, but not for those from another.

It is also important to note that this obstruction is not identically zero, in general. We remark that in the case when the isomorphism of cohomology groups is not given, if a global section is not critical, then the class $[\Xi_V \rfloor \eta]$ is an obstruction for any other homotopic section to be a critical section [28].

Acknowledgement
Research supported by Department of Mathematics, University of Torino. One of us (E. W.) has been partially supported by Department of Mathematics, University of Ostrava.

References

[1] E. Bessel-Hagen: Über die Erhaltungssätze der Elektrodynamik, *Math. Ann.* **84** (1921) 258–276.
[2] A. Borowiec, M. Ferraris, M. Francaviglia, M. Palese: Conservation laws for non-global Lagrangians, *Univ. Iagel. Acta Math.* **41** (2003) 319–331.
[3] G.E. Bredon: *Sheaf Theory*, McGraw–Hill (New York, 1967).
[4] M. Ferraris, M. Francaviglia, M. Palese, E. Winterroth: Canonical connections in gauge-natural field theories, *Int. J. Geom. Methods Mod. Phys.* 5 (6) (2008) 973–988.
[5] M. Ferraris, M. Francaviglia, M. Palese, E. Winterroth: Gauge-natural Noether currents and connection fields, *Int. J. Geom. Methods Mod. Phys.* **8**(1) (2011) 177–185.
[6] M. Ferraris, M. Palese, E. Winterroth: Local variational problems and conservation laws, *Diff. Geom. Appl* **29** (2011) S80–S85.

[7] M. Francaviglia, M. Palese, R. Vitolo: Symmetries in finite order variational sequences, *Czech. Math. J.* **52(127)** (1) (2002) 197–213.
[8] M. Francaviglia, M. Palese, E. Winterroth: Second variational derivative of gauge-natural invariant Lagrangians and conservation laws, in *Differential geometry and its applications*, Matfyzpress, Prague (2005), 591–604.
[9] M. Francaviglia, M. Palese, E. Winterroth: Locally variational invariant field equations and global currents: Chern-Simons theories, *Commun. Math.* **20** (1) (2012)13–22.
[10] M. Francaviglia, M. Palese, E. Winterroth: Variationally equivalent problems and variations of Noether currents, *Int. J. Geom. Meth. Mod. Phys.* **10**(1) (2013) art. no. 1220024.
[11] M. Francaviglia, M. Palese, E. Winterroth: Field equations or conservation laws? in *Physics, Astronomy and Engineering. Critical Problems in The History of Science*, R. Pisano *et al.* eds., The Scientia Socialis Press, Siauliai (2013).
[12] M. Krbek, J. Musilová: Representation of the variational sequence by differential forms, *Acta Appl. Math.***88** (2) (2005) 177–199.
[13] D. Krupka: Variational Sequences on Finite Order Jet Spaces, *Proc. Diff. Geom. Appl.*; J. Janyška, D. Krupka eds., World Sci. (Singapore, 1990) 236–254.
[14] D. Krupka: Global variational theory in fibred spaces, in *Handbook of global analysis*, 773–836, 1215, Elsevier Sci. B. V., Amsterdam, 2008.
[15] D. Krupka, O. Krupková, D. Saunders: The Cartan form and its generalizations in the calculus of variations, *Int. J. Geom. Methods Mod. Phys.***7** (4) (2010) 631–654.
[16] D. Krupka, J. Šedĕnková: Variational sequences and Lepage forms, in *Differential geometry and its applications*, 617–627, Matfyzpress, Prague, 2005.
[17] O. Krupková: Lepage forms in the calculus of variations, in *Variations, geometry and physics*, 27–55, Nova Sci. Publ., New York, 2009.
[18] Th.H.J. Lepage: Sur les champ geodesiques du Calcul de Variations, I, II, *Bull. Acad. Roy. Belg., Cl. Sci.* **22** (1936) 716–729, 1036–1046.
[19] J. Musilová, M. Lenc: Lepage forms in variational theories: from Lepage's idea to the variational sequence, in *Variations, geometry and physics*, 3–26, Nova Sci. Publ., New York, 2009.
[20] E. Noether: Invariante Variationsprobleme, *Nachr. Ges. Wiss. Gött., Math. Phys. Kl.* **II** (1918) 235–257.
[21] M. Palese: Variational derivatives and the representation of variational sequences by differential forms, in preparation, Int. Conf. on Diff. Geom. Appl., Brno 2013.
[22] M. Palese, E. Winterroth: Covariant gauge-natural conservation laws, *Rep. Math. Phys.* **54** (3) (2004) 349–364.
[23] M. Palese, E. Winterroth: Global Generalized Bianchi Identities for Invariant Variational Problems on Gauge-natural Bundles, *Arch. Math. (Brno)* **41** (3) (2005) 289–310.
[24] M. Palese, E. Winterroth: On the relation between the Jacobi morphism and the Hessian in gauge-natural field theories, *Theoret. and Math. Phys.* **152** (2) (2007) 1191–1200.
[25] M. Palese, E. Winterroth: Symmetries of Helmholtz forms and globally variational dynamical forms, *Jour. Phys. Conf. Series* **343** (2012) art. no. 012129.
[26] M. Palese, E. Winterroth: Higgs fields on spinor gauge-natural bundles, *Jour. Phys. Conf. Series* **411(1)** (2013) art. no. 012025.
[27] M. Palese, E. Winterroth, E. Garrone: Second variational derivative of local variational problems and conservation laws, *Arch. Math. (Brno)* **47**(5) (2011) 395–403.
[28] E. Winterroth: Some remarks on topological obstructions in the calculus of variations, in preparation, Int. Conf. on Diff. Geom. Appl., Brno 2013.

Generating functions for tensor product decomposition

Jan Fuksa[1,2], **Severin Pošta**[1]

[1] Department of mathematics, Faculty of nuclear sciences and physical engineering, Czech Technical University in Prague, Trojanova 13, 120 00 Prague, Czech Republic

[2] Bogoliubov Laboratory of Theoretical Physics, JINR, Joliot-Curie 6, 141980 Dubna, Moscow region, Russia

E-mail: `fuksajan@fjfi.cvut.cz`, `severin.posta@fjfi.cvut.cz`

Abstract. The paper deals with the tensor product decomposition problem. Tensor product decompositions are of great importance in the quantum physics. A short outline of the state of the art for the of semisimple Lie groups is mentioned. The generality of generating functions is used to solve tensor products. The corresponding generating function is rational. The feature of this technique lies in the fact that the decompositions of all tensor products of all irreducible representations are solved simultaneously. Obtaining the generating function is a difficult task in general. We propose some changes to an algorithm using Patera-Sharp character generators to find this generating function, which simplifies the whole problem to simple operations over rational functions.

1. Introduction

Generating functions have been usefully applied in many areas of mathematics and mathematical physics as powerful and general tools. In the representation theory of continuous and discrete groups, they provide techniques superior to other existing ones. The distinguishing feature is that they solve an infinity of problems of a given type at the same time.

One important application of generating functions is computation of restricted representations of a group. Restriction forms a representation of a subgroup from a representation of the whole group. Often, the restricted representation is simpler to understand. Rules for decomposing the restriction of an irreducible representation of the group into irreducible representations of the subgroup are of great importance in physics. For example, in the case of explicit symmetry breaking, the symmetry group of the problem is reduced from the whole group to one of its subgroups. In quantum mechanics, this reduction in symmetry appears as a splitting of degenerate states into multiplets, as in the Stark or Zeeman effect.

Branching rules for classical groups are well known. Generating functions have been used to decompose characters of semisimple Lie groups obtained by restriction from overgroups or by tensor products of irreducible representations. [6, 1] In the paper of A. Cohen and G. M. Ruitenburg [2] is proved that these generating functions are always rational, and an example of derivation of the generating function for restricted representations of $\boldsymbol{G}_2$ to $\boldsymbol{A}_2$ is provided. Nevertheless, the proposed algorithm, which is efficient to compute branching rules, turns out to be rather ponderous to decompose tensor products of representations.

We propose some change to a more suitable algorithm to compute tensor products decompositions. The algorithm is based on character generators [7]

$$X(A) = \sum_{\lambda \in \Lambda^+} \operatorname{ch} V(\lambda) \cdot A^\lambda, \tag{1}$$

the generating functions of characters of representations. The distinguishing feature of our algorithm lies in the fact that we do not use generating functions only to solve some recurrence, but we propose an operation over character generators to obtain a generating function for tensor products decompositions. This "*generatingfunctionologist*" [8] approach turns out to be simple and straightforward.

The paper is divided into three sections. In the first section, basics of semisimple Lie groups and their representations are summarized; general properties of tensor product decompositions are mentioned. In the second section, a generating function for tensor product decompositions is outlined. In the last section, the method to obtain the generating function is demonstrated on the example of $SU(3)$.

2. Tensor product decomposition

Let $\boldsymbol{G}$ be a semisimple connected complex Lie group with the Lie algebra $\mathfrak{g}$ of rank n. Let $\boldsymbol{T} \subset \boldsymbol{B}$ is a maximal torus of $\boldsymbol{G}$ with the Lie algebra $\mathfrak{t}$, where $\boldsymbol{B}$ is a fixed Borel subgroup of $\boldsymbol{G}$. Let $W = W_{\boldsymbol{G}}$ be the associated Weyl group. Let $\Lambda = \Lambda(\boldsymbol{T})$ denote the character group of $\boldsymbol{T}$, i.e., the group of all the algebraic group morphisms $\boldsymbol{T} \mapsto \mathbb{C}$. The Weyl group W acts on Λ. There is a straightforward identification between elements of Λ and weights $\lambda \in \mathfrak{t}^*$, which make a lattice in $\mathfrak{t}^*$. Therefore, we will identify Λ and the lattice of the weights in $\mathfrak{t}^*$. Let $R = R_{\mathfrak{g}} \subset \mathfrak{t}^*$ be the set of roots of $\mathfrak{g}$ with respect to the Cartan subalgebra $\mathfrak{t}$, and let R^+ be the set of positive roots. Let $\Delta = \{\alpha_1, \dots, \alpha_n\} \subset R^+$ be the set of simple roots, $\{\alpha_1^\vee, \dots, \alpha_n^\vee\} \subset \mathfrak{t}$ the corresponding simple coroots. For any $1 \leq j \leq n$, the fundamental weight $\omega_j \in \Lambda$ is defined by

$$\omega_j(\alpha_i^\vee) = \delta_{j,i}, \ \forall\, 1 \leq i \leq n. \tag{2}$$

An element $\lambda \in \Lambda$ is called dominant if $\lambda(\alpha_i^\vee) \geq 0$ for all the simple coroots $\alpha_i^\vee$. Let Λ^+ denote the set of all the dominant characters. There is an isomorphism between the set of isomorphism classes of irreducible finite-dimensional representations of $\boldsymbol{G}$ and Λ^+ via the highest weight of the irreducible representation. For $\lambda \in \Lambda^+$, we denote by $V(\lambda)$ the corresponding irreducible representation of the highest weight λ. The W-orbit of any $\lambda \in \Lambda$ contains a unique element in Λ^+. We also have the shifted action of W on Λ via $w * \lambda = w(\lambda + \rho) - \rho$, where ρ is the half sum of positive roots.

Let $\mathbb{Z}[\Lambda(\boldsymbol{G})] = \{e^\lambda \mid \lambda \in \Lambda(\boldsymbol{G})\}$ be a ring of formal exponentials and their linear combinations with coefficients in the ring $\mathbb{Z}$ with the multiplication rule $e^\lambda \cdot e^\mu = e^{\lambda+\mu}$, for a group $\boldsymbol{G}$.

An algebraic character $\operatorname{ch} V(\lambda)$, $\lambda \in \Lambda^+$, of a compact Lie group $\boldsymbol{G}$ can be obtained by the Weyl character formula

$$\operatorname{ch} V(\lambda) = \frac{\sum_{w \in W} \operatorname{sgn}(w) \cdot e^{w(\lambda+\rho)}}{\sum_{w \in W} \operatorname{sgn}(w) \cdot e^{w(\rho)}}, \tag{3}$$

where the elements e^λ are from the ring $\mathbb{Z}[\Lambda(\boldsymbol{G})]$.

For a semisimple connected complex Lie group $\boldsymbol{G}$ with the Lie algebra $\mathfrak{g}$, the irreducible finite-dimensional representations of $\boldsymbol{G}$ are parametrized by the set Λ^+ of dominant characters of $\boldsymbol{T}$, where $\boldsymbol{T}$ is a maximal torus of $\boldsymbol{G}$ with the Lie algebra $\mathfrak{t}$. By the complete reducibility, for any $\lambda, \mu \in \Lambda^+$, the tensor product $V(\lambda) \otimes V(\mu)$ decomposes as

$$V(\lambda) \otimes V(\mu) = \bigoplus_{\nu \in \Lambda^+} m_{\lambda,\mu}^\nu V(\nu), \tag{4}$$

where $m^{\nu}_{\lambda,\mu}$ are the tensor product multiplicities, which denote the multiplicity of $V(\nu)$ in the tensor product $V(\lambda) \otimes V(\mu)$. We say that $V(\nu)$ occurs in $V(\lambda) \otimes V(\mu)$ if $m^{\nu}_{\lambda,\mu} > 0$.

One of the major goals of the tensor product decomposition is to determine all the components of $V(\lambda) \otimes V(\mu)$ together with their multiplicities. In general, this is a very difficult task. [5]

3. Generating function for tensor products decomposition

3.1. *Restricted representations*

Let $\boldsymbol{G}$ be a semisimple connected complex Lie group and $\boldsymbol{H}$ its connected subgroup. Let $\boldsymbol{T}$ and $\boldsymbol{S}$ are maximal tori in $\boldsymbol{G}$ resp. $\boldsymbol{H}$, such that $\boldsymbol{S} \subset \boldsymbol{T}$. The straightforward algorithm to compute a decomposition $(V(\mu)|V(\lambda) \downarrow \boldsymbol{H})$ of the restricted representation $V(\lambda)$ of $\boldsymbol{G}$ into irreducible representations $V(\mu)$ of the subgroup $\boldsymbol{H}$ is to determine the set of all weights of the $\boldsymbol{G}$-module $V(\lambda)$, then to compute their restrictions to $\boldsymbol{S}$, and then to decompose them into $\boldsymbol{H}$-modules.

Let $r : \Lambda(\boldsymbol{G}) \mapsto \Lambda(\boldsymbol{H})$ be a linear restriction of the weights of $\boldsymbol{T}$ to the weights of $\boldsymbol{S}$. By choosing appropriate Borel subgroups, we may assume that $r(\alpha) \notin R^-_{\boldsymbol{H}}$ for $\alpha \in R^+_{\boldsymbol{G}}$. Let $R_{\boldsymbol{0}} = \{\alpha \in R_{\boldsymbol{G}} \,|\, r(\alpha) = 0\}$ and $R^+_{\boldsymbol{0}} = R_{\boldsymbol{0}} \cap R^+_{\boldsymbol{G}}$. Let $W_{\boldsymbol{0}}$ be the subgroup of $W_{\boldsymbol{G}}$ generated by $R_{\boldsymbol{0}}$. Let $W = W_{\boldsymbol{G}}/W_{\boldsymbol{0}}$; each coset of W has a unique representative in $W_{\boldsymbol{G}}$ of a minimal length.

Let $A = r(R^+_{\boldsymbol{G}} \backslash \{0\})$, let L be the lattice of non-negative integral linear combinations of A. The Kostant's partition function p_A on L is defined in the following way:[4] for $\beta \in L$, let $p_A(\beta)$ be the number of all different linear combinations of elements of A giving the element β, i.e., if $a = |A|$, then $p_A(\beta) = |\{(k_1, \ldots, k_a) \,|\, \beta = \sum_{i=1}^{a} k_i \alpha_i, k_i \in \mathbb{Z}\}|$, where $|M|$ is the cardinality of M.

For $\lambda \in \Lambda^+(\boldsymbol{G})$ and $\mu \in \Lambda^+(\boldsymbol{H})$ holds Kostant's formula[3]

$$(V(\mu)|V(\lambda) \downarrow \boldsymbol{H}) = \sum_{w \in W} \det(w) \cdot D_{\boldsymbol{H}}(w(\lambda + \rho_{\boldsymbol{G}})) \cdot p_A[r(w(\lambda + \rho_{\boldsymbol{G}})) - (\mu + r(\lambda + \rho_{\boldsymbol{G}}))], \quad (5)$$

where

$$D_{\boldsymbol{H}}(\nu) = \prod_{\alpha \in R^+_{\boldsymbol{H}}} \frac{(\nu, \alpha)}{(\rho_{\boldsymbol{H}}, \alpha)}, \quad \text{for } \nu \in \Lambda^+(\boldsymbol{H}),$$

is the Weyl dimensional polynomial. The corresponding generating function $P^{\boldsymbol{G}}_{\boldsymbol{H}}$ for restricted representations is obtained[2] by multiplying (5) by A^{λ} from the ring $\mathbb{Z}[\Lambda^+(\boldsymbol{G})]$, $X^{\mu} \in \mathbb{Z}[\Lambda^+(\boldsymbol{H})]$ and by summing over all $\lambda \in \Lambda^+(\boldsymbol{G})$, $\mu \in \Lambda^+(\boldsymbol{H})$.

$$P^{\boldsymbol{G}}_{\boldsymbol{H}}(A, X) = \sum_{\lambda \in \Lambda^+(\boldsymbol{G})} \sum_{\mu \in \Lambda^+(\boldsymbol{H})} (V(\mu)|V(\lambda) \downarrow \boldsymbol{H}) A^{\lambda} X^{\mu}. \quad (6)$$

3.2. *Tensor product of representations as restricted representation*

To compute the decomposition of a tensor product of two irreducible representations $V(\lambda) \otimes V(\mu)$ of a Lie group $\boldsymbol{H}$ it is possible to consider representations of the group $\boldsymbol{H}$ as restricted representations of the diagonally embedded subgroup isomorphic to $\boldsymbol{H}$ with respect to the overgroup $\boldsymbol{G} = \boldsymbol{H} \times \boldsymbol{H}$. We can identify $\Lambda^+(\boldsymbol{G})$ with $\Lambda^+(\boldsymbol{H}) \times \Lambda^+(\boldsymbol{H})$. The tensor product multiplicities for the Lie group $\boldsymbol{H}$ then fulfill the equality $m^{\nu}_{\lambda,\mu} = (V(\nu)|\, V(\lambda) \otimes V(\mu) \downarrow \boldsymbol{H})$, where $\mu, \nu, \lambda \in \Lambda^+(\boldsymbol{H})$.

Let $\mathfrak{g}$ and $\mathfrak{h}$ are the Lie algebras of $\boldsymbol{G}$ resp. $\boldsymbol{H}$, let $\mathfrak{s}$ resp. $\mathfrak{t}$ are the corresponding Cartan subalgebras. We construct the linear restriction r in the following way. Let $\pi : \mathfrak{h} \mapsto \mathfrak{g}$ be the projection of $\mathfrak{h}$ into $\mathfrak{g}$, which is identical on $\mathfrak{h}$. It follows that π is identical on $\mathfrak{t}$ as the projection into $\mathfrak{s}$. The restriction $r : \mathfrak{s}^* \mapsto \mathfrak{t}^*$ is defined by $r(\xi) = \xi \circ \pi$ for $\xi \in \mathfrak{s}^*$. It holds that $r : \Lambda(\boldsymbol{G}) \mapsto \Lambda(\boldsymbol{H})$.

Following the equation (6), the generating function $P_{\boldsymbol{H}}^{\boldsymbol{H}\times\boldsymbol{H}}$ can be obtained by multiplying the multiplicities $m_{\lambda,\mu}^{\nu}$ by $A^{\lambda}, B^{\mu}, X^{\nu} \in \mathbb{Z}[\Lambda(\boldsymbol{H})]$ and by summing over all $\mu, \nu, \lambda \in \Lambda(\boldsymbol{H})$.

$$P_{\boldsymbol{H}}^{\boldsymbol{H}\times\boldsymbol{H}}(A,B,X) = \sum_{\lambda,\mu,\nu\in\Lambda^{+}(\boldsymbol{H})} (V(\nu)|V(\lambda)\otimes V(\mu)\downarrow \boldsymbol{H})A^{\lambda}B^{\mu}X^{\nu}, \tag{7}$$

where the multiplicity $(V(\nu)|V(\lambda)\otimes V(\mu))$ is computed via Kostant's formula (5).

4. An algorithm for tensor products decomposition

The above mentioned algorithm[2] to obtain a generating function for tensor products decomposition, in fact, use an advantage of generating functions only partially. The essence lies in Kostant's formula (5), which is a recurrence relation in the weights of Λ^{+}. The authors use the generating functions method to solve this recurrence.

We use an algorithm which employs the Patera-Sharp character generators (1). The distinguishing feature of our algorithm lies in the fact that we do not use the generating function only to solve some recurrence, but we use an operation over character generators, which are generating functions themselves, to produce a new generating function for the tensor products decomposition.

Let $\boldsymbol{G}$ be a semisimple connected complex Lie group with the Lie algebra $\mathfrak{g}$. As mentioned above, the irreducible finite-dimensional representations of $\boldsymbol{G}$ are parametrized by the set Λ^{+} of the dominant characters of a maximal torus $\boldsymbol{T}$. Let $\lambda, \mu \in \Lambda^{+}$; by the complete reducibility theorem (4) and orthogonality relations, we have the following relation for the characters of the corresponding representations

$$\operatorname{ch} V(\lambda)\cdot \operatorname{ch} V(\mu) = \sum_{\nu\in\Lambda^{+}} m_{\lambda,\mu}^{\nu} \operatorname{ch} V(\nu). \tag{8}$$

The equation (8) is the key for the following considerations. The characters on the right hand side of (8) can be computed by Weyl character formula (3). Nevertheless, all the characters are completely determined by the highest weight $\nu \in \Lambda^{+}$ of the corresponding representation $V(\nu)$. Therefore, all the terms on the right hand side of (8) can be simply substituted by the expression e^{ν} without an information loss on the presence of the representation $V(\nu)$.

Using Weyl character formula (3), the expression e^{ν} can be obtained from the character $\operatorname{ch} V(\nu)$, easily. First of all, the character $\operatorname{ch} V(\nu)$ is to be multiplied by the denominator of the Weyl character $\sum_{w\in W}\operatorname{sgn}(w)e^{w(\rho)}$, then by $e^{-\rho}$, where ρ is the half sum of the positive roots of $\boldsymbol{H}$. At last, the annihilating operator $\mathfrak{D}$,

$$\mathfrak{D}\left(\sum_{i,j,\dots,k=-\infty}^{\infty} a_{i,j,\dots,k}x^{i}y^{j}\dots z^{k}\right) = \sum_{i,j,\dots,k=0}^{\infty} a_{i,j,\dots,k}x^{i}y^{j}\dots z^{k},$$

which annihilates all the terms with negative powers, is to be applied.

If the whole sequence of the steps is applied to equation (8), obtain

$$\mathfrak{D}\left[\left(e^{-\rho}\sum_{w\in W}\operatorname{sgn}(w)e^{w(\rho)}\right)\operatorname{ch} V(\lambda)\cdot \operatorname{ch} V(\mu)\right] = \sum_{\nu\in\Lambda^{+}} m_{\lambda,\mu}^{\nu} e^{\nu}. \tag{9}$$

If equation (9) is multiplied by $A^{\lambda}, B^{\mu} \in \mathbb{Z}[\Lambda(\boldsymbol{H})]$ and summed over all $\lambda, \mu \in \Lambda^{+}$, the generating function for the tensor product decomposition $P_{\boldsymbol{H}}^{\boldsymbol{H}\times\boldsymbol{H}}$ is obtained.

$$\mathfrak{D}\left[\left(e^{-\delta}\sum_{w\in W}\operatorname{sgn}(w)e^{w(\delta)}\right)X(A)\cdot X(B)\right] = \sum_{\lambda,\mu\in\Lambda^{+}}\sum_{\nu\in\Lambda^{+}} m_{\lambda,\mu}^{\nu}e^{\nu}A^{\lambda}B^{\mu}. \tag{10}$$

The object on the right hand side of (10) is the generating function $P_{\boldsymbol{H}}^{\boldsymbol{H}\times\boldsymbol{H}}$; on the left hand side appear the character generators $X(A)$ and $X(B)$.

As can be seen, the tensor products decomposition generating function can be obtained by simple calculations over the characters generators. Let us mention two examples for the groups $SU(2)$ and $SU(3)$, respectively.

4.1. SU(2) case

The weight lattice of the group $SU(2)$ is $\Lambda = \mathbb{Z}\omega$, where ω is the single fundamental weight. Then the character generator for $SU(2)$ is of the form

$$X(A,x) = \sum_{m=0}^{\infty} \operatorname{ch} V(m\omega) A^{m\omega} = \sum_{n=0}^{\infty} \frac{x^{m+1} - x^{-m-1}}{x - x^{-1}} A^m = \frac{1}{(1-Ax)(1-Ax^{-1})}, \tag{11}$$

where we denote $x = e^{\omega}$ and $A = A^{\omega}$, for the elements of the ring $\mathbb{Z}[\Lambda]$, for simplicity.

The generating function for tensor product decomposition $P_{SU(2)}^{SU(2)\times SU(2)}$ is obtained simple by multiplying $X(A,x)X(B,x)$ by $x^{-1}(x - x^{-1})$, which corresponds to $e^{-\rho}\sum_{w\in W}\operatorname{sgn}(w)e^{w(\rho)}$, and by applying the operator $\mathfrak{D}$.

$$P_{SU(2)}^{SU(2)\times SU(2)}(A,B,x) = \mathfrak{D}\left[x^{-1}(x-x^{-1})X(A,x)X(B,x)\right] = \frac{1}{(1-Ax)(1-AB)(1-Bx)}. \tag{12}$$

4.2. SU(3) case

The weight lattice of $SU(3)$ is $\Lambda = \mathbb{Z}\omega_1 + \mathbb{Z}\omega_2$, where $\{\omega_1, \omega_2\}$ are the fundamental weights. The character generator is computed

$$\begin{aligned} X(A,B,x,y) = \sum_{m,n=0}^{\infty} \operatorname{ch} V(m\omega_1 + n\omega_2) A^{m\omega_1} B^{n\omega_2} = \\ \frac{x^2y^2(1-AB)}{(1-Ax)(1-By)(y-A)(x-B)(x-Ay)(y-Bx)}, \end{aligned} \tag{13}$$

where again $x = e^{\omega_1}, y = e^{\omega_2}$ and $A = A^{\omega_1}, B = B^{\omega_2}$, for simplicity.

The generating function for tensor product decomposition $P_{SU(3)}^{SU(3)\times SU(3)}$ is obtained simple by multiplying $X(A,B,x,y)X(R,S,x,y)$ by the term

$$\left(x^{-1}y^{-1}\right)\left(xy - x^{-1}y^2 - x^2y^{-1} - x^{-1}y^{-1} + x^{-2}y + xy^{-2}\right), \tag{14}$$

which corresponds to $e^{-\rho}\sum_{w\in W}\operatorname{sgn}(w)e^{w(\rho)}$, and by applying the operator $\mathfrak{D}$. We obtain

$$\begin{aligned} P_{SU(3)}^{SU(3)\times SU(3)}(A,B,R,S,x,y) = \\ \frac{1-ABRSxy}{(AS-1)(BR-1)(Ax-1)(By-1)(Rx-1)(Sy-1)(ARy-1)(BSx-1)}. \end{aligned} \tag{15}$$

Let us describe this process in more detail how to obtain (15). Let us denote the product $X(A,B,x,y)X(R,S,x,y)$ multiplied by factor (14) by $g(x)$, i. e.

$$g(x) = \frac{xy(AB-1)(RS-1)\left(x^4y - x^3y^3 - x^3 + xy^4 + xy - y^3\right)}{d(x)},$$

where

$$d(x) = (Ax-1)(A-y)(B-x)(By-1)(Rx-1)(R-y)\times \\ (x-S)(Sy-1)(x-Ay)(Bx-y)(x-Ry)(Sx-y).$$

We now perform a partial fractions decomposition of g, obtaining

$$g(x) = -\frac{A^2\left(ARSy^3-Ay^3-RSy+y\right)}{(A-R)(AS-1)(Ax-1)(By-1)(R-y)(Sy-1)(Ay-B)(ARy-1)(Ay-S)} - \\ \frac{y\left(A^2RSy^2-A^2y^2-ARS+A\right)}{(A-R)(AS-1)(By-1)(R-y)(Sy-1)(Ay-B)(ARy-1)(Ay-S)(x-Ay)} - \\ \frac{B^2RSy-B^2y-BRSy^3+By^3}{(A-y)(BR-1)(B-S)(x-B)(R-y)(Sy-1)(B-Ay)(B-Ry)(BS-y)} + \\ \frac{B^2y\left(BRS-B-RSy^2+y^2\right)}{(y-A)(BR-1)(B-S)(R-y)(Sy-1)(B-Ay)(B-Ry)(BS-y)(Bx-y)} + \\ \frac{R^2\left(ABRy^3-ABy-Ry^3+y\right)}{(A-R)(A-y)(BR-1)(By-1)(Rx-1)(Sy-1)(ARy-1)(Ry-B)(Ry-S)} + \\ \frac{y\left(ABR^2y^2-ABR-R^2y^2+R\right)}{(A-R)(A-y)(BR-1)(By-1)(Sy-1)(ARy-1)(Ry-B)(Ry-S)(x-Ry)} - \\ \frac{ABS^2y-ABSy^3-S^2y+Sy^3}{(AS-1)(A-y)(B-S)(By-1)(y-R)(x-S)(S-Ay)(BS-y)(S-Ry)} - \\ \frac{S^2y\left(ABS-ABy^2-S+y^2\right)}{(AS-1)(A-y)(B-S)(By-1)(y-R)(S-Ay)(BS-y)(S-Ry)(Sx-y)}.$$

We now look at the terms which can be thrown away immediatelly. The first one, for example, cannot be thrown away because it contains the term $(Ax-1)$ in the denominator and this leads to positive powers of x. The second one can be thrown away, because the term $(x-Ay)$ can lead to negative powers of x. Similar way we examine all terms, obtaining the result consisting of four terms only, denoted by $\tilde{g}(y)$:

$$\tilde{g}(y) = -\frac{A^2\left(ARSy^3-Ay^3-RSy+y\right)}{(A-R)(AS-1)(Ax-1)(By-1)(R-y)(Sy-1)(Ay-B)(ARy-1)(Ay-S)} + \\ \frac{B^2y\left(BRS-B-RSy^2+y^2\right)}{(y-A)(BR-1)(B-S)(R-y)(Sy-1)(B-Ay)(B-Ry)(BS-y)(Bx-y)} + \\ \frac{R^2\left(ABRy^3-ABy-Ry^3+y\right)}{(A-R)(A-y)(BR-1)(By-1)(Rx-1)(Sy-1)(ARy-1)(Ry-B)(Ry-S)} - \\ \frac{S^2y\left(ABS-ABy^2-S+y^2\right)}{(AS-1)(A-y)(B-S)(By-1)(y-R)(S-Ay)(BS-y)(S-Ry)(Sx-y)}.$$

Now comes the decomposition to the partial fractions concerning the variable y. Generally, such decomposition is possible via solving system of linear equations. We obtain the result consisting of 18 terms:

$$
\begin{aligned}
\tilde{g}(y) = &\frac{R^2A^3}{(A-R)(AR-B)(BR-1)(AR-S)(AS-1)(Ax-1)(ARy-1)}+\\
&\frac{RA^2}{(A-R)(AR-B)(BR-1)(AR-S)(AS-1)(Ax-1)(y-R)}-\\
&\frac{R^3A^2}{(A-R)(AR-B)(BR-1)(AR-S)(AS-1)(Rx-1)(ARy-1)}+\\
&\frac{S^2A^2}{(A-R)(AR-S)(S-B)(AS-1)(A-BS)(Ax-1)(Sy-1)}-\\
&\frac{R^2A}{(A-R)(AR-B)(BR-1)(AR-S)(AS-1)(Rx-1)(y-A)}-\\
&\frac{S^2A}{(A-R)(B-S)(AR-S)(AS-1)(A-BS)(A-Sx)(y-A)}+\\
&\frac{AB^2RS-AB^2}{(A-R)(AR-B)(BR-1)(B-S)(AS-1)(A-BS)(A-Bx)(y-A)}-\\
&\frac{B^2R}{(A-R)(AR-B)(BR-1)(B-S)(R-BS)(R-Bx)(y-R)}+\\
&\frac{ABRS^2-RS^2}{(A-R)(BR-1)(B-S)(AR-S)(AS-1)(R-BS)(R-Sx)(y-R)}-\\
&\frac{BS}{(BR-1)(B-S)(AS-1)(BS-A)(BS-R)(B-x)(y-BS)}+\\
&\frac{BS}{(BR-1)(B-S)(AS-1)(BS-A)(BS-R)(S-x)(y-BS)}-\\
&\frac{B(RS-1)\left(Bx^3-x\right)}{(BR-1)(B-S)(S-x)(Ax-1)(Bx-A)(Bx-R)(Rx-1)(BSx-1)(y-Bx)}+\\
&\frac{(AB-1)S\left(Sx^3-x\right)}{(B-S)(AS-1)(B-x)(Ax-1)(Rx-1)(Sx-A)(Sx-R)(BSx-1)(y-Sx)}+\\
&\frac{B^2\left(A^2RS-A^2\right)}{(A-R)(AR-B)(BR-1)(B-S)(AS-1)(A-BS)(Ax-1)(By-1)}-\\
&\frac{B^2R^2}{(A-R)(AR-B)(BR-1)(B-S)(R-BS)(Rx-1)(By-1)}+\\
&\frac{B^3S^2}{(BR-1)(B-S)(AS-1)(BS-A)(BS-R)(BSx-1)(By-1)}+\\
&\frac{\left(ABR^2-R^2\right)S^2}{(A-R)(BR-1)(B-S)(AR-S)(AS-1)(R-BS)(Rx-1)(Sy-1)}-\\
&\frac{B^2S^3}{(BR-1)(B-S)(AS-1)(BS-A)(BS-R)(BSx-1)(Sy-1)}.
\end{aligned}
$$

We apply the same procedure to the above sum. The first term cannot be thrown away because of the $(ARy-1)$ in the denominator. The second one is thrown because of $(y-R)$, etc. Finally, we get sum of eight terms, namely

$$
\begin{aligned}
g_1 = & \frac{A^3R^2}{(A-R)(AS-1)(Ax-1)(BR-1)(AR-B)(AR-S)(ARy-1)}+ \\
& \frac{B^2\left(A^2RS-A^2\right)}{(A-R)(AS-1)(Ax-1)(BR-1)(B-S)(By-1)(AR-B)(A-BS)}- \\
& \frac{A^2R^3}{(A-R)(AS-1)(BR-1)(Rx-1)(AR-B)(AR-S)(ARy-1)}+ \\
& \frac{A^2S^2}{(A-R)(AS-1)(Ax-1)(S-B)(Sy-1)(A-BS)(AR-S)}+ \\
& \frac{B^3S^2}{(AS-1)(BR-1)(B-S)(By-1)(BS-A)(BS-R)(BSx-1)}- \\
& \frac{B^2R^2}{(A-R)(BR-1)(B-S)(By-1)(Rx-1)(AR-B)(R-BS)}- \\
& \frac{B^2S^3}{(AS-1)(BR-1)(B-S)(Sy-1)(BS-A)(BS-R)(BSx-1)}+ \\
& \frac{S^2\left(ABR^2-R^2\right)}{(A-R)(AS-1)(BR-1)(B-S)(Rx-1)(Sy-1)(AR-S)(R-BS)}.
\end{aligned}
$$

Summing these terms up, we get the desired result (15).

5. Conclusions

The feature of the obtained generating functions is the generality, i.e., they solve all the tensor products decomposition problems for a given group at the same time. In other words, "they are equivalent to the table of all tensor product decompositions". [7]

For example, the generating function $P_{SU(3)}^{SU(3)\times SU(3)}$ for the group $SU(3)$ computed in subsection 4.2 has the following meaning. The symbols A, B resp. R, S correspond to the first resp. second factorrepresentation. A term A^aB^b corresponds to the representation $V(a\omega_1+b\omega_2)$, a term R^rS^s corresponds to the representation $V(r\omega_1+s\omega_2)$, a term x^my^n corresponds to the representation $V(m\omega_1+n\omega_2)$. Then, the presence of a term $A^aB^bR^rS^sx^my^n$ in the expansion of $G(A,B,R,S,x,y)$ into a series implies that the tensor product of the representations $V(a\omega_1+b\omega_2)$ and $V(r\omega_1+s\omega_2)$ contains the representation $V(m,n)$, i.e.,

$$V(a\omega_1+b\omega_2)\otimes V(r\omega_1+s\omega_2)\supset V(m\omega_1+n\omega_2).$$

Using the Patera-Sharp character generators, we can simply obtain the generating functions for the tensor products decomposition for all semisimple Lie groups. These character generators are always rational functions.[2] Therefore, the algorithm used in the section 4 involves only operations over rational functions. The advantage of the algorithm against other methods is in its simultaneous simplicity and generality.

An unsolved task, which should be studied in the future, is the action of the annihilating operator $\mathfrak{D}$. Generally, it is not obvious, how it acts on a general rational function. This should be the direction of the following investigations.

Acknowledgments

JF thanks CTU for support via Project SGS12/198/OHK4/3T/14. SP acknowledges support of P201/10/1509, project of the Grant Agency of the Czech Republic.

References

[1] M. R. Bremner, R. V. Moody, and J. Patera. *Tables of dominant weight multiplicities for representations of simple Lie algebras.* Marcel Dekker, New York, 1985.

[2] A. M. Cohen and G. C. M. Ruitenburg. Generating functions and Lie groups. In *Computational aspects of Lie group representations and related topics: Proc. Comp. Alg. Sem. at CWI.* Amsterdam, 1991.

[3] G. Heckman. Projections of orbits and asymptotic behavior of multiplicities for compact connencted lie groups. *Inventiones Math.*, 67, 1982.

[4] J. E. Humphreys. *Introduction to Lie Algebras and Representation Theory.* Springer-Verlag, New York, 1972.

[5] S. Kumar. Tensor product decomposition. In *Proc. Int. Cong. Math.* Hyderabad, India, 2010.

[6] W. G. McKay and J. Patera. *Tables of dimensions, indices, and branching rules for representations of simple Lie algebras.* Marcel Dekker, New York, 1981.

[7] J. Patera and R. T. Sharp. Generating functions for characters of group representations and their applications. *Lecture Notes in Physics*, 94, 1979.

[8] H. S. Wilf. *Generatingfunctionology.* Academic Press, Waltham, 1994.

Action-angle variables for spherical mechanics related to near horizon extremal Myers–Perry black hole

Anton Galajinsky[1], **Armen Nersessian**[2] **and Armen Saghatelian**[2]

[1]Tomsk Polytechnic University, Lenin Ave. 30, 634050 Tomsk, Russia

[2] Yerevan State University, 1 Alex Manoogian St., Yerevan, 0025, Armenia

E-mail: galajinsky@tpu.ru, arnerses@ysu.am, armen.saghatelian@ysu.am

Abstract. The action-angle formulation for the spherical part of the conformal mechanics describing a massive relativistic particle moving near the horizon of an extremal rotating black hole in arbitrary dimension is presented for the special case that all rotation parameters are equal.

An extremal rotating black hole in arbitrary dimension exhibits the conformal symmetry $SO(2,1)$ in the near horizon limit (see, e.g., [1] and references therein). Because to each Killing vector of a background geometry there corresponds the first integral of the geodesic equations, a model of a massive relativistic particle propagating near the horizon of an extremal rotating black hole is automatically conformal invariant. Conformal mechanics originating from the near horizon extremal black hole geometry has been extensively investigated in the past [2]–[16] with a particular emphasis on its dynamics, a relation to nonrelativistic conformal mechanics, and the construction of supersymmetric extensions. Quite recently, based on an earlier work [17], such a conformal mechanics has been used to build new superintegrable models [18, 19]. In particular, in Ref. [19] a maximally superintegrable spherical mechanics associated with the black hole in $d = 2n + 1$ dimensions has been constructed. It was also demonstrated that its counterpart related to the black hole in $d = 2(n + 1)$ dimensions lacks for only one integral of motion to be maximally superintegrable.

The goal of this note is to extend the analysis in [19] and to construct the action-angle variables. There are several reasons for this study. The action–angle formulation of an integrable system gives a comprehensive geometric description of its dynamics and provides a useful means for developing the perturbation theory [20]. The use of the action-angle variables suggests a criterion of the (non)equivalence of two integrable systems. The variables provide a systematic way to reveal hidden symmetries and to construct new non–trivial examples of maximally superintegrable models (see, e.g., [21]). Worth mentioning also is that the formulation in terms of the action-angle variables is a base for the (semiclassical) quantization, which implies imposing on the action variables the Bohr–Sommerfeld quantization conditions.

Because only integrable systems with a compact phase space admit the action–angle formulation, the full conformal mechanics associated with the near horizon geometry of an extremal rotating black hole is not amenable to such a description. However, one can separate the radial (noncompact) motion from the angular variables dynamics, which is compact and can be considered on its own. Note that it is in this way that the action–angle variables were constructed recently for the Calogero model [22] and for some variants of the conformal mechanics related to the near horizon extremal black holes [14, 15, 16]. The present work extends the list of such examples and also illuminates the origin of the superintegrability of the models constructed recently in [19]. Below we first discuss the model originating from a black hole geometry in $d = 2n + 1$ dimensions and then turn to the case of $d = 2(n + 1)$.

The Hamiltonian of a spherical mechanics related to the extremal rotating black hole in $d = 2n + 1$ is a kind of matryoshka doll [19]

$$\mathcal{I}_0 = F_{n-1}, \tag{1}$$

where F_{n-1} is derived from the recurrence relation

$$F_i = p_{\theta_i}^2 + \frac{g_{i+1}^2}{\cos^2\theta_i} + \frac{F_{i-1}}{\sin^2\theta_i}. \tag{2}$$

Here (θ_i, p_{θ_i}), with $i = 1, \ldots, n-1$, constitute the canonical pairs, g_i are coupling constants, and $F_0 = g_1^2$. As is obvious from (1) and (2), the functionally independent integrals of motion in involution F_i ensure the integrability of (1). Let us demonstrate the maximal superintegrability (and in fact exact solvability) of this system by constructing the action–angle variables. Following the standard procedure [20], one has to write down the generating function

$$S_0^{odd}(F_i, |g_i|, \theta_i) = \sum_{i=1}^{n-1} \int p_{\theta_i}(F_1, \ldots, F_{n-1}, \theta_i) \mathrm{d}\theta_i, \tag{3}$$

where $p_{\theta_i}(F_1, \ldots, F_{n-1}, \theta_i)$ are to be expressed from (2). For the action variables one gets (for the details of evaluation of the integrals see, e.g., Ref. [21])

$$I_i = \frac{1}{2\pi} \oint \mathrm{d}\theta_i \left[\sqrt{F_i - \frac{F_{i-1}}{\sin^2\theta_i} - \frac{g_{i+1}^2}{\cos^2\theta_i}}\right] = \frac{1}{2}(\sqrt{F_i} - \sqrt{F_{i-1}} - |g_{i+1}|), \tag{4}$$

with $i = 1, \ldots, n-1$. Inverting these expression, one finds

$$F_i = \left(2\sum_{k=1}^{i} I_k + \sum_{k=1}^{i+1} |g_k|\right)^2, \tag{5}$$

while the angle variables read

$$\Phi_i^{odd} = \frac{\partial S}{\partial I_i} = \sum_{k=i}^{n-1} \arcsin X_k + 4 \sum_{k=i+1}^{n-1} \arctan Y_k, \tag{6}$$

where we denoted

$$X_k = \frac{\left(F_k + F_{k-1} - g_{k+1}^2\right) - 2F_k \sin^2\theta_k}{\sqrt{\left(-F_k + F_{k-1} - g_{k+1}^2\right)^2 - 4F_k g_{k+1}^2}}, \tag{7}$$

$$Y_k = \frac{\left(F_k + F_{k-1} - g_{k+1}^2\right)\sqrt{F_k \sin^2\theta_k \cos^2\theta_k - F_{k-1}\cos^2\theta_k - g_{k+1}^2 \sin^2\theta_k}}{\sqrt{F_{k-1}}\left((F_k + F_{k-1} - g_{k+1}^2) - 2F_k \sin^2\theta_k\right)} - \frac{\sin^2\theta_k \sqrt{F_k\left(F_k + F_{k-1} - g_{k+1}^2\right)^2 - F_k^2 F_{k-1}}}{\sqrt{F_{k-1}}\left((F_k + F_{k-1} - g_{k+1}^2) - 2F_k \sin^2\theta_k\right)}. \tag{8}$$

Thus, being rewritten in the action–angle variables, the Hamiltonian has the following form:

$$\mathcal{I}_0 = \left(2\sum_{k=1}^{n-1} I_k + \sum_{k=1}^{n} |g_k| \right)^2 . \tag{9}$$

Up to the shift of the action variables, it coincides with the Hamiltonian of a free particle moving on the $(n-1)$–dimensional sphere [21]. So the two systems differ only by the range of $\sum_i I_i$ ($[0,\infty)$ and $[\sum_{k=1}^{n}|g_k|,\infty)$, respectively). The former model thus possesses $SO(n)$ symmetry and is obviously maximally superintegrable.

The action–angle formulation allows one to immediately construct functionally independent hidden constants of the motion. Because the evolution of the angle variables is governed by the equation (cf. [21, 23])

$$\frac{d\Phi_i}{dt} = 2\left(2\sum_{k=1}^{n-1} I_k + \sum_{k=1}^{n} |g_k| \right), \tag{10}$$

the expressions $\cos(\Phi_i - \Phi_j + \mathrm{const})$ determine (globally defined) constants of the motion $n-2$ of which are functionally independent

$$G_i = \cos\left(\Phi_i - \Phi_{i+1}\right) = \frac{\sqrt{1 - X_i^2(1 - Y_{i+1}^2)} - 2X_i Y_{i+1}}{1 + Y_{i+1}^2}, \tag{11}$$

with $i = 1, \ldots, n-2$. Hence, the system (1) with $(n-1)$ configuration space degrees of freedom possesses $2n-3$ functionally independent constants of the motion and, as thus, is maximally superintegrable. That it is exactly solvable follows from the fact that the Hamiltonian is expressed via the action variables in terms of elementary functions only.

Now let us turn to the case of $d = 2(n+1)$. The Hamiltonian of a spherical mechanics related to the extremal rotating black hole in $d = 2(n+1)$ dimensions reads [19]

$$\mathcal{I}_0 = 2np_{\theta_n}^2 + \nu\sin^2\theta_n + \left(\frac{2n}{\sin^2\theta_n} - 2n + 1 \right) F_{n-1}, \tag{12}$$

where F_{n-1} is given in (2). As we have seen above, F_{n-1} is the Hamiltonian of a particle moving on $\mathcal{S}^{n-1}$. This sector provides $2(n-1)-1$ functionally independent integrals of motion. Because (12) involves one more canonical pair (θ_n, p_{θ_n}) and only one extra integral of motion (the Hamiltonian (12) itself), the full theory lacks for only one integral of motion to be maximally superintegrable.

Let us construct the action-angle variables for this system. One starts with the generating function

$$S_0^{even} = \sum_{i=1}^{n} \int p_{\theta_i}(\mathcal{I}_0, F_1, \ldots, F_{n-1}, \theta_i)\mathrm{d}\theta_i = \int p_{\theta_n}(\mathcal{I}_0, F_{n-1}, \theta_n)d\theta_n + S_0^{odd}, \tag{13}$$

where S_0^{odd} is defined in (3) and p_{θ_n} is to be derived from (12). The action variables $I_1, \ldots I_{n-1}$ coincide with those in (4), while I_n reads

$$I_n = \sqrt{\frac{-a^-\nu}{8n}} a^+ \mathcal{F}_1\left(\frac{1}{2}, 1, -\frac{1}{2}, 2, a^+, \frac{a^+}{a^-} \right), \tag{14}$$

where $\mathcal{F}_1$ is the Appell's first hypergeometric series[1] and

$$a^{\pm} = 1 - \frac{\mathcal{I}_0}{2\nu} - \frac{2n-1}{2\nu}F_{n-1} \pm \sqrt{\left(1 - \frac{\mathcal{I}_0}{2\nu} - \frac{2n-1}{2\nu}F_{n-1}\right)^2 + \frac{\mathcal{I}_0}{\nu} - \frac{F_{n-1}}{\nu} - 1}. \tag{15}$$

[1] Here and in what follows we use the notations in [24, 25].

Inverting these expressions, one gets the Hamiltonian in terms of the action variables

$$\mathcal{I}_0 = \mathcal{I}_0\left(I_n, F_{n-1}\right), \quad F_{n-1} = \left(2\sum_{k=1}^{n-1} I_k + \sum_{k=1}^{n} |g_k|\right)^2 . \tag{16}$$

Obviously, this cannot be done in terms of elementary functions. We thus conclude that the system under consideration is integrable but not exactly solvable.

The angle variable conjugated to I_n reads

$$\Phi_n^{even} = \frac{\partial \mathcal{I}_0}{\partial I_n}\frac{1}{\sqrt{8\nu n a^+}}\mathcal{F}\left(\arcsin\sqrt{\frac{a^+}{a^+ - \cos^2\theta_n}}, 1 - \frac{a^-}{a^+}\right), \tag{17}$$

while the remaining angle variables are defined by the following expressions ($i = 1, \ldots, n-1$):

$$\begin{aligned}\Phi_i^{even} = \Phi_i^{odd} \quad &- \quad A\,\Pi\left(1 - \frac{1}{a^+}, \arcsin\sqrt{\frac{a^+}{a^+ - \cos^2\theta_n}}, 1 - \frac{a^-}{a^+}\right) + \\ &+ \quad B\,\mathcal{F}\left(\arcsin\sqrt{\frac{a^+}{a^+ - \cos^2\theta_n}}, 1 - \frac{a^-}{a^+}\right), \end{aligned} \tag{18}$$

where Φ_i^{odd} are given in (6), $\mathcal{F}(\phi|m)$ is the elliptic integral of the first kind, $\Pi(n;\phi|m)$ is the elliptic integral of the third kind, and

$$A = \sqrt{\frac{8nF_{n-1}}{\nu}}\frac{1}{\sqrt{a^+}(a^+ - 1)}, \qquad B = A + \frac{\sqrt{2F_{n-1}}}{\sqrt{n\nu a^+}}\left(\frac{\partial \mathcal{I}_0}{\partial F_{n-1}} + 2n - 1\right). \tag{19}$$

Since the Hamiltonian of our system is of the form (16), the ratio of the effective frequencies $\omega_1 = \partial\mathcal{I}/\partial I_n$ and $\omega_2 = \partial\mathcal{I}/\partial F_{n-1}$ is not a rational number but a function of the action variables (for a similar situation see [21, 23]). Hence, although $(\omega_2\Phi_n - \omega_1\Phi_i)$ commutes with the Hamiltonian $\mathcal{I}_0$, it is not a periodic function. All hidden symmetries of our model are thus exhausted by (11). The system with n configuration space degrees of freedom thus has $n + (n-2) = 2n - 2$ constants of the motion and lacks for only one constant of the motion to be maximally superintegrable.

Acknowledgments

This work was supported by the RFBR grant 13-02-90602-Arm, MSE Program "Nauka" under the grant 1.604.2011 (A.G.), the Armenian State Committee of Science under the grant 13RF-018 (A.N.,A.S.), by the ANSEF Grant hepth-3267 (A.S.) and Volkswagenstiftung under the contracts 86 260 (A.S.) and I/84 496 (A.N.).

References

[1] Lu H, Mei J and Pope C N 2009 *JHEP* **0904** 054, arXiv:0811.2225.
[2] Claus P, Derix M, Kallosh R, Kumar J, Townsend P K and Van Proeyen A 1998 *Phys. Rev. Lett.* **81** 4553, hep-th/9804177.
[3] De Azcárraga J A, Izquierdo J M, Pérez Bueno J C and Townsend P K 1999 *Phys. Rev.* D **59** 084015, hep-th/9810230.
[4] Cacciatori S, Klemm D and Zanon D 2000 *Class. Qauntum Grav.* **17** 1731, hep-th/9910065.
[5] Clement G and Gal'tsov D 2001 *Nucl. Phys.* B **619** 741, hep-th/0105237.
[6] Bellucci S, Galajinsky A, Ivanov E and Krivonos S 2003 *Phys. Lett.* B **555** 99, hep-th/0212204.
[7] Leiva C and Plyushchay M 2003 *Annals Phys.* **307** 372, hep-th/0301244.

[8] Ivanov E, Krivonos S and Niederle J 2004 *Nucl. Phys.* B **677** 485, hep-th/0210196.
[9] Anabalon A, Gomis J and Kamimura K, Zanelli J 2006 *JHEP* **0610** 068, hep-th/0607124.
[10] Galajinsky A 2008 *Phys. Rev.* D **78** 044014, arXiv:0806.1629.
[11] Galajinsky A 2010 *JHEP* **1011** 126, arXiv:1009.2341.
[12] Galajinsky A and Orekhov K 2011 *Nucl. Phys.* B **850** 339, arXiv:1103.1047.
[13] Bellucci S and Krivonos S 2011 *JHEP* **1110** 014, arXiv:1106.4453.
[14] Galajinsky A and Nersessian A 2011 *JHEP* **1111** 135, arXiv:1108.3394.
[15] Bellucci S, Nersessian A and Yeghikyan V 2012 *Mod. Phys. Lett.* A **27** 1250191, arXiv:1112.4713.
[16] Saghatelian A 2012 *Class. Quant. Grav.* **29** 245018, arXiv:1205.6270.
[17] Hakobyan T, Krivonos S, Lechtenfeld O and Nersessian A 2010 *Phys. Lett.* A **374** 801, arXiv:0908.3290
[18] Galajinsky A 2013 *Phys. Rev.* D **87** 024023, arXiv:1209.5034.
[19] Galajinsky A, Nersessian A and Saghatelian A 2013 *JHEP* **1306** 002, arXiv:1303.4901.
[20] Arnold V I 1973 *Mathematical methods in classical mechanics*, Nauka Publ., Moscow.
[21] Hakobyan T, Lechtenfeld O, Nersessian A, Saghatelian A and Yeghikyan V 2012 *Phys. Lett.* A **376** 679, arXiv:1108.5189.
[22] Lechtenfeld O, Nersessian A and Yeghikyan V 2010 *Phys. Lett.* A **374** 4647, arXiv:1005.0464.
[23] Gonera C 2012 *Phys. Lett.* A **376** 2341, arXiv:1010.2915.
[24] Dwight H B 1961 *Tables of integrals and other mathematical data.*, 4th edition, The Macmillan Company, N.Y.,
[25] Prudnikov A P, Brychkov Yu A and Mar'ichev O I 1981 *Integrals and series* (in Russian), Nauka Publ., Moscow.

Journal of Physics: Conference Series **474** (2013) 012020 doi:10.1088/1742-6596/474/1/012020

Functional relations and the Yang-Baxter algebra

Wellington Galleas

Institute for Theoretical Physics and Spinoza Institute, Utrecht University, Leuvenlaan 4, 3584 CE Utrecht, The Netherlands

E-mail: w.galleas@uu.nl

Abstract. Functional equations methods are a fundamental part of the theory of Exactly Solvable Models in Statistical Mechanics and they are intimately connected with Baxter's concept of commuting transfer matrices. This concept has culminated in the celebrated Yang-Baxter equation which plays a fundamental role for the construction of quantum integrable systems and also for obtaining their exact solution. Here I shall discuss a proposal that has been put forward in the past years, in which the Yang-Baxter algebra is viewed as a source of functional equations describing quantities of physical interest. For instance, this method has been successfully applied for the description of the spectrum of open spin chains, partition functions of elliptic models with domain wall boundaries and scalar product of Bethe vectors. Further applications of this method are also discussed.

1. Introduction

Exact solutions have played an important role for the development of physical theories and assumptions and their contributions can be seen in a variety of contexts. For instance, Onsager's solution of the two-dimensional Ising model [1] not only showed that the formalism of Statistical Mechanics was indeed able to describe phase transitions, but also unveiled that the critical behavior of the Ising model specific heat was not included in Landau's theory of critical exponents [2]. In a different context, the exact solution of the one-dimensional Heisenberg spin chain [3] was also of fundamental importance for elucidating the value of the spin carried by a spin wave [4].

Bethe's celebrated solution of the isotropic Heisenberg chain [3] consists of a fundamental stone of the modern theory of quantum integrable systems and its influence can be seen in several areas ranging from Quantum Field Theory [5,6] to Combinatorics [7]. The hypothesis employed by Bethe for the model wave function is known nowadays as 'Bethe ansatz' and it became a standard tool in the theory of quantum integrable systems. On the other hand, Onsager's solution of the two-dimensional Ising model was based on Kramers and Wannier transfer matrix technique [8,9] which did not have any previous connection with Bethe ansatz. Nevertheless, the transfer matrix technique was later on recognized as a fundamental ingredient for the establishment of integrability in the sense of Baxter [10].

Within this scenario Baxter's concept of integrability appeared as an analogous of Liouville's classical concept where now the transfer matrix was playing the role of generating function of quantities in involution [10]. More precisely, in Baxter's framework a family of mutually commuting operators is obtained as a consequence of transfer matrices which commute for different values of their parameters. In their turn, these commutative transfer matrices are built directly from solutions of the Yang-Baxter equation.

The importance of the Yang-Baxter equation was only better understood with the proposal of the Quantum Inverse Scattering Method [11, 12]. This method unified the transfer matrix approach, the Yang-Baxter equation and the Bethe ansatz employed to solve a variety of one-dimensional quantum many-body systems. Besides that, the Quantum Inverse Scattering Method, or QISM for short, put in evidence the so called Yang-Baxter algebra which latter on led to the notion of Quantum Groups [13].

The Yang-Baxter algebra plays a fundamental role within the QISM and it is one of the main ingredients for the construction of exact eigenvectors of transfer matrices of two-dimensional lattice systems and hamiltonians of one-dimensional quantum many-body systems. However, the applications of the Yang-Baxter algebra are not limited to that and alternative ways of exploring the Yang-Baxter algebra are also known in the literature. For instance, it can be used to build solutions of the Knizhnik-Zamolodchikov equation [14] in the sense of [15, 16].

More recently, the Yang-Baxter algebra was also shown to be capable of rendering functional equations describing quantities such as the spectrum of spin chains and partition functions of vertex models [17–22]. Here we aim to discuss this latter possibility.

This article is organized as follows. In Section 2 we introduce definitions which will be relevant throughout this paper and also present the lattice systems we shall consider by means of this algebraic-functional approach. In Section 3 we illustrate how the Yang-Baxter algebra can be converted into functional equations and, in particular, we derive functional relations describing the partition function of the elliptic Eight-Vertex-SOS model with domain wall boundaries and scalar products of Bethe vectors. The solutions of the aforementioned functional equations are also presented in Section 3, and in Section 4 we unveil a family of partial differential equations underlying our functional relations. Concluding remarks are then discussed in Section 5.

2. Yang-Baxter relations and lattice systems

Lattice systems of Statistical Mechanics have a long history and remarkable examples share the property of being exactly solvable [23]. The most prominent examples, such as the Ising model and Eight-Vertex model, have been solved in two-dimensions and this choice of dimensionality undoubtedly grants them special properties. For instance, in [24, 25] Smirnov proved that the scaling limit of the critical site percolation on a two-dimensional triangular lattice is conformally invariant. The importance of this proof can be seen in two ways: from the mathematical perspective Smirnov's proof introduced the concept of 'discrete' harmonic functions. On the other hand, this proof provides a solid ground for the CFT (Conformal Field Theory) methods employed by Cardy in [26].

The concept of exact solvability seems to be intrinsically dependent on the method we are employing. However, it is nowadays well accepted that Baxter's commuting transfer matrices approach [10] plays a fundamental role for two-dimensional lattice systems within a variety of methods. For instance, the requirement of commuting transfer matrices leads us to the Yang-Baxter equation/algebra [11, 12] and also their dynamical counterparts [27, 28]. Those algebraic relations constitute the foundations of the algebraic Bethe ansatz [12] and, as firstly demonstrated in [17], they are also able to describe spectral problems in terms of functional equations.

Vertex and Solid-on-Solid models are some important examples of exactly solvable lattice systems and both of them admit an operatorial description in terms of generators of the Yang-Baxter algebra and its dynamical version. Here we will be mainly interested in the so called Eight-Vertex-SOS and Six-Vertex models, and for that it is enough to present only the dynamical Yang-Baxter equation/algebra while the standard Yang-Baxter relations will be obtained as a particular limit.

Dynamical Yang-Baxter equation. Let $\mathfrak{g}$ be a finite dimensional Lie algebra over $\mathbb{C}$ and $\mathfrak{h} \subset \mathfrak{g}$ be an abelian Lie subalgebra. Also, let $\mathbb{V} = \bigoplus_{\phi \in \mathfrak{h}^*} \mathbb{V}[\phi]$ with $\mathbb{V}[\phi] = \{v \in \mathbb{V} \mid hv = \phi(h)v \text{ for } h \in \mathfrak{h}\}$ be a diagonalizable $\mathfrak{h}$-module. Then for $\lambda_j, \gamma, \theta \in \mathbb{C}$ the dynamical Yang-Baxter equation reads

$$\begin{aligned} &\mathcal{R}_{12}(\lambda_1 - \lambda_2, \theta - \gamma h_3)\mathcal{R}_{13}(\lambda_1 - \lambda_3, \theta)\mathcal{R}_{23}(\lambda_2 - \lambda_3, \theta - \gamma h_1) = \\ &\qquad \mathcal{R}_{23}(\lambda_2 - \lambda_3, \theta)\mathcal{R}_{13}(\lambda_1 - \lambda_3, \theta - \gamma h_2)\mathcal{R}_{12}(\lambda_1 - \lambda_2, \theta) \; . \end{aligned} \tag{2.1}$$

Eq. (2.1) is a relation for an operator $\mathcal{R}_{ij} \; : \; \mathbb{C} \times \mathfrak{h}^* \mapsto \mathrm{End}(\mathbb{V}_i \otimes \mathbb{V}_j)$ where $\mathbb{V}_i$ $(i = 1, 2, 3)$ are finite dimensional diagonalizable $\mathfrak{h}$-modules. In this way we have (2.1) defined in $\mathrm{End}(\mathbb{V}_1 \otimes \mathbb{V}_2 \otimes \mathbb{V}_3)$ with tensor products being understood as

$$\mathcal{R}_{12}(\lambda, \theta - \gamma h_3)(v_1 \otimes v_2 \otimes v_3) = (\mathcal{R}_{12}(\lambda, \theta - \gamma\phi)(v_1 \otimes v_2)) \otimes v_3 \; . \tag{2.2}$$

The term ϕ in (2.2) corresponds to the weight of v_3 while the remaining elements $\mathcal{R}_{13}$ and $\mathcal{R}_{23}$ are computed by analogy. As far as the solutions of (2.1) are concerned, it is currently well understood the importance of the elliptic quantum groups $E_{p,q}[\mathfrak{g}]$ for their characterization. Here we shall restrict ourselves to the case $\mathfrak{g} \simeq \mathfrak{sl}(2)$, and in that case we consider $\mathfrak{h}$ as the $\mathfrak{sl}(2)$ Cartan subalgebra while $\mathbb{V} \cong \mathbb{C}^2$. Thus $h = \mathrm{diag}(1, -1)$ and we have the explicit solution

$$\mathcal{R}(\lambda, \theta) = \begin{pmatrix} a_+(\lambda, \theta) & 0 & 0 & 0 \\ 0 & b_+(\lambda, \theta) & c_+(\lambda, \theta) & 0 \\ 0 & c_-(\lambda, \theta) & b_-(\lambda, \theta) & 0 \\ 0 & 0 & 0 & a_-(\lambda, \theta) \end{pmatrix} \qquad \begin{aligned} a_\pm(\lambda, \theta) &= f(\lambda + \gamma) \\ b_\pm(\lambda, \theta) &= f(\lambda)\frac{f(\theta \mp \gamma)}{f(\theta)} \\ c_\pm(\lambda, \theta) &= f(\gamma)\frac{f(\theta \mp \lambda)}{f(\theta)} \end{aligned} \; . \tag{2.3}$$

The function f in (2.3) is essentially a Jacobi Theta-function. More precisely we have $f(\lambda) = \Theta_1(\mathrm{i}\lambda, \tau)/2$, according to the conventions of [29], and the dependence of f with the elliptic nome τ is omitted for convenience.

Dynamical Yang-Baxter algebra. Let $\mathbb{V}_a \cong \mathbb{V}$, $\mathbb{V}_{\mathcal{Q}} \cong \mathbb{V}^{\otimes L}$ and consider an operator $\mathcal{T}_a \in \mathrm{End}(\mathbb{V}_a \otimes \mathbb{V}_{\mathcal{Q}})$ for an arbitrary integer L. Then the dynamical Yang-Baxter equation (2.1) ensures the associativity of the relation

$$\begin{aligned} &\mathcal{R}_{ab}(\lambda_1 - \lambda_2, \theta - \gamma \mathrm{H})\mathcal{T}_a(\lambda_1, \theta)\mathcal{T}_b(\lambda_2, \theta - \gamma\hat{h}_a) = \\ &\qquad \mathcal{T}_b(\lambda_2, \theta)\mathcal{T}_a(\lambda_1, \theta - \gamma\hat{h}_b)\mathcal{R}_{ab}(\lambda_1 - \lambda_2, \theta) \; , \end{aligned} \tag{2.4}$$

for $\mathcal{R}$-matrices satisfying the weight zero condition, i.e. $[\mathcal{R}, h \otimes 1 + 1 \otimes h] = 0$. As usual we define $h_i \in \mathbb{V}_{\mathcal{Q}}$ as h acting on the i-th node of the tensor product space $\mathbb{V}_{\mathcal{Q}}$ while $\mathrm{H} = \sum_{i=1}^{L} h_i$. The generator H can be identified with the $\mathfrak{sl}(2)$ Cartan generator acting on $\mathbb{V}_{\mathcal{Q}}$ and one can also show that

$$\mathcal{T}_a(\lambda, \theta) = \overrightarrow{\prod_{1 \leq i \leq L}} \mathcal{R}_{ai}(\lambda - \mu_i, \hat{\theta}_i) \tag{2.5}$$

is a representation of (2.4) with $\hat{\theta}_i = \theta - \gamma \sum_{k=i+1}^{L} h_k$ and arbitrary parameters $\mu_i \in \mathbb{C}$. The operator $\mathcal{T}$ is usually denominated dynamical monodromy matrix, or simply monodromy matrix, and it consists of a matrix in the space $\mathbb{V}_a$ whose entries are operators in the space $\mathbb{V}_{\mathcal{Q}}$. Thus for the $E_{p,q}[\mathfrak{sl}(2)]$ solution (2.3), our monodromy matrix can be recasted as

$$\mathcal{T}(\lambda, \theta) = \begin{pmatrix} \mathcal{A}(\lambda, \theta) & \mathcal{B}(\lambda, \theta) \\ \mathcal{C}(\lambda, \theta) & \mathcal{D}(\lambda, \theta) \end{pmatrix} \tag{2.6}$$

where $\mathcal{A}, \mathcal{B}, \mathcal{C}, \mathcal{D} \in \mathrm{End}((\mathbb{C}^2)^{\otimes L})$.

Remark 1. *An important limit of (2.3) is the limit $p = e^{i\pi\tau} \to 0$ where the elliptic Theta-functions degenerate into hyperbolic ones. Furthermore, if we also consider the limit $\theta \to \infty$ we are left with the standard $\mathcal{R}$-matrix and Yang-Baxter relations of the six-vertex model [23].*

Integrable lattice systems. Exact solvability of two-dimensional lattice systems can be achieved from certain conditions of integrability in analogy to the theory of integrable differential equations. This condition is fulfilled by commuting transfer matrices which is assured by local equivalence transformations satisfied by the model statistical weights [23]. For vertex models we can encode the statistical weight of a given vertex configuration as the entry of a certain matrix $\mathcal{R}$. In this way the aforementioned equivalence transformation requires this $\mathcal{R}$-matrix to satisfy the celebrated Yang-Baxter equation. For Solid-on-Solid models, or SOS for short, the condition of integrability in the sense of Baxter requires the model statistical weights to satisfy the so called Hexagon identity [30]. This condition was shown in [27] to be directly related to the dynamical Yang-Baxter equation (2.1), and in what follows we shall briefly describe some important examples of integrable SOS and vertex lattice systems.

(i) *Eight-Vertex-SOS model with domain wall boundaries:* This model consists of a two-dimensional lattice system defined on a square lattice. It is built from the juxtaposition of plaquettes where we associate a set of state variables to the corners of each plaquette in order to characterize its allowed configurations. As far as boundary conditions are concerned, the case of domain wall boundaries consists of the assumption that the plaquettes at the border are fixed at a particular configuration. This model has been previously considered in [31–34] and here we shall adopt the conventions of [20, 21]. In this way the partition function of the Eight-Vertex-SOS model with domain wall boundaries can be written as

$$\mathcal{Z}_\theta = \langle \bar{0} | \overrightarrow{\prod_{1 \leq j \leq L}} \mathcal{B}(\lambda_j, \theta + j\gamma) | 0 \rangle , \tag{2.7}$$

where the operators $\mathcal{B}(\lambda, \theta)$ are defined through the relations (2.6), (2.3) and (2.5). In their turn the vectors $|0\rangle$ and $|\bar{0}\rangle$ are explicitly given by

$$|0\rangle = \begin{pmatrix} 1 \\ 0 \end{pmatrix}^{\otimes L} \qquad |\bar{0}\rangle = \begin{pmatrix} 0 \\ 1 \end{pmatrix}^{\otimes L} . \tag{2.8}$$

(ii) *Scalar product of Bethe vectors:* The evaluation of partition functions of lattice systems such as vertex models with periodic boundary conditions can be conveniently translated into an eigenvalue problem for an operator usually denominated transfer matrix [8, 9]. The diagonalization of transfer matrices for integrable vertex models can be performed, for instance, through the Quantum Inverse Scattering Method [12]. Within that method, Bethe vectors arise as an ansatz capable of determining transfer matrices exact eigenvectors. In a slightly different context, scalar products of Bethe vectors can also be regarded as the partition function of a vertex model with special boundary conditions [35,36], and this is the interpretation we shall pursue here. As it was stressed out in Remark 1, the standard six-vertex model relations can be obtained from (2.3) in a particular limit. In that limit we have $\mathcal{A}(\lambda,\theta) \rightarrow A(\lambda)$, $\mathcal{B}(\lambda,\theta) \rightarrow B(\lambda)$, $\mathcal{C}(\lambda,\theta) \rightarrow C(\lambda)$, $\mathcal{D}(\lambda,\theta) \rightarrow D(\lambda)$ and the scalar product of Bethe vectors S_n then reads

$$S_n = \langle 0| \overleftarrow{\prod_{1\leq i\leq n}} C(\lambda_i^C) \overrightarrow{\prod_{1\leq i\leq n}} B(\lambda_i^B) |0\rangle \ . \tag{2.9}$$

In formulae (2.9) we have considered the description employed in [36] while the vector $|0\rangle$ has been defined in (2.8).

From (2.7) and (2.9) we can see that the above defined partition functions are given as the expected value of a product of generators of the Yang-Baxter algebra and its dynamical counterpart. In what follows we shall demonstrate how the algebraic relations (2.4) can be explored yielding functional equations determining the aforementioned quantities.

3. Yang-Baxter algebra and functional relations

In order to illustrate how the Yang-Baxter algebra can be employed to derive functional relations, we shall consider the six-vertex model limit of (2.1), (2.3) and (2.4) for the sake of simplicity. In particular, the relation (2.4) then encodes a total of sixteen commutation rules involving the set of generators $\mathcal{M}(\lambda) = \{A, B, C, D\}(\lambda)$ evaluated at different values of the spectral parameter λ.

From the perspective of Quantum Field Theory one can regard the operators $A(\lambda)$ and $D(\lambda)$ as diagonal fields while $B(\lambda)$ and $C(\lambda)$ plays the role of creation and annihilation fields. As far as we are concerned with an eigenvalue problem involving the set of generators $\mathcal{M}(\lambda)$, i.e. the diagonalization of a transfer matrix [37,12], the framework of Quantum Field Theory is quite appealing and it seems natural to build the corresponding eigenvectors as elements of a Fock space. In particular, this approach is encouraged by the structure of the commutation relations in (2.4). Those commutation rules constitute one of the corner stones of the algebraic Bethe ansatz [12] and for instance let us single out the following one

$$A(\lambda_1)B(\lambda_2) = \frac{a(\lambda_2-\lambda_1)}{b(\lambda_2-\lambda_1)} B(\lambda_2)A(\lambda_1) - \frac{c(\lambda_2-\lambda_1)}{b(\lambda_2-\lambda_1)} B(\lambda_1)A(\lambda_2) \ , \tag{3.1}$$

where $a(\lambda) = \sinh(\lambda+\gamma)$, $b(\lambda) = \sinh(\lambda)$ and $c(\lambda) = \sinh(\gamma)$. Within the framework of the algebraic Bethe ansatz we usually regard (3.1) as a relation between a diagonal field and a creation field. Nevertheless, this is not the only way one can explore relations of type (3.1), and in what follows we shall see they can also be projected as a functional relation.

Definition 1. *Let $\pi : End(\mathbb{V}_{\mathcal{Q}}) \times End(\mathbb{V}_{\mathcal{Q}}) \mapsto \mathbb{C}$ be a continuous and bi-additive map. Due to (3.1), or more generally (2.4), it is convenient to specialize the map π to π_2 defined as*

$$\pi_2 : \qquad \mathcal{M}(\lambda) \times \mathcal{M}(\mu) \mapsto \mathbb{C}[\lambda^{\pm 1}, \mu^{\pm 1}] \ . \tag{3.2}$$

The 2-tuple $(\xi_1, \xi_2) \ : \ \xi_1 \in \mathcal{M}(\lambda)$, $\xi_2 \in \mathcal{M}(\mu)$ originated from the Cartesian product $\mathcal{M}(\lambda) \times \mathcal{M}(\mu)$ is then simply understood as the matrix product $\xi_1 \xi_2$. In other words the map π_2 associates a two-variable complex function to any quadratic form in (2.4).

The map π_2 defined in (3.2) is able to associate a functional relation to any commutation rule contained in (2.4). For instance, the map (3.2) applied on (3.1) yields the relation

$$b(\lambda_2 - \lambda_1) f(\lambda_1, \lambda_2) = a(\lambda_2 - \lambda_1) \bar{f}(\lambda_2, \lambda_1) - c(\lambda_2 - \lambda_1) \bar{f}(\lambda_1, \lambda_2) \tag{3.3}$$

where $f(\lambda_1, \lambda_2) = \pi_2(A(\lambda_1)B(\lambda_2))$ and $\bar{f}(\lambda_1, \lambda_2) = \pi_2(B(\lambda_1)A(\lambda_2))$.

The study of functional equations has a long history, see for instance the monograph [38], and they play a remarkable role in Statistical Mechanics [23] and Conformal Field Theory [39]. Within those contexts they appear intimately related to Baxter's concept of commuting transfer matrices [10] and among prominent examples we have Baxter's $T - Q$ relation [40], inversion relation [41], analytical Bethe ansatz [42] and Y-system [43,44]. Also, it is worth mentioning the quantum Knizhnik-Zamolodchikov equation [45] which describes form factors and correlation functions in integrable field theories [46]. Here we intend to demonstrate that partition functions of integrable lattice models, such as (2.7) and (2.9), can also be described by functional equations. Interestingly, the functional equations describing those partition functions follow directly from the Yang-Baxter algebra within the lines above discussed. In order to show that we first need to generalize the Definition 1 in the following way.

Definition 2. *Let n be an integer. Then we define the n-additive continuous map π_n as*

$$\pi_n : \qquad \mathcal{M}(\lambda_1) \times \mathcal{M}(\lambda_2) \times \cdots \times \mathcal{M}(\lambda_n) \mapsto \mathbb{C}[\lambda_1^{\pm 1}, \lambda_2^{\pm 1}, \ldots, \lambda_n^{\pm 1}] \,. \tag{3.4}$$

Similarly to (3.2) the n-tuple $(\xi_1, \xi_2, \ldots, \xi_n) : \xi_i \in \mathcal{M}(\lambda_i)$ is understood as the non-commutative product $\overrightarrow{\prod\limits_{1 \leq i \leq n}} \xi_i$. In other words, the map π_n associates a n-variable complex function to a product of n generators of the Yang-Baxter algebra.

Realization of π_n. A simple choice of realization of π_n is the scalar product with vectors $|\psi\rangle\,, |\psi'\rangle \in \mathbb{V}_{\mathcal{Q}}$. More precisely, we can readily see that

$$\pi_n(\mathcal{F}) = \langle \psi' | \, \mathcal{F} \, | \psi \rangle \tag{3.5}$$

is a realization of π_n for any element $\mathcal{F} \in \mathcal{M}(\lambda_1) \times \mathcal{M}(\lambda_2) \times \cdots \times \mathcal{M}(\lambda_n)$. At this stage $|\psi\rangle$ and $|\psi'\rangle$ are arbitrary vectors but it will become clear that particular choices can render interesting functional equations describing quantities such as (2.7) and (2.9).

3.1. Functional equation for $\mathcal{Z}_\theta$

In this section we aim to show how a functional equation for the partition function $\mathcal{Z}_\theta$ can be derived within the lines discussed in Section 3. For that the first step is to find appropriate vectors $|\psi\rangle$ and $|\psi'\rangle$ in order to employ the realization (3.5). In addition to that it is also important to consider suitable elements $\mathcal{F}$ such that we end up with an equation capable of determining the desired partition function. Although there is no precise recipe for selecting $|\psi\rangle$, $|\psi'\rangle$ and $\mathcal{F}$, we shall see that some properties of the elements entering in the definitions (2.7) and (2.8) can help us to sort that out.

Highest weight vectors. The vectors $|0\rangle$ and $|\bar{0}\rangle$ defined in (2.8) are $\mathfrak{sl}(2)$ highest and lowest weight vectors. They satisfy the following properties:

$$
\begin{aligned}
&\mathcal{A}(\lambda,\theta)\,|\bar{0}\rangle = \frac{f(\theta-\gamma)}{f(\theta+(L-1)\gamma)}\prod_{j=1}^{L} f(\lambda-\mu_j)\,|\bar{0}\rangle \qquad && \mathcal{A}(\lambda,\theta)\,|0\rangle = \prod_{j=1}^{L} f(\lambda-\mu_j+\gamma)\,|0\rangle \\
&\mathcal{D}(\lambda,\theta)\,|0\rangle = \frac{f(\theta+\gamma)}{f(\theta-(L-1)\gamma)}\prod_{j=1}^{L} f(\lambda-\mu_j)\,|0\rangle \qquad && \mathcal{D}(\lambda,\theta)\,|\bar{0}\rangle = \prod_{j=1}^{L} f(\lambda-\mu_j+\gamma)\,|\bar{0}\rangle \\
&\mathcal{C}(\lambda,\theta)\,|0\rangle = 0 && \mathcal{B}(\lambda,\theta)\,|\bar{0}\rangle = 0
\end{aligned} \tag{3.6}
$$

$$
\begin{aligned}
&\langle\bar{0}|\,\mathcal{A}(\lambda,\theta) = \frac{f(\theta-\gamma)}{f(\theta+(L-1)\gamma)}\prod_{j=1}^{L} f(\lambda-\mu_j)\,\langle\bar{0}| \qquad && \langle 0|\,\mathcal{A}(\lambda,\theta) = \prod_{j=1}^{L} f(\lambda-\mu_j+\gamma)\,\langle 0| \\
&\langle 0|\,\mathcal{D}(\lambda,\theta) = \frac{f(\theta+\gamma)}{f(\theta-(L-1)\gamma)}\prod_{j=1}^{L} f(\lambda-\mu_j)\,\langle 0| \qquad && \langle\bar{0}|\,\mathcal{D}(\lambda,\theta) = \prod_{j=1}^{L} f(\lambda-\mu_j+\gamma)\,\langle\bar{0}| \\
&\langle\bar{0}|\,\mathcal{C}(\lambda,\theta) = 0 && \langle 0|\,\mathcal{B}(\lambda,\theta) = 0
\end{aligned} \tag{3.7}
$$

The expressions (3.6) and (3.7) follow from the definitions (2.5), (2.3) and the highest/lowest weight property of $|0\rangle$ and $|\bar{0}\rangle$.

Yang-Baxter relations of degree n. The relations arising from the dynamical Yang-Baxter algebra (2.4) involve the set of generators $\mathcal{M}(\lambda,\theta)=\{\mathcal{A},\mathcal{B},\mathcal{C},\mathcal{D}\}(\lambda,\theta)$ and H. Although some entries of (2.4) contain products of the form $\mathcal{M}(\lambda_1,\theta_1)\times\mathcal{M}(\lambda_2,\theta_2)\times f(\mathrm{H})$, there are still commutation rules with all terms in $\mathcal{M}(\lambda_1,\theta_1)\times\mathcal{M}(\lambda_2,\theta_2)$. Those are the relations that will be explored here and among them we have the following ones

$$
\begin{aligned}
\mathcal{B}(\lambda_1,\theta)\mathcal{B}(\lambda_2,\theta+\gamma) &= \mathcal{B}(\lambda_2,\theta)\mathcal{B}(\lambda_1,\theta+\gamma) \\
\mathcal{A}(\lambda_1,\theta+\gamma)\mathcal{B}(\lambda_2,\theta) &= \frac{f(\lambda_2-\lambda_1+\gamma)}{f(\lambda_2-\lambda_1)}\frac{f(\theta+\gamma)}{f(\theta+2\gamma)}\mathcal{B}(\lambda_2,\theta+\gamma)\mathcal{A}(\lambda_1,\theta+2\gamma) \\
&- \frac{f(\theta+\gamma-\lambda_2+\lambda_1)}{f(\lambda_2-\lambda_1)}\frac{f(\gamma)}{f(\theta+2\gamma)}\mathcal{B}(\lambda_1,\theta+\gamma)\mathcal{A}(\lambda_2,\theta+2\gamma)\,.
\end{aligned} \tag{3.8}
$$

Both expressions in (3.8) are quadratic and their repeated use is able to provide relations of degree n for a subset of elements $\mathcal{F}\subset\mathcal{W}_n=\mathcal{M}(\lambda_0,\theta_0)\times\mathcal{M}(\lambda_1,\theta_1)\times\cdots\times\mathcal{M}(\lambda_{n-1},\theta_{n-1})$. More precisely, the iteration of (3.8) yields the following relation of degree $n+1$,

$$
\begin{aligned}
&\mathcal{A}(\lambda_0,\theta+\gamma)Y_{\theta-\gamma}(\lambda_1,\dots,\lambda_n) = \\
&\frac{f(\theta+\gamma)}{f(\theta+(n+1)\gamma)}\prod_{j=1}^{n}\frac{f(\lambda_j-\lambda_0+\gamma)}{f(\lambda_j-\lambda_0)}Y_{\theta}(\lambda_1,\dots,\lambda_n)\mathcal{A}(\lambda_0,\theta+(n+1)\gamma) \\
&-\sum_{i=1}^{n}\frac{f(\theta+\gamma-\lambda_i+\lambda_0)}{f(\theta+(n+1)\gamma)}\frac{f(\gamma)}{f(\lambda_i-\lambda_0)}\prod_{\substack{j=1\\ j\neq i}}^{n}\frac{f(\lambda_j-\lambda_i+\gamma)}{f(\lambda_j-\lambda_i)}\times \\
&\qquad Y_{\theta}(\lambda_0,\lambda_1,\dots,\lambda_{i-1},\lambda_{i+1},\dots,\lambda_n)\mathcal{A}(\lambda_i,\theta+(n+1)\gamma)\,,
\end{aligned} \tag{3.9}
$$

where $Y_\theta(\lambda_1,\dots,\lambda_n)=\overrightarrow{\prod_{1\leq j\leq n}}B(\lambda_j,\theta+j\gamma)$.

Building π_n for $\mathcal{Z}_\theta$. The relations (3.6) and (3.7) suggest the prescription $|\psi\rangle = |0\rangle$ and $|\psi'\rangle = |\bar{0}\rangle$. As we shall see this particular choice of π_n is able to generate a functional equation for the partition function $\mathcal{Z}_\theta$ from the algebraic relation (3.9).

Taking into account the discussion of Section 3 we then set $n = L$ in the relation (3.9). Next we consider the action of the map π_{L+1} on (3.9) and by doing so we find only terms of the form

$$\pi_{L+1}(\mathcal{A}(\lambda_0, \theta+\gamma) Y_{\theta-\gamma}(\lambda_1, \dots, \lambda_L)) \tag{3.10}$$

and

$$\pi_{L+1}(Y_\theta(\lambda_0, \lambda_1, \dots, \lambda_{i-1}, \lambda_{i+1}, \dots, \lambda_L) \mathcal{A}(\lambda_i, \theta + (L+1)\gamma)) \ . \tag{3.11}$$

Interestingly, the relations (3.6) and (3.7) obtained as a consequence of the highest/lowest weight property of $|0\rangle$ and $|\bar{0}\rangle$ give rise to a map $\pi_{L+1} \mapsto \pi_L$. More precisely we have

$$\pi_{L+1}(\mathcal{A}(\lambda_0, \theta+\gamma) Y_{\theta-\gamma}(\lambda_1, \dots, \lambda_L)) = \frac{f(\theta)}{f(\theta+L\gamma)} \prod_{j=1}^{L} f(\lambda_0 - \mu_j) \ \pi_L(Y_{\theta-\gamma}(\lambda_1, \dots, \lambda_L)) \tag{3.12}$$

and

$$\pi_{L+1}(Y_\theta(\lambda_0, \lambda_1, \dots, \lambda_{i-1}, \lambda_{i+1}, \dots, \lambda_L) \mathcal{A}(\lambda_i, \theta + (L+1)\gamma)) = \prod_{j=1}^{L} f(\lambda_i - \mu_j) \ \pi_L(Y_\theta(\lambda_0, \lambda_1, \dots, \lambda_{i-1}, \lambda_{i+1}, \dots, \lambda_L)) \ . \tag{3.13}$$

The partition function (2.7) can now be promptly identified with $\pi_L(Y_\theta(\lambda_1, \dots, \lambda_L))$. In this way the action of π_{L+1} on (3.9), in addition to the relations (3.12) and (3.13), leaves us with the following functional equation for $\mathcal{Z}_\theta$,

$$M_0 \ \mathcal{Z}_{\theta-\gamma}(\lambda_1, \dots, \lambda_L) + \sum_{i=0}^{L} N_i \ \mathcal{Z}_\theta(\lambda_0, \dots, \lambda_{i-1}, \lambda_{i+1}, \dots, \lambda_L) = 0 \ . \tag{3.14}$$

The structure of the coefficients M_0 and N_i is a direct consequence of the dynamical Yang-Baxter algebra relations (2.4) and the highest weight properties (3.6) and (3.7). For convenience we shall postpone presenting their explicit form. It is also worth remarking that (3.14) made its first appearance in [21] and due to the fact that the operators $\mathcal{B}$ satisfy the relation $\mathcal{B}(\lambda_1, \theta)\mathcal{B}(\lambda_2, \theta+\gamma) = \mathcal{B}(\lambda_2, \theta)\mathcal{B}(\lambda_1, \theta+\gamma)$, as given by (3.8), the ordering of the arguments of $\mathcal{Z}_\theta$ in (3.14) is indeed arbitrary. This property leads us to the following lemma.

Lemma 1 (Symmetric function). *The partition function $\mathcal{Z}_\theta$ is symmetric under the permutation of its variables, i.e.*

$$\mathcal{Z}_\theta(\dots, \lambda_i, \dots, \lambda_j, \dots) = \mathcal{Z}_\theta(\dots, \lambda_j, \dots, \lambda_i, \dots) \ . \tag{3.15}$$

Proof. This property follows directly from the commutation relation (3.8) or from the functional equation (3.14) as demonstrated in [21]. □

Remark 2. *Due to (3.15) we can safely employ the notation $\mathcal{Z}_\theta(\lambda_1, \dots, \lambda_L) = \mathcal{Z}_\theta(X^{1,L})$ where $X^{i,j} = \{\lambda_k \ : \ i \leq k \leq j\}$.*

Taking into account Remark 2, it is also convenient to introduce the set $X_k^{i,j} = X^{i,j} \backslash \{\lambda_k\}$ in such a way that (3.14) can be simply recasted as

$$M_0 \; \mathcal{Z}_{\theta-\gamma}(X^{1,L}) + \sum_{i=0}^{L} N_i \; \mathcal{Z}_{\theta}(X_i^{0,L}) = 0 \; . \tag{3.16}$$

In their turn the coefficients M_0 and N_i explicitly read

$$\begin{aligned} M_0 &= \frac{f(\theta)}{f(\theta + L\gamma)} \prod_{j=1}^{L} f(\lambda_0 - \mu_j) \\ N_i &= -\frac{f(\theta + \gamma + \lambda_0 - \lambda_i)}{f(\theta + (L+1)\gamma)} \frac{f(\gamma)}{f(\lambda_0 - \lambda_i + \gamma)} \prod_{j=1}^{L} f(\lambda_i - \mu_j + \gamma) \prod_{\lambda \in X_i^{0,L}} \frac{f(\lambda - \lambda_i + \gamma)}{f(\lambda - \lambda_i)} \; . \end{aligned} \tag{3.17}$$

Some further remarks are important at this stage. The reader familiar with the theory of Knizhnik-Zamolodchikov (KZ) equations can notice that (3.16) exhibits a structure which resembles that of the classical KZ equation [14]. For instance, the first term of (3.16) consists of the partition function with shifted variable θ which could be thought of as an analogous of the derivative. The second term of (3.16) consists of a linear combination of partition functions with a given variable λ_i in the argument being replaced by a variable λ_0. This variable replacement can be regarded as the action of an operator which can be considered as a sort of 'hamiltonian'. Although KZ equations are vector equations while here we are dealing with a scalar equation, we can see that both terms of (3.16) have a counterpart in the KZ theory. Furthermore, we shall find that the solution of (3.16) also resembles solutions of KZ-like equations [47].

Solution. The functional relation (3.16) consists of an equation for the partition function $\mathcal{Z}_\theta(X^{1,L})$ over the set of variables $X^{0,L}$. Thus we have one more variable than is required to describe the partition function itself. This feature is typical of functional equations such as the d'Alembert equation [38], but had not appeared previously in the functional relations describing Exactly Solvable Models to the best of our knowledge. This extra variable can be set at will in order to help us with the resolution of (3.16), and this approach is the basis of the method considered in [20] and [21]. Moreover, Eq. (3.16) also exhibits some special properties providing some guidance through the steps required to obtain its solution. These properties are as follows:

- *Scale invariance*: Eq. (3.16) is invariant under the symmetry transformation $\mathcal{Z}_\theta(X^{1,L}) \mapsto \alpha \mathcal{Z}_\theta(X^{1,L})$ where $\alpha \in \mathbb{C}$ is independent of θ and λ_j. This property tells us that (3.16) is only able to determine the partition function up to an overall multiplicative factor independent of the variables θ and λ_j. In this way the full determination of the partition function will require we know the precise value of $\mathcal{Z}_\theta$ for a particular choice of the aforementioned variables.
- *Linearity*: Eq. (3.16) is linear and as usual this implies that if $\mathcal{Z}_\theta^{(1)}$ and $\mathcal{Z}_\theta^{(2)}$ are solutions, then the linear combination $\mathcal{Z}_\theta = \mathcal{Z}_\theta^{(1)} + \mathcal{Z}_\theta^{(2)}$ is also a solution. As far as the determination of (2.7) is concerned, this property is telling us we need to establish an uniqueness criterium in order to characterize the desired partition function.

Here we do not intend to give a detailed description of the method developed to solve (3.16). Nevertheless, we can still comment on how the properties of scale invariance and linearity have been employed. In [21] we have shown that the partition function (2.7) can be explicitly computed in the limit $(\theta, \lambda_j) \rightarrow \infty$. This result can then be used to completely fix the overall multiplicative factor which is not constrained by (3.16). Concerning the issue raised by the linearity of (3.16), we have also shown in [21] that the desired solution consists of a higher order Theta-function [48]. As such, it is uniquely characterized by its zeroes up to an overall factor.

As a matter of fact, unveiling special zeroes of $\mathcal{Z}_\theta$ plays an important role for the resolution of (3.16) and we have found the following solution in [21],

$$\mathcal{Z}_\theta(X^{1,L}) = [f'(0)f(\gamma)]^L$$
$$\oint \cdots \oint \prod_{j=1}^{L} \frac{\mathrm{d}w_j}{2\mathrm{i}\pi} \frac{\prod_{j>i}^{L} f(w_j - w_i + \gamma) f(w_j - w_i)}{\prod_{i,j=1}^{L} f(w_i - \lambda_j)} \prod_{j=1}^{L} \frac{f(\theta + j\gamma - w_j + \mu_j)}{f(\theta + j\gamma)}$$
$$\times \prod_{j<i}^{L} f(\mu_j - w_i) \prod_{j>i}^{L} f(w_i - \mu_j + \gamma) \; . \tag{3.18}$$

Formulae (3.18) is given in terms of a multiple contour integral whose integration contour for each variable w_j encloses all variables in the set $X^{1,L}$. It is also worth remarking that similar multiple contour integrals also emerge as solutions of the KZ equation [47] and its q-deformed version [49].

3.2. *Functional equation for S_n*

The partition function (2.7) is not the only quantity which satisfy a functional equation such as the one described in Section 3.1. Similar equations can also be derived for scalar products of Bethe vectors as we shall demonstrate. Although the case of domain wall boundary conditions was introduced in [35] as a building block of scalar products, here we shall not follow that approach but instead consider scalar products defined by (2.9) as an independent quantity.

The derivation of (3.14) required two main ingredients: the construction of a suitable realization of π_n and the derivation of appropriate Yang-Baxter relations of degree n. In order to apply the same methodology for scalar products S_n we then first need to consider the commutation relations from (2.4) in the six-vertex model limit as discussed in Remark 1. In what follows we present the ones that will be required,

$$\begin{aligned} A(\lambda_1)B(\lambda_2) &= \frac{a(\lambda_2 - \lambda_1)}{b(\lambda_2 - \lambda_1)} B(\lambda_2)A(\lambda_1) - \frac{c(\lambda_2 - \lambda_1)}{b(\lambda_2 - \lambda_1)} B(\lambda_1)A(\lambda_2) \\ C(\lambda_1)A(\lambda_2) &= \frac{a(\lambda_1 - \lambda_2)}{b(\lambda_1 - \lambda_2)} A(\lambda_2)C(\lambda_1) - \frac{c(\lambda_1 - \lambda_2)}{b(\lambda_1 - \lambda_2)} A(\lambda_1)C(\lambda_2) \end{aligned} \tag{3.19}$$

$$\begin{aligned} D(\lambda_1)B(\lambda_2) &= \frac{a(\lambda_1 - \lambda_2)}{b(\lambda_1 - \lambda_2)} B(\lambda_2)D(\lambda_1) - \frac{c(\lambda_1 - \lambda_2)}{b(\lambda_1 - \lambda_2)} B(\lambda_1)D(\lambda_2) \\ C(\lambda_1)D(\lambda_2) &= \frac{a(\lambda_2 - \lambda_1)}{b(\lambda_2 - \lambda_1)} D(\lambda_2)C(\lambda_1) - \frac{c(\lambda_2 - \lambda_1)}{b(\lambda_2 - \lambda_1)} D(\lambda_1)C(\lambda_2) \end{aligned} \tag{3.20}$$

$$\begin{aligned} B(\lambda)B(\mu) &= B(\mu)B(\lambda) \\ C(\lambda)C(\mu) &= C(\mu)C(\lambda) \; . \end{aligned} \tag{3.21}$$

Yang-Baxter relation of degree n. The relations (3.19), (3.20) and (3.21) are a subset of the commutation rules contained in (2.4) in the six-vertex model limit and, as such, they are relations of degree 2 according to the discussion of Section 3.1. In order to describe the scalar product S_n we need an appropriate relation of degree n which can be obtained by the repeated use of (3.19)-(3.21). The direct inspection of the commutation relations (3.19)-(3.21) suggests us to consider the quantity

$$T_A = \overleftarrow{\prod_{1 \leq i \leq n}} C(\lambda_i^C)\, A(\lambda_0) \overrightarrow{\prod_{1 \leq i \leq n}} B(\lambda_i^B)\,, \tag{3.22}$$

which can be analyzed in at least two different ways through the relations (3.19) and (3.21). Here we shall restrict our discussion to the following ways of evaluating T_A. Firstly, we can move the operator $A(\lambda_0)$ in (3.22) all the way to the right through all the string of operators B with the help of the first relation in (3.19). Alternatively, we can also move the operator $A(\lambda_0)$ to the left by making use of the second relation in (3.19). These two different ways of evaluating the same quantity yields the following Yang-Baxter relation of order $2n+1$,

$$\begin{aligned}
&\prod_{i=1}^{n} \frac{a(\lambda_i^C - \lambda_0)}{b(\lambda_i^C - \lambda_0)} A(\lambda_0) \overleftarrow{\prod_{1 \leq i \leq n}} C(\lambda_i^C) \overrightarrow{\prod_{1 \leq i \leq n}} B(\lambda_i^B) \\
&- \sum_{i=1}^{n} \frac{c(\lambda_i^C - \lambda_0)}{b(\lambda_i^C - \lambda_0)} \prod_{\substack{j=1 \\ j \neq i}}^{n} \frac{a(\lambda_j^C - \lambda_i^C)}{b(\lambda_j^C - \lambda_i^C)} A(\lambda_i^C) \overleftarrow{\prod_{\substack{0 \leq j \leq n \\ j \neq i}}} C(\lambda_j^C) \overrightarrow{\prod_{1 \leq j \leq n}} B(\lambda_j^B) = \\
&\prod_{i=1}^{n} \frac{a(\lambda_i^B - \lambda_0)}{b(\lambda_i^B - \lambda_0)} \overleftarrow{\prod_{1 \leq i \leq n}} C(\lambda_i^C) \overrightarrow{\prod_{1 \leq i \leq n}} B(\lambda_i^B)\, A(\lambda_0) \\
&- \sum_{i=1}^{n} \frac{c(\lambda_i^B - \lambda_0)}{b(\lambda_i^B - \lambda_0)} \prod_{\substack{j=1 \\ j \neq i}}^{n} \frac{a(\lambda_j^B - \lambda_i^B)}{b(\lambda_j^B - \lambda_i^B)} \overleftarrow{\prod_{1 \leq j \leq n}} C(\lambda_j^C) \overrightarrow{\prod_{\substack{0 \leq j \leq n \\ j \neq i}}} B(\lambda_j^B)\, A(\lambda_i^B)\,.
\end{aligned} \tag{3.23}$$

It is important to stress here that the derivation of (3.23) also makes explicit use of the relations (3.21). The expression (3.23) will be left at rest for a while and we shall focus on another quantity. For instance, instead of (3.22) we could have performed the same analysis starting with

$$T_D = \overleftarrow{\prod_{1 \leq i \leq n}} C(\lambda_i^C)\, D(\lambda_0) \overrightarrow{\prod_{1 \leq i \leq n}} B(\lambda_i^B)\,. \tag{3.24}$$

In that case we need to consider the relations (3.20) and (3.21), and we end up with the following identity,

$$\begin{aligned}
&\prod_{i=1}^{n} \frac{a(\lambda_0 - \lambda_i^C)}{b(\lambda_0 - \lambda_i^C)} D(\lambda_0) \overleftarrow{\prod_{1 \leq i \leq n}} C(\lambda_i^C) \overrightarrow{\prod_{1 \leq i \leq n}} B(\lambda_i^B) \\
&- \sum_{i=1}^{n} \frac{c(\lambda_0 - \lambda_i^C)}{b(\lambda_0 - \lambda_i^C)} \prod_{\substack{j=1 \\ j \neq i}}^{n} \frac{a(\lambda_i^C - \lambda_j^C)}{b(\lambda_i^C - \lambda_j^C)} D(\lambda_i^C) \overleftarrow{\prod_{\substack{0 \leq j \leq n \\ j \neq i}}} C(\lambda_j^C) \overrightarrow{\prod_{1 \leq j \leq n}} B(\lambda_j^B) = \\
&\prod_{i=1}^{n} \frac{a(\lambda_0 - \lambda_i^B)}{b(\lambda_0 - \lambda_i^B)} \overleftarrow{\prod_{1 \leq i \leq n}} C(\lambda_i^C) \overrightarrow{\prod_{1 \leq i \leq n}} B(\lambda_i^B)\, D(\lambda_0) \\
&- \sum_{i=1}^{n} \frac{c(\lambda_0 - \lambda_i^B)}{b(\lambda_0 - \lambda_i^B)} \prod_{\substack{j=1 \\ j \neq i}}^{n} \frac{a(\lambda_i^B - \lambda_j^B)}{b(\lambda_i^B - \lambda_j^B)} \overleftarrow{\prod_{1 \leq j \leq n}} C(\lambda_j^C) \overrightarrow{\prod_{\substack{0 \leq j \leq n \\ j \neq i}}} B(\lambda_j^B)\, D(\lambda_i^B)\,.
\end{aligned} \tag{3.25}$$

Both expressions (3.23) and (3.25) consist of Yang-Baxter relations of order $2n+1$ and they can be converted into functional equations for S_n with a proper choice of π_{2n+1}.

The map π_m for S_n. Taking into account the relations (3.6), (3.7), (3.23) and (3.25) we can readily see that the choice $|\psi\rangle = |\psi'\rangle = |0\rangle$ for the realization (3.5) gives rise to functional relations for the scalar product S_n. In order to see that we apply the map π_{2n+1} on (3.23) and (3.25). By doing so we only find terms of the form

$$\pi_{2n+1}(A(v_0^A) \overleftarrow{\prod_{1\leq i\leq n}} C(v_i^C) \overrightarrow{\prod_{1\leq i\leq n}} B(v_i^B)) \quad , \quad \pi_{2n+1}(\overleftarrow{\prod_{1\leq i\leq n}} C(v_i^C) \overrightarrow{\prod_{1\leq i\leq n}} B(v_i^B)\, A(\bar{v}_0^A))$$
$$\pi_{2n+1}(D(v_0^D) \overleftarrow{\prod_{1\leq i\leq n}} C(v_i^C) \overrightarrow{\prod_{1\leq i\leq n}} B(v_i^B)) \quad , \quad \pi_{2n+1}(\overleftarrow{\prod_{1\leq i\leq n}} C(v_i^C) \overrightarrow{\prod_{1\leq i\leq n}} B(v_i^B)\, D(\bar{v}_0^D)) \quad (3.26)$$

with parameters $v_0^A, v_0^D, v_i^C \in \{\lambda_0, \lambda_1^C, \dots, \lambda_n^C\}$ and $\bar{v}_0^A, \bar{v}_0^D, v_i^B \in \{\lambda_0, \lambda_1^B, \dots, \lambda_n^B\}$.

Similarly to the case discussed in Section 3.1, here we also have a map $\pi_{2n+1} \mapsto \pi_{2n}$ induced by the highest weight property of $|0\rangle$. More precisely, from (3.6) we obtain

$$\begin{aligned}
\pi_{2n+1}(\overleftarrow{\prod_{1\leq i\leq n}} C(v_i^C) \overrightarrow{\prod_{1\leq i\leq n}} B(v_i^B)\, A(\bar{v}_0^A)) &= \\
\prod_{j=1}^{L} a(\bar{v}_0^A - \mu_j) &\quad \pi_{2n}(\overleftarrow{\prod_{1\leq i\leq n}} C(v_i^C) \overrightarrow{\prod_{1\leq i\leq n}} B(v_i^B)) \\
\pi_{2n+1}(\overleftarrow{\prod_{1\leq i\leq n}} C(v_i^C) \overrightarrow{\prod_{1\leq i\leq n}} B(v_i^B)\, D(\bar{v}_0^A)) &= \\
\prod_{j=1}^{L} b(\bar{v}_0^A - \mu_j) &\quad \pi_{2n}(\overleftarrow{\prod_{1\leq i\leq n}} C(v_i^C) \overrightarrow{\prod_{1\leq i\leq n}} B(v_i^B)) .
\end{aligned} \quad (3.27)$$

On the other hand, the property (3.7) yields

$$\begin{aligned}
\pi_{2n+1}(A(v_0^A) \overleftarrow{\prod_{1\leq i\leq n}} C(v_i^C) \overrightarrow{\prod_{1\leq i\leq n}} B(v_i^B)) &= \\
\prod_{j=1}^{L} a(v_0^A - \mu_j) &\quad \pi_{2n}(\overleftarrow{\prod_{1\leq i\leq n}} C(v_i^C) \overrightarrow{\prod_{1\leq i\leq n}} B(v_i^B)) \\
\pi_{2n+1}(D(v_0^D) \overleftarrow{\prod_{1\leq i\leq n}} C(v_i^C) \overrightarrow{\prod_{1\leq i\leq n}} B(v_i^B)) &= \\
\prod_{j=1}^{L} b(v_0^D - \mu_j) &\quad \pi_{2n}(\overleftarrow{\prod_{1\leq i\leq n}} C(v_i^C) \overrightarrow{\prod_{1\leq i\leq n}} B(v_i^B)) .
\end{aligned} \quad (3.28)$$

The terms $\pi_{2n}(\overleftarrow{\prod_{1\leq i\leq n}} C(v_i^C) \overrightarrow{\prod_{1\leq i\leq n}} B(v_i^B))$ can now be identified with the scalar products S_n as defined in (2.9). In this way the map π_m above discussed, together with the relations (3.23),

(3.25), (3.27) and (3.28), leaves us with the following functional equations,

$$
\begin{aligned}
J_0 S_n(X^{1,n}|Y^{1,n}) &+ \sum_{i=1}^{n} K_i^{(B)} S_n(X_i^{0,n}|Y^{1,n}) + \sum_{i=1}^{n} K_i^{(C)} S_n(X^{1,n}|Y_i^{0,n}) = 0 \\
\tilde{J}_0 S_n(X^{1,n}|Y^{1,n}) &+ \sum_{i=1}^{n} \tilde{K}_i^{(B)} S_n(X_i^{0,n}|Y^{1,n}) + \sum_{i=1}^{n} \tilde{K}_i^{(C)} S_n(X^{1,n}|Y_i^{0,n}) = 0 \, .
\end{aligned}
\tag{3.29}
$$

In their turn the coefficients appearing in (3.29) can be conveniently written as

$$
\begin{aligned}
J_0 &= \prod_{j=1}^{L} a(\lambda_0 - \mu_j) \left[\prod_{i=1}^{n} \frac{a(\lambda_i^C - \lambda_0)}{b(\lambda_i^C - \lambda_0)} - \prod_{i=1}^{n} \frac{a(\lambda_i^B - \lambda_0)}{b(\lambda_i^B - \lambda_0)} \right] \\
K_i^{(B,C)} &= \alpha_{B,C} \frac{c(\lambda_i^{B,C} - \lambda_0)}{b(\lambda_i^{B,C} - \lambda_0)} \prod_{j=1}^{L} a(\lambda_i^{B,C} - \mu_j) \prod_{\substack{j=1 \\ j \neq i}}^{n} \frac{a(\lambda_j^{B,C} - \lambda_i^{B,C})}{b(\lambda_j^{B,C} - \lambda_i^{B,C})} \, ,
\end{aligned}
\tag{3.30}
$$

and

$$
\begin{aligned}
\tilde{J}_0 &= \prod_{j=1}^{L} b(\lambda_0 - \mu_j) \left[\prod_{i=1}^{n} \frac{a(\lambda_0 - \lambda_i^C)}{b(\lambda_0 - \lambda_i^C)} - \prod_{i=1}^{n} \frac{a(\lambda_0 - \lambda_i^B)}{b(\lambda_0 - \lambda_i^B)} \right] \\
\tilde{K}_i^{(B,C)} &= \alpha_{B,C} \frac{c(\lambda_0 - \lambda_i^{B,C})}{b(\lambda_0 - \lambda_i^{B,C})} \prod_{j=1}^{L} b(\lambda_i^{B,C} - \mu_j) \prod_{\substack{j=1 \\ j \neq i}}^{n} \frac{a(\lambda_i^{B,C} - \lambda_j^{B,C})}{b(\lambda_i^{B,C} - \lambda_j^{B,C})} \, .
\end{aligned}
\tag{3.31}
$$

where $\alpha_B = 1$ and $\alpha_C = -1$.

The Remark 2 of Section 3.1 can be immediately extended to the case of scalar products, and in (3.29) we have employed the notation

$$
S_n(\lambda_1^B, \dots, \lambda_n^B | \lambda_1^C, \dots, \lambda_n^C) = S_n(X^{1,n}|Y^{1,n}) \tag{3.32}
$$

where $X^{i,j} = \{\lambda_k^B \ : \ i \leq k \leq j\}$ and $Y^{i,j} = \{\lambda_k^C \ : \ i \leq k \leq j\}$. Moreover, we have also considered the definitions $X_k^{i,j} = X^{i,j} \backslash \{\lambda_k^B\}$ and $Y_k^{i,j} = Y^{i,j} \backslash \{\lambda_k^C\}$. This possibility is granted by the following lemma.

Lemma 2 (Doubly symmetric function). *The scalar product $S_n(\lambda_1^B, \dots, \lambda_n^B | \lambda_1^C, \dots, \lambda_n^C)$ is symmetric under the permutation of variables $\lambda_i^B \leftrightarrow \lambda_j^B$ and $\lambda_i^C \leftrightarrow \lambda_j^C$ performed independently. More precisely we have*

$$
S_n(\dots, \lambda_i^B, \dots, \lambda_j^B, \dots | \lambda_1^C, \dots, \lambda_n^C) = S_n(\dots, \lambda_j^B, \dots, \lambda_i^B, \dots | \lambda_1^C, \dots, \lambda_n^C) \tag{3.33}
$$

and

$$
S_n(\lambda_1^C, \dots, \lambda_n^C | \dots, \lambda_i^B, \dots, \lambda_j^B, \dots) = S_n(\lambda_1^C, \dots, \lambda_n^C | \dots, \lambda_j^B, \dots, \lambda_i^B, \dots) \, . \tag{3.34}
$$

Proof. The proof follows directly from the commutation rules (3.21). Alternatively, one can demonstrate it from the analysis of (3.29) as performed in [22]. □

Solution. The same discussion of Section 3.1 concerning the resolution of the functional equation (3.14) is also valid for the set of equations (3.29). For instance, we can readily see that each equation in (3.29) is scale invariant and linear. Nevertheless, there is one main difference concerning (3.29) which is the fact that here we have two equations instead of only one. This might suggest that one of the equations is redundant but the direct inspection of our system of equations reveals that this is not the case. In fact, the scalar products we are interested consist of certain polynomials and the use of a polynomial ansatz for solving (3.29) shows that only one equation is not able to fix all the coefficients. On the other hand, the simultaneous resolution of both equations indeed fix the coefficients up to an overall multiplicative factor.

Although the process of solving the system of equations (3.29) is more involving than that for the single equation (3.14), the same methodology still applies. The solution of (3.29) was firstly obtained in [22] and here we restrict ourselves to presenting only the final expression. The scalar product S_n is then given by,

$$S_n(X^{1,n}|Y^{1,n}) = \oint \dots \oint \prod_{i=1}^{n} \frac{\mathrm{d}w_i}{2\mathrm{i}\pi} \frac{\mathrm{d}\bar{w}_i}{2\mathrm{i}\pi} \frac{H(w_1,\dots,w_n|\bar{w}_1,\dots,\bar{w}_n)}{\prod_{i,j=1}^{n} b(w_i-\lambda_j^C) b(\bar{w}_i-\lambda_j^B)} , \tag{3.35}$$

where the function H explicitly reads,

$$H(w_1,\dots,w_n|\bar{w}_1,\dots,\bar{w}_n) = \\ (-1)^{Ln+\frac{n(n+1)}{2}} c^{2n} \frac{\prod_{j>i}^{n} b(w_i-w_j)^2 b(\bar{w}_i-\bar{w}_j)^2 a(w_j-\mu_i) a(\bar{w}_j-\mu_i)}{\prod_{i=1}^{n} b(w_i-\mu_i) b(\bar{w}_i-\mu_i)} \prod_{i=1}^{n} R_i^{-1} \Lambda_i , \tag{3.36}$$

with functions R_i and Λ_i given by

$$\begin{aligned} R_i &= \prod_{k=i}^{n} \frac{a(w_k-\mu_i)}{b(w_k-\mu_i)} - \prod_{k=i}^{n} \frac{a(\bar{w}_k-\mu_i)}{b(\bar{w}_k-\mu_i)} \\ \Lambda_i &= \prod_{k=i}^{L} a(\bar{w}_i-\mu_k) b(\mu_k-w_i) \prod_{k=i+1}^{n} \frac{a(w_i-w_k)}{b(w_i-w_k)} \frac{a(\bar{w}_k-\bar{w}_i)}{b(\bar{w}_k-\bar{w}_i)} \\ &\quad - \prod_{k=i}^{L} a(w_i-\mu_k) b(\mu_k-\bar{w}_i) \prod_{k=i+1}^{n} \frac{a(w_k-w_i)}{b(w_k-w_i)} \frac{a(\bar{w}_i-\bar{w}_k)}{b(\bar{w}_i-\bar{w}_k)} . \end{aligned} \tag{3.37}$$

Formulae (3.35) is commonly denominated off-shell scalar product as it is valid for arbitrary complex parameters λ_i^B and λ_i^C. In its turn, when the variables λ_i^B are constrained by Bethe ansatz equations, see [37] for instance, the function S_n receives the name on-shell scalar product and the analysis of (3.29) in that case has also been performed in [22].

4. Partial differential equations

The functional equations (3.14) and (3.29) contain a very rich structure which is not apparent at first sight. In order to illustrate how these hidden structures emerge let us consider Eq. (3.14) in the standard six-vertex model limit. In that case we have $\mathcal{Z}_\theta \rightarrow Z$ and are left with the following equation,

$$\bar{M}_0 \, Z(X^{1,L}) + \sum_{i=1}^{L} \bar{N}_i \, Z(X_i^{0,L}) = 0 , \tag{4.1}$$

with coefficients $\bar{M}_0$ and $\bar{N}_i$ given by

$$\begin{aligned}
\bar{M}_0 &= \prod_{j=1}^{L} b(\lambda_0 - \mu_j) - \prod_{j=1}^{L} a(\lambda_0 - \mu_j) \prod_{j=1}^{L} \frac{a(\lambda_j - \lambda_0)}{b(\lambda_j - \lambda_0)} \\
\bar{N}_i &= \frac{c(\lambda_i - \lambda_0)}{b(\lambda_i - \lambda_0)} \prod_{j=1}^{L} a(\lambda_i - \mu_j) \prod_{\substack{j=1 \\ j \neq i}}^{L} \frac{a(\lambda_j - \lambda_i)}{b(\lambda_j - \lambda_i)} \, .
\end{aligned} \tag{4.2}$$

In what follows we intend to demonstrate that (4.1) encodes a family of partial differential equations and for that we need to introduce some extra definitions and lemmas.

Definition 3. *Let f be a complex valued function $f(z) \in \mathbb{C}[z]$ and $z = (z_1, \dots, z_n) \in \mathbb{C}^n$. Then for $\alpha \notin \{1, 2, \dots, n\}$ we define the operator D_i^α as*

$$D_i^\alpha : \qquad f(z_1, \dots, z_i, \dots, z_n) \mapsto f(z_1, \dots, z_\alpha, \dots, z_n) \, . \tag{4.3}$$

The operator D_i^α essentially replaces the variable z_i by z_α. It is worth mentioning that D_i^α had been previously introduced in [19].

Lemma 3 (Differential realization)**.** *The operator D_i^α admits the realization*

$$D_i^\alpha = \sum_{k=0}^{m} \frac{(z_\alpha - z_i)^k}{k!} \frac{\partial^k}{\partial z_i^k} \tag{4.4}$$

when its action is restricted to the ring of multivariate polynomials of degree m.

Proof. Let $f = f(z_1, z_2, \dots, z_n)$ and $\mathbb{K}^m[z_1, z_2, \dots, z_n]$ be the ring of polynomials in $z_1, \dots, z_n$ with degree m. The ring $\mathbb{K}^m[z_1, z_2, \dots, z_n]$ shall be simply denoted as $\mathbb{K}^m[z]$ and the condition $f \in \mathbb{K}^m[z]$ implies

$$\frac{\partial^k f}{\partial z_i^k} = 0 \qquad \text{if} \quad k > m \, . \tag{4.5}$$

Next we consider the Taylor expansion of f in the variable z_i around the point z_α. Due to (4.5) the expansion is truncated and convergent. Thus we have,

$$\begin{aligned}
f &= f(\dots, z_{i-1}, z_\alpha, z_{i+1}, \dots) + \left.\frac{\partial f}{\partial z_i}\right|_{i=\alpha} (z_i - z_\alpha) \\
&+ \frac{1}{2} \left.\frac{\partial^2 f}{\partial z_i^2}\right|_{i=\alpha} (z_i - z_\alpha)^2 + \dots + \frac{1}{m!} \left.\frac{\partial^m f}{\partial z_i^m}\right|_{i=\alpha} (z_i - z_\alpha)^m \, .
\end{aligned} \tag{4.6}$$

The expression (4.6) holds for indexes $i \in \{1, 2, \dots, n\}$, and as long as $\alpha \notin \{1, 2, \dots, n\}$ we can write

$$\left.\frac{\partial^k f}{\partial z_i^k}\right|_{i=\alpha} = \frac{\partial^k}{\partial z_\alpha^k} f(\dots, z_{i-1}, z_\alpha, z_{i+1}, \dots) \, . \tag{4.7}$$

In this way formulas (4.6) and (4.7) yields the relation

$$f(\dots, z_{i-1}, z_i, z_{i+1}, \dots) = \left[\sum_{k=0}^{m} \frac{(z_i - z_\alpha)^k}{k!} \frac{\partial^k}{\partial z_\alpha^k} \right] f(\dots, z_{i-1}, z_\alpha, z_{i+1}, \dots) \, . \tag{4.8}$$

The term inside the brackets in (4.8) performs the operation (4.3) and we thus obtain the differential realization (4.4). □

Next we notice the functional equation (4.1) can be written in terms of operators D_i^{α}. For that we only need to consider $n = L$ and $z_i = \lambda_i$. By doing so (4.1) becomes $\mathfrak{L}(\lambda_0) Z(X^{1,L}) = 0$ where

$$\mathfrak{L}(\lambda_0) = \bar{M}_0 + \sum_{i=1}^{L} \bar{N}_i \, D_i^0 \, . \tag{4.9}$$

Some remarks are required at this stage. For instance, although the functional equation (4.1) depends on the set of variables $X^{0,L}$, the use of Definition 3 localizes the whole dependence with the variable λ_0 in the operator $\mathfrak{L}$. It is also important to stress here that we can not immediately use the differential realization (4.4) in (4.9) since it is valid only for functions in $\mathbb{K}^m[z]$. Nevertheless, in what follows we shall discuss how (4.4) can be adjusted for $Z(X^{1,L})$.

Lemma 4 (Polynomial structure). *In terms of variables $x_i = e^{2\lambda_i}$ the partition function $Z(X^{1,L})$ is of the form*

$$Z(X^{1,L}) = \prod_{j=1}^{L} x_j^{\frac{1-L}{2}} \, \bar{Z}(x_1, \dots, x_L) \, , \tag{4.10}$$

where $\bar{Z}(x_1, \dots, x_L)$ is a polynomial of degree $L-1$ in each variable x_i separately.

Proof. A detailed proof can be found in [35] and [18]. □

Lemma 4 is telling us that $Z(X^{1,L})$ consists of a multivariate polynomial up to an overall multiplicative factor when the appropriate variable is considered. As a matter of fact we have $\bar{Z}(x_1, \dots, x_L) \in \mathbb{K}^{L-1}[x]$ and therefore the realization (4.4) can be employed for $\bar{Z}$. Due to that it is convenient to define the functions

$$\check{M}_0 = \bar{M}_0 \prod_{j=1}^{L} x_j^{\frac{1-L}{2}} \qquad \text{and} \qquad \check{N}_i = \bar{N}_i \prod_{\substack{j=0 \\ j \neq i}}^{L} x_j^{\frac{1-L}{2}} \tag{4.11}$$

in such a way that (4.1) becomes $\bar{\mathfrak{L}}(x_0) \bar{Z}(x_1, \dots, x_L) = 0$ with

$$\bar{\mathfrak{L}}(x_0) = \check{M}_0 + \sum_{i=1}^{L} \check{N}_i \, D_i^0 \, . \tag{4.12}$$

Now the formulae (4.4) can be substituted into (4.12) [1] leaving us with the expression

$$\bar{\mathfrak{L}}(x_0) = \sum_{k=0}^{L-2} x_0^k \, \Omega_k \, . \tag{4.13}$$

The coefficients Ω_k are differential operators whilst $\bar{\mathfrak{L}}(x_0)$ is a polynomial of degree $L-2$ in the variable x_0. In this way the equation $\bar{\mathfrak{L}}(x_0) \bar{Z}(x_1, \dots, x_L) = 0$ needs to be independently satisfied for each power in x_0 which leaves us with the following family of partial differential equations,

$$\Omega_k \, \bar{Z}(x_1, \dots, x_L) = 0 \qquad 0 \leq k \leq L-2 \, . \tag{4.14}$$

Although the explicit form of the operators Ω_k can be written down for any length L, it is usually given by cumbersome expressions for most of the indexes k. Fortunately the situation

[1] Here we need to set $z_i = x_i$ due to the change of variables discussed in Lemma 4.

is different for $k = L-2$ and the leading term coefficient Ω_{L-2} exhibits a compact expression given by

$$\Omega_{L-2} = \sum_{i=1}^{L} \bar{a}(x_i, y_i) - \frac{q^{2(1-L)}}{(L-1)!} \sum_{i=1}^{L} \prod_{j=1}^{L} \bar{a}(x_i, y_j) \prod_{\substack{j=1 \\ j \neq i}}^{L} \frac{\bar{a}(x_j, x_i)}{\bar{b}(x_j, x_i)} \frac{\partial^{L-1}}{\partial x_i^{L-1}} . \quad (4.15)$$

The expression (4.15) takes into account the further conventions $q = e^{\gamma}$, $y_i = e^{2\mu_i}$ and the remaining functions are then defined as $\bar{a}(x,y) = xq^2 - y$ and $\bar{b}(x,y) = x - y$.

From (4.15) we can see that the operator Ω_{L-2} exhibits some very interesting characteristics. For instance, it naturally decomposes into two kinds of terms and it is tempting to interpret it as the hamiltonian of a many-body system. Although Ω_{L-2} contains higher order derivatives, the first term in the RHS of (4.15) could be thought of as 'potential energy' while the second term can be regarded as 'kinetic energy'. The most obvious problem with this interpretation is that the interaction factor $\prod_{j \neq i}^{L} \frac{\bar{a}(x_j,x_i)}{\bar{b}(x_j,x_i)}$ appears in the 'kinetic energy' term and it is not clear if a change of variables could have this issue properly fixed. Nevertheless, taking into account this analogy it is sensible to consider the eigenvalue problem for the operator (4.15), i.e. $\Omega_{L-2}\Psi = \Lambda\Psi$. In this way the partition function $\bar{Z}$ can be regarded as the null eigenvalue wave function associated with Ω_{L-2}.

5. Concluding remarks

In this article we have described a mechanism allowing to extract functional equations satisfied by certain partition functions of two-dimensional lattice models directly from the Yang-Baxter algebra. More precisely, we have applied this method for the elliptic Eight-Vertex-SOS model with domain wall boundaries [21] and for scalar products of Bethe vectors [22]. For those systems we have obtained functional relations satisfied by their partition functions whose solution are then given in terms of multiple contour integrals.

The class of functional equations we describe here share some similarities, as far as their structure is concerned, with the classical Knizhnik-Zamolodchikov equation [14] as discussed in Section 3.1. This similarity seems to extend to their solutions as multiple contour integrals are also convenient to describe solutions of KZ equations [47].

Although classical KZ equations consist of a system of partial differential equations, whilst here we are dealing with functional equations, in Section 4 we have also demonstrated that there is a family of partial differential equations underlying our functional relations originated from the Yang-Baxter algebra. Interestingly, one member of this family exhibits a structure which resembles that of a generalized Schrödinger equation for a quantum many-body hamiltonian.

Concerning this algebraic-functional approach, it is fair to say that this method is still under development as there are still many opened questions. For instance, motivated by the similarities shared with the classical KZ equation, one could ask if there is an analogous of the whole theory of KZ equations [47, 49] for the functional equations presented here. Moreover, as far as the computation of physical quantities are concerned, one important problem is the evaluation of the model free-energy per site in the thermodynamical limit from the multiple contour integrals we have obtained. This analysis would not only give us access to the model physical properties but also help us to understand the influence of boundary conditions for vertex and SOS models [50, 51].

Acknowledgments

The author is supported by the Netherlands Organization for Scientific Research (NWO) under the VICI grant 680-47-602. The work of W. Galleas is also part of the ERC Advanced grant research programme No. 246974, *"Supersymmetry: a window to non-perturbative physics"*.

References

[1] Onsager, L 1944 *Phys. Rev.* **65** 117–149
[2] Ma S 2000 *Modern Theory of Critical Phenomena* Advanced book classics (Perseus)
[3] Bethe H 1931 *Zeitschrift für Physik* 225–226
[4] Faddeev, L D and Takhtajan, L A 1981 *Phys. Lett. A* **85** 375–377
[5] Zamolodchikov, A B 1990 *Nucl. Phys. B* **342** 695–720
[6] Minahan J A and Zarembo K 2003 *JHEP*
[7] Brak R and Galleas W 2013 *Lett. Math. Phys.* **103** 1261–1272
[8] Kramers H A and Wannier G H 1941 *Phys. Rev.* **60** 252
[9] Kramers H A and Wannier G H 1941 *Phys. Rev.* **60** 263
[10] Baxter R J 1971 *Phys. Rev. Lett.* **26** 832
[11] Sklyanin E K, Takhtadzhyan L A and Faddeev L D 1979 *Theor. Math. Phys.* **40** 688
[12] Takhtadzhyan L A and Faddeev L D 1979 *Russ. Math. Surv.* **34** 11
[13] Chari V and Pressley A 1995 *A Guide to Quantum Groups* (Cambridge University Press)
[14] Knizhnik, V G and Zamolodchikov, A B 1984 *Nucl. Phys. B* **247** 83–103
[15] Babujian, H M 1993 *J. Phys. A: Math. and Gen.* **26** 6981–6990
[16] Babujian, H M and Flume, R 1994 *Mod. Phys. Lett. A* **9** 2029–2039
[17] Galleas W 2008 *Nucl. Phys. B* **790** 524–542
[18] Galleas W 2010 *J. Stat. Mech.* P06008
[19] Galleas W 2011 *J. Stat. Mech.* P01013
[20] Galleas W 2012 *Nucl. Phys. B* **858** 117–141 (*Preprint* `math-ph/1111.6683`)
[21] Galleas W 2013 *Nucl. Phys. B* **867** 855–871
[22] Galleas, W (*Preprint* `arXiv: 1211.7342`)
[23] Baxter R J 2007 *Exactly Solved Models in Statistical Mechanics* (Mineola, New York: Dover Publications, Inc.) ISBN 978-0-486-46271-4; 0-486-46271-4
[24] Smirnov S 2001 *Comptes rendus de L'academie des sciences serie I - Mathematique* **333** 239–244
[25] Smirnov S 2001 *Pre-print* (*Preprint* `arXiv: math.PR/1211.3968`)
[26] Cardy J L 1992 *J. Phys. A - Math. and Gen.* **25** L201–L206
[27] Felder G 1995 *Proceedings of the International Congress of Mathematicians* **1** 1247
[28] Felder G 1996 *Nucl. Phys. B* **480** 485
[29] Whittaker E T and Watson G N 1927 *A Course of Modern Analysis* 4th ed (Cambridge University Press)
[30] D'Ariano G M, Montorsi A and Rasetti M G 1985 *Integrable Systems in Statistical Mechanics* (World Scientific)
[31] Rosengren H 2009 *Adv. Appl. Math.* **43** 137
[32] Pakuliak S, Rubtsov V and Silantyev A 2008 *J. Phys. A* **41** 295204
[33] Yang W L and Zhang Y Z 2009 *J. Math. Phys.* **50** 083518
[34] Razumov A G and Stroganov Y G 2009 *Theor. Math. Phys.* **161** 1325
[35] Korepin V E 1982 *Commun. Math. Phys.* **86**(3) 391–418
[36] de Gier J, Galleas W and Sorrell M 2011 (*Preprint* `hep-th/1111.3712`)
[37] Korepin V E, Bogoliubov N M and Izergin A G 1993 *Quantum inverse scattering method and correlation functions* (Cambridge University Press)
[38] Aczél, J 1984 *Functional equations: History, Applications and Theory* (D. Reidel Publishing Company)
[39] Bazhanov V V, Lukyanov S L and Zamolodchikov A B 1997 *Comm. Math. Phys.* **190** 247–278
[40] Baxter R J 1972 *Ann. Phys.* **70** 193
[41] Stroganov Y G 1982 *Unpublished thesis*
[42] Reshetikhin N Y 1987 *Lett. Math. Phys.* **14** 235–246
[43] Kuniba A, Nakanishi T and Suzuki J 1994 *Int. J. Mod. Phys. A* **9** 5215–5266
[44] Kuniba A, Nakanishi T and Suzuki J 2011 *J. Phys. A - Math. and Theor.* **44**
[45] Frenkel I B and Reshetikhin N Y 1992 *Comm. Math. Phys.* **146** 1–60
[46] Smirnov, F A 1992 *Form factors in completely integrable models of quantum field theory* (Advanced Series in Mathematical Physics. 14. Singapore: World Scientific,. xi, 208 p.)

[47] Varchenko, A N 2003 *Special Functions, KZ Type Equations, and Representation Theory* (American Mathematical Society)
[48] Weber H 1891 *Elliptische Functionen und algebraische Zahlen* 4th ed (Braunschweig: Friedrich Vieweg und Sohn)
[49] Etingof, P I and Frenkel, I B and Kirillov, A A 1998 *Lectures on Representation Theory and Knizhnik-Zamolodchikov Equations* (American Mathematical Society)
[50] Rosengren H 2011 *Adv. Appl. Math.* **46** 481
[51] Korepin V E and Zinn-Justin P 2000 *J. Phys. A: Math. Gen.* **33** 7053

Discretization of four types of Weyl group orbit functions

Jiří Hrivnák[1]

[1] Department of Physics, Faculty of Nuclear Sciences and Physical Engineering, Czech Technical University in Prague, Břehová 7, 115 19 Prague 1, Czech Republic

E-mail: jiri.hrivnak@fjfi.cvut.cz

Abstract. The discrete Fourier calculus of the four families of special functions, called $C-$, $S-$, S^s- and S^l-functions, is summarized. Functions from each of the four families of special functions are discretely orthogonal over a certain finite set of points. The generalizations of discrete cosine and sine transforms of one variable — the discrete S^s- and S^l-transforms of the group F_4 — are considered in detail required for their exploitation in discrete Fourier spectral methods. The continuous interpolations, induced by the discrete expansions, are presented.

1. Introduction

The standard discrete Fourier analysis of one real variable is known to be a very valuable tool in mathematics, physics and elsewhere [3, 13, 14]. Various approaches can be followed to formulate this analysis in higher dimensions. Here we summarize the approach based on Weyl groups – the one dimensional case as well as its straightforward Cartesian product generalization are then a special case of this calculus. Four types of the multidimensional generalizations of cosine and sine functions, which are available for the root systems with two different lengths of roots and form the core of this approach, appeared in various stages of development of the Lie theory [2, 4, 9, 10, 11]. Their discrete Fourier calculus, formulated explicitly and in full generality, has been developed only recently [5, 6, 7]. Summarizing this calculus and adding a specific example of the group F_4 is the main purpose of this paper.

Consider the Lie algebra of the compact simply connected simple Lie group G of rank n. The set of simple roots of G is denoted by $\Delta = \{\alpha_1, \dots, \alpha_n\}$ and forms a basis of the Euclidean vector space $\mathbb{R}^n$ equipped with the standard scalar product $\langle\,,\,\rangle$. Only such simple algebras which have two different lengths of roots are considered, namely B_n, $n \geq 3$, C_n, $n \geq 2$, G_2 and F_4. For these algebras the set of simple roots consists of short simple roots Δ_s and long simple roots Δ_l. The following quantities can be deduced from the entire root system Δ by standard methods: the Cartan matrix C, the highest root $\xi \equiv -\alpha_0 = m_1\alpha_1 + \dots + m_n\alpha_n$, the root lattice $Q = \mathbb{Z}\alpha_1 + \dots + \mathbb{Z}\alpha_n$, the $\mathbb{Z}$-dual lattice $P^\vee = \mathbb{Z}\omega_1^\vee + \dots + \mathbb{Z}\omega_n^\vee$, the dual root lattice $Q^\vee = \mathbb{Z}\alpha_1^\vee + \dots + \mathbb{Z}\alpha_n^\vee$, where $\alpha_i^\vee = 2\alpha_i/\langle\alpha_i, \alpha_i\rangle$ and the weight lattice $P = \mathbb{Z}\omega_1 + \dots + \mathbb{Z}\omega_n$. The set of vectors $\{\alpha_1^\vee, \dots, \alpha_n^\vee\}$ also forms a root system called the dual root system $\Delta^\vee$ of G. This dual root system determines the highest dual root $\eta \equiv -\alpha_0^\vee = m_1^\vee\alpha_1^\vee + \dots + m_n^\vee\alpha_n^\vee$.

The reflections r_α, $\alpha \in \Delta$ are given as reflections in $(n-1)$-dimensional 'mirrors' orthogonal to the simple roots α and intersecting at the origin. Similarly are defined the reflections r_ξ, r_η of the highest root ξ and the highest dual root r_η. The Weyl group W is generated by n

reflections r_α, $\alpha \in \Delta$ and the affine Weyl group W^{aff} of G is defined as a semidirect product $W^{\mathrm{aff}} = Q^\vee \rtimes W$. The affine Weyl group is generated by n reflections r_α and the affine reflection r_0, which composes of the reflection r_ξ and the shift by $\frac{2\xi}{\langle \xi, \xi \rangle}$. The fundamental region $F \subset \mathbb{R}^n$ of W^{aff} consists of precisely one point from each W^{aff}–orbit. It can be chosen as the n-simplex [8] of the form $F = \left\{0, \frac{\omega_1^\vee}{m_1}, \ldots, \frac{\omega_n^\vee}{m_n}\right\}_\kappa$.

The formulation of the discrete Fourier calculus requires careful analysis of the boundaries of the fundamental domain F – for this purpose short and long boundaries are introduced. The points of F which are stabilized by some r_α, $\alpha \in \Delta_s$ are collected in the short boundary H^s of F. The points of F which are stabilized by some r_α, $\alpha \in \Delta_l$ or r_0 are collected in the long boundary H^l of F. The short and the long fundamental domains F^s and F^l are defined in [7] as

$$F^s = F \setminus H^s, \quad F^l = F \setminus H^l.$$

The dual affine Weyl group $\widehat{W}^{\mathrm{aff}}$ of G is defined similarly as a semidirect product $\widehat{W}^{\mathrm{aff}} = Q \rtimes W$. The dual affine Weyl group is generated by n reflections r_α and the affine reflection $r_0^\vee$, which composes of the reflection r_η and the shift by $\frac{2\eta}{\langle \eta, \eta \rangle}$. The simplex $F^\vee \subset \mathbb{R}^n$ of the form $F^\vee = \left\{0, \frac{\omega_1}{m_1^\vee}, \ldots, \frac{\omega_n}{m_n^\vee}\right\}_\kappa$ forms the fundamental region of $\widehat{W}^{\mathrm{aff}}$. The points of $F^\vee$ which are stabilized by some r_α, $\alpha \in \Delta_s$ or $r_0^\vee$ are collected in the short dual boundary $H^{s\vee}$ of $F^\vee$. The points of $F^\vee$ which are stabilized by some r_α, $\alpha \in \Delta_l$ are collected in the long dual boundary $H^{l\vee}$ of $F^\vee$. The short and the long dual fundamental domains $F^{s\vee}$ and $F^{l\vee}$ are defined [7] as

$$F^{s\vee} = F^\vee \setminus H^{s\vee}, \quad F^{l\vee} = F^\vee \setminus H^{l\vee}.$$

2. Discrete Fourier calculus of $C-$, $S-$, S^s- and S^l-functions

Considering a weight $b \in P$, the normalized $C-$functions and the $S-$functions are given by

$$\Phi_b(a) = \sum_{w \in W} e^{2\pi i \langle wb, a \rangle}, \quad \varphi_b(a) = \sum_{w \in W} (\det w) e^{2\pi i \langle wb, a \rangle}.$$

The family of $C-$functions and the family of $S-$function could be taken as the generalization of cosine and sine functions, respectively. Two other families, which carry mixed properties of sines and cosines, can also be singled out from the continuum of Heckman-Opdam polynomials [4]. Two 'sign' homomorphisms $\sigma^s, \sigma^l : W \to \{\pm 1\}$ are defined [11] by their values on the generating reflections r_α, $\alpha \in \Delta$ of W,

$$\sigma^s(r_\alpha) = \begin{cases} 1, & \alpha \in \Delta_l \\ -1, & \alpha \in \Delta_s \end{cases}$$

$$\sigma^l(r_\alpha) = \begin{cases} 1, & \alpha \in \Delta_s \\ -1, & \alpha \in \Delta_l. \end{cases}$$

These sign homomorphisms σ^s and σ^l determine the S^s-functions and the S^l-functions

$$\varphi_b^s(x) = \sum_{w \in W} \sigma^s(w)\, e^{2\pi i \langle wb, x \rangle}, \quad \varphi_b^l(x) = \sum_{w \in W} \sigma^l(w)\, e^{2\pi i \langle wb, x \rangle}.$$

Contours plots of some lowest S^s- and S^l-functions of the group G_2 are depicted in Figures 1 and 2.

For the detailed review of $C-$functions see [9]; the $S-$functions, which are well-known from the character functions, are reviewed in [10]. The character functions, which appear in the

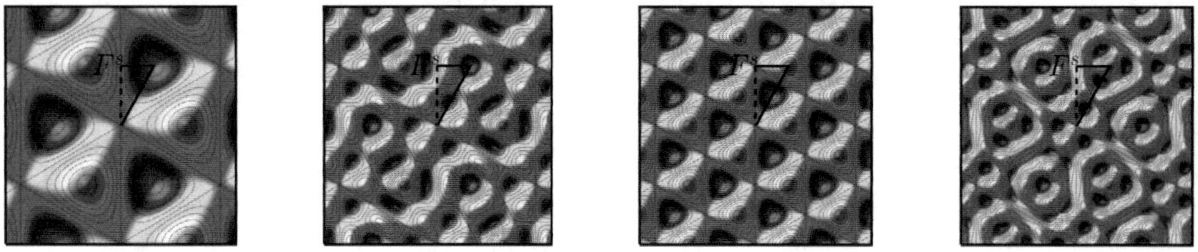

$\varphi^s_{(0,1)}$ $\varphi^s_{(1,1)}$ $\varphi^s_{(0,2)}$ $\varphi^s_{(1,2)}$

Figure 1. The contour plots of S^s- functions of G_2. The black triangle, with the dashed border H^s excluded, forms the fundamental domain F^s.

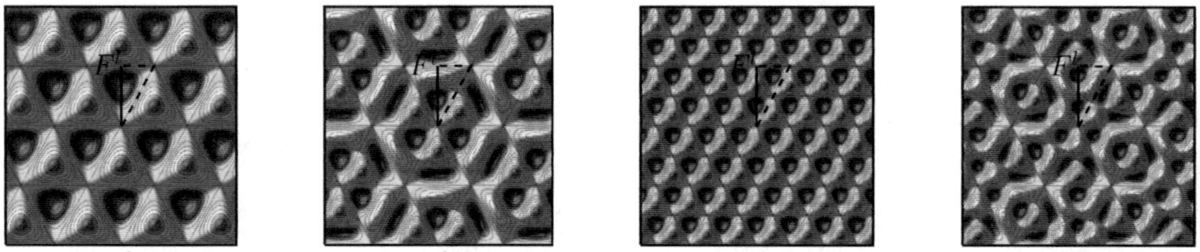

$\varphi^l_{(1,0)}$ $\varphi^l_{(1,1)}$ $\varphi^l_{(2,0)}$ $\varphi^l_{(1,2)}$

Figure 2. The contour plots of S^l- functions of G_2. The black triangle, with the dashed border H^l excluded, forms the fundamental domain F^l.

Weyl character formula, are ratios of the general $S-$functions and the lowest possible one. Similar ratios for the S^s- and S^l-functions lead to short and long character functions. Note that $C-$functions together with the three character functions are special cases of the Jacobi polynomials generalized to root systems [4].

Next, building the discrete Fourier calculus, we consider the values of $C-$, $S-$, S^s- and S^l-functions on certain finite grids inside the regions F, F°, F^s and F^l. Given an arbitrary $M \in \mathbb{N}$, the quotient group $\frac{1}{M}P^\vee/Q^\vee$ forms a W-invariant finite group. We define the finite grids F_M, $\widetilde{F}_M$, F^s_M and F^l_M as such elements from $\frac{1}{M}P^\vee/Q^\vee$ which have representative points in F, F°, F^s and F^l, respectively:

$$F_M = \frac{1}{M}P^\vee/Q^\vee \cap F, \quad \widetilde{F}_M = \frac{1}{M}P^\vee/Q^\vee \cap F^\circ,$$
$$F^s_M = \frac{1}{M}P^\vee/Q^\vee \cap F^s, \quad F^l_M = \frac{1}{M}P^\vee/Q^\vee \cap F^l.$$

The group $\frac{1}{6}P^\vee/Q^\vee$ and the grids F_6, $\widetilde{F}_6$, F^s_6 and F^l_6 of G_2, together with the roots $\alpha, \alpha^\vee$ and the weights $\omega, \omega^\vee$, are depicted in Figure 3. The counting formulas for the numbers of points in the grids F_M, $\widetilde{F}_M$, F^s_M and F^l_M are for all cases derived in [5, 7].

When $C-$, $S-$, S^s- and S^l-functions are sampled on the grids F_M, $\widetilde{F}_M$, F^s_M and F^l_M they can be labeled only by the labels from the finite subsets of P. As weights in the subsets Λ_M, $\widetilde{\Lambda}_M$, Λ^s_M and Λ^l_M, only the following elements of the W-invariant finite group P/MQ are taken

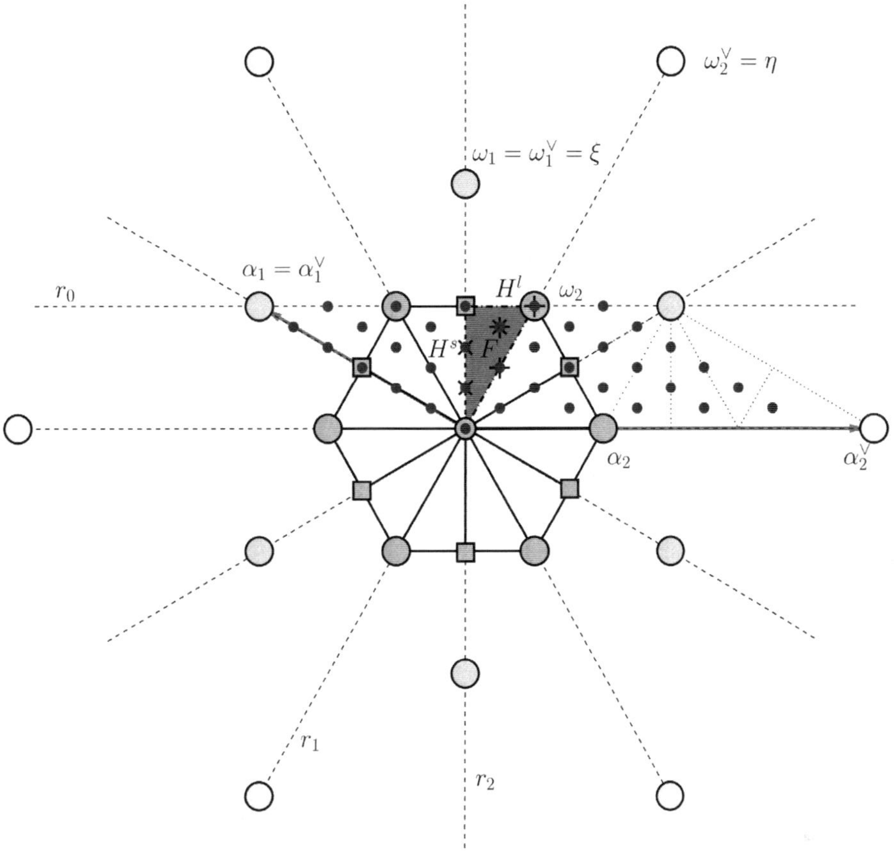

Figure 3. The fundamental regions F, F^s and F^l of G_2. The fundamental domain F is depicted as the red triangle containing its borders H^s and H^l, which are drawn as the thick dashed line and dot-and-dashed lines, respectively. The coset representatives of $\frac{1}{6}P^\vee/Q^\vee$ are depicted as 36 black dots. The three representatives belonging to F_6^s and F_6^l are crossed with '+' and '×', respectively. The dashed lines represent 'mirrors' r_0, r_1 and r_2.

to label the functions [5, 7]:

$$\Lambda_M = MF^\vee \cap P/MQ, \quad \widetilde{\Lambda}_M = M(F^\vee)^\circ \cap P/MQ,$$
$$\Lambda_M^s = MF^{s\vee} \cap P/MQ, \quad \Lambda_M^l = MF^{l\vee} \cap P/MQ.$$

Moreover, it is shown in [5, 7] that the numbers of weights in the sets Λ_M, $\widetilde{\Lambda}_M$, Λ_M^s and Λ_M^l coincide with the numbers of points in the sets F_M, $\widetilde{F}_M$, F_M^s and F_M^l for all cases of G, i.e.

$$|\Lambda_M| = |F_M|, \quad |\widetilde{\Lambda}_M| = |\widetilde{F}_M|,$$
$$|\Lambda_M^s| = |F_M^s|, \quad |\Lambda_M^l| = |F_M^l|.$$

The order of an W-orbit of a point $x \in \frac{1}{M}P^\vee/Q^\vee$ is denoted by $\varepsilon(x)$; the order of a W-stabilizer of a weight $\lambda \in P/MQ$ is denoted by $h_\lambda^\vee$. The discrete orthogonality of $C-$ and $S-$functions is proven in [5] and extended to the discrete orthogonality of S^s- and S^l-functions in [7]. These results from [5, 7] can be summarized as follows.

Theorem 2.1. *For $\lambda, \lambda' \in \Lambda_M$ it holds that*

$$\sum_{x \in F_M} \varepsilon(x) \Phi_\lambda(x) \overline{\Phi_{\lambda'}(x)} = c\,|W|\, M^n h_\lambda^\vee \delta_{\lambda,\lambda'}$$

and for $\lambda, \lambda' \in \widetilde{\Lambda}_M$ it holds that

$$\sum_{x \in \widetilde{F}_M} \varphi_\lambda(x) \overline{\varphi_{\lambda'}(x)} = c M^n \delta_{\lambda,\lambda'}.$$

For $\lambda, \lambda' \in \Lambda_M^s$ it holds that

$$\sum_{x \in F_M^s} \varepsilon(x) \varphi_\lambda^s(x) \overline{\varphi_{\lambda'}^s(x)} = c\,|W|\, M^n h_\lambda^\vee \delta_{\lambda,\lambda'} \tag{1}$$

and for $\lambda, \lambda' \in \Lambda_M^l$ it holds that

$$\sum_{x \in F_M^l} \varepsilon(x) \varphi_\lambda^l(x) \overline{\varphi_{\lambda'}^l(x)} = c\,|W|\, M^n h_\lambda^\vee \delta_{\lambda,\lambda'}, \tag{2}$$

where c is the determinant of the Cartan matrix, $|W|$ is the number of elements of the Weyl group W and n is the rank of G.

3. The group F_4

The highest root ξ and the highest dual root η of the exceptional group F_4 are determined by the formulas

$$\xi = 2\alpha_1 + 3\alpha_2 + 4\alpha_3 + 2\alpha_4, \quad \eta = 2\alpha_1 + 3\alpha_2 + 4\alpha_3 + 2\alpha_4.$$

The Weyl group of F_4 has 1152 elements [1, 8]. Discretization of the $C-$ and $S-$ functions of F_4 is detailed in [6]. Here we focus on discretization of the S^s- and S^l-functions of F_4. Taking the variables $u_0^s, u_1^s, u_2^s \in \mathbb{Z}^{\geq 0}$ and $u_3^s, u_4^s \in \mathbb{N}$, the grid F_M^s is given by

$$F_M^s = \left\{ \frac{u_1^s}{M}\omega_1^\vee + \cdots + \frac{u_4^s}{M}\omega_4^\vee \,\middle|\, u_0^s + 2u_1^s + 3u_2^s + 4u_3^s + 2u_4^s = M \right\}$$

and for the numbers of its elements it holds [7]:

$$\begin{aligned}
|F_{12k}^s| &= 18k^4 - k^2, \\
|F_{12k+1}^s| &= 18k^4 + 6k^3 - \frac{5}{2}k^2 - \frac{1}{2}k, \\
|F_{12k+2}^s| &= 18k^4 + 12k^3 + 2k^2, \\
|F_{12k+3}^s| &= 18k^4 + 18k^3 + \frac{7}{2}k^2 - \frac{1}{2}k \\
|F_{12k+4}^s| &= 18k^4 + 24k^3 + 11k^2 + 2k, \\
|F_{12k+5}^s| &= 18k^4 + 30k^3 + \frac{31}{2}k^2 + \frac{5}{2}k
\end{aligned}$$

$$
\begin{aligned}
|F^s_{12k+6}| &= 18k^4 + 36k^3 + 26k^2 + 8k + 1,\\
|F^s_{12k+7}| &= 18k^4 + 42k^3 + \frac{67}{2}k^2 + \frac{21}{2}k + 1\\
|F^s_{12k+8}| &= 18k^4 + 48k^3 + 47k^2 + 20k + 3,\\
|F^s_{12k+9}| &= 18k^4 + 54k^3 + \frac{115}{2}k^2 + \frac{51}{2}k + 4\\
|F^s_{12k+10}| &= 18k^4 + 60k^3 + 74k^2 + 40k + 8,\\
|F^s_{12k+11}| &= 18k^4 + 66k^3 + \frac{175}{2}k^2 + \frac{99}{2}k + 10.
\end{aligned}
$$

Taking the variables $u^l_3, u^l_4 \in \mathbb{Z}^{\geq 0}$ and $u^l_0, u^l_1, u^l_2 \in \mathbb{N}$, the grid F^l_M is given by

$$F^l_M = \left\{ \frac{u^l_1}{M}\omega^\vee_1 + \cdots + \frac{u^l_4}{M}\omega^\vee_4 \;\middle|\; u^l_0 + 2u^l_1 + 3u^l_2 + 4u^l_3 + 2u^l_4 = M \right\}$$

and for the numbers of its elements it holds $|F^l_M| = |F^s_M|$.

Taking the variables $t^s_1, t^s_2 \in \mathbb{Z}^{\geq 0}$ and $t^s_0, t^s_3, t^s_4 \in \mathbb{N}$, the grid Λ^s_M is determined by

$$\Lambda^s_M = \left\{ t^s_1\omega_1 + \cdots + t^s_4\omega_4 \;\middle|\; t^s_0 + 2t^s_1 + 4t^s_2 + 3t^s_3 + 2t^s_4 = M \right\}.$$

Taking the variables $t^l_0, t^l_3, t^l_4 \in \mathbb{Z}^{\geq 0}$ and $t^l_1, t^l_2 \in \mathbb{N}$, the grid Λ^l_M is determined by

$$\Lambda^l_M = \left\{ t^l_1\omega_1 + \cdots + t^l_4\omega_4 \;\middle|\; t^l_0 + 2t^l_1 + 4t^l_2 + 3t^l_3 + 2t^l_4 = M \right\}.$$

The discrete orthogonality relations of S^s- and S^l- functions, which hold for any two functions φ^s_λ, $\varphi^s_{\lambda'}$ labeled by $\lambda, \lambda' \in \Lambda^s_M$, and φ^l_λ, $\varphi^l_{\lambda'}$ labeled by $\lambda, \lambda' \in \Lambda^l_M$, are of the form (1) and (2), respectively. The coefficients $\varepsilon(x)$ and $h^\vee_\lambda$ are listed in Table 1 in [6].

4. Discrete S^s- and S^l-transforms of F_4

For a given function sampled on F^s_M or F^l_M of F_4, we define continous interpolating functions I^s_M, I^l_M

$$
\begin{aligned}
I^s_M(x) &= \sum_{\lambda\in\Lambda^s_M} c^s_\lambda \varphi^s_\lambda(x), \quad x \in \mathbb{R}^4\\
I^l_M(x) &= \sum_{\lambda\in\Lambda^l_M} c^l_\lambda \varphi^l_\lambda(x), \quad x \in \mathbb{R}^4.
\end{aligned}
$$

These interpolating functions I^s_M and I^l_M are defined as linear combinations of basis functions $\varphi^s_\lambda, \varphi^l_\lambda$ with expansion coefficients c^s_λ, c^l_λ whose values need to be determined from the conditions

$$
\begin{aligned}
I^s_M(x) &= f(x), \quad x \in F^s_M\\
I^l_M(x) &= f(x), \quad x \in F^l_M.
\end{aligned}
$$

The formulas for calculation of c^s_λ and c^l_λ, which can be also viewed as discrete S^s- and S^l-transforms, are of the explicit form:

$$c^s_\lambda = (1152\, M^4 h^\vee_\lambda)^{-1} \sum_{x\in F^s_M} \varepsilon(x) f(x) \overline{\varphi^s_\lambda(x)} \tag{3}$$

$$c^l_\lambda = (1152\, M^4 h^\vee_\lambda)^{-1} \sum_{x\in F^l_M} \varepsilon(x) f(x) \overline{\varphi^l_\lambda(x)}. \tag{4}$$

5. Concluding remarks

- The products of two $C-$, $S-$, S^s- and S^l-functions with the affine Weyl group of the same type and the same arguments $x \in \mathbb{R}^n$ and any two weights $\lambda, \lambda' \in P$, decompose into the sum of $C-$functions,

$$\Phi_\lambda(x) \cdot \Phi_{\lambda'}(x) = \sum_{w \in W} \Phi_{\lambda + w\lambda'}(x), \quad \varphi_\lambda(x) \cdot \varphi_{\lambda'}(x) = \sum_{w \in W} \det(w) \Phi_{\lambda + w\lambda'}(x)$$

$$\varphi^s_\lambda(x) \cdot \varphi^s_{\lambda'}(x) = \sum_{w \in W} \sigma^s(w) \Phi_{\lambda + w\lambda'}(x), \quad \varphi^l_\lambda(x) \cdot \varphi^l_{\lambda'}(x) = \sum_{w \in W} \sigma^l(w) \Phi_{\lambda + w\lambda'}(x).$$

- To any function $f : F \to \mathbb{C}$ one can assign two functional series $\{I^s_M\}_{M=1}^\infty$, $\{I^l_M\}_{M=1}^\infty$ with the coefficients calculated from (3), (4). Similar functional series can be constructed using $C-$ and $S-$functions [6]. The form of suitable conditions on f, which would be similar to the conditions on one-variable functions in [14] and would guarantee convergence of these functional series, poses an open problem.
- The standard relation between the sine and cosine functions of one variable and Chebyshev polynomials of the first and second kind is exploited in various applications as polynomial interpolation and cubature formulas. The discrete orthogonality of the sine and cosine functions and corresponding discrete cosine transform (DCT) and discrete sine transform (DST) play a crucial role in these methods. Since analogous relations hold for the $C-$, $S-$, S^s- and S^l-functions and corresponding Heckman-Opdam polynomials, a similar variety of methods is to be expected for these polynomials of several variables [12].

Acknowledgments

JH gratefully acknowledges the support of this work by RVO68407700 and by the Ministry of Education of Czech Republic from the project MSM6840770039.

References

[1] A. Bjorner, F. Brenti, *Combinatorics of Coxeter groups,* Graduate Texts in Mathematics, **231** (2005) Springer, New York
[2] N. Bourbaki, *Groupes et algèbres de Lie,* Chapiters IV, V, VI, Hermann, Paris 1968
[3] V. Britanak, P. Yip, K. Rao, *Discrete cosine and sine transforms. General properties, fast algorithms and integer approximations.* Elsevier, Amsterdam, (2007).
[4] G. J. Heckman, E. M. Opdam, *Root systems and hypergeometric functions. I.* Compositio Math. **64** no. 3, (1987), 329–352
[5] J. Hrivnák, J. Patera, *On discretization of tori of compact simple Lie groups,* J. Phys. A: Math. Theor. **42** (2009) 385208, arXiv:0905.2395
[6] J. Hrivnák, J. Patera, *Discretization of Tori of Exceptional Compact Simple Lie Groups* , AIP Conf. Proc. **1191** (2009) 110–115
[7] J. Hrivnák, L. Motlochová, J. Patera, *On discretization of tori of compact simple Lie groups II*, J. Phys. A: Math. Theor. **45** (2012) 255201; arXiv:1206.0240
[8] J. E. Humphreys, *Reflection groups and Coxeter groups*, Cambridge Studies in Advanced Mathematics, **29** (1990) Cambridge University Press, Cambridge
[9] A. U. Klimyk, J. Patera, *Orbit functions,* SIGMA (Symmetry, Integrability and Geometry: Methods and Applications) **2** (2006), 006, math-ph/0601037
[10] A. U. Klimyk, J. Patera, *Antisymmetric orbit functions,* SIGMA (Symmetry, Integrability and Geometry: Methods and Applications) **3** (2007), 023, math-ph/0702040v1
[11] R. V. Moody, L. Motlochová, and J. Patera, *New families of Weyl group orbit functions,* arXiv:1202.4415
[12] R. V. Moody, J. Patera, *Cubature formulae for orthogonal polynomials in terms of elements of finite order of compact simple Lie groups,* Advances of Applied Math. **47** (2011) 509–535
[13] G. Strang, *The discrete cosine transform,* SIAM Review, **41** (1999) 135-147.
[14] P. Tolstov, *Fourier series.* Dover Publications, Inc., New York, 1976

q-Virasoro algebra at root of unity limit and 2d-4d connection

H Itoyama[1,2], T Oota[2] and R Yoshioka[2]

[1] *Department of Mathematics and Physics, Graduate School of Science, Osaka City University*
[2] *Osaka City University Advanced Mathematical Institute (OCAMI)*
3-3-138, Sugimoto, Sumiyoshi-ku, Osaka, 558-8585, Japan

E-mail: itoyama@sci.osaka-cu.ac.jp, toota@sci.osaka-cu.ac.jp, yoshioka@sci.osaka-cu.ac.jp

Abstract. We propose a limiting procedure in which, starting from the q-lifted version (or K-theoretic five dimensional version) of the (W)AGT conjecture to be assumed, the Virasoro/W block is generated in the r-th root of unity limit in q in the $2d$ side, while the same limit automatically generates the projection of the five dimensional instanton partition function onto that on the ALE space $\mathbb{R}^4/\mathbb{Z}_r$. This proceeding is based on arXiv:1308.2068

1. Introduction

Continuing attention has been paid to the correspondence between the two dimensional conformal block [1] and the instanton sum [2, 3] identified as the partition function of the four dimensional $\mathcal{N} = 2$ supersymmetric gauge theory. The both sides of this correspondence [4, 5] have already been intensively studied for more than a few years and a wealth of such examples has been found by now. One of the central tools for our study is the β-deformed matrix model controlling the integral representation of the conformal block [6, 7, 8, 9, 10, 11, 12, 13, 14] and the use [10] of the formula [15, 16] on multiple integrals. This general correspondence, on the other hand, has stayed as conjectures in most of the examples except the few ones [17, 18, 19, 20] and one of the next steps in the developments would be to obtain efficient understanding among these while avoiding making many conjectures.

We regard the correspondence between q-Virasoro [21, 22, 23, 24, 25, 26]/W block versus five dimensional instanton partition function as a parent one. We propose the following procedure [27] on the orbifolded examples of the correspondence [28, 29, 30, 31, 32, 33, 34, 35, 36, 37]:

(i) assume the q-lifted version (or K-theoretic five dimensional version) of the (W)AGT conjecture

(ii) introduce the limiting procedure $q \to \omega$, where ω is the r-th root of unity.

(iii) apply the same limiting procedure to $Z^{5\text{d}}_{\text{inst}}$, which automatically generates the instanton partition function on ALE space.

We emphasize that, through this limiting procedure (and the assumption (i)), the resulting $2d$ conformal block is guaranteed to agree with the corresponding instanton partition function on ALE space. Our procedure is illustrated in Figure 1.

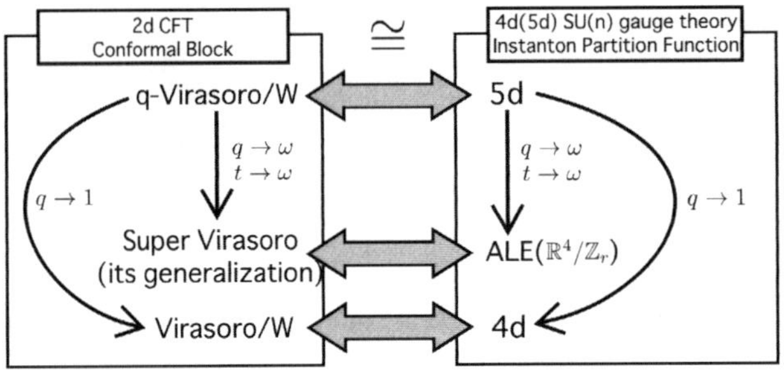

Figure 1. Schematic sketch of our procedure.

2. q-Virasoro algebra and root of unit limit

We introduce the q-Virasoro algebra which contains two parameters q and $t = q^\beta$ and consider the $q \to \omega$ limit.

2.1. q-Virasoro algebra

The defining relation of the q-deformed Virasoro algebra [21] is

$$f(z'/z)\mathcal{T}(z)\mathcal{T}(z') - f(z/z')\mathcal{T}(z')\mathcal{T}(z) = \frac{(1-q)(1-t^{-1})}{(1-p)}\Big[\delta(pz/z') - \delta(p^{-1}z/z')\Big], \tag{2.1}$$

with $p = q/t$. Here $\mathcal{T}(z)$ is the q-Virasoro operator and

$$f(z) = \exp\left(\sum_{n=1}^{\infty} \frac{1}{n}\frac{(1-q^n)(1-t^{-n})}{(1+p^n)} z^n\right), \qquad \delta(z) = \sum_{n\in\mathbb{Z}} z^n. \tag{2.2}$$

Using the q-deformed Heisenberg algebra defined by

$$\begin{aligned} [\alpha_n, \alpha_m] &= -\frac{1}{n}\frac{(1-q^n)(1-t^{-n})}{(1+p^n)}\delta_{n+m,0}, \qquad (n \neq 0), \\ [\alpha_n, Q] &= \delta_{n,0}, \end{aligned} \tag{2.3}$$

the q-Virasoro operator is realized as

$$\mathcal{T}(z) =: \exp\left(\sum_{n\neq 0} \alpha_n z^{-n}\right) : p^{1/2} q^{\sqrt{\beta}\alpha_0} + : \exp\left(-\sum_{n\neq 0} \alpha_n (pz)^{-n}\right) : p^{-1/2} q^{-\sqrt{\beta}\alpha_0}. \tag{2.4}$$

Let us introduce two q-deformed free bosons by

$$\widetilde{\varphi}^{(\pm)}(z) = \beta^{\pm 1/2} Q + 2\beta^{\pm 1/2}\alpha_0 \log z + \sum_{n\neq 0} \frac{(1+p^{-n})}{(1-\xi_\pm{}^n)}\alpha_n z^{-n}, \tag{2.5}$$

where $\xi_+ = q$, $\xi_- = t$ and define two kinds of deformed screening currents

$$S_\pm(z) =: \mathrm{e}^{\pm\widetilde{\varphi}^{(\pm)}(z)} : . \tag{2.6}$$

It can be checked that $S_\pm$ commute with the q-Virasoro generators up to total q- or t-derivative defined by

$$\frac{\mathrm{d}_{\xi_\pm}}{\mathrm{d}_{\xi_\pm} z} f(z) = \frac{f(z) - f(\xi_\pm z)}{(1-\xi_\pm)z}. \tag{2.7}$$

Deformed screening charges [25] are defined by the Jackson integral

$$Q^{+}_{[a,b]} = \int_a^b \mathrm{d}_q z\, S_+(z), \qquad Q^{-}_{[a,b]} = \int_a^b \mathrm{d}_t z\, S_-(z), \tag{2.8}$$

over a suitably chosen integral domain $[a,b]$.

2.2. $q \to -1$, $t \to -1$ limit

We consider a $q \to -1$ limit. Simultaneously, $t \to -1$ limit is taken. Let us realize this limit by

$$q = -\mathrm{e}^{-(1/\sqrt{\beta})h}, \qquad t = -\mathrm{e}^{-\sqrt{\beta}h}, \qquad p = q/t = \mathrm{e}^{Q_E h}, \qquad h \to +0, \tag{2.9}$$

where $Q_E = \sqrt{\beta} - 1/\sqrt{\beta}$. Choosing the branch of logarithms of q and t appropriately, we see that β takes the following rational value:

$$\beta = \frac{k_- + 1/2}{k_+ + 1/2} = \frac{2k_- + 1}{2k_+ + 1}, \tag{2.10}$$

where $k_\pm$ are non-negative integers.

First, we consider the $q \to -1$ limit of the q-bosons (2.5). Let us decompose them into "even" and "odd" parts:

$$\widetilde{\varphi}^{(\pm)}(z) = \widetilde{\varphi}^{(\pm)}_{\mathrm{even}}(z) + \widetilde{\varphi}^{(\pm)}_{\mathrm{odd}}(z), \tag{2.11}$$

where

$$\begin{aligned} \widetilde{\varphi}^{(\pm)}_{\mathrm{even}}(z) &:= \beta^{\pm 1/2} Q + \beta^{\pm 1/2}\alpha_0 \log(z^2) + \sum_{n\neq 0} \frac{1+p^{-2n}}{1-\xi_\pm{}^{2n}} \alpha_{2n} z^{-2n}, \\ \widetilde{\varphi}^{(\pm)}_{\mathrm{odd}}(z) &:= \sum_{n\in\mathbb{Z}} \frac{1+p^{-2n-1}}{1-\xi_\pm{}^{2n+1}} \alpha_{2n+1} z^{-2n-1}. \end{aligned} \tag{2.12}$$

In the $q \to -1$ limit of the q-bosons, we obtain two free bosons, $\phi(w)$ and $\varphi(w)$:

$$\begin{aligned} \widetilde{\varphi}^{(\pm)}_{\mathrm{even}}(z) &= \beta^{\pm 1/2}\phi(w) + O(h), \\ \widetilde{\varphi}^{(\pm)}_{\mathrm{odd}}(z) &= \varphi(w) + O(h), \end{aligned} \tag{2.13}$$

where $w = z^2$ and

$$\phi(w) = Q + a_0 \log w - \sum_{n\neq 0} \frac{a_n}{n} w^{-n}, \tag{2.14}$$

$$\varphi(w) = \sum_{n\in\mathbb{Z}} \frac{\tilde{a}_{n+1/2}}{n+1/2} w^{-n-1/2}, \tag{2.15}$$

with the following standard commutation relations of boson and twisted boson oscillators:

$$[a_n, a_m] = n\delta_{n+m,0}, \qquad [a_n, Q] = \delta_{n,0}, \tag{2.16}$$

$$[\tilde{a}_{n+1/2}, \tilde{a}_{-m-1/2}] = (n+1/2)\delta_{n,m}. \tag{2.17}$$

Next, we consider the $q \to -1$ limit of the screening currents (2.6). Using the limit of q-bosons (2.13), we obtain

$$\lim_{q\to -1} S_+(z) =: \mathrm{e}^{\sqrt{\beta}\phi(w)} \mathrm{e}^{\varphi(w)} :, \qquad \lim_{q\to -1} S_-(z) =: \mathrm{e}^{-(1/\sqrt{\beta})\phi(w)} \mathrm{e}^{-\varphi(w)} : . \tag{2.18}$$

We can construct the two fermion,

$$\psi(w) \equiv \frac{\mathrm{i}}{2\sqrt{2w}}\Big(: \mathrm{e}^{\varphi(w)} : - : \mathrm{e}^{-\varphi(w)} : \Big), \qquad \widehat{\psi}(w) \equiv \frac{1}{2\sqrt{2w}}\Big(: \mathrm{e}^{\varphi(w)} : + : \mathrm{e}^{-\varphi(w)} : \Big), \tag{2.19}$$

and see that $\psi(w)$ (resp. $\widehat{\psi}(w)$) is an NS fermion (resp. an R fermion) on the w-plane. The screening charge together with the regularization factor becomes

$$\lim_{q\to -1} \frac{\mathrm{i}}{\sqrt{2}}(1+q)Q^{+}_{[a,b]} = \int_{a^2}^{b^2} \mathrm{d}w\, \psi(w) : \mathrm{e}^{\sqrt{\beta}\phi(w)} :\equiv Q^{(+)}_{[a^2,b^2]}, \tag{2.20}$$

which is the screening charge for the superconformal block [38, 39]. Here we have ignored the subtlety due to the zero-mode and negative power terms. To be more precise, the "limit" (2.20) includes a kind of projection imposed by hand.

Let us choose a q-deformed Vertex operator $V_\alpha(z)$ as

$$V_\alpha(z) =: \mathrm{e}^{\Phi_\alpha(z)} :, \tag{2.21}$$

where

$$\Phi_\alpha(z) = \alpha\, Q + 2\alpha\, \alpha_0 \log z + \sum_{n\neq 0} \frac{q^{-n}(1-q^{2\sqrt{\beta}\alpha|n|})}{(1-q^{-|n|})(1-t^{-n})}\alpha_n z^{-n}. \tag{2.22}$$

We restrict the parameter α to take values corresponding to those of the primary fields of the minimal theories in the NS sector:

$$\alpha = \alpha_{r,s} = -\left(\frac{1-r}{2}\right)\frac{1}{\sqrt{\beta}} + \left(\frac{1-s}{2}\right)\sqrt{\beta}, \tag{2.23}$$

where

$$1 \le r \le 2k_-, \qquad 1 \le s \le 2k_+, \qquad r - s \in 2\mathbb{Z}. \tag{2.24}$$

The $q \to -1$ limit of the deformed vertex operator (2.21) is given by

$$\lim_{q\to -1} V_{\alpha_{r,s}}(z) =: \mathrm{e}^{\alpha_{r,s}\phi(w)} : \quad \text{for } L_{r,s} \text{ even}, \tag{2.25}$$

where

$$L_{r,s} \equiv (2k_+ + 1)\sqrt{\beta}\alpha_{r,s} = -k_+(1-r) + (1-s)k_- + \left(\frac{r-s}{2}\right) \in \mathbb{Z}. \tag{2.26}$$

Note that $: \mathrm{e}^{\alpha_{r,s}(w)} :$ is exactly equal to the Coulomb gas representation of the bosonic primary field in the NS sector with scaling dimension

$$\Delta_{\alpha_{r,s}} = \frac{1}{2}\alpha_{r,s}(\alpha_{r,s} - Q_E) = -\frac{1}{8}Q_E^2 + \frac{1}{8}\left(-\frac{r}{\sqrt{\beta}} + s\sqrt{\beta}\right)^2. \tag{2.27}$$

2.3. $q \to -1$ limit of q-Virasoro generators and $N = 1$ superconformal algebra

In this subsection, we consider directly the $q \to -1$ limit of the generating function $\mathcal{T}(z)$ of the q-Virasoro generators (2.4). In this limit, we can see that the fermionic currents

$$\begin{aligned} G(w) &= \psi(w)\partial\phi(w) + Q_E\, \partial\psi(w), \\ \widehat{G}(w) &= \widehat{\psi}(w)\partial\phi(w) + Q_E\, \partial\widehat{\psi}(w), \end{aligned} \tag{2.28}$$

appear [27]. Through OPE, we obtain the stress tensors $T(w)$ and $\hat{T}(w)$. Then, $T(w)$ and $G(w)$ (resp. $\hat{T}(w)$ and $\hat{G}(w)$) obey the $N=1$ superconformal algebra in the NS (resp. R) sector. The central charge is $c = \frac{3}{2}\hat{c}$ where

$$\hat{c} = 1 - 2Q_E^2 = 1 - 2\left(\sqrt{\beta} - \frac{1}{\sqrt{\beta}}\right)^2 = 1 - \frac{8(k_- - k_+)^2}{(2k_+ + 1)(2k_- + 1)}. \tag{2.29}$$

Recall that the $N=1$ minimal superconformal models have the central charge

$$\hat{c} = 1 - \frac{2(m' - m)^2}{mm'}. \tag{2.30}$$

Therefore, $\hat{c}$ (2.29) corresponds to an $N=1$ minimal model such that m and m' are both positive odd integers: $m = 2k_+ + 1$ and $m' = 2k_- + 1$. Without loss of generality, we can take $m' > m$, i.e., $k_- > k_+$. The unitary minimal models [40] occur when $k_- = k_+ + 1$ and $k_+ \geq 1$.

2.4. r-th root of unity limit

The limiting procedure is realized by

$$q = \mathrm{e}^{2\pi \mathrm{i}/r}\,\mathrm{e}^{-(1/\sqrt{\beta})h}, \qquad t = \mathrm{e}^{2\pi \mathrm{i}/r}\,\mathrm{e}^{-\sqrt{\beta}h}, \qquad p = q/t = \mathrm{e}^{Q_E h}, \qquad h \to +0. \tag{2.31}$$

Let us decompose the q-deformed free bosons (2.5) into two parts:

$$\widetilde{\varphi}^{(\pm)}(z) = \widetilde{\varphi}_0^{(\pm)}(z) + \widetilde{\varphi}_R^{(\pm)}(z), \tag{2.32}$$

where

$$\begin{aligned}
\widetilde{\varphi}_0^{(\pm)}(z) &= \beta^{\pm 1/2} Q + \frac{2}{r}\beta^{\pm 1/2}\alpha_0 \log z^r + \sum_{n\neq 0} \frac{(1 + p^{-nr})}{(1 - \xi_\pm^{nr})}\alpha_{nr}\, z^{-nr}, \\
\widetilde{\varphi}_R^{(\pm)}(z) &= \sum_{\ell=1}^{r-1}\sum_{n\in\mathbb{Z}} \frac{(1 + p^{-nr-\ell})}{1 - \xi_\pm^{nr+\ell}}\alpha_{nr+\ell}\, z^{-nr-\ell}.
\end{aligned} \tag{2.33}$$

In the $q \to \mathrm{e}^{2\pi \mathrm{i}/r}$ limit, we have

$$\lim_{h\to +0} \widetilde{\varphi}_0^{(\pm)}(z) = \sqrt{\frac{2}{r}}\,\beta^{\pm 1/2}\phi(w), \qquad \lim_{h\to +0} \widetilde{\varphi}_R^{(\pm)}(z) = \sqrt{\frac{2}{r}}\,\varphi(w), \tag{2.34}$$

where $w = z^r$ and

$$\phi(w) = Q_0 + a_0 \log w - \sum_{n\neq 0} \frac{a_n}{n} w^{-n}, \tag{2.35}$$

$$\varphi(w) = \sum_{\ell=1}^{r-1} \varphi^{(\ell)}(w), \qquad \varphi^{(\ell)}(w) = \sum_{n\subset\mathbb{Z}} \frac{\tilde{a}_{n+\ell/r}}{n + \ell/r} w^{-n-\ell/r}. \tag{2.36}$$

The oscillators obey the following commutation relations:

$$[a_m, a_n] = m\delta_{m+n,0}, \qquad [a_n, Q_0] = \delta_{n,0}, \tag{2.37}$$

$$[\tilde{a}_{n+\ell/r}, \tilde{a}_{-m-\ell'/r}] = (n + \ell/r)\delta_{m,n}\delta_{\ell,\ell'}. \tag{2.38}$$

The screening currents (2.6) turn into

$$\lim_{q\to e^{2\pi i/r}} S_\pm(z) =: \exp\left(\pm\sqrt{\frac{2}{r}}\beta^{\pm1/2}\phi(w)\right) :: \exp\left(\pm\sqrt{\frac{2}{r}}\varphi(w)\right) : . \tag{2.39}$$

Using the vertex operators $e^{\pm\sqrt{2/r}\,\varphi(w)}$, we can introduce the following fields

$$\Psi_\pm^{(\ell)}(w) := \frac{1}{r^{3/2}w^{1-(1/r)}}\sum_{k=0}^{r-1} e^{2\pi i k\ell/r} : \exp\left(\pm\sqrt{\frac{2}{r}}\varphi(e^{2\pi i k}w)\right) : . \tag{2.40}$$

Note that for $w_1 \to w_2$,

$$\langle 0|\Psi_\pm^{(\ell_1)}(w_1)\Psi_\mp^{(\ell_2)}(w_2)|0\rangle = \frac{\delta_{\ell_1+\ell_2,r}}{(w_1-w_2)^{2(1-(1/r))}}\left\{1+O\Big((w_1-w_2)^2\Big)\right\}. \tag{2.41}$$

These fields $\Psi_\pm^{(\ell)}(w)$ are analogue of the parafermion current with scaling dimension $\Delta_1 = 1-(1/r)$.

3. $SU(n)$ conformal/W block and q-deformation

The q-deformed W algebra at roots of unity itself will not be exploited here for our study of the q-lifted version of (W)AGT conjecture. For the conformal block, there is a simple recipe for the q-lift (q-deformation) without explicitly treating the q-W algebra.

3.1. integral representation of the SU(n) Vir/W block

In this subsection, we review the conformal block of W_n algebra. Let us consider the conformal field theory with the central charge c

$$c = (n-1)\big(1-n(n+1)Q_E^2\big), \tag{3.1}$$

associated with the $A_{n-1} = \mathfrak{sl}(n)$ Lie algebra. Let $\mathfrak{h}$ be the Cartan subalgebra of A_{n-1} and $\mathfrak{h}^*$ its dual.

The four-point conformal block of the chiral vertex operators can be expressed in terms of free fields. Let $\phi(z)$ be an $\mathfrak{h}$-valued free chiral boson with correlation functions:

$$\begin{aligned}\langle\phi_a(z)\phi_b(w)\rangle = (e_a,e_b)\log(z-w), \quad \phi_a(z) = \langle e_a,\phi(z)\rangle,\\ e_a\in\mathfrak{h}^*:\ \text{a simple root of } A_{n-1}, \quad a=1,\cdots,n-1.\end{aligned} \tag{3.2}$$

Here $(\cdot,\cdot)$ is the symmetric bilinear form on $\mathfrak{h}^*$ and $\langle\cdot,\cdot\rangle$ is the natural pairing between $\mathfrak{h}^*$ and $\mathfrak{h}$. With $(e_a,e_a)=2$, $C_{ab} := (e_a,e_b)$ is the Cartan matrix of the A_{n-1} algebra.

The free-field representation of the conformal block is then given by

$$\begin{aligned}&\mathcal{F}(c,\Delta_I,\Delta_i|\Lambda)\\ &= \left\langle V_{(1/\sqrt{\beta})\alpha_1}(0)V_{(1/\sqrt{\beta})\alpha_2}(\Lambda)V_{(1/\sqrt{\beta})\alpha_3}(1)V_{(1/\sqrt{\beta})\alpha_4}(\infty)\prod_{a=1}^{n-1}\mathcal{Q}_a^{\,N_a}\widetilde{\mathcal{Q}}_a^{\,\widetilde{N}_a}\right\rangle,\end{aligned} \tag{3.3}$$

with $\alpha_i\in\mathfrak{h}^*$ $(i=1,2,3,4)$. The vertex operator is defined by $V_\alpha(z) =: e^{\langle\alpha,\phi(z)\rangle} :$ $(\alpha\in\mathfrak{h}^*)$ and two kinds of screening charges are inserted:

$$\mathcal{Q}_a = \int_0^\Lambda dz\, V_{\sqrt{\beta}e_a}(z), \qquad \widetilde{\mathcal{Q}}_a = \int_1^\infty dz\, V_{\sqrt{\beta}e_a}(z). \tag{3.4}$$

The four points are set to $z_1 = 0$, $z_2 = \Lambda$, $z_3 = 1$, $z_4 = \infty$. Hence Λ is the cross ratio.

By using the Wick's theorem and changing the integration variables from $z_I^{(a)}$ ($I = 1, 2, \ldots, N_a + \widetilde{N}_a$) to $x_I^{(a)}$ ($I = 1, 2, \ldots, N_a$) and $y_J^{(a)}$ ($J = 1, 2, \ldots, \widetilde{N}_a$), defined by

$$\Lambda x_I^{(a)} := z_I^{(a)}, \qquad \frac{1}{y_J^{(a)}} := z_{N_a+J}^{(a)}, \tag{3.5}$$

we obtain the following Selberg type multiple integral [10]:

$$\mathcal{F} = \Lambda^{\Delta_I - \Delta_1 - \Delta_2}(1-\Lambda)^{(\alpha_2,\alpha_3)/\beta} Z_S(\Lambda), \tag{3.6}$$

where

$$\begin{aligned} Z_S(\Lambda) := \int_{[0,1]^N} \mathrm{d}x \left(\Delta_{A_{n-1}}(x)\right)^\beta \prod_{a=1}^{n-1}\prod_{I=1}^{N_a} (x_I^{(a)})^{(\alpha_1,e_a)}(1-x_I^{(a)})^{(\alpha_2,e_a)} \\ \times \int_{[0,1]^{\widetilde{N}}} \mathrm{d}y \left(\Delta_{A_{n-1}}(y)\right)^\beta \prod_{b=1}^{n-1}\prod_{J=1}^{N_b} (y_J^{(b)})^{(\alpha_4,e_b)}(1-y_J^{(b)})^{(\alpha_3,e_b)}\, F(x,y|\Lambda). \end{aligned} \tag{3.7}$$

Here $N := \sum_a N_a$, $\widetilde{N} := \sum_a \widetilde{N}_a$,

$$F(x,y|\Lambda) := \prod_{a=1}^{n-1}\left\{\prod_{I=1}^{N_a}(1-\Lambda x_I^{(a)})^{(\alpha_3,e_a)} \prod_{J=1}^{\widetilde{N}_a}(1-\Lambda y_J^{(a)})^{(\alpha_2,e_a)}\right\} \prod_{a=1}^{n-1}\prod_{b=1}^{n-1}\prod_{I=1}^{N_a}\prod_{J=1}^{\widetilde{N}_b}(1-\Lambda x_I^{(a)} y_J^{(b)})^{\beta C_{ab}}, \tag{3.8}$$

$$\mathrm{d}x = \prod_{a=1}^{n-1}\prod_{I=1}^{N_a}\mathrm{d}x_I^{(a)}, \qquad \mathrm{d}y = \prod_{a=1}^{n-1}\prod_{J=1}^{\widetilde{N}_a}\mathrm{d}y_J^{(a)}, \tag{3.9}$$

$$\Delta_{A_{n-1}}(x) = \prod_{a=1}^{n-1}\prod_{I<J}^{N_a}|x_I^{(a)} - x_J^{(a)}|^2 \prod_{b=1}^{n-2}\prod_{I=1}^{N_b}\prod_{J=1}^{N_{b+1}}|x_J^{(b+1)} - x_I^{(b)}|^{-1}, \tag{3.10}$$

$$\Delta_{A_{n-1}}(y) = \prod_{a=1}^{n-1}\prod_{I<J}^{\widetilde{N}_a}|y_I^{(a)} - y_J^{(a)}|^2 \prod_{b=1}^{n-2}\prod_{I=1}^{\widetilde{N}_b}\prod_{J=1}^{\widetilde{N}_{b+1}}|y_J^{(b+1)} - y_I^{(b)}|^{-1}. \tag{3.11}$$

Let us rewrite (3.7) as

$$Z_S(\Lambda) = Z_s(0)\left\langle\left\langle F(x,y|\Lambda)\right\rangle_+\right\rangle_- . \tag{3.12}$$

Here $\langle\cdots\rangle_+$ (resp. $\langle\cdots\rangle_-$) is the average over the A_{n-1} Selberg integral normalized as $\langle 1\rangle_\pm = 1$.

The AGT relation implies that $Z_S(\Lambda)/Z_S(0)$ is equal to the instanton part of the Nekrasov partition function of $\mathcal{N}=2$ $SU(n)$ gauge theory with $N_f = 2n$ fundamental matters.

3.2. q-deformation

In order to study the five-dimensional version of AGT conjecture, we need a q-deformed conformal block. There is a simple recipe [42, 47] to obtain a q-deformation of the Selberg-type multiple integral which is given by the following replacements:

$$\int_0^1 \mathrm{d}x_I \;\to\; \int_0^1 \mathrm{d}_q x_I, \tag{3.13}$$

and

$$(1-x_I)^{\beta_2-1} \rightarrow \prod_{k=0}^{\beta_2-1}(1-q^k x_I),$$
$$\prod_{1\le I<J\le n}(x_I-x_J)^{2\beta} \rightarrow \prod_{1\le I\neq J\le n}\prod_{k=0}^{\beta-1}(x_I-q^k x_J). \tag{3.14}$$

By this replacement, we obtain the q-deformation of the conformal block (3.12):

$$Z_S^{(q)} = \left\langle\left\langle \prod_{a=1}^{n-1}\left\{\prod_{I=1}^{N_a}\prod_{i=0}^{v_{a-}-1}(1-\Lambda x_I^{(a)}q^i)\prod_{J=1}^{\tilde{N}_a}\prod_{j=0}^{v_{a+}-1}(1-\Lambda y_J^{(a)}q^j)\right\}\times\right.\right.$$
$$\left.\left.\times\prod_{a,b=1}^{n-1}\prod_{\ell=0}^{\beta-1}\prod_{I=1}^{N_a}\prod_{J=1}^{\tilde{N}_a}(1-\Lambda x_I^{(a)}y_J^{(b)}q^\ell)^{C_{ab}}\right\rangle_{N+,q}\right\rangle_{\tilde{N}-,q}, \tag{3.15}$$

where

$$\left\langle f(x)\right\rangle_{N\pm,q} = \frac{1}{S_{N,q}}\left(\prod_{I=1}^{N}\int_0^1 \mathrm{d}_q x_I\right)\prod_{I=1}^{N}x_I^{u_{a\pm}}\prod_{i=1}^{v_{a\pm}-1}(1-x_I q^i)\prod_{1\le I\neq J\le N}\prod_{i=1}^{\beta-1}(x_I-q^i x_J)f(x) \tag{3.16}$$

Rearranging the integrand of eq.(3.15), we obtain

$$Z_S^{(q)} = \langle\langle I_S^{(q)}\rangle_{N+,q}\rangle_{\tilde{N}-,q}, \tag{3.17}$$

where

$$I_S^{(q)} = \exp\left\{-\sum_{k=1}^{\infty}[\beta]_{q^k}\frac{\Lambda^k}{k}\sum_{a=1}^{n}\left[\left(r_k^{(a)}+\frac{[v_{a+}]'_{q^k}}{[\beta]_{q^k}}\right)\left(\tilde{r}_k^{(a)}+\frac{[v_{a-}]'_{q^k}}{[\beta]_{q^k}}\right)-\frac{[v_{a+}]'_{q^k}}{[\beta]_{q^k}}\frac{[v_{a-}]'_{q^k}}{[\beta]_{q^k}}\right]\right\}. \tag{3.18}$$

Here we have introduced the q-number

$$[a]_q := \frac{1-q^a}{1-q}, \tag{3.19}$$

and

$$[v_{a-}]'_{q^k} := -\sum_{s=1}^{a-1}[v_{s-}]_{q^k}, \qquad [v_{(n-a)+}]'_{q^k} := \sum_{s=1}^{a}[v_{(n-s)+}]_{q^k},$$

with

$$[v_{1-}]'_{q^k} = [v_{n+}]'_{q^k} = 0.$$

Let λ be a partition, λ' be its conjugate and $a_\lambda(s)$ (resp. $\ell_\lambda(s)$) be the arm length (resp. leg length) at $s=(i,j)\in\lambda$:

$$a_\lambda(s) = a_\lambda(i,j) = \lambda_i - j, \qquad \ell_\lambda(s) = \ell_\lambda(i,j) = \lambda'_j - i. \tag{3.20}$$

Using the Cauchy identity for the Macdonald polynomials $P_\lambda(x;q,t)$,

$$\sum_\lambda \frac{C_\lambda}{C'_\lambda} P_\lambda(x;q,t) P_\lambda(y;q,t) = \exp\left\{\sum_{k=1}^{\infty} [\beta]_{q^k} \frac{1}{k} p_k \tilde{p}_k\right\}, \tag{3.21}$$

where

$$\frac{C_\lambda}{C'_\lambda} = \prod_{s\in\lambda} \frac{[a_\lambda(s)+\beta\ell_\lambda(s)+1]_q}{[a_\lambda(s)+\beta\ell_\lambda(s)+\beta]_q}, \tag{3.22}$$

we obtain a Λ-expansion of the q-deformed conformal block into a basis given by products of the Macdonald polynomials:

$$Z_S^{(q)} = \sum_{k=0}^{\infty} \Lambda^k \sum_{\substack{\vec{Y} \\ |\vec{Y}|=k}} \prod_{a=1}^{n} \frac{C_{Y_a}}{C'_{Y_a}} \left\langle \prod_{a=1}^{n} P_{Y_a}\left(-r_k^{(a)} - \frac{[v_{a+}]'_{q^k}}{[\beta]_{q^k}}\right)\right\rangle_{N+} \times$$
$$\times \left\langle \prod_{a=1}^{n} P_{Y_a}\left(\tilde{r}_k^{(a)} + \frac{[v_{a-}]'_{q^k}}{[\beta]_{q^k}}\right)\right\rangle_{\tilde{N}-}. \tag{3.23}$$

According to AGT conjecture at q-lifted case, this q-deformed conformal block is identified with the five dimensional instanton partition function. In Section 2, we have seen that the deformed screening charges and vertex operators reduce to those of the superconformal model and its generalization. We will see that the ALE instanton partition function can be obtained from the five dimensional instanton partition function by taking the root of unity limit in next section. Therefore, once the dictionary is found, AGT conjecture at q-lifted case provides the equality between the block and ALE instanton partition function.

4. ALE ($\mathbb{R}^4/\mathbb{Z}_r$) instanton partition function

4.1. brief review of 4d SU(n) instanton partition function on ALE

The instanton partition function of four dimensional $\mathcal{N}=2$ supersymmetric $SU(n)$ gauge theory on $\mathbb{R}^4$ (with Ω-deformation) can be calculated by the method of localization. The fixed points of the torus action $U(1)^2\times U(1)^n$ is labeled by an n-tuple of Young diagrams $\vec{Y}=(Y_\alpha)_{\alpha=1,2,\cdots,n}$. For the fixed points corresponding to the k-instanton, $|\vec{Y}|=\sum_\alpha |Y_\alpha| = k$. Here we denote by $|Y_\alpha|$ the number of boxes carried by Y_α.

Suppose that the torus action is generated by $(\epsilon_1,\epsilon_2,a_1,\cdots,a_n)$. Then the weight of an individual box $(i,j)\in Y_\alpha$ of Young diagram is given by $a_\alpha+(i-1)\epsilon_1+(j-1)\epsilon_2$. Here $\epsilon_1=\sqrt{\beta}g_s$ and $\epsilon_2=-g_s/\sqrt{\beta}$. Therefore,

$$g_s^2 = -\epsilon_1\epsilon_2, \qquad \beta = -\frac{\epsilon_1}{\epsilon_2}. \tag{4.1}$$

Let us consider the instanton on $\mathbb{R}^4/\mathbb{Z}_r$. The $\mathbb{Z}_r$ action is [45, 48]

$$\epsilon_1 \to \epsilon_1 - \frac{2\pi i}{r}, \quad \epsilon_2 \to \epsilon_2 + \frac{2\pi i}{r}, \quad a_\alpha \to a_\alpha + q_\alpha \frac{2\pi i}{r}. \tag{4.2}$$

From the weight associated with the torus action, we see that $\mathbb{Z}_r$-charges are assigned to each Young diagram and its boxes. The charge of each box (i,j) of the Young diagram Y_α is

$$q_{\alpha,(i,j)} = q_\alpha - (i-1) + (j-1) \pmod r, \qquad \alpha=1,2,\cdots,n, \tag{4.3}$$

and q_α is regarded as the $\mathbb{Z}_r$-charge of each Young diagram Y_α.

4.2. more on the labeling of ALE instantons
Let us consider the case where the 1st Chern class vanishes. This condition leads to

$$n_\ell - 2k_\ell + k_{\ell+1} + k_{\ell-1} = 0. \tag{4.4}$$

Here n_ℓ is the number of Young diagram $\{Y_\alpha\}$ such that $\mathbb{Z}_r$-charge $q_\alpha = \ell$ and k_ℓ is the total number of the boxes such that the $\mathbb{Z}_r$-charge $q_{\alpha,(i,j)} = \ell$. Of course they satisfy $n = \sum_\ell n_\ell$ and $k = \sum_\ell k_\ell$. When (4.4) is satisfied, the 2nd Chern number is k/r and the ALE partition function is schematically written as

$$Z^{\mathbb{R}^4/\mathbb{Z}_r} = \sum_{k=0}^{\infty} (\Lambda')^{k/r} \sum_{|\vec{Y}|=k} \mathcal{A}_{\vec{Y}} \Big|_{\substack{\text{1st Chern}=0 \\ \mathbb{Z}_r\text{charge of each factor} = 0}} . \tag{4.5}$$

Here $\mathcal{A}_{\vec{Y}}$ represents the contribution of the fixed point $\vec{Y}$.

4.3. limiting procedure from 5d instanton partition function
The five dimensional instanton partition function is given by given by [46, 25]

$$Z^{\mathbb{R}^5} = \sum_{k=0}^{\infty} \widetilde{\Lambda}^k \sum_{|\vec{Y}|=k} \widetilde{\mathcal{A}}_{\vec{Y}}, \tag{4.6}$$

$$\widetilde{\mathcal{A}}_{\vec{Y}} = \frac{\prod_{\alpha=1}^{n} \prod_{k=1}^{n} f_{Y_\alpha}^{q+}(m_k + a_\alpha) f_{Y_\alpha}^{q-}(m_{k+n} + a_\alpha)}{\prod_{\alpha,\alpha'}^{n} g^q_{Y_\alpha Y_{\alpha'}}(a_\alpha - a_{\alpha'})}, \tag{4.7}$$

$$g^q_{YW}(x) = \prod_{(i,j)\in Y} [x + \beta \ell_Y(i,j) + a_W(i,j) + \beta]_q [-x - \beta \ell_Y(i,j) - a_W(i,j) - 1]_q, \tag{4.8}$$

$$f_Y^{q\pm}(x) = \prod_{(i,j)\in Y} [\pm x \mp \beta(i-1) \pm (j-1)]_q. \tag{4.9}$$

The parameter m_i $(i = 1, \cdots, 2n)$ is related to the five dimensional fundamental mass m_i^{5d} by

$$m_i = m_i^{5d} + \frac{1}{2}(1-\beta). \tag{4.10}$$

Eq. (4.2) can be read off as the shift in q as well as in t [49]

$$q = \mathrm{e}^{\epsilon_2} \to \omega\, q, \qquad t = \mathrm{e}^{-\epsilon_1} \to \omega\, t, \qquad q^{\frac{a_\alpha}{\epsilon_2}} = \mathrm{e}^{a_\alpha} \to \omega^{q_\alpha} q^{\frac{a_\alpha}{\epsilon_2}}, \qquad \omega = \mathrm{e}^{2\pi \mathrm{i}/r}. \tag{4.11}$$

Here we have rescaled $a_\alpha \to a_\alpha/\epsilon_2$. If we take the limit $q \to 1$ subsequently, we expect that this is equivalent to taking the $\mathbb{Z}_r$ orbifold projection on four dimensional space. On the other hand, this limit is equal to the root of unity limit of q and t and the five dimensional Nekrasov partition function reduces to that on $\mathbb{R}^4/\mathbb{Z}_r$. In what follows, let us realize this procedure as

$$q = \omega\, \mathrm{e}^{h\epsilon_2}, \quad t = \omega\, \mathrm{e}^{-h\epsilon_1}, \quad q^{\frac{a_\alpha}{\epsilon_2}} = \omega^{q_\alpha} \mathrm{e}^{h a_\alpha}, \quad q^{\frac{m_i}{\epsilon_2}} = \mathrm{e}^{h m_i}, \qquad h \to +0. \tag{4.12}$$

Here we have rescaled m_i by m_i/ϵ_2.

4.4. obtaining ALE instanton partition function

We have established the method of obtaining the $SU(n)$ instanton partition function on $\mathbb{R}^4/\mathbb{Z}_r$ from the five dimensional partition function. The appropriate factors to remove the extra coefficients which appear in the limiting process have been also obtained. In the following, $\widetilde{\mathcal{A}}^{(q_\alpha)}_{\vec{Y}}$ represents that each Young diagram Y_α is assigned the $\mathbb{Z}_r$-charge q_α. The results are as follows:

• $SU(2)$ and $r=2$ case

$$Z^{\mathbb{R}^4/\mathbb{Z}_2}_{SU(2)} = \sum_{k=0}^{\infty} (\Lambda')^k Z^{(2)}_k + \sum_{k=0}^{\infty} (\Lambda')^{k+1/2} Z^{(2)}_{k+1/2}, \tag{4.13}$$

where

$$Z^{(2)}_{k+1/2} := \lim_{h\to 0} \frac{h^2}{2^2} \sum_{|A|+|B|=2k+1} \widetilde{\mathcal{A}}^{(1,1)}_{AB} \tag{4.14}$$

$$Z^{(2)}_{k} := \lim_{h\to 0} \sum_{|A|+|B|=2k} \widetilde{\mathcal{A}}^{(0,0)}_{AB}. \tag{4.15}$$

• $SU(2)$ and general r case

$$Z^{\mathbb{R}^4/\mathbb{Z}_r}_{SU(2)} = \sum_{q_a=0}^{\lfloor \frac{r}{2} \rfloor} \sum_{k=0}^{\infty} (\Lambda')^{k+\frac{q_a(r-q_a)}{r}} Z^{(r)}_{k+\frac{q_a(r-q_a)}{r}}, \tag{4.16}$$

where

$$Z^{(r)}_{k+\frac{q_a(r-q_a)}{r}} := \lim_{h\to 0} \Xi_{q_a} \sum_{\substack{|A|+|B| \\ =rk+q_a(r-q_a)}} \widetilde{\mathcal{A}}^{(q_a,-q_a)}_{AB}. \tag{4.17}$$

Here

$$\Xi_0 = 1, \quad \Xi_1 = h^2 \frac{1}{(1-\omega)(1-\omega^{-1})}, \tag{4.18}$$

$$\Xi_i = h^{2i} \frac{\prod_{k=1}^{i-1}(1-\omega^k)^{2i-3k}(1-\omega^{-k})^{2i-3k}}{\prod_{l=1}^{i}(1-\omega^{i+l-1})(1-\omega^{-(i+l-1)})}, \qquad 2 \le i \le \left\lfloor \frac{r}{2} \right\rfloor. \tag{4.19}$$

• $SU(n)$ and general r case

$$Z^{\mathbb{R}^4/\mathbb{Z}_r}_{SU(n)} = \sum_{q_\alpha} \sum_{k=0}^{\infty} (\Lambda')^{k+\frac{d}{r}} Z^{(r)}_{k+\frac{d}{r}}, \tag{4.20}$$

where

$$Z^{(r)}_{k+\frac{d}{r}} := \lim_{h\to 0} \Xi_{q_\alpha} \sum_{\substack{|\vec{Y}| \\ =rk+d}} \widetilde{\mathcal{A}}^{(q_\alpha)}_{\vec{Y}}. \tag{4.21}$$

Here

$$\Xi_{q_\alpha} = h^{2\sum_{\alpha=1}^{n-1}\alpha q_\alpha} \xi_{q_\alpha}(\omega)\xi_{q_\alpha}(\omega^{-1}), \tag{4.22}$$

$$\xi_{q_\alpha}(\omega) = \prod_{a<b}^{n-1} \prod_{j=1}^{q_b-1} \frac{(1-\omega^j)^{2(q_b-j)}}{(1-\omega^{\sum_{i=a}^{b-1} q_i+j})^{q_b-j}(1-\omega^{\sum_{i=a}^{b-1} q_i+j})^{q_b-j}} \frac{1}{(1-\omega^{\sum_{i=a}^{b-1} q_i})^{q_b}}$$
$$\times \prod_{a=1}^{n-1} \prod_{j=1}^{q_a-1} \frac{(1-\omega^j)^{q_a-j}(1-\omega^{\sum_{i=a+1}^{n-1} q_i+j})^{q_a-j}}{(1-\omega^{\sum_{i=1}^{a-1} q_i+j})^{j}(1-\omega^{\sum_{i=1}^{n-1} q_i+j})^{q_a-j}} \frac{1}{(1-\omega^{\sum_{i=1}^{a} q_i})^{q_a}}$$
$$\times \prod_{a=1}^{n-1} \prod_{j=1}^{q_{a+1}+\cdots q_{n-1}} \frac{(1-\omega^j)^{q_a}}{(1-\omega^{\sum_{i=1}^{a} q_i+j})^{q_a}}$$
$$\times \prod_{a+2\leq b}^{n-1} \prod_{j=1}^{q_b-1} \frac{(1-\omega^j)^{j}(1-\omega^{\sum_{i=a+1}^{b-1} q_i+j})^{q_b-j}}{(1-\omega^{q_a-q_b+j})^{j}(1-\omega^{\sum_{i=k}^{l-1} q_i-j})^{j}} \prod_{j=0}^{q_{a+1}+\cdots+q_{b-1}-q_b} \frac{(1-\omega^{q_b+j})^{q_b}}{(1-\omega^{q_a+j})^{q_b}}. \quad (4.23)$$

References

[1] A. A. Belavin, A. M. Polyakov, and A. B. Zamolodchikov, Nucl. Phys. B **241**, 333-380 (1984).
[2] H. Nakajima, *Lectures on Hilbert Schemes of points on surfaces*, American Mathematical Society (1999).
[3] N. Nekrasov, Adv. Theor. Math. Phys. **7**, 831-864 (2004) [arXiv:hep-th/0206161].
[4] L. F. Alday, D. Gaiotto, and Y. Tachikawa, Lett. Math. Phys. **9**, 167-197 (2010) [arXiv:0906.3219 [hep-th]].
[5] N. Wyllard, JHEP **0911**, 002 (2009) [arXiv:0907.2189 [hep-th]].
[6] R. Dijkgraaf and C. Vafa, arXiv:0909.2453 [hep-th].
[7] H. Itoyama, K. Maruyoshi and T. Oota, Prog. Theor. Phys. **123**, 957-987 (2010) [arXiv:0911.4244 [hep-th]].
[8] A. Mironov, A. Morozov and Sh. Shakirov, JHEP **1002**, 030 (2010) [arXiv:0911.5721 [hep-th]].
[9] A. Mironov, A. Morozov and Sh. Shakirov, J. Mod. Phys. A **25**, 3173-3207 (2010) [arXiv:1001.0563 [hep-th]].
[10] H. Itoyama and T. Oota, Nucl. Phys. B **838**, 298-330 (2010) [arXiv:1003.2929 [hep-th]].
[11] A. Mironov, A. Morozov and And. Morozov, Nucl. Phys. B **843**, 534-557 (2011) [arXiv:1003.5752 [hep-th]].
[12] H. Itoyama, T. Oota and N. Yonezawa, Phys. Rev. D **82**, 085031 (2010) [arXiv:1008.1861 [hep-th]].
[13] T. Nishinaka and C. Rim, JHEP **1202**, 114 (2012) [arXiv:1112.3545 [hep-th]].
[14] F. Fucito, J. F. Morales and D. R. Pacifici, JHEP **1301**, 091 (2013) [arXiv:1210.3580 [hep-th]].
[15] J. Kaneko, Ann. scient. Éc. Norm. Sup. **29**, 583-637 (1996).
[16] K. W. J. Kadell, Adv. Math. **130**, 33-102 (1997).
[17] V. A. Fateev and A. V. Litvinov, JHEP **1002**, 014 (2010) [arXiv:0912.0504 [hep-th]].
[18] M. Hadasz, Z. Jaskólski and P. Suchanek, JHEP **1006**, 046 (2010) [arXiv:arXiv:1004.1841 [hep-th]].
[19] A. Mironov, A. Morozov and Sh. Shakirov, ' JHEP **1102**, 067 (2011) [arXiv:1012.3137 [hep-th]].
[20] A. Morozov and A. Smirnov, arXiv:1307.2576 [hep-th].
[21] J. Shiraishi, H. Kubo, H. Awata and S. Odake, Lett. Math. Phys. **38**, 33-51 (1996) [arXiv:q-alg/9507034].
[22] E. Frenkel and N. Reshetikhin, [arXiv:q-alg/9706023].
[23] H. Awata, H. Kubo, S. Odake and J. Shiraishi, Commun. Math. Phys. **179**, 401-416 (1996) [arXiv:q-alg/9508011].
[24] H. Awata, H. Kubo, Y. Morita, S. Odake and J. Shiraishi, Lett. Math. Phys. **41**, 65-78 (1997) [arXiv:q-alg/9604023].
[25] H. Awata and Y. Yamada, JHEP **1001**, 125 (2010) [arXiv:0910.4431 [hep-th]; Prog. Theor. Phys. **124**, 227-262 (2010) [arXiv:1004.5122 [hep-th]].
[26] P. Bouwknegt and K. Pilch, Commun. Math. Phys. **196**, 249-288 (1998) [arXiv:q-alg/9710026].
[27] H. Itoyama, T. Oota and R. Yoshioka, [arXiv:1308.2068 [hep-th]]
[28] V. Belavin and B. Feigin, JHEP **1107**, 079 (2011) [arXiv:1105.5800 [hep-th]].
[29] G. Bonelli, K. Maruyoshi and A. Tanzini, JHEP **1108**, 056 (2011) [arXiv:1106.2505 [hep-th]]; Lett. Math. Phys. **101**, 103-124 (2012) [arXiv:1107.4609 [hep-th]].
[30] A. Belavin, V. Belavin and M. Bershtein, JHEP **1109**, 117 (2011) [arXiv:1106.4001 [hep-th]].
[31] N. Wyllard, arXiv:1109.4264 [hep-th].
[32] M. N. Alfimov and G. M. Tarnopolsky, JHEP **1202**, 036 (2012) [arXiv:1110.5628 [hep-th]].
[33] Y. Ito, Nucl. Phys. B **861**, 387-402 (2012) [arXiv:1110.2176 [hep-th]].
[34] A. A. Belavin, M. A. Bershtein, B. L. Feigin, A. V. Litvinov and G. M. Tarnopolsky, Commun. Math. Phys. **319**, 269-301 (2013) [arXiv:1111.2803 [hep-th]].
[35] A. A. Belavin, M. A. Bershtein and G. M. Tarnopolsky, arXiv:1211.2788 [hep-th].
[36] M. N. Alfimov, A. A. Belavin and G. M. Tarnopolsky, arXiv:1306.3938 [hep-th].
[37] V. A. Belavin, hep-th/0611295.

[38] Y. Kitazawa, N. Ishibashi, A. Kato, K. Kobayashi, Y. Matsuo and S. Odake, Nucl. Phys. B **306**, 425-444 (1988).
[39] L. Alvarez-Gaumé and Ph. Zaugg, Annals Phys. **215**, 171-230 (1992) [hep-th/9109050].
[40] D. Friedan, Z. Qiu and S. H. Shenker, Phys. Lett. B **151**, 37-43 (1985).
[41] M. A. Bershadsky, V. G. Knizhnik and M. G. Teitelman, Phys. Lett. B **151**, 31-36 (1985).
[42] S. O. Warnaar, Ramanujan J. **10**, 237-268 (2005).
[43] P. J. Forrester and S. O. Warnaar, Bull. Amer. Math. Soc. **45**, 489-534 (2008) [arXiv:0710.3981 [math.CA]].
[44] S. O. Warnaar, Acta Math. **203**, 269-304 (2009) [arXiv:0708.1193 [math.CA]].
[45] P. Kronheimer and H. Nakajima, Math. Ann. **288**, 263-307 (1990).
[46] H. Awata and H. Kanno, Int. J. Mod. Phys. A **24**, 2253-2306 (2009) [arXiv:0805.0191 [hep-th]].
[47] A. Mironov, A. Morozov, Sh. Shakirov and A. Smirnov, Nucl. Phys. B **855**, 128-151 (2012) [arXiv:1105.0948 [hep-th]].
[48] F. Fucito, J. F. Morales and R. Poghossian, Nucl. Phys. B **703**, 518-536 (2004) [arXiv:hep-th/0406243].
[49] T. Kimura, JHEP **1109**, 015 (2011) [arXiv:1105.6091 [hep-th]].

New approach to duality-invariant nonlinear electrodynamics

E. Ivanov[1], **O. Lechtenfeld**[2], **B. Zupnik**[1]

[1] Bogoliubov Laboratory of Theoretical Physics, JINR, 141980, Dubna, Moscow Region, Russia

[2] Institut für Theoretische Physik and Riemann Center for Geometry and Physics, Leibniz Universität Hannover, Appelstraße 2, 30167 Hannover, Germany

E-mail: eivanov@theor.jinr.ru, zupnik@theor.jinr.ru, lechtenf@itp.uni-hannover.de

Abstract. We survey a new approach to the duality-invariant systems of nonlinear electrodynamics, based on introducing auxiliary bi-spinor fields. In this approach, the entire information about the given self-dual system is encoded in the $U(1)$ invariant interaction of the auxiliary fields, while the standard self-dual Lagrangians appear on shell as a result of eliminating auxiliary fields by their equations of motion. Starting from the simplest $U(1)$ duality, we show how this approach can be generalized to the $U(N)$ duality (with N independent Maxwell field strengths), as well as to self-dual systems of $\mathcal{N}=1$ supersymmetric electrodynamics. Also, it works perfectly for self-dual systems with higher derivatives in the action.

1. Motivations

$U(N)$ duality invariance is on-shell symmetry of a wide class of the nonlinear electrodynamics models including the renowned Born-Infeld theory [1]-[6]. It is a generalization of the well known $O(2)$ symmetry between the free Maxwell equations of motion (EOM) and Bianchi identities for $F_{mn} = \partial_m A_n - \partial_n A_m, \tilde{F}_{mn} = \frac{1}{2}\epsilon_{mnpq}F^{pq}$:

$$\text{EOM}: \ \partial^m F_{mn} = 0 \qquad \Longleftrightarrow \qquad \text{Bianchi}: \ \partial^m \tilde{F}_{mn} = 0\,, \tag{1}$$

$$\delta F_{mn} = -\omega \tilde{F}_{mn}, \qquad \delta \tilde{F}_{mn} = \omega F_{mn}\,. \tag{2}$$

In the nonlinear case with the general Lagrangian $L(F)$ the dual field strength is defined as $P_{mn} = -2\frac{\partial L(F)}{\partial F^{mn}}$, and eqs. (1), (2) generalize to

$$\text{EOM}: \ \partial^m P_{mn} = 0 \qquad \Longleftrightarrow \qquad \text{Bianchi}: \ \partial^m \tilde{F}_{mn} = 0\,, \tag{3}$$

$$\delta P_{mn} = -\omega \tilde{F}_{mn}\,, \qquad \delta \tilde{F}_{mn} = \omega P_{mn}\,. \tag{4}$$

Since P_{mn} is a function of F_{mn}, the self-consistency condition (the so called self-duality condition [1, 2]) should be also imposed:

$$P\tilde{P} + F\tilde{F} = 0\,. \tag{5}$$

In the case of N Maxwell strengths the $O(2)$ ($U(1)$) duality group is enhanced to $U(N)$.

Recently, there was a rebirth of interest in the duality-invariant theories and their superextensions [7]-[12], mainly due to the hypothesis that the generalized duality symmetries at quantum level can play the decisive role in proving the conjectured UV finiteness of $\mathcal{N}=8,4D$ supergravity. Keeping this in mind, new efficient general methods for treating duality invariant theories and their supersymmetric extensions are urgently needed.

A decade ago, two of us proposed a new general formulation of the duality invariant theories which exploits the tensorial (bispinor) auxiliary fields [13]-[15]. In this approach, the $U(N)$ duality is realized as the *linear* off-shell symmetry of the nonlinear interaction constructed out of the auxiliary fields. The self-duality constraint (5) is also *linearized*. In the $U(1)$ case the problem of restoring the nonlinear electrodynamics action by the auxiliary interaction is reduced to solving some *algebraic* equations. On the contrary, in the standard approach, while restoring the Lagrangian as a solution of the constraint (5), one needs to solve *differential* equations.

We realized that some methods recently invented for the systematic construction of various duality invariant systems in [9]-[11] are in fact completely equivalent to our 10-years old approach just mentioned. This motivated us to return to the original formulation in order to see how the latest developments in the duality area can be ascribed into its framework [16, 17]. Supersymmetrization of the approach with auxiliary bispinor fields was undertaken in [18, 19].

The present talk is a brief comprehensive review of all these issues.

2. The standard setting

We will make use of the bispinor formalism:

$$F_{mn} \Rightarrow (F_{\alpha\beta},\ \ \bar{F}_{\dot\alpha\dot\beta}), \quad \varphi := F^{\alpha\beta}F_{\alpha\beta}\,, \quad \bar\varphi := \bar{F}^{\dot\alpha\dot\beta}\bar{F}_{\dot\alpha\dot\beta}\,,$$
$$L(\varphi,\bar\varphi) = -\frac{1}{2}(\varphi+\bar\varphi) + L^{int}(\varphi,\bar\varphi) \tag{6}$$

In this notation, the EOM and Bianchi identities read:

$$\partial^{\dot\beta}_{\alpha}\bar{P}_{\dot\alpha\dot\beta}(F) - \partial^{\beta}_{\dot\alpha}P_{\alpha\beta}(F) = 0\,, \quad \partial^{\dot\beta}_{\alpha}\bar{F}_{\dot\alpha\dot\beta} - \partial^{\beta}_{\dot\alpha}F_{\alpha\beta} = 0\,, \quad P_{\alpha\beta} = i\frac{\partial L}{\partial F^{\alpha\beta}}\,. \tag{7}$$

The $O(2)$ duality transformations are realized as

$$\delta_\omega F_{\alpha\beta} = \omega P_{\alpha\beta}, \qquad \delta_\omega P_{\alpha\beta} = -\omega F_{\alpha\beta}\,. \tag{8}$$

The self-consistency constraint and the Gaillard-Zumino (GZ) representation for L are:

$$F_{\alpha\beta}F^{\alpha\beta} + P_{\alpha\beta}P^{\alpha\beta} - \text{c.c} = 0 \quad \Leftrightarrow \quad \varphi - \bar\varphi - 4\,[\varphi(L_\varphi)^2 - \bar\varphi(L_{\bar\varphi})^2] = 0\,,$$
$$L^{sd} = \frac{i}{2}(\bar{P}\bar{F} - PF) + I(\varphi,\bar\varphi)\,, \quad \delta_\omega I(\varphi,\bar\varphi) = 0\,. \tag{9}$$

The GZ representation (9) does not give any explicit recipe how to determine the unknown $U(1)$ duality-invariant function $I(\varphi,\bar\varphi)$ and so how to restore the whole lagrangian L^{sd}. Our approach with the auxiliary tensorial fields provides an answer to this key question.

3. Formulation with bispinor auxiliary fields

We introduce the auxiliary unconstrained fields $V_{\alpha\beta}$ and $\bar{V}_{\dot\alpha\dot\beta}$ and write the extended Lagrangian in the (F,V)-representation as

$$\mathcal{L}(V,F) = \mathcal{L}_2(V,F) + E(\nu,\bar\nu), \quad \mathcal{L}_2(V,F) = \frac{1}{2}(\varphi+\bar\varphi) + \nu + \bar\nu - 2\,(V\cdot F + \bar{V}\cdot\bar{F}), \tag{10}$$

where $\nu = V^2$, $\bar{\nu} = \bar{V}^2$. Here $\mathcal{L}_2(V, F)$ is the bilinear part only through which the Maxwell field strength enters the action and $E(\nu, \bar{\nu})$ is the nonlinear interaction involving only auxiliary fields.

Dynamical equations of motion

$$\partial_\alpha^{\dot\beta} \bar{P}_{\dot\alpha\dot\beta}(V, F) - \partial_{\dot\alpha}^{\beta} P_{\alpha\beta}(V, F) = 0\,, \quad P_{\alpha\beta}(F, V) = i\frac{\partial \mathcal{L}(V, F)}{\partial F^{\alpha\beta}} = i(F_{\alpha\beta} - 2V_{\alpha\beta}), \tag{11}$$

together with Bianchi identity, are covariant under $O(2)$ transformations

$$\delta V_{\alpha\beta} = -i\omega V_{\alpha\beta},\ \delta F_{\alpha\beta} = i\omega(F_{\alpha\beta} - 2V_{\alpha\beta}),\ \ \delta\nu = -2i\omega\nu\,. \tag{12}$$

The algebraic equation of motion for $V_{\alpha\beta}$,

$$F_{\alpha\beta} = V_{\alpha\beta} + \frac{1}{2}\frac{\partial E}{\partial V^{\alpha\beta}} = V_{\alpha\beta}(1 + E_\nu)\,, \tag{13}$$

is $O(2)$ covariant *if and only if* the proper constraint holds:

$$\nu E_\nu - \bar{\nu} E_{\bar{\nu}} = 0 \quad \Rightarrow \quad E(\nu, \bar{\nu}) = \mathcal{E}(a)\,, \quad a := \nu\bar{\nu}\,. \tag{14}$$

The meaning of this constraint is that $E(\nu, \bar{\nu})$ should be $O(2)$ invariant function of the auxiliary tensor variables:

$$\delta_\omega E = 2i\omega(\bar{\nu} E_{\bar{\nu}} - \nu E_\nu) = 0\,. \tag{15}$$

This is none other than the self-duality constraint in the new setting:

$$F^2 + P^2 - \bar{F}^2 - \bar{P}^2 = 0 \quad \Leftrightarrow \quad \nu E_\nu - \bar{\nu} E_{\bar{\nu}} = 0\,. \tag{16}$$

The auxiliary equation can be now written as

$$F_{\alpha\beta} - V_{\alpha\beta} = V_{\alpha\beta}\bar{\nu}\,\mathcal{E}'. \tag{17}$$

It serves to express $V_{\alpha\beta}, \bar{V}_{\dot\alpha\dot\beta}$ in terms of $F_{\alpha\beta}, \bar{F}_{\dot\alpha\dot\beta}$:

$$V_{\alpha\beta}(F) = F_{\alpha\beta} G(\varphi, \bar{\varphi}), \quad G(\varphi, \bar{\varphi}) = \frac{1}{2} - L_\varphi = (1 + \bar{\nu}\mathcal{E}_a)^{-1}\,. \tag{18}$$

After substituting these expressions back into $\mathcal{L}$ we obtain the corresponding self-dual Lagrangian $L^{sd}(\varphi, \bar{\varphi})$

$$L^{sd}(\varphi, \bar{\varphi}) = \mathcal{L}(V(F), F) = -\frac{1}{2}\frac{(\varphi + \bar{\varphi})(1 - a\mathcal{E}_a^2) + 8a^2\mathcal{E}_a^3}{1 + a\mathcal{E}_a^2} + \mathcal{E}(a)\,, \tag{19}$$

where a is related to $\varphi, \bar{\varphi}$ by the algebraic equation

$$(1 + a\mathcal{E}_a^2)^2\varphi\bar{\varphi} = a[(\varphi + \bar{\varphi})\mathcal{E}_a + (1 - a\mathcal{E}_a^2)^2]^2\,. \tag{20}$$

The invariant GZ function is now represented as $I(\varphi, \bar{\varphi}) = \mathcal{E}(a) - 2a\mathcal{E}_a$, $a = \nu(\varphi, \bar{\varphi})\bar{\nu}(\varphi, \bar{\varphi})$.

To summarize, *all* $O(2)$ duality-symmetric systems of nonlinear electrodynamics without derivatives on the field strengths are parametrized by the $O(2)$ invariant off-shell interaction $\mathcal{E}(a)$ which is a function of the real quartic combination of the auxiliary fields. This universality is the basic advantage of the approach with tensorial auxiliary fields. The problem of constructing $O(2)$ duality-symmetric systems is reduced to choosing one or another specific $\mathcal{E}(a)$. As distinct from other existing approaches to solving the self-duality constraint, our approach uses the algebraic equations instead of the differential ones and automatically yields the Lagrangians $L^{sd}(\varphi, \bar{\varphi})$ which are analytic at $\varphi = \bar{\varphi} = 0$.

After passing to the tensorial notation, our basic auxiliary field equation (proposed ten years ago) precisely coincides with what was recently called "nonlinear twisted self-duality constraint" [9]-[11], and $\mathcal{E}(a)$ with what is called there "duality invariant source of deformation".

4. Alternative auxiliary field representation

One of the advantages of our approach is the possibility to choose some other auxiliary field formulations which sometimes prove to be simpler technically. It is convenient, e.g., to deal with the auxiliary complex variables μ and $\bar{\mu}$ related to ν and $\bar{\nu}$ by Legendre transformation

$$\mu := E_\nu, \; E - \nu E_\nu - \bar{\nu} E_{\bar{\nu}} := H(\mu, \bar{\mu}), \quad \nu = -H_\mu, \; E = H - \mu H_\mu - \bar{\mu} H_{\bar{\mu}} \,. \tag{21}$$

Under the $O(2)$ duality $\delta_\omega \mu = 2i\omega\, \mu, \; \delta_\omega \bar{\mu} = -2i\omega\, \bar{\mu}$, so

$$H(\mu, \bar{\mu}) = I(b) \,, \qquad b = \mu\bar{\mu} \,.$$

The basic relations of this formalism are

$$L^{sd}(\varphi, \bar{\varphi}) = -\frac{1}{2}\,(\varphi + \bar{\varphi} + 4bI_b)\,\frac{1-b}{1+b} + I(b) \,,$$
$$(b+1)^2\, \varphi\bar{\varphi} = b\,[\varphi + \bar{\varphi} - I_b(b-1)^2]^2 \,. \tag{22}$$

These relations can be reproduced by eliminating μ and $\bar{\mu}$ as independent scalar auxiliary fields from the off-shell Lagrangian

$$\tilde{L}(\varphi, \mu) = \frac{\varphi(\mu - 1)}{2(1+\mu)} + \frac{\bar{\varphi}(\bar{\mu} - 1)}{2(1+\bar{\mu})} + I(\mu\bar{\mu}). \tag{23}$$

Yet there even exists a combined tensor - scalar auxiliary field "master" formulation which enables to establish relations between different equivalent off-shell descriptions of the same duality-invariant system:

$$\mathcal{L}(F, V, \mu) = \frac{1}{2}(F^2 + \bar{F}^2) - 2(VF) - (\bar{V}\bar{F}) + V^2(1+\mu) + (1+\bar{\mu})V^2 + I(\mu\bar{\mu}) \,. \tag{24}$$

5. Examples

I. Born - Infeld. This model has a simpler description in the μ (or b) representation:

$$I^{BI}(b) = \frac{2b}{b-1}, \qquad I^{BI}_b = -\frac{2}{(b-1)^2} \,. \tag{25}$$

The equation for computing b becomes quadratic:

$$\varphi\bar{\varphi}\, b^2 + [2\varphi\bar{\varphi} - (\varphi + \bar{\varphi} + 2)^2]\, b + \varphi\bar{\varphi} = 0 \quad \Rightarrow$$
$$b = \frac{4\varphi\bar{\varphi}}{[2(1+Q) + \varphi + \bar{\varphi}]^2} \,, \quad Q(\varphi) := \sqrt{1 + \varphi + \bar{\varphi} + (1/4)(\varphi - \bar{\varphi})^2} \,. \tag{26}$$

After substituting this solution into the general formula for $L^{sd}(\varphi, \bar{\varphi})$ the standard BI Lagrangian is recovered

$$L^{BI}(\varphi, \bar{\varphi}) = 1 - \sqrt{1 + \varphi + \bar{\varphi} + (1/4)(\varphi - \bar{\varphi})^2} \tag{27}$$

II. The simplest interaction (SI) model. This model is the simplest example of the auxiliary interaction generating the non-polynomial self-dual electromagnetic Lagrangian. The relevant interaction in both ν and μ representations are linear functions

$$\mathcal{E}^{SI}(a) = \frac{1}{2}\, a \,, \quad I^{SI}(b) = -2b \,, \quad a = \nu\bar{\nu}, \;\; b = \mu\bar{\mu} \,. \tag{28}$$

Despite such a simple off-shell form of the auxiliary interaction, it is difficult to find a closed on-shell form of the nonlinear Lagrangian $L^{SI}(\varphi,\bar\varphi)$, since the algebraic equations relating a (or b) to $\varphi,\bar\varphi$ are of the 5-th order. E.g.,

$$(b+1)^2\,\varphi\bar\varphi = b\,[\varphi+\bar\varphi+2(b-1)^2]^2\,.$$

Nevertheless, it is straightforward to solve these equations as infinite series in b or a and then to restore $L^{SI}(\varphi,\bar\varphi)$ to any order. Up to 10-th order in $F,\bar F$:

$$\begin{aligned} L^{SI}(\varphi,\bar\varphi) &= -\frac{1}{2}(\varphi+\bar\varphi)+e_1\varphi\bar\varphi-e_1^2(\varphi^2\bar\varphi+\varphi\bar\varphi^2)+e_1^3(\varphi^3\bar\varphi+\varphi\bar\varphi^3)\\ &\quad+4e_1^3\varphi^2\bar\varphi^2-e_1^4(\varphi^4\bar\varphi+\varphi\bar\varphi^4)-10e_1^4(\varphi^3\bar\varphi^2+\varphi^2\bar\varphi^3)+O(F^{12}). \end{aligned} \tag{29}$$

Here $e_1=\frac{1}{2}$.

6. Systems with higher derivatives

The nonlinear electromagnetic Lagrangians with higher derivatives are functions of the variables

$$F,\quad \partial_m F,\quad \partial_m\partial_n F,\quad \partial_m\partial_n\partial_r F\ldots$$

and their complex conjugates. The higher-derivative Lagrangians in the explicit form involve various scalar combinations of these variables, e.g.,

$$F^2,\ \ (\partial^m F\partial_m F),\ \ (\partial^m F^2\partial_m F^2),\ \ (F\Box^N F),\ldots.$$

It is known that the higher-derivative generalizations of the duality-invariant Lagrangians contain all orders of derivatives of $F_{\alpha\beta}$ and $F_{\dot\alpha\dot\beta}$ [9].

In our formulation the generalized self-dual Lagrangian is

$$\mathcal{L}(F,V,\partial V)=\mathcal{L}_2+\mathcal{E}(V,\partial V)\,, \tag{30}$$

where $\mathcal{L}_2$ is the same "free" bilinear part as before and $\mathcal{E}(V,\partial V)$ is the $O(2)$ invariant self-interaction which can now involve any Lorentz invariant combinations of the tensorial auxiliary fields and their derivatives. In order to avoid non-localities and ghosts, it is reasonable to assume that $\mathcal{E}(V,\partial V)$ contains no terms bilinear in $V,\bar V$, i.e. that the extra derivatives appear only at the interaction level.

The equations of motion for this Lagrangian contain the Lagrange derivative of $\mathcal{E}$

$$\partial^\alpha_{\dot\beta}(F-2V)_{\alpha\beta}+\partial^{\dot\alpha}_{\beta}(\bar F-2\bar V)_{\dot\alpha\dot\beta}=0\,,\quad F_{\alpha\beta}=V_{\alpha\beta}+\frac{1}{2}\frac{\Delta\mathcal{E}}{\Delta V^{\alpha\beta}}\,. \tag{31}$$

This set of equations together with the Bianchi identity for $F,\bar F$ is covariant under the $O(2)$ duality transformations, provided that $\mathcal{E}(V,\partial V)$ is $O(2)$ invariant,

$$\delta_\omega\,\mathcal{E}(V,\partial V)=0\,.$$

An analog of the condition (5) is the vanishing of the integral

$$\int d^4x[P^2(F,V)+F^2-\bar P^2(F,V)-\bar F^2]=0\,,\ P_{\alpha\beta}=i\frac{\Delta\mathcal{E}}{\Delta F^{\alpha\beta}}=i(F-2V)_{\alpha\beta}\,, \tag{32}$$

and this again amounts to the $O(2)$ invariance of $\mathcal{E}(V,\partial V)$.

Due to the property that the derivatives appear only in the interaction, one can solve the auxiliary field equations for $V_{\alpha\beta}, \bar{V}_{\dot{\alpha}\dot{\beta}}$ by recursions, like in the case without derivatives, and to finally obtain $L^{sd}(F, \partial F)$ as a series expansion to any order in derivatives and the field strengths.

Some examples. As the first example we consider

$$\mathcal{E}_{(2)} = \frac{1}{2}\nu\bar{\nu} + c\partial^m\nu\partial_m\bar{\nu}, \qquad \frac{\Delta\mathcal{E}_{(2)}}{\Delta V^{\alpha\beta}} = 2V_{\alpha\beta}[1 + \frac{1}{2}\bar{\nu} - c\Box\bar{\nu}]\,. \tag{33}$$

The auxiliary field equation is:

$$F_{\alpha\beta} = V_{\alpha\beta}(1 + \frac{1}{2}\bar{\nu} - c\Box\bar{\nu})\,. \tag{34}$$

Its perturbative solution is given by:

$$\begin{aligned}
V^{(1)}_{\alpha\beta} &= F_{\alpha\beta}\,, \\
V^{(3)}_{\alpha\beta} &= -F_{\alpha\beta}(\frac{1}{2}\bar{\varphi} - c\Box\bar{\varphi}), \\
V^{(5)}_{\alpha\beta} &= F_{\alpha\beta}\{\frac{1}{2}\bar{\varphi}(\frac{1}{2}\bar{\varphi} - c\Box\bar{\varphi}) + \varphi(\frac{1}{2}\bar{\varphi} - c\Box\bar{\varphi}) - c(\Box\bar{\varphi})(\frac{1}{2}\bar{\varphi} - c\Box\bar{\varphi}) \\
&\quad - 2c\Box[\bar{\varphi}(\frac{1}{2}\varphi - c\Box\varphi)]\}, \qquad \text{etc}\,,
\end{aligned}$$

$$\begin{aligned}
L(F, \partial^N F) &= -\frac{1}{2}(\varphi + \bar{\varphi}) + \frac{1}{2}\varphi\bar{\varphi} - \frac{1}{4}\varphi^2\bar{\varphi} - \frac{1}{4}\varphi\bar{\varphi}^2 + c\partial^m\varphi\partial_m\bar{\varphi} \\
&\quad + c\varphi\bar{\varphi}[(\Box\varphi) + (\Box\bar{\varphi})] - c^2\varphi(\Box\bar{\varphi})^2 - c^2\bar{\varphi}(\Box\varphi)^2 + O(F^8)\,.
\end{aligned} \tag{35}$$

This model can be regarded as the "minimal" higher-derivative deformation of the SI model (the latter is recovered at $c = 0$).

As the second example, we consider

$$\mathcal{E}_{(4)} = \gamma(\partial^m V\partial^n V)(\partial_m\bar{V}\partial_n\bar{V}), \tag{36}$$

where γ is a coupling constant and brackets denote traces with respect to the $SL(2, C)$ indices.

The auxiliary field equation reads:

$$F_{\alpha\beta} = V_{\alpha\beta} - \gamma\partial^m\left[\partial^n V_{\alpha\beta}(\partial_m\bar{V}\cdot\partial_n\bar{V})\right]. \tag{37}$$

The perturbative solution is given by:

$$V^{(1)}_{\alpha\beta} = F_{\alpha\beta}, \quad V^{(3)}_{\alpha\beta} = \gamma\partial^m\left[\partial^n F_{\alpha\beta}(\partial_m\bar{F}\cdot\partial_n\bar{F})\right], \quad \text{etc}\,.$$

The Lagrangian in the F-representation involves higher derivatives, starting from the sixth order in fields

$$L^{(6)} = (V^{(3)}V^{(3)}) - 2\gamma[(V^{(3)}\partial^n\partial^m F)(\partial_m\bar{F}\partial_n\bar{F}) + (V^{(3)}\partial^m F)\partial^n(\partial_m\bar{F}\partial_n\bar{F})] + \text{c.c.}\,. \tag{38}$$

7. $U(N)$ case

The standard setting. The starting point is the nonlinear Lagrangian with N abelian gauge field strengths

$$L(F^k,\bar F^l) = -\frac{1}{2}[(F^kF^k)+(\bar F^k\bar F^k)] + L^{int}(\varphi^{kl},\bar\varphi^{kl}), \qquad (39)$$
$$\varphi^{kl}=\varphi^{lk}=(F^kF^l), \quad \bar\varphi^{kl}=(\bar F^k\bar F^l).$$

It is chosen to be invariant off shell under $O(N)$ transformations

$$\delta_\xi F^k_{\alpha\beta}=\xi^{kl}F^l_{\alpha\beta}, \quad \delta_\xi \bar F^k_{\dot\alpha\dot\beta}=\xi^{kl}\bar F^k_{\dot\alpha\dot\beta}, \quad \xi^{kl}=-\xi^{lk}, \qquad (40)$$

where $\xi^{lk}=-\xi^{kl}$ are the corresponding group parameters. The nonlinear equations of motion

$$E^k_{\alpha\dot\alpha} := \partial^{\dot\beta}_\alpha \bar P^k_{\dot\alpha\dot\beta}(F) - \partial^\beta_{\dot\alpha}P^k_{\alpha\beta}(F)=0, \; P^k_{\alpha\beta}(F)=i\frac{\partial L}{\partial F^{k\alpha\beta}},$$

together with the Bianchi identities

$$B^k_{\alpha\dot\alpha}=\partial^{\dot\beta}_\alpha \bar F^k_{\dot\alpha\dot\beta}-\partial^\beta_{\dot\alpha}F^k_{\alpha\beta}=0,$$

are on-shell covariant under the $U(N)$ duality transformations

$$\delta_\eta F^k_{\alpha\beta}=\eta^{kl}P^l_{\alpha\beta}, \quad \delta_\eta P^k_{\alpha\beta}=-\eta^{kl}F^l_{\alpha\beta}, \; \eta^{kl}=\eta^{lk}, \qquad (41)$$

where $\eta^{kl}=\eta^{lk}$ are $\frac{1}{2}N(N+1)$ real parameters completing $O(N)$ to $U(N)$, provided that the appropriate generalized self-duality consistency conditions hold:

$$(P^kP^l)+(F^kF^l)-\text{c.c.}=0, \quad (F^kP^l)-(F^lP^k)-\text{c.c.}=0. \qquad (42)$$

The (F,V) representation. An $U(N)$ analog of the Lagrangian (10) is [17]

$$\mathcal{L}(F^k,V^k) = \mathcal{L}_2(F^k,V^k)+E(\nu^{kl},\bar\nu^{kl}), \quad \nu^{kl}=(V^kV^l), \quad \bar\nu^{kl}=(\bar V^k\bar V^l), \qquad (43)$$
$$\mathcal{L}_2(F^k,V^k) = \frac{1}{2}[(F^kF^k)+(\bar F^k\bar F^k)]-2[(F^kV^k)+(\bar F^k\bar V^k)] + (V^kV^k)+(\bar V^k\bar V^k). \qquad (44)$$

The $U(N)/O(N)$ duality transformations are implemented as

$$\delta_\eta F^k_{\alpha\beta}=\eta^{kl}P^l_{\alpha\beta}=i\eta^{kl}(F^l-2V^l)_{\alpha\beta}, \quad \delta_\eta P^k_{\alpha\beta}=-\eta^{kl}F^l_{\alpha\beta}. \qquad (45)$$

For the whole set of equations of motion to be $U(N)$ duality invariant, the interaction $E(\nu^{kl},\bar\nu^{kl})$ should be $U(N)$ invariant:

$$E(\nu^{kl},\bar\nu^{kl}) \Rightarrow \mathcal{E}(A_1,\ldots,A_N), \qquad A_1=\bar\nu^{kl}\nu^{lk},\ldots \qquad (46)$$

The algebraic equations of the $U(N)$ duality-invariant models are

$$(F^k-V^k)_{\alpha\beta}=\mathcal{E}^{kl}V^l_{\alpha\beta}, \quad (\bar F^k-\bar V^k)_{\dot\alpha\dot\beta}=\bar{\mathcal{E}}^{kl}\bar V^l_{\dot\alpha\dot\beta}, \quad \mathcal{E}^{kl}:=\frac{\partial\mathcal{E}}{\partial\nu^{kl}}. \qquad (47)$$

These equations of motion are equivalent to the general "nonlinear twisted self-duality constraints". Solving them, e.g., by recursions, we can restore the whole nonlinear $U(N)$ duality invariant action by the invariant interaction $\mathcal{E}(A_1,\ldots,A_N)$. The duality invariant actions with higher derivatives can be constructed by the $U(N)$ invariant interaction involving derivatives of the auxiliary tensorial fields, like in the $U(1)$ case.

8. Auxiliary superfields in $\mathcal{N}$=1 electrodynamics

The superfield action of nonlinear $\mathcal{N} = 1$ electrodynamics can be written as:

$$S(W) = \frac{1}{4}\int d^6\zeta W^2 + \frac{1}{4}\int d^6\bar{\zeta}\bar{W}^2 + \frac{1}{4}\int d^8z\, W^2\bar{W}^2\Lambda(w,\bar{w},y,\bar{y}), \tag{48}$$
$$w = \frac{1}{8}\bar{D}^2\bar{W}^2, \quad \bar{w} = \frac{1}{8}D^2W^2, \qquad y \equiv D^\alpha W_\alpha = \bar{D}_{\dot{\alpha}}\bar{W}^{\dot{\alpha}}.$$

Here, $W_\alpha(x,\theta_\alpha,\bar{\theta}_{\dot{\beta}}) = \frac{i}{2}(\sigma^m\bar{\sigma}^n)_\alpha^\beta F_{mn}\theta_\beta - \theta_\alpha D + \ldots$, is the spinor chiral $\mathcal{N} = 1$ Maxwell superfield strength ($\bar{D}_{\dot{\alpha}}W_\alpha = 0$, $D^\alpha W_\alpha = \bar{D}_{\dot{\alpha}}\bar{W}^{\dot{\alpha}}$).

The on-shell $U(1)$ duality rotations and the $\mathcal{N} = 1$ analog of the self-duality constraint read [3]

$$\delta W_\alpha = \omega M_\alpha(W,\bar{W}), \quad \delta M_\alpha = -\omega W_\alpha \quad M_\alpha := -2i\frac{\delta S}{\delta W^\alpha}, \tag{49}$$
$$\mathrm{Im}\int d^6\zeta\,(W^2 + M^2) = 0. \tag{50}$$

The $\mathcal{N} = 1$, $U(1)$ duality is the symmetry between the superfield equations of motion and Bianchi identity [3]:

$$D^\alpha M_\alpha - \bar{D}_{\dot{\alpha}}\bar{M}^{\dot{\alpha}} = 0 \quad \Leftrightarrow \quad D^\alpha W_\alpha - \bar{D}_{\dot{\alpha}}\bar{W}^{\dot{\alpha}} = 0. \tag{51}$$

How to supersymmetrize the bispinor formulation? The basic idea [18, 19] is to embed tensorial auxiliary fields into chiral auxiliary superfields

$$V_{\alpha\beta}(x) \Rightarrow U_\alpha(x,\theta,\bar{\theta}) = v_\alpha(x) + \theta^\beta V_{\alpha\beta}(x) + \ldots, \quad \bar{D}_{\dot{\gamma}}U_\alpha(x,\theta,\bar{\theta}) = 0 \tag{52}$$

(similarly, for the $\mathcal{N} = 2$ case).

We make the substitution $S(W) \to S(W,U)$, with [19]

$$\begin{aligned} S(W,U) &= \int d^6\zeta\left(UW - \frac{1}{2}U^2 - \frac{1}{4}W^2\right) + \text{c.c.} \\ &\quad + \frac{1}{4}\int d^8z\, U^2\bar{U}^2\, E(u,\bar{u},g,\bar{g}), \\ &\quad u = \frac{1}{8}\bar{D}^2\bar{U}^2, \quad \bar{u} = \frac{1}{8}D^2U^2, \quad g = D^\alpha U_\alpha. \end{aligned} \tag{53}$$

The duality-invariant $\mathcal{N} = 1$ systems amount to the special choice of the interaction in (53) as $U(1)$ invariant one

$$E_{inv} = \mathcal{F}(B,A,C) + \bar{\mathcal{F}}(\bar{B},A,C), \quad A := u\bar{u}, \ C := g\bar{g}, \ B := ug^2, \ \bar{B} := \bar{u}\bar{g}^2. \tag{54}$$

One can also define the M -representation of the self-dual $\mathcal{N} = 1$ systems as a generalization of the μ representation of the bosonic systems. The corresponding superfield action is as follows

$$S(W,U,M) = \int d^6\zeta\left(UW - \frac{1}{2}U^2 - \frac{1}{4}W^2\right) + \text{c.c.} + S_{int}(W,U,M), \tag{55}$$
$$S_{int}(W,U,M) = \frac{1}{4}\int d^8z\,\left[(U^2\bar{M} + \bar{U}^2M) + M\bar{M}\,J(m,\bar{m})\right]. \tag{56}$$

Here $m = \frac{1}{8}\bar{D}^2\bar{M}$, $\bar{m} = \frac{1}{8}D^2M$ and M is a complex general scalar $\mathcal{N} = 1$ superfield. The duality transformations are realized as

$$\delta M = 2i\omega M\,, \quad \delta\bar{M} = -2i\omega\bar{M}\,, \quad \delta\bar{m} = 2i\omega\bar{m}\,, \quad \delta m = 2i\omega m\,. \tag{57}$$

The duality invariant systems correspond to the choice

$$J(m,\bar{m}) = J_{inv}(B)\,, \quad B := m\bar{m}\,.$$

Example [19]. $\mathcal{N} = 1$ Born-Infeld theory:

$$J^{(BI)}_{inv}(B) = \frac{2}{B-1}\,. \tag{58}$$

This should be compared with the standard $W, \bar{W}$ representation of the same theory [20]

$$S^{(BI)}_{int} = \frac{1}{4}\int d^8z\, W^2\bar{W}^2\Lambda^{(BI)}(w,\bar{w})\,, \quad w = \frac{1}{8}\bar{D}^2\bar{W}^2\,, \; \bar{w} = \frac{1}{8}D^2W^2\,,$$

$$\Lambda^{(BI)}(w,\bar{w}) = \left[1 + \frac{1}{2}(w+\bar{w}) + \sqrt{1 + (w+\bar{w}) + \frac{1}{4}(w-\bar{w})^2}\,\right]^{-1}.$$

9. Summary and outlook

- *All* duality invariant systems of nonlinear electrodynamics (including those with higher derivatives) admit an off-shell formulation with the auxiliary bispinor (tensorial) fields. These fields are fully unconstrained off shell, there is no need to express them through any secondary gauge potentials, etc.
- The full information about the given duality invariant system is encoded in the $O(2)$ invariant interaction function which depends only on the auxiliary fields (or also on their derivatives) and can be chosen *at will*. In many cases it looks much simpler compared to the final action written in terms of the Maxwell field strengths.
- The renowned nonlinear self-duality constraint is *linearized* in the new formulation and becomes just the requirement of $O(2)$ invariance of the auxiliary interaction. The $O(2)$ (and, in fact, $U(N)$ [17]) duality transformations are linearly realized off shell.
- The basic algebraic equations eliminating the auxiliary tensor fields are equivalent to the recently employed *"nonlinear twisted self-duality constraints"*. In our approach this sort of conditions appear as *equations of motion* corresponding to the well defined off-shell Lagrangian.

10. Some further lines of development

- (a) The full construction and exploration of extensions to $\mathcal{N} = 1, \mathcal{N} = 2, \ldots$ supersymmetric duality systems, including the most interesting supersymmetric Born-Infeld theories, in both flat and supergravity backgrounds, as a further development of the study in [18, 19].
- (b) Adding, in a self-consistent way, scalar and other fields into the auxiliary tensorial field formulation: $U(1)$ duality group $\Rightarrow SL(2,R)$ [2]. One adds two scalar fields, axion and dilaton $\mathcal{S}_1$ and $\mathcal{S}_2$, which support a nonlinear realization of $SL(2,R)$, and properly generalizes the self-dual Lagrangian in the (F,V)-representation. The corresponding equations of motion together with Bianchi identity exhibit $SL(2,R)$ duality invariance. Similarly, $U(N)$ duality can be extended to $Sp(2N,R)$ via coupling to the coset $Sp(2N,R)/U(N)$ fields. Supersymmetric versions of this new setting for generalized self-duality can also be constructed.

- (c) The formulation presented suggests a *new look* at the duality invariant systems: in both the *classical* and the *quantum* cases *not* to eliminate the tensorial auxiliary fields by their equations of motion, but to deal with the off-shell actions *at all intermediate steps.* In many cases, the interaction looks much simpler when the auxiliary fields are retained in the action. In this connection, recall the off-shell superfield approach in supersymmetric theories, which in many cases radically facilitates the quantum calculations and unveils the intrinsic geometric properties of the corresponding theories without any need to pass on shell by eliminating the auxiliary fields (at least until the final step of calculating S matrix, etc). It is worth recalling that the tensorial auxiliary fields have originally appeared just within an off-shell superfield formulation, that of $\mathcal{N}=3$ supersymmetric Born-Infeld theory [13].

Acknowledgments
We acknowledge a support from a grant of the Heisenberg-Landau Program, from the RFBR grants Nr. 12-02-00517, Nr. 13-02-91330 and the grant DFG LE 838/12-1. E.I. thanks the organizers of the conference ISQS'2013 for inviting him to present this talk and for kind hospitality in Prague.

References

[1] Gaillard M K and Zumino B 1981 *Nucl. Phys.* B **193** 221
[2] Gibbons G W and Rasheed D A 1995 *Nucl. Phys.* B **454** 185 (*Preprint* hep-th/9506035)
[3] Kuzenko S M and Theisen S 2001 *Fortsch. Phys.* **49** 273 (*Preprint* hep-th/0007231)
[4] Hatsuda M *et al* 1999 *Nucl. Phys.* B **561** 341 (*Preprint* hep-th/9906103)
[5] Bekaert X and Cucu S 2001 *Nucl. Phys.* B **610** 433 (*Preprint* hep-th/0104048)
[6] Aschieri P *et al* 2008 *Riv. Nuovo Cim.* **31** 625 (*Preprint* arXiv:0807.4039 [hep-th])
[7] Bossard G *et al* 2010 *JHEP* **1012** 052 (*Preprint* arXiv:1007.5472 [hep-th])
[8] Kallosh R 2012 *JHEP* **1203** 083 (*Preprint* arXiv:1103.4115 [hep-th])
Kallosh R 2011 *JHEP* **1106** 073 (*Preprint* arXiv:1104.5480 [hep-th])
[9] Bossard G and Nicolai H 2011 *JHEP* **1108** 074 (*Preprint* arXiv:1105.1273 [hep-th])
[10] Carrasco J J M *et al* 2012 *Phys. Rev.* D **85** 025007 (*Preprint* arXiv:1108.4390 [hep-th])
[11] Chemissany W *et al* 2012 *Phys. Rev.* D **85** 046002 (*Preprint* arXiv:1112.0332 [hep-th])
[12] Aschieri P and Ferrara S 2013 *JHEP* **1305** 087 (*Preprint* arXiv:1302.4737 [hep-th])
[13] Ivanov E A and Zupnik B M 2001 *Nucl. Phys.* B **618** 3 (*Preprint* hep-th/0110074)
[14] Ivanov E A and Zupnik B M 2002 New representation for Lagrangians of self-dual nonlinear electrodynamics (*Preprint* hep-th/0202203)
[15] Ivanov E A and Zupnik B M 2004 *Yadern. Fiz.* **67** 2212 [*Phys. Atom. Nucl.* **67** 2188] (*Preprint* hep-th/0303192)
[16] Ivanov E A and Zupnik B M 2013 *Phys. Rev.* D **87** 065023 (*Preprint* arXiv:1212.6637 [hep-th])
[17] Ivanov E A and Zupnik B M 2013 Bispinor auxiliary fields in duality-invariant electrodynamics revisited: The $U(N)$ case (*Preprint* arXiv:1304.1366 [hep-th])
[18] Kuzenko S M 2013 *JHEP* **1303** 153 (*Preprint* arXiv:1301.5194 [hep-th])
[19] Ivanov E *et al* 2013 *JHEP* **1305** 133 (*Preprint* arXiv:1303.5962 [hep-th])
[20] Cecotti S and Ferrara S 1987 *Phys. Lett.* B **187** 335
Bagger J and Galperin A 1997 *Phys. Rev.* D **55** 1091 (*Preprint* hep-th/9608177)

The geometry of variations in Batalin–Vilkovisky formalism

Arthemy V. Kiselev

Johann Bernoulli Institute for Mathematics and Computer Science, University of Groningen, P.O. Box 407, 9700 AK Groningen, The Netherlands

E-mail: `A.V.Kiselev@rug.nl`

Abstract. We explain why no sources of divergence are built into the Batalin–Vilkovisky (BV) Laplacian, whence there is no need to postulate any *ad hoc* conventions such as "$\boldsymbol{\delta}(0) = 0$" and "$\log \boldsymbol{\delta}(0) = 0$" within BV-approach to quantisation of gauge systems. Remarkably, the geometry of iterated variations does not refer at all to the construction of Dirac's $\boldsymbol{\delta}$-function as a limit of smooth kernels. We illustrate the reasoning by re-deriving –but not just 'formally postulating'– the standard properties of BV-Laplacian and Schouten bracket and by verifying their basic inter-relations (e.g., cohomology preservation by gauge symmetries of the quantum master-equation).

Introduction

This is a paper about geometry of variations. We formulate definitions of the objects and structures which are cornerstones of Batalin–Vilkovisky formalism [5, 7, 20, 22, 55]. To confirm the intrinsic self-regularisation of BV-Laplacian, we explain why there are no divergencies in it (such excessive elements are traditionally encoded by using derivatives of Dirac's $\boldsymbol{\delta}$-distribution). Namely, we specify the geometry in which the following canonical inter-relations between the variational Schouten bracket $[\![\, ,]\!]$ and BV-Laplacian Δ are rigorously proven for any BV-functionals F, G, H:

$$[\![F, G \cdot H]\!] = [\![F, G]\!] \cdot H + (-)^{(|F|-1)\cdot|G|} G \cdot [\![F, H]\!], \tag{1a}$$

$$\Delta(F \cdot G) = \Delta F \cdot G + (-)^{|F|}[\![F, G]\!] + (-)^{|F|} F \cdot \Delta G, \tag{1b}$$

$$\Delta\bigl([\![F, G]\!]\bigr) = [\![\Delta F, G]\!] + (-)^{|F|-1}[\![F, \Delta G]\!], \tag{1c}$$

$$\Delta^2 = 0 \qquad \Longleftrightarrow \qquad \text{Jacobi}\bigl([\![\, ,]\!]\bigr) = 0. \tag{1d}$$

There is an immense literature on this subject's intrinsic difficulties and attempts of regularisation of apparent divergencies in it (e.g., see [12, 13, 25, 50, 51] vs [21]). While the BV-quantisation technique has advanced far from its sources [7, 8], it is still admitted that it lacks sound mathematical consistency ([22, §15] or [3, §3]). The calculus in this field is thus reduced to formal operation with expressions which are expected to render the theory's main objects and structures. Several *ad hoc* techniques for cancellation of divergencies, allowing one to strike through calculations and obtain meaningful results, are adopted by repetition; we briefly review the plurality of such tricks in what follows.

Our reasoning is independent from such conventional schemes for cancellation of infinities or from other practised roundabouts for regularisation of terms which are believed to be infinite (e.g., by erasing 'infinite constants' [11]). In particular, we do not pronounce the traditional password

$$\boldsymbol{\delta}(0) := 0 \tag{2}$$

which lets one enter the existing paradigm and use its quantum alchemistry for operation with what remains from Dirac's $\boldsymbol{\delta}$-distribution.[1] Our message is this: we do not propose to replace 'bad slogans' with 'good slogans,' which would mean that a choice of conventions is still left to the one who attempts regularisation in the BV-setup. Such deficiency would symptomise that the theory remains a formal procedure. We now focus on the true sources of known difficulties. By analysing the geometry of variations of functionals at a very basic level, we prove the absence of apparently divergent essences. The intrinsically regularised definitions of the BV-Laplacian Δ and Schouten bracket $[\![\,,\,]\!]$ are the main result of this paper.

The new understanding leaves intact but substantiates the bulk of results which have been obtained by using various *ad hoc* techniques (that is, explicitly or tacitly referring to the surreal equalities $\boldsymbol{\delta}(0) = 0$ and $\log \boldsymbol{\delta}(0) = 0$); we refer to a detailed review [3] for an account of early developments in BV-formalism. We do not aim at a reformulation or reproduction of any old or recent achievements, accomplishing here a different task.

In fact, we invent nothing new. It is the coupling of dual vector spaces which ensures the intrinsic self-regularisation of BV-Laplacian and validity of equalities (1), with (1c) in particular. Therefore, it would be redundant to start developing any brand-new formalism (cf. [51]); on the other hand, we *prove* properties (1) and not just *postulate* these assertions (cf. [21]).

We employ standard notions, constructions, and techniques from the geometry of jet spaces [28, 40, 45]. Because the geometry of BV-objects is essentially variational, it would be methodologically incomplete to handle them as if the space-time, that is, the base manifold in the bundles of physical fields, were just a point ([27, 48] or [39]). The language of jet spaces is extensively used in the study of BV-models, see [3, 6, 21, 43]: the bundles of jets of sections usually appear in such traditional contexts as calculation of symmetries or conservation laws. In this paper we apply these geometric techniques at a much more profound level and give rigorous definitions for BV-objects. Let us emphasize that we do not aim at extending one's ability to write more formulas according to a regularly emended system of accepted algorithms; we explicate the genuine nature of objects and their canonical matchings, not taking any formulas for quasi-definitions.

This paper is structured as follows. Containing a brief overview of traditional approaches to regularisation of the BV-formalism, this introduction concludes with a parable; the line of our reasoning is reminiscent to that of *Lettres persanes* by Montesquieu.

In section 1 we describe the true geometry of variations; we first reveal the correspondence between action functionals and infinitesimal shifts of classical trajectories or physical fields. An understanding of nontrivial mechanism of such matching achieved for *one* variation, the picture of *many* variations becomes clear. This approach resolves the obstructions for regularisation of iterated variations in BV-formalism; we remark that Dirac's $\boldsymbol{\delta}$-function does not appear in section 2 at all.[2]

In section 2.1 we recall in proper detail the standard construction of Batalin–Vilkovisky (BV) vector bundles with canonically conjugate pairs of ghost parity-even and odd variables.

[1] Another convention is $\log \boldsymbol{\delta}(0) = 0$; we show that natural counterparts of the true geometry of variations lead to this intuitive convention and simultaneously to (2) — none of the two being actually required.

[2] We refer to [19] for the theory of distributions. Let us specify that singular linear integral operators which emerge in the course of our reasoning will **not** be approached via parametric families of regular linear integral functionals with piecewise continuous or smooth kernels (in which context the notation "$\boldsymbol{\delta}(0)$" for Dirac's *function* is used in the literature).

Journal of Physics: Conference Series **474** (2013) 012024 doi:10.1088/1742-6596/474/1/012024

In this specific setup we analyse the construction of two distinct couplings of the BV-fibres' ghost parity-homogeneous vector subspaces with their respective duals. In particular, in section 2.2 we focus on the rule of signs which determines the anti-commutation of differential one-forms in the geometry at hand. Applying the geometric concept of iterated variations in section 2.3, we represent the left- and right variations of functionals in terms of left- or right-directed singular linear integral operators; this framework ensures the intrinsic regularisation of iterated variations. We then formulate in section 2.4 the definitions of BV-Laplacian Δ and variational Schouten bracket $[\![\,,\,]\!]$ (or *antibracket*). We show that these definitions are operational, amounting to natural, well-defined reconfigurations of the geometry (but not to any hand-made algorithms for cancellation of divergent terms; for those do not appear at all). Our main result, which is contained in section 2.5, is an explicit proof – that is, starting from basic principles – of relations (1). In other words, we neither postulate a validity of these properties nor elaborate a cunning syllogism the aim of which would be to convince why such assertions should hold provided that one knows when various (derivatives of) Dirac's $\boldsymbol{\delta}$-functions must be erased in the course of so arguable a reasoning.

For consistency, we first apply the above theory to a standard derivation of the quantum master-equation from the Schwinger–Dyson condition that essentially eliminates a dependence on the unphysical, ghost parity-odd dimensions (see section 3.1); we also recall here the construction of quantum BV-differential. The point is that neither divergencies nor *ad hoc* cancellations occur in the entire argument. On the same grounds we address in section 3.2 the quantum BV-cohomology preservation by infinitesimal gauge symmetries of the quantum master-equation. (We refer to [7, 8, 20, 22] and also [1, 37, 51] in this context; several methodological comments, which highlight our concept, are placed in section 3 along the lines of a well-known reasoning.)

The paper concludes with a statement that an intrinsic regularisation in the geometry of iterated variations relies on the principle of locality (which manifests also through causality). We argue that a logical complexity of geometric objects grows while they accumulate the (iterated) variations; a conversion of such composite-structure objects into maps which take physical field configurations to numbers entails a decrease of the complexity via a loss of information. Having motivated this claim in section 2, we prove that the logic of analytic reasonings may not be interrupted; for example, the right-hand side of (1c) is not assembled from the would-be constituent blocks ΔF and ΔG for which it is known in advance how they take field configurations to numbers whenever the functionals F and G are given.

The paper explicitly answers the question what variations are — in particular, what iterated variations are. Moreover, we tacitly describe a geometric mechanism which is responsible for the anti-commutation of differential one-forms; such mechanism ensures that the results of calculations match empiric data even if the exterior algebras of forms are introduced by hand. The roots of this principle are none other that the ordering of dual vector spaces which stem in the course of variations in models of nonlinear phenomena (this picture is addressed in section 2.2).

We illustrate our approach with elementary starting section 1 in which we inspect the matching of geometries –one for an action functional, the other for a field's test shift– in the course of derivation of Euler–Lagrange equation of motion in field theory. The second example on pp. 34–36 clarifies the idea specifically in the BV-setup of (anti)fields and (anti)ghosts. We thus provide a pattern for all types of calculations which involve the Schouten bracket and BV-Laplacian in any model.

Historical context: an overview

There is a class of significant papers in which the BV-formalism is developed under assumption that the space-time is a point. Indeed, such hypothesis is equivalent to an agreement that the only admissible sections of bundles over space-time are constant; this implies that even if their derivatives are nominally present in some formulas, they are always equal to zero. The calculus of variations then reduces to usual differential geometry on the bundles' fibres. It must be noted that publications containing the above assumption did contribute to the subject and in many cases guided its further development (we recall the respective comment in [51] and refer to [12, 22, 27, 39, 48, 53]). Moreover, the no-derivatives reduction sometimes allows one to jump at conclusions which are correct; an integration by parts over the base manifold M^n is restored –whenever possible– at the end of the day. Still this oversimplification is potentially dangerous because variational calculus of integral functionals conceptually exceeds any classical differential geometry on the fibres (see [33] for discussion and [28, 34]). In the variational setup, the objects and their properties become geometrically different from their analogues on usual manifolds even if the terminology is kept unchanged. Here we recall for example that variational multivectors do not split to wedge products of variational one-vectors and likewise, several Leibniz rules are irreparably lost but this can not be noticed when all derivatives equal zero. In fact, it is the abyss between classical geometry of manifolds and geometry of variations for jet spaces of maps of manifolds which motivated our earlier study [34]. Yet the misconception is still present in active research, e.g., see [4, 27, 39, 44].

The fact of incompleteness of such heuristic analogies from usual geometry of manifolds is signalled in [51]. Paradoxically, it is simultaneously not true that a solution of the regularisation problem for BV-Laplacian has no analogues in the case of ODE dynamics on manifolds. From section 1 below it is readily seen that good old techniques persist in the finite-dimensional ODE geometry at the level of standard linear algebra of dual vector spaces.[3]

The article [51] is a considerable step towards a solution of the regularisation problem in BV-formalism. A weighted, critical overview of various inconsistencies, *ad hoc* practices, and roundabouts is summed up there. The object of [51] was to formulate a self-contained analytic concept which would make the variational calculus of functionals free from divergencies and infinities. Still it remained unclear from [51] what the generality of underlying geometry is and why such self-consistent formalism should actually exist at the level of objects, i.e., beyond a mere ability to write formulas. In particular, it remained unnoticed that the main motivating example –namely, the canonical BV-setup– itself is the only class of geometries in which the technique is grounded.[4] A correctness but incompleteness of the approach in [51] means the following in practice. Whenever a theorist refers to the formalism of *loc. cit.*, Nature immediately creates a new, principally inobservable essence –a metric field which is denoted by $E(x_1, \ldots, x_n; \Gamma)$ in [51]– on top of the electromagnetic and weak gauge connections, as well as the fields for strong force, gravity, or any other gauge fields Γ. It is perhaps this methodological difficulty which hints us why the approach of [51] is considered "formal" by many experts; that conceptual paper remains scarcely known to a wider community.[5]

[3] On the other hand, the variational setup highlights the fundamental concept of a physical field as a system with degrees of freedom attached at every point of the space-time M^n; we focus on this aspect in what follows.

[4] The integration of closed algebra of gauge symmetries for the quantum master-equation to a group of transformations of the master-action $S^\hbar$ remains a separate problem, which is also addressed in [51]. Suppose that the standard cohomological obstructions to such integration vanish (see section 3.2 below), whence (i) all infinitesimal transformations of the functional $S^\hbar$ are exact, i.e., they are generated by odd ghost-parity elements F, and also (ii) such transformations can be extended from the master-action $S^\hbar$ to evolution of the observables $\mathcal{O}$. We remark that, unlike it is claimed in [51], neither of the two groups of functionals' transformations is induced by any well-defined change of BV-coordinates; of course, evolutionary vector fields are well-defined objects in that geometry and one could study them regardless of these functionals' transformations. We shall recall in section 3.2 the standard construction of automorphisms for quantum BV-cohomology groups; it illustrates our concept because the notion of quantum gauge symmetries explicitly refers to all basic properties of the BV-Laplacian and Schouten bracket, see (1) on p. 1.

[5] An attempt to interpret the formalism of [51] in terms of the language of PDE geometry (particularly, in the context of [41], see also [28, 40, 45]) was performed in [23] and published in abridged form in [24]. The construction of Schouten bracket in [23] relies on the notion of variational cotangent bundle [41] and on classical approach to

Journal of Physics: Conference Series **474** (2013) 012024 doi:10.1088/1742-6596/474/1/012024

To demystify the notion of a "metric field $E(x_1, \ldots, x_n; \Gamma)$," we describe in this paper an elementary geometric mechanism for the long-expected but still intuitively paradoxical analytic behaviour of variations. This mechanism implies that Nature is not obliged to respond to the needs of a theorist and create such multi-entry distributions upon request.

Another line of reasoning, which led to much progress in a revision of BV-structures and regularisation of divergences, was pursued in [12, 13]. We recall that the language of *loc. cit.* is functional analytic so that the theory's objects are viewed as (Dirac's) distributions (and heat kernels are implemented). According to [12, §1.8], the BV-Laplacian Δ which is used in physical theories is ill-defined because for a given action S over space-time M^n of positive dimension n the object ΔS involves a multiplication of singular distributions (and thus –a quotation from [12] continues– ΔS has the same kind of singularities as appear in one-loop Feynman diagrams). The regularisation technique proposed in [12, 13] stems from analysis of the distributions' limit behaviour as one approaches the "physical" structures by using regular ones.

The resolution to apparent difficulties is that there are several distinct geometric constructions which yield the same singular linear operators with support on the diagonal (in what follows we study in detail on which space such operators are defined).

We now discuss a peculiar, well-established domain, the very form of existence of which could be hardly believed in. In that theory, there is a serious lack of rigorous definitions for the most elementary objects; at the same time, there is a rapidly growing number of monumental reviews. Whereas the theory's difficulties are clearly inherited from a deficit of boring rigour at the initial stage, such hardships are proclaimed the theory's immanent components. At expert level it is mandatory to have a firm knowledge of the built-in difficulties and readily classify the descriptive objects which those apparent obstructions bring into the mathematical apparatus. (There is no firm guarantee that the (un)necessary objects really exist beyond written formulas.) The way of handling inconveniences largely amounts not to resolving them by a thorough study of their origins but to some *ad hoc* methods for hiding their presence. Doing research is thus substituted by practising a ritual.

However, the community of experts who mature in operation with formulas (a part of which are believed to express something objectively existing) maintains a considerable pluralism about a proper way to mask the symptoms of troubles:

- The radicals declare that undefined objects which seem to make trouble must be set equal to zero.
- The revisionist approach prescribes a postfactum erasing of not the entire objects (which are still undefined) but of undesirable elements in those objects' description.
- A diplomatic viewpoint is that there might be sources of trouble but their contribution to final results is suppressed as soon as the objects' desired properties are postulated (regardless of the actual presence or absence of such sources and one's ability to substantiate those properties).

For an external observer, this state-of-the-art could seem atypical for a consistent theory. Indeed, the reliability of its main pillar is a matter of irrational belief.

1. The geometry of variations

Let us first analyse the basic geometry of variations of functionals; by comprehending the full setup of a one-time variation, we shall understand the geometry of many. Specifically, in this

the theory of variations. On one hand, this ensures the validity of Jacobi identity for the bracket (see the *second* half of Eq. (1d) but *not* the first one). But on the other hand, we have showed by a counterexample in [35, §3] that the old approach fails to relate by (1c) the Schouten bracket to BV-Laplacian. In other words, the BV-Laplacian did not entirely generate the variational Schouten bracket, making only Eq. (1b) but not (1c) possible in that geometry (cf. [39]).

section we reveal the interrelation of bundles in the course of integration by parts; we also explain a rigorous construction of iterated variations.

The core of traditional difficulties in this domain is that a use of only fibre bundles π of physical fields, which are subjected to test shifts, is insufficient. We argue that the tangent bundles $T\pi$ to the bundles π may not be discarded (see Fig. 1). For identities (1) to hold one

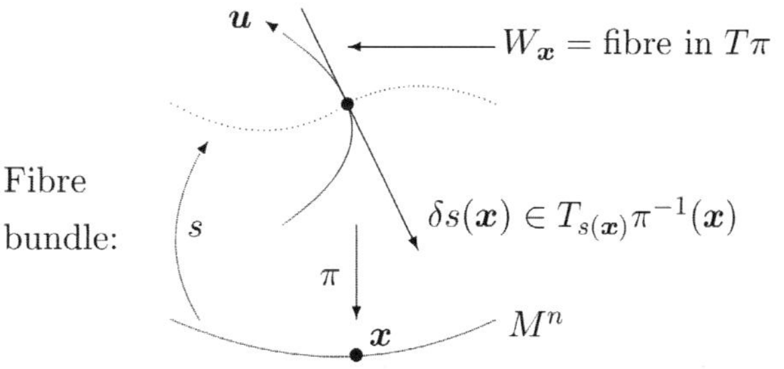

Figure 1. The fibre bundle π of fields s and vector bundle $T\pi$ of their variations δs.

must substantiate why higher-order variational derivatives are (graded-)permutable whenever one inspects the response of a given functional to shifts of its argument along several directions. To resolve the difficulties, we properly enlarge the space of functionals and adjust a description of the geometry for the functionals' variations: in fact, each variation brings its own copy of the base M^n into the picture (see Fig. 3 on p. 14).

1.1. Notation

We now fix some notations, in most cases matching that from [28] (for a more detailed exposition of these matters, see for example [28, 40, 45]).

Let $\pi\colon E \to M$ be a smooth fibre bundle[6] with m-dimensional fibres $\pi^{-1}(\boldsymbol{x})$ over points $\boldsymbol{x}$ of a smooth real oriented manifold M of dimension n; we assume that all mappings, including those which determine the smoothness class of manifolds, are infinitely smooth.

We let x^i denote local coordinates in a chart $U_\alpha \subseteq M^n$ and u_j be the fibre coordinates. We denote by $[\boldsymbol{u}]$ a differential dependence of the fibre variables (specifically in the BV-setup, a differential dependence $[\boldsymbol{q}]$ on physical fields and other ghost parity-even variables, and we denote by $[\boldsymbol{q}^\dagger]$ that of ghost parity-odd BV-variables).

Remark 1.1. We suppose that the initially given bundle π of physical fields is not graded. In what follows, starting with π, we shall construct new bundles whose fibres are endowed with the $\mathbb{Z}_2$-valued ghost parity gh($\cdot$). However, our reasoning remains valid for superbundles $\pi^{(0|1)}$ over supermanifolds $M^{(n_0|n_1)}$ ([10, 52]) and to a noncommutative setup of cyclic-invariant words (see [29, 32] and references therein), cf. Fig. 8 below.

We take the infinite jet space $\pi_\infty\colon J^\infty(\pi) \to M$ associated with this bundle [15, 45]; a point from the jet space is then $\theta = (x^i, \boldsymbol{u}, u^k_{x^i}, u^k_{x^ix^j}, \ldots, \boldsymbol{u}_\sigma, \ldots) \in J^\infty(\pi)$, where σ is a multi-index and we put $\boldsymbol{u}_\varnothing \equiv \boldsymbol{u}$. If $s \in \Gamma(\pi)$ is a section of π, or a *field*, we denote by $j^\infty(s)$ its infinite jet, which is a section $j^\infty(s) \in \Gamma(\pi_\infty)$. Its value at $\boldsymbol{x} \in M$ is $j^\infty_{\boldsymbol{x}}(s) = (x^i, s^\alpha(x), \frac{\partial s^\alpha}{\partial x^i}(x), \ldots, \frac{\partial^{|\sigma|}s^\alpha}{\partial x^\sigma}(x), \ldots) \in J^\infty(\pi)$.

We denote by $\mathcal{F}(\pi)$ the properly understood algebra of finite differential order smooth functions on the infinite jet space $J^\infty(\pi)$, see [28, 40] for details. The space of top-degree

[6] Vector bundles are primary examples but we do not actually use the linear vector space structure of their fibres so that π could be any smooth fibre bundle.

horizontal forms on $J^\infty(\pi)$ is denoted by $\overline{\Lambda}^n(\pi)$; let us also assume that at every $\boldsymbol{x} \in M$ a volume element $\mathrm{dvol}(\boldsymbol{x})$ is specified so that its pull-back under π_∞^* is an n-th degree form in $\overline{\Lambda}^n(\pi)$, cf. Remark 1.5 on p. 9.

The highest horizontal cohomology, i. e., the space of equivalence classes of n-forms from $\overline{\Lambda}^n(\pi)$ modulo the image of the horizontal exterior differential $\bar{\mathrm{d}}$ on $J^\infty(\pi)$, is denoted by $\overline{H}^n(\pi)$; the equivalence class of $\omega \in \overline{\Lambda}^n(\pi)$ is denoted by $\int \omega \in \overline{H}^n(\pi)$. We assume that sections $s \in \Gamma(\pi)$ are such that integration of functionals $\Gamma(\pi) \to \Bbbk$ by parts is allowed and does not result in any boundary terms (for example, the base manifold is closed, or the sections all have compact support, or decay sufficiently fast towards infinity, or are periodic).

1.2. Euler–Lagrange equations

A derivation of Euler–Lagrange equations $\mathcal{E}_{\mathrm{EL}}$ for a given action functional $S = \int \mathcal{L}(\boldsymbol{x}, [\boldsymbol{u}])\,\mathrm{dvol}(\boldsymbol{x})$ is a model example which illustrates the correlation of two geometries:[7] one for "trajectories" $s \in \Gamma(\pi)$ and the other for shifts δs. It is well known that the functional's response to a test shift δs of its argument $s \in \Gamma(\pi)$ is described by the formula [2, §12]

$$\frac{\mathrm{d}}{\mathrm{d}\varepsilon}\bigg|_{\varepsilon=0} S(s + \varepsilon \cdot \overleftarrow{\delta s}) = \int_M \mathrm{dvol}(\boldsymbol{x})\, \delta s(\boldsymbol{x}) \cdot \left.\frac{\overleftarrow{\delta}\mathcal{L}(\boldsymbol{x}, [\boldsymbol{u}])}{\delta \boldsymbol{u}}\right|_{j^\infty_{\boldsymbol{x}}(s)}. \tag{3}$$

We now claim that this one-step procedure is a correct consequence of definitions but itself not a definition of the functional's variation. The above formula conceals a longer, nontrivial reasoning of which the right-hand side in (3) is an implication — provided that the functional S will not be varied by using any other test shifts, i. e., if the correspondence $S \mapsto \mathcal{E}_{\mathrm{EL}}$ yields the object $\mathcal{E}_{\mathrm{EL}}$ of further study (cf. [2, §13]). Indeed, we notice that the left-hand side of (3) refers to *three* bundles (namely, the fibre bundle π for a section $s \in \Gamma(\pi)$ whose infinite jet is $j^\infty(s) \in \Gamma(\pi_\infty)$, the bundle π_∞ for the integral functional $S \in \overline{H}^n(\pi)$, and the tangent vector bundle $T\pi$ such that $\delta s \in \Gamma(T\pi)$ at the graph of s in π, see Fig. 1. (In what follows, a reference to attachment points $s(\boldsymbol{x}) \in \pi^{-1}(\boldsymbol{x})$ will always be implicit in the notation for δs: for a given section $s \in \Gamma(\pi)$, the base manifold M^n is the domain of definition for a test shift $\delta s(\boldsymbol{x}, s(\boldsymbol{x})) = \delta s(\boldsymbol{x})$ that takes values in $T_{s(\boldsymbol{x})}\pi^{-1}(\boldsymbol{x})$.) Let us figure out how the domains of definition for the sections s and δs merge to one copy of the manifold M^n over which an integration is performed in the right-hand side of (3). Strictly speaking, from (3) it is unclear whether the variational derivative,

$$\frac{\overleftarrow{\delta}\mathcal{L}(\boldsymbol{x}, [\boldsymbol{u}])}{\delta \boldsymbol{u}} = \sum_{|\sigma| \geq 0} \left(-\frac{\vec{\mathrm{d}}}{\mathrm{d}\boldsymbol{x}}\right)^\sigma \frac{\vec{\partial}\mathcal{L}(\boldsymbol{x}, [\boldsymbol{u}])}{\partial \boldsymbol{u}_\sigma},$$

stems from one (which would be false) or both (true!) copies of the base M.

To have a clear vision of the variations' geometry and by this avoid an appearance of phantoms in description, we now vary the action functional S at $s \in \Gamma(\pi)$ along $\delta s \in \Gamma(T\pi)$, commenting on each step we make. In fact, it suffices to figure out where the objects and structures at hand belong to — in particular, we should explain the nature of binary operation $\cdot$ in the right-hand side of conventional formula (3). The key idea is to understand what we are actually doing but not what we have got used to think we do in order to obtain an understandable result [2, §13]. The discovery is that this "multiplication of functions" is a shorthand notation for the

[7] An arrow over a variational derivative indicates the direction along which the shift δs is transported left- or rightmost. While the objects are non-graded commutative, this indication is not important. It becomes mandatory in the $\mathbb{Z}_2$-graded commutative setup (see section 2): likewise, the arrows are also mandatory and fix the direction of rotation for non-commutative cyclic words [29, 32, 36]; note that our formalism is extended verbatim to the variational calculus of such necklaces and their brackets.

canonically defined coupling between vectors and covectors from (co)tangent spaces $W_{s(\boldsymbol{x})}$ and $W^\dagger_{s(\boldsymbol{x})}$, respectively, at the points $s(\boldsymbol{x})$ of fibres $\pi^{-1}(\boldsymbol{x})$ in the bundle π.

To encode this linear-algebraic setup, let i, j run from 1 to $m = \dim(\pi^{-1}(\boldsymbol{x})) = \operatorname{rank}(T\pi)$ and take a local basis $\vec{e}_i(\boldsymbol{y})$ in the tangent spaces $W_{s(\boldsymbol{y})} = T_{s(\boldsymbol{y})}(\pi^{-1}(\boldsymbol{y}))$ at $s(\boldsymbol{y})$ over base points $\boldsymbol{y} \in M$. Introduce the dual basis $\vec{e}^{\,\dagger j}(\boldsymbol{x})$ in $W^\dagger_{s(\boldsymbol{x})}$ attached at $s(\boldsymbol{x})$ over $\boldsymbol{x} \in M$. By construction, this means that the value

$$\Big\langle \vec{e}_i(\boldsymbol{y}), \vec{e}^{\,\dagger j}(\boldsymbol{x}) \Big\rangle \tag{4}$$

is equal to the Kronecker symbol δ_i^j if and only if $\boldsymbol{x} = \boldsymbol{y}$ and the vector $\vec{e}_i(\boldsymbol{y}) \in W_{p_1}$ and covector $\vec{e}^{\,\dagger j}(\boldsymbol{x}) \in W^\dagger_{p_2}$ are attached at the same point $p_1 = p_2$ of the fibre $\pi^{-1}(\boldsymbol{x})$ over $\boldsymbol{x} = \boldsymbol{y} \in M$.

The *locality* of this coupling is an absolute geometric postulate: the coupling is not defined whenever $\boldsymbol{x} \neq \boldsymbol{y}$ or the values $p_1 = s_1(\boldsymbol{y})$ and $p_2 = s_2(\boldsymbol{x})$ of two local sections $s_1, s_2 \in \Gamma(\pi)$ are not equal at $\boldsymbol{x} = \boldsymbol{y}$. Physically speaking, the coupling is then not defined because there is no channel of information which would communicate the value $\delta s^i(\boldsymbol{y}) \cdot \vec{e}_i(\boldsymbol{y})$ of excitation of the physical field $s \in \Gamma(\pi)$ at a point $\boldsymbol{y} \in M$ to another point $\boldsymbol{x} \neq \boldsymbol{y}$ of the space-time M.

Remark 1.2. Let us remember that the definition of coupling between sections of (co)tangent bundles — i. e., (co)tangent to either a given manifold or a given bundle π which is the case here for Euler–Lagrange equations — forces the congruence $\{\boldsymbol{x} = \boldsymbol{y},\ s_1(\boldsymbol{y}) = s_2(\boldsymbol{x})\}$ of the (co)vectors' attachment points. We notice further that such congruence mechanism does not refer to any limiting procedure for smooth distributed kernels and regular linear operators on the space of (co)vector fields. Indeed, vectors couple with their duals at a given point regardless of any phantom limiting procedure which would grasp the (co)vector's values at any other points of the manifold.[8]

Remark 1.3. The coupling is a matching between test-shift vector fields which are tangent to the fibres of π and, on the other hand, with the elements of $\Gamma(T^*\pi)$ which are determined by the Lagrangian $\mathcal{L}$. This binary operation yields the singular integral operator $\int_M \mathrm{d}\boldsymbol{y}\, \langle \delta s^i(\boldsymbol{y})\vec{e}_i(\boldsymbol{y})|$ with support on the diagonal. Independently, the same operator can reappear as the limit in a parametric family of regular integral operators with smooth, distributed kernels. This shows that the same object is constructed by using several algorithms. Yet the analytic behaviour of the limit is determined not only by the limit itself but also by an algorithm how it is attained. Consequently, the object's analytic properties in the course of derivations could be (and actually, indeed they are) drastically different for different scenarios. This is the key point in a regularisation of the formalism; to achieve this goal, we properly identify the objects which are de facto handled.

Remark 1.4. Referring to a concept of locality of events, this definition of coupling $\langle\,,\,\rangle$ ensures a very interesting analytic behaviour of the value $\langle \vec{e}_i(\boldsymbol{y}), \vec{e}^{\,\dagger i}(\boldsymbol{x})\rangle$ of pairing for dual objects $\vec{e}_i(\boldsymbol{y})$ and $\vec{e}^{\,\dagger i}(\boldsymbol{x})$ at fixed i. Namely, this value is a constant scalar field which equals unit $1 \in \Bbbk$ at all points of the manifold M; the scalar field's partial derivatives with respect to x^j or y^k, $1 \leq j, k \leq n$, vanish identically. We shall use this property in what follows (see Remark 1.7 on p. 11). We also note that the logarithm of this coupling's unit value vanishes as well: $\log\langle \vec{e}_i(\boldsymbol{y}), \vec{e}^{\,\dagger i}(\boldsymbol{x})\rangle = 0$ whenever the coupling is well defined and $1 \leq i \leq m$.

Now let us return to the initial setup in context of Euler–Lagrange equation $\mathcal{E}_{\mathrm{EL}}$ and one-step correspondence $S \mapsto \mathcal{E}_{\mathrm{EL}}$, see Fig. 1. We have that $S = \int \mathcal{L}(\boldsymbol{x}, [\boldsymbol{u}])\, \mathrm{dvol}(\boldsymbol{x})$ is an integral functional; we let $s \in \Gamma(\pi)$ be a background section (e. g., a sought-for solution of the Euler–Lagrange stationary point equation $\delta S|_s = 0$) and $\delta s \in \Gamma(T\pi)$ be a test shift of s. The linear

[8] We recall that a similar, purely local geometric principle, not referring to the objects' values at non-coinciding points, works in the definition of Hirota's bilinear derivative.

term in a response of $S\colon \Gamma(\pi) \to \Bbbk$ to a shift of its argument s along δs is (cf.(14) on p. 20)

$$\frac{\mathrm{d}}{\mathrm{d}\varepsilon}\Big|_{\varepsilon=0} S(s + \varepsilon\,\overleftarrow{\delta s}) =$$

$$= \sum_{i,j}\sum_{|\sigma|\geq 0} \int_M \mathrm{d}\boldsymbol{y} \int_M \mathrm{dvol}(\boldsymbol{x}) \left\langle (\delta s^i)\left(\frac{\overleftarrow{\partial}}{\partial \boldsymbol{y}}\right)^{\sigma}(\boldsymbol{y})\,\vec{e}_i(\boldsymbol{y}), \vec{e}^{\,\dagger j}(\boldsymbol{x})\,\frac{\overrightarrow{\partial}\mathcal{L}(\boldsymbol{x},[\boldsymbol{u}])}{\partial u^j_\sigma}\Big|_{j^\infty_{\boldsymbol{x}}(s)}\right\rangle. \quad (5)$$

Remark 1.5. The rôles of two integral signs in (5) are different. Namely, the volume form $\mathrm{dvol}(\boldsymbol{x})$ at $\boldsymbol{x} \in M^n$ comes from the integral functional $S \in \overline{H}^n(\pi)$; should a formal choice of the volume form be different, the Euler–Lagrange equations would also change.[9] At the same time, the other integral sign $\int \mathrm{d}\boldsymbol{y}$ denotes the singular linear operator $\Gamma(T^*\pi) \to \Bbbk$ with support on the diagonal [19]; in fact, this notation means that a point $\boldsymbol{y}$ runs through the entire integration domain M.

1.3. Integration by parts

The most interesting things start to happen when one integrates by parts over the domain M^n of test shifts δs. (By default, we let the supports of local perturbations δs be such that no boundary terms appear in the course of integration by parts over M.)

For the sake of transparency let us first consider a model situation when there is just one derivative falling on δs at $\boldsymbol{y}$; all higher-order cases are processed recursively. By the definition of a (partial) derivative $\partial/\partial y^i$, we have that[10]

$$\int_M \mathrm{d}\boldsymbol{y} \int_M \mathrm{dvol}(\boldsymbol{x}) \left\langle (\delta s)\frac{\overleftarrow{\partial}}{\partial \boldsymbol{y}}(\boldsymbol{y})\,\vec{e}(\boldsymbol{y}), \vec{e}^{\,\dagger}(\boldsymbol{x})\,\frac{\overrightarrow{\partial}\mathcal{L}(\boldsymbol{x},[\boldsymbol{u}])}{\partial \boldsymbol{u}_\sigma}\Big|_{j^\infty_{\boldsymbol{x}}(s)}\right\rangle =$$

$$= -\int_M \mathrm{d}\boldsymbol{y} \int_M \mathrm{dvol}(\boldsymbol{x})\,\delta s(\boldsymbol{y})\frac{\overrightarrow{\partial}}{\partial \boldsymbol{y}}\left\{\langle \vec{e}(\boldsymbol{y}), \vec{e}^{\,\dagger}(\boldsymbol{x})\rangle\,\frac{\overrightarrow{\partial}\mathcal{L}(\boldsymbol{x},[\boldsymbol{u}])}{\partial \boldsymbol{u}_\sigma}\Big|_{j^\infty_{\boldsymbol{x}}(s)}\right\}.$$

By using a definition of the partial derivative which falls on the comultiple of δs, we obtain the

[9] There are natural classes of geometries in which the Lagrangian $\mathcal{L}(\boldsymbol{x},[\boldsymbol{u}])$ in the action S is a well-defined top-degree differential form, e. g., if the unknowns $\boldsymbol{u}$ are differential one-forms (we recall the Yang–Mills or Chern–Simons gauge theories in this context). Let us remember also that a construction of $\mathcal{L}$ could refer to a choice of volume form $\mathrm{dvol}(\boldsymbol{x})$ on M^n. For instance, such is the case when the Hodge structure $*$ is involved (the Yang–Mills Lagrangian yields an example: $\mathcal{L} \sim F_{\mu\nu} * F^{\mu\nu}$ in standard notation for the stress tensor). To avoid excessive case-study, we use a uniform notation thus writing $\mathrm{dvol}(\boldsymbol{x})$ explicitly.

We recall further that the integration measure $\mathrm{dvol}(\boldsymbol{x}, s(\boldsymbol{x})) = \sqrt{|\det(g_{\mu\nu}(\boldsymbol{x},s))|}\mathrm{d}\boldsymbol{x}$ is field-dependent by virtue of Einstein's general relativity equations which –in their right-hand sides– absorb the energy-momentum tensor of physical fields $s \in \Gamma(\pi)$. The volume element will be denoted by $\mathrm{dvol}(\boldsymbol{x})$ in order to emphasize that the space-time M^n is unique: Namely, field-dependent objects interact at its points only if the local geometry of underlying space-time is the same near $\boldsymbol{x} \in M^n$ for all objects (see Theorem 3 and Remark 2.11 on p. 29 for a realisation of this principle for the smooth manifold M^n endowed with metric tensor $g_{\mu\nu}$).

[10] In the definition of derivative, the calculation of length $|\Delta \boldsymbol{y}|$ in denominators refers to the standard Euclidean metric in the linear vector spaces which determine coordinate neighbourhoods near points of the manifold M at hand.

difference[11]

$$\stackrel{\text{def}}{=} -\int_M \mathrm{d}\boldsymbol{y} \int_M \mathrm{dvol}(\boldsymbol{x})\, \delta s(\boldsymbol{y}) \cdot \lim_{|\Delta \boldsymbol{y}| \to 0} \frac{1}{|\Delta \boldsymbol{y}|} \Big\{ \langle \vec{e}(\boldsymbol{y}+\Delta\boldsymbol{y}), \vec{e}(\boldsymbol{x}) \rangle \left. \frac{\overrightarrow{\partial} \mathcal{L}(\boldsymbol{x}, [\boldsymbol{u}])}{\partial \boldsymbol{u}_\sigma} \right|_{j^\infty_{\boldsymbol{x}}(s)} - \langle \vec{e}(\boldsymbol{y}), \vec{e}(\boldsymbol{x}) \rangle \left. \frac{\overrightarrow{\partial} \mathcal{L}(\boldsymbol{x}, [\boldsymbol{u}])}{\partial \boldsymbol{u}_\sigma} \right|_{j^\infty_{\boldsymbol{x}}(s)} \Big\}.$$

The locality postulate for coupling between (co)vectors $\vec{e}$ and $\vec{e}^\dagger$ forces the equality $\boldsymbol{y}+\Delta\boldsymbol{y} = \boldsymbol{x}$ in the minuend, which yields the two different points at which the restriction of Lagrangian $\mathcal{L}$ to the jet $j^\infty(s)$ of section $s \in \Gamma(\pi)$ is evaluated:

$$= -\int_M \mathrm{d}\boldsymbol{y} \int_M \mathrm{dvol}(\boldsymbol{x})\, \delta s(\boldsymbol{y}) \cdot \langle \vec{e}(\boldsymbol{y}), \vec{e}^\dagger(\boldsymbol{x}) \rangle \cdot \lim_{|\Delta \boldsymbol{y}| \to 0} \frac{1}{|\Delta \boldsymbol{y}|} \Big\{ \left. \frac{\overrightarrow{\partial} \mathcal{L}(\boldsymbol{x}, [\boldsymbol{u}])}{\partial \boldsymbol{u}_\sigma} \right|_{j^\infty_{\boldsymbol{x}+\Delta\boldsymbol{y}}(s)} - \left. \frac{\overrightarrow{\partial} \mathcal{L}(\boldsymbol{x}, [\boldsymbol{u}])}{\partial \boldsymbol{u}_\sigma} \right|_{j^\infty_{\boldsymbol{x}}(s)} \Big\}.$$

(Here we use the fact that the scalar product $\langle\, ,\, \rangle$, whenever defined, is the Kronecker symbol.) We continue the equality,

$$\stackrel{\text{def}}{=} -\int_M \mathrm{d}\boldsymbol{y} \int_M \mathrm{dvol}(\boldsymbol{x}) \left\langle \delta s(\boldsymbol{y})\, \vec{e}(\boldsymbol{y}), \vec{e}^\dagger(\boldsymbol{x}) \frac{\overrightarrow{\partial}}{\partial \boldsymbol{x}} \left(\left. \frac{\overrightarrow{\partial} \mathcal{L}(\boldsymbol{x}, [\boldsymbol{u}])}{\partial \boldsymbol{u}_\sigma} \right|_{j^\infty_{\boldsymbol{x}}(s)} \right) \right\rangle.$$

We finally recall that the total derivative $\mathrm{d}/\mathrm{d}\boldsymbol{x}$ is defined[12] via an application of $\partial/\partial\boldsymbol{x}$ to restriction to infinite jets $j^\infty(s)$ of sections s at base points $\boldsymbol{x}$. Therefore, the above expression is equal to

$$\stackrel{\text{def}}{=} \int_M \mathrm{d}\boldsymbol{y} \int_M \mathrm{dvol}(\boldsymbol{x}) \left\langle \delta s(\boldsymbol{y})\, \vec{e}(\boldsymbol{y}), \vec{e}^\dagger(\boldsymbol{x}) \left. \left(\left(-\frac{\vec{\mathrm{d}}}{\mathrm{d}\boldsymbol{x}} \right) \frac{\overrightarrow{\partial} \mathcal{L}(\boldsymbol{x}, [\boldsymbol{u}])}{\partial \boldsymbol{u}_\sigma} \right) \right|_{j^\infty_{\boldsymbol{x}}(s)} \right\rangle.$$

This shows that an integration by parts over the base M in the geometry of test shift δs reappears as integration by parts in the bundle where lives the background section $s \in \Gamma(\pi)$.

Repeating the integration by parts $|\sigma| \geq 0$ times in each term of the sum in (5), we obtain the expression

$$\sum_{i,j} \sum_{|\sigma| \geq 0} \int_M \mathrm{d}\boldsymbol{y} \int_M \mathrm{dvol}(\boldsymbol{x}) \left\langle \delta s^i(\boldsymbol{y})\, \vec{e}_i(\boldsymbol{y}), \vec{e}^{\dagger j}(\boldsymbol{x}) \left. \left(\left(-\frac{\vec{\mathrm{d}}}{\mathrm{d}\boldsymbol{x}} \right)^\sigma \frac{\overrightarrow{\partial} \mathcal{L}(\boldsymbol{x}, [\boldsymbol{u}])}{\partial u^j_\sigma} \right) \right|_{j^\infty_{\boldsymbol{x}}(s)} \right\rangle.$$

Let us recall once more that the coupling's support is the diagonal in $M \times M$, at points of which the value $\langle \vec{e}_i(\boldsymbol{y}), \vec{e}^{\dagger j}(\boldsymbol{x}) \rangle$ is the Kronecker symbol δ^j_i. Consequently, we arrive at

$$= \sum_{i,j} \sum_{|\sigma| \geq 0} \int_M \mathrm{dvol}(\boldsymbol{x})\, \delta s^i(\boldsymbol{x}) \cdot \left. \left(\left(-\frac{\vec{\mathrm{d}}}{\mathrm{d}\boldsymbol{x}} \right)^\sigma \frac{\overrightarrow{\partial} \mathcal{L}(\boldsymbol{x}, [\boldsymbol{u}])}{\partial u^i_\sigma} \right) \right|_{j^\infty_{\boldsymbol{x}}(s)}.$$

[11] Here and in the equalities below we suppress the indexes i running through $1, \ldots, m$ at $\delta s^i(\boldsymbol{y})$ and $\vec{e}_i(\boldsymbol{y})$ or $\vec{e}^{\dagger i}(\boldsymbol{x})$, or at u^i_σ in the derivative which acts on $\mathcal{L}$; we thus avoid an agglomeration of formulas.
[12] By definition, $(\vec{\mathrm{d}}f/\mathrm{d}x^i)\big|_{j^\infty(s)}(\boldsymbol{x}) = (\vec{\partial}/\partial x^i(f\big|_{j^\infty(s)}))(\boldsymbol{x})$ for differential functions f, see [28, 40, 45].

This is formula (3); it is familiar from any textbook on variational principles of classical mechanics (e. g., see [2, §12–13]).

A standard reasoning shows that, whenever a response of the functional's value $S(s)$ to a test shift of s along any direction δs vanishes, the Euler–Lagrange equation holds:

$$\left.\frac{\overleftarrow{\delta S}}{\delta \boldsymbol{u}}\right|_{j^\infty(s)} = 0. \tag{6}$$

Its left-hand side belongs to the space $\Gamma(T^*\pi)$ of sections of the cotangent bundle to π.

Remark 1.6. This conclusion tells us that traditional attempts of a brute-force labelling of equations in a given system (6) by using the unknowns $\boldsymbol{u}$ is not geometric. Indeed, the equations' left-hand sides are sections of a *vector* bundle, thus forming linear $\Bbbk$-vector spaces so that addition is well defined for the equations within a system. On the other hand, the fibres in the bundle π can be smooth manifolds (i. e., not necessarily being vector spaces) so that one may not add points of those fibres; for such operation is in general not defined at all. Even if π is a vector bundle, the fibres of which are endowed with linear vector space structure, the two structures are not related.

Remark 1.7. The integration by parts transforms a derivative $\partial/\partial \boldsymbol{y}$ along one copy of the base M to the minus derivative $-\partial/\partial \boldsymbol{x}$ along the other copy. This produces no visible effect on the mechanism which ensures a restriction onto the diagonal in $M \times M$, i. e., there appears no would-be third term in the Leibniz rule for the product which is defined only on the diagonal. A desperate prescription (2) was introduced in the literature in order to mimick this paradoxical analytic behaviour of the coupling between elements of dual bases.

1.4. Why are variations permutable ?

Having outlined the matching of geometries in the course of one sequence of integrations by parts for *one* fixed pair $M \times M \ni (\boldsymbol{y}, \boldsymbol{x})$ of copies of the base manifold, we emphasize that such integrations must be performed last, i. e., only when the objects at hand are finally viewed as maps $\Gamma(\pi) \to \Bbbk$.

Should one haste in absence of clear understanding of what is actually being done and for which purpose, further calculation of higher-order variations could predictably but uncontrollably lead to meaningless, manifestly erroneous conclusions (e. g., compare left- and right-hand sides in (7) below).

Namely, there exist integral functionals which determine equal maps $\Gamma(\pi) \to \Bbbk$ but, belonging to different spaces, behave differently in the course of variations, should one attempt any. We say that such functionals are *synonyms*; for instance, see Example 2.4 in the next section for a nontrivial synonym ΔG of the zero functional (cf. Fig. 2). Informally speaking, the composite structure objects with repeated integrals over products $M \times M \times \ldots \times M$ of the base retain a kind of memory of the way how they were obtained from primary objects such as the action S. Let us illustrate these claims.

Example 1.1. Let $\delta s_1 \in \Gamma(T\pi)$ be a test shift at $s \in \Gamma(\pi)$ for an integral functional $S = \int \mathcal{L}(\boldsymbol{x}, [\boldsymbol{u}])\,\mathrm{dvol}(\boldsymbol{x})$ with density $\mathcal{L}$ of positive differential order. (That is, we suppose that some positive-order derivatives are always present in densities of all representatives of the cohomology class $S \in \overline{H}^n(\pi)$; this assumption is not to any extent restrictive but it allows us to not take into account $\bar{\mathrm{d}}$-exact terms whose orders may not be bounded.) By using S, let us construct two new integral functionals. First, we set

$$F = \sum_i \sum_{|\sigma|\geq 0} \int \mathrm{dvol}(\boldsymbol{x})\, \delta s_1^i(\boldsymbol{x}) \cdot \left(-\frac{\mathrm{d}}{\mathrm{d}\boldsymbol{x}}\right)^\sigma \left(\frac{\overrightarrow{\partial} \mathcal{L}(\boldsymbol{x}, [\boldsymbol{u}])}{\partial u^i_\sigma}\right) \in \overline{H}^n(\pi),$$

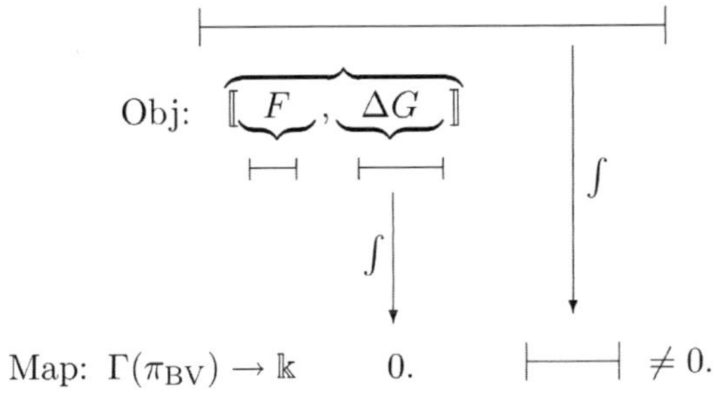

Figure 2. The synonyms ΔG of zero functional yield constant maps $0\colon \Gamma(\pi_{\text{BV}}) \to \Bbbk$ yet they can nontrivially contribute to larger structures such as $[\![F, \Delta G]\!]$, see Example 2.4 on p. 34.

so that the mapping $F\colon \Gamma(\pi) \to \Bbbk$ is defined at $s \in \Gamma(\pi)$ by restriction of the integrand to the jet $j^\infty(s)$ and then by actual integration over M.

Let the other functional $G \in \overline{H}^{2n}(\pi, T\pi)$ be such that its value at the same section $s \in \Gamma(\pi)$ is

$$G(s) = \sum_{i,j} \sum_{|\sigma| \geq 0} \int_M \mathrm{d}\boldsymbol{y} \int_M \mathrm{dvol}(\boldsymbol{x}) \left\langle (\delta s_1^i) \left(\frac{\overleftarrow{\partial}}{\partial \boldsymbol{y}} \right)^{\sigma} (\boldsymbol{y})\, \vec{e}(\boldsymbol{y}), \vec{e}^{\,\dagger j}(\boldsymbol{x})\, \frac{\overrightarrow{\partial} \mathcal{L}(\boldsymbol{x}, [\boldsymbol{u}])}{\partial u_\sigma^j} \bigg|_{j_{\boldsymbol{x}}^\infty(s)} \right\rangle.$$

From the previous section it is clear that F and G are indistinguishable as mappings to $\Bbbk$ for every $s \in \Gamma(\pi)$. Yet their variations, i. e., the responses to an extra shift $\delta s_2 \in \Gamma(T\pi)$, are different. Indeed, they are equal to, first,

$$\left(\frac{\mathrm{d}}{\mathrm{d}\varepsilon_2} \bigg|_{\varepsilon_2=0} F \right) (s + \varepsilon_2 \overleftarrow{\delta s_2}) = \sum_{i_1, i_2} \sum_{\substack{|\sigma_1| \geq 0 \\ |\sigma_2| \geq 0}} \int_M \mathrm{dvol}(\boldsymbol{x})\, \delta s_2^{i_2}(\boldsymbol{x}) \delta s_1^{i_1}(\boldsymbol{x}) \cdot$$
$$\cdot \left\{ \left(-\frac{\overrightarrow{\mathrm{d}}}{\mathrm{d}\boldsymbol{x}} \right)^{\sigma_2} \frac{\overrightarrow{\partial}}{\partial u_{\sigma_2}^{i_2}} \left(\left(-\frac{\overrightarrow{\mathrm{d}}}{\mathrm{d}\boldsymbol{x}} \right)^{\sigma_1} \frac{\overrightarrow{\partial} \mathcal{L}(\boldsymbol{x}, [\boldsymbol{u}])}{\partial u_{\sigma_1}^{i_1}} \right) \right\} \bigg|_{j_{\boldsymbol{x}}^\infty(s)}.$$

The above formula corresponds to a step-by-step calculation within a naïve approach to the geometry of variations. However, the genuine value of second variation of the integral functional S along δs_1 and then δs_2 at a section s is

$$\left(\frac{\mathrm{d}}{\mathrm{d}\varepsilon_2} \bigg|_{\varepsilon_2=0} G \right) (s + \varepsilon_2 \overleftarrow{\delta s_2}) = \sum_{\substack{i_1, i_2 \\ j_1, j_2}} \sum_{\substack{|\sigma_1| \geq 0 \\ |\sigma_2| \geq 0}} \int_M \mathrm{d}\boldsymbol{y}_2 \int_M \mathrm{d}\boldsymbol{y}_1 \int_M \mathrm{dvol}(\boldsymbol{x})$$
$$\left\{ (\delta s_2^{i_2}) \left(\frac{\overleftarrow{\partial}}{\partial \boldsymbol{y}_2} \right)^{\sigma_2} (\boldsymbol{y}_2)\, \langle \vec{e}_{i_2}(\boldsymbol{y}_2), \vec{e}^{\,\dagger j_2}(\boldsymbol{x}) \rangle \cdot (\delta s_1^{i_1}) \left(\frac{\overleftarrow{\partial}}{\partial \boldsymbol{y}_1} \right)^{\sigma_1} (\boldsymbol{y}_1)\, \langle \vec{e}_{i_1}(\boldsymbol{y}_1), \vec{e}^{\,\dagger j_1}(\boldsymbol{x}) \rangle \right\}$$
$$\cdot \frac{\overrightarrow{\partial}^2 \mathcal{L}(\boldsymbol{x}, [\boldsymbol{u}])}{\partial u_{\sigma_2}^{j_2} \partial u_{\sigma_1}^{j_1}} \bigg|_{j_{\boldsymbol{x}}^\infty(s)}.$$

The analytic distinction between the operators

$$\underbrace{\left(-\frac{\vec{\mathrm{d}}}{\mathrm{d}\boldsymbol{x}}\right)^{\sigma_2}\circ\frac{\vec{\partial}}{\partial u^{i_2}_{\sigma_2}}\circ\left(-\frac{\vec{\mathrm{d}}}{\mathrm{d}\boldsymbol{x}}\right)^{\sigma_1}\circ\frac{\vec{\partial}}{\partial u^{i_1}_{\sigma_1}}}_{\text{naïve approach}} \quad\text{and}\quad \underbrace{\left(-\frac{\vec{\mathrm{d}}}{\mathrm{d}\boldsymbol{x}}\right)^{\sigma_1\cup\sigma_2}\circ\frac{\vec{\partial}^2}{\partial u^{i_2}_{\sigma_2}\partial u^{i_1}_{\sigma_1}}}_{\text{geometric theory}} \tag{7}$$

reveals why in positive-order Lagrangian models it is forbidden to haste, which would imply that the derivatives along distinct copies of M for variations δs_1, …, δs_k are too early transformed to derivatives along the functional's own base. Such a conceptual error would repercuss with inexplicable, redundant terms in variations to-follow.

On the other hand, as soon as the product-bundle geometry of iterated variations is properly realized — so that all restrictions to the diagonals are postponed as late as possible, — the variations become (graded-)permutable.[13] Namely, denote by $|u^i|$, $1\leq i\leq m$, the overall $\mathbb{Z}_2$-valued parities of the fibre coordinates u^i; the ghost parity $\mathrm{gh}(u^i)$ or individual $\mathbb{Z}$- or $\mathbb{Z}_2$-valued gradings in the bundle π contribute additively to $|u^i|$ and then a residue modulo 2 is taken. Suppose that $\delta s_1=(\delta s_1^{i_1})$ and $\delta s_2=(\delta s_2^{i_2})$ are test shifts and $S=\int\mathcal{L}(\boldsymbol{x},[\boldsymbol{u}])\,\mathrm{dvol}(\boldsymbol{x})$ is an integral functional which maps a section $s\in\Gamma(\pi)$ to $\Bbbk$. Then, *after* the integrations by parts in the product-bundle geometry $\pi\times T\pi\times T\pi$ which is described above, there remains

$$\sum_{i_1,i_2}\sum_{\substack{|\sigma_1|\geq0\\|\sigma_2|\geq0}}\int_M \mathrm{dvol}(\boldsymbol{x})\,\delta s_2^{i_2}(\boldsymbol{x})\delta s_1^{i_1}(\boldsymbol{x})\left(\left(-\frac{\vec{\mathrm{d}}}{\mathrm{d}\boldsymbol{x}}\right)^{\sigma_1\cup\sigma_2}\frac{\vec{\partial}^2\mathcal{L}(\boldsymbol{x},[\boldsymbol{u}])}{\partial u^{i_2}_{\sigma_2}\partial u^{i_1}_{\sigma_1}}\right)\Bigg|_{j^\infty_{\boldsymbol{x}}(s)}$$

$$=\sum_{i_1,i_2}\sum_{\substack{|\sigma_1|\geq0\\|\sigma_2|\geq0}}(-)^{|u^{i_1}|\cdot|u^{i_2}|}\int_M \mathrm{dvol}(\boldsymbol{x})\,\delta s_1^{i_1}(\boldsymbol{x})\delta s_2^{i_2}(\boldsymbol{x})\left(\left(-\frac{\vec{\mathrm{d}}}{\mathrm{d}\boldsymbol{x}}\right)^{\sigma_1\cup\sigma_2}\frac{\vec{\partial}^2\mathcal{L}(\boldsymbol{x},[\boldsymbol{u}])}{\partial u^{i_1}_{\sigma_1}\partial u^{i_2}_{\sigma_2}}\right)\Bigg|_{j^\infty_{\boldsymbol{x}}(s)}.$$

Likewise, higher-order iterated variations with $k\geq2$ test shifts $\delta s_1,\ldots,\delta s_k$ are (graded-)permutable with the same rule of signs for permutations of order in which the (graded) partial derivatives $\vec{\partial}/\partial u^{i_1}_{\sigma_1},\ldots,\vec{\partial}/\partial u^{i_k}_{\sigma_k}$ fall from the left on the density $\mathcal{L}$ of the functional S. (A case of $\mathbb{Z}_2$-graded base manifold $M^{(n_0|n_1)}$ would bring more signs which are also captured in a standard way.)

Let there be $k\geq2$ variations $\delta s_1,\ldots,\delta s_k\in\Gamma(T\pi)$. We finally have that

$$\frac{\mathrm{d}}{\mathrm{d}\varepsilon_k}\Big|_{\varepsilon_k=0}\circ\ldots\circ\frac{\mathrm{d}}{\mathrm{d}\varepsilon_1}\Big|_{\varepsilon_1=0} S(s+\varepsilon_1\overleftarrow{\delta s}_1+\ldots+\varepsilon_k\overleftarrow{\delta s}_k)=$$

$$=\sum_{\substack{i_1,\ldots,i_k\\j_1,\ldots,j_k}}\sum_{\substack{|\sigma_1|\geq0\\\ldots\\|\sigma_k|\geq0}}\int_M \mathrm{d}\boldsymbol{y}_k\ldots\int_M\mathrm{d}\boldsymbol{y}_1\int_M\mathrm{dvol}(\boldsymbol{x})\cdot$$

$$\left\{(\delta s_k^{i_k})\left(\frac{\overleftarrow{\partial}}{\partial\boldsymbol{y}_k}\right)^{\sigma_k}(\boldsymbol{y}_k)\left\langle\vec{e}_{i_k}(\boldsymbol{y}_k),\vec{e}^{\,\dagger j_k}(\boldsymbol{x})\right\rangle\cdot\ldots\cdot(\delta s_1^{i_1})\left(\frac{\overleftarrow{\partial}}{\partial\boldsymbol{y}_1}\right)^{\sigma_1}(\boldsymbol{y}_1)\left\langle\vec{e}_{i_1}(\boldsymbol{y}_1),\vec{e}^{\,\dagger j_1}(\boldsymbol{x})\right\rangle\right\}$$

$$\cdot\frac{\vec{\partial}^k\mathcal{L}(\boldsymbol{x},[\boldsymbol{u}])}{\partial u^{j_k}_{\sigma_k}\ldots\partial u^{j_1}_{\sigma_1}}\Bigg|_{j^\infty_{\boldsymbol{x}}(s)}. \tag{8}$$

[13] An idea that iterated variations must be taken at nominally different points $\boldsymbol{x}$ and $\boldsymbol{y}$ has been in the air for a long time (let us refer to [38, §1] which contains due credits to E. Witten). A somewhat less obvious fact is that those different points belong to different copies of the manifold M in the product bundle $\pi\times T\pi\times\ldots\times T\pi$ over $M\times M\times\ldots\times M$.

Journal of Physics: Conference Series **474** (2013) 012024 doi:10.1088/1742-6596/474/1/012024

Whenever any $k-1$ variation(s) are fixed in the above formula, the co-multiple $|\rangle$ of the remaining, ℓth variation $\delta s_\ell = \langle \delta s_\ell^{i_\ell}(\boldsymbol{y}_k)\vec{e}_{i_\ell}(\boldsymbol{y}_\ell)|$ is an element of the cotangent vector space $T^*_{s(\boldsymbol{x})}\pi^{-1}(\boldsymbol{x}) = V^\dagger_{\boldsymbol{x}}$ at the point $s(\boldsymbol{x})$ in the fibre $\pi^{-1}(\boldsymbol{x})$ over a base point $\boldsymbol{x} \in M^n$.

Remark 1.8. The composite object in the left-hand side of equality (8) is an integral functional in the bundle $\pi \times T\pi \times \ldots \times T\pi$ which properly contains the geometry of k variations from $\Gamma(T\pi)$, see Fig. 3. This construction lives not on a Whitney sum $\pi \times_M T\pi \times_M \ldots \times_M T\pi$ over

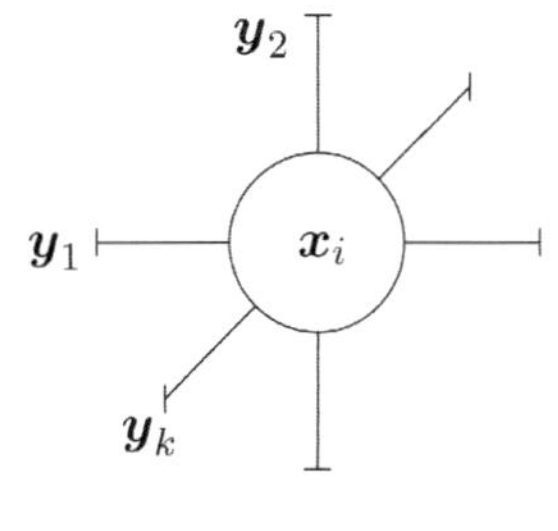

Figure 3. Each variation δs_1, ..., δs_k brings its own copy of the base $M^n \ni \boldsymbol{y}_\ell$ into the product bundle $\pi \times T\pi \times \ldots \times T\pi$ over $M \times M \times \ldots \times M$.

the base manifold M; that would force an untimely restriction to the diagonal in the product $M \times M \times \ldots \times M$ of bases and hence reproduce the old difficulties of the theory.

1.5. The spaces of functionals

The integral functionals $S \in \overline{H}^n(\pi)$, which we have been dealing with until now, are building blocks in a wider class of mappings $\Gamma(\pi) \to \Bbbk$. By viewing elements of $\Gamma(\pi)$ as "points" and functionals from $\overline{H}^n(\pi)$ as "elementary functions" (see [40] and references therein), we consider pointwise-defined (formal sums of) products of such maps, e. g., we let

$$(S_1 \cdot S_2)(s) \stackrel{\text{def}}{=} S_1(s) \cdot S_2(s)$$

for any two already defined functionals S_1 and S_2; the binary operation $\cdot$ for their values at $s \in \Gamma(\pi)$ is the usual multiplication of $\Bbbk$-numbers ($\Bbbk = \mathbb{R}$ or $\mathbb{C}$). By definition, we put

$$\overline{\mathfrak{N}}^n(\pi, T\pi) = \bigoplus_{\ell=1}^{+\infty} \bigotimes_{i=1}^{\ell}{}_{\Bbbk} \bigoplus_{k=0}^{+\infty} \overline{H}^{n(1+k)}(\pi \times \underbrace{T\pi \times \ldots \times T\pi}_{k \text{ variations}}).$$

This space contains the linear subspace of *local functionals*,

$$\overline{\mathfrak{M}}^n(\pi) = \bigoplus_{\ell=1}^{+\infty} \bigotimes_{i=1}^{\ell}{}_{\Bbbk} \overline{H}^n(\pi),$$

for instance, such as the standard weight factor $\exp(\frac{i}{\hbar}S^\hbar)$ in BV-models with quantum BV-action $S^\hbar$ (see section 3.2 below, cf. [9]). The larger space $\overline{\mathfrak{N}}^n(\pi, T\pi) \supsetneq \overline{\mathfrak{M}}^n(\pi)$ harbours local functionals *and* their variations of arbitrarily high order. The (products of) integral functionals in $\overline{\mathfrak{M}}^n(\pi) \supset \overline{H}^n(\pi)$ could be viewed as primary objects. In the course of variations, their descendants in $\overline{\mathfrak{N}}^n(\pi, T\pi)$ absorb new test shifts and retain the information about initial building blocks from $\overline{H}^n(\pi)$. This memory governs the analytic behaviour of descendants in operations such as calculation of the BV-Laplacian or taking the Schouten bracket; we also refer to sections 1.4 above and 3.1 in what follows. The composite structure of the bundle

$\pi \times T\pi \times \ldots \times T\pi$ is crucial whenever one wants to not only describe initial setup such as a given BV-model but to perform rigorous calculations in it, handling higher-order variations of objects (e. g., third-order variations occur in (1c) on p. 1, see also Example 2.4 on p. 34 below, — and the order is equal to four in property (1d) for the BV-Laplacian Δ to be a differential). The geometric approach to (graded-)permutable variations of functionals makes such calculations well-defined and proofs free from any *ad hoc* regularisation recipes.

2. The geometry of Batalin–Vilkovisky formalism

The geometry of variations which we analysed in the previous section was not specific to a bundle π of unknowns. In this section we first recall a construction of the BV-superbundle whose fibres are endowed with $\mathbb{Z}_2$-valued ghost parity. By definition, the BV-bundle $\boldsymbol{\pi}_{\text{BV}}^{(0|1)} = \pi_\infty^*(\boldsymbol{\zeta}_\infty^{(0|1)})$ is induced from the Whitney sum $\boldsymbol{\zeta}^{(0|1)} = \zeta_0 \times_M \zeta_1 \times_M \ldots \times_M \zeta_\lambda \times_M \Pi\widehat{\zeta_0} \times_M \Pi\widehat{\zeta_1} \times_M \ldots \times_M \Pi\widehat{\zeta_\lambda}$ of some $\mathbb{Z}_2$-graded vector bundles over M (in what follows we sum up the construction of $\zeta_0, \ldots, \zeta_\lambda$ and their parity-reversed duals $\Pi\widehat{\zeta_0}, \ldots, \Pi\widehat{\zeta_\lambda}$) by the infinite jet bundle $\pi_\infty \colon J^\infty(\pi) \to M$ associated with the smooth fibre bundle π of physical fields.[14]

2.1. The BV-zoo

Let a fibre bundle π of physical fields over the base manifold M^n be given and denote by ϕ the fibre coordinates in it. Suppose that

$$S_0 = \int \mathcal{L}_0(\boldsymbol{x}, [\phi]) \,\text{dvol}(\boldsymbol{x}) \in \overline{H}^n(\pi)$$

is the action of a field model under study. By using the theory and techniques from section 1 we know how one derives, via the stationary point condition $\overleftarrow{\delta S}\big|_s = 0$ at $s \in \Gamma(\pi)$ the Euler–Lagrange equations of motion $\mathcal{E}_{\text{EL}} = \{\overleftarrow{\delta S_0}/\delta\phi = 0\}$ whose left-hand sides belong to the $C^\infty(J^\infty(\pi))$-module of sections $P_0 = \Gamma(\pi_\infty^*(\zeta_0))$ for the cotangent bundle ζ_0 to π such that $\overleftarrow{\delta S_0}/\delta\phi|_{j^\infty(s)} \cdot \text{dvol}(\cdot) \in \Gamma(T^*\pi) \otimes_{C^\infty(M)} \Lambda^n(M)$ for any field configuration $s \in \Gamma(\pi)$.

We recall from Remark 1.6 that by following a misfortunate but long-established tradition it is the unknowns ϕ in π but not the global coordinates $\boldsymbol{F}$ in the fibre of cotangent bundle $T^*\pi$ to π which are used to parametrise the equations within Euler–Lagrange system $\mathcal{E}_{\text{EL}}$ at points of the graph of a section $\phi \in \Gamma(\pi)$.

If the model at hand is gauge-invariant, then it admits an off-shell differential dependence $\boldsymbol{\Phi}(\boldsymbol{x}, [\phi]; [\boldsymbol{F}]) \equiv 0 \in \Gamma((\pi_\infty \times_M \zeta_{0,\infty})^*(\zeta_1))$ between the left-hand sides $\boldsymbol{F}$ of equations $\mathcal{E}_{\text{EL}}$. We recall further that the dependence of Noether's identities $\boldsymbol{\Phi}$ on (the derivatives of) $\boldsymbol{F}$ is *linear* for Euler–Lagrange systems $\mathcal{E}_{\text{EL}}$; the generators $\boldsymbol{p}(\boldsymbol{x}, [\phi]) \in \widehat{P_1} = \Gamma(\pi_\infty^*(\widehat{\zeta_1}))$ of Noether's gauge symmetries for S_0 are sections of the bundle $\widehat{\zeta_1}$ which is induced from the dual to ζ_1 with respect to the top-degree horizontal form-valued coupling $\langle\, ,\, \rangle$. Indeed, if

$$0 \equiv \langle \boldsymbol{p}, \boldsymbol{\Phi}(\boldsymbol{x}, [\phi]; [\boldsymbol{F}]) \rangle$$

and $\boldsymbol{\Phi}$ is linear in $\boldsymbol{F}$ or its finite-order derivatives,

$$\boldsymbol{\Phi}(\boldsymbol{x}, [\phi]; [\boldsymbol{F}]) = \ell_{\boldsymbol{\Phi}}^{(\boldsymbol{F})}(\boldsymbol{F}) \equiv 0,$$

[14] A subtle point, which we reconsider in section 2.1 (see also Remark 1.6), is that the *fibre* bundle π is often identified with the *vector* bundle component ζ_0 in $\boldsymbol{\zeta}^{(0|1)}$. Nevertheless, it is the construction of induced bundle $\pi_\infty^*(\zeta_0 \times_M \ldots)$ by using which the physical fields and their derivatives are remembered by the Euler–Lagrange equations (referred to ζ_0), Noether's identities (in ζ_1), and higher geberations of syzygies from $\zeta_2, \ldots, \zeta_\lambda$ (if any).

Journal of Physics: Conference Series **474** (2013) 012024 doi:10.1088/1742-6596/474/1/012024

then an integration by parts yields that

$$0 \cong \left\langle (\ell_{\boldsymbol{\Phi}}^{(\boldsymbol{F})})^{\dagger}(\boldsymbol{p}), \delta S_0/\delta\phi \right\rangle \cong \vec{\partial}^{(\phi)}_{(\ell_{\boldsymbol{\Phi}}^{(\boldsymbol{F})})^{\dagger}(\boldsymbol{p})}(S_0).$$

This shows that the evolutionary vector field $\vec{\partial}^{(\phi)}_{A(\boldsymbol{p})}$ with $A = (\ell_{\boldsymbol{\Phi}}^{(\boldsymbol{F})})^{\dagger}$ and $\boldsymbol{p} = \boldsymbol{p}(\boldsymbol{x}, [\phi])$ is a Noether symmetry of the action S_0. By reading the above equalities backwards, one obtains the linear Noether relations $\boldsymbol{\Phi} = A^{\dagger}(\boldsymbol{F})$ between the Euler–Lagrange equations of motion.

Likewise, there could in principle appear higher generations of linear identities $\Psi_2(\boldsymbol{x}, [\phi], [\boldsymbol{F}]; [\boldsymbol{\Phi}]) \equiv 0$, $\ldots$, $\Psi_\lambda(\boldsymbol{x}, [\phi], [\boldsymbol{F}], [\boldsymbol{\Phi}], \ldots, [\Psi_{\lambda-2}]; [\Psi_{\lambda-1}]) \equiv 0$ which hold for all ϕ, sections $\boldsymbol{F}$ in ζ_0, and so on up to the coordinates $\Psi_{\lambda-2}$. Each ith generation of such identities arises with the respective vector bundle ζ_i with fibre dimension m_i; the total number of generations is bounded from above by a constant $\lambda \in \mathbb{N} \cup \{0\}$ due to Hilbert's theorem on syzygies [16]: $0 \leq i \leq \lambda \leq n$, where n is the dimension of base manifold M^n. For example, we have that $\lambda = 1$ for Yang–Mills theory, and $\lambda = 2$ for gravity over a fourfold M^4.

We denote by $\boldsymbol{F}$ (alas! at once identifying this global m-tuple in ζ_0 for the equations with the local field variables ϕ), and by $\boldsymbol{\gamma}^{\dagger}$, $\mathbf{c}^{\dagger}$, $\ldots$, $\mathbf{c}^{\dagger}_{\lambda}$ the global fibre coordinates in ζ_1 for Noether's identities, and so on up to ζ_λ, respectively (see Fig. 4).

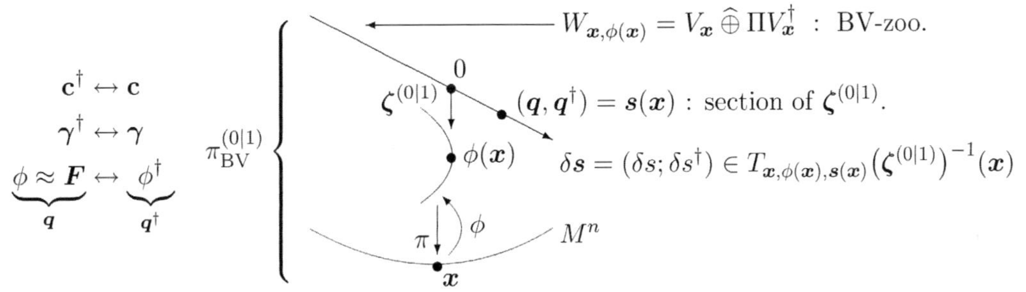

Figure 4. The fibre bundle π of physical fields ϕ, the bundle $\boldsymbol{\zeta}^{(0|1)}$ of BV-variables $(\boldsymbol{q}, \boldsymbol{q}^{\dagger})$, and the vector bundle $T\boldsymbol{\zeta}^{(0|1)}$ of their variations $\delta\boldsymbol{s} = (\delta s; \delta s^{\dagger})$.

In turn, each vector bundle ζ_0, $\ldots$, ζ_λ brings its $\langle\,,\,\rangle_i$-dual $\widehat{\zeta}_i$ into the picture. (Note that the equations $\overleftarrow{\delta S_0}\big|_s = 0$ upon $s \in \Gamma(\pi)$ for $S_0 = \int \mathcal{L}(\boldsymbol{x}, [\phi]) \cdot \mathrm{dvol}(\boldsymbol{x})$ and all equations' linear-differential descendants retain the volume form $\mathrm{dvol}(\boldsymbol{x})$ from the model's action S_0 at all points $\boldsymbol{x} \in M^n$.)

We now reverse the parity of linear vector space fibres in $\widehat{\zeta}_0$, $\ldots$, $\widehat{\zeta}_\lambda$ by introducing the $\mathbb{Z}_2$-valued ghost parity $\mathrm{gh}(\cdot)$ and considering the odd *neighbours* $\Pi\widehat{\zeta}_0$, $\ldots$, $\Pi\widehat{\zeta}_\lambda$ of the dual vector bundles (see [34, 52] and also Appendix A in [33] for discussion). Let us denote by $\phi^{\dagger}$, $\boldsymbol{\gamma}$, $\mathbf{c}$, $\ldots$, $\mathbf{c}_\lambda$ the ghost parity-odd global coordinates along linear vector space fibres in $\Pi\widehat{\zeta}_0$, $\ldots$, $\Pi\widehat{\zeta}_\lambda$, respectively. These variables' proper names are easily recognized from the standard notation: ϕ replacing $\boldsymbol{F}$ are the fields and $\phi^{\dagger}$ are odd-parity *antifields*, $\boldsymbol{\gamma}$ are the odd *ghosts* and $\boldsymbol{\gamma}^{\dagger}$ are the parity-even *antighosts*, whereas the canonically conjugate variables $\mathbf{c} \leftrightarrow \mathbf{c}^{\dagger}$, $\ldots$, $\mathbf{c}_\lambda \leftrightarrow \mathbf{c}^{\dagger}_\lambda$ are higher ghost-antighost pairs of opposite ghost parities (resp., odd and even). We denote by $\boldsymbol{q}$ the agglomeration of ghost parity-even variables and by $\boldsymbol{q}^{\dagger}$ their respective canonically conjugate parity-odd neighbours.[15]

[15] Consider Feynman's path integral $\int_{\Gamma(\zeta^0)} [D\boldsymbol{q}]\, \mathcal{O}([\boldsymbol{q}], [\boldsymbol{q}^{\dagger}])$ of an observable $\mathcal{O}$ over the space of ghost parity-even sections. The BV-Laplacian Δ is the tool which ensures the integral's effective independence from the unphysical ghost parity-odd variables $\boldsymbol{q}^{\dagger}$, see section 3.1.

Journal of Physics: Conference Series **474** (2013) 012024 doi:10.1088/1742-6596/474/1/012024

Remark 2.1. Let us emphasize that by using the word "parity" we always refer to the ghost parity $\text{gh}(\,\cdot\,)$ of objects.[16] In this paper we aim at understanding the geometry of variations so that the graded arithmetic and algebra of derivations play auxiliary rôles. However, as soon as the interaction of geometries is properly fixed, their extension to a $\mathbb{Z}_2$-graded setup of superbundle $\pi\colon E^{(m_0+n_0|m_1+n_1)} \to M^{(n_0|n_1)}$ of physical fields (possibly, over a base supermanifold $M^{(n_0|n_1)}$) makes no conceptual difficulty ([10], see also [22] and references therein). The theory then becomes bi-graded: it involves (i) the $\mathbb{Z}_2$-grading $|\cdot|$ in the ring of field coordinates, which echoes in the $\mathbb{Z}_2$-grading of Euler–Lagrange equations of motion, Noether identities, etc., (the model's action functional S_0 has even grading by default), and (ii) the ghost parity $\text{gh}(\cdot)$, see [52].

The $\mathbb{Z}_2$-grading $|\cdot|$ and the ghost parity $\text{gh}(\cdot)$ are independent from each other. We denote by $\boldsymbol{q} = \boldsymbol{q}^{(0|1)}$ the ghost parity-even BV-fibre variables, which are then grouped in even- and odd-grading components. Likewise, the ghost parity-odd BV-variables $\boldsymbol{q}^\dagger = (\boldsymbol{q}^\dagger)^{(0|1)}$ are arranged in exactly the same way. By construction, the values of $\mathbb{Z}_2$-gradings for canonically conjugate variables $(\boldsymbol{q}, \boldsymbol{q}^\dagger)$ coincide: we have that $|\boldsymbol{q}| = |\boldsymbol{q}^\dagger|$ and $\text{gh}(\boldsymbol{q}^\dagger) \equiv \text{gh}(\boldsymbol{q}) + 1 \mod 2$.

Next, we take the Whitney sum

$$\boldsymbol{\zeta}^{(0|1)} \stackrel{\text{def}}{=} \zeta_0 \times_M \zeta_1 \times_M \ldots \times_M \zeta_\lambda \times_M \Pi\widehat{\zeta_0} \times_M \Pi\widehat{\zeta_1} \times_M \ldots \times_M \Pi\widehat{\zeta_\lambda}$$

of the double set of dual bundles with opposite ghost parities of fibre coordinates. Finally, let us lift the Whitney sum of infinite jets of those bundles, putting it over the bundle of physical fields by using a pull-back under π_∞. We denote the resulting bundle over the total space $J^\infty(\pi) \to M$ by

$$\pi_{\text{BV}}^{(0|1)} = \pi_\infty^*\big(\boldsymbol{\zeta}_\infty^{(0|1)}\big).$$

The fibre $W_{\boldsymbol{x}} = V_{\boldsymbol{x}} \mathbin{\widehat{\oplus}} \Pi V_{\boldsymbol{x}}^\dagger$ of $\boldsymbol{\zeta}^{(0|1)}$ admits the canonical decomposition in two dual halves of opposite parities;[17] this is shown in Fig. 5.

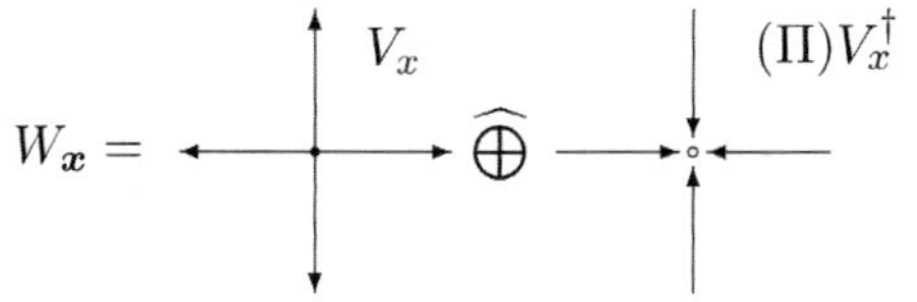

Figure 5. The BV-fibre is a direct sum of dual vector spaces; one is parity-even and the other is proclaimed ghost parity-odd.

Bearing in mind that the fields ϕ are artifically incorporated into the newly built fibre by ζ_0, we shall omit an ever-present reference to points $(\boldsymbol{x}, \phi(\boldsymbol{x}))$ of jets of sections of the initial bundle π when dealing with variations $\delta\boldsymbol{s} = (\delta s; \delta s^\dagger)$ for sections s of $\boldsymbol{\zeta}^{(0|1)}$ at $\phi(\boldsymbol{x})$, see Fig. 4.

[16] By construction, the ghost parities of canonically conjugate BV-variables are complementary modulo 2, that is, to each even-parity variable q there corresponds its odd-parity dual neighbour $q^\dagger$. Of course, there remains much freedom in a choice of the integer ghost numbers followed by the group homomorphism $(-)^{\text{gh}(\,\cdot\,)}\colon \mathbb{Z} \to \mathbb{Z}_2$. For example, let $(q, q^\dagger)$ be a pair of conjugate BV-variables; then one balances $\text{gh}(q) = \text{gh}(q^\dagger) \pm 1$ or $\text{gh}(q) = -\text{gh}(q^\dagger) \pm 1$, or by using any other integers such that one is even and the other is odd. Obviously any shift by an even integer (e.g., $\text{gh}(q) \mapsto -\text{gh}(q) = \text{gh}(q) - 2\cdot\text{gh}(q)$) does not alter any values in the parity group $\mathbb{Z}_2$; this is no more than another way to describe the same theory.

[17] To highlight this duality between ghost parity-even vector space $V_{\boldsymbol{x}}$ and ghost parity-odd subspace $\Pi V_{\boldsymbol{x}}^\dagger$ in $W_{\boldsymbol{x}}$, we use the notation $\widehat{\oplus}$ for their direct sum; whenever a coordinate in $V_{\boldsymbol{x}}$ is rescaled by const times, the respective conjugate variable in $\Pi V_{\boldsymbol{x}}^\dagger$ is transformed inverse-proportionally by const^{-1} times, see Remark 2.5 below.

2.2. *The signs convention in Nature*

The construction of canonically conjugate pairs of global coordinates $(\boldsymbol{q}, \boldsymbol{q}^\dagger)$ in the fibres $W_{\boldsymbol{x}} = V_{\boldsymbol{x}} \mathbin{\widehat{\oplus}} \Pi V_{\boldsymbol{x}}^\dagger$ refers to a choice of the smooth field of dual bases in the two subspaces of even and odd ghost parity. Suppose that $\vec{e}_i(\boldsymbol{x})$ is a frame in $V_{\boldsymbol{x}}$ and $\vec{e}^{\dagger i}(\boldsymbol{x})$ is its dual in $\Pi V_{\boldsymbol{x}}^\dagger$, where the index i runs from 1 to the total dimension of even- and odd-parity component in the fibre of $\boldsymbol{\zeta}^{(0|1)}$; we denote by $N = m + m_1 + \ldots + m_\lambda$ each of the two dimensions so that the fibre of the Whitney sum $\boldsymbol{\zeta}^{(0|1)}$ has superdimension $(N|N)$.

Let us recall that it is the parity of coordinates $\boldsymbol{q}^\dagger$ but not of the vectors $\vec{e}^{\dagger i}$ in a basis which is reversed by the operation Π. The odd-parity component in the vector bundle $\boldsymbol{\zeta}^{(0|1)}$ is topologically indistinguishable from $\widehat{\zeta}_0 \times_M \ldots \times_M \widehat{\zeta}_\lambda$ but the rules become new for arithmetic in the algebra of coordinate functions on the total space. Therefore, we let the notation $\vec{e}^{\dagger i}(\boldsymbol{x})$ be identical for the same bases in $V_{\boldsymbol{x}}^\dagger$ and $\Pi V_{\boldsymbol{x}}^\dagger$.

Remark 2.2. The presence of *two* dual vector spaces, $V_{\boldsymbol{x}}$ and $(\Pi)V_{\boldsymbol{x}}^\dagger$, standardly implies that there are *two* couplings,

$$\langle\,,\,\rangle\colon V_{\boldsymbol{x}} \times (\Pi)V_{\boldsymbol{x}}^\dagger \to \Bbbk \quad \text{and} \quad \langle\,,\,\rangle\colon (\Pi)V_{\boldsymbol{x}}^\dagger \times V_{\boldsymbol{x}} \to \Bbbk; \tag{9}$$

we denote both operations in the same way because the order of arguments uniquely determines the choice. Let us remember also that it is not the linear vector space fibres of the superbundle $\boldsymbol{\zeta}^{(0|1)}$ over the bundle π of physical fields but it is the tangent spaces

$$T_{(\boldsymbol{x},\phi(\boldsymbol{x}),s(\boldsymbol{x}))}\left(V_{\boldsymbol{x}} \mathbin{\widehat{\oplus}} \Pi V_{\boldsymbol{x}}^\dagger\right) \cong V_{\boldsymbol{x}} \mathbin{\widehat{\oplus}} \Pi V_{\boldsymbol{x}}^\dagger$$

to those fibres which harbour the variations $\delta\boldsymbol{s} = (\delta s; \delta s^\dagger)$ of sections s of the BV-bundle.

A reason to study the geometry of variations in tangent spaces to the fibres is clear from section 1. In fact, although we have substantiated in section 2.1 that Euler–Lagrange equations and their descendants do form linear vector spaces, this structure is incidental for the BV-formalism while Feynman path integration is not yet begun. The guiding geometric principle is that linear vector spaces appear only in the course of inspection of functionals' responses to infinitesimal test shifts of their arguments.

Couplings (9) are defined only if the linear vector spaces $V_{\boldsymbol{x}} \ni \delta s(\boldsymbol{x})$ and $\Pi V_{\boldsymbol{x}}^\dagger \ni \delta s^\dagger(\boldsymbol{x})$ are located over the same point $\boldsymbol{x} \in M^n$ of the base manifold, and over it they are attached as the two components of tangent space $T_{s(\boldsymbol{x})}\big(\boldsymbol{\zeta}^{(0|1)}\big)^{-1}(\boldsymbol{x}, \phi(\boldsymbol{x}))$, at the same point $s(\boldsymbol{x}) = s\,(\boldsymbol{x}, \phi(\boldsymbol{x}))$ of fibre in the superbundle $\boldsymbol{\zeta}^{(0|1)}$ over a point $(\boldsymbol{x}, \phi(\boldsymbol{x}))$ of the total space for the bundle π of physical fields (see Fig. 4).

A distinction between the vector space $V_{\boldsymbol{x}}$ and its parity-reversed dual nontrivially determines the couplings' values whenever they are defined. Namely, each of the two finite-dimensional vector spaces is reflexive,

$$\Big((V_{\boldsymbol{x}})^\dagger\Big)^\dagger \cong V_{\boldsymbol{x}} \quad \text{and} \quad \Big((\Pi V_{\boldsymbol{x}}^\dagger)^\dagger\Big)^\dagger \cong \Pi V_{\boldsymbol{x}}^\dagger, \tag{10}$$

but these isomorphisms are not always identity mappings. We have that

$$\Big\langle \vec{e}_i(\boldsymbol{x}), \vec{e}^{\dagger j}(\boldsymbol{x}) \Big\rangle = \boldsymbol{\delta}_i^j \quad \text{yet} \quad \Big\langle \vec{e}^{\dagger j}(\boldsymbol{x}), \vec{e}_i(\boldsymbol{x}), \Big\rangle = -\boldsymbol{\delta}_i^j, \tag{11}$$

where $\boldsymbol{\delta}_i^j$ is the Kronecker symbol whose value is the unit iff $i = j$ and which is set equal to zero otherwise.

Remark 2.3. We claim that this mechanism is responsible, in particular, for the skew-symmetry of various Poisson brackets (e. g., of the parity-odd Schouten bracket). Let us emphasize that this is a principle of order between geometric objects; the concept is not restricted to the BV-setup which we study here. Actually, Eq. (11) is the fundamental reason for differential 1-forms to anticommute[18] (in the class of geometries for which a coupling is defined between the linear vector spaces of co-multiples under the wedge product $\wedge$; for instance, such is the case of the Helmholtz criterion $\psi = \delta S/\delta \boldsymbol{q} \Leftrightarrow \vec{\ell}^{(\boldsymbol{q})}_{\psi} = \big(\vec{\ell}^{(\boldsymbol{q})}_{\psi}\big)^{\dagger}$ for images of the variational derivative [28, 45]). Physically speaking, the binary count by "a vector space," "not the former, hence its dual," and "not the dual, but the initial space's image under central symmetry" builds on the notion of order and realizes the law of the excluded middle.

2.3. Left- and right-variations via operators

Suppose that

$$S = \int \mathcal{L}(\boldsymbol{x}, [\boldsymbol{q}], [\boldsymbol{q}^{\dagger}])\, \mathrm{dvol}(\boldsymbol{x})$$

is an integral functional $\Gamma(\pi_{\mathrm{BV}}) \to \Bbbk$. Let us focus on the correspondence between test shifts $\delta\boldsymbol{s} = (\delta s; \delta s^{\dagger}) = \delta s^{i} \cdot \vec{e}_{i} + \delta s^{\dagger}_{i} \cdot \vec{e}^{\,\dagger i}$ of BV-fields $s \in \Gamma(\pi_{\mathrm{BV}})$ and, on the other hand, left- or right-acting linear singular integral operators $\overleftarrow{\delta\boldsymbol{s}}$ and $\overrightarrow{\delta\boldsymbol{s}}$ which yield the functional's responses to shifts of its argument $\boldsymbol{s}$. By definition, we put

$$\overrightarrow{\delta\boldsymbol{s}} = \int_{M} \mathrm{d}\boldsymbol{y} \Big\{ (\delta s^{i}) \left(\frac{\overleftarrow{\partial}}{\partial \boldsymbol{y}}\right)^{\sigma} (\boldsymbol{y}) \cdot \Big\langle (\vec{e}^{\,\dagger i})^{\dagger}(\boldsymbol{y}), \vec{e}^{\,\dagger j}(\cdot) \Big\rangle \frac{\overrightarrow{\partial}}{\partial q^{j}_{\sigma}} + \\ + (\delta s^{\dagger}_{i}) \left(\frac{\overleftarrow{\partial}}{\partial \boldsymbol{y}}\right)^{\sigma} (\boldsymbol{y}) \cdot \Big\langle (\vec{e}_{i})^{\dagger}(\boldsymbol{y}), \vec{e}_{j}(\cdot) \Big\rangle \frac{\overrightarrow{\partial}}{\partial q^{\dagger}_{j,\sigma}} \Big\} \qquad (12a)$$

and

$$\overleftarrow{\delta\boldsymbol{s}} = \int_{M} \mathrm{d}\boldsymbol{y} \Big\{ \frac{\overleftarrow{\partial}}{\partial q^{j}_{\sigma}} \Big\langle \vec{e}^{\,\dagger j}(\cdot), {}^{\dagger}(\vec{e}^{\,\dagger i})(\boldsymbol{y}) \Big\rangle \left(\frac{\overrightarrow{\partial}}{\partial \boldsymbol{y}}\right)^{\sigma} (\delta s^{i})(\boldsymbol{y}) + \\ + \frac{\overleftarrow{\partial}}{\partial q^{\dagger}_{j,\sigma}} \Big\langle \vec{e}_{j}(\cdot), {}^{\dagger}(\vec{e}_{i})(\boldsymbol{y}) \Big\rangle \left(\frac{\overrightarrow{\partial}}{\partial \boldsymbol{y}}\right)^{\sigma} (\delta s^{\dagger}_{i})(\boldsymbol{y}) \Big\}. \qquad (12b)$$

The above formulas for directed operators $\overrightarrow{\delta\boldsymbol{s}}$ and $\overleftarrow{\delta\boldsymbol{s}}$ contain new notation $(\vec{e}_{i})^{\dagger}$, $(\vec{e}^{\,\dagger i})^{\dagger}$ and ${}^{\dagger}(\vec{e}_{i})$, ${}^{\dagger}(\vec{e}^{\,\dagger i})$, also referring to an important sign convention which fully determines those adjoint objects. Namely, let us agree that over every $\boldsymbol{x} \in M^{n}$ the covectors

$$\vec{e}^{\,\dagger i}(\boldsymbol{x}) \left(\frac{\overrightarrow{\partial}}{\partial q^{i}_{\sigma}} \mathcal{L}(\boldsymbol{x}, [\boldsymbol{q}], [\boldsymbol{q}^{\dagger}]) \right)\Bigg|_{j^{\infty}_{\boldsymbol{x}}(\boldsymbol{s})} + \vec{e}_{i}(\boldsymbol{x}) \left(\frac{\overrightarrow{\partial}}{\partial q^{\dagger}_{i,\sigma}} \mathcal{L}(\boldsymbol{x}, [\boldsymbol{q}], [\boldsymbol{q}^{\dagger}]) \right)\Bigg|_{j^{\infty}_{\boldsymbol{x}}(\boldsymbol{s})}$$

and

$$\left(\mathcal{L}(\boldsymbol{x}, [\boldsymbol{q}], [\boldsymbol{q}^{\dagger}]) \frac{\overleftarrow{\partial}}{\partial q^{i}_{\sigma}} \right)\Bigg|_{j^{\infty}_{\boldsymbol{x}}(\boldsymbol{s})} \vec{e}^{\,\dagger i}(\boldsymbol{x}) + \left(\mathcal{L}(\boldsymbol{x}, [\boldsymbol{q}], [\boldsymbol{q}^{\dagger}]) \frac{\overleftarrow{\partial}}{\partial q^{\dagger}_{i,\sigma}} \right)\Bigg|_{j^{\infty}_{\boldsymbol{x}}(\boldsymbol{s})} \vec{e}_{i}(\boldsymbol{x})$$

[18] That is, this argument reveals why a mathematical axiom that differential forms do anticommute in the course of calculations leads to verifiable and relevant theoretic predictions which match experimental data.

are expanded in the cotangent space $T^*_{\boldsymbol{s}(\boldsymbol{x})}W_{\boldsymbol{x}} \cong V^\dagger_{\boldsymbol{x}} \,\widehat{\oplus}\, (TV^\dagger_{\boldsymbol{x}})^\dagger$ with respect to the original basis $(+\vec{e}^{\,\dagger i}, +\vec{e}_i)$; note the signs (any other convention here would nohow alter the theory's content but it would (in)appropriately modify the signs in (13) below). The normalization of left- and right-adjoint objects $(\vec{e}_i)^\dagger$, $(\vec{e}^{\,\dagger i})^\dagger$ and ${}^\dagger(\vec{e}_i)$, ${}^\dagger(\vec{e}^{\,\dagger i})$ is immediate under assumption that the couplings' equations yield (5) and then (3) after integration by parts — no extra sign factors appear in those formulas. This requirement determines the table

$$\begin{aligned} (\vec{e}^{\,\dagger i})^\dagger &= \ \vec{e}_i, & {}^\dagger(\vec{e}^{\,\dagger i}) &= -\vec{e}_i, \\ (\vec{e}_i)^\dagger &= -\vec{e}^{\,\dagger i}, & {}^\dagger(\vec{e}_i) &= \ \vec{e}^{\,\dagger i}, \end{aligned} \tag{13}$$

so that the following defining relations hold:

$$\Big\langle (\vec{e}_i)^\dagger, \vec{e}_i \Big\rangle = \Big\langle \vec{e}_i, {}^\dagger(\vec{e}_i) \Big\rangle = \Big\langle (\vec{e}^{\,\dagger i})^\dagger, \vec{e}^{\,\dagger i} \Big\rangle = \Big\langle \vec{e}^{\,\dagger i}, {}^\dagger(\vec{e}^{\,\dagger i}) \Big\rangle = +1.$$

Let us notice that the left- and right-acting operation $\dagger$ provides the analogue of left and right $\langle\,,\,\rangle$-dual in this ordered world; the first column in (13) determines a clockwise rotation in the oriented plane spanned by $\vec{e}_i \prec \vec{e}^{\,\dagger i}$, whereas taking the adjoints ${}^\dagger(\cdot)\colon \vec{e}_i \mapsto \vec{e}^{\,\dagger i}$ and $\vec{e}^{\,\dagger i} \mapsto -\vec{e}_i$ induces the counterclockwise rotation in that plane as shown in Fig. 6.

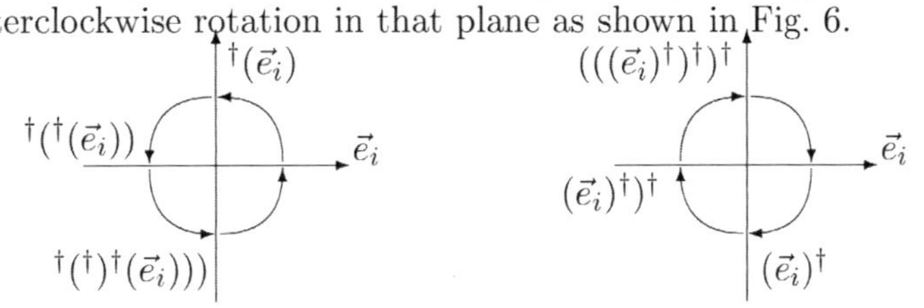

Figure 6. The orientation $\vec{e}_i \prec \vec{e}^{\,\dagger i}$ and configuration of the left- and right- duals with respect to the couplings $\langle\,,\,\rangle$.

Example 2.1. Identities (13) show up in the directed variations $\overleftarrow{\delta S}\Big|_s^{\delta\boldsymbol{s}} = \overrightarrow{\delta\boldsymbol{s}}(S)(s)$ and $\overrightarrow{\delta S}\Big|_s^{\delta\boldsymbol{s}} = (S)\overleftarrow{\delta\boldsymbol{s}}(s)$ of an integral functional $S = \int \mathcal{L}(\boldsymbol{x}, [\boldsymbol{q}], [\boldsymbol{q}^\dagger]) \cdot \mathrm{dvol}(\boldsymbol{x})$. Namely, we have that

$$\begin{aligned} \overleftarrow{\delta S}\Big|_s^{(\delta s, \delta s^\dagger)} &= \\ = \int_M \mathrm{d}\boldsymbol{y} \int_M \mathrm{dvol}(\boldsymbol{x}) \Bigg\{ &(\delta s^i) \left(\frac{\overleftarrow{\partial}}{\partial \boldsymbol{y}}\right)^\sigma (\boldsymbol{y}) \Big\langle \vec{e}_i(\boldsymbol{y}), \vec{e}^{\,\dagger j}(\boldsymbol{x}) \Big\rangle \left(\frac{\overrightarrow{\partial}}{\partial q^j_\sigma} \mathcal{L}(\boldsymbol{x}, [\boldsymbol{q}], [\boldsymbol{q}^\dagger]) \right)\Bigg|_{j^\infty_{\boldsymbol{x}}(s)} + \\ &+ (\delta s^\dagger_i) \left(\frac{\overleftarrow{\partial}}{\partial \boldsymbol{y}}\right)^\sigma (\boldsymbol{y}) \Big\langle -\vec{e}^{\,\dagger i}(\boldsymbol{y}), \vec{e}_j(\boldsymbol{x}) \Big\rangle \left(\frac{\overrightarrow{\partial}}{\partial q^\dagger_{j,\sigma}} \mathcal{L}(\boldsymbol{x}, [\boldsymbol{q}], [\boldsymbol{q}^\dagger]) \right)\Bigg|_{j^\infty_{\boldsymbol{x}}(s)} \Bigg\} \end{aligned} \tag{14a}$$

and

$$\begin{aligned} \overrightarrow{\delta S}\Big|_s^{(\delta s, \delta s^\dagger)} &= \\ = \int_M \mathrm{d}\boldsymbol{y} \int_M \mathrm{dvol}(\boldsymbol{x}) \Bigg\{ &\left(\mathcal{L}(\boldsymbol{x}, [\boldsymbol{q}], [\boldsymbol{q}^\dagger]) \frac{\overleftarrow{\partial}}{\partial q^j_\sigma} \right)\Bigg|_{j^\infty_{\boldsymbol{x}}(s)} \Big\langle \vec{e}^{\,\dagger j}(\boldsymbol{x}), -\vec{e}_i(\boldsymbol{y}) \Big\rangle \left(\frac{\overrightarrow{\partial}}{\partial \boldsymbol{y}}\right)^\sigma (\delta s^i)(\boldsymbol{y}) + \\ &+ \left(\mathcal{L}(\boldsymbol{x}, [\boldsymbol{q}], [\boldsymbol{q}^\dagger]) \frac{\overleftarrow{\partial}}{\partial q^\dagger_{j,\sigma}} \right)\Bigg|_{j^\infty_{\boldsymbol{x}}(s)} \Big\langle \vec{e}_j(\boldsymbol{x}), \vec{e}^{\,\dagger i}(\boldsymbol{y}) \Big\rangle \left(\frac{\overrightarrow{\partial}}{\partial \boldsymbol{y}}\right)^\sigma (\delta s^\dagger_i)(\boldsymbol{y}) \Bigg\}. \end{aligned} \tag{14b}$$

The operators $\overrightarrow{\delta\boldsymbol{s}}$ and $\overleftarrow{\delta\boldsymbol{s}}$ act via ghost-parity graded Leibniz' rule on formal products of integral functionals (and on their images under other infinitesimal variation operators as well), so that the two operators are defined on the entire space $\overline{\mathfrak{N}}^n(\pi_{\text{BV}}, T\pi_{\text{BV}})$, see section 2.4.2 below.

Remark 2.4. A reversion $\overleftarrow{\delta\boldsymbol{s}} \rightleftarrows \overrightarrow{\delta\boldsymbol{s}}$ of the direction along which such an operator acts means that the initially given operator (for definition, let it be $\overleftarrow{\delta\boldsymbol{s}}$ which acts to the left) is *destroyed* and in its place the other, opposite-direction operator is created (here it would be $\overrightarrow{\delta\boldsymbol{s}}$). Note that the variation $\delta\boldsymbol{s} \in \Gamma(T\pi)$ itself stays unchanged; it is the two realizations of this object via $\overleftarrow{\delta\boldsymbol{s}}$ and then via $\overrightarrow{\delta\boldsymbol{s}}$ which differ. (This concept of test shifts as primary geometric objects which contain information about the operators will be essential in Definition 2 of the variational Schouten bracket.)

Remark 2.5. The postulate of duality between $\vec{e}_i(\boldsymbol{x})$ and $\vec{e}^{\dagger i}(\boldsymbol{x})$ correlates their transformation laws under dilations: a rescaling $\vec{e}_i \mapsto \text{const}\cdot\vec{e}_i$ with $\text{const} \in \Bbbk \setminus \{0\}$ determines the inverse-proportional mapping $\vec{e}^{\dagger i} \mapsto \text{const}^{-1}\cdot\vec{e}^{\dagger i}$ of respective dual vectors. (Likewise, the coordinates in $V_{\boldsymbol{x}}$ and $\Pi V_{\boldsymbol{x}}^{\dagger}$ are then rescaled by $q^i \mapsto \text{const}^{-1}\cdot q^i$ and $q_i^{\dagger} \mapsto \text{const}\cdot q_i^{\dagger}$. respectively.)

Consider a variation $\delta\boldsymbol{s} = (\delta s; \delta s^{\dagger}) \in \Gamma(T\boldsymbol{\zeta}^{(0|1)})$ of a BV-section $s \in \Gamma(\boldsymbol{\zeta}^{(0|1)})$ over a given field configuration $\phi \in \Gamma(\pi)$ in the BV-bundle $\boldsymbol{\pi}_{\text{BV}}^{(0|1)}$. The infinitesimal variation vectors $\delta\boldsymbol{s} = (\delta s; \delta s^{\dagger})$ can be naturally split to ghost parity-homogeneous components:

$$\delta\boldsymbol{s} = (\delta s; 0) + (0; \delta s^{\dagger}). \tag{15}$$

Here we explicitly use the linear vector space structure in fibres of the tangent bundle $T\boldsymbol{\zeta}^{(0|1)}$. Let us recall that the two homogeneous variations

$$\delta s(\boldsymbol{x}) = \delta s^i(\boldsymbol{x})\cdot\vec{e}_i(\boldsymbol{x}) \quad \text{and} \quad \delta s^{\dagger}(\boldsymbol{x}) = \delta s_i^{\dagger}(\boldsymbol{x})\cdot\vec{e}^{\dagger i}(\boldsymbol{x})$$

in the right-hand side of (15) are the canonically dual to each other.

Moreover, by Remark 2.5 it is then possible to have δs and $\delta s^{\dagger}$ normalized, for every i running from 1 to the dimension N, by the equalities

$$\delta s^i(\boldsymbol{x})\cdot\delta s_i^{\dagger}(\boldsymbol{x}) \equiv +1 \tag{16}$$

at every $\boldsymbol{x} \in M^n$ where the smooth fields of dual bases $\vec{e}_i$ and $\vec{e}^{\dagger i}$ are defined for the section s. From now on, let us deal only with such *normalized* variations. This implies that the coupling of these geometric objects are "invisible" but still the order in which the co-multiples δs and $\delta s^{\dagger}$ occur in (11) does determine the signs in various formulas (e. g., in the definition of Schouten bracket, see p. 28 below).

2.4. Definitions of the BV-Laplacian and Schouten bracket

We now combine the geometry of graded-permutable iterated variations, which we explored in section 1 and which absorbs a new copy of the underlying base manifold M^n for each new infinitesimal test shift $\delta\boldsymbol{s}(\boldsymbol{x}) \in T_{s(\boldsymbol{x})}W_{\boldsymbol{x}}$ of the functionals' arguments at $\boldsymbol{x} \in M^n$, with the algebra of two couplings (9) between ghost parity-homogeneous halves of infinitesimal variations in the BV-setup $T_{s(\boldsymbol{x})}W_{\boldsymbol{x}} \cong V_{\boldsymbol{x}} \hat{\oplus} \Pi V_{\boldsymbol{x}}^{\dagger}$; the absolute locality of such coupling events is a fundamental principle.

To avoid an agglomeration of formulas and to match the notation with that in section 1, we omit an explicit reference to field configuration $\{\phi(\boldsymbol{x}),\ \boldsymbol{x} \in M^n\}$, indicating only the base points $\boldsymbol{x} \in M^n$. We also denote by π_{BV} the composite-structure superbundle over M^n (see Fig. 4) so that the notation for the vector bundle of BV-sections' infinitesimal variations is $T\pi_{\text{BV}}$. However, let us remember that only the linear BV-fibre variables $(\boldsymbol{q}, \boldsymbol{q}^\dagger)$ but not the physical fields ϕ are subjected to variations at points $s(\boldsymbol{x}) \in \big(\boldsymbol{\zeta}^{(0|1)}\big)^{-1}(\boldsymbol{x}, s(\boldsymbol{x}))$ over $(\boldsymbol{x}, \phi(\boldsymbol{x})) \in \pi^{-1}(\boldsymbol{x})$. A brute force labelling of Euler–Lagrange equations by the respective unknowns is an act of will by the one who writes formulas but it is not a prescription from the model's geometry.

This section contains rigorous, self-regularizing definitions of the BV-Laplacian and Schouten bracket for integral functionals from $\overline{H}^n(\pi_{\mathrm{BV}}) \subsetneq \overline{\mathfrak{M}}^n(\pi_{\mathrm{BV}}) \subsetneq \overline{\mathfrak{N}}^n(\pi_{\mathrm{BV}}, T\pi_{\mathrm{BV}})$. We shall extend the definition to the space $\overline{\mathfrak{N}}^n(\pi_{\mathrm{BV}}, T\pi_{\mathrm{BV}})$ of products of integral functionals, possibly with earlier-absorbed variations, in the subsequent sections of this paper. We then establish the main properties of these structures and prove relations between them. We note that the definitions which we give here are operational: each of them is a surgery for the couplings and their reconfiguration algorithm. (The locality postulate ensures the restrictions onto diagonals in the product $M \times \ldots \times M$ so that those recombinations make sense at every point of M.)

2.4.1. The BV-Laplacian Δ Let us first introduce some shorthand notation. Let $F = \int f(\boldsymbol{x}, [\boldsymbol{q}], [\boldsymbol{q}^\dagger]) \cdot \mathrm{dvol}(\boldsymbol{x})$ be an integral functional and $\delta\boldsymbol{s} = (\delta s; 0) + (0; \delta s^\dagger)$ be a variation's splitting in two ghost parity-homogeneous variations. From section 1 we know that each of the two is referred to its own copy of the base: let it be $\delta s(\boldsymbol{y}_1)$ and $\delta s^\dagger(\boldsymbol{y}_2)$ so that formula (5) defines the response of F to an infinitesimal shift of its argument along each of the two directions.

Definition 1. Let $\delta\boldsymbol{s} \in \Gamma(T\pi_{\mathrm{BV}})$ be a test shift normalized by (16) and then split to the sum $(\delta s; 0) + (0; \delta s^\dagger)$ of ghost parity-homogeneous, $\langle\,,\,\rangle$-dual halves. The BV-*Laplacian* is the linear operator $\Delta\colon \overline{H}^n(\pi_{\mathrm{BV}}) \to \overline{\mathfrak{N}}^n(\pi_{\mathrm{BV}}, T\pi_{\mathrm{BV}})$; for a ghost parity-homogeneous integral functional $F \in \overline{H}^n(\pi)$ and its argument s, the operator Δ is an algorithm for reconfiguration of couplings in the second variation

$$\frac{\mathrm{d}}{\mathrm{d}\varepsilon}\bigg|_{\varepsilon=0} \frac{\mathrm{d}}{\mathrm{d}\varepsilon^\dagger}\bigg|_{\varepsilon^\dagger=0} F(s + \varepsilon \cdot \overleftarrow{\delta s} + \varepsilon^\dagger \cdot \overleftarrow{\delta s}^\dagger) = \sum_{\substack{i_1,i_2\\ j_1,j_2}} \sum_{\substack{|\sigma_1|\geq 0\\ |\sigma_2|\geq 0}} \int_M \mathrm{d}\boldsymbol{y}_1 \int_M \mathrm{d}\boldsymbol{y}_2 \int_M \mathrm{dvol}(\boldsymbol{x})$$

$$\left\{\begin{array}{l} (\delta s^{i_1}) \left(\frac{\overleftarrow{\partial}}{\partial \boldsymbol{y}_1}\right)^{\sigma_1} (\boldsymbol{y}_1) \langle\ \vec{e}_{i_1}(\boldsymbol{y}_1), \vec{e}^{\,\dagger j_1}(\boldsymbol{x})\rangle \hookleftarrow \\ \hookrightarrow (\delta s^\dagger_{i_2}) \left(\frac{\overleftarrow{\partial}}{\partial \boldsymbol{y}_2}\right)^{\sigma_2} (\boldsymbol{y}_2) \langle -\vec{e}^{\,\dagger i_2}(\boldsymbol{y}_2), \vec{e}_{j_2}(\boldsymbol{x})\rangle \end{array}\right\} \cdot \frac{\overrightarrow{\partial}^2 f(\boldsymbol{x}, [\boldsymbol{q}], [\boldsymbol{q}^\dagger])}{\partial q^{j_1}_{\sigma_1} \partial q^\dagger_{j_2,\sigma_2}}\bigg|_{j^\infty_{\boldsymbol{x}}(s)}.$$

This second variation's integrand contains the couplings $\langle\,,\,\rangle$:

$$T_{s(\boldsymbol{y}_1)} V_{\boldsymbol{y}_1} \times T^*_{s(\boldsymbol{x})}(\Pi) V_{\boldsymbol{x}} \to \Bbbk \quad \text{and} \quad T_{s(\boldsymbol{y}_2)} \Pi V^\dagger_{\boldsymbol{y}_2} \times T^*_{s(\boldsymbol{x})}(\Pi) V^\dagger_{\boldsymbol{x}} \to \Bbbk$$

which are defined only if the attachment points coincide for these (co)vectors; an optional presence of the parity reversion operator indicates a possibility of having ghost parity-odd functional F.

At the moment when the object ΔF under construction – or a larger object of which ΔF is an element, see (1c) – is evaluated at a section $s \in \Gamma(\pi_{\mathrm{BV}})$, the integrations by parts carry the derivatives away from the variations' components: $\overleftarrow{\partial}/\partial \boldsymbol{y}_i \mapsto \overrightarrow{\partial}/\partial \boldsymbol{y}_i$ as explained in section 1.3. The third step in definition of Δ acting on F is a surgery algorithm for an on-the-diagonal reattachment of the couplings, see Figure 7. In other words, *after* the integration by parts the

$$\begin{array}{cccc} \langle 1| & & |3\rangle & \\ & \langle 2| & & |4\rangle \end{array} \quad \longmapsto \quad \begin{array}{cccc} \langle 1| & & \langle 3| & \\ & |2\rangle & & |4\rangle \end{array}$$

Figure 7. The on-the-diagonal coupling of variations versus taking the trace of bi-linear form.

surgery yields the following:

$$(\Delta F)\Big|_{s}^{\delta s} = \sum_{\substack{i_1,i_2\\ j_1,j_2}} \sum_{\substack{|\sigma_1|\geq 0\\ |\sigma_2|\geq 0}} \int_M \mathrm{d}\boldsymbol{y}_1 \int_M \mathrm{d}\boldsymbol{y}_2 \int_M \mathrm{dvol}(\boldsymbol{x})\cdot$$
$$\cdot\left\{ \delta s^{i_1}(\boldsymbol{y}_1)\left(-\frac{\overrightarrow{\partial}}{\partial \boldsymbol{y}_1}\right)^{\sigma_1} \underbrace{\langle \vec{e}_{i_1}(\boldsymbol{y}_1), -\vec{e}^{\dagger i_2}(\boldsymbol{y}_2)\rangle}_{-1} \cdot \delta s^{\dagger}_{i_2}(\boldsymbol{y}_2)\left(-\frac{\overrightarrow{\partial}}{\partial \boldsymbol{y}_2}\right)^{\sigma_2}\right\}\cdot$$
$$\cdot\left\{ \underbrace{\langle \vec{e}^{\dagger j_1}(\boldsymbol{x}), \vec{e}_{j_2}(\boldsymbol{x})\rangle}_{-1} \cdot \frac{\vec{\partial}^2 f(\boldsymbol{x},[\boldsymbol{q}],[\boldsymbol{q}^\dagger])}{\partial q^{j_1}_{\sigma_1}\partial q^{\dagger}_{j_2,\sigma_2}}\Big|_{j^\infty_{\boldsymbol{x}}(s)}\right\}. \quad (17)$$

Note that the left-to-right order in $\langle \vec{e}_{i_1}(\boldsymbol{y}_1), \vec{e}^{\dagger j_1}(\boldsymbol{x})\rangle \cdot \langle -\vec{e}^{\dagger i_2}(\boldsymbol{y}_2), \vec{e}_{j_2}(\boldsymbol{x})\rangle$ is preserved by the respective couplings' arguments in $\langle \vec{e}_{i_1}(\boldsymbol{y}_1), -\vec{e}^{\dagger i_2}(\boldsymbol{y}_2)\rangle \cdot \langle \vec{e}^{\dagger j_1}(\boldsymbol{x}), \vec{e}_{j_2}(\boldsymbol{x})\rangle$, cf. Fig. 7.

Remark 2.6. Until the moment when the integrations by parts are performed in ΔF, the derivatives $\partial/\partial\boldsymbol{y}_1$ and $\partial/\partial\boldsymbol{y}_2$ refer to different copies of the manifold M^n in the base $M^n\times M^n\times M^n$ of the product bundle $\pi_{\mathrm{BV}}\times T\pi_{\mathrm{BV}}\times T\pi_{\mathrm{BV}}$. This implies that the two variations of F in the definition of Δ are graded-permutable between each other and with all other variations falling on $f(\boldsymbol{x},[\boldsymbol{q}],[\boldsymbol{q}^\dagger])$ whenever ΔF is a constituent element of a larger object (e. g., see (1c–1d) on p. 1).

Remark 2.7. To keep track of multiple copies of the base M^n for functionals and variations (here $\boldsymbol{x}\in M^n$, $\boldsymbol{y}_1\in M^n$, $\boldsymbol{y}_2\in M^n$) in the course of integration by parts (see section 1.3), we indicate the respective variations' bases by explicitly writing $\boldsymbol{q}(\boldsymbol{y}_1)$ and $\boldsymbol{q}^\dagger(\boldsymbol{y}_2)$ in the denominators *and* we denote by $\partial/\partial\boldsymbol{y}_1$ and $\partial/\partial\boldsymbol{y}_2$ the derivatives which now fall on the functional's density $f(\boldsymbol{x},[\boldsymbol{q}],[\boldsymbol{q}^\dagger])$ — for instance, we do so in Example 2.4 on p. 34 below. Namely, we put

$$\frac{\overleftarrow{\delta} f(\boldsymbol{x},[\boldsymbol{q}],[\boldsymbol{q}^\dagger])}{\delta q^\alpha(\boldsymbol{y}_1)}\Big|_{j^\infty_{\boldsymbol{x}}(s)} = \sum_{|\sigma_1|\geqslant 0}\left(-\frac{\vec{\partial}}{\partial\boldsymbol{y}_1}\right)^{\sigma_1}\left(\frac{\vec{\partial} f(x,[\boldsymbol{q}],[\boldsymbol{q}^\dagger])}{\partial q^\alpha_{\sigma_1}}\right)\Big|_{j^\infty_{\boldsymbol{x}}(s)} = \tag{18a}$$
$$= \sum_{|\sigma_1|\geq 0}\left(\left(-\frac{\vec{\mathrm{d}}}{\mathrm{d}\boldsymbol{y}_1}\right)^{\sigma_1}\frac{\vec{\partial} f(\boldsymbol{x},[\boldsymbol{q}],[\boldsymbol{q}^\dagger]}{\partial q^\alpha_{\sigma_1}}\right)\Big|_{j^\infty_{\boldsymbol{x}}(s)}$$

and

$$\frac{\overleftarrow{\delta} f(\boldsymbol{x},[\boldsymbol{q}],[\boldsymbol{q}^\dagger])}{\delta q^\dagger_\beta(\boldsymbol{y}_2)}\Big|_{j^\infty_{\boldsymbol{x}}(s)} = \sum_{|\sigma_2|\geqslant 0}\left(-\frac{\vec{\partial}}{\partial\boldsymbol{y}_2}\right)^{\sigma_2}\left(\frac{\vec{\partial} f(\boldsymbol{x},[\boldsymbol{q}],[\boldsymbol{q}^\dagger])}{\partial q^\dagger_{\beta,\sigma_2}}\right)\Big|_{j^\infty_{\boldsymbol{x}}(s)} = \tag{18b}$$
$$= \sum_{|\sigma_2|\geq 0}\left(\left(-\frac{\vec{\mathrm{d}}}{\mathrm{d}\boldsymbol{y}_2}\right)^{\sigma_2}\frac{\vec{\partial} f(\boldsymbol{x},[\boldsymbol{q}],[\boldsymbol{q}^\dagger]}{\partial q^\dagger_{\beta,\sigma_2}}\right)\Big|_{j^\infty_{\boldsymbol{x}}(s)}$$

for the ghost parity-homogeneous components of variational derivative. At every point $(\boldsymbol{x},\phi(\boldsymbol{x}),s(\boldsymbol{x}))$ of the total space for the bundle π_{BV}, and for a given functional F which is assumed ghost parity-homogeneous, we have that

$$\frac{\overleftarrow{\delta} f(\boldsymbol{x},[\boldsymbol{q}],[\boldsymbol{q}^\dagger])}{\delta q^\alpha(\boldsymbol{y}_1)}\Big|_{j^\infty_{\boldsymbol{x}}(s)} \in T^*_{s(\boldsymbol{x})}(\Pi)V_{\boldsymbol{x}} \qquad\text{and}\qquad \frac{\overleftarrow{\delta} f(\boldsymbol{x},[\boldsymbol{q}],[\boldsymbol{q}^\dagger])}{\delta q^\dagger_\beta(\boldsymbol{y}_2)}\Big|_{j^\infty_{\boldsymbol{x}}(s)} \in T^*_{s(\boldsymbol{x})}(\Pi)V^\dagger_{\boldsymbol{x}}.$$

Let us remember that an attribution of denominators to $\boldsymbol{y}_1$ or $\boldsymbol{y}_2$ is a matter of notation in (18); whenever happening, everything happens at $\boldsymbol{x}\in M^n$.

Journal of Physics: Conference Series **474** (2013) 012024 doi:10.1088/1742-6596/474/1/012024

Lemma 1. The BV-Laplacian Δ is independent of a choice of the variation $\delta\boldsymbol{s}$ normalized by (16).

Indeed, whenever the integrations by parts are performed, products (16) of the dual components are always the same at all points of the intersection of their domains of definition.[19] We illustrate the definition of BV-Laplacian Δ by using Fig. 8; let us notice that it properly

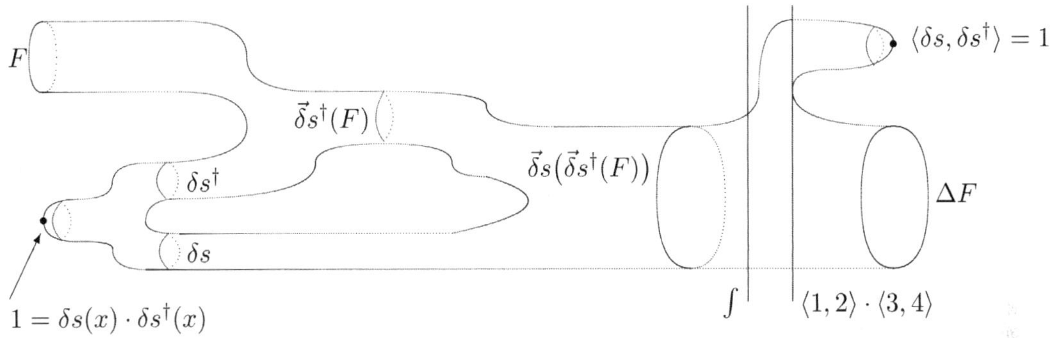

Figure 8. A variational update of the cyclic wor(l)d from [36]: the (anti)words δs and $\delta s^\dagger$ are pasted into a necklace F according to the graded Leibniz rule. Then they annihilate in such a way that the respective loose ends of the string join, the cyclic order of gems preserved; this yields ΔF.

renders the assertion of Lemma 1 in a wider, noncommutative setup of [36] and [29, 32] (see Remark 1.1 on p. 6).

Corollary 2. In particular, we obtain the equality for immediate numeric value of ΔF at s. Namely, we have that

$$(\Delta F)(s) = \sum_{i_1,i_2} \sum_{\substack{|\sigma_1|\geq 0\\ |\sigma_2|\geq 0}} \int_M \mathrm{dvol}(\boldsymbol{x})$$
$$\left\{ \delta s^{i_1}(\boldsymbol{y}_1) \cdot \boldsymbol{\delta}^{i_2}_{i_1} \cdot \delta s^\dagger_{i_2}(\boldsymbol{y}_2) \cdot \left(\left(-\frac{\vec{\mathrm{d}}}{\mathrm{d}\boldsymbol{x}} \right)^{\sigma_1 \cup \sigma_2} \frac{\vec{\partial}^2 f(\boldsymbol{x}, [\boldsymbol{q}], [\boldsymbol{q}^\dagger])}{\partial q^{i_1}_{\sigma_1} \partial q^\dagger_{i_2,\sigma_2}} \right) \Bigg|_{j^\infty_{\boldsymbol{x}}(s)} \right\} \Bigg|_{\substack{\boldsymbol{y}_1 = \boldsymbol{x}\\ \boldsymbol{y}_2 = \boldsymbol{x}}} \in \Bbbk.$$

By taking one sum containing Kronecker's $\boldsymbol{\delta}$-symbol, one arrives at a conventional formula with a summation over the diagonal:

$$(\Delta F)(s) = \sum_{i=1}^{N} \sum_{\substack{|\sigma_1|\geq 0\\ |\sigma_2|\geq 0}} \int_M \mathrm{dvol}(\boldsymbol{x}) \left(\left(-\frac{\vec{\mathrm{d}}}{\mathrm{d}\boldsymbol{x}} \right)^{\sigma_1 \cup \sigma_2} \frac{\vec{\partial}^2 f(\boldsymbol{x}, [\boldsymbol{q}], [\boldsymbol{q}^\dagger])}{\partial q^i_{\sigma_1} \partial q^\dagger_{i,\sigma_2}} \right) \Bigg|_{j^\infty_{\boldsymbol{x}(s)}} \stackrel{\text{def}}{=}$$
$$\stackrel{\text{def}}{=} \sum_{i=1}^{N} \int_M \mathrm{dvol}(\boldsymbol{x}) \, \frac{\overleftarrow{\delta^2} f(\boldsymbol{x}, [\boldsymbol{q}], [\boldsymbol{q}^\dagger])}{\delta q^i \delta q^\dagger_i}. \quad (19)$$

We refer to footnote 13 on p. 13 in this context.

[19] The assertion of Lemma 1 extends to the variational Schouten bracket, which is a derivative structure with respect to the BV-Laplacian (see Definition 2 on p. 28). Moreover, the independence of a specific choice of variations implies that their coefficients $(\delta s_1, \delta s_1^\dagger)$ and $(\delta s_2, \delta s_2^\dagger)$, which are built into Δ and $[\![\,,\,]\!]$, can be swapped, not altering an object that contains these test shifts $\delta \boldsymbol{s}_1$ and $\delta \boldsymbol{s}_2$ (see the proof of Lemma 5 on p. 31).

Remark 2.8. The conventional formula

$$\left.\frac{\overleftarrow{\delta^2} f(\boldsymbol{x},[\boldsymbol{q}],[\boldsymbol{q}^\dagger])}{\delta\boldsymbol{q}(\boldsymbol{y}_1)\delta\boldsymbol{q}^\dagger(\boldsymbol{y}_2)}\right|_{\substack{\boldsymbol{y}_1=\boldsymbol{x}\\ \boldsymbol{y}_2=\boldsymbol{x}}}$$

itself is not the definition of a density of the BV-Laplacian ΔF for an integral functional $F=\int f(\boldsymbol{x},[\boldsymbol{q}],[\boldsymbol{q}^\dagger])\cdot \mathrm{dvol}(\boldsymbol{x})$. Not containing any built-in sources of divergence, the geometric definition and its implication (19) yield identical results only when one calculates the numeric value $(\Delta F)(s)\in\Bbbk$ — but not earlier: structurally different objects (17) and (19) belong to non-isomorphic spaces (so that the former contains more information then the latter), and their analytic behaviour is also different, see Example 2.4 on p. 34.

The following two examples are quoted from [35]; they show that the structure Δ defined above coincides – but only in the simplest situation– with the one which is intuitively known from the literature. We refer to the main Example 2.4 on p. 34 which illustrates the multiple-base geometry in a logically more complex situation of (1c).

Example 2.2. Take a compact, semisimple Lie group G with Lie algebra $\mathfrak{g}$ and consider the corresponding Yang–Mills theory. Write A_i^a for the (coordinate expression of) the gauge potential A – a lower index i because A is a one-form on the base manifold (i. e., a covector), and an upper index a because A is a vector in the Lie algebra $\mathfrak{g}$ of the Lie group G. Defining the field strength $\mathcal{F}$ by $\mathcal{F}_{ij}^a=\partial_i A_j^a-\partial_j A_i^a+f_{bc}^a A_i^b A_j^c$ where f_{bc}^a are the structure constants of the Lie algebra $\mathfrak{g}$, the Yang–Mills action is[20]

$$S_{\mathrm{YM}}=\tfrac{1}{4}\int \mathcal{F}_{ij}^a\mathcal{F}^{a,ij}\,\mathrm{d}^4x,$$

and the full BV-action S_{BV} is[21]

$$S_{\mathrm{BV}}=S_{\mathrm{YM}}+\int A_a^{i\dagger}\bigl(\tfrac{\mathrm{d}}{\mathrm{d}x^i}\gamma^a+f_{bc}^aA_i^b\gamma^c\bigr)\,\mathrm{d}^4x-\tfrac{1}{2}\int f_{ab}^c\gamma^a\gamma^b\gamma_c^\dagger\,\mathrm{d}^4x.$$

Let us calculate the BV-Laplacian of this functional. By Corollary 2, the only terms which survive in $\Delta(S_{\mathrm{BV}})$ are those which contain both A and $A^\dagger$, or both γ and $\gamma^\dagger$. Therefore,

$$\begin{aligned}\Delta(S_{\mathrm{BV}})&=\int\left(\frac{\overleftarrow{\delta}}{\delta A_j^d}\frac{\overleftarrow{\delta}}{\delta A_d^{j\dagger}}(f_{bc}^aA_a^{i\dagger}A_i^b\gamma^c)-\frac{1}{2}\frac{\overleftarrow{\delta}}{\delta\gamma_d^\dagger}\frac{\overleftarrow{\delta}}{\delta\gamma^d}(f_{ab}^c\gamma^a\gamma^b\gamma_c^\dagger)\right)\mathrm{d}^4x\\&=\int\left(\frac{\overleftarrow{\delta}}{\delta A_j^d}(f_{bc}^dA_j^b\gamma^c)-\frac{1}{2}\frac{\overleftarrow{\delta}}{\delta\gamma_d^\dagger}(f_{db}^c\gamma^b\gamma_c^\dagger-f_{ad}^c\gamma^a\gamma_c^\dagger)\right)\mathrm{d}^4x\\&=\int\Bigl(f_{dc}^d\gamma^c-\tfrac{1}{2}\bigl(f_{db}^d\gamma^b-f_{ad}^d\gamma^a\bigr)\Bigr)\,\mathrm{d}^4x=0.\end{aligned}$$

Let us note also that, since the BV-action S_{BV} is by construction such that the horizontal cohomology class of $[\![S_{\mathrm{BV}},S_{\mathrm{BV}}]\!]$ is zero, as one easily checks by using Definition 2 below, the functional S_{BV} satisfies quantum master-equation (40) tautologically: both sides are, by independent calculations, equal to zero — should one inspect those values at any section s of the BV-bundle.

[20] The action functional S_{YM} is referred to Minkowski flat coordinates such that $\mathrm{dvol}(\boldsymbol{x})=\sqrt{|-1|}\,\mathrm{d}^4x$ in the weak gauge field limit.
[21] We denote by $A_a^{i\dagger}$ the parity-odd *antifields*, by γ^a the odd *ghosts*, and by $\gamma_a^\dagger$ the parity-even *antighosts*.

Example 2.3. Consider the nonlinear Poisson sigma model introduced in [11]. Since its fields are not all purely even, we have to generalize all of our reasoning so far to a $\mathbb{Z}_2$-graded setup — which is, as noted in Remark 2.1, tedious but straightforward. A verification that $\Delta(S_{\text{CF}})(s) = 0$ for the BV-action S_{CF} of this model and a section s of the respective BV-bundle would, up to minor differences in conventions and notations, proceed just as it does in that paper itself, in section 3.2 thereof — except that no infinite constants or Dirac's $\boldsymbol{\delta}$-function appear.

Remark 2.9. The BV-Laplacian Δ is extended by using Leibniz' rule from the space $\overline{H}^n(\pi_{\text{BV}})$ of building blocks in $\overline{\mathfrak{M}}^n(\pi_{\text{BV}})$ to the space $\overline{\mathfrak{N}}^n(\pi_{\text{BV}}, T\pi_{\text{BV}})$, see Theorem 3 on p. 29. The couplings' (re)attachment algorithm then results in formula (1b) on p. 1, which is taken as a *definition* of the variational Schouten bracket $[\![\,,\,]\!]$, see [39]. In turn, that structure's extention from $\overline{H}^n(\pi_{\text{BV}}) \times \overline{H}^n(\pi_{\text{BV}})$ to $\overline{\mathfrak{N}}^n(\pi_{\text{BV}}, T\pi_{\text{BV}}) \times \overline{\mathfrak{N}}^n(\pi_{\text{BV}}, T\pi_{\text{BV}})$ is immediate (see Theorem 4 below).

The correspondence between Δ and $[\![\,,\,]\!]$ is furthered to an equivalence between the property $\Delta^2 = 0$ of BV-Laplacian to be a differential and, on the other hand, Jacobi's identity for the variational Schouten bracket. We emphasize that the latter *can be* verified within the old approach [41] to geometry of variations. (We refer to [32] for a proof; its crucial idea is that with evolutionary vector fields it does not matter under "whose" total derivatives, $\mathrm{d}/\mathrm{d}\boldsymbol{x}$ or $\mathrm{d}/\mathrm{d}\boldsymbol{y}_i$, such fields dive.) Nevertheless, the traditional paradigm fails to reveal that the operator Δ is a differential because of a necessity to have the variations graded-permutable and for that, to distinguish between the functionals' and variations' domains of definition. Our geometric approach resolves that obstruction and ensures the validity of identities (1c) and (1d) (see Theorems 6 and 8, respectively).

2.4.2. The variational Schouten bracket $[\![\,,\,]\!]$ The parity-odd Laplacian Δ is the parent object[22] which induces the variational Schouten bracket $[\![\,,\,]\!]$. Namely, the bracket appears in the course of that operator's extension from the space $\overline{H}^n(\pi_{BV}) \ni F$ to the space $\overline{\mathfrak{N}}^n(\pi_{\text{BV}}, T\pi_{\text{BV}}) \supseteq \overline{\mathfrak{M}}^n(\pi_{\text{BV}})$ of local functionals $F_1 \cdot \ldots \cdot F_\ell$ (it is possible that F_i's already contain some normalized variations).

A distinction between *left* and *right* in the directed operators $\overleftarrow{\delta\boldsymbol{s}}$ and $\overrightarrow{\delta\boldsymbol{s}}$, the orientation $\vec{e}_i \prec \vec{e}^{\,\dagger i}$ in the composite BV-fibres $W_{\boldsymbol{x}} \cong V_{\boldsymbol{x}} \mathbin{\widehat{\oplus}} \Pi V_{\boldsymbol{x}}^\dagger$ equipped with two couplings (9), and the ordering of variations $\delta\boldsymbol{s}_1, \ldots, \delta\boldsymbol{s}_k$ specify the logic of operational Definition 2, which is given in this section.

Remark 2.10. For the sake of brevity, we extend the BV-Laplacian Δ from the space $\overline{H}^n(\pi_{\text{BV}})$ of integral functionals $F_1, \ldots, F_\ell$ to the space $\overline{\mathfrak{M}}^n(\pi_{\text{BV}})$ of local functionals such as $F_1 \cdot \ldots \cdot F_\ell$, the factors of which do not explicitly contain any built-in variations. To further this extension verbatim onto the full space $\overline{\mathfrak{N}}^n(\pi_{\text{BV}}, T\pi_{\text{BV}}) \supsetneq \overline{\mathfrak{M}}^n(\pi_{\text{BV}})$, one must remember that it is forbidden to break the order in which the directed variation operators $\overrightarrow{\delta\boldsymbol{s}_k}$ and $\overleftarrow{\delta\boldsymbol{s}_k}$ appear in the (ordered collection of) objects at hand. (Such concept is illustrated by the third term in (20) below.)

Likewise, we extend Δ to products of just two factors; in the case of arbitrary number $\ell \geq 2$ of building blocks $F_1, \ldots, F_\ell$ one proceeds inductively by using the ghost parity-graded Leibniz rule, then extending Δ onto the vector space $\overline{\mathfrak{N}}^n(\pi_{\text{BV}}, T\pi_{\text{BV}})$ by linearity.

Let $F = \int f(\boldsymbol{x}_1, [\boldsymbol{q}], [\boldsymbol{q}^\dagger])\,\mathrm{dvol}(\boldsymbol{x}_1)$ and $G = \int g(\boldsymbol{x}_2, [\boldsymbol{q}], [\boldsymbol{q}^\dagger])\,\mathrm{dvol}(\boldsymbol{x}_2)$ be integral functionals $\Gamma(\pi_{\text{BV}}) \to \Bbbk$ and let $\delta\boldsymbol{s} = (\delta s; \delta s^\dagger)$ be a normalized test shift of their product's argument $s \in \Gamma(\pi_{\text{BV}})$. We now define the operator Δ acting on the element $F \cdot G$ at $\boldsymbol{s}$ by variations first along $(0; \delta s^\dagger)$ and then along $(\delta s; 0)$.

[22] In particular, the definition of BV-Laplacian logically precedes the construction of Schouten bracket in BV-formalism (although such parity-odd variational Poisson bracket is often introduced through postulated formula (25) in the context of Hamiltonian dynamics and infinite-dimensional completely integrable systems [14, 18, 26, 41, 42]). Indeed, the entire Schouten-bracket machinery of (quantum) BV-cohomology groups and their automorphisms, which we consider in secction 3.2, stems from quantum master-equation (40), see p. 41.

According to (14), the object to start with is

$$\int_M \mathrm{d}\boldsymbol{y}_1 \int_M \mathrm{d}\boldsymbol{y}_2 \sum_{\substack{i_1,i_2\\ j_1,j_2}} \sum_{\substack{|\sigma_1|\geq 0\\ |\sigma_2|\geq 0}} \Bigl\{ (\delta s^{i_1}) \Bigl(\frac{\overleftarrow{\partial}}{\partial \boldsymbol{y}_1}\Bigr)^{\sigma_1} (\boldsymbol{y}_1) \bigl\langle \vec{e}_{i_1}(\boldsymbol{y}_1), \vec{e}^{\,\dagger j_1}(\cdot)\bigr\rangle \frac{\overrightarrow{\partial}}{\partial q^{j_1}_{\sigma_1}} \circ$$
$$\circ (\delta s^\dagger_{i_2}) \Bigl(\frac{\overleftarrow{\partial}}{\partial \boldsymbol{y}_2}\Bigr)^{\sigma_2} (\boldsymbol{y}_2) \bigl\langle -\vec{e}^{\,\dagger i_2}(\boldsymbol{y}_2), \vec{e}_{j_2}(\cdot)\bigr\rangle \frac{\overrightarrow{\partial}}{\partial q^\dagger_{j_2,\sigma_2}} \Bigr\}$$
$$\Bigl(\int f(\boldsymbol{x}_1, [\boldsymbol{q}], [\boldsymbol{q}^\dagger])\,\mathrm{dvol}(\boldsymbol{x}_1) \cdot \int g(\boldsymbol{x}_2, [\boldsymbol{q}], [\boldsymbol{q}^\dagger])\,\mathrm{dvol}(\boldsymbol{x}_2) \Bigr)(s).$$

Their order preserved, the directed operators $\overrightarrow{\delta s}$ and $\overrightarrow{\delta s}^\dagger$ spread over the two factors F and G by the binomial formula because of the Leibniz rule for graded derivations $\overrightarrow{\partial}/\partial q^{j_1}_{\sigma_1}$ and $\overrightarrow{\partial}/\partial q^\dagger_{j_2,\sigma_2}$. Note that whenever the ghost parity-odd object $\overrightarrow{\partial}/\partial q^\dagger_{j_2,\sigma_2}$ overtakes the density f of ghost parity $\mathrm{gh}(F)$, there appears an overall sign factor $(-)^{\mathrm{gh}(F)}$. We thus obtain

$$(\overrightarrow{\delta s} \circ \overrightarrow{\delta s}^\dagger)(F) + (-)^{\mathrm{gh}(F)} \overrightarrow{\delta s}(F)\, \overrightarrow{\delta s}^\dagger(G) + \overrightarrow{\delta s} \xrightarrow{\cdot\,\overrightarrow{\delta s}^\dagger(F)\cdot} G$$
$$+ (-1)^{\mathrm{gh}(F)} F \cdot (\overrightarrow{\delta s} \circ \overrightarrow{\delta s}^\dagger)(G). \quad (20)$$

The next step is to push right through F its single variations in the middle two terms of the above expression. This yields the equality

$$= (\overrightarrow{\delta s} \circ \overrightarrow{\delta s}^\dagger)(F) \cdot G + (-)^{\mathrm{gh}(F)} \Bigl\{ (F)\overleftarrow{\delta s} \cdot \overrightarrow{\delta s}^\dagger(G) + \overrightarrow{\delta s} \xrightarrow{\cdot (F)\overleftarrow{\delta s}^\dagger\cdot} G \Bigr\}$$
$$+ (-)^{\mathrm{gh}(F)} F \cdot (\overrightarrow{\delta s} \circ \overrightarrow{\delta s}^\dagger)(G). \quad (21)$$

We emphasize that the operators $\overleftarrow{\delta s}$ and $\overleftarrow{\delta s}^\dagger$ in the variations $(F)\overleftarrow{\delta s} = \overrightarrow{\delta_{\boldsymbol{q}} F}$ and $(F)\overleftarrow{\delta s}^\dagger = \overrightarrow{\delta_{\boldsymbol{q}^\dagger} F}$ are temporarily redirected to the left so that the middle terms in (21) are $(-)^{\mathrm{gh}(F)}$ times

$$\overrightarrow{\delta_{\boldsymbol{q}} F} \cdot \overleftarrow{\delta_{\boldsymbol{q}^\dagger} F} + \overleftarrow{\delta_{\boldsymbol{q}}} \xrightarrow{\overrightarrow{\delta_{\boldsymbol{q}^\dagger} F}\cdot} \overrightarrow{G}\,; \quad (22)$$

this is the input datum for a traditional definition of the variational Schouten bracket (e. g., see [11] vs [39]). Let us remember that the BV-fibres orientation $\boldsymbol{q} \prec \boldsymbol{q}^\dagger$ expressed by (9) is built into the last term of (22) even if it is written as follows,

$$(F)\overleftarrow{\delta_{\boldsymbol{q}}} \cdot \overrightarrow{\delta_{\boldsymbol{q}^\dagger}}(G) + (F)\overleftarrow{\delta_{\boldsymbol{q}^\dagger}} \cdot \overrightarrow{\delta_{\boldsymbol{q}}}(G).$$

Should this be the notation for input, one then usually proclaims that "differential 1-forms anticommute" so that $\langle \delta\boldsymbol{q}^\dagger \wedge \delta\boldsymbol{q}\rangle = -\langle \delta\boldsymbol{q} \wedge \delta\boldsymbol{q}^\dagger\rangle = -1$ in $[\![F, G]\!] = \langle \overrightarrow{\delta F} \wedge \overleftarrow{\delta G}\rangle$.

We now are almost in a position to (re)configure the couplings in the four terms of (21). The first term will of course become $\Delta F \cdot G$, and the last will provide $(-)^{\mathrm{gh}(F)} F \cdot \Delta G$; one is here allowed to integrate by parts (as explained in section 1.3) in order to shake the derivatives off δs^{i_1} and $\delta s^\dagger_{i_2}$ prior to evaluation of couplings in the resulting object's numeric value at its argument s. Yet there remains one more logical step to be done with (22): let us reverse back $\overleftarrow{\delta s} \mapsto \overrightarrow{\delta s}$ and $\overleftarrow{\delta s}^\dagger \mapsto \overrightarrow{\delta s}^\dagger$ so that on one hand, the vertical differentials fall on F but on the other hand, the normalization of the basis which stands near $\delta s^i(\boldsymbol{y}_1)$ and $\delta s^\dagger_i(\boldsymbol{y}_2)$ is the *first* not second column in (13). This yields the following integrand of $(-)^{\mathrm{gh}(F)}\int_M \mathrm{d}\boldsymbol{y}_1 \int_M \mathrm{d}\boldsymbol{y}_2 \int_M \mathrm{dvol}(\boldsymbol{x}_1) \int_M \mathrm{dvol}(\boldsymbol{x}_2)$, with a summation over i_1, i_2, j_1, j_2, and $|\sigma_1| \geq 0,\ |\sigma_2| \geq 0$,

$$\left(f(\boldsymbol{x}_1, [\boldsymbol{q}], [\boldsymbol{q}^\dagger]) \frac{\overleftarrow{\partial}}{\partial q^{j_1}_{\sigma_1}} \right)\Bigg|_{j^\infty_{\boldsymbol{x}_1}(s)} \langle \vec{e}^{\,\dagger j_1}(\boldsymbol{x}_1), +\vec{e}_{i_1}(\boldsymbol{y}_1)\rangle \left(\frac{\overrightarrow{\partial}}{\partial \boldsymbol{y}_1}\right)^{\sigma_1} (\delta s^{i_1})(\boldsymbol{y}_1)\cdot \qquad (23)$$

$$\cdot (\delta s^\dagger_{i_2}) \left(\frac{\overleftarrow{\partial}}{\partial \boldsymbol{y}_2}\right)^{\sigma_2} (\boldsymbol{y}_2) \langle -\vec{e}^{\,\dagger i_2}(\boldsymbol{y}_2), \vec{e}_{j_2}(\boldsymbol{x}_2)\rangle \left(\frac{\overrightarrow{\partial}}{\partial q^\dagger_{j_2,\sigma_2}} g(\boldsymbol{x}_2, [\boldsymbol{q}], [\boldsymbol{q}^\dagger]) \right)\Bigg|_{j^\infty_{\boldsymbol{x}_2}(s)} +$$

$$+ \left(f(\boldsymbol{x}_1, [\boldsymbol{q}], [\boldsymbol{q}^\dagger]) \frac{\overleftarrow{\partial}}{\partial q^\dagger_{j_1,\sigma_1}} \right)\Bigg|_{j^\infty_{\boldsymbol{x}_1}(s)} \cdot$$

$$\left\{ \begin{array}{llll} & (\delta s^{i_1}) \left(\frac{\overleftarrow{\partial}}{\partial \boldsymbol{y}_1}\right)^{\sigma_2} (\boldsymbol{y}_1) \langle \vec{e}_{i_1}(\boldsymbol{y}_1)| & & |\vec{e}^{\,\dagger j_2}(\boldsymbol{x}_2)\rangle \\ \langle \vec{e}_{j_1}(\boldsymbol{x}_1)| & & |-\vec{e}^{\,\dagger i_2}(\boldsymbol{y}_2)\rangle \left(\frac{\overrightarrow{\partial}}{\partial \boldsymbol{y}_2}\right)^{\sigma_1} (\delta s^\dagger_{i_2})(\boldsymbol{y}_2)\rangle & \end{array} \right\}.$$

$$\left(\frac{\overrightarrow{\partial}}{\partial q^{j_2}_{\sigma_2}} g(\boldsymbol{x}_2, [\boldsymbol{q}], [\boldsymbol{q}^\dagger]) \right)\Bigg|_{j^\infty_{\boldsymbol{x}_2}(s)}.$$

The integrations by parts are performed and couplings are reconfigured at the end of the day in exactly same manner as it has been done in Definition 1; let us recall that we now define the BV-Laplacian on a larger space. Namely, the variations couple with the dual variations whereas the differentials of functionals' densities attach to each other.

Definition 2. The *variational Schouten bracket* of two integral functionals

$$F = \int f(\boldsymbol{x}_1, [\boldsymbol{q}], [\boldsymbol{q}^\dagger] \cdot \mathrm{dvol}(\boldsymbol{x}_1) \qquad \text{and} \qquad G = \int g(\boldsymbol{x}_2, [\boldsymbol{q}], [\boldsymbol{q}^\dagger] \cdot \mathrm{dvol}(\boldsymbol{x}_2)$$

is the on-the-diagonal couplings surgery which, by using a normalized test shift $\delta\boldsymbol{s} = (\delta s; 0) + (0; \delta s^\dagger) \in \Gamma(T\pi_{\mathrm{BV}})$, yields the functional from $\overline{\mathfrak{N}}^n(\pi_{\mathrm{BV}}, T\pi_{\mathrm{BV}})$ whose construction at a BV-section $s \in \Gamma(\pi_{\mathrm{BV}})$ is[23]

$$\int_M \mathrm{d}\boldsymbol{y}_1 \int_M \mathrm{d}\boldsymbol{y}_2 \int_M \mathrm{d}\boldsymbol{x}_1 \int_M \mathrm{dvol}(\boldsymbol{x}_2) \Bigg[\left(f(\boldsymbol{x}_1, [\boldsymbol{q}], [\boldsymbol{q}^\dagger]) \frac{\overleftarrow{\partial}}{\partial q^{j_1}_{\sigma_1}} \right)\Bigg|_{j^\infty_{\boldsymbol{x}_1}(s)}$$

$$\left\{ \begin{array}{l} \left(-\frac{\overleftarrow{\partial}}{\partial \boldsymbol{y}_1}\right)^{\sigma_1} \delta s^{i_1}(\boldsymbol{y}_1) \overbrace{\langle \vec{e}_{i_1}(\boldsymbol{y}_1), -\vec{e}^{\,\dagger i_2}(\boldsymbol{y}_2)\rangle}^{-1} \delta s^\dagger_{i_2}(\boldsymbol{y}_2) \left(-\frac{\overrightarrow{\partial}}{\partial \boldsymbol{y}_2}\right)^{\sigma_2} \\ \underbrace{\langle \vec{e}^{\,\dagger j_1}(\boldsymbol{x}_1)| \qquad\qquad , \qquad\qquad |\vec{e}_{j_2}(\boldsymbol{x}_2)\rangle}_{-1} \end{array} \right\} \left(\frac{\overrightarrow{\partial}}{\partial q^\dagger_{j_2,\sigma_2}} g(\boldsymbol{x}_2, [\boldsymbol{q}], [\boldsymbol{q}^\dagger]) \right)\Bigg|_{j^\infty_{\boldsymbol{x}_2}(s)}$$

[23] Note that the directions of $\partial/\partial\boldsymbol{y}_i$ are reversed so that the minus signs appear. We emphasize that, prior to the evaluation of reconfigured couplings, the (co)vectors at $\boldsymbol{x}_j$ channel the partial derivatives to f or g according to the couplings' old arrangement.

$$
+\left(f(\boldsymbol{x}_1,[\boldsymbol{q}],[\boldsymbol{q}^\dagger])\frac{\overleftarrow{\partial}}{\partial q^\dagger_{j_1,\sigma_1}}\right)\bigg|_{j^\infty_{\boldsymbol{x}_1}(s)}
$$

$$
\underbrace{\left\{\begin{matrix} \delta s^{i_1}(\boldsymbol{y}_1)\left(-\frac{\overrightarrow{\partial}}{\partial\boldsymbol{y}_1}\right)^{\sigma_2}\overbrace{\langle\vec{e}_{i_1}(\boldsymbol{y}_1),(-\vec{e}^{\,\dagger i_2})(\boldsymbol{y}_2)\rangle}^{-1}\cdot\left(-\frac{\overleftarrow{\partial}}{\partial\boldsymbol{y}_2}\right)^{\sigma_1}\delta s^\dagger_{i_2}(\boldsymbol{y}_2) \\ \langle\vec{e}_{j_1}(\boldsymbol{x}_1)| \qquad\qquad , \qquad\qquad |\vec{e}^{\,\dagger j_2}(\boldsymbol{x}_2)\rangle \end{matrix}\right\}}_{+1}
$$

$$
\left(\frac{\overrightarrow{\partial}}{\partial q^{j_2}_{\sigma_2}}g(\boldsymbol{x}_1,[\boldsymbol{q}],[\boldsymbol{q}^\dagger])\right)\bigg|_{j^\infty_{\boldsymbol{x}_2}(s)}\Bigg].
$$

Note that the inner couplings between variations provide a restriction to the diagonal $\boldsymbol{y}_1=\boldsymbol{y}_2$ and yield the singular integral operators which then act to the right via multiplication by -1 only if $\boldsymbol{y}_2=\boldsymbol{x}_2$ and $\boldsymbol{y}_1=\boldsymbol{x}_1$, respectively. The outer coupling then furnishes the main diagonal $\boldsymbol{x}_1=\boldsymbol{y}_1=\boldsymbol{y}_2=\boldsymbol{x}_2$, restricting the objects further to the same BV-fibre point in the total space of the BV-bundle. This reveals why over each point of the base M^n the (derivatives of the) densities f and g are restricted to the infinite jet of the same section s; this also means that, since the moment when the couplings are reconfigured, the volume element $\mathrm{dvol}(\boldsymbol{x}_1)$ is discarded because appears a new singular linear integral operator with a standard sign $\int \mathrm{d}\boldsymbol{x}_1$.

We conclude the reasoning and sum up the definitions and notations in the following theorem.

Theorem 3. *The BV-Laplacian Δ is the linear operator*

$$
\overline{\mathfrak{N}}^n(\pi_{\mathrm{BV}},T\pi_{\mathrm{BV}})\times\overline{\mathfrak{N}}^n(\pi_{\mathrm{BV}},T\pi_{\mathrm{BV}})\to\overline{\mathfrak{N}}^n(\pi_{\mathrm{BV}},T\pi_{\mathrm{BV}})
$$

which acts on products of functionals F and $G\in\overline{\mathfrak{N}}^n(\pi_{\mathrm{BV}},T\pi_{\mathrm{BV}})$ by the rule

$$
\Delta(F\cdot G)=\Delta(F)\cdot G+(-)^{\mathrm{gh}(F)}[\![F,G]\!]+(-)^{\mathrm{gh}(F)}F\cdot\Delta G. \tag{24}
$$

The variational Schouten bracket $[\![\,,\,]\!]$ measures the deviation for the BV-Laplacian Δ from being a derivation.

- *After integration by parts, Definition 2 implies the renouned coordinate formula*

$$
[\![F,G]\!]=\int\mathrm{dvol}(\boldsymbol{x})\left(\frac{\overrightarrow{\delta}f(\boldsymbol{x},[\boldsymbol{q}],[\boldsymbol{q}^\dagger])}{\delta\boldsymbol{q}}\cdot\frac{\overleftarrow{\delta}g(\boldsymbol{x},[\boldsymbol{q}],[\boldsymbol{q}^\dagger])}{\delta\boldsymbol{q}^\dagger}-\right.
$$

$$
\left.-\frac{\overrightarrow{\delta}f(\boldsymbol{x},[\boldsymbol{q}],[\boldsymbol{q}^\dagger])}{\delta\boldsymbol{q}^\dagger}\cdot\frac{\overleftarrow{\delta}g(\boldsymbol{x},[\boldsymbol{q}],[\boldsymbol{q}^\dagger])}{\delta\boldsymbol{q}}\right). \tag{25}
$$

Remark 2.11. Let us recall from Remark 1.5 that the building blocks of local functionals are encoded by equivalence classes of their densities, whereas the underlying integration manifold M^n is endowed with the field-dependent volume element $\mathrm{dvol}(\boldsymbol{x},\phi)$. The variational Schouten bracket transforms two given integral functionals F and G into $[\![F,G]\!]$. For every configuration of physical fields $\phi\in\Gamma(\pi)$, the integration measure is the same in F, G, and $[\![F,G]\!]$. This is because the couplings are local over points $\big(\boldsymbol{x},\phi(\boldsymbol{x})\big)$ in the total space of the bundle π of physical fields, see Remark 2.2 on p. 18; the equality of local sections ϕ at which all (derivatives of) functionals' densities are evaluated ensures the equality of metric tensor elements $g_{\mu\nu}$ in all functionals by virtue of Einstein's general relativity equations.

The operational definition of the *antibracket* $[\![\,,\,]\!]$ determines the way how this structure acts on the square $\overline{\mathfrak{N}}^n(\pi_{\mathrm{BV}},T\pi_{\mathrm{BV}})\times\overline{\mathfrak{N}}^n(\pi_{\mathrm{BV}},T\pi_{\mathrm{BV}})$ of entire space $\overline{\mathfrak{N}}^n(\pi_{\mathrm{BV}},T\pi_{\mathrm{BV}})$ containing formal products of functionals.

Theorem 4. *Let F, G, and $H \in \overline{\mathfrak{N}}^n(\pi_{\rm BV}, T\pi_{\rm BV})$ be ghost parity-homogeneous functionals. The variational Schouten bracket $[\![\,,\,]\!]\colon \overline{\mathfrak{N}}^n(\pi_{\rm BV}, T\pi_{\rm BV}) \times \overline{\mathfrak{N}}^n(\pi_{\rm BV}, T\pi_{\rm BV}) \to \overline{\mathfrak{N}}^n(\pi_{\rm BV}, T\pi_{\rm BV})$ has the following properties:*

(i) The value of $[\![\,,\,]\!]$ at two arguments F and $G \cdot H$ is

$$[\![F, G \cdot H]\!] = [\![F, G]\!] \cdot H + (-)^{(\mathrm{gh}(F)-1)\,\mathrm{gh}(G)} G \cdot [\![F, H]\!]. \tag{26}$$

This formula recursively extends to products of arbitrary finite number of factors in the second argument.

(ii) The bracket $[\![\,,\,]\!]$ is shifted-graded skew-symmetric:

$$[\![F, G]\!] = -(-)^{(\mathrm{gh}(F)-1)\cdot(\mathrm{gh}(G)-1)} [\![G, F]\!], \tag{27}$$

which extends $[\![\,,\,]\!]$ to products of arbitrary finite number of factors taken as its first argument in (26).

(iii) The bracket $[\![\,,\,]\!]$ satisfies the shifted-graded Jacobi identity

$$(-)^{(\mathrm{gh}(F)-1)(\mathrm{gh}(H)-1)} [\![F, [\![G, H]\!]]\!] + (-)^{(\mathrm{gh}(F)-1)(\mathrm{gh}(G)-1)} [\![G, [\![H, F]\!]]\!] + \\ + (-)^{(\mathrm{gh}(G)-1)(\mathrm{gh}(H)-1)} [\![H, [\![F, G]\!]]\!] = 0, \tag{28}$$

which stems from graded Leibniz rule (36) for evolutionary vector fields $\boldsymbol{Q}^F$ defined by the rule $\boldsymbol{Q}^F(\cdot) \cong [\![F, \cdot]\!]$ (here the equivalence up to integration by parts is denoted by $\cong$).

Finally, the variational Schouten bracket extends by linearity to formal sums of elements from $\overline{\mathfrak{N}}^n(\pi_{\rm BV}, T\pi_{\rm BV})$.

Proof. The bilinearity of $[\![\,,\,]\!]$ is obvious. It is also clear that the terms in $[\![F, G\cdot H]\!]$ are grouped in two parts: those in which the ghost-parity graded derivations $\overrightarrow{\partial}/\partial \boldsymbol{q}^\dagger$ act on G and those for H; the former do not contribute with any extra sign factors whereas the latter do — in a way which depends on the parity gh(G). This means that $[\![F, G\cdot H]\!] = [\![F, G]\!] \cdot H + \ldots$; to grasp the sign in front of the term which has been omitted, let us swap the graded multiples G and H. We have that $G \cdot H = (-)^{\mathrm{gh}(G)\,\mathrm{gh}(H)} H \cdot G$, whence $[\![F, G\cdot H]\!] = (-)^{\mathrm{gh}(G)\,\mathrm{gh}(H)} [\![F, H]\!] \cdot G + \cdots$. By recalling that $\mathrm{gh}([\![F, H]\!]) = \mathrm{gh}(F) + \mathrm{gh}(H) - 1$, we conclude that

$$[\![F, G \cdot H]\!] = [\![F, G]\!] \cdot H + (-)^{\mathrm{gh}(G)\,\mathrm{gh}(H)} (-)^{(\mathrm{gh}(F)+\mathrm{gh}(H)-1)\cdot\mathrm{gh}(G)} G \cdot [\![F, H]\!],$$

which yields formula (26).

Proving (27) amounts to a count of signs whenever the bracket $[\![F, G]\!]$ of an ordered pair of ghost parity-graded objects is virtually transformed into $[\![G, F]\!]$. By using the rule of signs for odd-parity coordinates, $q^\dagger_{\alpha,\sigma} \cdot q^\dagger_{\beta,\tau} = -q^\dagger_{\beta,\tau} \cdot q^\dagger_{\alpha,\sigma}$, we first note that

$$\frac{\overrightarrow{\partial}}{\partial q^\dagger_{j_2,\sigma_2}} g(\boldsymbol{x}_2, [\boldsymbol{q}], [\boldsymbol{q}^\dagger]) = (-)^{\mathrm{gh}(G)-1} \Bigl(g(\boldsymbol{x}_2, [\boldsymbol{q}], [\boldsymbol{q}^\dagger]) \Bigr) \frac{\overleftarrow{\partial}}{\partial q^\dagger_{j_2,\sigma_2}},$$

with a similar formula for the left- and right-acting graded derivative of f. By swapping the (variational) derivatives of the densities f and g, we gain the signs $(-)^{\mathrm{gh}(F)\cdot(\mathrm{gh}(G)-1)}$ and $(-)^{(\mathrm{gh}(F)-1)\cdot\mathrm{gh}(G)}$ for the respective terms in (23) on p. 28. Combined together, the two steps accumulate equal factors $(-)^{(\mathrm{gh}(F)+1)\cdot(\mathrm{gh}(G)-1)} = (-)^{(\mathrm{gh}(F)-1)\cdot(\mathrm{gh}(G)+1)} = (-)^{(\mathrm{gh}(F)-1)\cdot(\mathrm{gh}(G)-1)}$. Thirdly, by comparing $(-)^{(\mathrm{gh}(F)-1)\cdot(\mathrm{gh}(G)-1)} [\![F, G]\!]$ –in which the derivatives of f and g are interchanged and the derivations' directions are reversed– with $[\![G, F]\!]$, we conclude that the reconfiguration of couplings in the second term in (23) for the former expression yields *minus* the first term in $[\![G, F]\!]$. Likewise, the couplings reattachment in the first term of such (23) produces minus the second term in $[\![G, F]\!]$. This is because the (co)vectors in the differentials of densities remain unswapped, now going in the 'wrong' order.

We now refer to [32, Proposition 3] for a proof of property (iii) in a wider, non-commutative setup of cyclic words (cf. [29, 36, 46]). It is remarkable that the reasoning persists within a naïve theory of variations, not referring to our main idea that each test shift brings its own copy of the base M^n into the picture. A key point in the proof is that the rule $\boldsymbol{Q}^F(\cdot) \cong [\![F, \cdot]\!]$ naturally associates with functionals F the evolutionary fields $\boldsymbol{Q}^F$ on the infinite jet superbundles at hand, and with *evolutionary* vector fields it does not matter under 'whose" total derivatives such fields dive, obeying their defining property $[\boldsymbol{Q}^F, \overrightarrow{\mathrm{d}}/\mathrm{d}\boldsymbol{x}] = 0$ (i. e., any integrations by parts, which transform the derivatives $\overrightarrow{\partial}/\partial \boldsymbol{y}_i$ falling on test shifts into total derivatives $\overrightarrow{\mathrm{d}}/\mathrm{d}\boldsymbol{x}$ falling on the functionals' densities, do not mar the outcome even if one attempts to perform such integrations ahead of time). □

2.5. Main result: the proof of properties (1c–1d)

We are ready to *prove* the main interrelations between the BV-Laplacian Δ and variational Schouten bracket $[\![\,,\,]\!]$. Let us recall that either a validity of these properties was postulated (see [21]) or an *ad hoc* regularization technique was formally employed in the literature in order to mask the seemingly present divergencies (which are actually not there), cf. [22, §15].

Let us fix the terms. In what follows we refer to building blocks from $\overline{H}^n(\pi_{\mathrm{BV}})$ and their descendants – containing reconfigured variations – from $\overline{H}^{n(1+k)}(\pi_{\mathrm{BV}} \times T\pi_{\mathrm{BV}} \times \ldots \times T\pi_{\mathrm{BV}})$ as *integral functionals*. Such objects will be used for bases of inductive proofs of Lemmas 5 and 7. We then extend the properties (1c) and $\Delta^2 = 0$ to the space $\overline{\mathfrak{N}}^n(\pi_{\mathrm{BV}}, T\pi_{\mathrm{BV}}) \supseteq \overline{\mathfrak{M}}^n(\pi_{\mathrm{BV}})$ of *local functionals*, that is, of formal sums of products of (varied descendants of) building blocks.

Lemma 5. Let $F \in \overline{H}^{n(1+k)}\big(\pi_{\mathrm{BV}} \times T\pi_{\mathrm{BV}} \times \ldots \times T\pi_{\mathrm{BV}}\big)$ and $G \in \overline{H}^{n(1+\ell)}\big(\pi_{\mathrm{BV}} \times T\pi_{\mathrm{BV}} \times \ldots \times T\pi_{\mathrm{BV}}\big)$ be two integral functionals; here $k, \ell \geqslant 0$. Then

$$\Delta\big([\![F, G]\!]\big) = [\![\Delta F, G]\!] + (-)^{\mathrm{gh}(F)-1}[\![F, \Delta G]\!]. \tag{29}$$

Proof. The key idea is that the structures Δ and $[\![\,,\,]\!]$ yield equivalence classes of integral functionals which, after an integration by parts at the end of the day, are *independent* of a choice of the built-in test shifts normalized by (16). Consequently, the composite structure $\Delta([\![\cdot, \cdot]\!])$ does not change under swapping $\delta s_1^\alpha \rightleftarrows \delta s_2^\beta$, $\delta s^\dagger_{1,\alpha} \rightleftarrows \delta s^\dagger_{2,\beta}$ of the respective variations $\delta \boldsymbol{s}_1$ and $\delta \boldsymbol{s}_2$ in Δ and $[\![\,,\,]\!]$. Hence the terms which are skew-symmetric under such exchange necessarily vanish.

For the sake of clarity, let us assume that $F = \int f(\boldsymbol{x}_1, [\boldsymbol{q}], [\boldsymbol{q}^\dagger])\,\mathrm{dvol}(\boldsymbol{x}_1)$ and $G = \int g(\boldsymbol{x}_2, [\boldsymbol{q}], [\boldsymbol{q}^\dagger])\,\mathrm{dvol}(\boldsymbol{x}_2)$ are just building blocks from the cohomology group $\overline{H}^n(\pi_{\mathrm{BV}})$; this simplification is legitimate because new variations which come from Δ and $[\![\,,\,]\!]$ do not interfere with any other test shifts if those are already absorbed by the densities f and g. Suppose that $\delta \boldsymbol{s}_1$ and $\delta \boldsymbol{s}_2$ are two normalized variations of a section $s \in \Gamma(\pi_{\mathrm{BV}})$. By definition, we have that[24]

$$\Delta\left([\![F, G]\!]\right)(s) = \int_M \mathrm{d}\boldsymbol{z}_1 \int_M \mathrm{d}\boldsymbol{z}_2 \int_M \mathrm{d}\boldsymbol{y}_1 \int_M \mathrm{d}\boldsymbol{y}_2 \int_M \mathrm{d}\boldsymbol{x}_1 \int_M \mathrm{dvol}(\boldsymbol{x}_2)\, \cdot$$
$$\Big\{ (\delta s_1^\alpha) \Big(\frac{\overleftarrow{\partial}}{\partial \boldsymbol{z}_1}\Big)^{\sigma_1} (\boldsymbol{z}_1) \Big\langle \vec{e}_\alpha(\boldsymbol{z}_1), -\vec{e}^{\,\dagger\alpha}(\boldsymbol{z}_2) \Big\rangle (\delta s^\dagger_{1,\alpha}) \Big(\frac{\overleftarrow{\partial}}{\partial \boldsymbol{z}_2}\Big)^{\sigma_2} (\boldsymbol{z}_2) \langle \vec{e}^{\,\dagger\alpha}(\cdot), \vec{e}_\alpha(\cdot)\rangle \frac{\overrightarrow{\partial}}{\partial q^\alpha_{\sigma_1}} \frac{\overrightarrow{\partial}}{\partial q^\dagger_{\alpha,\sigma_2}}$$

[24] To keep track of their origin, we let the directed derivatives $\partial/\partial \boldsymbol{y}_i$ or $\partial/\partial \boldsymbol{z}_j$ remain falling on the respective coefficients in $\delta \boldsymbol{s}_1$ and $\delta \boldsymbol{s}_2$; the integration by parts is performed in a standard way prior to the reconfigurations which are shown in the formula.

$$\Bigg[f(\boldsymbol{x}_1.[\boldsymbol{q}],[\boldsymbol{q}^\dagger])\frac{\overleftarrow{\partial}}{\partial q^\beta_{\tau_1}}\,\underline{\langle\vec{e}^{\dagger\beta}(\boldsymbol{x}_1)|}\Big\langle\Big(\frac{\overrightarrow{\partial}}{\partial\boldsymbol{y}_1}\Big)^{\tau_1}(\delta s_2^\beta)(\boldsymbol{y}_1)\,\vec{e}_\beta(\boldsymbol{y}_1),$$
$$-\vec{e}^{\dagger.\beta}(\boldsymbol{y}_2)\,(\delta s^\dagger_{2,\beta})\Big(\frac{\overleftarrow{\partial}}{\partial\boldsymbol{y}_2}\Big)^{\tau_2}(\boldsymbol{y}_2)\Big\rangle\,\underline{|\vec{e}_\beta(\boldsymbol{x}_2)\rangle}\,\frac{\overrightarrow{\partial}}{\partial q^\dagger_{\beta,\tau_2}}g(\boldsymbol{x}_2,[\boldsymbol{q}],[\boldsymbol{q}^\dagger])+$$
$$+f(\boldsymbol{x}_1,[\boldsymbol{q}],[\boldsymbol{q}^\dagger])\frac{\overleftarrow{\partial}}{\partial q^\dagger_{\beta,\tau_2}}\,\underline{\langle\vec{e}_\beta(\boldsymbol{x}_1)|}\Big\langle(\delta s_2^\beta)\Big(\frac{\overleftarrow{\partial}}{\partial\boldsymbol{y}_1}\Big)^{\tau_1}(\boldsymbol{y}_1)\,\vec{e}_\beta(\boldsymbol{y}_1),$$
$$-\vec{e}^{\dagger\beta}(\boldsymbol{y}_2)\Big(\frac{\overrightarrow{\partial}}{\partial\boldsymbol{y}_2}\Big)^{\tau_2}(\delta s^\dagger_{2,\beta})(\boldsymbol{y}_2)\Big\rangle\,\underline{|\vec{e}^{\dagger\beta}(\boldsymbol{x}_2)\rangle}\,\frac{\overrightarrow{\partial}}{\partial q^\beta_{\tau_1}}g(\boldsymbol{x}_2,[\boldsymbol{q}],[\boldsymbol{q}^\dagger])\Bigg]\Bigg\}\Bigg|^{j^\infty(s)}_{\boldsymbol{x}_i=\boldsymbol{y}_j=\boldsymbol{z}_k}.$$

The partial derivatives $\overrightarrow{\partial}/\partial q^\alpha_{\sigma_1}\circ\overrightarrow{\partial}/\partial q^\dagger_{\alpha,\sigma_2}$ are distributed between the arguments f and g by the graded Leibniz rule. Whenever *none* of the two operators overtakes the density of F, the reconfiguration yields $[\![\Delta F,G]\!](\boldsymbol{s})$. Likewise, if *both* derivatives indexed by α overtake F and an old derivative that fell on g, then we obtain $(-)^{\mathrm{gh}(F)-1}[\![F,\Delta G]\!](\boldsymbol{s})$, which is the second term in the right-hand side of (29). We claim that the remaining four terms cancel out by virtue of independence of Δ and $[\![\,,\,]\!]$ from a choice of normalized variations. To prove this claim, we consecutively inspect the behaviour of those four terms under a swap $\delta\boldsymbol{s}_1\rightleftarrows\delta\boldsymbol{s}_2$ of coefficients in the normalized test shifts.

The first and second terms sum up to the difference

$$\Big\langle(\delta s_1^\alpha)\Big(\frac{\overleftarrow{\partial}}{\partial\boldsymbol{z}_1}\Big)^{\sigma_1}(\boldsymbol{z}_1)\,\overbrace{\vec{e}_\alpha(\boldsymbol{z}_1),(-\vec{e}^{\dagger\alpha})(\boldsymbol{z}_2)}^{-1}(\delta s^\dagger_{1,\alpha})\Big(\frac{\overleftarrow{\partial}}{\partial\boldsymbol{z}_2}\Big)^{\sigma_2}(\boldsymbol{z}_2)\Big\rangle\,\underbrace{\underline{\langle\vec{e}^{\dagger\alpha}(\boldsymbol{x}_2),\vec{e}_\alpha(\boldsymbol{x}_1)\rangle}}_{-1}\cdot$$
$$\cdot\frac{\overrightarrow{\partial}}{\partial q^\dagger_{\alpha,\sigma_2}}f(\boldsymbol{x}_1,[\boldsymbol{q}],[\boldsymbol{q}^\dagger])\frac{\overleftarrow{\partial}}{\partial q^\beta_{\tau_1}}\,\underline{\langle\vec{e}^{\dagger\beta}(\boldsymbol{x}_1)|}\Big\langle\Big(\frac{\overrightarrow{\partial}}{\partial\boldsymbol{y}_1}\Big)^{\tau_1}(\delta s_2^\beta)(\boldsymbol{y}_1)\,\overbrace{\vec{e}_\beta(\boldsymbol{y}_1),(-\vec{e}^{\dagger\beta})(\boldsymbol{y}_2)}^{-1}$$
$$(\delta s^\dagger_{2,\beta})\Big(\frac{\overleftarrow{\partial}}{\partial\boldsymbol{y}_2}\Big)^{\tau_2}(\boldsymbol{y}_2)\Big\rangle\,\underbrace{\underline{|\vec{e}_\beta(\boldsymbol{x}_2)\rangle}}_{-1}\,\frac{\overrightarrow{\partial}}{\partial q^\alpha_{\sigma_1}}\frac{\overrightarrow{\partial}}{\partial q^\dagger_{\beta,\tau_2}}g(\boldsymbol{x}_2,[\boldsymbol{q}],[\boldsymbol{q}^\dagger])+$$
$$+\Big\langle(\delta s_1^\alpha)\Big(\frac{\overleftarrow{\partial}}{\partial\boldsymbol{z}_1}\Big)^{\sigma_1}(\boldsymbol{z}_1)\,\overbrace{\vec{e}_\alpha(\boldsymbol{z}_1),(-\vec{e}^{\dagger\alpha})(\boldsymbol{z}_2)}^{-1}(\delta s^\dagger_{1,\alpha})\Big(\frac{\overleftarrow{\partial}}{\partial\boldsymbol{z}_2}\Big)^{\sigma_2}(\boldsymbol{z}_2)\Big\rangle\,\underbrace{\underline{\langle\vec{e}^{\dagger\alpha}(\boldsymbol{x}_1),\vec{e}_\alpha(\boldsymbol{x}_2)\rangle}}_{-1}\cdot$$
$$\cdot(-)^{\mathrm{gh}(F)-1}\frac{\overrightarrow{\partial}}{\partial q^\alpha_{\sigma_1}}f(\boldsymbol{x}_1,[\boldsymbol{q}],[\boldsymbol{q}^\dagger])\frac{\overleftarrow{\partial}}{\partial q^\dagger_{\beta,\tau_2}}\,\underline{\langle\vec{e}_\beta(\boldsymbol{x}_1)|}\Big\langle(\delta s_2^\beta)\Big(\frac{\overleftarrow{\partial}}{\partial\boldsymbol{y}_1}\Big)^{\tau_1}(\boldsymbol{y}_1)\,\overbrace{\vec{e}_\beta(\boldsymbol{y}_1),(-\vec{e}^{\dagger\beta})(\boldsymbol{y}_2)}^{-1}$$
$$\Big(\frac{\overrightarrow{\partial}}{\partial\boldsymbol{y}_2}\Big)^{\tau_2}(\delta s^\dagger_{2,\beta})(\boldsymbol{y}_2)\Big\rangle\,\underbrace{\underline{|\vec{e}^{\dagger\beta}(\boldsymbol{x}_2)\rangle}}_{+1}\,\frac{\overrightarrow{\partial}}{\partial q^\dagger_{\alpha,\sigma_2}}\frac{\overrightarrow{\partial}}{\partial q^\beta_{\tau_1}}g(\boldsymbol{x}_2,[\boldsymbol{q}],[\boldsymbol{q}^\dagger]).\quad(30)$$

Recalling that

$$f(\boldsymbol{x}_1,[\boldsymbol{q}],[\boldsymbol{q}^\dagger])\frac{\overleftarrow{\partial}}{\partial q^\dagger_{\beta,\tau_2}}=(-)^{\mathrm{gh}(F)-1}\frac{\overrightarrow{\partial}}{\partial q^\dagger_{\beta,\tau_2}}f(\boldsymbol{x}_1,[\boldsymbol{q}],[\boldsymbol{q}^\dagger]),$$

let us swap the derivations which fall on f from the left and right; this eliminates the sign $(-)^{\mathrm{gh}(F)-1}$. We proceed likewise for g and then transport the variations $\delta\boldsymbol{s}_1$ and $\delta\boldsymbol{s}_2$, exchanging their places (and their rôles with respect to Δ and $[\![\,,\,]\!]$). The second term in formula (30) becomes

$$\Bigl\langle (\delta s_2^\beta) \Bigl(\frac{\overleftarrow{\partial}}{\partial \boldsymbol{y}_1}\Bigr)^{\tau_1} (\boldsymbol{y}_1)\, \overbrace{\vec{e}_\beta(\boldsymbol{y}_1), (-\vec{e}^{\,\dagger\beta})(\boldsymbol{y}_2)}^{-1} \Bigl(\frac{\overrightarrow{\partial}}{\partial \boldsymbol{y}_2}\Bigr)^{\tau_2} (\delta s^\dagger_{2,\beta})(\boldsymbol{y}_2) \Bigr\rangle \underbrace{\langle \vec{e}_\beta(\boldsymbol{x}_1), \vec{e}^{\,\dagger\beta}(\boldsymbol{x}_2)\rangle}_{+1}$$

$$\cdot \frac{\overrightarrow{\partial}}{\partial q^\dagger_{\beta,\tau_2}} f(\boldsymbol{x}_1, [\boldsymbol{q}], [\boldsymbol{q}^\dagger]) \frac{\overleftarrow{\partial}}{\partial q^\alpha_{\sigma_1}} \underline{\langle \vec{e}^{\,\dagger\alpha}(\boldsymbol{x}_1) |} \Bigl\langle \Bigl(\frac{\overrightarrow{\partial}}{\partial \boldsymbol{z}_1}\Bigr)^{\sigma_1} (\delta s_1^\alpha)(\boldsymbol{z}_1)\, \overbrace{\vec{e}_\alpha(\boldsymbol{z}_1), (-\vec{e}^{\,\dagger\alpha})(\boldsymbol{z}_2)}^{-1} (\delta s^\dagger_{1,\alpha}) \Bigl(\frac{\overleftarrow{\partial}}{\partial \boldsymbol{z}_2}\Bigr) (\boldsymbol{z}_2) \Bigr\rangle$$

$$\underbrace{| \vec{e}_\alpha(\boldsymbol{x}_2)\rangle}_{-1} \frac{\overrightarrow{\partial}}{\partial q^\beta_{\tau_1}} \frac{\overrightarrow{\partial}}{\partial q^\dagger_{\alpha,\sigma_2}} g(\boldsymbol{x}_2, [\boldsymbol{q}], [\boldsymbol{q}^\dagger]).$$

It is now readily seen that the first term in (30) and this equivalent expression of its second term are opposite to each other. Indeed, relabel the summation indexes $\alpha \rightleftarrows \beta$, $\sigma \rightleftarrows \tau$ so that $\delta s_1^\alpha \rightleftarrows \delta s_2^\beta$, $\delta s^\dagger_{1,\alpha} \rightleftarrows \delta s^\dagger_{2,\beta}$, and swap the copies of base manifold M^n by $\boldsymbol{y} \rightleftarrows \boldsymbol{z}$. Due to the second factors in the products $(-1)\cdot(-1)\cdot(-1)\cdot(-1) = +1$ versus $(-1)\cdot(+1)\cdot(-1)\cdot(-1) = -1$, the two terms in (30) cancel out after the integration by parts and evaluation of the couplings in view of (16).

Next, the integrand of $\Delta([\![F, G]\!])(s)$ contains a restriction to the infinite jet $j^\infty(s)$ of the third term, which is

$$\Bigl\langle (\delta s_1^\alpha) \Bigl(\frac{\overleftarrow{\partial}}{\partial \boldsymbol{z}_1}\Bigr)^{\sigma_1} (\boldsymbol{z}_1)\, \overbrace{\vec{e}_\alpha(\boldsymbol{z}_1), (-\vec{e}^{\,\dagger\alpha})(\boldsymbol{z}_2)}^{-1} (\delta s^\dagger_{1,\alpha}) \Bigl(\frac{\overleftarrow{\partial}}{\partial \boldsymbol{z}_2}\Bigr)^{\sigma_2} (\boldsymbol{z}_2) \Bigr\rangle \underbrace{\langle \vec{e}^{\,\dagger\alpha}(\boldsymbol{x}_2), \vec{e}_\alpha(\boldsymbol{x}_1)\rangle}_{-1}$$

$$\cdot \frac{\overrightarrow{\partial}}{\partial q^\dagger_{\alpha,\sigma_2}} \Bigl(f(\boldsymbol{x}_1, [\boldsymbol{q}], [\boldsymbol{q}^\dagger]) \frac{\overleftarrow{\partial}}{\partial q^\dagger_{\beta,\tau_2}} \Bigr)$$

$$\underline{\langle \vec{e}_\beta(\boldsymbol{x}_1) |} \Bigl\langle (\delta s_2^\beta) \Bigl(\frac{\overleftarrow{\partial}}{\partial \boldsymbol{y}_1}\Bigr)^{\tau_1} (\boldsymbol{y}_1)\, \overbrace{\vec{e}_\beta(\boldsymbol{y}_1), (-\vec{e}^{\,\dagger\beta})(\boldsymbol{y}_2)}^{-1} \Bigl(\frac{\overrightarrow{\partial}}{\partial \boldsymbol{y}_2}\Bigr)^{\tau_2} (\delta s^\dagger_{2,\beta})(\boldsymbol{y}_2) \Bigr\rangle \underbrace{| \vec{e}^{\,\dagger\beta}(\boldsymbol{x}_2)\rangle}_{+1}$$

$$\frac{\overrightarrow{\partial}}{\partial q^\alpha_{\sigma_1}} \frac{\overrightarrow{\partial}}{\partial q^\beta_{\tau_1}} g(\boldsymbol{x}_2, [\boldsymbol{q}], [\boldsymbol{q}^\dagger]).$$

Let the summation indexes be relabelled as above: $\alpha \rightleftarrows \beta$, $\sigma \rightleftarrows \tau$, and $\delta s_1^\alpha \rightleftarrows \delta s_2^\beta$, $\delta s^\dagger_{1,\alpha} \rightleftarrows \delta s^\dagger_{2,\beta}$ on top of $\boldsymbol{y} \rightleftarrows \boldsymbol{z}$. The transformation of graded derivations falling from the left and right on f is then

$$\frac{\overrightarrow{\partial}}{\partial q^\dagger_{\alpha,\sigma_2}} \Bigl(f \frac{\overleftarrow{\partial}}{\partial q^\dagger_{\beta,\tau_2}} \Bigr) \longmapsto \frac{\overrightarrow{\partial}}{\partial q^\dagger_{\beta,\tau_2}} \Bigl(f \frac{\overleftarrow{\partial}}{\partial q^\dagger_{\alpha,\sigma_2}} \Bigr) = \frac{\overrightarrow{\partial}}{\partial q^\dagger_{\beta,\tau_2}} \Bigl((-)^{\mathrm{gh}(F)-1} \frac{\overrightarrow{\partial}}{\partial q^\dagger_{\alpha,\sigma_2}} f \Bigr) =$$
$$= (-)^{\mathrm{gh}(F)-2} \cdot (-)^{\mathrm{gh}(F)-1} \Bigl(\frac{\overrightarrow{\partial}}{\partial q^\dagger_{\alpha,\sigma_2}} f \Bigr) \frac{\overleftarrow{\partial}}{\partial q^\dagger_{\beta,\tau_2}} = - \frac{\overrightarrow{\partial}}{\partial q^\dagger_{\alpha,\sigma_2}} \Bigl(f \frac{\overleftarrow{\partial}}{\partial q^\dagger_{\beta,\tau_2}} \Bigr).$$

This minus sign shows that the third term as it was written initially, and the newly produced one in which the variations $\delta \boldsymbol{s}_1$ and $\delta \boldsymbol{s}_2$ are interchanged have opposite signs. At the same time, these integral functionals must be equal to each other due to independence of Δ and $[\![\,,\,]\!]$ of a choice of the test shifts. Therefore, each of those expressions vanishes.

The fourth term is processed analogously; its integrand is

$$(-)^{\mathrm{gh}(F)}\left\langle (\delta s_1^{\alpha})\left(\frac{\overleftarrow{\partial}}{\partial \boldsymbol{z}_1}\right)^{\sigma_1}(\boldsymbol{z}_1)\,\overbrace{\vec{e}_{\alpha}(\boldsymbol{z}_1),(-\vec{e}^{\dagger\alpha})(\boldsymbol{z}_2)}^{-1}(\delta s^{\dagger}_{1,\alpha})\left(\frac{\overleftarrow{\partial}}{\partial \boldsymbol{z}_2}\right)^{\sigma_2}(\boldsymbol{z}_2)\right\rangle\cdot\underbrace{\langle\vec{e}^{\dagger\alpha}(\boldsymbol{x}_1),\vec{e}_{\alpha}(\boldsymbol{x}_2)\rangle}_{-1}$$

$$\cdot\frac{\overrightarrow{\partial}}{\partial q^{\alpha}_{\sigma_1}}f(\boldsymbol{x}_1,[\boldsymbol{q}],[\boldsymbol{q}^{\dagger}])\frac{\overleftarrow{\partial}}{\partial q^{\beta}_{\tau_1}}$$

$$\underline{\langle\vec{e}^{\dagger\beta}(\boldsymbol{x}_1)|}\left\langle\left(\frac{\overleftarrow{\partial}}{\partial \boldsymbol{y}_1}\right)^{\tau_1}(\delta s_2^{\beta})(\boldsymbol{y}_1)\,\overbrace{\vec{e}_{\beta}(\boldsymbol{y}_1),(-\vec{e}^{\dagger\beta})(\boldsymbol{y}_2)}^{-1}(\delta s^{\dagger}_{2,\beta})\left(\frac{\overleftarrow{\partial}}{\partial \boldsymbol{y}_2}\right)^{\tau_2}(\boldsymbol{y}_2)\right\rangle\underbrace{|\vec{e}_{\beta}(\boldsymbol{x}_2)\rangle}_{-1}$$

$$\frac{\overrightarrow{\partial}}{\partial q^{\dagger}_{\alpha,\sigma_2}}\frac{\overrightarrow{\partial}}{\partial q^{\dagger}_{\beta,\tau_2}}g(\boldsymbol{x}_2,[\boldsymbol{q}],[\boldsymbol{q}^{\dagger}]).$$

The very same procedure of two variations interchange and relabelling restores an almost identical expression in which, however, the parity-odd derivations go in the inverse order $\overrightarrow{\partial}/\partial q^{\dagger}_{\beta,\tau_2}\circ\overrightarrow{\partial}/\partial q^{\dagger}_{\alpha,\sigma_2}$. Equal to minus itself, the fourth term vanishes. This concludes the proof. □

The following example illustrates the assertion of Lemma 5 (but not a technique of its proof which itself accompanies Lemma 1). We use the convention from Remark 2.7, denoting by $\mathrm{d}/\mathrm{d}\boldsymbol{y}_i$ or $\mathrm{d}/\mathrm{d}\boldsymbol{z}_j$ the total derivatives which act on the functionals' densities at points $\boldsymbol{x}_k$; this keeps track of those derivatives origin and lets us indicate the couplings' values as they appear after the integrations by parts, contributing only with sign factors ± 1. For the sake of brevity we do not write the (co)vectors $\vec{e}_i$ and $\vec{e}^{\dagger i}$ in the formulas below, referring to the proofs in preceding sections. Likewise, we do not indicate the base point congruences that occur due to the absolute locality of couplings.

An overall comment to Example 2.4 below is that, fully aware of the goal which is to calculate $\Delta([\![F,G]\!])$ or, respectively, $[\![\Delta F,G]\!]$ and $[\![F,\Delta G]\!]$, we do not interrupt the logic of our reasoning by attempting to view the intermediate objects $[\![F,G]\!]$ or ΔF and ΔG as mappings $\Gamma(\pi_{\mathrm{BV}})\to\Bbbk$, cf. Corollary 2 on p. 24. Such mappings would not be elements of the structures which stand in the left- and right-hand sides of the identity under examination. The slogan is that a step-by-step evaluation is illegal; derivations of the end-product from input data must not be interrupted at half-way.

We also emphasize that the example below is a prototype reasoning which is equally well applicable to any other arguments F and G in (29); a choice of the functionals is here not specific to any model. The point is that equality (29) holds and does not require any manual regularization.

Example 2.4. Consider the integral functionals

$$F=\int q^{\dagger}qq_{x_1x_1}\,\mathrm{d}x_1\quad\text{and}\quad G=\int q^{\dagger}_{x_2x_2}\cos q\,\mathrm{d}x_2.$$

Let us show that equality (29) is satisfied for F and G, that is,

$$\Delta([\![F,G]\!])=[\![\Delta F,G]\!]+[\![F,\Delta G]\!],\qquad \mathrm{gh}(F)=1,\tag{31}$$

in the frames of product-bundle geometry of variations and operational definitions of the BV-Laplacian Δ and variational Schouten bracket $[\![\,,\,]\!]$.

We have

$$[\![F,G]\!] = \iiiint dx_1 dx_2 dy_1 dy_2 \Big\langle \underbrace{\Big(q^\dagger q_{xx} + \tfrac{d^2}{dy_1^2}(q^\dagger q)\Big)}_{x_1} \cdot \underbrace{\langle \delta s_2(y_1), \delta s_2^\dagger(y_2)\rangle}_{+1} \cdot \tfrac{d^2}{dy_2^2} \underbrace{(\cos q)}_{x_2} \Big\rangle$$

$$+ \iiiint dx_1 dx_2 dy_1 dy_2 \Big\langle \underbrace{(q q_{xx})}_{x_1} \cdot \underbrace{\langle \delta s_2^\dagger(y_2), \delta s_2(y_1)\rangle}_{-1} \cdot \underbrace{(-q_{xx}^\dagger \sin q)}_{x_2} \Big\rangle.$$

Therefore, one side of the expected equality is

$$\Delta([\![F,G]\!]) = \int dz_1 \int dz_2 \int dx_1 \int dx_2 \int dy_1 \int dy_2 \underbrace{\langle \delta s_1(z_1), \delta s_1^\dagger(z_2)\rangle}_{+1} \cdot \underbrace{\langle \delta s_2(y_1), \delta s_2^\dagger(y_2)\rangle}_{+1} \cdot$$

$$\cdot \Big\langle \tfrac{d^2}{dz_1^2} \underbrace{(1)}_{x_1} \cdot \tfrac{d^2}{dy_2^2} \underbrace{(\cos q)}_{x_2} + \underline{\underbrace{q_{xx}}_{x_1} \cdot \tfrac{d^2}{dy_2^2} \underbrace{(-\sin q)}_{x_2}} + \tfrac{d^2}{dy_1^2} \underbrace{(1)}_{x_1} \cdot \tfrac{d^2}{dy_2^2} \underbrace{(\cos q)}_{x_2} + \underline{\underline{\tfrac{d^2}{dy_1^2} \underbrace{(q)}_{x_1} \cdot \tfrac{d^2}{dy_2^2} \underbrace{(-\sin q)}_{x_2}}} \Big\rangle$$

$$+ \int dz_1 \int dz_2 \int dx_1 \int dx_2 \int dy_1 \int dy_2 \underbrace{\langle \delta s_1(z_1), \delta s_1^\dagger(z_2)\rangle}_{+1} \cdot \underbrace{\langle \delta s_2^\dagger(y_2), \delta s_2(y_1)\rangle}_{-1} \cdot$$

$$\cdot \Big\langle \underline{\underbrace{q_{xx}}_{x_1} \cdot \tfrac{d^2}{dz_2^2} \underbrace{(-\sin q)}_{x_2}} + \underline{\underline{\tfrac{d^2}{dz_1^2} \underbrace{(q)}_{x_1} \cdot \tfrac{d^2}{dz_2^2} \underbrace{(-\sin q)}_{x_2}}} + \underbrace{(q q_{xx})}_{x_1} \cdot \tfrac{d^2}{dz_2^2} \underbrace{(-\cos q)}_{x_2} \Big\rangle .$$

The respective pairs of underlined terms cancel out and there remains only

$$= \int \cdots \int dz_1\, dz_2\, dx_1\, dx_2\, dy_1\, dy_2 \underbrace{\langle \delta s_1(z_1), \delta s_1^\dagger(z_2)\rangle}_{+1} \cdot \Big\langle \underbrace{(q q_{xx})}_{x_1} \cdot \tfrac{d^2}{dz_2^2} \underbrace{(-\cos q)}_{x_2} \Big\rangle \cdot \underbrace{\langle \delta s_2^\dagger(y_2), \delta s_2(y_1)\rangle}_{-1}. \quad (32)$$

On the other hand, we obtain that

$$\Delta F = \iiint dz_1 dz_2 dx_1 \underbrace{\langle \delta s_1(z_1), \delta s_1^\dagger(z_2)\rangle}_{+1} \cdot \Big\langle \underbrace{q_{xx}}_{x_1} + \tfrac{d^2}{dz_1^2} \underbrace{(q)}_{x_1} \Big\rangle,$$

which yields

$$[\![\Delta F, G]\!] = \int dz_1 \int dz_2 \int dx_1 \int dx_2 \int dy_1 \int dy_2 \underbrace{\langle \delta s_1(z_1), \delta s_1^\dagger(z_2)\rangle}_{+1} \cdot$$

$$\Big\langle \Big(\tfrac{d^2}{dy_1^2} \underbrace{(1)}_{x_1} + \tfrac{d^2}{dz_1^2} \underbrace{(1)}_{x_1} \Big) \cdot \tfrac{d^2}{dy_2^2} \underbrace{(\cos q)}_{x_2} \Big\rangle \cdot \underbrace{\langle \delta s_2(y_1), \delta s_2^\dagger(y_2)\rangle}_{+1} = 0.$$

From the fact that the other BV-Laplacian,

$$\Delta G = \iiint dz_1\, dz_2\, dx_2 \underbrace{\langle \delta s_1(z_1), \delta s_1^\dagger(z_2)\rangle}_{+1} \cdot \Big\langle \tfrac{d^2}{dz_2^2} \underbrace{(-\sin q)}_{x_2} \Big\rangle,$$

does not contain $q^\dagger$ so that the first half of the Schouten bracket $[\![F, \Delta G]\!]$ drops out, we deduce that

$$[\![F, \Delta G]\!] = \int \cdots \int dz_1\, dz_2\, dx_1\, dx_2\, dy_1\, dy_2\, \underbrace{\langle \delta s_1(z_1), \delta s_1^\dagger(z_2)\rangle}_{+1} \cdot \langle \underbrace{(qq_{xx})}_{x_1} \cdot \frac{d^2}{dz_2^2} \underbrace{(-\cos q)}_{x_2} \rangle \cdot \underbrace{\langle \delta s_2^\dagger(y_2), \delta s_2(y_1)\rangle}_{-1}. \quad (33)$$

Consequently, the two sides of (31), namely, $\Delta([\![F, G]\!])$ expressed by (32) and $[\![\Delta F, G]\!] + [\![F, \Delta G]\!]$ accumulated in (33), match perfectly for the functionals F and G at hand.

Theorem 6. *Let $F, G \in \overline{\mathfrak{N}}^n(\pi_{\mathrm{BV}}, T\pi_{\mathrm{BV}})$ be two functionals. The Batalin–Vilkovisky Laplacian Δ satisfies the relation*

$$\Delta([\![F, G]\!]) = [\![\Delta F, G]\!] + (-)^{\mathrm{gh}(F)-1}[\![F, \Delta G]\!]. \quad (29)$$

In other words, the operator Δ is a graded derivation of the variational Schouten bracket $[\![\,,]\!]$.

Proof. We prove this by induction over the number of building blocks in each argument of the Schouten bracket in the left hand side of (29). If F and G both belong to $\overline{H}^*(\pi_{\mathrm{BV}} \times T\pi_{\mathrm{BV}} \times \ldots \times T\pi_{\mathrm{BV}})$, then Lemma 5 states the assertion, which is the base of induction. To make an inductive step, without loss of generality let us assume that the second argument of $[\![\,,]\!]$ in (29) is a product of two elements from $\overline{\mathfrak{N}}^n(\pi_{\mathrm{BV}}, T\pi_{\mathrm{BV}})$, each of them containing less multiples from $\overline{H}^*(\pi_{\mathrm{BV}} \times T\pi_{\mathrm{BV}} \times \ldots T\pi_{\mathrm{BV}})$ than the product. Denote such factors by G and H and recall that by Theorem 4,

$$[\![F, G \cdot H]\!] = [\![F, G]\!] \cdot H + (-)^{(\mathrm{gh}(F)-1)\cdot \mathrm{gh}(G)} G \cdot [\![F, H]\!].$$

Therefore, using Theorem 3 we have that

$$\begin{aligned}
&\Delta([\![F, G \cdot H]\!]) \\
&= \Delta([\![F, G]\!]) \cdot H + (-)^{\mathrm{gh}(F)+\mathrm{gh}(G)-1}[\![[\![F, G]\!], H]\!] + (-)^{\mathrm{gh}(F)+\mathrm{gh}(G)-1}[\![F, G]\!] \cdot \Delta H \\
&\quad + (-)^{(\mathrm{gh}(F)-1)\,\mathrm{gh}(G)} \Big(\Delta G \cdot [\![F, H]\!] + (-)^{\mathrm{gh}(G)}[\![G, [\![F, H]\!]]\!] + (-)^{\mathrm{gh}(G)} G \cdot \Delta([\![F, H]\!])\Big).
\end{aligned}$$

Using the inductive hypothesis in the first and last terms of the right-hand side in the above formula, we continue the equality and obtain

$$\begin{aligned}
&= [\![\Delta F, G]\!] \cdot H + (-)^{\mathrm{gh}(F)-1}[\![F, \Delta G]\!] \cdot H + (-)^{\mathrm{gh}(F)+\mathrm{gh}(G)-1}[\![[\![F, G]\!], H]\!] \\
&\quad + (-)^{\mathrm{gh}(F)\,\mathrm{gh}(G)}[\![G, [\![F, H]\!]]\!] + (-)^{\mathrm{gh}(F)+\mathrm{gh}(G)-1}[\![F, G]\!] \cdot \Delta H \\
&\quad + (-)^{(\mathrm{gh}(F)-1)\,\mathrm{gh}(G)} \Delta G \cdot [\![F, H]\!] + (-)^{\mathrm{gh}(F)\,\mathrm{gh}(G)} G \cdot [\![\Delta F, H]\!] \\
&\quad + (-)^{\mathrm{gh}(F)\,\mathrm{gh}(G)+\mathrm{gh}(F)-1} G \cdot [\![F, \Delta H]\!].
\end{aligned} \quad (34)$$

On the other hand, let us expand the formula

$$[\![\Delta F, G \cdot H]\!] + (-)^{\mathrm{gh}(F)-1}[\![F, \Delta(G \cdot H)]\!],$$

Journal of Physics: Conference Series **474** (2013) 012024 doi:10.1088/1742-6596/474/1/012024

which is the right hand side of (29) in the inductive claim. We obtain

$$
\begin{aligned}
&= [\![\Delta F, G]\!]\cdot H + (-)^{(\mathrm{gh}(\Delta F)-1)\,\mathrm{gh}(G)} G\cdot[\![\Delta F, H]\!] \\
&\quad + (-)^{\mathrm{gh}(F)-1}[\![F,\ \Delta G\cdot H + (-)^{\mathrm{gh}(G)}[\![G,H]\!] + (-)^{\mathrm{gh}(G)} G\cdot\Delta H\ ]\!] \\
&= [\![\Delta F, G]\!]\cdot H + (-)^{\mathrm{gh}(F)\,\mathrm{gh}(G)} G\cdot[\![\Delta F, H]\!] + (-)^{\mathrm{gh}(F)-1}[\![F,\Delta G]\!]\cdot H \\
&\quad + (-)^{\mathrm{gh}(F)-1}(-)^{(\mathrm{gh}(F)-1)(\mathrm{gh}(G)-1)}\Delta G\cdot[\![F,H]\!] + (-)^{\mathrm{gh}(F)-1}(-)^{\mathrm{gh}(G)}[\![F,[\![G,H]\!]]\!] \\
&\quad + (-)^{\mathrm{gh}(F)-1}(-)^{\mathrm{gh}(G)}[\![F,G]\!]\cdot\Delta H + (-)^{\mathrm{gh}(F)-1}(-)^{\mathrm{gh}(G)}(-)^{(\mathrm{gh}(F)-1)\,\mathrm{gh}(G)} G\cdot[\![F,\Delta H]\!].
\end{aligned}
\tag{35}
$$

Comparing (35) with (34), which was derived from the inductive hypothesis, we see that all terms match except for

$$(-)^{\mathrm{gh}(F)+\mathrm{gh}(G)-1}[\![[\![F,G]\!],H]\!] + (-)^{\mathrm{gh}(F)\,\mathrm{gh}(G)}[\![G,[\![F,H]\!]]\!]$$

from (34) versus

$$(-)^{\mathrm{gh}(F)+\mathrm{gh}(G)-1}[\![F,[\![G,H]\!]]\!]$$

from (35). However, these three terms constitute Jacobi's identity (28) for the variational Schouten bracket. Namely, we have that (cf. [32])

$$[\![F,[\![G,H]\!]]\!] = [\![[\![F,G]\!],H]\!] + (-)^{(\mathrm{gh}(F)-1)(\mathrm{gh}(G)-1)}[\![G,[\![F,H]\!]]\!], \tag{36}$$

so that by multiplying both sides of the identity by $(-)^{\mathrm{gh}(F)+\mathrm{gh}(G)-1}$, we fully balance (34) and (35). This completes the inductive step and concludes the proof. □

Lemma 7. The linear operator

$$\Delta\colon \overline{H}^{n(1+k)}\big(\pi_{\mathrm{BV}}\times T\pi_{\mathrm{BV}}\times\ldots\times T\pi_{\mathrm{BV}}\big) \longrightarrow \overline{H}^{n(2+k)}\big(\pi_{\mathrm{BV}}\times T\pi_{\mathrm{BV}}\times\ldots\times T\pi_{\mathrm{BV}}\big)$$

is a differential for every $k\geqslant 0$.

The proof of Lemma 7 is conceptually close to the second and third steps in the proof of Lemma 5. Namely, two normalized variations are swapped in an integral functional within the image of Δ^2, which yields an indistinguishable result of opposite sign.

Proof. Let $\delta\boldsymbol{s}_1$ and $\delta\boldsymbol{s}_2$ be normalized test shifts of a section $s\in\Gamma(\pi_{\mathrm{BV}})$, and let $H = \int h(\boldsymbol{x},[\boldsymbol{q}],[\boldsymbol{q}^\dagger])\cdot\mathrm{dvol}(\boldsymbol{x})$ be an integral functional. (It suffices to consider a simplified picture $H\in\overline{H}^n(\pi_{\mathrm{BV}})$, not taking into account any built-in variations in the construction of H.) By definition, we have that

$$
\begin{aligned}
\Delta(\Delta H)(s) &= \int_M \mathrm{d}\boldsymbol{z}_1\int_M \mathrm{d}\boldsymbol{z}_2\int_M \mathrm{d}\boldsymbol{y}_1\int_M \mathrm{d}\boldsymbol{y}_2\int_M \mathrm{dvol}(\boldsymbol{x})\cdot \\
&\cdot\Big\{\Big\langle(\delta s_1^\alpha)\Big(\frac{\overleftarrow{\partial}}{\partial\boldsymbol{z}_1}\Big)^{\sigma_1}(\boldsymbol{z}_1)\,\overbrace{\vec{e}_\alpha(\boldsymbol{z}_1),(-\vec{e}^{\dagger\alpha})(\boldsymbol{z}_2)}^{-1}(\delta s^\dagger_{1,\alpha})\Big(\frac{\overleftarrow{\partial}}{\partial\boldsymbol{z}_2}\Big)^{\sigma_2}(\boldsymbol{z}_2)\Big\rangle\underbrace{\big\langle\vec{e}^{\dagger\alpha}(\boldsymbol{x}),\vec{e}_\alpha(\boldsymbol{x})\big\rangle}_{-1} \\
&\Big\langle(\delta s_2^\beta)\Big(\frac{\overleftarrow{\partial}}{\partial\boldsymbol{y}_1}\Big)^{\tau_1}(\boldsymbol{y}_1)\,\overbrace{\vec{e}_\beta(\boldsymbol{y}_1),(-\vec{e}^{\dagger\beta})(\boldsymbol{y}_2)}^{-1}(\delta s^\dagger_{2,\beta})\Big(\frac{\overleftarrow{\partial}}{\partial\boldsymbol{y}_2}\Big)^{\tau_2}(\boldsymbol{y}_2)\Big\rangle\underbrace{\big\langle\vec{e}^{\dagger\beta}(\boldsymbol{x}),\vec{e}_\beta(\boldsymbol{x})\big\rangle}_{-1} \\
&\qquad\frac{\overrightarrow{\partial}}{\partial q^\alpha_{\sigma_1}}\frac{\overrightarrow{\partial}}{\partial q^\dagger_{\alpha,\sigma_2}}\frac{\overrightarrow{\partial}}{\partial q^\beta_{\tau_1}}\frac{\overrightarrow{\partial}}{\partial q^\dagger_{\beta,\tau_2}}h(\boldsymbol{x},[\boldsymbol{q}],[\boldsymbol{q}^\dagger])\Big\}\Big|_{j^\infty_{\boldsymbol{x}}(s)}.
\end{aligned}
$$

By exchanging the integrand's upper two lines and then relabelling $\alpha \rightleftarrows \beta$, $\sigma \rightleftarrows \tau$ so that $\delta s_1^\alpha \rightleftarrows \delta s_2^\beta$ and $\delta s^\dagger_{1,\alpha} \rightleftarrows \delta s^\dagger_{2,\beta}$, and by swapping the reference $\boldsymbol{y} \rightleftarrows \boldsymbol{z}$ to copies of the base manifold M^n, we almost recover the initial expression (which should be the case), yet the order in which the parity-odd partial derivatives follow is inverse,

$$\frac{\vec{\partial}}{\partial q^\dagger_{\alpha,\sigma_2}} \circ \frac{\vec{\partial}}{\partial q^\dagger_{\beta,\tau_2}} \longmapsto \frac{\vec{\partial}}{\partial q^\dagger_{\beta,\tau_2}} \circ \frac{\vec{\partial}}{\partial q^\dagger_{\alpha,\sigma_2}} = -\frac{\vec{\partial}}{\partial q^\dagger_{\alpha,\sigma_2}} \circ \frac{\vec{\partial}}{\partial q^\dagger_{\beta,\tau_2}}.$$

Therefore the integrand of functional $\Delta^2 H$ vanishes, which proves the assertion. □

Theorem 8. *The Batalin–Vilkovisky Laplacian Δ is a differential: for all $H \in \overline{\mathfrak{N}}^n(\pi_{\mathrm{BV}}, T\pi_{\mathrm{BV}})$ we have*

$$\Delta^2(H) = 0.$$

Proof. We prove Theorem 8 by induction over the number of building blocks from $\overline{H}^*(\pi_{\mathrm{BV}} \times T\pi_{\mathrm{BV}} \times \ldots \times T\pi_{\mathrm{BV}})$ in the argument $H \in \overline{\mathfrak{N}}^n(\pi_{\mathrm{BV}}, T\pi_{\mathrm{BV}})$ of Δ^2. If $H \in \overline{H}^*(\pi_{\mathrm{BV}} \times T\pi_{\mathrm{BV}} \times \ldots \times T\pi_{\mathrm{BV}})$ itself is an integral functional, then by Lemma 7 there remains nothing to prove. Suppose now that $H = F \cdot G$ for some $F, G \in \overline{\mathfrak{N}}^n(\pi_{\mathrm{BV}}, T\pi_{\mathrm{BV}})$. Then Theorem 3 yields that

$$\Delta^2(F \cdot G) = \Delta\left(\Delta F \cdot G + (-)^{\mathrm{gh}(F)} [\![F, G]\!] + (-)^{\mathrm{gh}(F)} F \cdot \Delta G\right).$$

Using Theorem 3 again and also Theorem 6, we continue the equality:

$$\begin{aligned} &= \Delta^2 F \cdot G + (-)^{\mathrm{gh}(\Delta F)} [\![\Delta F, G]\!] + (-)^{\mathrm{gh}(\Delta F)} \Delta F \cdot \Delta G \\ &\quad + (-)^{\mathrm{gh}(F)} [\![\Delta F, G]\!] + (-)^{\mathrm{gh}(F)} (-)^{\mathrm{gh}(F)-1} [\![F, \Delta G]\!] \\ &\quad + (-)^{\mathrm{gh}(F)} \Delta F \cdot \Delta G + (-)^{\mathrm{gh}(F)} (-)^{\mathrm{gh}(F)} [\![F, \Delta G]\!] + (-)^{\mathrm{gh}(F)} (-)^{\mathrm{gh}(F)} F \cdot \Delta^2 G. \end{aligned}$$

By the inductive hypothesis, the first and last terms in the above formula vanish; taking into account that $\mathrm{gh}(\Delta F) = \mathrm{gh}(F) - 1$ in $\mathbb{Z}_2$, the terms with $\Delta F \cdot \Delta G$ cancel against each other, as do the terms containing $[\![\Delta F, G]\!]$ and $[\![F, \Delta G]\!]$. The proof is complete. □

3. The quantum master-equation

3.1. The Laplace equation

In this section we inspect the conditions upon functionals $F \in \overline{\mathfrak{N}}^n(\pi_{\mathrm{BV}}, T\pi_{\mathrm{BV}})$ under which the Feynman path integrals $\int_{\Gamma(\boldsymbol{\zeta}^0)} [Ds]\, F([s], [s^\dagger])$ are (infinitesimally) independent of the unphysical anti-objects $s^\dagger \in \Gamma(\boldsymbol{\zeta}^1)$. The derivation of such a condition (see equation (39) below) relies on an extra assumption of the translation invariance of a measure in the path integral. It must be noted, however, that we do not define Feynman's integral here and do not introduce that measure which essentially depends on the agreement about the classes of 'admissible' sections $\Gamma(\pi)$ or $\Gamma(\boldsymbol{\zeta}^{(0|1)})$. Consequently, our reasoning is to some extent heuristic.

The basics of path integration, which we recall here for consistency, are standard: they illustrate how the geometry of the BV-Laplacian works in practice. We draw the experts' attention only to the fact that in our notation Ψ is not the gauge fixing fermion $\boldsymbol{\Psi}$ such that the odd-component's section $s^\dagger \in \Gamma(\boldsymbol{\zeta}^1)$ is the restriction of $\delta\boldsymbol{\Psi}/\delta q$ to the jet of a section for $\boldsymbol{\zeta}^0$; instead, we let Ψ determine the infinitesimal shift $\dot{q}^\dagger = \delta\Psi/\delta q$ of coordinates along the fibre's parity-odd half. We also note that the preservation of parity is not mandatory here and thus an even-parity $\Psi \in \overline{H}^n(\boldsymbol{\zeta}^0) \hookrightarrow \overline{H}^n(\pi_{\mathrm{BV}})$ is a legitimate choice.

Let $F = \int f(\boldsymbol{x}, \boldsymbol{q}, \boldsymbol{q}^\dagger)\, \mathrm{dvol}(\boldsymbol{x}) \in \overline{H}^n(\pi_{\mathrm{BV}})$ be a functional; here and in what follows we proceed over the building blocks of elements from $\overline{\mathfrak{N}}^n(\pi_{\mathrm{BV}}, T\pi_{\mathrm{BV}})$ by the graded Leibniz rule. Let $\Psi = \int \psi(\boldsymbol{y},\boldsymbol{q})\,\mathrm{dvol}(\boldsymbol{y}) \in \overline{H}^n(\boldsymbol{\zeta}^0) \hookrightarrow \overline{H}^n(\pi_{\mathrm{BV}})$ be an integral functional which, by assumption, is constant along ghost parity-odd variables: $\Psi(s^\alpha, s^\dagger_\beta) = \Psi(s^\alpha, t^\dagger_\beta)$ for any sections $\{s^\alpha\} \in \Gamma(\boldsymbol{\zeta}^0)$ and $\{s^\dagger_\beta\}, \{t^\dagger_\beta\} \in \Gamma(\boldsymbol{\zeta}^1)$. We investigate under which conditions the path integral $\int_{\Gamma(\boldsymbol{\zeta}^0)}[Ds^\alpha]\,F(s^\alpha, s^\dagger_\beta)\colon \Gamma(\boldsymbol{\zeta}^1) \to \Bbbk$ is infinitesimally independent of a choice of the anti-objects:

$$\frac{\mathrm{d}}{\mathrm{d}\varepsilon^\dagger}\Big|_{\varepsilon^\dagger=0} \int_{\Gamma(\boldsymbol{\zeta}^0)} [Ds^\alpha]\, F\Big(s^\alpha, s^\dagger_\beta + \varepsilon^\dagger\, \frac{\vec{\delta}\psi}{\delta q^\beta}\Big|_{s^\alpha}\Big) = 0 \quad \text{for all } s^\dagger \in \Gamma(\boldsymbol{\zeta}^1). \tag{37}$$

Note that this formula makes sense because the bundles $\boldsymbol{\zeta}^0$ and $\boldsymbol{\zeta}^1$ are dual so that a variational covector in the geometry of $\boldsymbol{\zeta}^0$ acts as a shift vector in the geometry of $\boldsymbol{\zeta}^1$. The left-hand side of (37) equals

$$\int_{\Gamma(\boldsymbol{\zeta}^0)} [Ds^\alpha] \int_M \mathrm{dvol}(\boldsymbol{x})\, \frac{\overrightarrow{\delta\psi}}{\delta q^\beta}(\boldsymbol{x},\boldsymbol{q})\big|_{j^\infty_{\boldsymbol{x}}(s^\alpha)} \cdot \frac{\overleftarrow{\delta f}}{\delta q^\dagger_\beta}(\boldsymbol{x},\boldsymbol{q},\boldsymbol{q}^\dagger)\big|_{j^\infty_{\boldsymbol{x}}(s^\alpha, s^\dagger_\gamma)}, \qquad s^\dagger \in \Gamma(\boldsymbol{\zeta}^1).$$

Take any auxiliary section $\delta\boldsymbol{s} = (\delta s^\alpha, \delta s^\dagger_\beta) \in \Gamma\big(T\boldsymbol{\zeta}^{(0|1)}\big)$ normalized by $\delta s^\alpha(x) \cdot \delta s^\dagger_\alpha(x) \equiv 1$ at every $\boldsymbol{x} \in M^n$ for each $\alpha = 1, \ldots, m + m_1 + \cdots + m_\lambda = N$ and blow up the scalar integrand to a pointwise contraction of dual object taking their values in the fibres $T_{(\boldsymbol{x},\phi(\boldsymbol{x}),s(\boldsymbol{x}))}V_{\boldsymbol{x}}$ and $T_{(\boldsymbol{x},\phi(\boldsymbol{x}),s^\dagger(\boldsymbol{x}))}\Pi V^\dagger_{\boldsymbol{x}}$ of $T(\pi_{\mathrm{BV}})$ over $\phi(x)$: for $s = (s^\alpha, s^\dagger_\beta)$ we have

$$\begin{aligned}
&\int_M \mathrm{dvol}(\boldsymbol{x}) \left(\frac{\overrightarrow{\delta\psi}}{\delta q^\alpha} \cdot \frac{\overleftarrow{\delta f}}{\delta q^\dagger_\alpha}\right)\Bigg|_{j^\infty_{\boldsymbol{x}}(s)} \\
&\quad = \int_M \mathrm{dvol}(\boldsymbol{x}_1) \int_M \mathrm{dvol}(\boldsymbol{x}_2) \int_M \mathrm{d}\boldsymbol{y}_1 \int_M \mathrm{d}\boldsymbol{y}_2 \left(\psi(\boldsymbol{x}_1,\boldsymbol{q}) \frac{\overleftarrow{\partial}}{\partial q^{j_1}_{\sigma_1}}\right)\Bigg|_{j^\infty_{\boldsymbol{x}_1}(s)} \\
&\quad \cdot \underline{\langle \vec{e}^{\,\dagger j_1}(\boldsymbol{x}_1)|}\, \Big\langle \Big(\frac{\overleftarrow{\partial}}{\partial \boldsymbol{y}_1}\Big)^{\sigma_1} \delta s^{i_1}(\boldsymbol{y}_1)\, \vec{e}_{i_1}(\boldsymbol{y}_1), -\vec{e}^{\,\dagger i_2}(\boldsymbol{y}_2)\, \delta s^\dagger_{i_2}(\boldsymbol{y}_2) \Big(\frac{\overrightarrow{\partial}}{\partial \boldsymbol{y}_2}\Big)^{\sigma_2} \Big\rangle\, \underline{|\vec{e}_{j_2}(\boldsymbol{x}_2)\rangle} \\
&\qquad\qquad \cdot \left(\frac{\overrightarrow{\partial}}{\partial q^\dagger_{j_2,\sigma_2}} f(\boldsymbol{x}_2,\boldsymbol{q},\boldsymbol{q}^\dagger)\right)\Bigg|_{j^\infty_{\boldsymbol{x}_2}(s)}.
\end{aligned}$$

In fact, the integrand refers to a definition of the evolutionary vector field $\boldsymbol{Q}^\Psi$ such that $\boldsymbol{Q}^\Psi(F) \cong [\![\Psi, F]\!]$ modulo integration by parts in the building blocks of F, cf. [32]. Due to a special choice of the dependence of Ψ on s only, this is indeed the Schouten bracket $[\![\Psi, F]\!]$.

To rephrase the indifference of the path integral to a choice of Ψ in terms of an equation upon the functional F alone, we perform integration by parts in Feynman's integral. For this we employ the translation invariance $[Ds] = [D(s - \mu \cdot \delta s)]$ of the functional measure.

Lemma 9. Let $H = \int h(\boldsymbol{x},\boldsymbol{q},\boldsymbol{q}^\dagger)\,\mathrm{dvol}(\boldsymbol{x}) \in \overline{H}^n(\pi_{\mathrm{BV}}) \subset \overline{\mathfrak{N}}^n(\pi_{\mathrm{BV}}, T\pi_{\mathrm{BV}})$ be an integral functional and $\delta s \in \Gamma(T\boldsymbol{\zeta}^0) \hookrightarrow \Gamma\big(T\boldsymbol{\zeta}^{(0|1)}\big)$ be a shift. Then we have that

$$\int_{\Gamma(\boldsymbol{\zeta}^0)} [Ds^\alpha] \int_M \mathrm{dvol}(\boldsymbol{x})\, \delta s^\nu(\boldsymbol{x}) \cdot \frac{\overleftarrow{\delta h}}{\delta q^\nu}\Big|_{j^\infty_{\boldsymbol{x}}(s^\alpha, s^\dagger_\beta)} = 0,$$

where the section $s^\dagger \in \Gamma(\boldsymbol{\zeta}^1)$ is a parameter.

Proof. Indeed,

$$0 = \frac{\mathrm{d}}{\mathrm{d}\mu}\bigg|_{\mu=0} \int_{\Gamma(\boldsymbol{\zeta}^0)} [Ds^\alpha]\, H(s^\alpha, s^\dagger_\beta),$$

because the integral contains no parameter $\mu \in \Bbbk$. We continue the equality:

$$= \frac{\mathrm{d}}{\mathrm{d}\mu}\bigg|_{\mu=0} \int_{\Gamma(\boldsymbol{\zeta}^0)} [D(s^\alpha - \mu\,\delta s^\alpha)]\, H(s^\alpha, s^\dagger_\beta)$$
$$= \frac{\mathrm{d}}{\mathrm{d}\mu}\bigg|_{\mu=0} \int_{\Gamma(\boldsymbol{\zeta}^0)} [Ds^\alpha]\, H(s^\alpha + \mu\,\delta s^\alpha, s^\dagger_\beta) = \int_{\Gamma(\boldsymbol{\zeta}^0)} [Ds^\alpha]\, \frac{\mathrm{d}}{\mathrm{d}\mu}\bigg|_{\mu=0} H(s^\alpha + \mu\,\delta s^\alpha, s^\dagger_\beta),$$

which yields the helpful formula in the lemma's assertion. □

Returning to the functionals Ψ and F and denoting $G(s) := \frac{\mathrm{d}}{\mathrm{d}\ell}\big|_{\ell=0} F(s + \ell \cdot \overleftarrow{\delta s}^\dagger)$, we use the Leibniz rule for the derivative of $H = \Psi \cdot G$:

$$\frac{\mathrm{d}}{\mathrm{d}\mu}\bigg|_{\mu=0} (\Psi \cdot G)(s + \mu \cdot \overleftarrow{\delta s}) = \frac{\mathrm{d}}{\mathrm{d}\mu}\bigg|_{\mu=0} (\Psi)(s + \mu \cdot \overleftarrow{\delta s}) \cdot G(s) + \Psi(s) \cdot \frac{\mathrm{d}}{\mathrm{d}\mu}\bigg|_{\mu=0} (G)(s + \mu \cdot \overleftarrow{\delta s}).$$

Because the path integral over $[Ds^\alpha]$ of the entire expression vanishes by Lemma 9 in which we were ready to proceed by the Leibniz rule over building blocks, we infer that the path integrals of the two terms are opposite. Now take the traces over indexes in both variations. The integral of the first term equals the initial expression for the path integral containing F, i. e., the left-hand side of equation (37). Consequently, if

$$\int_{\Gamma(\boldsymbol{\zeta}^0)} [Ds^\alpha]\, \Psi(s^\alpha) \cdot \Delta F(s^\alpha, s^\dagger_\beta) = 0 \tag{38}$$

for $\{s^\dagger_\beta\} \in \Gamma(\boldsymbol{\zeta}^1)$ and for all $\Psi \in \overline{H}^n(\boldsymbol{\zeta}^0) \hookrightarrow \overline{H}^n(\pi_{\text{BV}})$, then the path integral over F is infinitesimally independent of a section $\{s^\dagger_\beta\} \in \Gamma(\boldsymbol{\zeta}^1)$.

The condition

$$\Delta F = 0 \tag{39}$$

is sufficient for equation (38), and therefore equation (37), to hold. By specifying a class $\Gamma(\pi_{\text{BV}})$ of admissible sections of the BV-bundle for a concrete field model, and endowing that space of sections with a suitable metric, one could reinstate a path integral analogue of the main lemma in the calculus of variations and then argue that the condition $\Delta F = 0$ is also necessary.

Summarizing, whenever equation (39) holds, one can assign arbitrary admissible values to the odd-parity coordinates; for example, one can let $s^\dagger_\beta(\boldsymbol{x}) = \delta\boldsymbol{\psi}/\delta q^\beta\big|_{j^\infty_{\boldsymbol{x}}(s^\alpha)}$ for a gauge-fixing integral $\boldsymbol{\Psi} = \int \boldsymbol{\psi}(\boldsymbol{x}, \boldsymbol{q})\, \mathrm{dvol}(\boldsymbol{x}) \in \overline{H}^n(\boldsymbol{\zeta}^0)$. This choice is reminiscent of the substitution principle, see [32] and [45].

Laplace's equation (39) ensures the infinitesimal independence from non-physical anti-objects for path integrals of functionals over physical fields – not only in the classical BV-geometry of the bundle π_{BV}, but also in the quantum setup, whenever all objects are tensored with formal power series $\Bbbk[[\hbar, \hbar^{-1}]]$ in the Planck constant $\hbar$. It is accepted that each quantum field $s^\hbar$ contributes to the expectation value of a functional $\mathcal{O}^\hbar$ with the factor $\exp(\boldsymbol{i}S^\hbar(s^\hbar)/\hbar)$, where $S^\hbar$ is the quantum BV-action of the model. Solutions $\mathcal{O}^\hbar$ of the equation $\Delta\big(\mathcal{O}^\hbar \cdot \exp(\boldsymbol{i}S^\hbar/\hbar)\big) = 0$ are the *observables*. In particular, the postulate that the unit $1\colon s^\hbar \mapsto 1 \in \Bbbk$ is averaged to unit by the Feynman integral of $1 \cdot \exp(\boldsymbol{i}S^\hbar(s^\hbar)/\hbar)$ over the space of quantum fields $s^\hbar$ normalizes the integration measure and constrains the quantum BV-action by the quantum master-equation (see, e.g., [7, 8, 20, 22, 54]).

Proposition 10. Let $S^\hbar$ be the even quantum BV-action (i. e., let it have a density that has an even number of ghost parity-odd coordinates in each of its terms). If the identity

$$\Delta\left(\exp(\tfrac{\boldsymbol{i}}{\hbar}S^\hbar)\right) = 0$$

holds, then $S^\hbar$ satisfies the quantum master-equation:

$$\boldsymbol{i}\hbar\,\Delta S^\hbar = \tfrac{1}{2}[\![S^\hbar, S^\hbar]\!]. \tag{40}$$

Proposition 11. If an even functional $\mathcal{O}$ and the quantum BV-action $S^\hbar$ are such that $\Delta\bigl(\mathcal{O}\exp(\boldsymbol{i}S^\hbar/\hbar)\bigr) = 0$ and $\Delta\bigl(\exp(\boldsymbol{i}S^\hbar/\hbar)\bigr) = 0$ hold, respectively, then $\mathcal{O}$ satisfies

$$\Omega^\hbar(\mathcal{O}) := -\boldsymbol{i}\hbar\,\Delta\mathcal{O} + [\![S^\hbar, \mathcal{O}]\!] = 0. \tag{41}$$

We quote the standard proofs of Propositions 10 and 11 from [35] in Appendix Appendix A — yet now we gain a deeper insight on a construction of the quantum BV-differential $\Omega^\hbar$.

Remark 3.1. A practical way to fix the signs which arise in the BV-Laplacian and Schouten bracket from the ghost parity and a grading in the case of a superbundle $\pi\colon E^{(m_0+n_0|m_1+n_1)} \to M^{(n_0|n_1)}$ of superfields is by a re-derivation of the Laplace equation $\Delta(\mathcal{O}\exp(\frac{\boldsymbol{i}}{\hbar}S^\hbar)) = 0$ upon an observable $\mathcal{O}$ starting from the Schwinger–Dyson condition,

$$\vec{\partial}^{(\boldsymbol{q}^\dagger)}_{\vec{\delta}\Psi/\delta\boldsymbol{q}}\left(\int [D\boldsymbol{q}]\,\mathcal{O}([\boldsymbol{q}],[\boldsymbol{q}^\dagger])\exp\left(\tfrac{\boldsymbol{i}}{\hbar}S^\hbar([\boldsymbol{q}],[\boldsymbol{q}^\dagger])\right)\right) = 0, \tag{42}$$

which postulates the Feynman path integral's independence of the non-physical BV-coordinates $\boldsymbol{q}^\dagger$ with odd ghost parity. Note that the measure in the path integral involves only ghost parity-even objects (whatever be their $\mathbb{Z}_2$-grading).

Theorem 12. *Let* $\mathcal{O} \in \overline{\mathfrak{N}}^n(\pi_{\text{BV}}, T\pi_{\text{BV}})$ *be a functional and let the even functional* $S^\hbar \in \overline{H}^n(\pi_{\text{BV}})$ *satisfy quantum master-equation* (40)*. Then the operator* $\Omega^\hbar$*, defined in* (41)*, squares to zero*:

$$(\Omega^\hbar)^2(\mathcal{O}) = 0.$$

Proof. We calculate, using Theorem 6,

$$\begin{aligned}(\Omega^\hbar)^2(\mathcal{O}) &= [\![S^\hbar, [\![S^\hbar, \mathcal{O}]\!] - \boldsymbol{i}\hbar\,\Delta\mathcal{O}]\!] - \boldsymbol{i}\hbar\,\Delta\bigl([\![S^\hbar, \mathcal{O}]\!] - \boldsymbol{i}\hbar\,\Delta\mathcal{O}\bigr)\\ &= [\![S^\hbar, [\![S^\hbar, \mathcal{O}]\!]]\!] - \boldsymbol{i}\hbar\,[\![S^\hbar, \Delta\mathcal{O}]\!] - \boldsymbol{i}\hbar\,[\![\Delta S^\hbar, \mathcal{O}]\!] + \boldsymbol{i}\hbar\,[\![S^\hbar, \Delta\mathcal{O}]\!] + (\boldsymbol{i}\hbar)^2\Delta^2\mathcal{O}.\end{aligned}$$

The last term vanishes identically by Theorem 8, while the second term cancels against the fourth term. Using Jacobi's identity (28) for the Schouten bracket on the first term, we obtain:

$$(\Omega^\hbar)^2(\mathcal{O}) = -\boldsymbol{i}\hbar\,[\![\Delta S^\hbar, \mathcal{O}]\!] + \tfrac{1}{2}[\![[\![S^\hbar, S^\hbar]\!], \mathcal{O}]\!] = [\![-\boldsymbol{i}\hbar\,\Delta S^\hbar + \tfrac{1}{2}[\![S^\hbar, S^\hbar]\!], \mathcal{O}]\!].$$

Now is the crucial moment in the entire proof. By the logic of our reasoning's objective, the theorem's claim is that the operator $(\Omega^\hbar)^2$ yields zero whenever acting on a functional $\mathcal{O}$. We accordingly transform the variational Schouten bracket of two terms to the operator realization,

$$\cong \vec{\boldsymbol{Q}}^{-\boldsymbol{i}\hbar\,\Delta S^\hbar + \frac{1}{2}[\![S^\hbar, S^\hbar]\!]}(\mathcal{O}),$$

with the evolutionary derivation now acting on the argument. Let us emphasize that a transition from the variational Schouten bracket –which increases the number of bases $M \times \ldots \times M$ by construction– to the evolutionary vector field chops a multiplication of geometries by uniquely fixing the field's generating section.[25] But by our initial assumption, this generating section is zero by virtue of (40). Therefore the image of $\mathcal{O}$ under such map vanishes, which proves the assertion. □

[25] It might happen otherwise that a co-multiple of $\mathcal{O}$ under $[\![\,,\,]\!]$ looks like zero as a map of the space $\Gamma(\pi_{\text{BV}})$ yet the bracket with it could still be nonzero, see, e. g., ΔG on p. 36 in Example 2.4.

3.2. Gauge automorphisms of quantum BV-cohomology groups

By using the quantum BV-differential $\Omega^\hbar$, let us construct a closed algebra of infinitesimal gauge symmetries for the quantum master-equation (40).

Proposition 13. Let $F \in \overline{\mathfrak{N}^n}(\pi_{\text{BV}}, T\pi_{\text{BV}})$ be an arbitrary odd-parity functional and $S^\hbar$ the quantum master-action satisfying (40). Then the infinitesimal shift of the functional $S^\hbar$,

$$\dot{S}^\hbar = \Omega^\hbar(F) \quad\Longleftrightarrow\quad S^\hbar \mapsto S^\hbar(\varepsilon) = S^\hbar + \varepsilon \cdot \Omega^\hbar(F) + \bar{o}(\varepsilon),\ \varepsilon \in \mathbb{R}, \tag{43}$$

is a symmetry of (40) so that $\Delta\left(\exp\left(\frac{\boldsymbol{i}}{\hbar}S^\hbar(\varepsilon)\right)\right) = \bar{o}(\varepsilon)$ in Peano's notation.

• The algebra of infinitesimal gauge symmetries (43) of the quantum master-equation is closed,

$$\left(\frac{\mathrm{d}}{\mathrm{d}\varepsilon_1} \circ \frac{\mathrm{d}}{\mathrm{d}\varepsilon_2} - \frac{\mathrm{d}}{\mathrm{d}\varepsilon_2} \circ \frac{\mathrm{d}}{\mathrm{d}\varepsilon_1}\right)\bigg|_{\varepsilon_i=0} (S^\hbar) = \Omega^\hbar([\![F_1, F_2]\!]), \tag{44}$$

i.e., the commutator of two even-parity symmetries with respective generators F_1 and F_2 is the infinitesimal gauge symmetry whose generator is the odd Poisson bracket of F_1 and F_2.

Remark 3.2. The odd-parity generators $F_i \in \overline{\mathfrak{N}^n}(\pi_{\text{BV}}, T\pi_{\text{BV}})$ never evolve in the course of a transformation which is induced by any generator F_j on the quantum BV-action functional $S^\hbar$.

Proof. Assuming a smooth dependence of $S^\hbar(\varepsilon)$ on ε, we obtain that[26]

$$\frac{\mathrm{d}}{\mathrm{d}\varepsilon}\Delta\left(\exp\left(\tfrac{\boldsymbol{i}}{\hbar}S^\hbar(\varepsilon)\right)\right) = \tfrac{\boldsymbol{i}}{\hbar}\dot{S}^\hbar \cdot \Delta\left(\exp\left(\tfrac{\boldsymbol{i}}{\hbar}S^\hbar(\varepsilon)\right)\right) + \Omega^\hbar(\dot{S}^\hbar)\cdot\exp\left(\tfrac{\boldsymbol{i}}{\hbar}S^\hbar(\varepsilon)\right).$$

Because $(\Omega^\hbar)^2 = 0$ by Theorem 12, for $\dot{S}^\hbar$ to be an infinitesimal symmetry of the equation $\Delta\left(\exp\left(\frac{\boldsymbol{i}}{\hbar}S^\hbar\right)\right) = 0$ it is sufficient that $\dot{S}^\hbar = \Omega^\hbar(F)$ for some odd-parity functional F.

Second, let

$$\frac{\mathrm{d}}{\mathrm{d}\varepsilon_i}(S^\hbar) = -\boldsymbol{i}\hbar\,\Delta F_i + [\![S^\hbar, F_i]\!] \qquad \text{for } i = 1, 2, \qquad \varepsilon_i \in \mathbb{R},$$

and postulate that $\frac{\mathrm{d}}{\mathrm{d}\varepsilon_i}(F_j) \equiv 0$ for all i and j. Then commutator (44) of even-parity infinitesimal transformations (43) generated by the functionals F_1 and F_2 is

$$\begin{aligned}
&[\![-\boldsymbol{i}\hbar\,\Delta F_1 + [\![S^\hbar, F_1]\!], F_2]\!] - [\![-\boldsymbol{i}\hbar\,\Delta F_2 + [\![S^\hbar, F_2]\!], F_1]\!] \\
&\qquad = -\boldsymbol{i}\hbar\,\left([\![\Delta F_1, F_2]\!] - [\![\Delta F_2, F_1]\!]\right) + \Big([\![[\![S^\hbar, F_1]\!], F_2]\!] - [\![[\![S^\hbar, F_2]\!], F_1]\!]\Big).
\end{aligned}$$

Because F_1 has odd parity, we swap the factors in $-[\![\Delta F_2, F_1]\!] = [\![F_1, \Delta F_2]\!]$; likewise, $+[\![F_1, [\![S^\hbar, F_2]\!]]\!]$ is the last term in the above expression. From our main Theorem 6 and by Jacobi identity (28) we conclude that the commutator is equal to

$$-\boldsymbol{i}\hbar\,\Delta([\![F_1, F_2]\!]) + [\![S^\hbar, [\![F_1, F_2]\!]]\!] = \Omega^\hbar([\![F_1, F_2]\!]),$$

that is, the Schouten bracket of F_1 and F_2 is the new gauge symmetry generator. □

[26] This proof is standard: it originates from the cohomological deformation theory for solutions of the Maurer–Cartan equation (e. g., of (40)), see [37].

Remark 3.3. (cf. [51, §5]). The transformation $\exp\left(\frac{\boldsymbol{i}}{\hbar}S^\hbar\right) \mapsto \exp\left(\frac{\boldsymbol{i}}{\hbar}S^\hbar(\varepsilon)\right)$ for a finite $\varepsilon \in \mathbb{R}$ is determined by the operator $\exp(\varepsilon[\Delta, F])$, where $[\ ,\ ]$ is the anticommutator of two odd-parity objects. Indeed, by Theorem 3 we have that

$$\begin{aligned}
&\Delta\bigl(F \cdot \exp\left(\tfrac{\boldsymbol{i}}{\hbar}S^\hbar\right)\bigr) + F \cdot \Delta\bigl(\exp\left(\tfrac{\boldsymbol{i}}{\hbar}S^\hbar\right)\bigr) \\
&\quad = \Delta F \cdot \exp\left(\tfrac{\boldsymbol{i}}{\hbar}S^\hbar\right) - [\![F, \exp\left(\tfrac{\boldsymbol{i}}{\hbar}S^\hbar\right)]\!] - F \cdot \Delta\left(\exp\left(\tfrac{\boldsymbol{i}}{\hbar}S^\hbar\right)\right) + F \cdot \Delta\left(\exp\left(\tfrac{\boldsymbol{i}}{\hbar}S^\hbar\right)\right) \\
&\quad = \tfrac{\boldsymbol{i}}{\hbar}(-\boldsymbol{i}\hbar\,\Delta F + [\![S^\hbar, F]\!]) \cdot \exp\left(\tfrac{\boldsymbol{i}}{\hbar}S^\hbar\right) = \tfrac{\boldsymbol{i}}{\hbar}\dot{S}^\hbar \cdot \exp\left(\tfrac{\boldsymbol{i}}{\hbar}S^\hbar\right) = \frac{\mathrm{d}}{\mathrm{d}\varepsilon}\Big|_{\varepsilon=0}\left(\exp\left(\tfrac{\boldsymbol{i}}{\hbar}S^\hbar\right)\right).
\end{aligned}$$

Note that the Schouten bracket acts on $\exp\left(\frac{\boldsymbol{i}}{\hbar}S^\hbar\right)$ by the Leibniz rule (see Theorem 4) and we then use the equality $-[\![F, \frac{\boldsymbol{i}}{\hbar}S^\hbar]\!] = \frac{\boldsymbol{i}}{\hbar}[\![S^\hbar, F]\!]$ which holds by Theorem 4 again.

Let us now regard the full quantum BV-action as the generating functional for ghost parity-even observables $\mathcal{O}$, see [54].

Lemma 14. There are no observables $\mathcal{O}$, other than the identically zero functional, which would be ghost parity-odd.

Proof. Indeed, Eq. (42) implies that the path integral

$$I = \int_{\Gamma(\varsigma^0)} [D\boldsymbol{q}]\, \mathcal{O}([\boldsymbol{q}], [\boldsymbol{q}^\dagger]) \exp\left(\tfrac{\boldsymbol{i}}{\hbar}S^\hbar([\boldsymbol{q}], [\boldsymbol{q}^\dagger])\right)$$

over the space of ghost parity-even BV-section components is effectively independent of the ghost parity-odd BV-variables $\boldsymbol{q}^\dagger$. Notice further that the ghost parity $\mathrm{gh}(I)$ of this constant function $I([\boldsymbol{q}^\dagger])$ is equal to that of $\mathcal{O}$; the quantum master-action $S^\hbar$ is parity-even. Under a (speculative) assumption that an observable $\mathcal{O}$ could be ghost parity-odd, we obtain an odd parity constant. Unless a possibility of their existence is postulated by brute force, this odd-parity constant must be equal to zero, whence the ghost parity-odd functional $\mathcal{O} \in \overline{H^n}(\pi_{\mathrm{BV}}) \subseteq \overline{\mathfrak{N}^n}(\pi_{\mathrm{BV}}, T\pi_{\mathrm{BV}})$ itself is zero. □

In what follows we accept for transparency that there is no grading in the initial geometry of physical fields, i.e., for sections of the bundle $\pi\colon E^{n+m} \to M^n$. Let us focus on the standard cohomological approach to quantum BV-models and to their gauge symmetries (cf. [37]).

Lemma 15. Suppose that an infinitesimal shift $S^\hbar \mapsto S^\hbar + \lambda \cdot \mathcal{O} + \bar{o}(\lambda)$ of the quantum BV-action by using an even-parity functional $\mathcal{O} \in \overline{H^n}(\pi_{\mathrm{BV}}) \subseteq \overline{\mathfrak{N}^n}(\pi_{\mathrm{BV}}, T\pi_{\mathrm{BV}})$ does not destroy the quantum master-equation,

$$\frac{\mathrm{d}}{\mathrm{d}\lambda}\Big|_{\lambda=0} \Delta\left(\exp\left(\tfrac{\boldsymbol{i}}{\hbar}(S^\hbar + \lambda \cdot \mathcal{O})\right)\right) = 0.$$

Then the *observable* $\mathcal{O}$ is $\Omega^\hbar$-closed: $-\boldsymbol{i}\hbar\,\Delta\mathcal{O} + [\![S^\hbar, \mathcal{O}]\!] = 0$.

Proof. The proof literally repeats that of Proposition 13. □

For a given odd-parity functional $F \in \overline{H^n}(\pi_{\mathrm{BV}})$, we organize the infinitesimal shift (43) of the master-functional $S^\hbar$ as follows:

$$\begin{aligned}
\dot{S}^\hbar &= -\boldsymbol{i}\hbar\,\Delta(F) + [\![S^\hbar, F]\!], \\
\dot{\mathcal{O}} &= [\![\mathcal{O}, F]\!].
\end{aligned}$$

Note that, unless one has that $\Delta F = 0$ incidentally, the transformation of the integral *functional* $S^\hbar$ is not induced by any infinitesimal transformation of the BV-*variables*, that is, by an evolutionary vector field on the horizontal infinite jet space at hand. No earlier than the transformation law $S^\hbar \mapsto S^\hbar(\varepsilon)$ is postulated, it becomes an act of will to think that the functional F is the generator of parity-preserving evolutionary vector field $\overleftarrow{Q}^F = \overrightarrow{Q}^F$ acting on the BV-variables so that $\dot{\mathcal{O}} \cong \overrightarrow{Q}^F(\mathcal{O})$ for all observables $\mathcal{O}$.

Furthermore, let us extend the deformation $\mathcal{O} \mapsto \mathcal{O}(\varepsilon)$ of even-parity cocycles $\mathcal{O} \in \ker \Omega^\hbar$ to the space of odd-parity functionals $\xi \in \overline{H^n}(\pi_{\text{BV}}) \subseteq \overline{\mathfrak{N}^n}(\pi_{\text{BV}}, T\pi_{\text{BV}})$ which produce the coboundaries $\Omega^\hbar(\xi)$. Namely, we postulate that

$$\dot{\xi} = [\![\xi, F]\!]$$

for all such functionals ξ; here we denote by the dot over ξ its velocity in the course of the transformation generated by a given F. Let us remember however that the law for evolution of the odd-parity functionals ξ which produce the $\Omega^\hbar$-coboundaries is different from our earlier postulate (see Proposition 13) that the odd-parity generators F_i of gauge symmetries do not evolve: $dF_i/d\varepsilon_j \equiv 0$ or, in shorthand notation,

$$\dot{F} \equiv 0. \tag{45}$$

We claim that under these hypotheses, the structure of quantum BV-cohomology group remains intact in the course of gauge symmetry transformations of the quantum master-action, $S^\hbar \mapsto S^\hbar(\varepsilon)$, even though the quantum BV-differential is modified, $\Omega^\hbar \mapsto \Omega^\hbar(\varepsilon)$, and the cocycles and coboundaries are also deformed.

Theorem 16. *An infinitesimal shift of the quantum BV-cohomology classes induced by* (43), (45), *and*

$$\begin{aligned} \dot{\mathcal{O}} &= [\![\mathcal{O}, F]\!], && \mathcal{O} \in \ker \Omega^\hbar, \\ \dot{\xi} &= [\![\xi, F]\!], && \xi \in \overline{H^n}(\pi_{\text{BV}}) \subseteq \overline{\mathfrak{N}^n}(\pi_{\text{BV}}, T\pi_{\text{BV}}),\ \xi \text{ odd}, \end{aligned}$$

yields an isomorphism of the $\Omega^\hbar$-cohomology group: under such mapping, every $\Omega^\hbar$-closed, even-parity $\Omega^\hbar$-cocycle $\mathcal{O}$ becomes $\Omega^\hbar(\varepsilon)$-closed, whereas the transformation of an even-parity coboundary $\Omega^\hbar(\xi)$ produces an $\Omega^\hbar(\varepsilon)$-coboundary: $(\Omega^\hbar(\xi))(\varepsilon) = \Omega^\hbar(\varepsilon)\bigl(\xi(\varepsilon)\bigr)$.

Proof. Let $\mathcal{O} \in \ker \Omega^\hbar$ be an even-parity observable and F an odd-parity generator of gauge transformation. Consider the equation $\Omega^\hbar(\varepsilon)(\mathcal{O}(\varepsilon)) = 0$ which states that the transformed functional $\mathcal{O}(\varepsilon)$ remains a coboundary. The term which is proportional to ε in this equation's left-hand side is equal to

$$\frac{\mathrm{d}}{\mathrm{d}\varepsilon}\Big|_{\varepsilon=0} \Bigl(-i\hbar\,\Delta\mathcal{O}(\varepsilon) + [\![S^\hbar(\varepsilon), \mathcal{O}(\varepsilon)]\!]\Bigr) = \Omega^\hbar(\dot{\mathcal{O}}) + [\![\dot{S}^\hbar, \mathcal{O}]\!] = \Omega^\hbar([\![\mathcal{O}, F]\!]) + [\![\Omega^\hbar(F), \mathcal{O}]\!];$$

recalling once again that $\Omega^\hbar = -i\hbar\,\Delta + [\![S^\hbar, \cdot\,]\!]$, we continue the equality

$$= -i\hbar\,\Delta([\![\mathcal{O}, F]\!]) + [\![S^\hbar, [\![\mathcal{O}, F]\!]]\!] + [\![-i\hbar\,\Delta F + [\![S^\hbar, F]\!], \mathcal{O}]\!].$$

Now by Theorem 6 we obtain that, the observable $\mathcal{O}$ being parity-even,

$$\begin{aligned} &= -i\hbar\,[\![\Delta\mathcal{O}, F]\!] + i\hbar\,[\![\mathcal{O}, \Delta F]\!] + [\![S^\hbar, [\![\mathcal{O}, F]\!]]\!] - i\hbar\,[\![\Delta F, \mathcal{O}]\!] + [\![\mathcal{O}, [\![S^\hbar, F]\!]]\!] = \\ &\qquad = [\![\Omega^\hbar(\mathcal{O}), F]\!] \cong -\vec{\boldsymbol{Q}}^F\bigl(\Omega^\hbar(\mathcal{O})\bigr) = 0, \end{aligned}$$

because $[\![S^\hbar,[\![\mathcal{O},F]\!]]\!] = [\![[\![S^\hbar,\mathcal{O}]\!],F]\!] - [\![\mathcal{O},[\![S^\hbar,F]\!]]\!]$ by Jacobi identity (28), because we are inspecting the ε-linear term in the operator $\Omega^\hbar(\varepsilon)\circ\bigl(\varepsilon=0\longmapsto\varepsilon\neq 0\bigr)$ applied to $\mathcal{O}$, and $\mathcal{O}$ is an $\Omega^\hbar$-cocycle. Therefore, the zero initial condition $\Omega^\hbar(\mathcal{O})=0$ evolves at zero velocity to the $\Omega^\hbar(\varepsilon)$-cocycle equation $\Omega^\hbar(\varepsilon)\bigl(\mathcal{O}(\varepsilon)\bigr)=0$ upon $\mathcal{O}(\varepsilon)$.

Likewise, let $\Omega^\hbar(\xi)$ be a coboundary for some odd-parity functional ξ which evolves by $\dot\xi=[\![\xi,F]\!]$. Then the even-parity observable $\Omega^\hbar(\xi)\in\ker\Omega^\hbar$ evolves as fast as $[\![\Omega^\hbar(\xi),F]\!]$ but simultaneously we have that the mapping $\Omega^\hbar$ and its argument ξ change. We claim that the two evolutions match so that $(\Omega^\hbar(\xi))(\varepsilon)$ is $\Omega^\hbar(\varepsilon)$-exact. Indeed, we have that

$$\frac{\mathrm{d}}{\mathrm{d}\varepsilon}\Big|_{\varepsilon=0}\bigl(\Omega^\hbar(\varepsilon)(\xi(\varepsilon))\bigr)=\Omega^\hbar([\![\xi,F]\!])+[\![\Omega^\hbar(F),\xi]\!]$$
$$=-i\hbar\,[\![\Delta\xi,F]\!]\underline{-i\hbar\,[\![\xi,\Delta F]\!]}+[\![S^\hbar,[\![\xi,F]\!]]\!]+[\![\underline{-i\hbar\,\Delta F}+[\![S^\hbar,F]\!],\underline{\xi}]\!];$$

by cancelling out the underlined Schouten brackets and then using the Jacobi identity we obtain

$$=[\![-i\hbar\,\Delta\xi,F]\!]+[\![[\![S^\hbar,\xi]\!],F]\!]+[\![\xi,[\![S^\hbar,F]\!]]\!]-[\![\xi,[\![S^\hbar,F]\!]]\!]=[\![\Omega^\hbar(\xi),F]\!],$$

which proves our claim.

Summarizing, we see that gauge symmetries of the quantum master-equation induce automorphisms of the $\Omega^\hbar$-cohomology group. $\square$

We conclude that it would be conceptually incorrect to say that the infinitesimal gauge transformations of all functionals in a quantum BV-model are induced by a canonical transformation, determined by the evolutionary vector field $\overrightarrow{Q}^F$ acting on the BV-variables. Let us remember that the even-parity quantum master-action $S^\hbar\in\overline{H^n}(\pi_{\text{BV}})$ and its descendants, the observables $\mathcal{O}$ evolve by

$$\dot S^\hbar=-i\hbar\,\Delta F+[\![S^\hbar,F]\!]=\Omega^\hbar(F),\qquad F\in\overline{H^n}(\pi_{\text{BV}})\subseteq\overline{\mathfrak{N}^n}(\pi_{\text{BV}},T\pi_{\text{BV}}),\quad F\text{ odd},$$

and

$$\dot{\mathcal{O}}=[\![\mathcal{O},F]\!].$$

We note that the evolution of the generating functional $S^\hbar_{\text{BV}}$ is **not** determined by a vector field on the space of BV-variables. Likewise, we recall that the odd-parity arguments ξ of $\Omega^\hbar$ for the coboundaries $\Omega^\hbar(\xi)\sim 0$ do evolve,

$$\dot\xi=[\![\xi,F]\!],$$

whereas the generators F of gauge symmetries for (40) never change: symbolically,

$$\dot F=0$$

(see Eq. (45) above). In fact, one may think that each F determines a parity-preserving evolutionary vector field $\overrightarrow{Q}^F$ on the space of BV-variables, but it is not the objects $\overrightarrow{Q}^F$ but the full systems of four distinct evolution equations which encode the deformation of respective functionals. Neither the functionals' attribution to the space of building blocks $\overline{H^n}(\pi_{\text{BV}})\ni S^\hbar$, $\mathcal{O}$, F nor a functional's parity, $\mathrm{gh}(S^\hbar)=\mathrm{gh}(\mathcal{O})$ and $\mathrm{gh}(F)=\mathrm{gh}(\xi)$, completely determines their individual transformation laws.

Remark 3.4. The supports of test shifts $\delta\boldsymbol{s}$ can be arbitrarily small[27] and they can be chosen in such a way that all boundary terms vanish in the course of integration by parts within equivalence classes from the horizontal cohomology groups $\overline{H}^{n(1+k)}(\pi_{\text{BV}} \times T\pi_{\text{BV}} \times \ldots \times T\pi_{\text{BV}})$. Let us note also that these integrations by parts (see section 1.3) transport the derivatives from one copy of the base manifold M^n to another copy; this reasoning stays local with respect to base points $\boldsymbol{x}$ and local volume elements $\text{dvol}(\boldsymbol{x})$ because the geometric mechanism of locality yields the diagonal in powers of the base manifold. However, an integration by parts in functionals from $\overline{H}^n(\pi_{\text{BV}})$ is a different issue. In fact, it refers to the topology of M^n or to a choice of the class $\Gamma(\pi_{\text{BV}})$ of admissible sections (so that there appear no boundary terms as well). Let us recall that the only place where such global, de Rham cohomology aspect is explicitly used is the proof of Jacobi's identity for the variational Schouten bracket (see [32]). In turn, Theorems 8 and 12 relate these properties of the bracket $[\![\,,\,]\!]$ to cohomology generators $\Delta^2 = 0$ and $(\Omega^\hbar)^2 = 0$. (The converse is also true: Jacobi's identity for $[\![\,,\,]\!]$ stems from $\Delta^2 = 0$.) This motivates why the de Rham and quantum BV-cohomologies are interrelated (cf. [5]).

Conclusion

Mathematical models are designed for description of phenomena of Nature; a construction of the models' objects is not the same as their evaluation at given configurations of the models, which would associate $\Bbbk$-numbers to physical fields $\phi \in \Gamma(\pi)$ in terms of such objects. Namely, consider an Euler–Lagrange model whose primary element is the action functional $S\colon \Gamma(\pi) \to \Bbbk$. By definition, derivative objects are obtained from S by using natural operations such as $\vec{\delta}$ or $[\![\,,\,]\!]$ and Δ. The derivative objects' geometric complexity is greater than that of S because they absorb the domains of definition for test shifts $\delta s_1, \ldots, \delta s_k$ of field configurations. We emphasize that such composite structure objects do not yet become maps $\Gamma(\pi) \to \Bbbk$ which would suit well for their evaluation at sections $\boldsymbol{s} \in \Gamma(\pi)$ yielding $\Bbbk$-numbers. The intermediate objects can rather be used as arguments of $[\![\,,\,]\!]$ or Δ in a construction of larger, logically and geometrically more complex objects; we illustrate by Fig. 9 the expansion of analytic structures and their shrinking

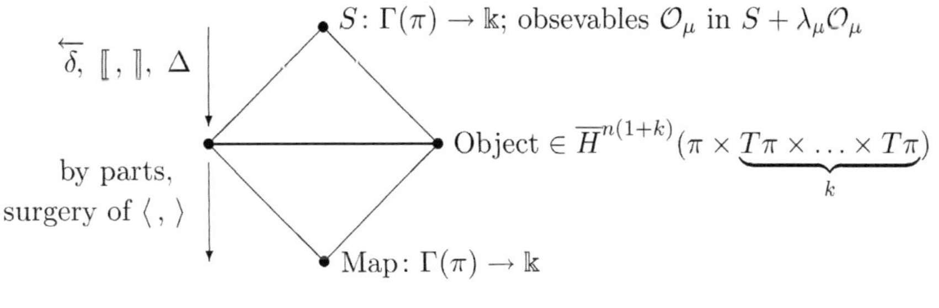

Figure 9. The action S as a generator of observables, building blocks of derivative objects as horizontal cohomology classes in products of bundles over $M \times M \times \ldots \times M$, and resulting mappings as the objects' contractions over Whitney's sum of bundles.

in the course of integration by parts and multiplication of normalized test shifts in reconfigured couplings. Indeed, the derivative objects become multi-linear maps with respect to k-tuples of the variations $\delta s_1, \ldots, \delta s_k \in \Gamma(T\pi)$ only when the integrations by parts carry all derivatives away from the test shifts, channelling the derivations to densities of the object's constituent blocks such as the Lagrangian in the action functional. A surgery of couplings then contracts

[27] We recall that the smoothness class of variations $\delta\boldsymbol{s}$ is determined by smoothness of the frame fields $\vec{e}_i(\boldsymbol{x})$, $\vec{e}^{\,\dagger i}(\boldsymbol{x})$ and coefficient functions $\delta s^i(\boldsymbol{x})$, $\delta s^\dagger_i(\boldsymbol{x})$.

the values of normalized test shifts by virtue of (16) at every point of the base manifold. This is how maps $\Gamma(\pi) \to \Bbbk$ are obtained.

We conclude that a calculation of composite-structure object may not be interrupted ahead of time. Otherwise speaking, every calculation stretches from its input data to the end value at $\boldsymbol{s}$; independently existing values at $\boldsymbol{s}$ for the resulting object's constituent elements not always contribute to the sought-for value of the large structure (e. g., consider (1c) on p. 1 and Example 2.4 on p. 34 and try to calculate consecutively the objects ΔF, ΔG, and their Schouten brackets with G and F, respectively, for that example's functionals F and G). Summarizing, it is illegal to construct composite objects step by step, redundantly inspecting the elements' values at field configurations. One must not deviate from a way towards the appointed end of logical reasoning.

In fact, it is us but not Nature who calculates (e. g., the left-hand sides of equations of motion): Nature neither calculates nor evaluates; for there is no built-in mechanism for doing that.[28] This implies that there is no ever-growing logical complexity in a description of the Universe; the flow of local, observer-dependent time does not require any perpetual increase of the number $k \geqslant 0$ of factors in the product-bundle location of objects over $k+1$ copies of the space-time. Conversely, there always remains a unique copy of the space-time for all local functionals.

The space-time geometry of information transfer is very restrictive: its pointwise locality of events of couplings between dual objects is an absolute principle; by weakening this hypothesis one could create a source of difficulties through causality violation. Consequently, a count of space-time points where the couplings with a given (co)vector occur makes the formalism of singular linear integral operators truly adequate in mathematical models of physical phenomena.[29]

We finally remark that the product-base approach of bundles $\pi \times T\pi \times \ldots \times T\pi$ over $M \times M \times \ldots \times M$ to the geometry of variations highlights the concept of physical field as infinite-dimensional system with degrees of freedom which are attached at every point of space-time. The locality principle for (co)vector interaction is the mechanism which distinguishes between space-time points with respect to its (non)Hausdorff topology.

Discussion
Let us finally address two logical aspects of the geometry of variations.

Linear vector space structures Nature is essentially nonlinear; for there is no mechanism which would realize – under a uniform time bound – an arbitrarily large number of replications of an object. This is tautological for those physical fields ϕ which take values in spaces without any linear structure. Moreover, even if there is a brute force labelling of Euler–Lagrange equations by using the fields ϕ, a linear vector space pattern of the equations of motion is not utilized (the same is true for the equations' descendants such as the antifields $\phi^\dagger$ or (anti)ghosts). Indeed, it

[28] The probabilistic approach to evolution of Nature suggests that maxima of transition (and correlation) functions concentrate near the zero loci of such deterministic equations' left-hand sides. At the same time, Noether symmetries of the action S are abundant in the models. Not referring to any actual transformation of a system's components, such symmetries reflect the model's geometry. The analytic machinery of self-regularizing structures yields the invariants – e. g., cohomology classes as in section 3.2 – which constrain the probabilistic laws of evolution.

[29] We recall from Remark 1.5 on p. 9 that the volume elements $\mathrm{dvol}\big(\boldsymbol{x},\phi(\boldsymbol{x})\big) = \sqrt{|\det(g_{\mu\nu})|}\,\mathrm{d}\boldsymbol{x}$ are present in the building blocks of composite-structure objects.Let us note further that an association of the weight factors $\mathrm{dvol}(\boldsymbol{x})$ with point $\boldsymbol{x} \in M^n$ is intrinsically related to the structure of space-time M^n as topological manifold (cf. [31]). It is readily seen that a discrete tiling of space-time converts the integrations over a measure on it to weighted sums over the points which mark the quantum domains. This links the concept with loop quantum gravity (see e. g. [17, 47, 49]).

is only their the *tangent* spaces whose linear structure is used, in particular, in order to split the variations in ghost parity-homogeneous components. Objects are linearized only in the course of variations under infinitesimal test shifts. For example, this determines the distinction between finite offsets $\Delta\boldsymbol{x}$ so that $(\boldsymbol{x}, \boldsymbol{x}+\Delta\boldsymbol{x}) \in M \times M$ and infinitesimal test shifts $\delta\boldsymbol{x}|_{\boldsymbol{x}} \in T_{\boldsymbol{x}}M$ which are mapped to the number field $\Bbbk$ by covectors $\mathrm{d}\boldsymbol{x}|_{\boldsymbol{x}} \in T^*_{\boldsymbol{x}}M$.

Annual reproduction rate for interspecimen breeding of cats and whales An immediate comment on the title of this paragraph is as follows. One could proclaim that the annual reproduction rate for interspecimen breeding of – without loss of generality – cats and whales is equal to zero for a given year. Alternatively, one should understand that such events never happen (not that a given year brought no brood).

This grotesque illustration works equally well for the (co)tangent spaces to fibres of the BV-zoo or, in broad terms, for a definition of Kronecker's symbol $\boldsymbol{\delta}_i^j$ by zero whenever the indices $i \neq j$ do not match so that the couplings in (11) do not eventuate. We argue that, on top of the absolute pointwise locality for couplings (9), a superficial definition of $\langle\,,\,\rangle$ by zero for mismatching elements $\vec{e}_i$ and $\vec{e}^{\dagger j}$ of dual bases is a mere act of will; in reality those evaluations do not occur. Consequently, the geometry dictates that

$$\log\left\langle \vec{e}_i(\boldsymbol{x}), {}^{\dagger}(\vec{e}_j)(\boldsymbol{x})\right\rangle = \log 1 = 0 \quad \text{and} \quad \log\left\langle \vec{e}^{\dagger j}(\boldsymbol{x}), {}^{\dagger}(\vec{e}^{\dagger i})(\boldsymbol{x})\right\rangle = \log 1 = 0.$$

Combined with the geometric locality principle (4) realized by singular linear integral operators (12), this argument finally resolves the paradoxical, *ad hoc* conventions $\boldsymbol{\delta}(0) = 0$ and $\log\boldsymbol{\delta}(0) = 0$ for Dirac's distribution.

Acknowledgments
The author thanks the Organizing committee of XXI International conference 'Integrable systems & quantum symmetries' (June 11 – 16, 2013; CVUT Prague, Czech Republic) for cooperation and warm atmosphere during the meeting. These notes follow the lecture course which was read by the author in October 2013 at the Taras Shevchenko National University and Bogolyubov Institute for Theoretical Physics in Kiev, Ukraine; the author is grateful to BITP for hospitality. The author thanks M. A. Vasiliev and A. G. Nikitin for helpful discussions and constructive criticisms.

This research was supported in part by JBI RUG project 103511 (Groningen). A part of this research was done while the author was visiting at the IHÉS (Bures-sur-Yvette); the financial support and hospitality of this institution are gratefully acknowledged.

Appendix A. Proof of Propositions 10 and 11

We need the following two lemmas.

Lemma 17. Let $F \in \overline{H}^n(\pi_{\mathrm{BV}})$ be an even integral functional, let $G \in \overline{\mathfrak{N}}^n(\pi_{\mathrm{BV}}, T\pi_{\mathrm{BV}})$ be another functional, and let $n \in \mathbb{N}_{\geq 1}$. Then

$$[\![G, F^n]\!] = n[\![G, F]\!]F^{n-1}.$$

Proof. We use induction on Theorem 4. Note that all signs vanish since F is even, meaning that whenever F is multiplied with any other integral functional, the factors may be freely swapped without this resulting in minus signs. For $n = 1$ the statement is trivial. Suppose the formula holds for some $n \in \mathbb{N}_{>1}$, then $[\![G, F^{n+1}]\!] =$

$$[\![G, F \cdot F^n]\!] = [\![G, F]\!] \cdot F^n + F \cdot [\![G, F^n]\!] = [\![G, F]\!] \cdot F^n + nF \cdot [\![G, F]\!]F^{n-1} = (n+1)[\![G, F]\!] \cdot F^n,$$

so that the statement also holds for $n + 1$. □

Lemma 18. Let $F \in \overline{H}^n(\pi_{\mathrm{BV}})$ be an even integral functional, and let $n \in \mathbb{N}_{\geq 2}$. Then

$$\Delta(F^n) = n(\Delta F) \cdot F^{n-1} + \tfrac{1}{2}n(n-1)[\![F, F]\!] \cdot F^{n-2}.$$

Proof. We use induction and the previous lemma. For $n = 2$ the formula clearly holds by Theorem 3. Suppose that it holds for some $n \in \mathbb{N}_{>2}$, then

$$\begin{aligned}\Delta(F^{n+1}) &= \Delta(F \cdot F^n) = (\Delta F) \cdot F^n + [\![F, F^n]\!] + F \cdot \Delta(F^n) \\ &= (\Delta F) \cdot F^n + n[\![F, F]\!] \cdot F^{n-1} + F \cdot n(\Delta F)F^{n-1} + \tfrac{1}{2}n(n-1)F \cdot [\![F, F]\!]F^{n-2} \\ &= (n+1)(\Delta F) \cdot F^n + \tfrac{1}{2}(n+1)n\,[\![F, F]\!] \cdot F^{n-1},\end{aligned}$$

so that the statement also holds for $n + 1$. □

Proof of Proposition 10. For convenience, we denote $F = \frac{i}{\hbar}S^\hbar$. Then

$$\begin{aligned}0 = \Delta(\exp F) &= \Delta\left(\sum_{n=0}^{\infty} \frac{1}{n!}F^n\right) = \sum_{n=0}^{\infty} \frac{1}{n!}\Delta(F^n) \\ &= \sum_{n=1}^{\infty} \frac{n}{n!}(\Delta F) \cdot F^{n-1} + \sum_{n=2}^{\infty} \frac{1}{2n!}n(n-1)[\![F, F]\!] \cdot F^{n-2} \\ &= (\Delta F) \cdot \sum_{n=1}^{\infty} \frac{1}{(n-1)!}F^{n-1} + \frac{1}{2}[\![F, F]\!] \cdot \sum_{n=2}^{\infty} \frac{1}{(n-2)!}F^{n-2} \\ &= \left(\Delta F + \tfrac{1}{2}[\![F, F]\!]\right) \cdot \exp F = \left(\frac{i}{\hbar}\Delta S^\hbar - \frac{1}{2\hbar^2}[\![S^\hbar, S^\hbar]\!]\right) \cdot \exp\left(\tfrac{i}{\hbar}S^\hbar\right),\end{aligned}$$

from which the result follows. □

Proof of Proposition 11 (*cf. Proposition* 13 *on p.* 42). Again, let us set $F = \frac{i}{\hbar}S^\hbar$. We first calculate, using Lemma 17,

$$[\![\mathcal{O}, \exp F]\!] = \sum_{n=0}^{\infty} \frac{1}{n!}[\![\mathcal{O}, F^n]\!] = \sum_{n=1}^{\infty} \frac{n}{n!}[\![\mathcal{O}, F]\!]F^{n-1} = [\![\mathcal{O}, F]\!] \exp F.$$

Then

$$\begin{aligned}0 = \Delta(\mathcal{O} \exp F) &= (\Delta\mathcal{O}) \exp F + [\![\mathcal{O}, \exp F]\!] + \mathcal{O} \cdot \Delta(\exp F) \\ &= \left(\Delta\mathcal{O} + [\![\mathcal{O}, F]\!]\right) \exp F = \left(\Delta\mathcal{O} + \tfrac{i}{\hbar}[\![\mathcal{O}, S^\hbar]\!]\right) \exp\left(\tfrac{i}{\hbar}S^\hbar\right),\end{aligned}$$

from which the assertion follows. □

References

[1] Alexandrov M, Schwarz A, Zaboronsky O and Kontsevich M 1997 *Int. J. Modern Phys.* A vol **12** 1405–29

[2] Arnol'd V I 1996 *Mathematical methods of classical mechanics* (Grad. Texts in Math. vol **60**) (New York: Springer–Verlag)

[3] Barnich G 2000 *Classical and quantum aspects of the extended antifield formalism* (These d'agregation ULB *Proc. spring school 'QFT and Hamiltonian systems'* (Calimanesti, Romania) pp 1–94 (*Preprint* hep-th/0011120)

[4] Barnich G 2010 (AIP Conf. Proc. vol **1307**) *XXIX Workshop on Geometric Methods in Physics* (Białowieża, Poland) 7–18 (*Preprint* math-ph/1010.0899)

[5] Barnich G, Brandt F and Henneaux M 1995 *Commun. Math. Phys.* vol **174** 57–91, 93–116 (*Preprint* hep-th/9405109)
[6] Barnich G, Brandt F and Henneaux M 2000 *Phys. Rep.* **338** 439–569 (*Preprint* hep-th/0002245)
[7] Batalin I and Vilkovisky G 1981 *Phys. Lett.* B vol **102** 27–31
Batalin I A and Vilkovisky G A 1983 *Phys. Rev.* D vol **29** 2567–82
[8] Becchi C, Rouet A and Stora R 1976 *Ann. Phys.* **98** 287–321
Tyutin I V 1975 Gauge invariance in field theory and statistical mechanics *Preprint* Lebedev FIAN no 39
[9] Beilinson A and Drinfeld V 2004 *Chiral algebras* (AMS Colloq. Publ. vol **51**) (Providence, RI: AMS)
[10] Berezin F A 1987 *Introduction to superanalysis* (Math. Phys. & Appl. Math. vol **9**) (Dordrecht: D. Reidel Publ. Co.)
[11] Cattaneo A S and Felder G 2000 *Commun. Math. Phys.* vol **212** 591–611
[12] Costello K J 2007 Renormalisation and the Batalin–Vilkovisky formalism (*Preprint* 0706.1533)
[13] Costello K 2011 *Renormalization and effective field theory* (Math. Surveys and Monographs vol **170**) (Providence, RI: AMS)
[14] Dorfman I Ya 1993 *Dirac structures* (J. Whiley & Sons)
[15] Ehresmann C 1953 *Geometrie Differentielle, Colloq. Inter. du Centre Nat. de la Recherche Scientifique (Strasbourg)* 97–110
[16] Eisenbud D 2005 *The geometry of syzygies. A second course in commutative algebra and algebraic geometry* (Grad. Texts in Math. vol **229**) (New York: Springer–Verlag)
[17] Gambini R and Pullin J 1996 *Loops, knots, gauge theories and quantum gravity* (Cambridge Monographs on Math. Phys.) (Cambridge: Cambridge Univ. Press)
[18] Gel'fand I M and Dorfman I Ja 1981 *Functional Anal. Appl.* vol **14** 223–6
[19] Gel'fand I M and Shilov G E 1964 *Generalized functions* vol **1** Properties and operations (NY–London: Academic Press)
Gel'fand I M and Shilov G E 1968 *Generalized functions* vol **2** Spaces of fundamental and generalized functions (NY–London: Academic Press)
[20] Gitman D M and Tyutin I V 1990 *Quantization of fields with constraints* (Springer Ser. Nucl. Part. Phys.) (Berlin: Springer-Verlag)
[21] Gomis J, París J and Samuel S 1995 *Phys. Rep.* vol **259** 1–145 (*Preprint* hep-th/9412228)
[22] Henneaux M and Teitelboim C 1992 *Quantization of gauge systems* (Princeton, NJ: Princeton Univ. Press)
[23] Igonin S, Verbovetsky A and Vitolo R 2002 On the formalism of local variational differential operators *Memorandum* 1641 (Faculty of Math. Sci., Univ. Twente, Enschede, The Netherlands)
[24] Igonin S, Verbovetsky A and Vitolo R 2004 *Symmetry in nonlinear mathematical physics (Kiev, 2003) Pr. Inst. Mat. Nats. Akad. Nauk Ukr. Mat. Zastos.* vol **50** 1335–42
[25] de Jonghe F, Siebelink R and Troost W 1993 Hiding anomalies *Phys. Lett.* B vol **306** 295–300
[26] Kersten P, Krasil'shchik I and Verbovetsky A 2004 *J. Geom. Phys.* vol **50** 273–302
[27] Khudaverdian H M and Voronov Th Th 2013 *J. Geom. Phys.* vol **64** 31–53
[28] Kiselev A V 2012 The twelve lectures in the (non)commutative geometry of differential equations *Preprint* IHÉS/M/12/13
[29] Kiselev A V 2012 *Physics of Particles and Nuclei* vol **43** 663–5 (*Preprint* 1112.5784 [math-ph])
[30] Kiselev A V 2012 *J. Phys.: Conf. Series* **343** 012058, 1–20 (*Preprint* 1111.3272)
[31] Kiselev A V 2013 *Proc. 6th Int. Workshop 'Group analysis of differential equations and integrable systems' (Protaras, Cyprus)* 111–26 (*Preprint* 1304.0234)
[32] Kiselev A V 2012 The calculus of multivectors on noncommutative jet spaces *Preprint* 1210.0726 [math.DG]
[33] Kiselev A V and Krutov A O 2013 Non-Abelian Lie algebroids over jet spaces *Preprint* 1305.4598 [math.DG]
[34] Kiselev A V and van de Leur J W 2011 *Theoret. and Mathem. Phys.* vol **167** 772–84 (*Preprint* 1006.4227 [math.DG])
[35] Kiselev A V and Ringers S 2013 On the geometry of the Batalin–Vilkovisky Laplacian *Preprint* 1302.4388 [math.DG]
[36] Kontsevich M 1993 *The Gel'fand Mathematical Seminars, 1990-1992* eds L Corwin, I Gelfand and J Lepowsky (Boston MA: Birkhäuser) 173–87
[37] Kontsevich M and Soibelman Y 2009 *Homological mirror symmetry. New developments and perspectives* (Lect. Notes in Phys. vol **757**) eds A Kapustin, M Kreuzer and K-G Schlesinger (Berlin: Springer) 153–219
[38] Kontsevich M and Vishik S 1994 Determinants of elliptic pseudo-differential operators *Preprint* hep-th/9404046
[39] Kosmann-Schwarzbach Y 2008 *SIGMA* vol **4**, 5 pp 1–30 (*Preprint* 0710.3098)
[40] Krasil'shchik I S and Vinogradov A M eds 1999 *Symmetries and conservation laws for differential equations of mathematical physics* (Providence, RI: AMS)

[41] Kupershmidt B A 1980 *Geometric methods in mathematical physics* (Lecture Notes in Math. vol **775**) *Proc. NSF–CBMS Conf. (Univ. Lowell, Mass.)* (Berlin: Springer) 162–218
[42] Magri F 1978 *J. Math. Phys.* vol **19** 1156–62
[43] McCloud P 1994 *Class. Quant. Grav.* vol **11** 567–87
[44] Merkulov S and Willwacher Th 2010 Grothendieck–Teichmüller and Batalin–Vilkovisky *Preprint* 1012.2467 [math.QA]
[45] Olver P J 1993 *Applications of Lie groups to differential equations* (Grad. Texts in Math. vol **107**) 2nd ed (New York: Springer–Verlag)
[46] Olver P J and Sokolov V V 1998 *Comm. Math. Phys.* vol **193** 245–68
[47] Rovelli C 2004 *Quantum gravity* (Cambridge Monographs on Math. Phys.) (Cambridge: Cambridge Univ. Press)
[48] Schwarz A 1993 *Commun. Math. Phys.* vol **155** 249–60 (*Preprint* hep-th/9205088)
[49] Thiemann Th 2007 *Modern canonical quantum general relativity* (Cambridge Monographs on Math. Phys.) (Cambridge: Cambridge Univ. Press)
[50] Troost W, van Nieuwenhuizen P and van Proeyen A 1990 *Nucl. Phys.* B vol **333** 727–70
[51] Voronov B L, Tyutin I V and Shakhverdiev Sh S 1999 *Theoret. and Mathem. Phys.* vol **120** 1026–44 (*Preprint* hep-th/9904215)
[52] Voronov T 2002 *Quantization, Poisson brackets, and beyond* (Contemp. Math. vol **315**) ed T Voronov (Providence, RI: AMS) pp 131–68 (*Preprint* math.DG/0105237)
[53] Witten E 1990 *Modern Phys. Lett.* A **5** 487–94
[54] Zinn-Justin J 1993 *Quantum field theory and critical phenomena* (Int. Ser. of Monographs on Phys. vol **85**) 2nd ed (Oxford: Oxford Univ. Press; New York: The Clarendon Press)
[55] Zinn-Justin J 1975 *Trends in Elementary Particle Theory* (Lect. Notes in Phys. vol **37**) eds H Rollnick and K. Dietz (Berlin: Springer) pp 2–39
Zinn-Justin J 1976 *Méthodes en théorie des champs / Methods in field theory* (École d'Été de Physique Théorique, Session XXVIII, tenue à Les Houches, 28 Juillet – 6 Septembre 1975) eds R Balian and J Zinn-Justin (Amsterdam: North-Holland Publ. Co.)

Journal of Physics: Conference Series **474** (2013) 012025 doi:10.1088/1742-6596/474/1/012025

Extended Supersymmetries and $2+1$ Dimensional Supersymmetric Chern Simons Theories[1]

V K Oikonomou
Max Planck Institute for Mathematics in the Sciences
Inselstrasse 22, 04103 Leipzig, Germany
E-mail: `voiko@physics.auth.gr`

Abstract. We study $N=2$ supersymmetric Chern-Simons Higgs models in (2+1)-dimensions. As we will demonstrate, an extended supersymmetric quantum mechanics algebras underlies the fermionic zero modes quantum system and the zero modes corresponding to bosonic fluctuations. These two algebras, in turn, combine to give an $N=4$ extended 1-dimensional supersymmetric algebra with central charge

Gauge field theories in $(2+1)$-dimensions [1, 2] are particularly interesting, since the Chern-Simons [3–5] term can be consistently incorporated in the theoretical framework. This term drastically alters the long distance behavior of the theory and modifies the allowed solitonic solutions. Specifically, the topologically stable solitonic solutions have electric charge [6–8]. For some Higgs potentials, the minimum energy vortex solutions satisfy the Bogomol'nyi equations [1,9–11]. These self-dual systems have static multisoliton solutions which are localized in different points in space, or superimposed at one point.

In reference to supersymmetric extensions of $(2+1)$-dimensional gauged Chern-Simons models, the fermionic zero modes represent the degeneracy of the solitonic states, while the zero modes of the bosonic fluctuations determine the collective coordinates which describe the solitons and their small velocity kinematics. As is well known [2], in $N=2$ spacetime supersymmetric models, the fermionic zero modes are directly related to the zero modes of bosonic fluctuations. In this presentation (which is based on [15]) we demonstrate that the zero modes of fermionic and bosonic fluctuations of Abelian gauge models having a global $N=2$ spacetime supersymmetry in $(2+1)$-dimensions, can separately constitute two $N=2$, $d=1$ supersymmetric quantum algebras [16–32], with the zero modes being the corresponding quantum Hilbert space vectors. Due to the $N=2$ global spacetime supersymmetry of the theory, the two one-dimensional $N=2$ supersymmetric quantum algebras combine to form a centrally extended $N=4$, $d=1$ supersymmetric quantum algebra. Consequently, the quantum system of fermionic zero modes and of bosonic fluctuations zero modes, constitute an quantum system with centrally extended supersymmetry. This provides us with important information about the bosonic fluctuations subsystem and particularly for the quantum theory of the vortices. This is because the bosonic fluctuations correspond to collective coordinates describing the vortices positions and slow velocity kinematics.

We shall present the models in the limiting case $\kappa=0$, with κ the coupling of the Chern-Simons

[1] Prepared for the Conference Integrable Systems and Quantum Symmetries June 2013 Prague

term. This case, corresponds to the supersymmetric extension of the Landau-Ginzburg model. Therefore, the results could provide us with interesting information, in reference to the quantum Hall effect and to the high-temperature superconductivity [12–14].

The model that we will extensively use in this presentation, following reference [2, 15], are described by the following $N = 2$ spacetime supersymmetric model Lagrangian,

$$\begin{aligned}\mathcal{L} = &-\frac{1}{4}F^{\mu\nu}_{\mu\nu} + \frac{1}{4}\kappa\epsilon^{\mu\nu\lambda}F_{\mu\nu}A_\lambda - |D_\mu\phi|^2 - \frac{1}{2}(\partial_\mu N)^2 \\ &-\frac{1}{2}(e|\phi|^2 + \kappa N - eu^2)^2 - e^2N^2|\phi|^2 \\ &i\bar{\psi}\gamma^\mu D_\mu\psi + i\bar{\chi}\gamma^\mu\partial_\mu\chi + \kappa\bar{\chi}\chi + i\sqrt{2}e(\bar{\psi}\chi\phi - \bar{\chi}\psi\phi^*) + eN\bar{\psi}\psi\end{aligned} \tag{1}$$

where $D_\mu = \partial_\mu - ieA_\mu$ and the gamma matrices satisfy the relation $\gamma^\mu\gamma^\nu = -\eta^{\mu\nu} - i\epsilon^{\mu\nu\lambda}\gamma_\lambda$. The fields ψ, χ are two component spinors, with ψ being a Weyl charged fermionic field and χ a neutral complex Weyl two component spinor. We turn our focus on the fermionic part of the $N = 2$ Lagrangian in order to find the fermionic zero modes, in the background of self dual vortices. The fermionic equations of motion corresponding to Lagrangian (1), are:

$$\begin{aligned}&\gamma^i D_i\psi + ie(\gamma^0A^0 - N)\psi - \sqrt{2}e\phi\chi = 0 \\ &\gamma^i\partial_i\chi - i\kappa\chi + \sqrt{2}e\phi^*\psi = 0\end{aligned} \tag{2}$$

Making the following conventions:

$$\gamma^0 = \sigma_3,\ \gamma^1 = i\sigma_2,\ \gamma^2 = i\sigma_1 \tag{3}$$

and by setting:

$$\psi = \begin{matrix}\psi_\uparrow \\ \psi_\downarrow\end{matrix},\ \chi = \begin{matrix}\chi_\uparrow \\ \chi_\downarrow\end{matrix} \tag{4}$$

and moreover assuming positive values of the flux, the equations of motion for the fermions can be cast in the following form:

$$\begin{aligned}&(D_1 + iD_2)\psi_\downarrow - \sqrt{2}e\phi\chi_\uparrow = 0 \\ &(\partial_1 - i\partial_2)\chi_\uparrow + i\kappa\chi_\downarrow - \sqrt{2}\phi^*\psi_\downarrow = 0 \\ &(D_1 - iD_2)\psi_\uparrow + 2ieA^0\psi_\downarrow + \sqrt{2}e\phi\chi_\downarrow = 0 \\ &(\partial_1 + i\partial_2)\chi_\downarrow - i\kappa\chi_\uparrow + \sqrt{2}e\phi^*\psi_\uparrow = 0\end{aligned} \tag{5}$$

Now, in the $\kappa = 0$ case, the fermionic equations of motion (5) become,

$$\begin{aligned}&(D_1 + iD_2)\psi_\downarrow - \sqrt{2}e\phi\chi_\uparrow = 0 \\ &(\partial_1 - i\partial_2)\chi_\uparrow - \sqrt{2}e\phi^*\psi_\downarrow = 0 \\ &(D_1 - iD_2)\psi_\uparrow + \sqrt{2}e\phi\chi_\downarrow = 0 \\ &(\partial_1 + i\partial_2)\chi_\downarrow + \sqrt{2}e\phi^*\psi_\uparrow = 0\end{aligned} \tag{6}$$

The last two equations of relation (6) have no localized fermionic solutions, apart from some trivial solutions [2]. Conversely, the first two equations of relation (6), have $2n$ solutions, with n the vorticity number [2]. We can form an operator $\mathcal{D}_{LG}$, corresponding to the first two equations of (6),

$$\mathcal{D}_{LG} = \begin{matrix} D_1 + iD_2 & -\sqrt{2}e\phi \\ -\sqrt{2}\phi^* & \partial_1 - i\partial_2\end{matrix} \tag{7}$$

which acts on the vector:

$$|\Psi_{LG}\rangle = \begin{pmatrix} \psi_{\downarrow} \\ \chi_{\uparrow} \end{pmatrix} . \tag{8}$$

Therefore, the first two equations of (6) can be written as:

$$\mathcal{D}_{LG}|\Psi_{LG}\rangle = 0 \tag{9}$$

The solutions of the above equation are the zero modes of the operator $\mathcal{D}_{LG}$, but since the first two equations of (6) have $2n$ normalized solutions, we may conclude that:

$$\dim \ker \mathcal{D}_{LG} = 2n \tag{10}$$

Furthermore, the adjoint of the operator $\mathcal{D}_{LG}$, namely $\mathcal{D}^{\dagger}_{LG}$, is equal to:

$$\mathcal{D}^{\dagger}_{LG} = \begin{pmatrix} D_1 - iD_2 & \sqrt{2}e\phi \\ \sqrt{2}\phi^* & \partial_1 + i\partial_2 \end{pmatrix} \tag{11}$$

and acts on the vector:

$$|\Psi'_{LG}\rangle = \begin{pmatrix} \psi_{\uparrow} \\ \chi_{\downarrow} \end{pmatrix} . \tag{12}$$

The zero modes of the operator $\mathcal{D}^{\dagger}_{LG}$, are identical to the solutions of the last two equations of (6), with the obvious replacement $e \to -e$. Since the corresponding pair of equations has no normalized solutions, the kernel of the operator $\mathcal{D}^{\dagger}_{LG}$ is null:

$$\dim \ker \mathcal{D}^{\dagger}_{LG} = 0 \tag{13}$$

The fermionic system in the self dual vortices background, has an unbroken $N = 2$, $d = 1$ supersymmetry. The supercharges of this $N = 2$, $d = 1$ SUSY algebra are equal to:

$$\mathcal{Q}_{LG} = \begin{pmatrix} 0 & \mathcal{D}_{LG} \\ 0 & 0 \end{pmatrix}, \quad \mathcal{Q}^{\dagger}_{LG} = \begin{pmatrix} 0 & 0 \\ \mathcal{D}^{\dagger}_{LG} & 0 \end{pmatrix} \tag{14}$$

while the quantum Hamiltonian of the quantum system is:

$$\mathcal{H}_{LG} = \begin{pmatrix} \mathcal{D}_{LG}\mathcal{D}^{\dagger}_{LG} & 0 \\ 0 & \mathcal{D}^{\dagger}_{LG}\mathcal{D}_{LG} \end{pmatrix} \tag{15}$$

These elements of the algebra, satisfy the $d = 1$ SUSY algebra:

$$\{\mathcal{Q}_{LG}, \mathcal{Q}^{\dagger}_{LG}\} = \mathcal{H}_{LG} \, , \mathcal{Q}^{2}_{LG} = 0, \; {\mathcal{Q}^{\dagger}_{LG}}^{2} = 0 \tag{16}$$

The Hilbert space of the supersymmetric quantum mechanical system, $\mathcal{H}_{LG}$ is Z_2 graded by the operator $\mathcal{W}$, the Witten parity, an involution operator which commutes with the total Hamiltonian,

$$[\mathcal{W}, \mathcal{H}_{LG}] = 0 \tag{17}$$

and anti-commutes with the supercharges,

$$\{\mathcal{W}, \mathcal{Q}_{LG}\} = \{\mathcal{W}, \mathcal{Q}^{\dagger}_{LG}\} = 0 \tag{18}$$

In addition, the operator $\mathcal{W}$ satisfies the following identity,

$$\mathcal{W}^2 = 1 \tag{19}$$

The Witten parity $\mathcal{W}$, spans the total Hilbert space into Z_2 equivalent subspaces. Therefore, the total Hilbert space of the quantum system can be written as:

$$\mathcal{H} = \mathcal{H}^+ \oplus \mathcal{H}^- \tag{20}$$

with the vectors that belong to the two subspaces $\mathcal{H}^\pm$, classified according to their Witten parity, to even and odd parity states, that is:

$$\mathcal{H}^\pm = \mathcal{P}^\pm \mathcal{H} = \{|\psi\rangle : \mathcal{W}|\psi\rangle = \pm|\psi\rangle\} \tag{21}$$

Moreover, the Hamiltonians of the Z_2 graded spaces are equal to:

$$\mathcal{H}_+ = \mathcal{D}_{LG}\,\mathcal{D}_{LG}^\dagger, \quad \mathcal{H}_- = \mathcal{D}_{LG}^\dagger\,\mathcal{D}_{LG} \tag{22}$$

The operator $\mathcal{W}$ can be represented in the following matrix form:

$$\mathcal{W} = \begin{matrix} 1 & 0 \\ 0 & -1 \end{matrix} \tag{23}$$

The eigenstates of the operator $\mathcal{P}^\pm$, namely $|\psi^\pm\rangle$, satisfy the following relation:

$$P^\pm|\psi^\pm\rangle = \pm|\psi^\pm\rangle \tag{24}$$

We call them positive and negative parity eigenstates, with "parity" referring to the $P^\pm$ operator, which is nothing else but the Witten parity operator. Using the representation (23) for the Witten parity operator, the parity eigenstates can represented by the vectors,

$$|\psi^+\rangle = \begin{matrix} |\phi^+\rangle \\ 0 \end{matrix}, \quad |\psi^-\rangle = \begin{matrix} 0 \\ |\phi^-\rangle \end{matrix} \tag{25}$$

with $|\phi^\pm\rangle \,\epsilon\, \mathcal{H}^\pm$. We can write the fermionic states of the system in terms of the SUSY quantum algebra:

$$|\Psi_{LG}\rangle = |\phi^-\rangle = \begin{matrix} \psi_\downarrow \\ \chi_\uparrow \end{matrix}, \quad |\Psi'_{LG}\rangle = |\phi^+\rangle = \begin{matrix} \psi_\uparrow \\ \chi_\downarrow \end{matrix} \tag{26}$$

Thereby, the even and odd parity SUSY quantum states are:

$$|\psi^+\rangle = \begin{matrix} |\Psi'_{LG}\rangle \\ 0 \end{matrix}, \quad |\psi^-\rangle = \begin{matrix} 0 \\ |\Psi_{LG}\rangle \end{matrix} \tag{27}$$

upon which, the Hamiltonian and the supercharges act. Supersymmetry is considered unbroken if the Witten index is a non-zero integer. The Witten index for Fredholm operators is defined to be:

$$\Delta = n_- - n_+ \tag{28}$$

with $n_\pm$ the number of zero modes of $\mathcal{H}_\pm$ in the subspace $\mathcal{H}^\pm$, with the constraint that these are finitely many.
The case for which the Witten index is zero and also if $n_+ = n_- = 0$, then supersymmetry is broken. Conversely, if $n_+ = n_- \neq 0$, the system has still an unbroken supersymmetry.
The Witten index is directly related to the Fredholm index of the operator $\mathcal{D}_{LG}$, as can be seen in the following equations:

$$\Delta = \dim\ker\mathcal{H}_- - \dim\ker\mathcal{H}_+ = \dim\ker\mathcal{D}_{LG}^\dagger\mathcal{D}_{LG} - \dim\ker\mathcal{D}_{LG}\mathcal{D}_{LG}^\dagger = \tag{29}$$
$$\mathrm{ind}\mathcal{D}_{LG} = \dim\ker\mathcal{D}_{LG} - \dim\ker\mathcal{D}_{LG}^\dagger$$

Recalling the results of equations (10) and (13), the Witten index is equal to:

$$\Delta = -2n \tag{30}$$

Therefore, the fermionic system in the self-dual Landau-Ginzburg vortices background with $N = 2$ spacetime supersymmetry, has an unbroken $N = 2$, $d = 1$ supersymmetry. This could be owing to the fact that the initial system has an unbroken $N = 2$ spacetime supersymmetry, so the Hilbert space of the zero modes states retains a $N = 2$, $d = 1$ supersymmetric quantum algebra. However this is not true. Supersymmetric algebras with equal number of supercharges, but in different dimensions are a different concept. For example, the spacetime supersymmetric algebra is four dimensional while the SUSY quantum mechanics algebra is one dimensional. There is an obvious correlation between the two supersymmetries, since extended (with $N = 4, 6...$) supersymmetric quantum mechanics models can describe the dimensional reduction to one (temporal) dimension of N = 2 and N = 1 Super-Yang Mills models [34, 35]. However, the $N = 2$, $d = 1$ SUSY quantum mechanics supercharges do not generate spacetime supersymmetry. In addition, the supersymmetry in supersymmetric quantum mechanics doesn't relates fermions and bosons, but SUSY QM supercharges classify the Hilbert space of quantum states according to the group Z_2. Moreover, the supercharges generate transformations between the Witten parity eigenstates, which are two orthogonal eigenstates of the quantum Hamiltonian. Finally, the supersymmetric quantum mechanics algebra is not a spacetime supersymmetry, but a quantum algebra that some of the representations of the super-Poincare algebra obey.

The bosonic zero modes equations for $\kappa = 0$ and $A^0 = N = 0$ are cast in the following form:

$$\begin{aligned} &(D_1 + iD_2)\delta\phi - ie\phi(\delta A_1 + i\delta A_2) = 0 \\ &(\partial_1 - \partial_2)(\delta A_1 + i\delta A_2) + 2ie\phi^*\delta\phi = 0 \end{aligned} \tag{31}$$

Note that the above equations become identical to equations (6), if we substitute:

$$\psi_\downarrow = \delta\phi, \;\; \chi_\uparrow = \frac{i}{\sqrt{2}}(\delta A_1 + i\delta A_2) \tag{32}$$

Therefore, we conclude that the number of the zero modes corresponding to equation (31), is equal to $2n$. As in the fermionic case, an unbroken $N = 2$ SUSY quantum mechanical algebra underlies the bosonic system as well. Indeed, equations (31) can be written in the following form:

$$\mathcal{D}'_{LG}|\Phi_{LG}\rangle = 0 \tag{33}$$

with the operator $\mathcal{D}'_{LG}$ being equal to:

$$\mathcal{D}'_{LG} = \begin{pmatrix} D_1 + iD_2 & -\sqrt{2}e\phi \\ -\sqrt{2}\phi^* & \partial_1 - i\partial_2 \end{pmatrix} \tag{34}$$

and is considered to act on the vector:

$$|\Phi_{LG}\rangle = \begin{pmatrix} \delta\phi \\ \frac{i}{\sqrt{2}}(\delta A_1 + i\delta A_2) \end{pmatrix}. \tag{35}$$

Hence:

$$\dim \ker \mathcal{D}'_{LG} = 2n \tag{36}$$

As in the fermionic case,

$$\dim \ker {\mathcal{D}'_{LG}}^\dagger = 0 \tag{37}$$

Journal of Physics: Conference Series **474** (2013) 012025 doi:10.1088/1742-6596/474/1/012025

The operators $\mathcal{D}'_{LG}$ and $\mathcal{D}'_{LG}{}^{\dagger}$ are Fredholm, and therefore any which operator constructed from these, is also Fredholm. The supercharges and the Hamiltonian, that constitute the $N = 2$, $d = 1$ algebra in the bosonic case, are:

$$\mathcal{Q}'_{LG} = \begin{pmatrix} 0 & \mathcal{D}'_{LG} \\ 0 & 0 \end{pmatrix}, \quad \mathcal{Q}'^{\dagger}{}_{LG} = \begin{pmatrix} 0 & 0 \\ \mathcal{D}'^{\dagger}_{LG} & 0 \end{pmatrix}, \quad \mathcal{H}'_{LG} = \begin{pmatrix} \mathcal{D}_{LG}\mathcal{D}^{\dagger}_{LG} & 0 \\ 0 & \mathcal{D}^{\dagger}_{LG}\mathcal{D}_{LG} \end{pmatrix} \tag{38}$$

The elements of the algebra, satisfy the $d = 1$ SUSY algebra:

$$\{\mathcal{Q}'_{LG}, \mathcal{Q}'^{\dagger}_{LG}\} = \mathcal{H}'_{LG}\,, \mathcal{Q}^2_{LG} = 0, \;\; \mathcal{Q}^{\dagger}_{LG}{}^{2} = 0 \tag{39}$$

Supersymmetry is unbroken, since the corresponding Witten index Δ' is a non-zero integer, in this case too. Indeed:

$$\Delta' = -2n \tag{40}$$

Hence, we can see that an unbroken $N = 2$ SUSY quantum mechanics algebra underlies the bosonic fluctuations in the self dual Landau-Ginzburg vortex background. This algebra is identical to the fermionic SUSY quantum mechanics algebra. It is reasonable to ask if there is a more rich symmetry structure in the fermion-boson system, apart from these two $N = 2$ algebras. The answer is yes. Indeed, we can see this if we calculate the following commutation and anti-commutation relations:

$$\begin{aligned}
&\{\mathcal{Q}'_{LG}, \mathcal{Q}'_{LG}{}^{\dagger}\} = 2\mathcal{H},\ \{\mathcal{Q}_{LG}, \mathcal{Q}_{LG}{}^{\dagger}\} = 2\mathcal{H},\ \{\mathcal{Q}_{LG}, \mathcal{Q}_{LG}\} = 0,\ \{\mathcal{Q}'_{LG}, \mathcal{Q}'_{LG}\} = 0, \\
&\{\mathcal{Q}_{LG}, \mathcal{Q}'_{LG}{}^{\dagger}\} = \mathcal{Z},\ \{\mathcal{Q}'_{LG}, \mathcal{Q}_{LG}{}^{\dagger}\} = \mathcal{Z}, \\
&\{\mathcal{Q}'_{LG}{}^{\dagger}, \mathcal{Q}'_{LG}{}^{\dagger}\} = 0, \{\mathcal{Q}_{LG}{}^{\dagger}, \mathcal{Q}_{LG}{}^{\dagger}\} = 0,\ \{\mathcal{Q}_{LG}{}^{\dagger}, \mathcal{Q}'_{LG}{}^{\dagger}\} = 0,\ \{\mathcal{Q}_{LG}, \mathcal{Q}'_{LG}\} = 0 \\
&[\mathcal{Q}'_{LG}, \mathcal{Q}_{LG}] = 0, [\mathcal{Q}_{LG}{}^{\dagger}, \mathcal{Q}'_{LG}{}^{\dagger}] = 0,\ [\mathcal{Q}'_{LG}, \mathcal{Q}'_{LG}] = 0\,[\mathcal{Q}'_{LG}{}^{\dagger}, \mathcal{Q}'_{LG}{}^{\dagger}] = 0, \\
&[\mathcal{H}'_{LG}, \mathcal{Q}'_{LG}] = 0,\ [\mathcal{H}'_{LG}, \mathcal{Q}'_{LG}{}^{\dagger}] = 0,\ [\mathcal{H}_{LG}, \mathcal{Q}_{LG}{}^{\dagger}] = 0,\ [\mathcal{H}_{LG}, \mathcal{Q}_{LG}] = 0,
\end{aligned} \tag{41}$$

with $\mathcal{Z}$:

$$\mathcal{Z} = 2\mathcal{H}_{LG} = 2\mathcal{H}'_{LG} \tag{42}$$

The operator $\mathcal{Z}$ commutes with all the elements of the two algebras, namely the supercharges $\mathcal{Q}_{LG}, \mathcal{Q}'_{LG}$, their conjugates $\mathcal{Q}_{LG}{}^{\dagger}, \mathcal{Q}'_{LG}{}^{\dagger}$ and finally the Hamiltonians, $\mathcal{H} = \mathcal{H}_{LG} = \mathcal{H}'_{LG}$. The relations (41) describe an $N = 4$ supersymmetric quantum mechanics algebra with central charge $\mathcal{Z}$, which is the Hamiltonian of each $N = 2$ subsystem. In general, for an $N = 4$ algebra we have:

$$\begin{aligned}
&\{Q_i, Q^{\dagger}_j\} = 2\delta^j_i H + Z_{ij},\ \ i = 1, 2 \\
&\{Q_i, Q_j\} = 0,\ \ \{Q^{\dagger}_i, Q^{\dagger}_j\} = 0
\end{aligned} \tag{43}$$

The algebra (41) possesses two central charges which are $Z_{12} = Z_{21} = \mathcal{Z}$.

Obviously, such a algebraic structure is interesting, since the fermions and the fluctuations of the bosonic fields, that are connected with an $N = 2$ spacetime supersymmetric algebra, in $(2 + 1)$-dimensions, form the Hilbert space of an $N = 4$ $d = 1$ supersymmetric quantum mechanics algebra.

References

[1] Dunne G, *Self-Dual Chern-Simons Theories*, Springer, Berlin 1995

[2] Lee Bum-Hoon ,Lee C, Min H, *Phys.Rev.* D **45**, 4588 (1992)

[3] Jackiw R, Templeton S, *Phys.Rev.* D **23**, 2291 (1981)

[4] Schonfeld J, *Nucl. Phys.* .B **185** , 157 (1981)
[5] Deser S, Jackiw R, Templeton S, , *Phys. Rev. Lett.* **48**, 975 (1982)
[6] Paul S, Khare A, *Phys. Lett.* B **174** , 420 (1986)
[7] de Vega H, Schaposnik F, *Phys. Rev. Lett.* , **56** 2564 (1986)
[8] Jacobs L, Khare A, Kumar C, Paul S, *Int. J. Mod. Phys.* A, **6** , 3441 (1991)
[9] Grignani G, Nardelli G, *Phys. Rev.* D **43** , 1919 (1991)
[10] Jackiw R, Lee K, Weinberg E J, *Phys. Rev.* D **42**, 3488 (1990)
[11] Lee C, Min H, Rim C, *Phys. Rev.* D **43** , 4100 (1991)
[12] Oh P, Park Q-Han, *Phys. Lett.* B **400**, 157 (1997)
[13] Oh P,Park Q-Han, *Phys. Lett.* B **400**, 157 (1997)
[14] Fordy A P, Kulish P P, *Commun. Math. Phys.* **89** , 427 (1983)
[15] Oikonomou V K, *Nucl.Phys.* B **870**, 477 (2013)
[16] Witten E, *Nucl. Phys.* B **249**, 557 (1985)
[17] Junker G, *Supersymmetric Methods in Quantum and Statistical Physics*, Springer, 1996
[18] Combescure M, Gieres F, Kibler M, *J. Phys. A: Math. Gen.* **37**, 10385 (2004)
[19] Gangopadhyaya A, Mallow J V, Rasinariu C, *Supersymmetric Quantum Mechanics, An Introduction*, World Scientific
[20] Kirchberg A, Laenge J D, Wipf A, *Annal.Phys.* **315**, 467 (2005)
[21] Cooper F, Khare A, Sukhatme U, *Phys. Rept.* **251**, 267 (1995)
[22] Bougie J, Gangopadhyaya A, Mallow J, Rasinariu C, *Symmetry* **4**, 452 (2012)
[23] Thaller, Bernd *The Dirac Equation*, Springer 1992
[24] Tanaka T *Int. J. Mod. Phys.* A **27**, 1250102 (2012)
[25] Filho E D, Ricotta R M , *Mod.Phys.Lett.* A **4** , 2283 (1989)
[26] Ndimubandi J, *Mod. Phys. Lett.* A **8**, 429 (1993)
[27] Spiridonov V, *Mod. Phys. Lett.* A **7** , 1241 (1992)
[28] Durand S, *Mod. Phys. Lett.* A **8**, 1795 (1993)
[29] Kuznetsova Z , Toppan F , *Int. J. Mod. Phys.* A **23**, 3947 (2008)
[30] Gavin E J O, Fiedeldey H, Leeb H, Sofianos S A, *Int.J.Mod.Phys.* A **8**, 1437 (1993)
[31] Wosiek J, *Int. J. Mod.Phys.* A 20, 4484 (2005)
[32] Witten E, *Int.J.Mod.Phys.A* **10**, 1247 (1995)
[33] Galperin A S, Ivanov E A, Ogievetsky V I, Sokatchev E S,*Harmonic Superspace*, Cambridge Monographs on Mathematical Physics, 2007
[34] Delduc F, Ivanov E A, *Nucl. Phys.* B **855**, 815 (2012)
[35] Ivanov E, Lechtenfeld O, Sutulin A, *Nucl. Phys.* B **790**, 493 (2008)

Journal of Physics: Conference Series **474** (2013) 012026 doi:10.1088/1742-6596/474/1/012026

Quantum polydisk, quantum ball, and a q-analog of Poincaré's theorem

A Yu Pirkovskii

Faculty of Mathematics, National Research University Higher School of Economics, 7 Vavilova, 117312 Moscow, Russia

E-mail: aupirkovskii@hse.ru

Abstract. The classical Poincaré theorem (1907) asserts that the polydisk $\mathbb{D}^n$ and the ball $\mathbb{B}^n$ in $\mathbb{C}^n$ are not biholomorphically equivalent for $n \geq 2$. Equivalently, this means that the Fréchet algebras $\mathscr{O}(\mathbb{D}^n)$ and $\mathscr{O}(\mathbb{B}^n)$ of holomorphic functions are not topologically isomorphic. Our goal is to prove a noncommutative version of the above result. Given $q \in \mathbb{C} \setminus \{0\}$, we define two noncommutative power series algebras $\mathscr{O}_q(\mathbb{D}^n)$ and $\mathscr{O}_q(\mathbb{B}^n)$, which can be viewed as q-analogs of $\mathscr{O}(\mathbb{D}^n)$ and $\mathscr{O}(\mathbb{B}^n)$, respectively. Both $\mathscr{O}_q(\mathbb{D}^n)$ and $\mathscr{O}_q(\mathbb{B}^n)$ are the completions of the algebraic quantum affine space $\mathscr{O}_q^{\text{reg}}(\mathbb{C}^n)$ w.r.t. certain families of seminorms. In the case where $0 < q < 1$, the algebra $\mathscr{O}_q(\mathbb{B}^n)$ admits an equivalent definition related to L. L. Vaksman's algebra $C_q(\bar{\mathbb{B}}^n)$ of continuous functions on the closed quantum ball. We show that both $\mathscr{O}_q(\mathbb{D}^n)$ and $\mathscr{O}_q(\mathbb{B}^n)$ can be interpreted as Fréchet algebra deformations (in a suitable sense) of $\mathscr{O}(\mathbb{D}^n)$ and $\mathscr{O}(\mathbb{B}^n)$, respectively. Our main result is that $\mathscr{O}_q(\mathbb{D}^n)$ and $\mathscr{O}_q(\mathbb{B}^n)$ are not isomorphic if $n \geq 2$ and $|q| = 1$, but are isomorphic if $|q| \neq 1$.

1. Introduction

Noncommutative geometry is a vast and rapidly growing subject consisting of a number of different branches (noncommutative algebraic geometry, noncommutative differential geometry, noncommutative topology, noncommutative measure theory, etc.). Each of these branches has its own objects of study and its own methods. Nevertheless, all of them share the common unifying "philosophy" that some classical constructions and results known from various fields of geometry and topology can be successfully applied to noncommutative objects, which, at the first glance, have nothing to do with geometry.

The subject of the present paper can be characterized as "noncommutative complex analysis", or "noncommutative complex analytic geometry". At the moment, this theory is much less developed than any of the above-mentioned parts of noncommutative geometry. However, a number of important results have been obtained in this field during the last decade. First of all, let us mention the tremendous work done by L. L. Vaksman's school (see, e.g., [63, 68–71] and references therein), which eventually resulted in the creation of the general theory of quantum bounded symmetric domains. A more operator-theoretic aspect of the subject is reflected in the papers by K. R. Davidson, D. R. Pitts, E. G. Katsoulis, C. Ramsey, and O. Shalit [7–12], F. H. Szafraniec [62], and G. Popescu [42–51]. An algebraic point of view is adopted by A. Polishchuk and A. Schwarz [38–41], M. Khalkhali, G. Landi, W. D. van Suijlekom, and A. Moatadelro [19–21], E. Beggs and S. P. Smith [3], R. Ó Buachalla [28, 29]. All this shows that noncommutative complex analysis and noncommutative complex geometry will hopefully reach their bloom period fairly soon.

Our primary goal is to prove a noncommutative version of a classical theorem by Poincaré [37], which asserts that the polydisk $\mathbb{D}^n$ and the ball $\mathbb{B}^n$ in $\mathbb{C}^n$ are not biholomorphically equivalent for $n \geq 2$. This result is often mentioned in textbooks as one of the first results in function theory of several complex variables (see, e.g., [27, 54]). To understand how a noncommutative version of Poincaré's theorem should look like, let us recall O. Forster's important result [14] whose informal meaning is that all essential information about a domain of holomorphy D in $\mathbb{C}^n$ is contained in the algebra $\mathscr{O}(D)$ of holomorphic functions on D. To be more precise, Forster's theorem states that the functor

$$\Big\{ \text{Stein spaces} \Big\} \to \Big\{ \text{Fréchet algebras} \Big\},$$
$$(X, \mathscr{O}_X) \mapsto \mathscr{O}(X),$$

is fully faithful. As a consequence, two domains of holomorphy D_1 and D_2 in $\mathbb{C}^n$ are biholomorphically equivalent if and only if the algebras $\mathscr{O}(D_1)$ and $\mathscr{O}(D_2)$ are topologically isomorphic. Thus Poincaré's theorem is equivalent to the assertion that the algebras $\mathscr{O}(\mathbb{D}^n)$ and $\mathscr{O}(\mathbb{B}^n)$ are not topologically isomorphic for $n \geq 2$.

Now we can explain our plan in more detail. Let $\mathbb{C}^\times = \mathbb{C} \setminus \{0\}$. For each $q \in \mathbb{C}^\times$ we define q-analogs of the algebras $\mathscr{O}(\mathbb{D}^n)$ and $\mathscr{O}(\mathbb{B}^n)$ to be the completions of the algebraic quantum affine space $\mathscr{O}_q^{\mathrm{reg}}(\mathbb{C}^n)$ with respect to certain families of seminorms. Both the resulting algebras $\mathscr{O}_q(\mathbb{D}^n)$ and $\mathscr{O}_q(\mathbb{B}^n)$ were introduced earlier in [35] (see also [36] for a more detailed treatment of $\mathscr{O}_q(\mathbb{D}^n)$), but the definition of $\mathscr{O}_q(\mathbb{B}^n)$ was given only for $0 < q < 1$. Here we propose a different approach to $\mathscr{O}_q(\mathbb{B}^n)$ which makes sense for all $q \in \mathbb{C}^\times$ and which is equivalent to the approach of [35] in the case where $0 < q < 1$. To justify our definitions, we show that the Fréchet algebra families $\{\mathscr{O}_q(\mathbb{D}^n) : q \in \mathbb{C}^\times\}$ and $\{\mathscr{O}_q(\mathbb{B}^n) : q \in \mathbb{C}^\times\}$ can be arranged into Fréchet algebra bundles over $\mathbb{C}^\times$, generalizing thereby our earlier result from [35]. Our main result, i.e., a q-analog of Poincaré's theorem, states that $\mathscr{O}_q(\mathbb{D}^n)$ and $\mathscr{O}_q(\mathbb{B}^n)$ are not topologically isomorphic if $n \geq 2$ and $|q| = 1$. On the other hand, we show that they are topologically isomorphic if $|q| \neq 1$.

This paper is mostly a survey. The proofs are either sketched or omitted. We plan to present the details elsewhere.

2. Preliminaries

We shall work over the field $\mathbb{C}$ of complex numbers. All algebras are assumed to be associative and unital, and all algebra homomorphisms are assumed to be unital (i.e., to preserve identity elements). By a *Fréchet algebra* we mean a complete metrizable locally convex algebra (i.e., a topological algebra whose underlying space is a Fréchet space). A *locally m-convex algebra* [25] is a topological algebra A whose topology can be defined by a family of submultiplicative seminorms (i.e., seminorms $\|\cdot\|$ satisfying $\|ab\| \leq \|a\|\|b\|$ for all $a, b \in A$). A complete locally m-convex algebra is called an *Arens-Michael algebra* [17]. The algebra of holomorphic functions on a complex manifold X will be denoted by $\mathscr{O}(X)$. Recall that $\mathscr{O}(X)$ is a Fréchet-Arens-Michael algebra with respect to the topology of uniform convergence on compact subsets of X.

3. Quantum affine space

Let $q \in \mathbb{C}^\times$. Recall that the algebra $\mathscr{O}_q^{\mathrm{reg}}(\mathbb{C}^n)$ *of regular functions on the quantum affine n-space* (see, e.g., [6]) is generated by n elements $x_1, \ldots, x_n$ subject to the relations $x_i x_j = q x_j x_i$ for all $i < j$. If $q = 1$, then $\mathscr{O}_q^{\mathrm{reg}}(\mathbb{C}^n)$ is nothing but the polynomial algebra $\mathbb{C}[x_1, \ldots, x_n] = \mathscr{O}^{\mathrm{reg}}(\mathbb{C}^n)$. Of course, $\mathscr{O}_q^{\mathrm{reg}}(\mathbb{C}^n)$ is noncommutative unless $q = 1$, but the monomials $x^k = x_1^{k_1} \cdots x_n^{k_n}$ ($k \in \mathbb{Z}_+^n$) still form a basis of $\mathscr{O}_q^{\mathrm{reg}}(\mathbb{C}^n)$. Thus $\mathscr{O}_q^{\mathrm{reg}}(\mathbb{C}^n)$ may be viewed as a "deformed" polynomial algebra.

The algebras $\mathscr{O}_q(\mathbb{D}^n_r)$ and $\mathscr{O}_q(\mathbb{B}^n_r)$ that we are going to define will be certain completions of $\mathscr{O}^{\mathrm{reg}}_q(\mathbb{C}^n)$. There are many nonequivalent ways to complete this algebra, but, among all the completions, there is a universal one. Recall that the *Arens-Michael envelope*, $\widehat{A}$, of an algebra A is the completion of A with respect to the family of all submultiplicative seminorms on A. The Arens-Michael envelope has the universal property that, for each Arens-Michael algebra B, there is a 1-1 correspondence

$$\mathrm{Hom}_{\mathsf{Alg}}(A, B) \cong \mathrm{Hom}_{\mathsf{AM}}(\widehat{A}, B),$$

where Alg is the category of algebras and AM is the category of Arens-Michael algebras. Moreover, the assignment $A \mapsto \widehat{A}$ is a functor from Alg to AM, and this functor is left adjoint to the forgetful functor $\mathsf{AM} \to \mathsf{Alg}$.

Arens-Michael envelopes were introduced by J. L. Taylor [64] under the name of "completed locally m-convex envelopes". Now it is customary to call them "Arens-Michael envelopes", following the terminology suggested by A. Ya. Helemskii [17]. As was observed in [33], the Arens-Michael envelope of a finitely generated algebra is a nuclear Fréchet algebra.

Example 3.1 ([65]). If $A = \mathbb{C}[x_1, \dots, x_n]$ is the polynomial algebra, then $\widehat{A} = \mathscr{O}(\mathbb{C}^n)$, the algebra of entire functions on $\mathbb{C}^n$.

Example 3.2 ([34]). If $(X, \mathscr{O}^{\mathrm{reg}}_X)$ is an affine scheme of finite type over $\mathbb{C}$, and if $A = \mathscr{O}^{\mathrm{reg}}(X)$, then $\widehat{A} = \mathscr{O}(X_h)$, where $(X_h, \mathscr{O}_{X_h})$ is the complex space associated to $(X, \mathscr{O}^{\mathrm{reg}}_X)$ (cf. [16, Appendix B]).

Using these examples as a motivation, we defined [34] the algebra $\mathscr{O}_q(\mathbb{C}^n)$ of *holomorphic functions on the quantum affine n-space* to be the Arens-Michael envelope of $\mathscr{O}^{\mathrm{reg}}_q(\mathbb{C}^n)$. This algebra can also be described explicitly as follows. Define a function $w_q \colon \mathbb{Z}^n_+ \to \mathbb{R}_+$ by

$$w_q(k) = \begin{cases} 1 & \text{if } |q| \ge 1, \\ |q|^{\sum_{i<j} k_i k_j} & \text{if } |q| < 1. \end{cases}$$

As was shown in [34], we have

$$\mathscr{O}_q(\mathbb{C}^n) = \Big\{ a = \sum_{k \in \mathbb{Z}^n_+} c_k x^k : \|a\|_\rho = \sum_{k \in \mathbb{Z}^n_+} |c_k| w_q(k) \rho^{|k|} < \infty \ \forall \rho > 0 \Big\}, \tag{1}$$

where $|k| = k_1 + \dots + k_n$ for $k = (k_1, \dots, k_n) \in \mathbb{Z}^n_+$. The topology on $\mathscr{O}_q(\mathbb{C}^n)$ is given by the norms $\|\cdot\|_\rho$ $(\rho > 0)$. Moreover, each norm $\|\cdot\|_\rho$ is submultiplicative.

4. Quantum polydisk and quantum ball

The explicit construction (1) of $\mathscr{O}_q(\mathbb{C}^n)$ leads naturally to the following definition.

Definition 4.1 ([35, 36]). Let $q \in \mathbb{C}^\times$, and let $r > 0$. We define the *algebra of holomorphic functions on the quantum n-polydisk of radius* $r \in (0, +\infty]$ by

$$\mathscr{O}_q(\mathbb{D}^n_r) = \Big\{ a = \sum_{k \in \mathbb{Z}^n_+} c_k x^k : \|a\|_\rho = \sum_{k \in \mathbb{Z}^n_+} |c_k| w_q(k) \rho^{|k|} < \infty \ \forall \rho \in (0, r) \Big\}. \tag{2}$$

The multiplication on $\mathscr{O}_q(\mathbb{D}^n_r)$ is uniquely determined by $x_i x_j = q x_j x_i$ $(i < j)$.

It follows from the above discussion that $\mathscr{O}_q(\mathbb{D}^n_r)$ is a Fréchet-Arens-Michael algebra with respect to the topology determined by the submultiplicative norms $\|\cdot\|_\rho$ $(\rho \in (0, r))$. It is a simple exercise to show that, if $q = 1$, then $\mathscr{O}_q(\mathbb{D}^n_r)$ is topologically isomorphic to the algebra $\mathscr{O}(\mathbb{D}^n_r)$ of holomorphic functions on the polydisk

$$\mathbb{D}^n_r = \{z = (z_1, \ldots, z_n) \in \mathbb{C}^n : \max_{1\le i\le n} |z_i| < r\}.$$

If $r = \infty$, then we clearly have $\mathscr{O}_q(\mathbb{D}^n_r) = \mathscr{O}_q(\mathbb{C}^n)$.

The definition of the algebra of holomorphic functions on the quantum ball is less straightforward. It is based on a theorem by L. A. Aizenberg and B. S. Mityagin [1]. Recall that a domain $D \subset \mathbb{C}^n$ is a *complete Reinhardt domain* if for each $z = (z_1, \ldots, z_n) \in D$ and each $(\lambda_1, \ldots, \lambda_n) \in \mathbb{C}^n$ satisfying $|\lambda_i| \le 1$ $(i = 1, \ldots, n)$ we have $(\lambda_1 z_1, \ldots, \lambda_n z_n) \in D$. Clearly, the polydisk $\mathbb{D}^n_r$ and the ball

$$\mathbb{B}^n_r = \Big\{z = (z_1, \ldots, z_n) \in \mathbb{C}^n : \sum_{i=1}^n |z_i|^2 < r^2\Big\}$$

are complete Reinhardt domains.

Given a complete bounded Reinhardt domain $D \subset \mathbb{C}^n$, let

$$b_k(D) = \sup_{z\in D} |z^k| \qquad (k \in \mathbb{Z}^n_+).$$

Aizenberg and Mityagin proved that there exists a topological isomorphism

$$\mathscr{O}(D) \cong \Big\{f = \sum_{k\in\mathbb{Z}^n_+} c_k z^k : \|f\|_s = \sum_{k\in\mathbb{Z}^n_+} |c_k| b_k(D) s^{|k|} < \infty \;\forall s \in (0,1)\Big\}.$$

Explicitly, the above isomorphism takes each function $f \in \mathscr{O}(D)$ to its Taylor expansion at 0.

We clearly have $b_k(\mathbb{D}^n_r) = r^{|k|}$. An explicit calculation involving Lagrange's multipliers shows that

$$b_k(\mathbb{B}^n_r) = \left(\frac{k^k}{|k|^{|k|}}\right)^{1/2} r^{|k|}.$$

Therefore

$$\mathscr{O}(\mathbb{B}^n_r) \cong \Big\{f = \sum_{k\in\mathbb{Z}^n_+} c_k z^k : \|f\|_\rho = \sum_{k\in\mathbb{Z}^n_+} |c_k| \left(\frac{k^k}{|k|^{|k|}}\right)^{1/2} \rho^{|k|} < \infty \;\forall \rho \in (0,r)\Big\}. \tag{3}$$

Now we have to quantize the above algebra. To this end, it will be convenient to represent $\mathscr{O}(\mathbb{B}^n_r)$ in a slightly different way.

Proposition 4.2. *There exists a topological isomorphism*

$$\mathscr{O}(\mathbb{B}^n_r) \cong \Big\{f = \sum_{k\in\mathbb{Z}^n_+} c_k z^k : \|f\|_\rho = \sum_{k\in\mathbb{Z}^n_+} |c_k| \left(\frac{k!}{|k|!}\right)^{1/2} \rho^{|k|} < \infty \;\forall \rho \in (0,r)\Big\}. \tag{4}$$

The proof is a simple application of Stirling's formula.

The power series representation (4) of $\mathscr{O}(\mathbb{B}^n_r)$ is more convenient to us than (3) because there is a standard way to quantize the factorial. For each $k \in \mathbb{N}$, let

$$[k]_q = 1 + q + \cdots + q^{k-1}; \quad [k]_q! = [1]_q [2]_q \cdots [k]_q.$$

It is also convenient to let $[0]_q! = 1$. Finally, given $k = (k_1, \ldots, k_n) \in \mathbb{Z}^n_+$, we let $[k]_q! = [k_1]_q! \cdots [k_n]_q!$.

Definition 4.3. The *space of holomorphic functions on the quantum n-ball* of radius $r \in (0, +\infty]$ is

$$\mathscr{O}_q(\mathbb{B}^n_r) = \left\{ a = \sum_{k \in \mathbb{Z}^n_+} c_k x^k : \|a\|_\rho = \sum_{k \in \mathbb{Z}^n_+} |c_k| \left(\frac{[k]_{|q|^{-2}}!}{\big[|k|\big]_{|q|^{-2}}!} \right)^{1/2} \rho^{|k|} < \infty \ \forall \rho \in (0, r) \right\}. \tag{5}$$

It is immediate from the definition that $\mathscr{O}_q(\mathbb{B}^n_r)$ is a Fréchet space with respect to the topology determined by the norms $\|\cdot\|_\rho$ $(\rho \in (0, r))$. However, it is not obvious at all whether it is an algebra. In fact it is:

Proposition 4.4. *The Fréchet space $\mathscr{O}_q(\mathbb{B}^n_r)$ is an Arens-Michael algebra with respect to the multiplication uniquely determined by $x_i x_j = q x_j x_i \quad (i < j)$. Moreover, each norm $\|\cdot\|_\rho$ given by (5) is submultiplicative.*

The proof is based on a q-analog of the Chu-Vandermonde formula (see, e.g., [23, 2.1.2, Proposition 3]).

If $q = 1$, then Proposition 4.2 implies that $\mathscr{O}_q(\mathbb{B}^n_r) \cong \mathscr{O}(\mathbb{B}^n_r)$. It can also be shown that, if $r = \infty$, then $\mathscr{O}_q(\mathbb{B}^n_r) = \mathscr{O}_q(\mathbb{C}^n)$ (although this is not that obvious as in the case of the quantum polydisk algebra).

Proposition 4.5. *For each $q \in \mathbb{C}^\times$ and each $r \in (0, +\infty]$ there exist Fréchet algebra isomorphisms*

$$\mathscr{O}_q(\mathbb{D}^n_r) \cong \mathscr{O}_{q^{-1}}(\mathbb{D}^n_r), \quad x_i \mapsto x_{n-i};$$
$$\mathscr{O}_q(\mathbb{B}^n_r) \cong \mathscr{O}_{q^{-1}}(\mathbb{B}^n_r), \quad x_i \mapsto x_{n-i}.$$

The idea of the proof of Proposition 4.5 is as follows. Clearly, there is an algebra isomorphism $\tau \colon \mathscr{O}_q^{\mathrm{reg}}(\mathbb{C}^n) \to \mathscr{O}_{q^{-1}}^{\mathrm{reg}}(\mathbb{C}^n)$ taking each x_i to x_{n-i}. An explicit calculation shows that τ is isometric with respect to each norm $\|\cdot\|_\rho$, both on $\mathscr{O}_q(\mathbb{D}^n_r)$ and $\mathscr{O}_q(\mathbb{B}^n_r)$. The rest is clear.

5. Quantum ball à la Vaksman

In the special case where $0 < q < 1$, $\mathscr{O}_q(\mathbb{B}^n_r)$ is closely related to L. L. Vaksman's q-analog of $A(\bar{\mathbb{B}}^n)$, the algebra of functions holomorphic on the open unit ball $\mathbb{B}^n = \mathbb{B}^n_1$ and continuous on the closed ball $\bar{\mathbb{B}}^n$ [70]. Let us recall how Vaksman's algebra is defined. Assume that $0 < q < 1$, and let $\mathrm{Pol}_q(\mathbb{C}^n)$ denote the $*$-algebra generated (as a $*$-algebra) by n elements $x_1, \ldots, x_n$ subject to the relations

$$\begin{aligned} x_i x_j &= q x_j x_i \quad (i < j); \\ x_i^* x_j &= q x_j x_i^* \quad (i \neq j); \\ x_i^* x_i &= q^2 x_i x_i^* + (1 - q^2)\Big(1 - \sum_{k > i} x_k x_k^*\Big). \end{aligned} \tag{6}$$

Clearly, for $q = 1$ we have $\mathrm{Pol}_q(\mathbb{C}^n) = \mathrm{Pol}(\mathbb{C}^n)$, where $\mathrm{Pol}(\mathbb{C}^n)$ is the algebra of polynomials in the complex coordinates $z_1, \ldots, z_n$ on $\mathbb{C}^n$ and their complex conjugates $\bar{z}_1, \ldots, \bar{z}_n$. Observe that the subalgebra of $\mathrm{Pol}_q(\mathbb{C}^n)$ generated (as an algebra) by $x_1, \ldots, x_n$ is exactly $\mathscr{O}_q^{\mathrm{reg}}(\mathbb{C}^n)$. The algebra $\mathrm{Pol}_q(\mathbb{C}^n)$ was introduced by W. Pusz and S. L. Woronowicz [53], although they used different $*$-generators $a_1, \ldots, a_n$ given by $a_i = (1 - q^2)^{-1/2} x_i^*$. Relations (6) divided by $1 - q^2$ and written in terms of the a_i's are called the "twisted canonical commutation relations", and the algebra $A_q = \mathrm{Pol}_q(\mathbb{C}^n)$ defined in terms of the a_i's is sometimes called the "quantum Weyl algebra" (see, e.g., [2, 18, 23, 72]). Note that, while $\mathrm{Pol}_q(\mathbb{C}^n)$ becomes $\mathrm{Pol}(\mathbb{C}^n)$ for $q = 1$, A_q becomes the Weyl algebra. The idea to use the generators x_i instead of the a_i's and to consider $\mathrm{Pol}_q(\mathbb{C}^n)$ as a q-analog of $\mathrm{Pol}(\mathbb{C}^n)$ is probably due to Vaksman [67]; the one-dimensional case was considered in [22]. The algebra $\mathrm{Pol}_q(\mathbb{C}^n)$ serves as a basic example in the general theory of quantum bounded symmetric domains developed by Vaksman and his collaborators (see [69,71] and references therein).

Now let H be a Hilbert space with an orthonormal basis $\{e_k : k \in \mathbb{Z}_+^n\}$. The algebra of bounded linear operators on H will be denoted by $\mathscr{B}(H)$. Following [53], for each $k = (k_1, \ldots, k_n) \in \mathbb{Z}_+^n$ we will write $|k_1, \ldots, k_n\rangle$ for e_k. As was proved by Pusz and Woronowicz [53], there exists a faithful irreducible $*$-representation $\pi\colon \mathrm{Pol}_q(\mathbb{C}^n) \to \mathscr{B}(H)$ uniquely determined by

$$\pi(x_j)e_k = \sqrt{1-q^2}\sqrt{[k_j+1]_{q^2}}\, q^{\sum_{i>j} k_i} |k_1, \ldots, k_j+1, \ldots, k_n\rangle$$
$$(j = 1, \ldots, n,\ k = (k_1, \ldots, k_n) \in \mathbb{Z}_+^n).$$

The completion of $\mathrm{Pol}_q(\mathbb{C}^n)$ with respect to the operator norm $\|a\|_{\mathrm{op}} = \|\pi(a)\|$ is denoted by $C_q(\bar{\mathbb{B}}^n)$ and is called the *algebra of continuous functions on the closed quantum ball* [70]; see also [52, 53]. The closure of $\mathscr{O}_q^{\mathrm{reg}}(\mathbb{C}^n)$ in $C_q(\bar{\mathbb{B}}^n)$ is denoted by $A_q(\bar{\mathbb{B}}^n)$ [70]; this is a natural q-analog of the algebra $A(\bar{\mathbb{B}}^n)$, which consists of those continuous functions on $\bar{\mathbb{B}}^n$ that are holomorphic on $\mathbb{B}^n$.

For each $\rho > 0$, let γ_ρ be the automorphism of $\mathscr{O}_q^{\mathrm{reg}}(\mathbb{C}^n)$ uniquely determined by $\gamma_\rho(x_i) = \rho x_i$ $(i = 1, \ldots, n)$. Define a norm $\|\cdot\|_\rho^\infty$ on $\mathscr{O}_q^{\mathrm{reg}}(\mathbb{C}^n)$ by

$$\|a\|_\rho^\infty = \|\gamma_\rho(a)\|_{\mathrm{op}} \qquad (a \in \mathscr{O}_q^{\mathrm{reg}}(\mathbb{C}^n)).$$

According to Vaksman's point of view, $\|\cdot\|_{\mathrm{op}}$ is a natural q-analog of the supremum norm over $\bar{\mathbb{B}}^n$. Therefore our $\|\cdot\|_\rho^\infty$ is a q-analog of the supremum norm over $\bar{\mathbb{B}}_\rho^n$. It is well known that $\mathscr{O}^{\mathrm{reg}}(\mathbb{C}^n) = \mathbb{C}[z_1, \ldots, z_n]$ is dense in $\mathscr{O}(\mathbb{D}_r^n)$; in other words, the completion of $\mathscr{O}^{\mathrm{reg}}(\mathbb{C}^n)$ with respect to the family $\{\|\cdot\|_\rho^\infty : 0 < \rho < r\}$ of norms is topologically isomorphic to $\mathscr{O}(\mathbb{B}_r^n)$. This result has the following q-analog.

Theorem 5.1. *For each $q \in (0,1)$ and each $r \in (0,+\infty]$, the completion of $\mathscr{O}_q^{\mathrm{reg}}(\mathbb{C}^n)$ with respect to the family $\{\|\cdot\|_\rho^\infty : 0 < \rho < r\}$ of norms is topologically isomorphic to $\mathscr{O}_q(\mathbb{B}_r^n)$.*

Thus we see that our definition of $\mathscr{O}_q(\mathbb{B}_r^n)$ is consistent with the definition given in [35].

6. Free polydisk and free ball

Let $F_n = \mathbb{C}\langle\zeta_1, \ldots, \zeta_n\rangle$ denote the free algebra on n generators $\zeta_1, \ldots, \zeta_n$. Clearly, $\mathscr{O}_q^{\mathrm{reg}}(\mathbb{C}^n)$ is nothing but the quotient of F_n modulo the two-sided ideal generated by the elements $\zeta_i\zeta_j - q\zeta_j\zeta_i$ $(i < j)$. Our next goal is to represent the algebras $\mathscr{O}_q(\mathbb{D}_r^n)$ and $\mathscr{O}_q(\mathbb{B}_r^n)$ in a similar way, i.e., as quotients of certain "analytic analogs" of F_n. This will enable us to interpret $\mathscr{O}_q(\mathbb{D}_r^n)$ and $\mathscr{O}_q(\mathbb{B}_r^n)$ as "deformations" (in a suitable sense) of $\mathscr{O}(\mathbb{D}_r^n)$ and $\mathscr{O}(\mathbb{B}_r^n)$, respectively.

Let us introduce some notation. For each $d \in \mathbb{Z}_+$, let $W_{n,d} = \{1, \ldots, n\}^d$, and let $W_n = \bigsqcup_{d\in\mathbb{Z}_+} W_{n,d}$. Thus a typical element of W_n is a d-tuple $\alpha = (\alpha_1, \ldots, \alpha_d)$ of arbitrary length $d \in \mathbb{Z}_+$, where $\alpha_j \in \{1, \ldots, n\}$ for all j. Given $\alpha \in W_{n,d} \subset W_n$, we let $|\alpha| = d$. The only element of $W_{n,0}$ will be denoted by $*$. For each $\alpha = (\alpha_1, \ldots, \alpha_d) \in W_n$ with $d > 0$, let $\zeta_\alpha = \zeta_{\alpha_1} \cdots \zeta_{\alpha_d} \in F_n$. It is also convenient to set $\zeta_* = 1 \in F_n$. The family $\{\zeta_\alpha : \alpha \in W_n\}$ of all words in $\zeta_1, \ldots, \zeta_n$ is the standard vector space basis of F_n.

Recall from [36] (see also [35]) that each family $(A_i)_{i\in I}$ of Arens-Michael algebras has a coproduct $\widehat{*}_{i\in I} A_i$ in the category AM of Arens-Michael algebras. The algebra $\widehat{*}_{i\in I} A_i$ is called the *Arens-Michael free product* of the A_i's. By definition [35, 36], the *algebra of holomorphic functions on the free n-polydisk of radius* $r \in (0,+\infty]$ is

$$\mathscr{F}(\mathbb{D}_r^n) = \mathscr{O}(\mathbb{D}_r) \mathbin{\widehat{*}} \cdots \mathbin{\widehat{*}} \mathscr{O}(\mathbb{D}_r).$$

The algebra $\mathscr{F}(\mathbb{D}_r^n)$ can also be described more explicitly as follows. Let $\mathfrak{F}_n$ denote the algebra of all free formal series $a = \sum_{\alpha \in W_n} c_\alpha \zeta_\alpha$ (where $c_\alpha \in \mathbb{C}$) with the obvious multiplication. In other words, $\mathfrak{F}_n = \varprojlim_d F_n/I^d$, where I is the ideal of F_n generated by $\zeta_1, \dots, \zeta_n$. Given $d \geq 2$ and $\alpha = (\alpha_1, \dots, \alpha_d) \in W_{n,d}$, let $s(\alpha)$ denote the cardinality of the set

$$\{i \in \{1, \dots, d-1\} : \alpha_i \neq \alpha_{i+1}\}.$$

If $d \in \{0, 1\}$, we let $s(\alpha) = d - 1$. By [36, Proposition 7.8], we have

$$\mathscr{F}(\mathbb{D}_r^n) = \Big\{a = \sum_{\alpha \in W_n} c_\alpha \zeta_\alpha \in \mathfrak{F}_n : \|a\|_{\rho,\tau} = \sum_{\alpha \in W_n} |c_\alpha| \rho^{|\alpha|} \tau^{s(\alpha)+1} < \infty \ \forall \rho \in (0,r), \ \forall \tau \geq 1\Big\}. \tag{7}$$

The topology on $\mathscr{F}(\mathbb{D}_r^n)$ is given by the norms $\|\cdot\|_{\rho,\tau}$ ($\rho \in (0,r)$, $\tau \geq 1$), and the multiplication is given by concatenation. Moreover, each norm $\|\cdot\|_{\rho,\tau}$ is submultiplicative.

Another natural candidate for the algebra of holomorphic functions on the free polydisk was introduced by J. L. Taylor [65, 66]. By definition,

$$\mathscr{F}^T(\mathbb{D}_r^n) = \Big\{a = \sum_{\alpha \in W_n} c_\alpha \zeta_\alpha \in \mathfrak{F}_n : \|a\|_\rho = \sum_{\alpha \in W_n} |c_\alpha| \rho^{|\alpha|} < \infty \ \forall \rho \in (0,r)\Big\}. \tag{8}$$

It is easy to see that $\mathscr{F}(\mathbb{D}_r^n) \subset \mathscr{F}^T(\mathbb{D}_r^n)$, and that the inclusion of $\mathscr{F}(\mathbb{D}_r^n)$ into $\mathscr{F}^T(\mathbb{D}_r^n)$ is continuous. On the other hand, $\mathscr{F}(\mathbb{D}_r^n) \neq \mathscr{F}^T(\mathbb{D}_r^n)$ unless $n = 1$ or $r = \infty$. Note also that, if $r = \infty$, then both $\mathscr{F}(\mathbb{D}_r^n)$ and $\mathscr{F}^T(\mathbb{D}_r^n)$ are topologically isomorphic to $\mathscr{F}(\mathbb{C}^n)$, the Arens-Michael envelope of F_n.

Theorem 6.1. *Let $q \in \mathbb{C}^\times$, $n \in \mathbb{N}$, and $r \in (0, +\infty]$.*

(i) *The algebra $\mathscr{O}_q(\mathbb{D}_r^n)$ is topologically isomorphic to the quotient of $\mathscr{F}(\mathbb{D}_r^n)$ modulo the closed two-sided ideal generated by the elements $\zeta_i\zeta_j - q\zeta_j\zeta_i$ ($i < j$). Moreover, for each $\rho \in (0,r)$ and each $\tau \geq 1$ the norm $\|\cdot\|_\rho$ on $\mathscr{O}_q(\mathbb{D}_r^n)$ given by (2) is equal to the quotient of the norm $\|\cdot\|_{\rho,\tau}$ on $\mathscr{F}(\mathbb{D}_r^n)$ given by (7).*

(ii) *The algebra $\mathscr{O}_q(\mathbb{D}_r^n)$ is topologically isomorphic to the quotient of $\mathscr{F}^T(\mathbb{D}_r^n)$ modulo the closed two-sided ideal generated by the elements $\zeta_i\zeta_j - q\zeta_j\zeta_i$ ($i < j$). Moreover, for each $\rho \in (0,r)$ the norm $\|\cdot\|_\rho$ on $\mathscr{O}_q(\mathbb{D}_r^n)$ given by (2) is equal to the quotient of the norm $\|\cdot\|_\rho$ on $\mathscr{F}^T(\mathbb{D}_r^n)$ given by (8).*

Part (i) of Theorem 6.1, except for the equality of the norms, was proved in [36, Theorem 7.13] in the more general multiparameter case. Part (ii) is new.

To formulate a similar result for $\mathscr{O}_q(\mathbb{B}_r^n)$, we need G. Popescu's algebra of "holomorphic functions on the free ball" [44]. Let H be a Hilbert space, and let $T = (T_1, \dots, T_n)$ be an n-tuple of bounded linear operators on H. Following [44], we identify T with the "row" operator acting from the Hilbert direct sum $H^n = H \oplus \cdots \oplus H$ to H. Thus we have $\|T\| = \|\sum_{i=1}^n T_i T_i^*\|^{1/2}$. For each $a = \sum_\alpha c_\alpha \zeta_\alpha \in \mathfrak{F}_n$, the *radius of convergence* $R(a) \in [0, +\infty]$ is given by the Cauchy-Hadamard-type formula

$$\frac{1}{R(a)} = \limsup_{d \to \infty} \Big(\sum_{|\alpha|=d} |c_\alpha|^2\Big)^{\frac{1}{2d}}.$$

By [44, Theorem 1.1], for each $T \in \mathscr{B}(H)^n$ such that $\|T\| < R(a)$, the series

$$\sum_{d=0}^{\infty} \Big(\sum_{|\alpha|=d} c_\alpha T_\alpha\Big) \tag{9}$$

converges in $\mathscr{B}(H)$ and, moreover, $\sum_d \| \sum_{|\alpha|=d} c_\alpha T_\alpha \| < \infty$. On the other hand, if H is infinite-dimensional, then for each $R' > R(a)$ there exists $T \in \mathscr{B}(H)^n$ with $\|T\| = R'$ such that the series (9) diverges. This "free operator analog" of the classical Hadamard lemma explains why the radius of convergence is so called.

By [44, Theorem 1.4], the collection of all $a \in \mathfrak{F}_n$ such that $R(a) \ge r$ is a subalgebra of $\mathfrak{F}_n$. We denote this algebra by $\mathscr{F}(\mathbb{B}^n_r)$ (Popescu uses the notation $Hol(\mathscr{B}(\mathcal{H})^n_r)$), and we call it the *algebra of holomorphic functions on the free n-ball of radius r* [35]. For each $a \in \mathscr{F}(\mathbb{B}^n_r)$, each Hilbert space H, and each $T \in \mathscr{B}(H)^n$ with $\|T\| < r$, the sum of the series (9) is denoted by $a(T)$. The map

$$\mathscr{F}(\mathbb{B}^n_r) \to \mathscr{B}(H), \qquad a \mapsto a(T),$$

is an algebra homomorphism.

Fix an infinite-dimensional Hilbert space $\mathcal{H}$, and, for each $\rho \in (0, r)$, define a seminorm $\| \cdot \|^P_\rho$ on $\mathscr{F}(\mathbb{B}^n_r)$ by

$$\|a\|^P_\rho = \sup\{\|a(T)\| : T \in \mathscr{B}(\mathcal{H})^n, \ \|T\| \le \rho\}.$$

As was observed in [44], $\| \cdot \|^P_\rho$ is in fact a norm on $\mathscr{F}(\mathbb{B}^n_r)$. This norm can be viewed as a "free operator analog" of the supremum norm over $\bar{\mathbb{B}}^n_\rho$. By [44, Theorem 5.6], $\mathscr{F}(\mathbb{B}^n_r)$ is a Fréchet algebra with respect to the topology determined by the family $\{\| \cdot \|^P_\rho : \rho \in (0, r)\}$ of norms.

For our purposes, a slightly different definition of $\mathscr{F}(\mathbb{B}^n_r)$ is needed. Consider the projection $p\colon W_n \to \mathbb{Z}^n_+$ given by

$$p(\alpha) = (p_1(\alpha), \dots, p_n(\alpha)), \quad p_j(\alpha) = |\alpha^{-1}(j)|.$$

Proposition 6.2. *There exists a topological isomorphism*

$$\mathscr{F}(\mathbb{B}^n_r) \cong \Big\{ a = \sum_{\alpha \in W_n} c_\alpha \zeta_\alpha \in \mathfrak{F}_n : \|a\|_\rho = \sum_{k \in \mathbb{Z}^n_+} \Big(\sum_{\alpha \in p^{-1}(k)} |c_\alpha|^2 \Big)^{1/2} \rho^{|k|} < \infty \ \forall \rho \in (0, r) \Big\}. \tag{10}$$

Moreover, each norm $\| \cdot \|_\rho$ given by (10) *is submultiplicative.*

Using Proposition 6.2, we obtain the following theorem, which extends our earlier result from [35].

Theorem 6.3. *For each $q \in \mathbb{C}^\times$, the algebra $\mathscr{O}_q(\mathbb{B}^n_r)$ is topologically isomorphic to the quotient of $\mathscr{F}(\mathbb{B}^n_r)$ modulo the closed two-sided ideal generated by the elements $\zeta_i\zeta_j - q\zeta_j\zeta_i$ $(i < j)$. Moreover, for each $\rho \in (0, r)$ the norm $\| \cdot \|_\rho$ on $\mathscr{O}_q(\mathbb{B}^n_r)$ given by* (5) *is equal to the quotient of the norm $\| \cdot \|_\rho$ on $\mathscr{F}(\mathbb{B}^n_r)$ given by* (10).

7. Fréchet algebra bundles

Now we can explain in which sense the algebras $\mathscr{O}_q(\mathbb{D}^n_r)$ and $\mathscr{O}_q(\mathbb{B}^n_r)$ are "deformations" of $\mathscr{O}(\mathbb{D}^n_r)$ and $\mathscr{O}(\mathbb{B}^n_r)$, respectively. Let us recall some definitions from [15] (in a slightly modified form). Suppose that X is a locally compact, Hausdorff topological space. By a *family of vector spaces* over X we mean a pair (E, p), where E is a set and $p\colon E \to X$ is a surjective map, together with a vector space structure on each fiber $E_x = p^{-1}(x)$ $(x \in X)$. As usual, we let

$$E \times_X E = \{(u, v) \in E \times E : p(u) = p(v)\}.$$

By a *prebundle of topological vector spaces* over X we mean a family (E, p) of vector spaces over X together with a topology on E such that p is continuous and open, the zero section $0\colon X \to E$ is continuous, and the operations

$$\begin{aligned} E \times_X E \to E, &\quad (u, v) \mapsto u + v, \\ \mathbb{C} \times E \to E, &\quad (\lambda, u) \mapsto \lambda u, \end{aligned} \tag{11}$$

are also continuous.

Let (E, p) be a family of vector spaces over X. By definition, a function $\|\cdot\|: E \to [0, +\infty)$ is a *seminorm* if the restriction of $\|\cdot\|$ to each fiber is a seminorm in the usual sense. A family $\mathscr{N} = \{\|\cdot\|_\lambda : \lambda \in \Lambda\}$ of seminorms on E is said to be *directed* if for each $\lambda, \mu \in \Lambda$ there exist $C > 0$ and $\nu \in \Lambda$ such that $\|\cdot\|_\lambda \le C\|\cdot\|_\nu$ and $\|\cdot\|_\mu \le C\|\cdot\|_\nu$. If $\mathscr{N} = \{\|\cdot\|_\lambda : \lambda \in \Lambda\}$ and $\mathscr{N}' = \{\|\cdot\|_\mu : \mu \in \Lambda'\}$ are two directed families of seminorms on E, then we say that $\mathscr{N}$ is *dominated* by $\mathscr{N}'$ (and write $\mathscr{N} \prec \mathscr{N}'$) if for each $\lambda \in \Lambda$ there exist $C > 0$ and $\mu \in \Lambda'$ such that $\|\cdot\|_\lambda \le C\|\cdot\|_\mu$. If $\mathscr{N} \prec \mathscr{N}'$ and $\mathscr{N}' \prec \mathscr{N}$, then we say that $\mathscr{N}$ and $\mathscr{N}'$ are *equivalent* and write $\mathscr{N} \sim \mathscr{N}'$.

Remark 7.1. A directed family $\mathscr{N} = \{\|\cdot\|_\lambda : \lambda \in \Lambda\}$ of seminorms on E determines a uniform structure $\mathscr{U}(\mathscr{N})$ on E whose basis consists of all sets of the form

$$\{(u, v) \in E \times_X E : \|u - v\|_\lambda < \varepsilon\} \qquad (\lambda \in \Lambda,\ \varepsilon > 0).$$

It is easy to see that $\mathscr{N} \prec \mathscr{N}'$ if and only if $\mathscr{U}(\mathscr{N}) \subset \mathscr{U}(\mathscr{N}')$, and consequently $\mathscr{N} \sim \mathscr{N}'$ if and only if $\mathscr{U}(\mathscr{N}) = \mathscr{U}(\mathscr{N}')$.

The following definition is essentially a locally convex version of the notion of a Banach bundle in the sense of J. M. G. Fell (see, e.g., [13]).

Definition 7.2. Let (E, p) be a prebundle of topological vector spaces over X, and let $\mathscr{N} = \{\|\cdot\|_\lambda : \lambda \in \Lambda\}$ be a directed family of seminorms on E. We say that $\mathscr{N}$ is *admissible* if for each $x \in X$ the sets

$$\{u \in E : p(u) \in U,\ \|u\|_\lambda < \varepsilon\} \qquad (\lambda \in \Lambda,\ \varepsilon > 0,\ U \subseteq X \text{ is an open neighborhood of } x)$$

form a base of open neighborhoods of $0 \in E_x$. By a *locally convex uniform structure* on (E, p) we mean the equivalence class of an admissible directed family of seminorms on E. By a *locally convex bundle* over X we mean a prebundle of topological vector spaces over X together with a locally convex uniform structure. A *locally convex algebra bundle* over X is a locally convex bundle (E, p) over X together with an algebra structure on each fiber E_x such that the map

$$m_E : E \times_X E \to E, \quad (u, v) \mapsto uv,$$

is continuous. A *Fréchet algebra bundle* over X is a locally convex algebra bundle (E, p) over X such that each fiber E_x is a Fréchet algebra.

The following result is an improvement of [35, Theorem 6.1].

Theorem 7.3. *Let $n \in \mathbb{N}$, and let $r \in (0, +\infty]$.*

(i) *There exists a Fréchet algebra bundle (D, p) over $\mathbb{C}^\times$ such that for each $q \in \mathbb{C}^\times$ we have $D_q \cong \mathscr{O}_q(\mathbb{D}_r^n)$.*

(ii) *There exists a Fréchet algebra bundle (B, p) over $\mathbb{C}^\times$ such that for each $q \in \mathbb{C}^\times$ we have $B_q \cong \mathscr{O}_q(\mathbb{B}_r^n)$.*

In fact, we constructed the bundles (D, p) and (B, p) already in [35, Theorem 6.1], but we did not know how the fibers of (B, p) look like unless $q \in (0, 1]$. The construction is as follows. Let z denote the complex coordinate on $\mathbb{C}^\times$, and let $\mathscr{O}_{\mathrm{def}}(\mathbb{D}_r^n)$ denote the quotient of $\mathscr{O}(\mathbb{C}^\times, \mathscr{F}(\mathbb{D}_r^n))$ modulo the closed two-sided ideal generated by the elements $x_i x_j - z x_j x_i$ $(i < j)$. The algebra $\mathscr{O}_{\mathrm{def}}(\mathbb{D}_r^n)$ is a Fréchet $\mathscr{O}(\mathbb{C}^\times)$-algebra in a canonical way (i.e., we have a continuous homomorphism from $\mathscr{O}(\mathbb{C}^\times)$ to the center of $\mathscr{O}_{\mathrm{def}}(\mathbb{D}_r^n)$). Given a Fréchet $\mathscr{O}(\mathbb{C}^\times)$-algebra R, we define the *fiber* of R over $q \in \mathbb{C}^\times$ to be $R_q = R/\overline{M_q R}$, where $M_q = \{f \in \mathscr{O}(\mathbb{C}^\times) : f(q) = 0\}$. Let now $E = \bigsqcup_{q \in \mathbb{C}^\times} R_q$, and let $p : E \to \mathbb{C}^\times$ take each R_q to q. There is a canonical way to topologize E in such a way that (E, p) becomes a Fréchet algebra bundle over $\mathbb{C}^\times$. Applying this construction to $R = \mathscr{O}_{\rm def}(\mathbb{D}_r^n)$, we obtain a Fréchet algebra bundle (D, p) over $\mathbb{C}^\times$ whose fibers are equal to those of $\mathscr{O}_{\rm def}(\mathbb{D}_r^n)$. On the other hand, Theorem 6.1 (i) implies that the fiber of $\mathscr{O}_{\rm def}(\mathbb{D}_r^n)$ over $q \in \mathbb{C}^\times$ is isomorphic to $\mathscr{O}_q(\mathbb{D}_r^n)$. We can also use $\mathscr{F}^T(\mathbb{D}_r^n)$ instead of $\mathscr{F}(\mathbb{D}_r^n)$ in the above construction; the deformation algebra $\mathscr{O}_{\rm def}(\mathbb{D}_r^n)$ and hence the bundle (D, p) will then be the same. By replacing $\mathscr{F}(\mathbb{D}_r^n)$ with $\mathscr{F}(\mathbb{B}_r^n)$ is the above construction, and by using Theorem 6.3 instead of Theorem 6.1, we obtain a Fréchet $\mathscr{O}(\mathbb{C}^\times)$-algebra $\mathscr{O}_{\rm def}(\mathbb{B}_r^n)$ and a Fréchet algebra bundle (B, p) whose fiber over $q \in \mathbb{C}^\times$ is isomorphic to $\mathscr{O}_q(\mathbb{B}_r^n)$.

Remark 7.4. We hope that Theorem 7.3 may be of interest from the viewpoint of "analytic" deformation theory of associative algebras. Recall that most papers and monographs on algebraic deformation theory deal with *formal* deformations. Roughly, this means that the deformed product of two elements of an algebra A is no longer an element of A, but is a formal power series with coefficients in A. To some extent, such an approach to deformation theory is dictated by convenience considerations. However, formal deformations are not entirely satisfactory from the point of view of physics. Indeed, only those deformations are physically interesting that are represented by convergent power series [4]. In other words, one should be able to substitute concrete values of Planck's constant into the deformed product. The first successful attempt to develop a theory satisfying this requirement was made by M. Rieffel [55–58]. In his approach, a deformation is a continuous field (or a bundle) of C^*-algebras endowed with an additional structure. See [59] for a recent survey.

For a number of reasons, it is natural to expect that there should be an approach intermediate between the formal and continuous (i.e., C^*-algebraic) deformation theories. Specifically, it is desirable to have a "differentiable" or "complex analytic" deformation theory. Perhaps the first attempt to develop such a theory was made by M. J. Pflaum and M. Schottenloher [32]. Important special cases of convergent deformed products were studied by H. Omori, Y. Maeda, N. Miyazaki, and A. Yoshioka [30, 31], and by S. Beiser, H. Römer, and S. Waldmann [4]. Quite recently, very interesting preprints by S. Beiser, G. Lechner, and S. Waldmann [5, 24] have appeared, in which a rather general approach to analytic deformation theory has been presented. On the other hand, so far nothing is known about deformations of algebras of holomorphic functions on classical domains. We do not try to give a general definition of a Fréchet algebra deformation here, but we believe that such a definition should be similar to that given by Rieffel in the C^*-algebra case. Also, we would like to note that our deformation algebras $\mathscr{O}_{\rm def}(\mathbb{D}_r^n)$ and $\mathscr{O}_{\rm def}(\mathbb{B}_r^n)$ are not topologically free (moreover, they are not topologically projective) over $\mathscr{O}(\mathbb{C}^\times)$, so they do not fit into the framework suggested in [32].

8. A q-analog of Poincaré's theorem

We now turn to our main question of whether $\mathscr{O}_q(\mathbb{D}_r^n)$ and $\mathscr{O}_q(\mathbb{B}_r^n)$ are topologically isomorphic. The answer is as follows.

Theorem 8.1. *Let $n \in \mathbb{N}$ and $r \in (0, +\infty]$.*
- (i) *If $n \geq 2$, $r < \infty$, and $|q| = 1$, then $\mathscr{O}_q(\mathbb{D}_r^n)$ and $\mathscr{O}_q(\mathbb{B}_r^n)$ are not topologically isomorphic.*
- (ii) *If $|q| \neq 1$, then $\mathscr{O}_q(\mathbb{D}_r^n)$ and $\mathscr{O}_q(\mathbb{B}_r^n)$ are topologically isomorphic (in fact, they are equal as power series algebras).*

The proof of part (ii) is elementary in the case where $|q| > 1$. The case where $|q| < 1$ is reduced to the previous one via Proposition 4.5.

The proof of part (i) is more involved and is based on an Arens-Michael algebra version of the joint spectral radius (see, e.g., [26]). Let A be an Arens-Michael algebra, and let $\{\|\cdot\|_\lambda : \lambda \in \Lambda\}$ be a directed defining family of submultiplicative seminorms on A. Given an n-tuple $a = (a_1, \dots, a_n) \in A^n$, we define the *joint ℓ^p-spectral radius* $r_p^A(a)$ by

$$\begin{aligned} r_p^A(a) &= \sup_{\lambda\in\Lambda} \lim_{d\to\infty} \Big(\sum_{\alpha\in W_{n,d}} \|a_\alpha\|_\lambda^p \Big)^{1/pd} \quad \text{for } 1 \le p < \infty; \\ r_\infty^A(a) &= \sup_{\lambda\in\Lambda} \lim_{d\to\infty} \Big(\sup_{\alpha\in W_{n,d}} \|a_\alpha\|_\lambda \Big)^{1/d}. \end{aligned} \tag{12}$$

In the case of Banach algebras, the joint ℓ^∞-spectral radius was introduced by G.-C. Rota and W. G. Strang [60]. The case $p < \infty$ was studied by A. Sołtysiak [61] for commuting n-tuples in a Banach algebra. The observation that a similar definition makes sense in the noncommutative case is probably due to V. Müller [26, C.32.2].

It is easy to show that $r_p^A(a)$ does not depend on the choice of a directed defining family of submultiplicative seminorms on A. Moreover, if $\varphi\colon A \to B$ is a continuous homomorphism of Arens-Michael algebras, then for each $a = (a_1, \dots, a_n) \in A^n$ we have $r_p^B(\varphi(a)) \le r_p^A(a)$, where $\varphi(a) = (\varphi(a_1), \dots, \varphi(a_n))$. In particular, if $\varphi\colon A \to B$ is a topological isomorphism, then $r_p^B(\varphi(a)) = r_p^A(a)$. In contrast to the Banach algebra case, it may happen that $r_p^A(a) = +\infty$. For example, if $A = \mathscr{O}(\mathbb{C})$ and $z \in A$ is the complex coordinate, then an easy computation shows that $r_p^A(z) = +\infty$ for all p.

The following example is crucial for our purposes. Let $|q| = 1$, and let $x = (x_1, \dots, x_n)$ denote the system of canonical generators of $\mathscr{O}_q(\mathbb{D}_r^n)$ or $\mathscr{O}_q(\mathbb{B}_r^n)$. Then we have

$$r_2^{\mathscr{O}_q(\mathbb{B}_r^n)}(x) = r \quad \text{and} \quad r_2^{\mathscr{O}_q(\mathbb{D}_r^n)}(x) = r\sqrt{n}. \tag{13}$$

Now we can explain the idea of the proof of Theorem 8.1 (i). Fix $q \in \mathbb{C}^\times$, $q \ne 1$, with $|q| = 1$. Let $A = \mathscr{O}_q(\mathbb{D}_r^n)$ and $B = \mathscr{O}_q(\mathbb{B}_r^n)$, and assume that $\varphi\colon B \to A$ is a topological isomorphism. For each $i = 1, \dots, n$, let $f_i = \varphi(x_i)$. A tedious but elementary algebraic argument shows that there exists a permutation $\sigma \in S_n$ such that

$$f_i = \lambda_i x_{\sigma(i)} + \text{terms of higher degree},$$

where $|\lambda_i| = 1$. Hence for each $\alpha = (\alpha_1, \dots, \alpha_d) \in W_n$ we have

$$f_\alpha = \lambda_\alpha x_{\sigma(\alpha)} + \text{terms of higher degree},$$

where $\sigma(\alpha) = (\sigma(\alpha_1), \dots, \sigma(\alpha_d))$ and $\lambda_\alpha = \lambda_{\alpha_1} \cdots \lambda_{\alpha_d}$. This implies that for each $\rho \in (0, r)$ we have $\|f_\alpha\|_\rho \ge \|x_{\sigma(\alpha)}\|_\rho = \|x_\alpha\|_\rho$, where $\|\cdot\|_\rho$ is the norm on $\mathscr{O}_q(\mathbb{D}_r^n)$ given by (2). Hence $r_2^A(f) \ge r_2^A(x)$. On the other hand, $r_2^A(f) = r_2^B(x)$, because φ is a topological isomorphism. Taking into account (13), we conclude that $r = r_2^B(x) \ge r_2^A(x) = r\sqrt{n}$, whence $n = 1$.

9. Open problems

We conclude the paper with a couple of open problems. The first problem was already mentioned in Section 5 and is inspired by the fact that the algebra $\mathscr{O}_q(\mathbb{B}_r^n)$ is defined for all $q \in \mathbb{C}^\times$.

Problem 9.1. *Is it possible to define Vaksman's algebras $C_q(\bar{\mathbb{B}}^n)$ and $A_q(\bar{\mathbb{B}}^n)$ in the case where $q \notin (0, 1]$? If yes, then can Theorem 5.1 be extended to this case?*

The second problem is related to the notion of an HFG algebra introduced in [35, 36]. Let $\mathscr{F}(\mathbb{C}^n)$ denote the Arens-Michael envelope of the free algebra F_n (see Section 6). A Fréchet algebra A is said to be *holomorphically finitely generated* (HFG for short) if A is isomorphic to a quotient of $\mathscr{F}(\mathbb{C}^n)$ for some n. There is also an "internal" definition given in terms of

J. L. Taylor's free functional calculus [65]. By [36, Theorem 3.22], a commutative Fréchet-Arens-Michael algebra is holomorphically finitely generated if and only if it is topologically isomorphic to $\mathscr{O}(X)$ for some Stein space $(X, \mathscr{O}_X)$ of finite embedding dimension. Together with Forster's theorem (see Section 1), this implies that the category of commutative HFG algebras is anti-equivalent to the category of Stein spaces of finite embedding dimension. There are many natural examples of noncommutative HFG algebras; see [36, Section 7]. For instance, $\mathscr{O}_q(\mathbb{D}_r^n)$ and $\mathscr{F}(\mathbb{D}_r^n)$ are HFG algebras. By Theorem 8.1 (ii), $\mathscr{O}_q(\mathbb{B}_r^n)$ is an HFG algebra provided that $|q| \neq 1$.

Problem 9.2. *Is $\mathscr{O}_q(\mathbb{B}_r^n)$ an HFG algebra in the case where $|q| = 1$, $q \neq 1$?*

Acknowledgments

This work was supported by the Russian Foundation for Basic Research (grant no. 12-01-00577). The author thanks A. Ya. Helemskii and D. Proskurin for helpful discussions.

References

[1] Aizenberg L A and Mityagin B S 1960 Spaces of analytic functions in polycircular domains (in Russian) *Sib. Mat. Zh.* **1** 153–170

[2] Alev J and Dumas F 1994 Sur le corps des fractions de certaines algèbres quantiques *J Algebra* **170** no 1 229–265

[3] Beggs E and Smith S P 2012 Noncommutative complex differential geometry *Preprint* arXiv:1209.3595 [math.AG]

[4] Beiser S, Römer H and Waldmann S 2007 Convergence of the Wick star product *Comm Math Phys* **272** no 1 25–52

[5] Beiser S and Waldmann S 2011 Fréchet algebraic deformation quantization of the Poincaré disk *Preprint* arXiv:1108.2004 [math.QA]

[6] Brown K A and Goodearl K R 2002 *Lectures on Algebraic Quantum Groups* (Basel: Birkhäuser Verlag)

[7] Davidson K R and Pitts D R 1998 The algebraic structure of non-commutative analytic Toeplitz algebras *Math Ann* **311** no 2 275–303.

[8] Davidson K R 2001 Free semigroup algebras. A survey *Systems, Approximation, Singular Integral Operators, and Related Topics (Bordeaux 2000), Oper Theory Adv Appl* **129** (Basel: Birkhäuser) pp 209–240

[9] Davidson K R and Popescu G 1998 Noncommutative disc algebras for semigroups *Canad J Math* **50** no 2 290–311

[10] Davidson K R and Katsoulis E G 2010 Biholomorphisms of the unit ball of $\mathbb{C}^n$ and semicrossed products *Operator theory live, Theta Ser Adv Math* **12** (Theta: Bucharest) pp 69–80

[11] Davidson K R and Katsoulis E G 2010 Dilating covariant representations of the non-commutative disc algebras. *J Funct Anal* **259** no 4 817–831

[12] Davidson K R Ramsey C and Shalit O 2011 The isomorphism problem for some universal operator algebras *Adv Math* **228** no 1 167–218

[13] Fell J M G and Doran R S 1988 *Representations of *-Algebras, Locally Compact Groups, and Banach *-Algebraic Bundles. Vol 1. Basic Representation Theory of Groups and Algebras* (Boston, MA: Academic Press)

[14] Forster O 1967 Zur Theorie der Steinschen Algebren und Moduln *Math Z* **97** 376–405

[15] Gierz G 1982 *Bundles of Topological Vector Spaces and Their Duality* (Berlin-New York: Springer-Verlag)

[16] Hartshorne R 1977 *Algebraic Geometry* (New York–Heidelberg: Springer-Verlag)

[17] Helemskii A Ya 1993 *Banach and Locally Convex Algebras* (New York: The Clarendon Press, Oxford University Press)

[18] Jordan D A 1995 A simple localization of the quantized Weyl algebra *J Algebra* **174** no 1 267–281

[19] Khalkhali M and Moatadelro A 2011 The homogeneous coordinate ring of the quantum projective plane *J Geom Phys* **61** no 1 276–289

[20] Khalkhali M, Landi G and van Suijlekom W D 2011 Holomorphic structures on the quantum projective line *Int Math Res Not IMRN* **4** 851–884

[21] Khalkhali M and Moatadelro A 2011 Noncommutative complex geometry of the quantum projective space *J Geom Phys* **61** no 12 2436–2452

[22] Klimek S and Lesniewski A 1993 A two-parameter quantum deformation of the unit disc *J Funct Anal* **115** no 1 1–23

[23] Klimyk A and Schmüdgen K 1997 *Quantum Groups and Their Representations* (Berlin: Springer-Verlag)

[24] Lechner G and Waldmann S 2011 Strict deformation quantization of locally convex algebras and modules *Preprint* arXiv:1109.5950 [math.QA]
[25] Michael E A 1952 Locally multiplicatively-convex topological algebras *Mem Amer Math Soc* **11**
[26] Müller V 2003 *Spectral Theory of Linear Operators and Spectral Systems in Banach Algebras* (Basel: Birkhäuser Verlag)
[27] Narasimhan R 1971 *Several Complex Variables* (Chicago and London: The Univ. of Chicago Press)
[28] Ó Buachalla R 2012 Quantum bundle description of quantum projective spaces *Comm Math Phys* **316** no 2 345–373
[29] Ó Buachalla R 2011 Noncommutative complex structures on quantum homogeneous spaces *Preprint* arXiv:1108.2374 [math.QA]
[30] Omori H, Maeda Y, Miyazaki N and Yoshioka A 2000 Deformation quantization of Fréchet-Poisson algebras: convergence of the Moyal product *Conférence Moshé Flato 1999, Vol II (Dijon), Math Phys Stud* **22** (Dordrecht: Kluwer Acad Publ) pp 233–245
[31] Omori H, Maeda Y, Miyazaki N and Yoshioka A 2007 Orderings and non-formal deformation quantization *Lett Math Phys* **82** no 2–3 153–175
[32] Pflaum M J and Schottenloher M 1998 Holomorphic deformation of Hopf algebras and applications to quantum groups *J Geom Phys* **28** no 1–2 31–44
[33] Pirkovskii A Yu 2006 Stably flat completions of universal enveloping algebras *Dissertationes Math (Rozprawy Math)* **441** 1–60
[34] Pirkovskii A Yu 2008 Arens-Michael envelopes, homological epimorphisms, and relatively quasi-free algebras *Trans Moscow Math Soc* 27–104
[35] Pirkovskii A Yu 2014 Noncommutative analogues of Stein spaces of finite embedding dimension *Algebraic Methods in Functional Analysis, Oper Theory Adv Appl* **233** (Basel: Birkhäuser/Springer) (*Preprint* arXiv:1204.4936 [math.FA])
[36] Pirkovskii A Yu 2014 Holomorphically finitely generated algebras *J Noncomm Geometry*, to appear (*Preprint* arXiv:1304.1991 [math.FA])
[37] Poincaré H 1907 Les fonctions analytiques de deux variables et la représentation conforme *Rend Circ Mat Palermo* **23** 185–220
[38] Polishchuk A and Schwarz A 2003 Categories of holomorphic vector bundles on noncommutative two-tori *Comm Math Phys* **236** no 1 135–159
[39] Polishchuk A 2004 Classification of holomorphic vector bundles on noncommutative two-tori *Doc Math* **9** 163–181
[40] Polishchuk A 2006 Analogues of the exponential map associated with complex structures on noncommutative two-tori *Pacific J Math* **226** no 1 153–178
[41] Polishchuk A 2007 Quasicoherent sheaves on complex noncommutative two-tori *Selecta Math (N.S.)* **13** no 1 137–173
[42] Popescu G 1996 Non-commutative disc algebras and their representations *Proc Amer Math Soc* **124** no 7 2137–2148
[43] Popescu G 1998 Noncommutative joint dilations and free product operator algebras *Pacific J Math* **186** no 1 111–140
[44] Popescu G 2006 Free holomorphic functions on the unit ball of $B(\mathcal{H})^n$ *J Funct Anal* **241** no 1 268–333
[45] Popescu G 2007 Operator theory on noncommutative varieties. II *Proc Amer Math Soc* **135** no 7 2151–2164
[46] Popescu G 2010 Free holomorphic automorphisms of the unit ball of $B(\mathcal{H})^n$ *J Reine Angew Math* **638** 119–168
[47] Popescu G 2010 Operator theory on noncommutative domains *Mem Amer Math Soc* **205** no 964
[48] Popescu G 2011 Free biholomorphic classification of noncommutative domains *Int Math Res Not IMRN* **4** 784–850
[49] Popescu G 2011 Joint similarity to operators in noncommutative varieties *Proc Lond Math Soc (3)* **103** no 2 331–370
[50] Popescu G 2012 Free biholomorphic functions and operator model theory *J Funct Anal* **262** no 7 3240–3308
[51] Popescu G 2013 Free biholomorphic functions and operator model theory, II *J Funct Anal* **265** no 5 786–836
[52] Proskurin D and Samoĭlenko Yu 2002 Stability of the C^*-algebra associated with twisted CCR *Algebr Represent Theory* **5** no 4 433–444
[53] Pusz W and Woronowicz S L 1989 Twisted second quantization *Rep Math Phys* **27** no 2 231–257
[54] Range R M 1986 *Holomorphic Functions and Integral Representations in Several Complex Variables* (New York: Springer-Verlag)
[55] Rieffel M A 1989 Deformation quantization of Heisenberg manifolds *Comm Math Phys* **122** no 4 531–562
[56] Rieffel M A 1990 Deformation quantization and operator algebras *Operator Theory: Operator Algebras and Applications, Part 1 (Durham, NH, 1988), Proc Sympos Pure Math* **51** Part 1 (Providence, RI: Amer

Math Soc) pp 411–423

[57] Rieffel M A 1990 Lie group convolution algebras as deformation quantizations of linear Poisson structures *Amer J Math* **112** no 4 657–685

[58] Rieffel M A 1993 Deformation quantization for actions of $\mathbb{R}^d$ *Mem Amer Math Soc* **506**

[59] Rieffel M A 1998 Questions on quantization *Operator Algebras and Operator Theory (Shanghai, 1997), Contemp Math* **228** (Providence, RI: Amer Math Soc) pp 315–326

[60] Rota G-C and Strang W G 1960 A note on the joint spectral radius *Indag Math* **22** 379–381

[61] Sołtysiak A 1993 On the joint spectral radii of commuting Banach algebra elements *Studia Math* **105** 93–99

[62] Szafraniec F H 2003 Multipliers in the reproducing kernel Hilbert space, subnormality and noncommutative complex analysis *Reproducing Kernel Spaces and Applications, Oper Theory Adv Appl* **143** (Basel: Birkhäuser) pp 313–331

[63] Sinel'shchikov S and Vaksman L 1998 On q-analogues of bounded symmetric domains and Dolbeault complexes *Math Phys Anal Geom* **1** no 1 75–100

[64] Taylor J L 1972 Homology and cohomology for topological algebras *Adv Math* **9** 137–182

[65] Taylor J L 1972 A general framework for a multi-operator functional calculus *Adv Math* **9** 183–252

[66] Taylor J L 1973 Functions of several noncommuting variables *Bull Amer Math Soc* **79** 1–34

[67] Vaksman L L 1995 Integral intertwining operators and quantum homogeneous spaces *Theoret and Math Phys* **105** no 3 1476–1483

[68] Vaksman L L and Shklyarov D L 1997 Integral representations of functions in the quantum disk. I (in Russian) *Mat Fiz Anal Geom* **4** no 3 286–308

[69] *Lectures on q-analogues of Cartan domains and associated Harish-Chandra modules* ed L Vaksman, Kharkov, Ukraine, 2001 (*Preprint* arXiv:math.QA/0109198)

[70] Vaksman L 2003 The maximum principle for "holomorphic functions" in the quantum ball (in Russian) *Mat Fiz Anal Geom* **10** no 1 12–28

[71] Vaksman L L 2010 *Quantum Bounded Symmetric Domains* (Providence, RI: American Mathematical Society)

[72] Wess J and Zumino B 1990 Covariant differential calculus on the quantum hyperplane *Recent Advances in Field Theory (Annecy-le-Vieux, 1990) Nuclear Phys B Proc Suppl* **18B** 302–312

Journal of Physics: Conference Series **474** (2013) 012027 doi:10.1088/1742-6596/474/1/012027

Moments of two noncommutative random variables in terms of their joint quantum operators

Gabriela Popa[1] **and Aurel I. Stan**[2]

[1] Department of Physics, College of Arts and Sciences, 240 Elton Hall, Ohio University, Zanesville, OH 43701, USA

[2] Department of Mathematics, Ohio State University at Marion, 1465 Mount Vernon Avenue, Marion, OH 43302, USA

E-mail: popag@ohio.edu, stan.7@osu.edu

Abstract. The computation of the generating function of the joint moments of two non–commutative gaussiano–poissonian random variables, from the commutators of their quantum operators, is presented first. An example of such random variables is also included in the paper.

1. Introduction

In this paper we apply the algorithm for recovering the joint moments from the commutators of the quantum operators, and the first order moments, outlined in [9], to two noncommutative random variables X and Y. These are not classical random variables, but symmetric operators densely defined on a Hilbert space. We must also mention the fact that, the algorithm mentioned above was extended to q–commutators in [4], and the authors of that paper worked out the joint moments of both commutative and noncommutative random variables.
The multi–dimensional case is much harder than the one–dimensional one, and for this reason we restrict our attention to the dimension $d = 2$ in this paper.

The paper is structured as follows. In section 2 we present a little bit of the background concerning Noncommutative Probability and the joint quantum operators, generated by a finite family of noncommutative random variables. In section 3 we pose a two–dimensional commutator problem and apply the algorithm, from [9], to find a recursive formula for the joint moments of the two random variables involved in that commutator problem. In section 4, we use the recursive formula, obtained in the previous section, to compute the joint Laplace transform of the two random variables from section 3. Finally, in section 5, we show the existence of the operators involved in the commutator problem from section 3.

2. Background

We present in this section a basic background of Noncommutative Probability and joint quantum operators generated by a family of not necessary commutative random variables. The definition of a noncommutative probability space is taken from [13], and the facts about the joint quantum operators from [10].

Let $(H, \langle \cdot, \cdot \rangle)$ be a Hilbert space over the field of real numbers $\mathbb{R}$, and let $X_1, X_2, \ldots, X_d$ be d symmetric densely defined linear operators on H. We denote by $\mathcal{A}$ the unital algebra generated by $X_1, X_2, \ldots, X_d$. It is clear that $\mathcal{A}$ is in fact, the space of all operators of the form $f(X_1, X_2, \ldots, X_d)$, where f is a polynomial of d noncommutative variables $x_1, x_2, \ldots, x_d$. We make the assumption that there exists a nonzero element ϕ of H, such that ϕ belongs to the domain of g, for any $g \in \mathcal{A}$. We can normalize ϕ and assume that:

$$\parallel \phi \parallel \; = \; 1, \tag{1}$$

where $\| \cdot \|$ denotes the norm of the Hilbert space H. We fix ϕ and call it the *vacuum vector.*

Definition 2.1 *We call the pair* $(\mathcal{A}, \phi)$ *a* probability space supported by H. *We call every element* g *of* $\mathcal{A}$, *a* random variable, *and define its* expectation *as:*

$$E[g] \; := \; \langle g\phi, \phi \rangle. \tag{2}$$

We also define the following equivalence relation (see [10]):

Definition 2.2 *Let* $(\mathcal{A}, \phi)$ *and* $(\mathcal{A}', \phi')$ *be two probability spaces supported by two Hilbert spaces* H *and* H', *and let* E *and* E' *denote their expectations. Let* $X_1, X_2, \ldots, X_d$ *be operators from* $\mathcal{A}$, *and* $X_1', X_2', \ldots, X_d'$ *operators from* $\mathcal{A}'$. *We say that the random vectors* $(X_1, X_2, \ldots, X_d)$ *and* $(X_1', X_2', \ldots, X_d')$ *are* moment equal *and denote this fact by* $(X_1, X_2, \ldots, X_d) \equiv (X_1', X_2', \ldots, X_d')$, *if for any polynomial* $p(x_1, x_2, \ldots, x_d)$ *of* d *noncommutative variables, we have:*

$$E\left[p\left(X_1, X_2, \ldots, X_d\right)\right] = E'\left[p\left(X_1', X_2', \ldots, X_d'\right)\right]. \tag{3}$$

We define the subspace F, of H, as:

$$F \; := \; \{g\phi \mid g \in \mathcal{A}\}. \tag{4}$$

For any non–negative integer n, we define the space F_n as the set of all vectors, of H, of the form $f(X_1, X_2, \ldots, X_d)\phi$, where f is a polynomial of total degree less than or equal to n. We observe that, since F_n is a finite–dimensional subspace of H, F_n is a closed subspace of H, for all $n \geq 0$. We have:

$$F_0 \subset F_1 \subset F_2 \subset \cdots \subset F \subset H \tag{5}$$

and

$$F \; = \; \cup_{n=0}^{\infty} F_n. \tag{6}$$

We define $G_0 := F_0$, and for all $n \geq 1$, $G_n := F_n \ominus F_{n-1}$, that means, G_n is the orthogonal complement of F_{n-1} into F_n. For any $n \geq 0$, we call G_n the *homogenous chaos space of order* n generated by $X_1, X_2, \ldots, X_d$. We also define the space:

$$\mathcal{H} \; := \; \oplus_{n=0}^{\infty} G_n,$$

and call $\mathcal{H}$ the *chaos space* generated by $X_1, X_2, \ldots, X_d$. It is not hard to see that $\mathcal{H}$ is the closure of the space $\mathcal{A}\phi := \{g\phi \mid g \in \mathcal{A}\}$ in H.

We have the following lemma:

Lemma 2.3 *For any* $i \in \{1, 2, \ldots, d\}$ *and any non–negative integer* n:

$$X_i G_n \;\perp\; G_k, \tag{7}$$

for all $k \neq n-1$, n, *and* $n+1$, *where* $\perp$ *means "orthogonal to".*

We present a proof of the above lemma, that follows exactly the proof of the same lemma, in the commutative case, from [1].

Proof. Let $i \in \{1, 2, \ldots, d\}$ be fixed.
If $k \notin \{n-1, n, n+1\}$, then either $k \geq n+2$ or $k \leq n-2$.

Case 1: If $k \geq n+2$, then let $u \in G_n$ and $v \in G_k$. Since $u \in G_n$, there exists f a polynomial of degree at most n, such that:

$$u = f(X_1, X_2, \ldots, X_d)\phi. \tag{8}$$

Since the polynomial $x_i f(x_1, x_2, \ldots, x_d)$ has degree at most $n+1$, we have:

$$\begin{aligned} X_i u &= X_i f(X_1, X_2, \ldots, X_d)\phi \\ &\in F_{n+1}. \end{aligned}$$

Because $n+1 < k$, we have $G_k \perp F_{n+1}$, and so:

$$v \perp X_i u.$$

Case 2: If $k \leq n-2$, then let $u \in G_n$ and $v \in G_k$. We can see as before that $X_i v \in F_{k+1}$. Using first the fact that X_i is a symmetric operator, and second that $G_n \perp F_{k+1}$ since $n > k+1$, we obtain:

$$\begin{aligned} \langle X_i u, v \rangle &= \langle u, X_i v \rangle \\ &= 0. \end{aligned}$$

□

To ease our notation, we write: $X = (X_1, X_2, \ldots, X_d)$ and $f(X) = f(X_1, X_2, \ldots, X_d)$ for every polynomial f of d variables.
Let n be a fixed nonnegative integer, and $f(X)\phi \in G_n$. Then, according to the previous lemma, for any $i \in \{1, 2, \ldots, d\}$:

$$X_i f(X)\phi \in G_{n-1} \oplus G_n \oplus G_{n+1}. \tag{9}$$

Thus, there exist and are unique $f_{i,n-1}(X)\phi \in G_{n-1}$, $f_{i,n}(X)\phi \in G_n$, and $f_{i,n+1}(X)\phi \in G_{n+1}$, such that:

$$X_i f(X)\phi = f_{i,n-1}(X)\phi + f_{i,n}(X)\phi + f_{i,n+1}(X)\phi. \tag{10}$$

We define the following three families of operators:

$$D_n^-(i) : G_n \to G_{n-1}, \tag{11}$$

defined as:

$$D_n^-(i) f(X)\phi := f_{i,n-1}(X)\phi. \tag{12}$$

Since f has degree n, while $f_{i,n-1}$ has degree $n-1$, we see that $D_n^-(i)$ decreases the degree of a homogenous polynomial by 1 unit, and so we call $D_n^-(i)$ an *annihilation* operator.

$$D_n^0(i) : G_n \to G_n, \tag{13}$$

defined as:

$$D_n^0(i)f(X)\phi \ := \ f_{i,n}(X)\phi. \tag{14}$$

Since f has degree n, and $f_{i,n}$ has degree n, too, we see that $D_n^0(i)$ preserves the degree of a homogenous polynomial, and so we call $D_n^0(i)$ a *preservation* operator.

$$D_n^+(i) : G_n \to G_{n+1}, \tag{15}$$

defined as:

$$D_n^+(i)f(X)\phi \ := \ f_{i,n+1}(X)\phi. \tag{16}$$

Since f has degree n, while $f_{i,n+1}$ has degree $n+1$, we see that $D_n^+(i)$ increases the degree of a homogenous polynomial by 1 unit, and so we call $D_n^+(i)$ a *creation* operator.
If we restrict the domains of the operators X_1, X_2, $\dots$, X_d to the space G_n, then we can see from (10) that, for all $i \in \{1, 2, \dots, d\}$, we have:

$$X_i|_{G_n} \ = \ D_n^-(i) + D_n^0(i) + D_n^+(i). \tag{17}$$

We have defined, so far, the annihilation, preservation, and creation operators only separately on each homogenous chaos space G_n, for $n \geq 0$. We can extend this partial definition, in a linear way, to the whole space $F = \mathcal{A}\phi$. More precisely, if f is a polynomial of degree k, where k is a nonnegative integer, then $f(X)\phi \in F_k$, and we know that:

$$F_k \ = \ G_0 \oplus G_1 \oplus \cdots \oplus G_k. \tag{18}$$

Thus, there exist and are unique $f_0(X)\phi \in G_0$, $f_1(X)\phi \in G_1$, $\dots$, $f_k(X)\phi \in G_k$, such that:

$$f(X)\phi \ = \ f_0(X)\phi + f_1(X)\phi + \cdots + f_k(X)\phi. \tag{19}$$

We define:

$$a^-(i)f(X)\phi \ := \ D_0^-(i)f_0(X)\phi + \cdots + D_k^-(i)f_k(X)\phi, \tag{20}$$

$$a^0(i)f(X)\phi \ := \ D_0^0(i)f_0(X)\phi + \cdots + D_k^0(i)f_k(X)\phi, \tag{21}$$

and

$$a^+(i)f(X)\phi \ := \ D_0^+(i)f_0(X)\phi + \cdots + D_k^+(i)f_k(X)\phi. \tag{22}$$

We call $a^-(i)$ an *annihilation* operator, $a^0(i)$ a *preservation* operator, and $a^+(i)$ a *creation* operator.
We also call the operators $\{a^-(i)\}_{1\leq i\leq d}$, $\{a^0(i)\}_{1\leq i\leq d}$, and $\{a^+(i)\}_{1\leq i\leq d}$, the *minimal joint quantum operators* generated by the family $\{X_i\}_{1\leq i\leq d}$. The domain of the minimal joint quantum operators is understood to be the space F.
Restricting the domain of each X_i, for $i \in \{1, 2, \dots, d\}$, to the space F, we have now:

$$X_i \ = \ a^-(i) + a^0(i) + a^+(i), \tag{23}$$

for all $i \in \{1, 2, \dots, d\}$.
For each $1 \leq i \leq d$, we call the equality (23), the *minimal quantum decomposition* of X_i.

One should not forget that, for any $i \in \{1, 2, \ldots, d\}$ and $n \geq 0$, $a^-(i) : G_n \to G_{n-1}$, $a^0(i) : G_n \to G_n$, and $a^+(i) : G_n \to G_{n+1}$, where $G_{-1} := \{0\}$ is the null space.
If for each $n \geq 0$, we denote by P_n the orthogonal projection of $\mathcal{H}$ onto its closed subspace F_n, and restrict the domain of P_n to F, then it is not hard to see that:

$$a^-(i) = \sum_{n=1}^{\infty} P_{n-1}X_iP_n, \tag{24}$$

$$a^0(i) = \sum_{n=0}^{\infty} P_nX_iP_n, \tag{25}$$

$$a^+(i) = \sum_{n=0}^{\infty} P_{n+1}X_iP_n, \tag{26}$$

for all $i \in \{1, 2, \ldots, d\}$. We can see from these three formulas that, since X_i is symmetric, for all u and v in F, and all $i \in \{1, 2, \ldots, d\}$, we have:

$$\langle a^+(i)u, v\rangle = \langle u, a^-(i)v\rangle, \tag{27}$$

$$\langle a^0(i)u, v\rangle = \langle u, a^0(i)v\rangle. \tag{28}$$

If U and V are two operators, then we define their commutator $[U, V]$ as:

$$[U, V] := UV - VU. \tag{29}$$

It was proven in [1], [2], and [10], that $X_1, X_2, \ldots, X_d$ commute among themselves if and only if the following conditions hold, for any $i, j \in \{1, 2, \ldots, d\}$:

$$\left[a^-(i), a^-(j)\right] = 0, \tag{30}$$

$$\left[a^-(i), a^0(j)\right] = \left[a^-(j), a^0(i)\right], \tag{31}$$

$$\left[a^-(i), a^+(j)\right] - \left[a^-(j), a^+(i)\right] = \left[a^0(j), a^0(i)\right]. \tag{32}$$

Example 2.4 *Let $X_1, X_2, \ldots, X_d$ be classic random variables defined on the same probability space $(\Omega, \mathcal{F}, P)$, having finite moments of any order. Let $H := L^2(\Omega, \mathcal{F}, P)$ and $\phi = 1$, i.e., the constant random variable equal to 1. We can view $X_1, X_2, \ldots, X_d$ as multiplication operators on the space $\mathcal{A}1 \subset H$, where $\mathcal{A}$ is the algebra of the random variables of the form $f(X_1, X_2, \ldots, X_d)$, where f is a polynomial of d variables. It is obvious that $X_iX_j = X_jX_i$, for all $1 \leq i < j \leq d$.*

The following definition is taken from [10].

Definition 2.5 *Let H be a Hilbert space, $(\mathcal{A}, \phi)$ a probability space supported by H, and $\{X_i\}_{1\leq i\leq d}$ elements of $\mathcal{A}$, that are symmetric operators. Let $\{a_{x_i}^-\}_{1\leq i\leq d}$, $\{a_{x_i}^0\}_{1\leq i\leq d}$, and $\{a_{x_i}^+\}_{1\leq i\leq d}$ be three families of linear operators, defined on subspaces of H, such that, ϕ belongs to the domain of $a_{x_{i_1}}^{\epsilon_1} a_{x_{i_2}}^{\epsilon_2} \cdots a_{x_{i_n}}^{\epsilon_n}$, for all $n \geq 1$, $(i_1, i_2, \ldots, i_n) \in \{1, 2, \ldots, d\}^n$, and $(\epsilon_1, \epsilon_2, \ldots, \epsilon_n) \in \{-, 0, +\}^n$. We say that these families of operators form* a joint quantum decomposition of $\{X_i\}_{1\leq i\leq d}$ relative to $\mathcal{A}$, *if the following conditions hold:*

$$X_i = a_{x_i}^- + a_{x_i}^0 + a_{x_i}^+, \tag{33}$$

$$\left(a_{x_i}^+\right)^* |\mathcal{A}\phi = a_{x_i}^-|\mathcal{A}\phi, \tag{34}$$

$$a_{x_i}^- H_n \subset H_{n-1}, \tag{35}$$

$$a_{x_i}^0 H_n \subset H_n, \tag{36}$$

for all $1 \leq i \leq d$ and $n \geq 0$, where $H_{-1} := \{0\}$, $H_0 := \mathbb{R}\phi$, and H_k is the vector space spanned by all vectors of the form $a^+_{x_{i_1}} a^+_{x_{i_2}} \cdots a^+_{x_{i_k}} \phi$, where $i_1, i_2, \ldots, i_k \in \{1, 2, \ldots, d\}$, for all $k \geq 1$. We call $a^-_{x_i}$ an annihilation *operator, $a^0_{x_i}$ a* preservation *operator, and $a^+_{x_i}$ a* creation *operator, for all $1 \leq i \leq d$.*

For all $1 \leq i \leq d$, the fact that $\left(a^+_{x_i}\right)^* |\mathcal{A}\phi = a^-_{x_i}|\mathcal{A}\phi$ is understood as:

$$\langle a^+_{x_i} u, v\rangle = \langle u, a^-_{x_i} v\rangle,$$

for all u and v in $\mathcal{A}\phi$, where $\langle \cdot, \cdot \rangle$ denotes the inner product of H. Because, for all $i \in \{1, 2, \ldots, d\}$, X_i is a symmetric operator, it follows from (34), that:

$$\left(a^0_{x_i}\right)^* |\mathcal{A}\phi = a^0_{x_i}|\mathcal{A}\phi. \tag{37}$$

The following lemma from [10] refers to the uniqueness of the joint quantum decomposition of a finite family of symmetric random variables.

Lemma 2.6 *Let $\{X_i\}_{1\leq i\leq d}$ be a family of symmetric random variables in a noncommutative probability space $(\mathcal{A}, \phi)$, and $\{a^-_{x_i}\}_{1\leq i\leq d}$, $\{a^0_{x_i}\}_{1\leq i\leq d}$, and $\{a^+_{x_i}\}_{1\leq i\leq d}$ a joint quantum decomposition of $\{X_i\}_{1\leq i\leq d}$ relative to $\mathcal{A}$. Let $\mathcal{A}'$ be the algebra generated by $\{X_i\}_{1\leq i\leq d}$, and $\{a^-(i)\}_{1\leq i\leq d}$, $\{a^0(i)\}_{1\leq i\leq d}$, and $\{a^+(i)\}_{1\leq i\leq d}$ be the minimal joint quantum decomposition of $\{X_i\}_{1\leq i\leq d}$. Then for any $i \in \{1, 2, \ldots, d\}$ and any $\epsilon \in \{-, 0, +\}$, we have:*

$$a^\epsilon_{x_i}|\mathcal{A}'\phi = a^\epsilon(i)|\mathcal{A}'\phi. \tag{38}$$

Moreover, if $\mathcal{A}''$ denotes the algebra generated by $\cup_{i=1}^d \{a^-_{x_i}, a^0_{x_i}, a^+_{x_i}\}$, then

$$\mathcal{A}''\phi = \mathcal{A}'\phi, \tag{39}$$

and $\{a^-_{x_i}\}_{1\leq i\leq d}$, $\{a^0_{x_i}\}_{1\leq i\leq d}$, and $\{a^+_{x_i}\}_{1\leq i\leq d}$ is a joint quantum decomposition of $\{X_i\}_{1\leq i\leq n}$ relative to $\mathcal{A}''$.

3. A commutator problem

Let us consider two symmetric "random variables" X and Y, whose joint quantum operators satisfy the following commutation relationships:

$$\left[a^-_x, a^-_y\right] = 0, \tag{1}$$
$$\left[a^-_x, a^+_x\right] = I, \tag{2}$$
$$\left[a^-_x, a^+_y\right] = bI, \tag{3}$$
$$\left[a^-_y, a^+_y\right] = I, \tag{4}$$
$$\left[a^-_x, a^0_x\right] = pa^-_x + p'a^-_y, \tag{5}$$
$$\left[a^-_x, a^0_y\right] = 0, \tag{6}$$
$$\left[a^-_y, a^0_x\right] = 0, \tag{7}$$
$$\left[a^-_y, a^0_y\right] = q'a_x + qa^-_y. \tag{8}$$

We assume that $E[X] = E[Y] = 0$.

Taking the adjoint in both sides of each of the relations (1), (3), (5), and (8), we obtain the following commutation relationships:

$$\begin{aligned} \left[a_y^+, a_x^+\right] &= 0, &(9)\\ \left[a_y^-, a_x^+\right] &= bI, &(10)\\ \left[a_x^0, a_x^+\right] &= pa_x^+ + p'a_y^+, &(11)\\ \left[a_y^0, a_y^+\right] &= q'a_x^+ + qa_y^+. &(12)\end{aligned}$$

From the Jacobi commutation identity:

$$\left[a_y^-, \left[a_x^0, a_x^+\right]\right] + \left[a_x^0, \left[a_x^+, a_y^-\right]\right] + \left[a_x^+, \left[a_y^-, a_x^0\right]\right] = 0,$$

we conclude that $p' = -pb$. Similarly, we can see that $q' = -qb$. Thus the commutation relationships (5) and (8) can be written as:

$$\begin{aligned} \left[a_x^-, a_x^0\right] &= p\left(a_x^- - ba_y^-\right), &(13)\\ \left[a_y^-, a_y^0\right] &= q\left(a_y^- - ba_x^-\right). &(14)\end{aligned}$$

We assume that $|b| \neq 1$, $b \neq 0$, $p \neq 0$, and $q \neq 0$. It follows that $XY \neq YX$, since if $XY = YX$, then we would have:

$$\begin{aligned} 0 &= [X, Y]\\ &= \left[a_x^- + a_x^0 + a_x^+, a_y^- + a_y^0 + a_y^+\right]\\ &= \sum_{(\epsilon_1,\epsilon_2)\in\{-,0,+\}^2} \left[a_x^{\epsilon_1}, a_y^{\epsilon_2}\right].\end{aligned}$$

In the above sum of commutators, we have $[a_x^{\epsilon_1}, a_y^{\epsilon_2}] + [a_x^{\epsilon_2}, a_y^{\epsilon_1}] = 0$, for all $\epsilon_1 \neq \epsilon_2$. Moreover, $[a_x^-, a_y^-] = [a_x^+, a_y^+] = 0$. Thus we obtain:

$$\begin{aligned} \left[a_x^0, a_y^0\right] &= [X, Y]\\ &= 0.\end{aligned}$$

Using now the Jacobi identity:

$$\left[a_x^-, \left[a_x^0, a_y^0\right]\right] + \left[a_x^0, \left[a_y^0, a_x^-\right]\right] + \left[a_y^0, \left[a_x^-, a_x^0\right]\right] = 0,$$

and the fact that $bpq \neq 0$, we conclude that $a_y^- = ba_x^-$. Taking the adjoint in both sides of this equality we get $a_y^+ = ba_x^+$. It follows now from (2) and (4) that $b^2 = 1$, which is not possible since $|b| \neq 1$. Therefore we are dealing with non–commutative random variables.

We will show at the end of the paper that, for each $b \in (-1, 1)$, $p \neq 0$, and $q \neq 0$, there is a model in which these commutation relationships are realized. The necessary condition $|b| \leq 1$ follows from the Schwarz inequality, as it will be explained below. We exclude the case $b = \pm 1$ from our discussion.

Since both X and Y are symmetric, using Schwarz inequality, we obtain:

$$\begin{aligned} |E[XY]| &= |\langle XY\phi, \phi\rangle|\\ &= |\langle Y\phi, X\phi\rangle|\\ &\leq (\langle Y\phi, Y\phi\rangle)^{1/2} (\langle X\phi, X\phi\rangle)^{1/2}\\ &= \left(\langle X^2\phi, \phi\rangle\right)^{1/2} \left(\langle Y^2, \phi\rangle\right)^{1/2}\\ &= \left(E\left[X^2\right]\right)^{1/2} \left(E\left[Y^2\right]\right)^{1/2}. &(15)\end{aligned}$$

Because $E[X] = E[Y] = 0$, we have $a_x^0\phi = a_y^0\phi = 0$. Since $a_x^-\phi = a_y^-\phi = 0$, we have:

$$\begin{aligned} E[XY] &= \langle (a_x^- + a_x^0 + a_x^+)(a_y^- + a_y^0 + a_y^+)\phi, \phi\rangle \\ &= \langle (a_x^- + a_x^0 + a_x^+)a_y^+\phi, \phi\rangle \\ &= \langle a_y^+\phi, (a_x^+ + a_x^0 + a_x^-)\phi\rangle \\ &= \langle a_y^+\phi, a_x^+\phi\rangle \\ &= \langle a_x^- a_y^+\phi, \phi\rangle \\ &= \langle (a_x^- a_y^+ - a_y^+ a_x^-)\phi, \phi\rangle \\ &= \langle [a_x^-, a_y^+]\phi, \phi\rangle \\ &= \langle bI\phi, \phi\rangle \\ &= b. \end{aligned} \tag{16}$$

Similarly, we have:

$$E[X^2] = 1 \tag{17}$$

and

$$E[Y^2] = 1. \tag{18}$$

It follows now from (15), (16), (17), and (18) that:

$$|b| \leq 1. \tag{19}$$

Inside the algebra $\mathcal{A}$, we can consider the bi–module $M(X, Y)$ (left module over the ring $\mathcal{P}(X)$ of all polynomials of X, and right module over the ring $\mathcal{P}(Y)$ of all polynomials of Y) generated by the identity operator I. That means, $M(X, Y)$ is the vector space spanned by the monomials of the form $\{X^mY^n\}_{m\geq 0, n\geq 0}$, in which all factors of X are to the left of all factors of Y. We show below a method for computing the expectation of all monomials from $M(X, Y)$.

Lemma 3.1 *If X and Y satisfy the commutation relationships (1)–(8), then for any m and n non–negative integers, with $m \geq 1$, we have:*

$$\begin{aligned} E[X^mY^n] &= (m-1)E[X^{m-2}Y^n] + nbE[X^{m-1}Y^{n-1}] \\ &\quad +(1-b^2)\binom{m-1}{2}pE[X^{m-3}Y^n] \\ &\quad +p^2 \sum_{0\leq j<i\leq m-2} \langle X^{m-3-j}(a_x^- - ba_y^-)X^jY^n\phi, \phi\rangle \\ &\quad -(m-1)bpqR_2, \end{aligned}$$

where

$$R_2 := \sum_{0\leq k\leq n-1} \langle X^{m-2}Y^{n-1-k}(a_y^- - ba_x^-)Y^k\phi, \phi\rangle.$$

Proof. Let m and n be two fixed non–negative integers, such that $m \geq 1$. We apply the commutator method from [9]. According to that method:

$$\begin{aligned} E[X^mY^n] &= \langle X^mY^n\phi, \phi\rangle \\ &= \langle (a_x^+ + a_x^0 + a_x^-)X^{m-1}Y^n\phi, \phi\rangle \\ &= \langle X^{m-1}Y^n\phi, a_x^-\phi\rangle + \langle X^{m-1}Y^n\phi, a_x^0\phi\rangle + \langle a_x^- X^{m-1}Y^n\phi, \phi\rangle \\ &= 0 + \langle X^{m-1}Y^n\phi, E[X]\phi\rangle + \langle a_x^- X^{m-1}Y^n\phi, \phi\rangle \\ &= \langle a_x^- X^{m-1}Y^n\phi, \phi\rangle, \end{aligned}$$

Journal of Physics: Conference Series **474** (2013) 012027 doi:10.1088/1742-6596/474/1/012027

since $E[X]=0$. We swap now a_x^- and $X^{m-1}Y^n$, using the product rule for commutators:

$$[A,B_1\cdots B_k] = \sum_{i=1}^{k} B_1\cdots B_{i-1}\,[A,B_i]\,B_{i+1}\cdots B_k,$$

for all operators A, B_1, …, B_k, and the fact that $a_x^-\phi=0$, and obtain:

$$\begin{aligned}
E[X^mY^n] &= \langle a_x^- X^{m-1}Y^n\phi,\phi\rangle \\
&= \langle X^{m-1}Y^n a_x^-\phi,\phi\rangle + \langle [a_x^-, X^{m-1}Y^n]\phi,\phi\rangle \\
&= \sum_{0\le i\le m-2} \langle X^{m-2-i}[a_x^-,X]X^iY^n\phi,\phi\rangle \\
&\quad + \sum_{0\le j\le n-1} \langle X^{m-1}Y^{n-1-j}[a_x^-,Y]Y^j\phi,\phi\rangle .
\end{aligned}$$

Let us observe now that:

$$\begin{aligned}
[a_x^-,X] &= [a_x^-, a_x^+ + a_x^0 + a_x^-] \\
&= [a_x^-,a_x^+] + [a_x^-,a_x^0] \\
&= I + p(a_x^- - ba_y^-)
\end{aligned}$$

and

$$\begin{aligned}
[a_x^-,Y] &= [a_x^-,a_y^+] + [a_x^-,a_y^0] + [a_x^-,a_y^-] \\
&= bI + 0 + 0 \\
&= bI.
\end{aligned}$$

Therefore, we have:

$$\begin{aligned}
E[X^mY^n] &= \sum_{0\le i\le m-2} \langle X^{m-2-i}\{I + p(a_x^- - ba_y^-)\}X^iY^n\phi,\phi\rangle \\
&\quad + \sum_{0\le j\le n-1} \langle X^{m-1}Y^{n-1-j}bIY^j\phi,\phi\rangle \\
&= (m-1)E[X^{m-2}Y^n] + nbE[X^{m-1}Y^{n-1}] \\
&\quad + p\sum_{0\le i\le m-2} \langle X^{m-2-i}(a_x^- - ba_y^-)X^iY^n\phi,\phi\rangle .
\end{aligned}$$

Let us swap now the operators $a_x^- - ba_y^-$ and X^iY^n, using again the product rule for commutators, and the fact that $(a_x^- - ba_y^-)\phi = 0$. We obtain:

$$\begin{aligned}
E[X^mY^n] &= (m-1)E[X^{m-2}Y^n] + nbE[X^{m-1}Y^{n-1}] \\
&\quad + p\sum_{0\le i\le m-2} \langle X^{m-2-i}(a_x^- - ba_y^-)X^iY^n\phi,\phi\rangle \\
&= (m-1)E[X^{m-2}Y^n] + nbE[X^{m-1}Y^{n-1}] \\
&\quad + p\sum_{0\le i\le m-2} \langle X^{m-2-i}[a_x^- - ba_y^-, X^iY^n]\phi,\phi\rangle \\
&= (m-1)E[X^{m-2}Y^n] + nbE[X^{m-1}Y^{n-1}] \\
&\quad + p\sum_{0\le j<i\le m-2} \langle X^{m-2-i}X^{i-1-j}[a_x^- - ba_y^-, X]X^jY^n\phi,\phi\rangle \\
&\quad + p\sum_{0\le i\le m-2}\sum_{0\le k\le n-1} \langle X^{m-2-i}X^iY^{n-1-k}[a_x^- - ba_y^-, Y]Y^k\phi,\phi\rangle .
\end{aligned}$$

Using the simple facts that $X^{m-2-i}X^{i-1-j} = X^{m-3-j}$ and $X^{m-2-i}X^i = X^{m-2}$, we obtain:

$$\begin{aligned} E[X^mY^n] &= (m-1)E[X^{m-2}Y^n] + nbE[X^{m-1}Y^{n-1}] \\ &\quad +p \sum_{0\leq j<i\leq m-2} \langle X^{m-3-j}[a_x^- - ba_y^-, X]X^jY^n\phi, \phi\rangle \\ &\quad +(m-1)p \sum_{0\leq k\leq n-1} \langle X^{m-2}Y^{n-1-k}[a_x^- - ba_y^-, Y]Y^k\phi, \phi\rangle. \end{aligned}$$

Let us observe that:

$$\begin{aligned} [a_x^- - ba_y^-, X] &= [a_x^-, X] - b[a_y^-, X] \\ &= I + p(a_x^- - ba_y^-) - b^2I \\ &= (1-b^2)I + p(a_x^- - ba_y^-) \end{aligned}$$

and

$$\begin{aligned} [a_x^- - ba_y^-, Y] &= [a_x^-, Y] - b[a_y^-, Y] \\ &= bI - b\{I + q(a_y^- - ba_x^-)\} \\ &= -bq(a_y^- - ba_x^-). \end{aligned}$$

Thus, we obtain:

$$\begin{aligned} E[X^mY^n] &= (m-1)E[X^{m-2}Y^n] + nbE[X^{m-1}Y^{n-1}] \\ &\quad +p \sum_{0\leq j<i\leq m-2} \langle X^{m-3-j}\{(1-b^2)I + p(a_x^- - ba_y^-)\}X^jY^n\phi, \phi\rangle \\ &\quad -(m-1)bpq \sum_{0\leq k\leq n-1} \langle X^{m-2}Y^{n-1-k}(a_y^- - ba_x^-)Y^k\phi, \phi\rangle \\ &= (m-1)E[X^{m-2}Y^n] + nbE[X^{m-1}Y^{n-1}] \\ &\quad +(1-b^2)\binom{m-1}{2}pE[X^{m-3}Y^n] \\ &\quad +p^2 \sum_{0\leq j<i\leq m-2} \langle X^{m-3-j}(a_x^- - ba_y^-)X^jY^n\phi, \phi\rangle \\ &\quad -(m-1)bpqR_2, \end{aligned}$$

where

$$R_2 := \sum_{0\leq k\leq n-1} \langle X^{m-2}Y^{n-1-k}(a_y^- - ba_x^-)Y^k\phi, \phi\rangle.$$

□

Lemma 3.2 *Using the notations from the Lemma 3.1, we have:*

$$R_2 = \frac{1-b^2}{q^2}E\left[X^{m-2}\left\{(Y+qI)^n - Y^n - nqY^{n-1}\right\}\right].$$

Proof. Swapping $a_y^- - ba_x^-$ and Y^k, we get:

$$\begin{aligned} R_2 &= \sum_{0\le k\le n-1} \langle X^{m-2}Y^{n-1-k}[a_y^- - ba_x^-, Y^k]\phi, \phi\rangle \\ &= \sum_{0\le l<k\le n-1} \langle X^{m-2}Y^{n-1-k}Y^{k-1-l}[a_y^- - ba_x^-, Y]Y^l\phi, \phi\rangle \\ &= \sum_{0\le l<k\le n-1} \langle X^{m-2}Y^{n-2-l}\{(1-b^2)I + q(a_y^- - ba_x^-)\}Y^l\phi, \phi\rangle \\ &= (1-b^2)\binom{n}{2}E[X^{m-2}Y^{n-2}] \\ &\quad +q \sum_{0\le l<k\le n-1} \langle X^{m-2}Y^{n-2-l}(a_y^- - ba_x^-)Y^l\phi, \phi\rangle. \end{aligned}$$

We repeat this procedure, swapping now $a_y^- - ba_x^-$ and Y^l, and so on, and obtain in the end that:

$$\begin{aligned} R_2 &= (1-b^2)\left\{\binom{n}{2}E[X^{m-2}Y^{n-2}] + \binom{n}{3}qE[X^{m-2}Y^{n-3}] + \cdots \right. \\ &\qquad \left. \cdots + \binom{n}{n}q^{n-2}E[X^{m-2}Y^0]\right\} \\ &= \frac{1-b^2}{q^2}E\left[X^{m-2}\left\{(Y+qI)^n - Y^n - nqY^{n-1}\right\}\right]. \end{aligned}$$

□

Lemma 3.3 *If X and Y satisfy the commutation relationships (1)–(8), then for any m and n non–negative integers, with $m \ge 1$, we have:*

$$\begin{aligned} &E[X^mY^n] \\ &= (m-1)E[X^{m-2}Y^n] + nbE[X^{m-1}Y^{n-1}] \\ &\quad + \frac{1-b^2}{p}E\left[\left\{(X+pI)^{m-1} - X^{m-1} - (m-1)pX^{m-2}\right\}Y^n\right] \\ &\quad - \frac{b(1-b^2)}{q}E\left[\left\{(X+pI)^{m-1} - X^{m-1}\right\}\left\{(Y+qI)^n - Y^n - nqY^{n-1}\right\}\right]. \end{aligned} \tag{20}$$

Proof. Using Lemmas 3.1 and 3.2, we get:

$$\begin{aligned} &E[X^mY^n] \\ &= (m-1)E[X^{m-2}Y^n] + nbE[X^{m-1}Y^{n-1}] \\ &\quad +(1-b^2)\binom{m-1}{2}pE[X^{m-3}Y^n] \\ &\quad +p^2 \sum_{0\le j<i\le m-2} \langle X^{m-3-j}(a_x^- - ba_y^-)X^jY^n\phi, \phi\rangle \\ &\quad - \frac{b(1-b^2)}{q}E\left[(m-1)pX^{m-2}\left\{(Y+qI)^n - Y^n - nqY^{n-1}\right\}\right]. \end{aligned}$$

We do now a similar procedure as in the proof of Lemma 3.1, by swapping first $(a_x^- - ba_y^-)$ and X^jY^n. In this way we arrive at a new sum R_3 which is similar to R_2. We compute R_3 in the same way that we calculated R_2 in the proof of Lemma 3.2, and so on, obtaining in the end that:

$$\begin{aligned}
& E[X^m Y^n] \\
= \;& (m-1)E[X^{m-2}Y^n] + nbE[X^{m-1}Y^{n-1}] \\
& +(1-b^2)\left\{\binom{m-1}{2}pE[X^{m-3}Y^n] + \binom{m-1}{3}p^2E[X^{m-4}Y^n] + \cdots \right. \\
& \qquad \left. \cdots + \binom{m-1}{m-1}p^{m-2}E[X^0Y^n]\right\} \\
& -\frac{b(1-b^2)}{q}E\left[\binom{m-1}{1}pX^{m-2}\left\{(Y+qI)^n - Y^n - nqY^{n-1}\right\}\right] \\
& -\frac{b(1-b^2)}{q}E\left[\binom{m-1}{2}p^2X^{m-3}\left\{(Y+qI)^n - Y^n - nqY^{n-1}\right\}\right] \\
& \ldots \\
& -\frac{b(1-b^2)}{q}E\left[\binom{m-1}{m-1}p^{m-1}X^0\left\{(Y+qI)^n - Y^n - nqY^{n-1}\right\}\right] \\
= \;& (m-1)E[X^{m-2}Y^n] + nbE[X^{m-1}Y^{n-1}] \\
& +\frac{1-b^2}{p}E\left[\left\{(X+pI)^{m-1} - X^{m-1} - (m-1)pX^{m-2}\right\}Y^n\right] \\
& -\frac{b(1-b^2)}{q}E\left[\left\{(X+pI)^{m-1} - X^{m-1}\right\}\left\{(Y+qI)^n - Y^n - nqY^{n-1}\right\}\right].
\end{aligned}$$

□

4. Joint left X – right Y Laplace transform of X and Y

In this section, we compute the joint Laplace transform of the random variables X and Y from the previous section. We start first by presenting some important bounds for the joint moments of X and Y.

Lemma 4.1 *Let $X = a_x^- + a_x^0 + a_x^+$ and $Y = a_y^0 + a_y^0 + a_y^+$ be joint quantum decompositions of two random variables, whose terms satisfy the commutation relationships (1)–(8), where b, p, p', q, and q' are real numbers, such that $|b| < 1$, $p' = -pb$, and $q' = -qb$. We assume that $E[X] = E[Y] = 0$. Then there exists a positive constant C, such that, for any m and n non–negative integers, we have:*

$$|E[X^m Y^n]| \;\leq\; Cm!n!. \tag{21}$$

Proof. Indeed, if we define, for all non–negative integers m and n,

$$C_{m,n} \;:=\; \max\{|E[X^iY^j]|/(i!j!) \mid 0 \leq i \leq m, 0 \leq j \leq n\}, \tag{22}$$

then the sequence $\{C_{m,n}\}_{m\geq 0, n\geq 0}$ is left and right increasing, in the sense that if we fix n, then it is increasing with respect to m, and vice–versa. We have:

$$\begin{aligned}
& \left|E\left[\left\{(X+pI)^{m-1} - X^{m-1} - (m-1)pX^{m-2}\right\}Y^n\right]\right| \\
\leq \;& (m-1)!n!\sum_{i=2}^{m-1}\frac{|E[X^{m-1-i}Y^n]|}{(m-1-i)!n!}\cdot\frac{p^i}{i!} \\
\leq \;& (m-1)!n!C_{m-3,n}\sum_{i=2}^{m-1}\frac{p^i}{i!} \\
\leq \;& \left(e^p - 1 - p\right)C_{m-3,n}(m-1)!n!.
\end{aligned}$$

Similarly, we have:

$$\begin{aligned} & \left|E\left[\{(X+pI)^{m-1}-X^{m-1}\}\{(Y+qI)^n-Y^n-nqY^{n-1}\}\right]\right| \\ \leq & \ (e^p-1)(e^q-1-q)\,C_{m-2,n-2}(m-1)!n!. \end{aligned}$$

It follows now from our recursive formula (20) that:

$$\begin{aligned} & |E[X^mY^n]| \\ \leq & \ (m-1)C_{m-2,n}(m-2)!n! + nbC_{m-1,n-1}(m-1)!(n-1)! \\ & + \frac{1-b^2}{|p|}(e^p-1-p)\,C_{m-3,n}(m-1)!n! \\ & + \frac{|b|(1-b^2)}{|q|}(e^p-1)(e^q-1-q)\,C_{m-2,n-2}(m-1)!n!. \end{aligned}$$

Since the sequence $\{C_{m,n}\}_{m\geq 0,n\geq 0}$ is left and right increasing, we conclude now that there is a constant K depending on b, p, and q, such that:

$$|E[X^mY^n]| \ \leq \ KC_{m-1,n}(m-1)!n!,$$

for all $m \geq 1$, $n \geq 0$. Dividing both sides of this inequality by $m!n!$, we obtain:

$$\frac{|E[X^mY^n]|}{m!n!} \ \leq \ \frac{K}{m}C_{m-1,n},$$

for all $m \geq 1$ and $n \geq 0$. Thus, we have:

$$\begin{aligned} C_{m,n} &= \max\left\{C_{m-1,n}, \left|E[X^mY^0]\right|/(m!0!), \ldots, \left|E[X^mY^n]\right|/(m!n!)\right\} \\ &\leq \max\left\{C_{m-1,n}, \frac{K}{m}C_{m-1,0}, \ldots, \frac{K}{m}C_{m-1,n}\right\} \\ &= \max\left\{C_{m-1,n}, \frac{K}{m}C_{m-1,n}\right\} \\ &= C_{m-1,n}, \end{aligned}$$

for all $m \geq \lfloor K \rfloor + 1$ and $n \geq 0$. On the other hand we have $C_{m,n} \geq C_{m-1,n}$, for all $m \geq 1$ and all $n \geq 0$. Thus, the sequences $\{C_{m,0}\}_{m\geq 1}$, $\{C_{m,1}\}_{m\geq 1}$, $\ldots$ are uniformly stationary, in the sense that there is natural number m_0, that is the same for all $n \geq 0$, such that $C_{m_0,n} = C_{m_0+1,n} = C_{m_0+2,n} = \cdots$.

On the other hand, since X and Y are symmetric operators, we have:

$$\begin{aligned} E[Y^nX^m] &= \langle Y^nX^m\phi, \phi\rangle \\ &= \langle \phi, X^mY^n\phi\rangle \\ &= E[X^mY^n]. \end{aligned}$$

Thus a similar argument will imply now that there exists an $n_0 \geq 1$, such that for all $m \geq 0$, $C_{m,n_0} = C_{m,n_0+1} = C_{m,n_0+2} = \cdots$. Since the sequence $\{C_{m,n}\}_{m\geq 0,n\geq 0}$ is left and right increasing, we conclude now that, for all $m \geq 0$ and all $n \geq 0$, we have $C_{m,n} \leq C_{m_0,n_0}$. Thus, for all $m \geq 0$ and $n \geq 0$, $|E[X^mY^n]| \leq C_{m_0,n_0}m!n!$. □

We are now ready to compute the joint Laplace transform of X and Y.

Theorem 4.2 *Let $X = a_x^- + a_x^0 + a_x^+$ and $Y = a_y^0 + a_y^0 + a_y^+$ be joint quantum decompositions of two random variables, whose terms satisfy the commutation relationships (1)–(8), where b, p, p', q, and q' are real numbers, such that $|b| < 1$, $p' = -pb$, and $q' = -qb$. We assume that $E[X] = E[Y] = 0$. Then, for all $(s,\, t) \in (-1,\, 1)^2$, the series:*

$$E\left[e^{sX}e^{tY}\right] := \sum_{m\geq 0}\sum_{n\geq 0}\frac{s^m t^n E[X^m Y^n]}{m!n!} \tag{23}$$

is absolutely convergent, and the joint left X–right Y Laplace transform *is:*

$$\begin{aligned} E\left[e^{sX}e^{tY}\right] &= e^{\frac{s^2+2bst+t^2}{2}} \\ &\times e^{\frac{1-b^2}{p^2}\left(e^{ps}-1-ps-\frac{1}{2}p^2s^2\right)} e^{\frac{1-b^2}{q^2}\left(e^{qt}-1-qt-\frac{1}{2}q^2t^2\right)} \\ &\times e^{-\frac{b(1-b^2)}{pq}\left(e^{ps}-1-ps\right)\left(e^{qt}-1-qt\right)}. \end{aligned} \tag{24}$$

If $p = 0$ or $q = 0$, then formula (24) makes sense by taking the limit as $p \to 0$ or $q \to 0$ in its right–hand side. Moreover, there exist two commuting random variables U and V, such that $(X,\, Y) \equiv (U,\, V)$, if and only if $bpq = 0$. If $bpq = 0$, then there exist an invertible 2×2 matrix T, a vector $(u,\, v)$ in $\mathbb{R}^2$, and two independent classic random variables X' and Y', that are either Gaussian or Poisson, such that $(X,\, Y) \equiv (X',\, Y')T + (u,\, v)$.

Proof. It follows from Lemma 4.1 that the series $\sum_{m\geq 0, n\geq 0} s^m t^n E[X^m Y^n]/m!n!$ is absolutely convergent for all s and t complex numbers, such that $|s| < 1$ and $|t| < 1$. We denote this series by $E[e^{sX}e^{tY}]$, to be in agreement with the classic commutative case. Let $\varphi(s,\, t) := E[e^{sX}e^{tY}]$. It follows from Weierstrass M–test that the double series, from the definition of $\varphi(s,\, t)$, can be differentiated term by term, with respect to both s and t, for $|s| < 1$ and $|t| < 1$. If we multiply both sides of the recursive relation (20) by $s^{m-1}t!/[(m-1)!n!]$ and sum up the resulting relations from $m = 1$ to ∞, and from $n = 0$ to ∞, then we obtain:

$$\begin{aligned} \frac{\partial\varphi}{\partial s}(s,t) &= s\varphi(s,t) + bt\varphi(s,t) \\ &+\frac{1-b^2}{p}(e^{ps}-1-ps)\varphi(s,t) \\ &-\frac{b(1-b^2)}{q}(e^{ps}-1)(e^{qt}-1-qt)\varphi(s,t). \end{aligned} \tag{25}$$

In particular, for $t = 0$, we conclude that the function $g(s) := E[e^{sX}]$ satisfies the differential equation:

$$\frac{dg}{ds}(s) = sg(s) + \frac{1-b^2}{p}\left(e^{ps}-1-ps\right)g(s).$$

Since $g(0) = 1$, we must have:

$$g(s) = e^{\frac{s^2}{2}+\frac{1-b^2}{p^2}\left(e^{ps}-1-ps-\frac{p^2s^2}{2}\right)}.$$

Similarly, we can see that:

$$E[e^{tY}] = e^{\frac{t^2}{2}+\frac{1-b^2}{q^2}\left(e^{qt}-1-qt-\frac{q^2t^2}{2}\right)}.$$

Solving now the differential equation, in the variable s, (25), we obtain:

$$\begin{aligned}\varphi(s,t) &= K(t)e^{\frac{s^2}{2}}e^{bst}e^{\frac{1-b^2}{p^2}\left(e^{ps}-1-ps-\frac{p^2s^2}{2}\right)}\\ &\quad\times e^{-\frac{b(1-b^2)}{pq}(e^{ps}-1-ps)(e^{qt}-1-qt)}.\end{aligned}$$

Setting $s=0$ in this equation and using the fact that $\varphi(0,t)=E[e^{tY}]$, we conclude that:

$$K(t) = e^{\frac{t^2}{2}+\frac{1-b^2}{q^2}\left(e^{qt}-1-qt-\frac{q^2t^2}{2}\right)}.$$

Thus, we have:

$$\begin{aligned}E[e^{sX}e^{tY}] &= e^{\frac{t^2}{2}+\frac{1-b^2}{q^2}(e^{qt}-1-qt-\frac{q^2t^2}{2})}\\ &\quad\times e^{\frac{s^2}{2}}e^{bst}e^{\frac{1-b^2}{p^2}(e^{ps}-1-ps-\frac{p^2s^2}{2})}\\ &\quad\times e^{-\frac{b(1-b^2)}{pq}(e^{ps}-1-ps)(e^{qt}-1-qt)}\\ &= e^{\frac{s^2+2bst+t^2}{2}}\\ &\quad\times e^{\frac{1-b^2}{p^2}(e^{ps}-1-ps-\frac{p^2s^2}{2})}\\ &\quad\times e^{\frac{1-b^2}{q^2}(e^{qt}-1-qt-\frac{q^2t^2}{2})}\\ &\quad\times e^{-\frac{b(1-b^2)}{pq}(e^{ps}-1-ps)(e^{qt}-1-qt)}.\end{aligned}$$

□

5. The existence of the operators X and Y

We close the paper by showing that, for any $|b|<1$, and any real numbers p and q, we can construct two random variables satisfying the commutation relationships from the previous theorem. Let b, p, and q be fixed, such that $|b|<1$. Let U and V be two jointly Gaussian classic (commutative) random variables, with mean 0, variance 1, and covariance b, defined on the same probability space $(\Omega,\mathcal{F},P)$. We have $U=a_u^-+a_u^+$ and $V=a_v^-+a_v^+$, $[a_u^-,a_u^+]=[a_v^-,a_v^+]=I$ and $[a_u^-,a_v^+]=[a_v^-,a_u^+]=E[UV]I=bI$. See [5], [6], and [8]. Let $\mathcal{A}$ be the unital algebra generated by a_u^-, a_u^+, a_v^-, and a_v^+. This is an algebra of operators that are densely defined on the Hilbert space $H:=L^2(\Omega,\sigma(U,V),P)$, where $\sigma(U,V)$ represents the smallest sub–sigma–algebra of $\mathcal{F}$, with respect to which both U and V are measurable functions. Inside the algebra $\mathcal{A}$, let us consider the following operators:

$$X := a_u^- + \left[\alpha_1 a_u^+a_u^- + \beta_1(a_u^+a_v^- + a_v^+a_u^-) + \gamma_1 a_v^+a_v^-\right] + a_u^+ \tag{26}$$

and

$$Y := a_v^- + \left[\alpha_2 a_v^+a_v^- + \beta_2(a_v^+a_u^- + a_u^+a_v^-) + \gamma_2 a_u^+a_u^-\right] + a_v^+, \tag{27}$$

where α_i, β_i, and γ_i, $1\le i\le 2$, are real constants that will be determined such that X and Y satisfy the commutation relationships (1)–(8). The coefficients of $a_x^+a_y^-$ and $a_y^+a_x^-$ are the same in the definitions of X and Y, β_1 and β_2, respectively, to ensure that X and Y are both symmetric operators. Let $\phi:=1\in H$, be the unit vector with respect to which the noncommutative expectation will be computed. The terms of a joint quantum decomposition of X and Y (relative to $\mathcal{A}$), are:

$$\begin{aligned}
a_x^- &= a_u^-, \\
a_x^0 &= \alpha_1 a_u^+ a_u^- + \beta_1(a_u^+ a_v^- + a_v^+ a_u^-) + \gamma_1 a_v^+ a_v^-, \\
a_x^+ &= a_u^+, \\
a_y^- &= a_v^-, \\
a_y^0 &= \alpha_2 a_v^+ a_v^- + \beta_2(a_v^+ a_u^- + a_u^+ a_v^-) + \gamma_2 a_u^+ a_u^-, \\
a_y^+ &= a_v^+.
\end{aligned}$$

It is now clear that (1), (2), (3), (4) hold. Also, because in the formulas of a_x^0 and a_y^0, the annihilation operators a_u^- and a_v^- are always to the right of the creation operators a_u^+ and a_v^+, we conclude that the conditions $E[X] = 0$ and $E[Y] = 0$ hold. To compute the commutators $[a_x^-, a_x^0]$, $[a_x^-, a_y^0]$, $[a_y^-, a_x^0]$, and $[a_y^-, a_y^0]$, we apply again the product rule for commutators. For example, one commutator involved in the computation of $[a_x^-, a_x^0]$ is $[a_u^-, a_v^+ a_u^-]$, which can be computed as follows:

$$\begin{aligned}
[a_u^-, a_v^+ a_u^-] &= [a_u^-, a_v^+]a_u^- + a_v^+[a_u^-, a_u^-] \\
&= bIa_u^- + 0 \\
&= ba_x^-.
\end{aligned}$$

Doing all these types of commutators, we obtain:

$$\begin{aligned}
[a_x^-, a_x^0] &= (\alpha_1 + b\beta_1)a_x^- + (\beta_1 + b\gamma_1)a_y^-, \\
[a_x^-, a_y^0] &= (\gamma_2 + b\beta_2)a_x^- + (\beta_2 + b\alpha_2)a_y^-, \\
[a_y^-, a_x^0] &= (\beta_1 + b\alpha_1)a_x^- + (\gamma_1 + b\beta_1)a_y^-, \\
[a_y^-, a_y^0] &= (\beta_2 + b\gamma_2)a_x^- + (\alpha_2 + b\beta_2)a_y^-.
\end{aligned}$$

If we want the commutation relationships (5), (6), (7), and (8) to hold, then identifying the coefficients of a_x^- and a_y^-, we should have:

$$\begin{aligned}
\alpha_1 + b\beta_1 &= p, \\
\beta_1 + b\gamma_1 &= -pb, \\
\gamma_2 + b\beta_2 &= 0, \\
\beta_2 + b\alpha_2 &= 0, \\
\beta_1 + b\alpha_1 &= 0, \\
\gamma_1 + b\beta_1 &= 0, \\
\beta_2 + b\gamma_2 &= -qb, \\
\alpha_2 + b\beta_2 &= q.
\end{aligned}$$

This is a system of eight equations and six unknowns, that can be split up into two sub–systems of four equations and three unknowns each. Each of these two sub–systems is consistent and has a unique solution:

$$(\alpha_1, \beta_1, \gamma_1) = \frac{p}{1-b^2}(1, -b, b^2)$$

and

$$(\alpha_2, \beta_2, \gamma_2) = \frac{q}{1-b^2}(1, -b, b^2).$$

Thus we have proven the existence of two noncommutative random variables X and Y, for which the commutation relationships (1)–(8) and the condition $E[X] = E[Y] = 0$ hold.

The left X–right Y Laplace transform $E[e^{sX}e^{tY}]$ is a generating function only for the joint moments of the form $E[X^m Y^n]$, $m \geq 0$, $n \geq 0$. Because X and Y do not commute, we also need to compute other Laplace transforms like $E[e^{s_1 X}e^{t_1 Y}e^{s_2 X}e^{t_2 Y}]$, $E[e^{s_1 X}e^{t_1 Y}e^{s_2 X}e^{t_2 Y}e^{s_3 X}e^{t_3 Y}]$ and so on, to generate the other joint moments like $E[X^{m_1}Y^{n_1}X^{m_2}Y^{n_2}]$, $E[X^{m_1}Y^{n_1}X^{m_2}Y^{n_2}X^{m_3}Y^{n_3}]$, ... Our commutator method can be applied to compute these other Laplace transforms, but the calculations are more complicated.

References

[1] Accardi L, Kuo H-H and Stan A I 2004 Characterization of probability measures through the canonically associated interacting Fock spaces *Infin. Dimens. Anal. Quantum Probab. Relat. Top.* **7** No 4 485-505
[2] Accardi L, Kuo H-H and Stan A I 2007 Moments and commutators of probability measures *Infin. Dimens. Anal. Quantum Probab. Relat. Top.* **10** No 4 591-612
[3] Chihara T S 1978 *An Introduction to Orthogonal Polynomials* (New York: Gordon & Breach)
[4] Drożdżewicz K and Matysiak W 2011 Moments and q-commutators of noncommutative random vectors *Infin. Dimens. Anal. Quantum Probab. Relat. Top.* **14** No 4 629-45
[5] Janson S 1997 *Gaussian Hilbert Spaces* Cambridge University Press
[6] Kuo H-H 1996 *White Noise Distribution Theory* (Boca Raton, FL: CRC Press)
[7] Meixner J Orthogonale 1934 Polynomsysteme mit einer besonderen Gestalt der erzeugenden Funktion *J. London Math. Soc.* **9** 6-13
[8] Obata N 1994 *White Noise Calculus and Fock Space* (Berlin Heidelberg: Springer-Verlag)
[9] Stan A I and Whitaker J J 2009 A study of probability measures through commutators J. Theoret. Probab. **22** No 1 123-45
[10] Stan A I 2011 Two-dimensional Meixner random vectors of class $\mathcal{M}_L$ J. Theoret. Probab. **24** No 1 39-65
[11] Sunder V S 1986 *An invitation to von Neumann Algebras* Springer–Verlag (New York: Springer-Verlag)
[12] Szegö M 1975 *Orthogonal Polynomials* Coll. Publ. **23** Amer. Math. Soc.
[13] Voiculescu D V, Dykema KJ and Nica A 1982 *Free Random Variables* **1** (Providence, RI: CRM Series, Amer. Math. Soc.)

Hamiltonian structure of an operator valued extension of Super KdV equations

Alvaro Restuccia[1], **Adrián Sotomayor**[2]

[1] Physics Department , Universidad de Antofagasta, Antofagasta, Chile
[1] Physics Department, Universidad Simón Bolívar, Caracas, Venezuela
[2] Departament of Matemathics, Universidad de Antofagasta, Antofagasta, Chile

E-mail: arestu@usb.ve,asotomayor@uantof.cl

Abstract. An extension of the super Korteweg-de Vries integrable system in terms of operator valued functions is obtained. In particular the extension contains the $N = 1$ Super KdV and coupled systems with functions valued on a symplectic space. We introduce a Miura transformation for the extended system and obtain its hamiltonian structure. We also obtain an extended Gardner transformation which allows to find an infinite number of conserved quantities of the extended system.

1. Introduction

The $N = 1$ supersymmetric extension of Korteweg-de Vries equation obtained in [1] is equivalent to the equations obtained in [2] by reduction from the super Kadomtsev-Petviashvili hierarchy. The equations are described by a system of nonlinear coupled partial differential equations for fields which take values on the even and odd parts of a Grassmann algebra. The system posses a sequence of infinite local as well as non-local conserved quantities [1, 3, 4, 5]. Later on, also supersymmetric extensions with more than one supersymmetric generator were obtained [6, 7, 8, 9, 10].

Coupled partial differential equations in terms of real commuting fields may be obtained from these supersymmetric models by expanding the even and odd fields in terms of a basis of the Grassmann algebra and separating the equations corresponding to each generator. Families of new solutions [11, 12] to super KdV were obtained using such bosonization procedure [13].

The bosonization approach becomes non-trivial when one considers the hamiltonian formulation, the Poisson structure on the unconstrained phase space and the Poisson structure on the constrained phase space given by the Dirac brackets. In fact, the formulation in terms of even variables introduces antisymmetric Poisson brackets while the one in terms of odd variables introduces symmetric Poisson brackets and the equivalence has to be shown explicitly. In the case we will consider in this paper both approaches becomes completely equivalent.

In this paper we will consider a very general algebraic structure for the fields describing the nonlinear systems. The systems are coupled partial differential equations extending the KdV equation. In particular our KdV extensions include the supersymmetric $N = 1$ SKdV system as well as deformations of it. It also describe systems whose odd part satisfies a symplectic bracket structure.

We will obtain the hamiltonian structure of these extended KdV systems. The Poisson structure on the constrained submanifold of phase space will be given in terms of the Dirac brackets defined for constrained systems [14, 15, 16, 17]. The extended KdV systems we will consider have an infinite sequence of conserved quantities which may be obtained via a Gardner transformation as was done in [18, 13].

2. Extension of KdV equation, hamiltonian structure and infinite sequence of conserved quantities

We consider an associative algebra of even and odd elements. The even elements belong to a commutative algebra $\mathcal{P}$ with unit while the odd ones belong to $\mathcal{Q}$ which satisfies

$$\begin{aligned} &\mathcal{Q}\mathcal{P} \subset \mathcal{Q} \\ &[\mathcal{Q},\mathcal{P}] = 0 \\ &[\mathcal{Q},\mathcal{Q}] \subset \mathcal{P}, \end{aligned} \tag{1}$$

and for any $q \in \mathcal{Q}$ there always exists $\hat{q} \in \mathcal{Q}$ such that $[q,\hat{q}] \neq 0$.

We introduce a Lagrangian formulated in terms of fields w and η valued on $\mathcal{P}$ and $\mathcal{Q}$ respectively. The Lagrangian depends on a real parameter λ. The Lagrangian is

$$\mathcal{L} = \frac{1}{2}\dot{w}w' + \frac{\lambda}{2}\left[\dot{\eta},\eta\right] - \frac{1}{2}(w'')^2 - \frac{1}{2}(w')^4 - \frac{1}{2}\lambda^2\left[\eta,\eta'\right]^2 - \frac{\lambda}{2}\left[\eta'',\eta'\right] - \frac{3}{2}\lambda(w')^2\left[\eta',\eta\right]. \tag{2}$$

We notice that if λ is positive one may redefine $\eta \to \lambda\eta$ and reduce the Lagrangian to the case $\lambda = +1$. If λ is negative one may reduce to the case $\lambda = -1$.

The associated field equations are

$$\dot{v} + v''' - 6v^2v' - 3\lambda\left(v\left[\eta',\eta\right]\right)' = 0 \tag{3}$$

$$\dot{\eta} + \eta''' - 3v^2\eta' - 3vv'\eta + \lambda\eta'\left[\eta,\eta'\right] + \frac{1}{2}\eta\lambda\left[\eta,\eta''\right] = 0 \tag{4}$$

where $v \equiv w'$.

In particular if $\mathcal{P}$ and $\mathcal{Q}$ generates a Grassmann algebra, the latter two terms on the second equations are zero.

There is a hamiltonian and a corresponding Poisson structure associated to this Lagrangian.

We denote by p and μ the conjugate momenta associated to w and η respectively.

We then obtain from the definition of p and μ the following primary constraints

$$\begin{aligned} \phi &\equiv p - \tfrac{1}{2}v = 0 \\ \psi &\equiv \mu - \tfrac{\lambda}{2}\eta = 0. \end{aligned} \tag{5}$$

It turns out that these are the only constraints of the theory. They are second class constraints. In fact, they satisfy the following Poisson bracket relations

$$\begin{aligned} \{\phi(x),\phi(y)\}_{PB} &= -\partial_x\delta(x,y) \\ \{\phi(x),\psi(y)\}_{PB} &= 0 \\ \{\psi(x),\psi(y)\}_{PB} &= -\frac{1}{\lambda}\delta(x,y). \end{aligned}$$

The hamiltonian density associated to the Lagrangian (2) may be obtained via a Legendre transformation and is expressed as

$$\mathcal{H} = \frac{1}{2}(v')^2 + \frac{1}{2}v^4 + \frac{1}{2}\lambda^2\left[\eta,\eta'\right]^2 + \frac{1}{2}\lambda\left[\eta'',\eta'\right] + \frac{3}{2}\lambda v^2\left[\eta',\eta\right] \tag{6}$$

subject to the constraints $\phi = 0, \psi = 0$.

The canonical field equations

$$\begin{aligned} \dot{v} &= \{v, H\}_{DB} \\ \dot{\eta} &= \{\eta, H\}_{DB} \end{aligned} \tag{7}$$

where $H = \langle \mathcal{H} \rangle_x$ is the integral on $\mathbb{R}$, exactly agree with the Lagrangian field equations (3),(4) as it should be.

We notice that if $\mathcal{P}$ and $\mathcal{Q}$ generate a Grassmann algebra H reduces to the hamiltonian of the $N = 1$ supersymmetric KdV equations.

In fact, in the case of a Grassmann algebra the term $[\eta, \eta']^2$ in $\mathcal{H}$ becomes zero and we may perform a Miura transformation

$$\begin{aligned} u &= v' + v^2 - \lambda\,[\eta, \eta'] \\ \xi &= \eta' + v\eta \end{aligned} \tag{8}$$

to obtain

$$\mathcal{H} = \frac{1}{2}u^2 + \frac{1}{2}\lambda\,[\xi', \xi] = \frac{1}{2}u^2 + \lambda\xi'\xi.$$

The canonical equations then reduce to the system

$$\begin{aligned} u_t &= -u''' + 6uu' - 6\lambda\xi\xi'' \\ \xi_t &= -\xi''' + 3(u\xi)', \end{aligned} \tag{9}$$

which is invariant under the supersymmetric transformation with odd parameter ϵ,

$$\begin{aligned} \delta_\epsilon u &= 2\epsilon\lambda\xi' \\ \delta_\epsilon \xi &= \epsilon u. \end{aligned} \tag{10}$$

(9) is a parametric Susy KdV equation, for $\lambda = 1$ it gives the $N = 1$ Super KdV equation.

If instead of considering a Grassmann algebra, we consider an associative algebra satisfying in addition to (1) the condition $\mathcal{Q}\mathcal{Q} \subset \mathcal{P}$, which implies $[\mathcal{Q}, \mathcal{Q}] \subset \mathcal{P}$, then the hamiltonian density reduces to

$$\mathcal{H} = \frac{1}{2}(v')^2 + \frac{1}{2}v^4 + \frac{1}{2}\lambda\,[\eta'', \eta'] + \frac{3}{2}\lambda v^2\,[\eta', \eta]\,.$$

After using the generalized Miura transformation given by (8) we obtain

$$\mathcal{H} = \frac{1}{2}u^2 + \frac{\lambda}{2}[\xi', \xi]$$

and the canonical field equations

$$\begin{aligned} u_t &= -u''' + 6uu' + 3\lambda[\xi'', \xi] \\ \xi_t &= -\xi''' + 3(u\xi)', \end{aligned} \tag{11}$$

which are invariant under the transformations

$$\begin{aligned} \delta_\epsilon u &= \lambda[\epsilon, \xi'] \\ \delta_\epsilon \xi &= \epsilon u. \end{aligned} \tag{12}$$

Moreover, the system (11) under assumption (1) has an infinite sequence of local conserved quantities for any value of λ. This property may be proven by using a Gardner transformation as was done for the case $\lambda = 1$ in [13]. In fact, we have

$$\begin{aligned} z_t &= (-z'' + 3z^2 + 3\lambda\,[\sigma',\sigma])' + \epsilon^2(2z^3 + 3\lambda z[\sigma',\sigma])' \\ \sigma_t &= (-\sigma'' + 3z\sigma)' + \epsilon^2 3(z^2\sigma' + zz'\sigma + \lambda\sigma'[\sigma',\sigma]), \end{aligned} \tag{13}$$

$$\begin{aligned} u &= z + \epsilon z' + \epsilon^2(z^2 + \lambda[\sigma',\sigma]) \\ \xi &= \sigma + \epsilon\sigma' + \epsilon^2 z\sigma, \end{aligned} \tag{14}$$

where (14) and (15) are the Gardner equations and associated Gardner transformations respectively.

After simplifying by crossing out derivatives in x and using the inverse Gardner transformation, the first four nontrivial conserved quantities for the operator-extended KdV system (13) are:

$$\begin{aligned} H_0 &= \int u dx \\ H_2 &= \int \left(u^2 + \lambda[\xi',\xi]\right) dx \\ H_4 &= \int \left(2u^3 + (u')^2 + 4\lambda u\,[\xi',\xi] + \lambda\,[\xi'',\xi']\right) dx \\ H_6 &= \int \Big(5u^4 + 10u(u')^2 + (u'')^2 + 15\lambda u^2\,[\xi',\xi] - 2\lambda u\,[\xi'',\xi'] \\ &- 8\lambda u\,[\xi''',\xi] + 3\lambda^2[\xi',\xi]^2 + \lambda\,[\xi''',\xi'']\Big)\, dx. \end{aligned} \tag{15}$$

3. Conclusions

We introduced an operatorial extension for Korteweg-de Vries equation which contains as particular cases several systems with the property of having a sequence of infinite local conserved quantities. In particular it contains the $N = 1$ super KdV system. We obtained the hamiltonian structure of such extension. The existence of an infinite sequence of conserved quantities for the operatorial extension was shown using a generalized Gardner transformation.

Acknowledgments

A. R. and A. S. are partially supported by Project Fondecyt 1121103, Chile.

References

[1] P. Mathieu, J. Math. Phys. 29, 2499 (1988).
[2] Yu. I. Manin, A. O. Radul, Commun. Math. Phys. 98, 65 (1985).
[3] P. Dargis, P. Mathieu, Phys. Lett. A 176, 67 (1993).
[4] S. Andrea, A. Restuccia, A. Sotomayor, J. Math. Phys. 46, 103517 (2005).
[5] S. Andrea, A. Restuccia and A. Sotomayor, Phys. Lett. A 376, 245251 (2012).
[6] P. Labelle and P. Mathieu, J. Math. Phys. 32, 923 (1991).
[7] S. Bellucci, E. Ivanov and S. Krivonos, J. Math. Phys. 34, 3087 (1993).
[8] F. Delduc and E. Ivanov, Phys. Lett. B 309, 312-319 (1993).
[9] F. Delduc and E. Ivanov and S. Krivonos, J. Math. Phys. 37, 1356 (1996).
[10] Z. Popowicz, Phys. Lett. B 459, 150-158 (1999).
[11] X. N. Gao and S. Y. Lou, Phys. Lett. B 707, 209 (2012).
[12] X. N. Gao, S. Y. Lou and X. Y. Tang, JHEP 05, 029 (2013).
[13] S. Andrea, A. Restuccia and A. Sotomayor, J. Math. Phys. 42, 2625 (2001).
[14] P. A. M. Dirac, "Lectures on Quantum Mechanics", Belfer Graduate School Monograph Series No.2, Yeshiva University, New York, (1964).
[15] Y. Nutku, J. Math. Phys. 25 (6) , June (1984).
[16] G. W. Kentwell, J. Math. Phys. 29, 46 (1988).

[17] A. Restuccia and A. Sotomayor, "The Hamiltonian structure of a coupled system derived from a supersymmetric breaking of Super KdV equations", arXiv:1306.0528 (2013).
[18] R. M. Miura, C. S. Gardner, and M. D. Kruskal, J. Math. Phys. 9, 1204 (1968).

An efficient iterative method for solving Zakharov-Kuznetsov Equation

Masoud Saravi[1], **Ali Nikkar**[2]
[1] Department of Mathematics, Islamic Azad University, Nour Branch, Nour 731, Iran
[2] Faculty of Civil Engineering, University of Tabriz, Tabriz, Iran

E-mail: msaravi2001@yahoo.com

Abstract. In this paper, we apply new modified of Variational Iteration Method (VIM-II) which is a kind of analytical approximate method then, use it to solve Zakharov-Kuznetsov (ZK) equation that governs the behavior of the weakly nonlinear ion-acoustic waves in plasma. Two cases of this equation are considered and the results are compared with those that obtained by Adomian Decomposition Method (ADM) and Variational Homotopy Perturbation Method (VHPM). The results illustrate that proposed technique yields a very rapid convergence of the solution as well as low computational effort.

1. Introduction

The ZK equation [1] is one of two well-studied canonical two-dimensional extensions of the Korteweg-de Vries equation [2]. The second one is the Kadomtsev-Petviashvilli (KP) equation [3]. The ZK equation is known to follow from the exact equations of potential water waves by the symmetry-preserving truncation at a certain order in wave steepness. This equation, being formulated in terms of nonlinear normal variables, has long been recognized as an indispensable tool for theoretical analysis of surface wave dynamics. However, its potential as the basis for the numerical modeling of wave evolution has not been adequately explored.

The ZK equation is valid in anisotropic setting which is exactly the case for rotating fluids where the differential latitudinal dependence of the rotation rate causes anisotropy between the meridional and the longitudinal directions. Moreover, in contrast to the KP equation the ZK equation supports stable lump solitary waves. This makes the ZK equation a very attractive model equation for the study of vortices in geophysical flows. The ZK equation was first derived for describing weakly nonlinear ion-acoustic waves in strongly magnetized lossless plasma in two dimensions [1]. Traveling waves are very important because various phenomena in nature such as vibration and soliton or self-reinforcing solitary waves are described by them.

Recently, advances were reported on producing and developments of analytical approaches for solving nonlinear engineering problems. Some of these well-known methods include homotopy perturbation [4-6], variational approach [7], differential ransformation [8], variational iteration [9], amplitude-frequency formulation [10] Adomian decomposition [11], and homotopy analysis method [12,13]. This influences researchers to consider these methods for finding analytical solution for ZK equation. Wazwaz [14] used extended tanh method for analytic treatment of the ZK equation, the modified ZK

[1] To whom any correspondence should be addressed.

equation, and the generalized forms of these equations. Huang [15] applied the polynomial expansion method to solve the coupled ZK equations. Zhao et al. [16] obtained numbers of solitary waves, periodic waves and kink waves using the theory of bifurcations of dynamical systems for the modified ZK equation. Inc [17] solved nonlinear dispersive ZK equation using the Adomian decomposition method, and Biazar et al. [18] applied the homotopy perturbation method to solve the ZK equation. VIM-II is a powerful analytical technique which utilizes Laplace Transform to decrease computational efforts and increase accuracy of the classical Variational Iteration Method. The method has been used by some authors in wide variety of scientific and engineering applications to solve different types of differential equations [19-20]. The most sensible advantages of VIM-II are using Laplace Transform and choosing initial conditions simply and easily to solve linear and nonlinear equations. In this paper we apply VIM-II to find analytical solutions for nonlinear ZK (m, n, k) equations in the following form:

$$u_t + a(u^m)_x + b(u^n)_{xxx} + c(u^k)_{yyx} = 0 \quad , \quad m,n,k \neq 0 \tag{1}$$

where a, b and c are arbitrary constants and m, n and k are integers.
The Effectiveness and convenience of the method is revealed in comparisons with the other solution techniques. The results reveal that these methods are very effective and convenient in predicting the solution of such problems, and it is predicted that the VIM-II can find a wide application in new engineering problems.

2. Basic concept of VIM-II

If we clarify the idea of the proposed method for solution of the nonlinear governing equation of Zakharov-Kuznetsov, the basic concept of Variational Iteration Method-II [21,22] is firstly treated. A general nonlinear equation of kth order is considered at the following form:

$$u^{(k)} + f(u', u'', \ldots, u^{(k)}) = 0 \tag{2}$$

The classical variational iteration algorithm is as follows:

$$u_{n+1}(t) = u_n(t) + \int_0^t \lambda(u_n^{(k)} + \tilde{f}_n)d\varepsilon \tag{3}$$

where λ is a general Lagrange multiplier. We apply Laplace transform to identify the Lagrange multiplier [23]. By using Laplace transform, we have:

$$s^k U(s) + \ell\{f(u, u', \ldots, u^{(k)})\} = 0 \tag{4}$$

$$U(s) = \frac{-\ell\{f(u, u', \ldots, u^{(k)})\}}{s^k}. \tag{5}$$

It is assumed that all required conditions in Laplace transform are zero. The inverse Laplace transform reads:

$$u(t) = (-1)^k \int_0^t \frac{(\varepsilon - t)^{k-1}}{(k-1)!} f(u(\varepsilon), u'(\varepsilon), \ldots, u^{(k)}(\varepsilon))d\varepsilon. \tag{6}$$

Hence, after identifying the Lagrange multiplier λ, the variational iteration algorithm-II [21,22] is constructed as follows:

$$u_{n+1}(t) = u_0(t) + (-1)^k \int_0^t \frac{(\varepsilon - t)^{k-1}}{(k-1)!} f(u_n(\varepsilon), u_n'(\varepsilon), \ldots, u_n^{(k)}(\varepsilon)) d\varepsilon. \tag{7}$$

The above equation is generally called as the variational iteration algorithm-II. In which, $u_0(t)$ is the initial solution. The initial values are usually used for selecting the zeroth approximation u_0. With u_0 determined, then several approximations u_n ,$n>0$, follow immediately. Consequently, the exact solution could be obtained as follows:

$$u(t) = \lim_{n\to\infty} u_n \tag{8}$$

In the next section, VIM-II is applied to solve nonlinear ZK (m, n, k) equation.

3. Implementation of VIM-II

In order to illustrate the effectiveness of the proposed method, we implement it to solve the ZK equation with different numerical examples.

3.1. Example 1

First, we consider the ZK (2. 2. 2) equation [24]:

$$u_t - (u^2)_x + \frac{1}{8}(u^2)_{xxx} + \frac{1}{8}(u^2)_{yyx} = 0 \tag{9}$$

with initial value problem of:

$$u(x,y,0) = \frac{4}{3}\xi \sinh^2\left(\frac{x+y}{2}\right) \tag{10}$$

where ζ is an arbitrary constant. We assume $\zeta = 1$. First, according to the method, by applying Laplace Transform to identify the Lagrange multiplier, we have:

$$\lambda = 1 \tag{11}$$

So, variational iteration algorithm-II is derived:

$$u_{n+1} = u_0 + \int_0^t ((u_n^2)_x - \frac{1}{8}(u_n^2)_{xxx} - \frac{1}{8}(u_n^2)_{yyx}) d\varepsilon \tag{12}$$

We start with an initial approximation $u_0(x,y,t) = \frac{4}{3}\sinh^2(\frac{x+y}{2})$ by the iteration formula (12), we can obtain the first components as follows:

$$\begin{aligned} u_1(x,y,t) &= \frac{4}{3}\sinh^2(\frac{x+y}{2}) + \frac{22}{9}\sinh^3(\frac{x+y}{2})\cosh(\frac{x+y}{2})t - \frac{1}{3}\cosh^4(\frac{x+y}{2})t \\ &-\frac{8}{3}\sinh^2(\frac{x+y}{2})\cosh^2(\frac{x+y}{2})t - \frac{5}{9}\sinh^4(\frac{x+y}{2}) - \frac{2}{3}\sinh(\frac{x+y}{2})\cosh^3(\frac{x+y}{2})t \end{aligned} \tag{13}$$

Proceeding in the same way, we can obtain the high order approximations. Comparison of exact solution [17] with the corresponding approximate solution of ZK (2,2,2) for fixed values of y=0.1 and t=0.001 using two iterations (n=2) is presented in Fig. 1. While in Fig. 2 the corresponding approximate solution is presented.

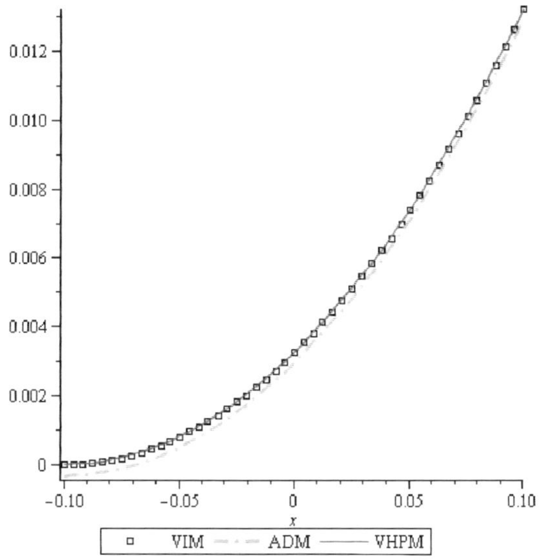

Figure 1. Comparison between the results of different solutions at y=0.1 and t=0.001

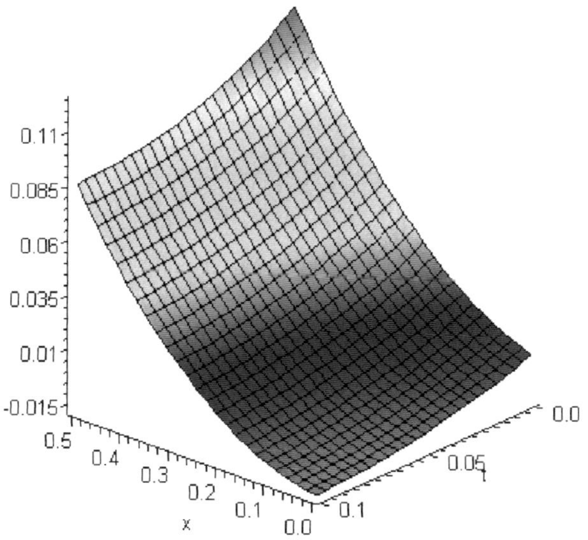

Figure 2. The surface shows the solution u(x,y,t) for ZK(2,2,2) with y=0.1

3.2. Example 2

Consider the ZK (3.3.3) equation [25]:

$$u_t + (u^3)_x + 2(u^3)_{xxx} + 2(u^3)_{yyx} = 0 \tag{14}$$

with initial condition:

$$u(x, y, 0) = \frac{3}{2}\zeta \sinh(\frac{x+y}{6}) \tag{15}$$

As mentioned in the previous example we assume $\zeta = 1$. According to the method, the Lagrange multiplier can therefore be simply identified as:

$$\lambda = 1 \tag{16}$$

and the following iteration formula can be obtained:

$$u_{n+1} = u_0 - \int_0^t ((u_n^3)_x + 2(u_n^3)_{xxx} + 2(u_n^3)_{yyx}) d\varepsilon \tag{17}$$

We start with an arbitrary initial approximation $u_0(x, y, t) = \frac{3}{2}\sinh(\frac{x+y}{6})$ that satisfies the initial condition and by using the iteration formula (17), we have the following successive approximation:

$$u_1(x, y, t) = \frac{3}{2}\sinh(\frac{x+y}{6}) - \frac{885}{512}\sinh^2(\frac{x+y}{6})\cosh(\frac{x+y}{6})t - \frac{5}{256}\sinh(\frac{x+y}{6})\cosh^2(\frac{x+y}{6})t - \frac{7}{1024}\sinh^3(\frac{x+y}{6})t - \frac{3}{256}\cosh^3(\frac{x+y}{6})t \tag{18}$$

Proceeding in the same way, we can obtain the high order approximations. Comparison of exact solution [17] with approximate solution of ZK (3,3,3) for fixed values of y =0.1 and t=0.001 using two iterations (n = 2) is presented in Fig. 3. While in Fig. 4 the corresponding approximate solution is presented.

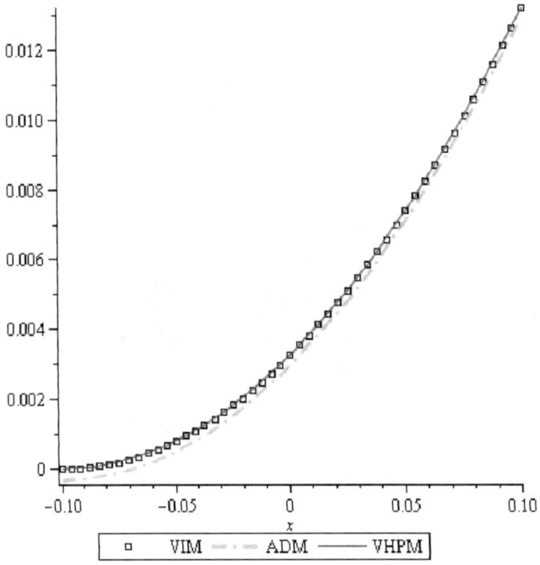

Figure 3. Comparison between the results of different solutions at y=0.1 and t=0.001

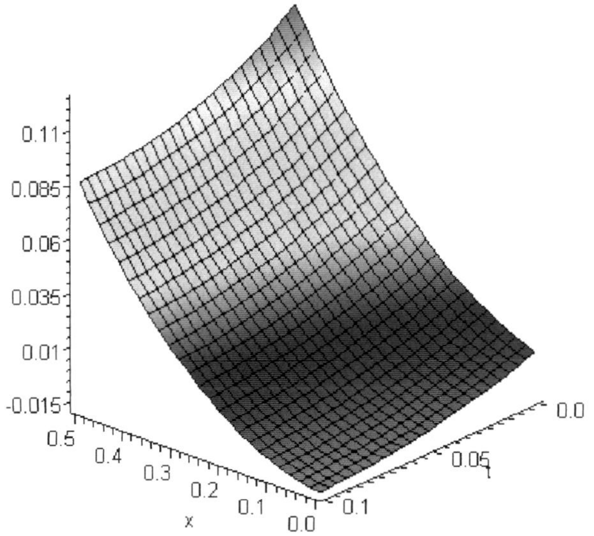

Figure 4. The surface shows the solution u(x,y,t) for ZK(3,3,3) with y=0.1

4. Conclusion

A novel approximate numerical solution for the ZK (m,n,k) equation is obtained by using VIM-II. Some illustrative examples are given to show the validity and simplicity of this method and results show that these schemes provide excellent approximations to the solution of this nonlinear problem with high accuracy. It is worth pointing out that this method presents a rapid convergence for the solutions.

5. References

[1] Zakharov V E, Kuznetsov E A, 1974 *Soviet Phys.* **39** 285
[2] Korteweg D D, De Vries D G, 1895 *Phil. Mag.* **39 (240)** 422
[3] Kadomtsev B B, Petviashvilli V I, 1970 *Phys. Dokl.* **15** 539
[4] Bararnia H, Ghasemi E, Soleimanikutanaei S, Barari A, Ganji D D, 2011 *J.of Porous Media* **14 (6)** 545
[5] Saravi M, Hermaan M, Ebarahimi khah H, 2013 *J. Theo. Appl. Phys.* **7** 1
[6] Rostamiyan Y, Fereidoon A, Davoudabadi M R, Yaghoobi H, Ganji D D, 2010 Math. *Comp. Appl.* **15 (5)** 816
[7] Bagheri S, Nikkar A, Ghaffarzadeh H, 2014 *Lat. Am. J. Sol. Str.* **11 (1)** 157
[8] Yaghoobi H, Torabi M, 2011 *Int. Comm. He. Mas. Trans.* **38 (6)** 815
[9] Nikkar A, Bagheri S, Saravi M, 2014 *Lat. Am. J. Sol. Str.* **11 (2)** 320
[10] Zhang H L, 2008 *Int. J. Non. Sci. Num. Sim.* **9 (3)** 297
[11] Sheikholeslami M, Ganji D D, Ashorynejad H R, Rokni H B, 2012 Appl. Math. Mech. **33 (1)** 25
[12] He J H, 2004 *Appl. Math. Comput.* **156 (2)** 527
[13] Khan Y, Taghipour R, Fallahian M, Nikkar A, 2012 *Neu. Compu. Appl.* doi 10.1007/s00521-012-1077-0
[14] Wazwaz A M, 2008 *Comm. in Non. Sci. Num. Sim.* **13** 1039
[15] Huang W, 2006 *Chaos Soli. Fractal.* **29(2)** 365
[16] Zhao X, Zhou H, Tang Y, Jia H, 2006 *App. Math. Compu.* **181 (1)** 634
[17] Inc M, 2007 *Chaos Soli. Fractal.* **33 (5)** 1783
[18] Biazar J, Badpeima F, Azimi F, 2009 *Com. Math. with App.* **58 (11)** 2391

[19] Nikkar A, Mighani Z, Saghebian S M, Nojabaei S B, Daie M, 2012 *Res. J. App. Sci. Eng. Tech.* **5 (1)** 296
[20] S. Ghasempour, Vahidi J, Nikkar A, Mighani M, 2012 *Res. J. App. Sci. Eng. Tech.* **5 (1)** 339
[21] He J H, 2012 *Abst. Appl. Analys.* Article ID 916793, doi:10.1155/2012/916793.
[22] He J H, 2011 *Phys. Let. A.* **375 (38)** 3362
[23] Hesameddin E, Latifizadeh H, 2009 *Int. J. Non. Sci. Num. Sim.* **10 (10)** (2009) 1365
[24] Hesam S, Nazemi A, Haghbin A, 2010 *Int. J. Eng. Nat. Sci.* **4 (4)** 235
[25] Matinfar M, Ghasemi M, Saeidy M, 2012 *Iran. J. Sci. Tech.* **36** 351

Poisson structure and stability analysis of a coupled system arising from the supersymmetric breaking of Super KdV

Adrián Sotomayor[1], **Alvaro Restuccia**[2]

[1] Departament of Mathematics, Universidad de Antofagasta, Antofagasta, Chile
[2] Physics Department, Universidad de Antofagasta, Antofagasta, Chile
[2] Physics Department, Universidad Simón Bolívar, Caracas, Venezuela

E-mail: asotomayor@uantof.cl,arestu@usb.ve

Abstract. The Poisson structure of a coupled system arising from a supersymmetric breaking of N=1 Super KdV equations is obtained. The supersymmetric breaking is implemented by introducing a Clifford algebra instead of a Grassmann algebra. The Poisson structure follows from the Dirac brackets obtained by the constraint analysis of the hamiltonian of the system. The coupled system has multisolitonic solutions. We show that the one soliton solutions are Liapunov stable.

1. Introduction

It is well known that the Korteweg-de Vries (KdV) equation has two hamiltonians and associated Poisson structures. The first hamiltonian structure was obtained directly in the framework of the field u satisfying the KdV equation. The second hamiltonian structure was obtained on the framework of the Miura equation (MKdV) and is related to the KdV equation via a Miura transformation. Although both hamiltonian structures may be formulated in terms of the u field, the origin of the second hamiltonian structure relays on the MKdV formulation. Many extensions of the KdV equation in terms of non-linear coupled partial differential equations for two or more commutative fields have been proposed in the literature [1, 2, 3, 4, 5, 6]. A particularly interesting case is the $N = 1$ super KdV system [7], which has only one local hamiltonian structure. In this case the fields take values on the even and odd part of a Grassmann algebra. When this algebra has finite number of generators, the supersymmetric system may be formulated directly in terms of the real coefficients of a field expansion in terms of the basis of the algebra. One ends up with a coupled system in terms of commutative real fields. The supersymmetric transformation may then be defined directly as a transformation on these real fields which leaves invariant the coupled system.

Recently using such bosonization approach [8] exact solutions of $N = 1$ SKdV equations were obtained [9, 10].

An important aspect of the analysis of the supersymmetric models is its behaviour under supersymmetric breaking conditions. We consider here a supersymmetry breaking implemented by replacing the Grassmann algebra by a Clifford one. This scheme has already been followed in several works, see for example [11].

We consider then a coupled system with fields valued on a Clifford algebra. We find the hamiltonian for such system as well as its Poisson structure. The hamiltonian formulation on phase space presents second class constraints. We then introduce Dirac brackets on the constrained submanifold to define the Poisson structure for such system. The hamiltonian of this system is bounded from below which ensures the a physical admissible content. The solutions of the system satisfy an a priori bound which suggests the existence of global solutions to the system.

Although the system has multisolitonic solutions it does not have an infinite number of local conserved quantities.

We finally consider the stability analysis of the one-soliton solutions of the extended KdV system arising from $N = 1$ SKdV under supersymmetric breaking conditions. Following the approach introduced by [12] and [13] we show that the one-soliton solutions are Liapunov stable, for details see [14].

2. Breaking supersymmetry on SKdV

The $N = 1$ SKdV equations have an infinite sequence of local conserved quantities [7] as well as an infinite sequence of non-local conserved quantities [15, 16, 17]. To break supersymmetry we consider fields u and ξ valued on a Clifford algebra instead of being Grassmann algebra valued. We thus take u to be a real valued field while ξ to be an expansion in terms of the generators $e_i, i = 1, \ldots$ of the Clifford algebra:

$$\xi = \sum_i \varphi_i e_i + \sum_{ij} \varphi_{ij} e_i e_j + \sum_{ijk} \varphi_{ijk} e_i e_j e_k + \cdots \tag{1}$$

where

$$e_i e_j + e_j e_i = -2\delta_{ij} \tag{2}$$

and $\varphi_i, \varphi_{ij}, \varphi_{ijk}, \ldots$ are real valued functions. We define $\bar{\xi} = \sum_{i=1}^{\infty} \varphi_i \bar{e}_i + \sum_{ij} \varphi_{ij} \bar{e}_j \bar{e}_i + \sum_{ijk} \varphi_{ijk} \bar{e}_k \bar{e}_j \bar{e}_i + \cdots$ where $\bar{e}_i = -e_i$. We denote as in superfield notation the body of the expansion those terms associated with the identity generator and the soul the remaining ones. Consequently the body of $\xi\bar{\xi}$, denoted by $\mathcal{P}(\xi\bar{\xi})$, is equal to $\Sigma_i \varphi_i^2 + \Sigma_{ij} \varphi_{ij}^2 + \Sigma_{ijk} \varphi_{ijk}^2 + \cdots$ In what follows, without loss of generality, we rewrite $\mathcal{P}(\xi\bar{\xi}) = \Sigma_i \varphi_i^2 + \Sigma_{ij} \varphi_{ij}^2 + \Sigma_{ijk} \varphi_{ijk}^2 + \cdots$ simply as $\mathcal{P}(\xi\bar{\xi}) = \Sigma_i \varphi_i^2$.

The resulting system after the breaking of supersymmetry becomes

$$\begin{array}{l} u_t = -u''' - uu' - \frac{1}{4}\left(\mathcal{P}\left(\xi\bar{\xi}\right)\right)' \\ \xi_t = -\xi''' - \frac{1}{2}(\xi u)'. \end{array} \tag{3}$$

This coupled KdV system arises from the following Lagrangian, expressed in terms of the fields w and η_1 related to u and φ_i by

$$u = w' \text{ and } \varphi_i = \eta_i', \tag{4}$$

$$S(w, \eta_i) \equiv \int dx dt \left[\frac{1}{2} w' \partial_t w + \frac{1}{6}(w')^3 - \frac{1}{2}(w'')^2 + \frac{1}{4} w' (\eta_i')^2 + \frac{1}{2} \eta_i' \partial_t \eta_i - \frac{1}{2}(\eta_i'')^2\right], \tag{5}$$

where a repeated index i implies summation on that index. Variations of the action $S(w, \eta_i)$ yields the corresponding field equations (3).

The hamiltonian and corresponding Poisson structure associated to the action (5) follows by introducing the conjugate momenta associated to w, η_i. We will denote them by p, σ_i respectively

$$\begin{array}{l} p := \frac{\partial \mathcal{L}}{\partial(\partial_t w)} = \frac{1}{2} w' = \frac{1}{2} u \\ \\ \sigma_i := \frac{\partial \mathcal{L}}{\partial(\partial_t \eta_i)} = \frac{1}{2} \eta_i' = \frac{1}{2} \varphi_i. \end{array} \tag{6}$$

(6) are primary constraints on the phase space. The hamiltonian may be obtained by performing a Legendre transformation.

We obtain

$$H = \int_{-\infty}^{+\infty} \left(-\frac{1}{3}u^3 - \frac{1}{2}u\mathcal{P}(\xi\bar{\xi}) + (u')^2 + \mathcal{P}(\xi'\bar{\xi}') \right) dx. \tag{7}$$

Following the Dirac approach for constrained systems it turns out that (6) are the only constraints on the phase space, and they are second class constraints. The Poisson structure on the constrained submanifold of phase space is then given by the Dirac brackets. They are given by

$$\begin{aligned} &\{u(x), u(y)\}_{DB} = \partial_x \delta(x,y), \\ &\{\varphi_i(x), \varphi_j(y)\}_{DB} = \delta_{ij}\partial_x \delta(x,y), \\ &\{u(x), \varphi_i(y)\}_{DB} = 0. \end{aligned} \tag{8}$$

We thus have obtained the hamiltonian H and the Poisson structure of the coupled KdV system (3). Besides the hamiltonian H there are other conserved quantities under the evolution determined by the coupled KdV system (3). The following are conserved quantities

$$\begin{aligned} &\hat{H}_{\frac{1}{2}} = \textstyle\int_{-\infty}^{+\infty} \xi dx, \\ &\hat{H}_1 = \textstyle\int_{-\infty}^{+\infty} u dx, \\ &V \equiv \hat{H}_3 = \textstyle\int_{-\infty}^{+\infty} \left(u^2 + \mathcal{P}(\xi\bar{\xi})\right) dx. \end{aligned} \tag{9}$$

It is interesting to notice that the non-local quantity

$$\int_{-\infty}^{\infty} \xi(x) \int_{-\infty}^{x} \xi(s) ds dx \tag{10}$$

is also conserved. Remarkably, this quantity in terms of a Grassmann valued ξ is also conserved for the $N = 1$ SKdV system.

The system (3) besides being not invariant under supersymmetry, by construction, it does not have an infinite sequence of local conserved quantities. It does have, however, multisolitonic solutions. We discuss now the stability properties of the one-soliton solutions.

3. Stability of the solitonic solutions

An important first step in the analysis is to show an a priori bound [18] for the solutions of the system (3).

We denote by $\|\ \|_{H_1}$ the Sobolev norm

$$\|(u,\xi)\|_{H_1}{}^2 = \int_{-\infty}^{+\infty} \left[\left(u^2 + \Sigma_{i=0}^{\infty}\varphi_i^2\right) + \left(u'^2 + \Sigma_{i=0}^{\infty}\varphi_i'^2\right) \right] dx. \tag{11}$$

We obtain from the expressions of V and H

$$\|(u,\xi)\|_{H_1}{}^2 \leq V + H + \frac{1}{2}\int_{-\infty}^{+\infty} |u| \left(u^2 + \mathcal{P}\left(\xi\bar{\xi}\right)\right) dx. \tag{12}$$

We may now use

$$\sup|u| \leq \frac{1}{\sqrt{2}}\|u\|_{H_1} \leq \frac{1}{\sqrt{2}}\|(u,\xi)\|_{H_1} \tag{13}$$

to get

$$\|(u,\xi)\|_{H_1}{}^2 \leq V + H + \frac{1}{2\sqrt{2}}V\|(u,\xi)\|_{H_1}. \tag{14}$$

It then follows

$$\|(u,\xi)\|_{H_1} \leq \frac{d+\sqrt{d^2+4e}}{2} \tag{15}$$

where $d = \frac{1}{2\sqrt{2}}V$ and $e = V + M$. We notice that $d^2 + 4e \geq 0$.

Given V and H from the initial data and a solution satisfying those initial conditions, then $\|(u,\xi)\|_{H_1}$ is bounded by (15) for all $t \geq 0$. This is an a priori bound and a strong evidence of the existence of the solution for $t \geq 0$.

We consider the stability of a solution $(\hat{u},\hat{\xi})$ of (3) in the Liapunov sense:

$(\hat{u},\hat{\xi})$ is stable if given ϵ there exists δ such that for any solution (u,ξ) of (3), satisfying at $t=0$

$$d_I\left[(u,\xi),\left(\hat{u},\hat{\xi}\right)\right] < \delta \tag{16}$$

then

$$d_{II}\left[(u,\xi),\left(\hat{u},\hat{\xi}\right)\right] < \epsilon \tag{17}$$

for all $t \geq 0$.

We consider first the stability of the ground state solution $\hat{u} = 0, \hat{\xi} = 0$. We take d_I and d_{II} to be the Sobolev norm $\|(u-\hat{u},\xi-\hat{\xi})\|_{H_1}$.

We get

$$V \leq \|(u,\xi)\|^2_{H_1}$$

and

$$H \leq \int_{-\infty}^{+\infty}\left(\frac{1}{2}|u|\left(u^2+\mathcal{P}\left(\xi\bar{\xi}\right)\right)+u'^2+\mathcal{P}\left(\xi'\bar{\xi}'\right)\right)dx \leq \frac{1}{2\sqrt{2}}\|(u,\xi)\|_{H_1}{}^3 + \|(u,\xi)\|_{H_1}{}^2.$$

It then follows from the a priori bound (15) the stability of the ground state $\hat{u} = 0, \hat{\xi} = 0$.

We now consider the stability of the one-soliton solution $\hat{u} = \phi, \hat{\xi} = 0$, where ϕ is the one-soliton solution of KdV which satisfies

$$\phi'' + \frac{1}{2}\phi^2 = \mathcal{C}\phi. \tag{18}$$

The proof of stability is based on estimates for the u field which are analogous to the one presented in [12, 13] while a new argument will be given for the ξ field. The distances we will use are

$$d_I\left[(u_1,\xi_1),(u_2,\xi_2)\right] = \|\left(u_1-u_2,\xi_1-\xi_2\right)\|_{H_1} \tag{19}$$

$$d_{II}\left[(u_1,\xi_1),(u_2,\xi_2)\right] = \inf_{\tau}\|\left(\tau u_1-u_2,\xi_1-\xi_2\right)\|_{H_1} \tag{20}$$

where τu_1 denotes the group of translations along the x-axis. d_{II} is a distance on a metric space obtained by identifying the translations of each $u \in H_1(\mathbb{R})$ [12]. d_{II} is related to a stability in the sense that a solution u remains close to $\hat{u} = \phi$ only in shape but not necessarily in position.

At $t = 0$ we assume that

$$d_I\left[(u,\xi),(\phi,0)\right] = \|\left(h,\xi\right)\|_{H_1} < \delta. \tag{21}$$

Using estimates as in [12, 13] one can show that

$$|\Delta H| \leq \left[\max\left(1,\mathcal{C}\right)+\frac{1}{3\sqrt{2}}\delta\right]\|\left(h,\xi\right)\|^2_{H_1} \tag{22}$$

where $h \equiv u - \phi$.

The next argument introduces besides the ideas in [12, 13], new estimates on ξ, see [14]. After several delicate estimates we obtain

$$\Delta H \geq \frac{1}{6} l \{ d_{II} \left[(u, \xi), (\phi, 0) \right] \}^2 \tag{23}$$

where $l = \min(1, \mathcal{C})$.

The upper and lower bounds for ΔH imply the stability of the one-soliton solution of the coupled KdV system.

4. Conclusions

By breaking the supersymmetry in the $N = 1$ SKdV equations we arrive to a coupled KdV system. We found its hamiltonian and associated Poisson structure. The hamiltonian is bounded from below and consequently has an admissible physical interpretation. This important property is then used to show the stability of the ground state solution as well as the one-soliton solutions of the coupled KdV system. We determine also an a priori bound giving a strong evidence of the existence of solutions for all $t \geq 0$.

The property of having a hamiltonian bounded from below is very important and it is shared with the hamiltonian of the super KdV equations. However, the stability of the solitonic solutions of $N = 1$ super KdV equations has, so far, not been proven in the literature.

Acknowledgments

A. S. and A. R. are partially supported by Project Fondecyt 1121103, Chile.

References

[1] R. Hirota and J. Satsuma, Phys. Lett. A 85, 407 (1981).
[2] M. Ito, Phys. Lett. 91 A, number 7, 335 (1982).
[3] S. Y. Lou, B. Tong, H. C. Hu and X. Y. Tang, J. Phys. A: Math. Gen. 39 513-527 (2006).
[4] J. R. Yang and J. J. Mao, Commun. Theor. Phys. 49, 22-26 (2008).
[5] J. R. Yang and J. J. Mao, Chinese Phys. B 17 4337 (2008).
[6] D. S. Wang, Appl. Math. and Comp. 216, 1349-1354 (2010).
[7] P. Mathieu, J. Math. Phys. 29, 2499 (1988).
[8] S. Andrea, A. Restuccia and A. Sotomayor, J. Math. Phys. 42, 2625 (2001).
[9] X. N. Gao and S. Y. Lou, Phys. Lett. B 707, 209 (2012).
[10] X. N. Gao, S. Y. Lou and X. Y. Tang, JHEP 05, 029 (2013).
[11] N. Seibert, JHEP 0306, 010 (2003).
[12] T. B. Benjamin, Proc. R. Soc. Lond. A. 328, 153-183 (1972).
[13] J. Bona, Proc. R. Soc. Lond. A. 344, 363-374 (1975).
[14] A. Restuccia and A. Sotomayor, arXiv:1207.5870 , accepted for publication in Boundary Value Problems (2013).
[15] P. Dargis and P. Mathieu, Phys. Lett. A 176, 67-74 (1993).
[16] S. Andrea, A. Restuccia and A. Sotomayor, J. Math. Phys. 46, 103517 (2005).
[17] S. Andrea, A. Restuccia and A. Sotomayor, Phys. Lett. A 376, 245-251 (2012).
[18] T. Tao, Bulletin (New Series) of the American Mathematical Society Volume 46, Number 1, Pages 1-33 (2009).

Journal of Physics: Conference Series **474** (2013) 012031 doi:10.1088/1742-6596/474/1/012031

CPT and Lorentz Invariance: Their Relation and Violation

Anca Tureanu
Department of Physics, University of Helsinki, P.O. Box 64, 00014 Helsinki
E-mail: anca.tureanu@helsinki.fi

Abstract. The CPT theorem in quantum field theory, its validity, breaking and consequences are reviewed. One can show that for CPT theorem to hold, Lorentz invariance is not always needed. Also one can have CPT violation, while there is Lorentz invariance. Field theoretical examples for both cases are given and mass differences between particle-antiparticle are discussed as well. Unambiguous tests of CPT violation, unrelated to the breaking of Lorentz invariance, are suggested.

1. Introduction

Lorentz symmetry and the CPT invariance are two of the most fundamental symmetries of Nature, whose violation has not yet been observed. While the Lorentz invariance is a continuous symmetry of space-time, the CPT involves the discrete space- and time-inversions, P, T, and the charge conjugation operation on the fields, C. Although the individual symmetries, C, P and T have been observed to be violated in various interactions, their combined product, CPT, remarkably remains still as an exact symmetry. The first proof of CPT theorem was given by Lüders and Pauli [1, 2] based on the Hamiltonian formulation of quantum field theory, which involves locality of the interaction, Lorentz invariance and Hermiticity of the Hamiltonian. Later on the theorem was proven by Jost [3] (see also [4–6]) within the axiomatic formulation of quantum field theory without reference to any specific form of interaction. This proof of CPT theorem relaxes the requirement of locality or *local commutativity* condition to the so-called *weak local commutativity*. Lorentz symmetry has been an essential ingredient of the proof, both in the Hamiltonian and in the axiomatic proofs.

A simple phenomenological classification of possible C, P, T, CP, PT, TC and CPT-violating effects is presented in [7]. For consequences of CPT and their experimental tests, as well as some theoretical considerations on the possibilities of violation of Lorentz invariance and CPT in the known interactions, we refer to [8–13] and references therein.

It is important to clarify the relation between the CPT and Lorentz invariance and in particular to see whether the violation of any of them implies the violation of the other. This issue has recently become a topical one due to the growing phenomenological importance of CPT violating scenarios, namely in neutrino physics as well as its cosmological and astrophysical consequences. Indeed, the relation between the CPT and Lorentz invariance has acquired a prominent place in nowadays particle physics with the attempts of explaining in a unified manner the contradictory results, "anomalies", in the interpretation of various neutrino physics experiments, without enlarging the neutrino sector. The idea was first suggested by Murayama and Yanagida [14] in the form of different masses for neutrino and antineutrino, based on phenomenological considerations. This proposal was formalized as a CPT-violating quantum field theory with a mass difference between neutrino and antineutrino in [15, 16]. The issue was taken up in relation with the Lorentz symmetry by Greenberg [17], the conclusion of Greenberg's analysis being that CPT violation implies violation of Lorentz invariance. In the following, we shall show that this claim is in general not valid. Moreover, a Lorentz-invariant model with mass splitting between particle and antiparticle has also been proposed.

2. Lorentz-invariance violating but CPT-invariant quantum field theories

During the last decade, we have learned that the violation of Lorentz invariance does not necessarily lead to the violation of the CPT theorem. The example comes from the quantum field theory on noncommutative space-time (NC QFT) with the canonical, Heisenberg-like, commutation relations for coordinate operators:

$$[x^\mu, x^\nu] = i\theta^{\mu\nu}, \tag{1}$$

with $\theta^{\mu\nu}$ an antisymmetric constant matrix.

In this case, by the nature of the above noncommutativity parameter $\theta^{\mu\nu}$ being a constant but not a tensor, Lorentz invariance is broken, but not the CPT symmetry [18–21]. Translational invariance is valid. In addition to the Lorentz invariance violation, such NC QFTs are nonlocal in the noncommuting coordinates. However, the Lorentz symmetry violation is of a very particular form, and invariance under the stability group of the matrix $\theta^{\mu\nu}$ is preserved under the so-called residual symmetry $SO(2) \times SO(1,1)$. This reduced symmetry is enough to prove the CPT theorem only for the scalar fields (for which the C operation is a simple Hermitian conjugation) on the noncommutative space-time (1) [19]. A full proof of the CPT theorem in Lorentz-violating noncommutative quantum field theory, however, could be achieved [20] only by using the twisted Poincaré symmetry [22, 23] which these theories possess. The twisted Poincaré invariance is a deformation of the Poincaré symmetry, considered as a Hopf algebra, a concept coming from the theory of quantum groups [24], as compared with the Lie algebra. The irreducible representations of twisted Poincaré are identical to those of the usual Poincaré algebra, i.e. labeled by the mass and spin of the particles. Therefore, the meaning of the charge conjugation has survived intact in the noncommutative quantum field theories. While parity and time reversal symmetries can be defined with any concept of space and time, the notion of charge conjugation has meaning only in the framework of Lorentz symmetry. Antiparticles are a consequence of special relativity. Particle and antiparticle are in the same irreducible representation of the Poincaré group. The CPT theorem is thus strongly connected to the Poincaré group representations, and not so much to the Lorentz symmetry, as the validity of the CPT theorem in the noncommutative space-time shows.

3. CPT-violating but Lorentz-invariant nonlocal model

In [25] was proposed a class of models which preserve Lorentz invariance while breaking the CPT symmetry through a (nonlocal) interaction. The latter attitude is taken as responsible for the violations of a symmetry, based on our experience that all the discrete, C, P and T invariances, as well as other symmetries, are broken in our description of Nature by means of interaction. We also know that nonlocal field theories appear, in general, as effective field theories of a larger theory.

Consider a field theory with the nonlocal interaction Hamiltonian of the type

$$\mathcal{H}_{int}(x) = \lambda \int d^4y \; \phi^*(x)\phi(x)\phi^*(x)\theta(x_0 - y_0)\theta((x-y)^2)\phi(y) + h.c., \tag{2}$$

where λ is a coupling constant with dimension appropriate for the Hamiltonian density, $\phi(x)$ is a Lorentz-scalar field and θ is the Heaviside step function, with values 0 or 1, for its negative and positive argument, respectively. The combination $\theta(x_0 - y_0)\theta((x-y)^2)$ in (2) ensures the Lorentz invariance, i.e. invariance under the proper orthochronous Lorentz transformations, since the order of the times x_0 and y_0 remains unchanged for time-like intervals, while for space-like distances the interaction vanishes. Also, the same combination makes the nonlocal interaction causal at the tree level, which dictates that there is no interaction when the fields are separated by space-like distances and thus there is a maximum speed of $c = 1$ for the propagation of information.

On the other hand, it is clear that C and P invariance are trivially satisfied in (2), while T invariance is broken due to the presence of $\theta(x_0 - y_0)$ in the integrand.

One can always insert into the Hamiltonian (2), without changing its symmetry properties, a weight function or form-factor $F((x-y)^2)$, for instance of a Gaussian type:

$$F = \exp\left(-\frac{(x-y)^2}{l^2}\right), \tag{3}$$

with l being a nonlocality length in the considered theory. Such a weight function would smear out the interaction and would guarantee the desired behaviour of the integrand in (2); in the limit of fundamental length $l \to 0$ in (3), the Hamiltonian (2) would correspond to a local, CPT- and Lorentz-invariant theory. The nonlocality length l could be looked upon as being a characteristic parameter relating the effective field theory to its parent one, for instance the radius of a compactified dimension when the parent theory is a higher-dimensional one. Furthermore, with such a weight function, the interaction vanishes at infinite $(x-y)^2$ separations and thus one can envisage the existence of in- and out-fields.

There exists a whole class of such CPT-violating, Lorentz invariant field theories involving different, scalar, spinor or higher-spin interacting fields [25]. Typical simplest examples are:

$$\mathcal{H}_{int}(x) = \lambda \int d^4y\, \phi_1^*(x)\phi_1(x)\theta(x_0 - y_0)\theta((x-y)^2)\phi_2(y) + h.c., \tag{4}$$

$$\mathcal{H}_{int}(x) = \lambda \int d^4y\, \bar{\psi}(x)\psi(x)\theta(x_0 - y_0)\theta((x-y)^2)\phi(y) + h.c., \tag{5}$$

$$\mathcal{H}_{int}(x) = \lambda \int d^4y\, \phi(x)\theta(x_0 - y_0)\theta((x-y)^2)\phi^2(y) + h.c. \tag{6}$$

The above Hamiltonians are nonlocal in time, and thus we expect difficulties in defining a unitary S-matrix in general. In addition, Marnelius [26] showed the breakdown of energy-momentum conservation in this class of theories if one naively applies canonical quantization. We suspect that this difficulty noted by Marnelius is associated with the absence of the well-defined canonical momentum and canonical quantization in theories nonlocal in time. To avoid this and related difficulties with canonical quantization, we use the path integral which is based on Schwinger's action principle, as explained in detail below. This path integral is quite general and has an advantage when applied to the analysis of a theory nonlocal in time: we do not start with the notion of canonical momentum, which is not generally defined in theories nonlocal in time. Instead, the canonical structure, should it exist, is fully extracted later from Green's functions defined in the path integral formalism by the Bjorken–Johnson–Low (BJL) prescription. The path integral thus defined is manifestly Lorentz-invariant and, in fact, Poincaré invariant. The energy-momentum conservation is thus ensured although the canonically quantized energy-momentum operator does not exist in general. Even the possible presence of Lorentz anomaly, i.e., quantum breaking of Lorentz symmetry, is reliably detected by the path integral.

4. Lagrangian model of fermion mass splitting

It has been shown [27] that, by including the CPT-violation only in the interaction term of a Lorentz-invariant Lagrangian, the equality of masses of particle and antiparticle remains valid. The question arises whether a Lorentz-invariant model can be constructed in which the masses of particle and antiparticle split. Obviously, besides CPT violation, such a model has necessarily to feature C and CP violation, since otherwise the antiparticle can always be defined by an operation containing C and ensuring the equality of masses.

In the present nonlocal formulation, we have a new possibility which is absent in a smooth nonlocal extension of the CPT-even local field theory. The term $i\mu\bar{\psi}(x)\psi(y)$ (to be precise, $i\mu\bar{\psi}(x)\psi(x)$) with a real μ does not appear in the local Lagrangian since it is canceled by its Hermitian conjugate. Also this term is CPT-odd. But in the present nonlocal theory one can consider the Hermitian combination

$$\int d^4x d^4y[\theta(x^0-y^0)-\theta(y^0-x^0)]\delta((x-y)^2-l^2)[i\mu\bar{\psi}(x)\psi(y)], \tag{7}$$

which is non-vanishing. Under CPT, we have $i\mu\bar{\psi}(x)\psi(y) \to -i\mu\bar{\psi}(-y)\psi(-x)$. By performing the change of integration variables $-x \to y$ and $-y \to x$, this combination is confirmed to be $CPT = -1$. In fact, we have the following transformation property of the operator part

$$\begin{aligned}
&\mathrm{C}: \ i\mu\bar{\psi}(x)\psi(y) \to i\mu\bar{\psi}(y)\psi(x), \\
&\mathrm{P}: \ i\mu\bar{\psi}(x^0,\vec{x})\psi(y^0,\vec{y}) \to i\mu\bar{\psi}(x^0,-\vec{x})\psi(y^0,-\vec{y}), \\
&\mathrm{T}: \ i\mu\bar{\psi}(x^0,\vec{x})\psi(y^0,\vec{y}) \to -i\mu\bar{\psi}(-x^0,\vec{x})\psi(-y^0,\vec{y}),
\end{aligned} \tag{8}$$

and thus the overall transformation property is $C=-1$, $P=1$, $T=1$. Namely, $C = CP = CPT = -1$.

It is thus interesting to examine a new action, proposed in [28],

$$\begin{aligned}
S &= \int d^4x\{\bar{\psi}(x)i\gamma^\mu\partial_\mu\psi(x) \quad m\bar{\psi}(x)\psi(x) \\
&\quad - \int d^4y[\theta(x^0-y^0)-\theta(y^0-x^0)]\delta((x-y)^2-l^2)[i\mu\bar{\psi}(x)\psi(y)]\},
\end{aligned} \tag{9}$$

which is Lorentz invariant and Hermitian. For the real parameter μ, the third term has $C = CP = CPT = -1$ and no symmetry to ensure the equality of particle and antiparticle masses.

The Dirac equation is replaced by

$$\begin{aligned}
&i\gamma^\mu\partial_\mu\psi(x) = m\psi(x) \\
&+i\mu\int d^4y[\theta(x^0-y^0)-\theta(y^0-x^0)]\delta((x-y)^2-l^2)\psi(y).
\end{aligned} \tag{10}$$

By inserting an ansatz for the possible solution

$$\psi(x) = e^{-ipx}U(p), \tag{11}$$

we have

$$\begin{aligned}
\not{p}U(p) &= mU(p) \\
&+ i\mu\int d^4y[\theta(x^0-y^0)-\theta(y^0-x^0)] \\
&\times \delta((x-y)^2-l^2)e^{-ip(y-x)}U(p) \\
&= mU(p) + i\mu[f_+(p)-f_-(p)]U(p),
\end{aligned} \tag{12}$$

where $f_{\pm}(p)$ is a Lorentz invariant form factor defined by

$$f_{\pm}(p) = \int d^4 z_1 e^{\pm ipz_1} \theta(z_1^0) \delta((z_1)^2 - l^2), \tag{13}$$

which are inequivalent for time-like p due to the factor $\theta(z_1^0)$. For time-like momentum p, one may choose a suitable Lorentz frame such that $\vec{p} = 0$ and

$$f_{\pm}(p^0) = 2\pi \int_0^{\infty} dz \frac{z^2 e^{\pm ip^0\sqrt{z^2+l^2}}}{\sqrt{z^2+l^2}}, \tag{14}$$

and for the space-like momentum p one may choose a suitable Lorentz frame such that $p^0 = 0$ and

$$f_{\pm}(\vec{p}) = \frac{2\pi}{|p|^2} \int_0^{\infty} dz\, z \frac{\sin z}{\sqrt{z^2 + (|p|l)^2}}, \tag{15}$$

which is analogous to the Fourier transform of the Coulomb potential and real. The expression $f_{\pm}(p)$ is mathematically related to the formula of the two-point Wightman function (for a free scalar field), which suggests that $f_{\pm}(p)$ is mathematically well-defined for $p \neq 0$ at least in the sense of distribution.

The (off-shell) propagator is defined by

$$\int d^4 x e^{ip(x-y)} \langle T^{\star} \psi(x) \bar{\psi}(y) \rangle = \frac{i}{\not{p} - m + i\epsilon - i\mu[f_+(p) - f_-(p)]}, \tag{16}$$

which is manifestly Lorentz covariant. Note that we use the $T^{\star}$-product for the path integral in accord with Schwinger's action principle, which is based on the equation of motion (10) with a source term added:

$$\langle 0, +\infty | 0, -\infty \rangle_J = \int \mathcal{D}\bar{\psi}\mathcal{D}\psi \exp i\Big\{ S + \int d^4 x \mathcal{L}_J] \Big\}, \tag{17}$$

where the action S is given in (9) and the source term is $\mathcal{L}_J = \bar{\psi}(x)\eta(x) + \bar{\eta}(x)\psi(x)$. The $T^{\star}$-product is quite different from the canonical T-product in the present nonlocal theory, and in fact the canonical quantization is not defined in the present theory. It is however important to note that the $T^{\star}$-product can reproduce all the results of the T-product, if the T-product is well-defined, by means of the Bjorken–Johnson–Low prescription [29]. In the present example, the presence of the sine-function in the denominator of the correlation function complicates this procedure, which is an indication of the absence of the canonical quantization of (9). We also emphasize that the analysis of the mass-splitting can be performed in terms of the exact solution of the (modified) free Dirac equation (10), which also defines the propagator in the present path integral prescription.

The propagator (16) is also an exact propagator for (9) in the sense of the propagator theory of relativistic quantum mechanics, and thus it could describe the particle and antiparticle propagation if one understands the antiparticle with negative energy propagating backward in time. However, if one attempts to describe the particle and antiparticle propagation with definite masses by pole approximation, for example, then the off-shell Lorentz covariance of the propagator (16) is lost.

For the space-like p, the extra term with μ in the denominator of the propagator (16) vanishes since $f_+(p) = f_-(p)$ for $p = (0, \vec{p})$, as can be easily seen from (15). Thus the propagator has poles only at the time-like momentum, and in this sense the present Hermitian action (9) does not allow a tachyon. By assuming a time-like p, we go to the frame where $\vec{p} = 0$. Then the eigenvalue equation is given by

$$p_0 = \gamma_0\{m + i\mu[f_+(p_0) - f_-(p_0)]\}, \tag{18}$$

namely,

$$p_0 = \gamma_0 \left[m - 4\pi\mu \int_0^\infty dz \frac{z^2 \sin[p_0\sqrt{z^2+l^2}]}{\sqrt{z^2+l^2}} \right], \tag{19}$$

where we used the explicit formula in (14). The solution p_0 of this equation (19) determines the possible mass eigenvalues.

This eigenvalue equation under $p_0 \to -p_0$ becomes:

$$-p_0 = \gamma_0 \left[m + 4\pi\mu \int_0^\infty dz \frac{z^2 \sin[p_0\sqrt{z^2+l^2}]}{\sqrt{z^2+l^2}} \right]. \tag{20}$$

By sandwiching this equation by γ_5, which is regarded as CPT operation, we have

$$-p_0 = \gamma_0 \left[-m - 4\pi\mu \int_0^\infty dz \frac{z^2 \sin[p_0\sqrt{z^2+l^2}]}{\sqrt{z^2+l^2}} \right], \tag{21}$$

i.e.,

$$p_0 = \gamma_0 \left[m + 4\pi\mu \int_0^\infty dz \frac{z^2 \sin[p_0\sqrt{z^2+l^2}]}{\sqrt{z^2+l^2}} \right], \tag{22}$$

which is not identical to the original equation in (19). In other words, if p_0 is the solution of the original equation, $-p_0$ cannot be the solution of the original equation except for $\mu = 0$. The last term in the Lagrangian (9) with $C = CP = CPT = -1$ splits the particle and antiparticle masses.

As a crude estimate of the mass splitting, one may assume $\mu \ll m$ and solve these equations iteratively. If the particle mass for (19) is chosen at

$$p_0 \simeq m - 4\pi\mu \int_0^\infty dz \frac{z^2 \sin[m\sqrt{z^2+l^2}]}{\sqrt{z^2+l^2}}, \tag{23}$$

then the antiparticle mass for (22) is estimated at

$$p_0 \simeq m + 4\pi\mu \int_0^\infty dz \frac{z^2 \sin[m\sqrt{z^2+l^2}]}{\sqrt{z^2+l^2}}. \tag{24}$$

Thus, the simple Lorentz invariant CPT violating Lagrangian model in (9), which produces the splitting of particle and antiparticle masses. The simple Lagrangian model will provide a useful theoretical laboratory when one investigates Lorentz invariant CPT violation effects.

The idea has been extended to the originally posed problem of the possible splitting of neutrino and antineutrino masses [30], assuming the neutrino masses to be predominantly Dirac-type in the Standard Model and without spoiling the $SU(2)_L \times U(1)$ gauge symmetry.

An attempt has been made [31] to incorporate the electromagnetic interaction in a Lorentz invariant but CPT violating nonlocal model with particle-antiparticle mass-splitting, which is regarded as a modified QED. The gauge invariance is maintained by the Schwinger non-integrable phase factor but the electromagnetic interaction breaks C, CP and CPT symmetries. To introduce the electromagnetic interaction in (9), we consider the simplest scheme (a modified QED):

$$\begin{aligned} S &= \int d^4x \Big\{ \bar\psi(x) i\gamma^\mu D_\mu \psi(x) - m\bar\psi(x)\psi(x) \\ &\quad - \int d^4y [\theta(x^0-y^0) - \theta(y^0-x^0)]\delta((x-y)^2 - l^2) i\mu\bar\psi(x) \exp\left[ie\int_y^x A_\mu(z)dz^\mu\right]\psi(y)\Big\} \\ &\quad -\frac{1}{4}\int d^4x F_{\mu\nu}(x)F^{\mu\nu}(x), \end{aligned} \tag{25}$$

with

$$D_\mu = \partial_\mu - ieA_\mu(x). \tag{26}$$

We added the Schwinger non-integrable phase factor,

$$\exp\left[ie\int_y^x A_\mu(z)dz^\mu\right], \tag{27}$$

to make the nonlocal term gauge invariant. This action is invariant under the gauge transformations

$$\begin{aligned} \psi(x) &\to e^{i\alpha(x)}\psi(x), \\ A_\mu(x) &\to A_\mu(x) + \frac{1}{e}\partial_\mu\alpha(x), \end{aligned} \tag{28}$$

and the C, CP and CPT transformation properties of each term in the action (25) are the same as in the theory without electromagnetic couplings.

It is natural to consider the non-integrable phase factor in (25) as an independent dynamical entity rather than a given external factor. In fact, Y. Nambu emphasized in many occasions the non-integrable phase factor as a manifestation of string-like objects appearing in the theory.

The proposal of [31] is to replace the non-integrable phase factor in (25) by a first quantized very massive particle propagation defined by the covariant path integral

$$\begin{aligned} &\exp\left[ie\int_y^x A_\mu(z)dz^\mu\right]\delta_{\alpha,\beta} \Rightarrow \\ &\int \mathcal{D}z^\mu \exp\left\{i\int_y^x \frac{1}{2}\left[(\dot z^\mu)^2 + M^2\right]d\tau + ie\int_y^x A_\mu(z)\frac{dz^\mu}{d\tau}d\tau\right\}\delta_{\alpha,\beta}, \end{aligned} \tag{29}$$

where the factor $\delta_{\alpha,\beta}$ contracts the spinor indices, an analogue of the Chan-Paton factor in string theory. In this way, the non-integrable phase factor becomes more dynamical and the flow of the charge is visualized in Feynman diagrams, although the second quantized particle and the first quantized particle appear in a mixed manner in Feynman diagrams. This use of a semi-static massive particle for the non-integrable phase factor is common in lattice gauge theory.

The full set of Ward–Takahashi identities, i.e. the relations among different Green's functions for the presented modified QED, can be derived formally in the path integral quantization, but the exact current becomes more involved due to the presence of the non-integrable phase factor in the full action [31]. An analysis of higher order effects in the electromagnetic coupling in the presence of the non-integrable phase factor even in the lowest order of the CPT-violation parameter μ is an interesting subject of future study.

5. Conclusion

Besides the fact that both CPT and Lorentz invariance are two most fundamental symmetries in physics, whose violations have not been hitherto observed, the relation between the two symmetries and their possible breaking are of considerable theoretical and experimental interest. Recent MINOS neutrino experiments with their favoured interpretation through a mass difference for muon neutrino and antineutrino have revived interest in CPT violation and its possible implication on Lorentz invariance breaking [14, 32–34].

It is an interesting question whether the CPT violation the models described above [25, 27, 28, 30, 31] could be a long distance effective description of some modified structure of space-time at short distances, for example.

A treatment of chiral gauge theory in the present scheme, which breaks C invariance, is in general a difficult but important future task to complete the analysis of the induced CPT violation in the present scheme. However, relaxing the quadratic CPT-violating term in the Lagrangian used in the present work as in (9), the question remains as of which symmetry is responsible for the equality of the masses of partice and antiparticle. By taking CPT violation to be due to only interaction (when C and CP are also violated), one can show that the equality of masses persists [35]. Thus, we can infer that as long as the quadratic part in the Lagrangian is not altered, the equality of the masses of particle and antiparticle is due to Lorentz invariance rather than to CPT [28, 35]. A further clarification of the basic mechanism which can ensure equal masses to the particle and anti-particle in the absence of C, CP and CPT symmetries is an interesting remaining issue. At the same time, if there exists a CPT violating interaction in Nature, for composite particles such as $p - \bar{p}$, $\pi^+ - \pi^-$, $K^0 - \bar{K}^0$, etc., as well as for bound states, such as hydrogen-antihydrogen, there are differences in their masses, due to different bound-state energies, in their total widths (life-times) and magnetic moments. Therefore, the experimental measurements on the latter characteristics of composite particles or bound systems would reveal the existence of a CPT-violating theory, although the theory is Lorentz invariant.

As for practical implications of CPT breaking in the modified QED [31], the search for the mass splitting of particle and antiparticle, just as the search for the neutrino antineutrino mass splitting in oscillation experiments [36], is interesting [37]. In the atomic transitions of the matter or antimatter systems, the frequency differences caused by the small mass difference between the electron and positron such as in (23) and (24) will be important. Other possibilities are to look for the possible small C and CP breaking in electromagnetic interactions other than those caused by weak interactions.

One problem for further study is whether the mechanism of inducing mass splitting between particle and antiparticle can be used to generate the baryon asymmetry of the Universe in equilibrium (with CPT violation, one of the three Sakharov conditions can be ignored) and what does such a case imply for the parameters of the model.

References

[1] G. Lüders, *On the Equivalence of Invariance under Time-Reversal and under Particle-Antiparticle Conjugation for Relativistic Field Theories*, Det. Kong. Danske Videnskabernes Selskab, Mat.-fys. Medd. **28** (5) (1954).
[2] W. Pauli, *Niels Bohr and the Development of Physics*, W. Pauli (ed.), Pergamon Press, New York, 1955.
[3] R. Jost, *A remark on the C.T.P. theorem*, Helv. Phys. Acta **30** (1957) 409.
[4] R. Jost, in *Theoretical Physics in the Twentieth Century*, Interscience, New York, 1960.
[5] R. F. Streater and A. S. Wightman, *PCT, Spin & Statistics, and All That*, Benjamin, New York, 1964.
[6] N. N. Bogoliubov, A. A. Logunov and I. T. Todorov, *Introduction to Axiomatic Quantum Field Theory*, Benjamin, Reading, 1975.
[7] L. B. Okun, *C, P, T are broken. Why not CPT?*, Talk given at 14th Rencontres de Blois: Matter - Anti-matter Asymmetry, Chateau de Blois, France, 17-22 Jun 2002 [arXiv:hep-ph/0210052].
[8] S. R. Coleman and S. L. Glashow, *High-Energy Tests of Lorentz Invariance*, Phys. Rev. D **59** (1999) 116008 [arXiv:hep-ph/9812418].
[9] D. Colladay and V. A. Kostelecky, *CPT violation and the standard model*, Phys. Rev. D **55** (1997) 6760

[arXiv:hep-ph/9703464]; *Lorentz-violating extension of the standard model*, Phys. Rev. D **58** (1998) 116002 [arXiv:hep-ph/9809521].
[10] V. A. Kostelecky and R. Lehnert, *Stability, causality, and Lorentz and CPT violation*, Phys. Rev. D **63** (2001) 065008 [arXiv:hep-th/0012060].
[11] V. A. Kostelecky, ed., *CPT and Lorentz Symmetry I-IV*, World Scientific, Singapore, 1999-2008; R. Bluhm, *Overview of the SME: Implications and phenomenology of Lorentz violation*, Lect. Notes Phys. **702** (2006) 191 [arXiv:hep-ph/0506054].
[12] A. G. Cohen and S. L. Glashow, *Very special relativity*, Phys. Rev. Lett. **97** (2006) 021601 [arXiv:hep-ph/0601236].
[13] Ralf Lehnert, *Tests of Lorentz symmetry*, Presented at 5th Patras Workshop on Axions, WIMPs and WISPs, Durham, England, United Kingdom, 13-17 Jul 2009. Published in "Durham 2009, Patras Workshop on Axions, WIMPs and WISPs" 171-174 [arXiv:1008.1746 [hep-ph]].
[14] H. Murayama and T. Yanagida, *LSND, SN1987A, and CPT violation,* Phys. Lett. **B 520** (2001) 263 [arXiv:hep-ph/0010178].
[15] G. Barenboim, L. Borissov, J. D. Lykken and A. Y. Smirnov, *Neutrinos as the messengers of CPT violation,* JHEP **0210** (2002) 001 [arXiv:hep-ph/0108199].
[16] G. Barenboim, L. Borissov and J. Lykken, *Neutrinos that violate CPT, and the experiments that love them,* Phys. Lett. **B 534** (2002) 106 [arXiv:hep-ph/0201080].
[17] O. W. Greenberg, *CPT violation implies violation of Lorentz invariance,* Phys. Rev. Lett. **89** (2002) 231602 [arXiv:hep-ph/0201258].
[18] M. Chaichian, K. Nishijima and A. Tureanu, *Spin statistics and CPT theorems in noncommutative field theory,* Phys. Lett. **B 568** (2003) 146 [arXiv:hep-th/0209006].
[19] L. Álvarez-Gaumé and M. A. Vázquez-Mozo, *General properties of noncommutative field theories,* Nucl. Phys. **B 668** (2003) 293 [arXiv:hep-th/0305093].
[20] M. Chaichian, M. N. Mnatsakanova, K. Nishijima, A. Tureanu and Yu. S. Vernov, *Towards an axiomatic formulation of noncommutative quantum field theory,* J. Math. Phys. (in print) [arXiv:hep-th/0402212].
[21] M. M. Sheikh-Jabbari, *C, P, and T invariance of noncommutative gauge theories,* Phys. Rev. Lett. **84** (2000) 5265 [arXiv:hep-th/0001167].
[22] M. Chaichian, P. P. Kulish, K. Nishijima and A. Tureanu, *On a Lorentz-invariant interpretation of noncommutative space-time and its implications on noncommutative QFT,* Phys. Lett. **B 604** (2004) 98 [arXiv:hep-th/0408069].
[23] M. Chaichian, P. Prešnajder and A. Tureanu, *New concept of relativistic invariance in NC space-time: Twisted Poincaré symmetry and its implications,* Phys. Rev. Lett. **94** (2005) 151602 [arXiv:hep-th/0409096v1].
[24] M. Chaichian and A. P. Demichev, *Introduction to Quantum Groups*, World Scientific, Singapore, 1996.
[25] M. Chaichian, A. D. Dolgov, V. A. Novikov and A. Tureanu, *CPT Violation Does Not Lead to Violation of Lorentz Invariance and Vice Versa,* Phys. Lett. B **699** (2011) 177 [arXiv:1103.0168 [hep-th]].
[26] R. Marnelius, *Can the S Matrix Be Defined in Relativistic Quantum Field Theories with Nonlocal Interaction?,* Phys. Rev. D **10** (1974) 3411.
[27] M. Chaichian, K. Fujikawa and A. Tureanu, *Lorentz invariant CPT violation,* Eur. Phys. J. C **73** (2013) 2349 [arXiv:1205.0152 [hep-th]].
[28] M. Chaichian, K. Fujikawa and A. Tureanu, *Lorentz invariant CPT violation: Particle and antiparticle mass splitting,* Phys. Lett. B **712** (2012) 115 [arXiv:1203.0267 [hep-th]].
[29] K. Fujikawa, *Path integral for space-time noncommutative field theory,* Phys. Rev. D **70** (2004) 085006 [hep-th/0406128].
[30] M. Chaichian, K. Fujikawa and A. Tureanu, *On neutrino masses via CPT violating Higgs interaction in the Standard Model,* Phys. Lett. B **718** (2012) 178 [arXiv:1208.5508 [hep-ph]].
[31] M. Chaichian, K. Fujikawa and A. Tureanu, *Electromagnetic interaction in theory with Lorentz invariant CPT violation,* Phys. Lett. B **718** (2013) 1500 [arXiv:1210.0208 [hep-th]].
[32] S. M. Bilenky, M. Freund, M. Lindner, T. Ohlsson and W. Winter, Phys. Rev. D **65**, 073024 (2002);
[33] G. Altarelli, *The Mystery of Neutrino Mixings,* arXiv:1111.6421 [hep-ph].
[34] P. Adamson *et al.* [MINOS Collaboration], *Improved search for muon-neutrino to electron-neutrino oscillations in MINOS,'* Phys. Rev. Lett. **107**, 181802 (2011); see, however, P. Adamson *et al.* [MINOS Collaboration], *An improved measurement of muon antineutrino disappearance in MINOS* [arXiv:1202.2772 [hep-ex]].
[35] M. Chaichian and A. Tureanu, in preparation.
[36] P. Adamson *et al.* [MINOS Collaboration], *Improved search for muon-neutrino to electron-neutrino oscillations in MINOS,* Phys. Rev. Lett. **107** (2011) 181802 [arXiv:1108.0015 [hep-ex]];
P. Adamson *et al.* [MINOS Collaboration], *An improved measurement of muon antineutrino disappearance*

in MINOS, Phys. Rev. Lett. **108** (2012) 191801 [arXiv:1202.2772 [hep-ex]];
P. Adamson *et al.* [MINOS Collaboration], *Search for Lorentz invariance and CPT violation with muon antineutrinos in the MINOS Near Detector,* Phys. Rev. D **85** (2012) 031101 [arXiv:1201.2631 [hep-ex]].

[37] T. Aaltonen *et al.* [CDF Collaboration], *Measurement of the mass difference between t and $\bar{t}$ quarks,* Phys. Rev. Lett. **106** (2011) 152001 [arXiv:1103.2782 [hep-ex]];
M. Hori *et al.* [ASACUSA Collaboration], *Two-photon laser spectroscopy of antiprotonic helium and the antiproton-to-electron mass ratio,* Nature **475** (2011) 484 [arXiv:1304.4330 [physics.atom-ph]].

Journal of Physics: Conference Series **474** (2013) 012032 doi:10.1088/1742-6596/474/1/012032

$U(1) \times SU(2)$ from the tangent bundle

Jose G. Vargas

PST Associates, 138 Promontory Rd, Columbia, SC 29209-1244, USA

E-mail: josegvargas@earthlink.net

Abstract. Élie Cartan developed modern differential geometry as theory of moving frames. Particles do not enter the equations of structure and thus play a less fundamental role. A Kaluza-Klein (KK) space without compactification brings particles into the core of geometry by making propertime (τ) the fifth dimension. (x^i, τ) emerges as the subspace for the quantum sector. This KK space does not make sense in SR, by virtue of non-orthogonality of τ to 3-space (actually), which brings a preferred frame to the fore. In contrast, propertime is perpendicular to 3-space in the "para-Lorentzian structure" with absolute time dilation (PL). Its (x^i, τ) subspace looks very much like (t, x^i) in SR. And its (t, x^i) sector is not made of orthogonal frames but does not cause contradictions with SR, which supporters of the thesis of conventionality of synchronizations have been claiming for many decades. In PL, the conjunction of the Clifford algebras of differential forms (Kaehler's) and of their valuedness gives rise to a commutative algebra of primitive idempotents that embodies the $U(1) \times SU(2)$ symmetry. $SU(3)$ also emerges in the process, but we do not deal with this issue beyond proposing the geometric palet of quarks.

1. The big picture

This paper is one in a series devoted to show that we do not need auxiliary bundles for the type of physics that goes under the name of Yang-Mills theory. There is room for that in a structure directly related to the tangent bundle by looking at the metric as a natural lifting condition relating spacetime and time-space-propertime.

A previous paper titled "Real Units Imaginary in Kaehler's Quantum Mechanics" is of the essence in the argument that we are about to present [1]. The main points that we made there were:

(a) Spin is an internal property only if one looks at particles independently of the field in which they are immersed. From the field's perspective, spin is at par with orbital angular momentum.

(b) The approach to rotations through Euclidean algebra (read Clifford algebra) allows one to geometrize the units imaginary that appear in relativistic quantum mechanics.

This geometrization leads nowhere if we think that "spacetime is everything there is to spacetime". Help comes to the rescue through the foundations of special relativity. Lest we are misunderstood, we state unequivocally that special relativity is here to stay, specially if we refer to its contents, abstraction made of the etymology of the term relativity. But its foundations may not be what have been believed to be so far. In the present paper, we fill a vacuum of related mathematical research. It has implications for Yang-Mills theory.

In a first part of the paper, we consider the foundations of differential geometry by interpreting the equations of structure of Euclidean space with the method of the moving frame. We find a vacuum here also. Dealing with it, we also fill the previously mentioned vacuum. Spacetime has to be replaced by time-space-propertime as the arena of the physics, a most natural Kaluza-Klein type of space, but without compactification. In the process, groups become ever more loosely connected with what appears to be most relevant physically for particle phenomenology. On the one hand, their Lie algebras play more direct role in any dynamics based on connections, since it is in those algebras that connections take their values. But the generators can alternatively be seen as living in a Clifford-algebra-valued Clifford algebra of differential forms. In this last structure, idempotents rather than generators may play the decisive role in determining what fundamental particles exist.

2. Introduction

2.1. Use of term tangent bundle

In a previous paper [1], we started what should become a long process of "classical geometrization" of quantum mechanics (QM). We now continue that process with the replacing of $U(1) \times SU(2)$ Yang-Mills symmetry with $U(1) \times SU(2)$ tangent bundle symmetry, thus our use of the term classical geometrization. In the interest of brevity and simplicity, tangent bundle means here not only the bundle of tangent vectors, but also other bundles directly related to it. Thus, for example, a bundle of orthonormal frames made of tangent vectors also comes under the umbrella of this denomination. More to the point, we also consider as "tangent bundle" anything directly related to the extension of the spacetime manifold with propertime, τ, to yield a 5-D Kaluza-Klein (KK) space, (t, x, τ). The dual vector of this fifth dimension "more than replaces" the 4-velocity of the paradigm.

Modern differential geometry is (or is equivalent to) a theory of moving frames [2], particles being absent from the equations of structure. Their motion is represented by the origins of a succession of frames, but this only becomes apparent in Finsler bundles [3] (See a summary of the argument after Eq. (3)). Giving particles a separate life independent of the frames may be viewed as a "mathematically overlooked issue", and as an approach different from other approaches to 5-D KK spaces in the literature. Physically, this type of structure is specially revealing (we would even say that it is only possible) in context of the specific preferred frame scenario to which we shall refer as para-Lorentzian (PL). It appears to have advantage over special relativity (SR). We proceed to explain why.

2.2. Of relativity and orthogonality

The subspace (x, τ) is a natural one for quantum physics, but not without problems when the spacetime subsector is the relativistic one. The reason is that the role of the 4-velocity is now taken by the unit vector dual to the propertime coordinate lines. It is not orthogonal to spacetime, its components being the components of the "old" 4-velocity. The (x, τ) frames are thus velocity-dependent. But quantum physics, like the classical one, must at least look relativistic, given how much relativity permeates a good functioning paradigm.

Paradoxically, the relativistic format (i.e. orthogonality, velocity independence) emerges in the (x, τ) subspace in PL context. As for the non-orthogonality of PL in the spacetime subspace, it is without consequence, at least for virtually any physical situation or experiment conceived so far. For an example of analysis of a recent experiment where that "indistinguishability" is explicitly shown, see our Opera paper [5]. Notice that we claimed that, if PL had to do with the unexpected discrepancy, we could only provide an explanation based on ad hoc assumption about an alternative hypothetical dynamics. Superluminal velocities needed not be involved; new physics, perhaps. One now believes that there was not any new physics and thus for an ad hoc assumption producing the 60 nanoseconds.

2.3. Indistinguishability and conventionalism

That indistinguishability has been believed or known for a long time by philosophers of science subscribing to the so called thesis of conventionality of synchronizations, though the type of argument leaves room for disagreement as to the limits of its applicability. See for instance the interesting analysis of the resonance experiment in a centrifuge by Maciel and Tiomno [6]. An alternative analysis of this experiment confirming that result was later provided by this author in co-authored paper with Torr [7].

Occasionally, some renowned physicist has also reached similar conclusions about that indistinguishability on more physical argument [4], though without a discussion of the limits of its applicability. The argument has to do with Einstein's synchronization, which is unavoidably implemented on clocks in a PL world, even if they stay put or are in slow motion in the laboratory frame.

The conclusion of indistinguishability is counter intuitive to, say, those involved in the dynamics of high energy accelerators. Some of them will vehemently dispute it In my doctoral dissertation, I had to suffer being caught between a supporter of the latter view and a conventionalist. It is fair to say that the limits of the thesis of conventionality of synchronizations remains a matter of controversy.

2.4. Substructure for special relativity

We posit that the relativistic-like behavior of the physics in both subspaces is a concomitant of the PL world, not of SR. The PL preferred frame structure makes possible velocity independence in the quantum sector (velocities may and will, however, be present in some role, but not in the form of the dynamical equations). It is thus more than just a competitor to SR. It may be the kinematical side of an underlying framework of a theory of which relativistic physics emerges as its facade.

In this new view, physics remains virtually what it appears to be —i.e. looking relativistic— except for being based on different foundations and except, perhaps, at the interface of the classical and quantum sectors, where a breakthrough may be very much needed. The Tajmar experiments may pertain to this interface, since they appear to suggest classical effects (specifically gravitational) of macroscopic quantum systems (superconductors) [8], [9], [10].

The proposed new scenario would concern only experts on very specialized areas of theoretical physics, the ones where a breakthrough is very much needed. It certainly would be theoretically interesting if $U(1)$ and $SU(2)$ could be viewed as spacetime related symmetries, provided that $su(3)$ could also be viewed in a similar light as $u(1)$ and $su(2)$. It will.

Since this paper touches sensitive points of modern physics, we shall let the mathematics speak, relegating physical issues for future papers. Readers reluctant to accept our results may always provide their own solution to the mathematical problem with which we launch our argument in the next section, and then draw their own consequences. Physicists who may feel the need to go deeper into these subjects and do not feel comfortable with the formal literature on differential geometry may wish to refer to our recent book [2]. Although it does not deal with Finsler geometry, readers can then go into a series of papers on the subject that I co-authored and which should then be easy reading [3], [11], [12], [13].

A different, less geometric approach to substructure for special relativity based on a preferred frame has been developed by Caban and Rembielinski [14]. The effort goes back to pioneering work by the last of these authors [15]. Their key word is "meta relativity" [16]. It emphasizes tachions, thus superluminal velocities (See for instance [17]). The driving idea is the same as in the series by this author, in frequent collaboration with D. G. Torr: there is more to structure in special relativity than meets the eye. But we do not see at this point a role for those velocities.

This paper is organized as follows. In section 3, we make the case for a Kaluza-Klein type of space directly related to the tangent bundle. Its development, which takes place in section

5, requires the justification and presentation of PL kinematics. In section 4, we show how this kinematics mimics SR. In section 6, we proceed to discuss the individual $U(1)$ and $SU(2)$ symmetries, as well as their coming together. In section 7, we start to consider electroweak stoichiometry. In section 8, we propose idempotents for the palet of quarks, but without yet going into any stoichiometry.

3. Kaluza-Klein space from tangent bundle

3.1. Motivation

In a paper that preceded his series on the theory of affine connections, Cartan [18] argued as follows. In Euclidean space, the change of the coordinates of a *fixed* point relative to a frame that experiences differential translation and rotation is given by

$$dx^i + x^j\omega_j^i + \omega^i = 0. \tag{1}$$

where ω^i represents the translation and $x^j\omega_j^i$ (with $\omega_{ij} + \omega_{ji} = 0$) the rotation. Exterior differentiating (1) and using (1) itself in the system resulting from the differentiation, one obtains

$$d\omega^i - \omega^j \wedge \omega_j^i + x^k(d\omega_k^i - \omega_k^j \wedge \omega_j^i) = 0. \tag{2}$$

Equation (2) is valid for all x, which results in the equations of structure of Euclidean space

$$d\omega^i - \omega^j \wedge \omega_j^i = 0, \quad d\omega_k^i - \omega_k^j \wedge \omega_j^i = 0, \quad \omega_{ij} + \omega_{ji} = 0. \tag{3}$$

This argument makes it clear that the equations of structure are about a geometry where points remain fixed and frames "move". Points are, so to speak, left out. We have explained through the Finslerian refibration of the frame bundle of spacetime that autoparallels are successions of origins of a succession of frames. The push-forward of the equation $d\mathbf{u} = 0$ to the Finsler bundle becomes $d\mathbf{e}_0 = 0$ or, equivalently, $\omega_0^i = 0$ [3]. As for extremals, this is a concept not connected with the theory of moving frames, except when dealing with the Levi-Civita connection. The extremals then become autoparallels and then also respond to the equation $d\mathbf{e}_0 = 0$ in the Finsler bundle.

3.2. Natural lifting condition for KK space

We address the issue of incorporating particles deeper into the geometry through a KK type of space. In the theory of moving frames, the standard quadratic forms known as metrics can be viewed as involving a double product:

$$d\mathbf{P}(\otimes,\cdot)d\mathbf{P} = dx^\mu \mathbf{e}_\mu(\otimes,\cdot)dx^\nu\mathbf{e}_\nu = \mathbf{e}_\mu\cdot\mathbf{e}_\nu dx^\mu \otimes dx^\nu = g_{\mu\nu}dx^\mu \otimes dx^\nu. \tag{4}$$

where $d\mathbf{P}$ is the translation element in the spacetime 4-dimensional manifold. The translation element of our KK space is [19] ($c = 1$)

$$d\wp = d\mathbf{P} + \mathbf{u}d\tau = \omega^\mu \mathbf{e}_\mu + \mathbf{u}d\tau. \tag{5}$$

The vector $\mathbf{u}$ is not contained in spacetime, since τ now is an independent coordinate. As for d, it is not here what one modernly calls the exterior derivative but part of a symbol. The same comment applies to $d\wp$.

In our KK space, (4) is replaced with

$$d\wp(\cdot,\cdot)d\wp = \omega^A \mathbf{e}_A(\cdot,\cdot)\omega^B\mathbf{e}_B = 0, \tag{6}$$

with $\omega^5 = d\tau$.In (6), two Clifford products are involved, respectively for tangent vector fields and differential forms viewed as integrands. This view of differential forms is very explicit in [20], [21]. Modernly, differential forms are defined as antisymmetric multilinear functions of vectors, but such will not be the case here.

If the spacetime dynamics is governed from the KK (t, x, τ) manifold, the latter's relevant curves are those that satisfy Eq. (6), which thus is a natural lifting condition. τ thus relates to the proper time assigned to those trajectories. Recall that, on curves, all differential forms are multiples (we do not mean integer multiples) of one of them, and thus of $d\tau$. But, in the (x^i, τ) submanifold, it is simply the evolution parameter of quantum systems.

3.3. *Clifford algebras and Kaluza-Klein structure*

Logically, we should view classical physics as being what it is because quantum physics is what it is. The sector (x, τ) should then determine what classical physics is like. But this is not actually so, due to historical accident.

He developed a exterior-interior calculus of differential forms for quantum mechanics. It does not resort to gamma matrices [21], [22]. We thoroughly rely on it given what we believe to be its enormous superiority over Dirac's.

Following Kähler, we shall use the signature $\eta = (-1, 1, 1, 1)$ for spacetime. In five dimensions, the metric will be postulated to be null, as per (6). The development of this equation yields

$$0 = \eta_{\mu\nu}\omega^\mu \cdot \omega^\nu + d\tau \cdot d\tau + (\omega^\mu \cdot d\tau)\mathbf{u} \cdot \mathbf{e}_\mu. \tag{7}$$

with

$$\mathbf{u} \cdot \mathbf{u} = 1, \qquad \mathbf{e}_0 \cdot \mathbf{e}_0 = -1. \tag{8}$$

If we ad hoc set

$$\omega^\mu \cdot d\tau = 0, \tag{9}$$

and specialize to $\omega^\mu = (dt, dx^i)$, the equation

$$0 = dt \cdot dt - \sum_i dx^i \cdot dx^i + d\tau \cdot d\tau \tag{10}$$

follows. We shall write it as

$$0 = dt^2 - \sum_i (dx^i)^2 - d\tau d\tau, \tag{11}$$

in parallel to writing in that way the metric tensor equation

$$0 = dt \otimes dt - \sum_i dx^i \otimes dx^i + d\tau \otimes d\tau, \tag{12}$$

which actually is an algebraic definition of the tensor $d\tau \otimes d\tau$.

The dot product in (10) has to be understood in a Clifford algebra context of differential forms, i.e. Kähler's algebra [21], [22]. We have, in the spacetime case,

$$\omega^\mu \vee \omega^\nu + \omega^\nu \vee \omega^\mu = 2\eta^{\mu\nu}. \tag{13}$$

In order to get to (13), we avoided the non-orthogonality of $\mathbf{u}$ and $\mathbf{e}_0$ in the relativistic case by virtue of (12). But we do not see how anything of theoretical interest could emerge under that non-orthogonality. But, if we consider the PL structure, a new theoretical scenario reveals itself. It can actually be viewed as the infrastructure that hides and paradoxically justifies that the quantum physics sector looks relativistic. Readers should start getting to the idea that relativistic and orthonormal need not always go together.

4. Preferred frame kinematics

4.1. Para-Lorentzian structure

Consider regions of spacetime as small as needed to be viewed as regions of affine spacetime, thus enjoying translational space symmetry and uniformity of time. Assume a frame Σ of isotropy, Einstein's preferred frame [23]. For motion along the x axis and under the assumption of homogeneity of spacetime and spatial isotropy in the preferred frame, the transformations must be of the form

$$x' = a \cdot (X - VT), \quad t' = hX + jT, \quad y' = eY, \quad z' = eZ. \tag{14}$$

Assume further that there is compliance with the interpretation by Robertson of the standard results attributed to the Michelson-Morley, Kennedy-Thorndike and Ives-Stilwell experiments. It implies [24]

$$MM : a = \gamma e, \qquad KT : j + hV = a\gamma^{-2}, \qquad IS : j + hV = \gamma^{-1}. \tag{15}$$

One readily checks that equality of the to and fro speeds of light requires

$$h + Vj = 0, \tag{16}$$

which, together with relations (15), makes (14) become the special Lorentz transformations.

Because of their relevance in applications of the KK theory of present interest, we shall consider the restriction of (14) by (16) only. One obtains:

$$x' = \gamma \cdot (X - VT), \qquad y' = Y, \qquad z' = Z, \qquad t' = hX + (\gamma^{-1} - Vh)T. \tag{17}$$

Clearly, the option of absolute simultaneity with time dilation with respect to the preferred frame in the ratio γ^{-1} is contained in this system of equations, specifically for $h = 0$:

$$x' = \gamma \cdot (X - VT), \qquad y' = Y, \qquad z' = Z, \qquad t' = \gamma^{-1}T. \tag{18}$$

We shall refer to this option as the para-Lorentzian option, or simply PL. If the velocity is not a constant, we should not write these equations but rather

$$\omega'^1 = \gamma \cdot (dX - VdT), \qquad \omega'^2 = dY, \qquad \omega'^3 = dZ, \qquad \omega'^0 = \gamma^{-1}dT. \tag{19}$$

Notice that the differentials of some of the right hand sides would not be zero if V were viewed as a coordinate and thus not as a constant, which is why we did not write on the left hand sides dx'^μ (since ddx'^μ is zero). In SR, the role of the Σ frame can be played by any Lorentzian frame, and we have

$$\omega^1 = \gamma \cdot (dX - VdT), \qquad \omega^2 = dY, \qquad \omega^3 = dZ, \qquad \omega^0 = \gamma \cdot (dT - VdX). \tag{20}$$

By eliminating (T, X^i) between (18) and the special Lorentz transformations, we get $t' = t + Vx, \quad x'^i = x^i$. Correspondingly, we have

$$t' = t + V_i x^i, \qquad x'^i = x^i \tag{21}$$

for arbitrary direction of the velocity. The relation between SR and PL frames is then given by

$$\mathbf{e}'_0 = \mathbf{e}_0, \qquad \mathbf{e}'_i = \mathbf{e}_i - V_i\mathbf{e}_0. \tag{22}$$

The ten dimensional set of PL frames is obtained by applying the usual rotations $SO(3)$ to the primed frames.

The PL metric may be obtained from the SR metric

$$ds^2 = (\omega^0)^2 - \sum_i (\omega^i)^2 = (\omega'^0 - V_i \omega'^i)^2 - \sum_i (\omega'^i)^2, \tag{23}$$

which, for mnemonic reasons, we shall write with the wrong notation

$$ds^2 = (dt' - V_i dx'^i)^2 - \sum_i (dx'^i)^2. \tag{24}$$

We may also use $\mathbf{e}'_\mu \cdot \mathbf{e}'_\nu = g'_{\mu\nu}$ and $\mathbf{e}_\mu \cdot \mathbf{e}_\nu = \eta_{\mu\nu}$.

4.2. Slow clock transport synchronization in SR and PL

We shall now look at slow clock transport synchronizations in a PL world. The argument about to be made was already made by Bohm decades ago [4]. Unfortunately, it is an argument that very few physicists appear to be aware of, including this author when he took his first steps as a researcher [25] (my blunder was that I overlooked that one factor in a key product went to infinity as the other factor went to zero). It helps keep a more open mind about the foundations of SR. We now provide an argument to the same effect, but where we take in PL the same steps as in SR. We start with the latter.

Let a clock move with a small velocity ϵ with respect to a frame $S(t, x)$ that in turn moves with respect to a frame $\Sigma(T, X)$ with velocity $\mathbf{V}$. The proper time τ of the moving clock will be given by

$$\tau \simeq T\sqrt{1 - (V + \epsilon)^2}; \quad T = \gamma_v \cdot (t + Vx), \quad x = \epsilon t, \tag{25}$$

with $\gamma_v = (1 - V^2)^{-1/2}$. Therefore, in first order,

$$\tau \simeq \frac{\sqrt{1 - (V + \epsilon)^2}}{\sqrt{1 - V^2}}(t + Vx) \simeq (1 - V\epsilon)t(1 + V\epsilon) \simeq t. \tag{26}$$

This means that, to within terms of order ϵ^2, the proper time of the clock moving with speed v with respect to S coincides with the relativistic time at the point in front of him (i.e. with coordinate $x = \epsilon t$).

The same computation in a PL world yields

$$\tau \simeq T\sqrt{1 - (V + \epsilon)^2}; \quad T = \gamma_v t', \quad x = x' = \epsilon t, \tag{27}$$

with $\gamma_v = (1 - V^2)^{-1/2}$. Therefore,

$$\tau \simeq \frac{\sqrt{1 - (V + \epsilon)^2}}{\sqrt{1 - V^2}} t' \simeq (1 - V\epsilon)(t + Vx) \simeq (1 - V\epsilon)t(1 + V\epsilon) \simeq t, \tag{28}$$

which shows that slow clock transport in a PL world performs a relativistic synchronization.

Actually, one need not perform this computation since $d\tau$ is in PL the same as in SR, namely $\gamma^{-1}T$. But the derivation just given makes our point more explicit. The bottom line is that, for most purposes, the kinematics of a PL world is unavoidably like in SR for most purposes. These considerations may not, however, apply if we had a Mossbauer experiment involving the comparison of frequencies of two identical clocks in a rotating disk, as this would not involve in any way a synchronization of clocks [6]. There is always room for doubt nevertheless, specially given the controversy surrounding rotating disks (Ehrenfest paradox).

5. Tangent bundle Kaluza-Klein spaces

5.1. SR, quantum mechanics and KK structure

The type of 5-dimensional space that is being introduced here has been overlooked, like spacetime was until the very early 20*th* century. Had it not been so, the subspace $(\mathbf{e}_i, \mathbf{u})$ rather than $(\mathbf{e}_0, \mathbf{e}_i)$ would have been the arena for the dynamics of quantum systems, and the hydrogen atom in particular. Said differently, propertime, not time, matters. It has its own representation rather than just been an ad hoc chosen laboratory time so that the hydrogen atom will be at rest in it.

On curves, and by virtue of $\mathbf{u}$ being dual to propertime, the dot products $\mathbf{u} \cdot \mathbf{e}_\mu$ $(= g_{4\mu})$ have to be interpreted as components U_μ of the 4-velocity in SR, thus not null. Hence $\mathbf{u}$ is not perpendicular to spacetime, and to 3-space in particular. We can certainly orthonormalize the vector basis but, in the normalized basis, the meaning of the unit vector would be lost. The relevant frame in this scenario for solving the hydrogen atom, then is $(\mathbf{e}_i, \mathbf{u})$, which is not orthonormal. This introduces ab initio dependence of the dynamics with respect to velocity the laboratory of the physics itself. Of course, this issue is sidestepped by "bringing the laboratory to the hydrogen atom", which is disguised as bringing the hydrogen atom to the laboratory.

It would be a waste of time to discuss the issue of the covariance of the Dirac equation as justification for the validity of, as we said, bringing the laboratory to the atom. Better to explore other avenues given that prominent physicists and mathematicians —even experts on the Dirac equation [26] (see his preface)— spontaneously admit that one does not quite understand this equation. The present author is totally convinced of the far greater naturalness and superiority of the Kähler equation[27], [22]. We shall use tangent vectors for the valuedness of the differential forms.

The negative definiteness of $-\sum_i (dx^i)^2 - d\tau d\tau$ should not be considered a counter argument (at least at this point) against this KK space. One could, for example, tinker with the signs of relevant equations in order to get the right results. Thus, for instance, the placement of the unit imaginary in the Kähler and Dirac equations is different and non-equivalent, but they give the same fine structure for the hydrogen atom [27], [22]. They also yield the same classical type electromagnetic Hamiltonian [28]. Furthermore, the unit imaginary will be replaced with members of real Clifford algebras [1]. We must wait for further work on this subject before understanding whether there is any problem here.

To summarize, if dealing with a quantum problem we had to choose between the subspaces (t, x) and (x, τ) because they yielded different results, it is clear that we should choose (x, τ) over (t, x), if we are serious about this KK option.

5.2. Para-Lorentzian KK structure and preferred frame

We proceed to consider the very revealing KK space for the PL spacetime. Let $\mathbf{E}_\mu$ denote a basis in the preference frame (up to a 3D rotation) and let $\mathbf{w}$ be the unit vector dual to propertime. Physically it represents in the KK space what in spacetime is the 4-velocity of a particle. We choose the symbol $\mathbf{w}$ where we previously chose the symbol $\mathbf{u}$, which one might now erroneously replace with the particle's velocity $\mathbf{u}'$ relative to S'. Time dilations are absolute and determined by speed with respect to the preferred frame.

In parallel to (4), we now consider

$$d\wp = d\mathbf{P} + d\tau \mathbf{w} = dX^\mu \mathbf{E}_\mu + d\tau \mathbf{w}. \tag{29}$$

We set

$$\mathbf{w} \cdot \mathbf{w} = -\mathbf{1}. \tag{30}$$

which yields

$$d\wp(\cdot, \cdot) d\wp = 0 = \eta_{\mu\nu} dX^\mu \cdot dX^\nu - d\tau \cdot d\tau + 2(dX^\mu \cdot d\tau)\mathbf{E}_\mu \cdot \mathbf{w}. \tag{31}$$

Clearly

$$\mathbf{E}_i \cdot \mathbf{w} = 0, \tag{32}$$

since, otherwise, we would not have invariance under rotations in the preferred frame. It is a tenet of post-SR preferred frame theories that we must have

$$0 = -dT^2 + \sum_i (dX^i)^2 + d\tau^2. \tag{33}$$

In order to get from (31) to (33), we need

$$(dX^\mu \cdot d\tau)(\mathbf{w} \cdot \mathbf{E}_\mu) = d\tau \cdot d\tau. \tag{34}$$

Using (32), we can write (34) as

$$(dT \cdot d\tau)(\mathbf{E}_0 \cdot \mathbf{w}) = d\tau \cdot d\tau. \tag{35}$$

The equation

$$d\tau = dT\ (1 - w^2)^{1/2} = dT\ \gamma_w^{-1}, \tag{36}$$

then allows to substitute dT in terms of dT in (35) and obtain

$$\mathbf{E}_0 \cdot \mathbf{w} = \gamma_w^{-1}. \tag{37}$$

5.3. Para-Lorentzian KK structure and moving frame

Assume that PL system S' moves with velocity of components $(V, 0, 0)$ with respect to the preferred frame. From

$$T\mathbf{E}_0 + X\mathbf{E}_1 = t'\mathbf{e}'_0 + x'\mathbf{e}'_1, \tag{38}$$

we get, using (18),

$$\mathbf{e}'_0 = \gamma(\mathbf{E}_0 + V\mathbf{E}_1), \quad \mathbf{e}'_1 = \gamma^{-1}\mathbf{E}_1, \tag{39}$$

where γ is $(1 - V^2)^{-1/2}$. The covariant form of (39) under rotations is

$$\mathbf{e}'_0 = \gamma(\mathbf{E}_0 + V^i\mathbf{E}_i), \quad \mathbf{e}'_i = \gamma^{-1}\mathbf{E}_i, \tag{40}$$

and, therefore,

$$\mathbf{e}'_i \cdot \mathbf{w} = 0. \tag{41}$$

Thus, regardless of the velocity $\mathbf{V}$ of the frame $(\mathbf{e}'_0,\mathbf{e}'_i)$, the subspace spanned by $(\mathbf{e}'_i)$ is perpendicular to the velocity $\mathbf{w}$ of particles with respect to the preferred frame. In other words, the sets $(\mathbf{e}'_i, \mathbf{w})$ constitute frames adapted to particles, as in Finsler bundles [3].

We further have, using the first of (40),

$$\mathbf{e}'_0 \cdot \mathbf{w} = \gamma(\mathbf{E}_0 \cdot \mathbf{w} + V^i\ \mathbf{E}_i \cdot \mathbf{w}) = \gamma\ \gamma_w^{-1}. \tag{42}$$

When a quantum system is at rest in the laboratory, $\gamma = \gamma_w$ and

$$\mathbf{e}'_i \cdot \mathbf{w} = 1. \tag{43}$$

The dx'^i's of a moving particle are then set to zero in the classical representation, in the sense that $\int dx'^i$ is zero. Hence, we have

$$0 = (dt'\mathbf{e}'_0 + d\tau\mathbf{w})(\cdot,\cdot)(dt'\mathbf{e}'_0 + d\tau\mathbf{w}) = \ - dt' \cdot dt' - d\tau \cdot d\tau + 2(dt' \cdot d\tau) \tag{44}$$

and, therefore,

$$dt' = d\tau \tag{45}$$

in this case. Furthermore, since

$$d\wp = d\tau(\mathbf{e}'_0 + \mathbf{w}), \tag{46}$$

the natural lifting condition implies $d\wp = 0$ and, therefore,

$$\mathbf{w} = -\mathbf{e}'_0. \tag{47}$$

These results speak of structural consistency. They will appear more natural when, in future papers, we are led to viewing this interplay of two algebras in the direction suggested in section 6.

Assume that a system S' moves with velocity $(V, 0, 0)$. The $\mathbf{e}'_i$ are not perpendicular among themselves since the process followed does not relate orthonormal 3D SR frames to orthonormal 3D PL frames. However, if the dot products are defined, one can always orthonormalize the 3-D basis. We emphasize that the crucial point here is the orthogonality of $\mathbf{w}$ and the subspace subtended by the $\mathbf{e}'_i$ regardless of what is the velocity $\mathbf{V}$ of the spacetime frame $(\mathbf{e}'_0, \mathbf{e}'_i)$ to which we attach a $(\mathbf{e}'_i)$ spatial basis.

To summarize, we have uncovered a time-space-propertime KK structure with particles represented with the help of frames but not by frames.

5.4. Non-SR, yet Lorentzian Kaluza-Klein structures

The type of KK structure that we have developed is inconsistent with the foundations of SR since it is inconsistent even with a Lorentzian structure with preferred frame, i.e. without SR. We proceed to prove it.

The Lorentz transformations do not imply SR [29]. Assume an ab initio preferred frame Σ. Assume that all frames are related to it by Lorentz transformations. If we hit without friction a body at rest in the preferred frame, it jumps by hypothesis to a Lorentz frame S_1 related to Σ by a Lorentz boost without rotation. Also from Σ, we hit without rotation an identical body in a similar manner but with different force and in a different direction. It falls to rest in Lorentz frame S_2. Let $\mathbf{v}$ be the relative velocity of S_2 with respect to S_1. If we now hit, again without friction, the body at S_1 so that it falls into S_2 and assume SR, this body reaches there with a Thomas rotation with respect to the one already there by boost from Σ. But a maker of worlds could in principle have created a world where the result of the blow from S_1 to S_2 is a Lorentz boost with rotation in such a way that the final state in S_2 is the same regardless of which of the two ways took as there. It is in such a context that we now proceed with the discussion.

$\mathbf{w}$ will again refer to the velocity of particles with respect to Σ. As in PL, we would choose $\mathbf{w} \cdot \mathbf{w}$ equal to -1. Let S be any other moving orthonormal frame that will play the role that S' played in the previous subsection. Equations (39) and (40) must now be replaced by Lorentz boots and Eq. (41) no longer applies. The same conclusion is reached if we had simply substituted the second of equations (15) in (41):

$$(\mathbf{e}_i - V_i\mathbf{e}_0) \cdot \mathbf{w} = 0. \tag{48}$$

We thus have orthonormality and, therefore, presumed be velocity dependence of the dynamics. The fine structure of the hydrogen atom would depend on the velocity with respect to the laboratory. We are not speaking here of the shifted spectrum of an atom by virtue of its motion with respect to the laboratory frame. We would be computing in the propertime scenario that this KK theory blindly assigns to it.

The standard relativistic approach to the hydrogen atom based on the $(\mathbf{e}_0, \mathbf{e}_i)$ subspace at zero velocity with respect to the laboratory yields the right spectrum (up to second quantization corrections). The KK scenario validates PL. The relativistic approach for the hydrogen atom is, therefore, mimicking the PL physics in the (x^i, τ) subspace, equivalently $\mathbf{e}'_i, \mathbf{w}$ subspace.

6. Partial geometric interpretation of $U(1) \times SU(2)$

We speak of partial interpretation because, going the full way with the interpretation, we fall into $SU(3)$. We want to follow the historical development —first electroweak, then $U(1) \times SU(2) \times SU(3)$— because that is the way in which it started to evolve in our own mind. In any case, going directly to $U(1) \times SU(2) \times SU(3)$ would be too much to digest at once. We leave it for a future paper, when this author will himself have a better grasp of the issues involved.

6.1. Rotations in a Clifford algebra and in its ideals

In Kähler's approach to quantum mechanics, the equation that supersedes Dirac's is called the Kähler equation. It is not about spinors but about a primordial field that is a member of an algebra. Let us refer to it as the Kähler algebra in order to distinguish it from other algebras to be considered. The Kähler algebra can be decomposed into ideals defined by symmetry properties. The field can correspondingly be decomposed into spinors.

The differential forms that are members of the Kähler algebra for the electromagnetic interaction are scalar valued. There are four basic mutually annulling primitive idempotents. They are associated with charge and chirality (or energy and spin if one considers the phase factors that multiply those idempotents). Not all symmetries of a system fit at the same time in the decomposition of the solutions. For example, only one component of spin fits in them even if there is spherical symmetry of the system. In Dirac's theory, this has to do with whether generators commute or not.

Because of scalar valuedness, Kähler had no alternative but to use the unit imaginary to form idempotents. This is totally unnecessary when the differential forms are Clifford valued. Rotations truly take place in the valuedness algebra, not in the algebra of scalar-valued differential forms [1]. Under a ϕ rotation around the unit vector $\mathbf{n}$, a member $\mathbf{A}$ of the valuedness (i.e. tangent) Clifford algebra transforms in the following way:

$$e^{-\frac{1}{2}\phi n^i \mathbf{a}_j \mathbf{a}_k} \,\mathbf{A}\, e^{\frac{1}{2}\phi n^i \mathbf{a}_j \mathbf{a}_k}, \tag{49}$$

with summation over the three cyclic permutations of $(1, 2, 3)$, the $(\mathbf{a}_i)$ constituting an orthonormal basis. Readers uninitiated on Clifford algebra be informed that $\mathbf{a}_j\mathbf{a}_k$ can be written as $i\mathbf{a}_i$ where the factor i is $\mathbf{a}_i\mathbf{a}_j\mathbf{a}_k$, of square minus one. The square of $\mathbf{a}_j\mathbf{a}_k$ also is minus one.

For simplicity take $\mathbf{n} = \mathbf{a}_3$. Then, (48) becomes

$$e^{-\frac{1}{2}\phi \mathbf{a}_1 \mathbf{a}_2} \,\mathbf{A}\, e^{\frac{1}{2}\phi \mathbf{a}_1 \mathbf{a}_2}. \tag{50}$$

Left spinors $\mathbf{L}$ of the same algebra, on the other hand, transform like

$$e^{-\frac{1}{2}\phi n^i \mathbf{a}_j \mathbf{a}_k}\, \mathbf{L}. \tag{51}$$

And, corresponding to (49),

$$e^{-\frac{1}{2}\phi \mathbf{a}_1 \mathbf{a}_2}\, \mathbf{L}. \tag{52}$$

Whether the rotation is represented by (48), i.e. $SO(3)$, or by (50), i.e. $SU(2)$, depends on the objects on which rotations act. It is then as legitimate to speak of rotations when dealing with $SU(2)$ as when dealing with $SO(3)$, even if the relation is 2 to 1.

6.2. $SU(2)$

Kähler's QM is more revealing than Dirac's regarding interactions beyond the electromagnetic one. The calculus on which he based his work follows long after his seminal work on exterior systems nowadays known as the Cartan-Kähler theory of differential equations. His here-crucial treatment of symmetries is what it is because of his replacement of Dirac's equation for spinors with an equation for a primordial field valued in some algebra. The spinors are present through the decomposition of the algebra into ideals.

He decomposed his Clifford algebra of differential forms through four commuting primitive idempotents, $\epsilon^{\pm}\tau^{*}$, where the asterisk means both signs independently of the sign in $\epsilon^{\pm}$. The $\tau^{\pm}$ and $\epsilon^{\pm}$ are

$$\tau^{\pm} \equiv \frac{1}{2}(1 \pm idxdy), \qquad \epsilon^{\pm} \equiv \frac{1}{2}(1 \mp idt). \tag{53}$$

The symmetry is around the z axis (actually by choice of axis). There are more pairs of mutually annulling primitive idempotents, say

$$\tau^{\pm}_{yz} \equiv \frac{1}{2}(1 \pm idydz), \qquad \tau^{\pm}_{zx} \equiv \frac{1}{2}(1 \pm idzdx), \tag{54}$$

but none of these commutes with $\tau^{\pm}$, to which we refer as $\tau^{\pm}_{xy}$ in the following.

With the geometrization of quantum mechanics that we have recently proposed [1], we make the replacement of

$$\tau^{\pm}_{ij} \equiv \frac{1}{2}(1 \pm idx^{i}dx^{j}) \tag{55}$$

and $\epsilon^{\pm}$ with

$$\boldsymbol{\tau}^{\pm}_{ij} \equiv \frac{1}{2}(1 \pm \mathbf{a}_i\mathbf{a}_j dx^{i}dx^{j}), \qquad \boldsymbol{\varepsilon}^{\pm} \equiv \frac{1}{2}(1 \mp \mathbf{w}d\tau), \tag{56}$$

where τ without indices is propertime. Clearly, no summation over repeated indices is understood.

All the six independent $\boldsymbol{\tau}^{\pm}_{ij}$ commute, which does not mean necessarily that they can be used as factors in hypothetical solutions with spherical symmetry. More than that would be needed, as explained elsewhere [28]. One may nevertheless replace the family $\epsilon^{\pm}\tau^{*}$ with the family

$$\boldsymbol{\varepsilon}^{\pm}\boldsymbol{\tau}^{*}_{yz}, \qquad \boldsymbol{\varepsilon}^{\pm}\boldsymbol{\tau}^{*}_{zx}, \qquad \boldsymbol{\varepsilon}^{\pm}\boldsymbol{\tau}^{*}_{xz}. \tag{57}$$

At this point, one may wonder what is the Lie product in the algebra that supposedly replaces the Lie algebra of the rotation group. The Clifford product would play that role when we multiplying two different factors. Define the natural order of the space bivectors and corresponding differential 2-forms as $\mathbf{a}_i\mathbf{a}_j$, $\mathbf{a}_j\mathbf{a}_k$, $\mathbf{a}_k\mathbf{a}_i$. Then:

$$(\mathbf{a}_i\mathbf{a}_j)(\mathbf{a}_j\mathbf{a}_k) = \mathbf{a}_i\mathbf{a}_k = -\mathbf{a}_k\mathbf{a}_i, \tag{58}$$

and, similarly,

$$(dx^{i}dx^{j})(dx^{j}dx^{k}) = dx^{i}dx^{k} = -dx^{k}dx^{i}. \tag{59}$$

The problem with this would appear to be that the product of $\mathbf{a}_i\mathbf{a}_j$ with itself (equivalently $dx^{i}dx^{j}$) does not yield zero. If need be, we recover an equivalent Lie algebra in both cases by defining a Lie product of $(\mathbf{a}_i\mathbf{a}_j)$ and $(\mathbf{a}_j\mathbf{a}_k)$ as

$$\frac{1}{2}[(\mathbf{a}_i\mathbf{a}_j)(\mathbf{a}_j\mathbf{a}_k) - (\mathbf{a}_j\mathbf{a}_k)(\mathbf{a}_i\mathbf{a}_j)], \tag{60}$$

and similarly for the $dx^{i}dx^{j}$.

Little by little it will become increasingly evident that the Lie products of the generators are irrelevant. What matters is the compatibility of idempotents, even if the algebra from which they are built is commutative.

6.3. $U(1)$ *and* $U(1) \times SU(2)$

Recall the opening mathematical argument of the paper. With the particle at rest, we obtained the equations of structure by moving the frame to which the particle is referred. The boosts of frames mimic the translation of the particles when, for instance, we consider this translation as being represented by autoparallels.

The translation of particles is now obviously represented by $\mathbf{w}$, but not in (x, τ) subspace, where it is multiplied by the differential $d\tau$ of the evolution parameter in the quantum sector τ. Hence, in this sector, we have to see the $U(1)$ symmetry as pertaining to this evolution. And the relation to electrodynamics is obvious by running the argument of the previous subsection in reverse, from the primitive idempotents (56) to the $\epsilon^{\pm}\tau^{*}$.

The $\varepsilon^{\pm}$ are then the idempotents associated with propertime translation symmetry in (x, τ) subspace. In the full (t, x, τ) space and through the natural lifting condition, τ represents the (propertime of a) particle rather than the evolution parameter for particle-less quantum mechanical field theory. [32] is then the Lie algebra for the one-parameter group of propertime time translation symmetry, $\mathbf{w}$ being the generator. But what matters is the associated idempotents (accompanied by the appropriate phase factor [28]).

At its deepest level, quantum physics is more directly concerned with its Lie algebra, which is where connections take their values. Thus the focussing on groups rather than on their Lie algebras may be specially misleading in the case of $U(1)$ because there is only one Lie algebra in dimension one. It does not matter whether the group is compact or not since the connection is a local concept.

Using rotations as an example, the point about different groups being locally equivalent was succinctly and authoritatively made by Donaldson and Kronheimer when they stated "*For the purposes of local differential geometry, SO(3) connections and SU(2) connections are completely equivalent*". and further down in the same paragraph: "*So there is not really much difference between working with the structure groups SU(2) and SO(3)* [32] (page 42, no italics in the original). Of course, there are global differences, but these may be totally irrelevant for a physicist if, as pointed out with a specific case by Cartan in correspondence with Einstein "... *every singularity-free solution of system (1), creates from the topological point of view,* the continuum in which it *exits*" [33] (p. 103, italics as in the original; the system (1) to which Cartan refers is an example he proposed to make a point to Einstein].

A too narrow focus on groups reflects a perspective of symmetries inherited from classical physics, where restrictions imposed by lack of commutativity of operators do not enter the theory. Rotations are very illuminating also to illustrate this point. A Kähler equation may have as input a spherically symmetric potential, but the solutions with angular momentum different from zero do not have spherical symmetry because of reasons that go beyond commutativity or not of the operators. Of the essence is the form of the output of the equation (spinors, in the case of particle solutions as opposed to preferred frame solutions) rather than the symmetry properties of the input (the electromagnetic potential), greater than those of the input.

But the everyday manifestation of classical electrodynamics —except for the trained eye who sees its effects in determining the properties of matter through quantum laws— is in the classical domain. In this domain, the equation of motion of particles is given by $d\mathbf{u} = \mathbf{0}$. This is the case not only for the theory of gravitation but also for electrodynamics, once we accept that it has to do with geometry in the tangent bundle [30], [31]. In turn, $\mathbf{u}$ and $\mathbf{w}$ are intimately related through the natural lifting condition, as we already saw. See argument that starts after Eq. (42) and goes all the way down to equation (46). So, propertime translation symmetry (i.e. the $U(1)$ group in the quantum sector) also is related to classical electrodynamics.

7. Limitation of "electroweak" stoichiometry

By the term stoichiometry, we mean here the balancing of idempotents in particle reactions. By the qualification electroweak, we mean not only the absence of quark considerations, but also the absence of vector bosons. These have lifetimes like third generation quarks, thus highly non-nucleonic. We have put weak in electro"weak" between inverted commas for the same reason as to why we spoke of "partial" interpretation in the title of the previous section.

The general stoichiometry of elementary particles is based on equations such as

$$1 = \tau^+ + \tau^= = \frac{1}{2}(1 + idxdy) + \frac{1}{2}(1 + idxdy). \tag{61}$$

$$1 = \epsilon^+ + \epsilon^- = \frac{1}{2}(1 - idt) + \frac{1}{2}(1 + idt). \tag{62}$$

$$1 = \epsilon^+\tau^+ + \epsilon^+\tau^- + \epsilon^-\tau^+ + \epsilon^-\tau^-. \tag{63}$$

All these equations are supposed to be multiplied by a solution of the primordial field on the left. One then brings dt and $dxdy$ factors in the primordial field to the right to be absorbed by the idempotents through

$$dt\ \epsilon^+ = \pm\epsilon^+, \qquad dxdy\ \tau^+ = \pm\tau^+. \tag{64}$$

Whether $\epsilon^-\tau^+$ or $\epsilon^-\tau^+ + \epsilon^-\tau^-(=\epsilon^-)$ represents an electron or not depends on what these idempotents are multiplied by in the end, i.e. when these operations have taken place. First, there has to be an appropriate phase factor. But that is not all. The factor that still remains must not contain t and dt in the case of the idempotents ϵ. The case of $\epsilon^-\tau^+$ and $\epsilon^-\tau^-$ follows along similar lines, but involving more variables and differentials.

According to this scheme of things, neutron decay,

$$n \to p + e + \bar{\nu}_e\ , \tag{65}$$

is represented by

$$\tau^+ = \epsilon^+ + \epsilon^- - \tau^-. \tag{66}$$

This can be rewritten as

$$0 = \epsilon^+ + \epsilon^- - (\tau^+ + \tau^-). \tag{67}$$

At this level of theory, one cannot even distinguish between the idempotents for neutrinos from those for neutrons since τ^- represents a neutrino. But, in another reaction, it could represent an antineutrino. One needs more idempotents to specify n and p, but this would be strong interaction stoichiometry.

The question then is, how does the electroweak case improve on the electromagnetic case in this regard. In the above scheme of things, one simply replaces $\tau^\pm$ and $\epsilon^\pm$ with

$$\boldsymbol{\tau}_{ij}^\pm \equiv \frac{1}{2}(1 \pm \mathbf{a}_i\mathbf{a}_j dx^i dx^j), \qquad \boldsymbol{\varepsilon}^\pm \equiv \frac{1}{2}(1 \mp \mathbf{w}d\tau). \tag{68}$$

Thus, equation (66) is now replaced with

$$\boldsymbol{\tau}_{xy}^+ = \boldsymbol{\varepsilon}^+ + \boldsymbol{\varepsilon}^- - \boldsymbol{\tau}_{xy}^-. \tag{69}$$

Very little has been gained, except that we also have

$$\boldsymbol{\tau}_{yz}^+ = \boldsymbol{\varepsilon}^+ + \boldsymbol{\varepsilon}^- - \boldsymbol{\tau}_{yz}^-, \qquad \boldsymbol{\tau}_{zx}^+ = \boldsymbol{\varepsilon}^+ + \boldsymbol{\varepsilon}^- - \boldsymbol{\tau}_{zx}^- \tag{70}$$

for neutron decay via the muon and tau channels.

One has to be careful with the handling of these equations before we deal with the strong interaction, at which point the different particles in the reaction will be represented by richer idempotent decompositions. One also must be careful not to mix equations of the types (65) and (66). One might be tempted to write the reactions

$$W^- \to e + \bar{\nu}_e, \qquad W^- \to \mu + \bar{\nu}_\mu, \qquad W^- \to \tau + \bar{\nu}_\tau \tag{71}$$

jointly as

$$W^- = e + \bar{\nu}_e = \mu + \bar{\nu}_\mu = \tau + \bar{\nu}_\tau, \tag{NO!}$$

This is clearly incorrect as a direct consequence of Eqs. (69)-(70).

But there is another very important reason. W^- is a boson. So, like the gammas, they must not appear in our equations. Bosons will be included in the primordial field, except that this field is more sophisticated in the electroweak case than in the electromagnetic case. In Eq. (63), the primordial electromagnetic field will be included in tis equation only after that equation is multiplied by it. so, we may say that, in that equation, it is represented by the unity. In the same way, we shall now have

$$\begin{aligned} 1 = {} & \boldsymbol{\varepsilon}^+\boldsymbol{\tau}^+_{yz} + \boldsymbol{\varepsilon}^+\boldsymbol{\tau}^-_{yz} + \boldsymbol{\varepsilon}^-\boldsymbol{\tau}^+_{yz} + \boldsymbol{\varepsilon}^-\boldsymbol{\tau}^-_{yz} + \\ & + \boldsymbol{\varepsilon}^+\boldsymbol{\tau}^+_{zx} + \boldsymbol{\varepsilon}^+\boldsymbol{\tau}^-_{zx} + \boldsymbol{\varepsilon}^-\boldsymbol{\tau}^+_{zx} + \boldsymbol{\varepsilon}^-\boldsymbol{\tau}^-_{zx} \\ & + \boldsymbol{\varepsilon}^+\boldsymbol{\tau}^+_{xy} + \boldsymbol{\varepsilon}^+\boldsymbol{\tau}^-_{xy} + \boldsymbol{\varepsilon}^-\boldsymbol{\tau}^+_{xy} + \boldsymbol{\varepsilon}^-\boldsymbol{\tau}^-_{xy}, \end{aligned} \tag{72}$$

the unit on the left representing the electroweak field. A more significant difference is that the photon is massless and the $W^\pm$ and Z are not.

8. Beyond electroweak stoichiometry: quarks

The chirality operators are associated with spin, thus with rotational symmetry, but nothing has been said of the relation of the axis of rotational symmetry to the motion of particles with respect to the preferred frame. It may make a difference whether something spins in the direction of the motion or perpendicular to it. That is an element to be taken into account in building $U(1) \times SU(2) \times SU(3)$ stoichiometry. It brings new factors into the idempotents that represent particles.

The three pairs $\boldsymbol{\tau}^\pm_{ij}$ are associated with spin/chirality in three orthogonal directions. One may think of quarks as organizations of the field that failed to be particles because they missed one of the elements needed to be so; they died while in gestation. "Failed" experiments of nature can be built by incorporating new idempotents as factors of the $\boldsymbol{\varepsilon}^\pm\boldsymbol{\tau}^*_{ij}$. We have specifically in mind translational symmetry idempotents

$$\chi_x \equiv \frac{1}{2}(1 + d\mathbf{x}), \qquad \chi_y \equiv \frac{1}{2}(1 + d\mathbf{y}), \qquad \chi_z \equiv \frac{1}{2}(1 + d\mathbf{z}), \tag{73}$$

since they are naturally correlated with rotational degrees of freedom and can be suggested by relevant velocities. The symbols $d\mathbf{x}$, $d\mathbf{y}$ and $d\mathbf{z}$ are just notation for $dx\mathbf{a}_1$, $dy\mathbf{a}_2$ and $dz\mathbf{a}_3$. They can participate similarly in the making of quarks.

Owing to reasons that we shall not enter at this point, one does not need a $\pm$ here because it is the relation between $\boldsymbol{\chi}$'s and $\boldsymbol{\tau}$'s that matters, and these are taken care of already with just the plus signs in the $\boldsymbol{\chi}$'s. We are also led to believe that idempotents (73) are associated with generations. Then, within each of these, color will be associated with rotational idempotents. But all that is for a future paper.

In order to see how (73) expands the algebra for any given generation, just notice that the subspace of our commutative algebra that $\boldsymbol{\varepsilon}^{\pm}\boldsymbol{\tau}^{*}_{xy}$ spans is the same one as generated by

$$1, \; d\tau\mathbf{w}, \; d\mathbf{x}d\mathbf{y}, \; d\mathbf{x}d\mathbf{y}d\tau\mathbf{w}. \tag{74}$$

In order to avoid distractions with a double use of the symbol tau, we do not write $d\tau\mathbf{w}$ as $d\boldsymbol{\tau}$.We can built copies of $SU(2)$ with these generators in many different ways. In each of the columns of the table that follows, the product of any two of them in cyclic order amount to the negative of the third one:

$$\begin{bmatrix} 1 & 1 & d\tau\mathbf{w} & \boldsymbol{\varepsilon}^{+} & \boldsymbol{\tau}^{+}_{xy} \\ d\mathbf{x}d\mathbf{y} & -d\tau\mathbf{w} & d\mathbf{x}d\mathbf{y} & d\mathbf{x}d\mathbf{y}\boldsymbol{\varepsilon}^{+} & -\boldsymbol{\tau}^{+}_{xy}d\tau\mathbf{w} \\ -d\mathbf{x}d\mathbf{y} & d\tau\mathbf{w} & -d\mathbf{x}d\mathbf{y}d\tau\mathbf{w} & -d\mathbf{x}d\mathbf{y}\boldsymbol{\varepsilon}^{+} & \boldsymbol{\tau}^{+}_{xy}d\tau\mathbf{w} \end{bmatrix}, \tag{75}$$

as per (58). The fourth and fifth columns result from multiplication of the first and second ones by ε^{+} and $\boldsymbol{\tau}^{+}_{xy}$ respectively. We might as well have multiplied by ε^{-} and $\boldsymbol{\tau}^{-}_{xy}$.

Let us now put together the contents of the previous two paragraphs. We multiply the first two rows of the three last columns by $d\mathbf{z}$ to get the following copies of $su(2)$:

$$\begin{bmatrix} d\mathbf{z}d\tau\mathbf{w} & d\mathbf{z}\boldsymbol{\varepsilon}^{+} & d\mathbf{z}\boldsymbol{\tau}^{+}_{xy} \\ d\mathbf{z}d\mathbf{x}d\mathbf{y} & d\mathbf{z}d\mathbf{x}d\mathbf{y}\boldsymbol{\varepsilon}^{+} & -d\mathbf{z}\boldsymbol{\tau}^{+}_{xy}d\tau\mathbf{w} \\ -d\mathbf{x}d\mathbf{y}d\tau\mathbf{w} & -d\mathbf{x}d\mathbf{y}\boldsymbol{\varepsilon}^{+} & \boldsymbol{\tau}^{+}_{xy}d\tau\mathbf{w} \end{bmatrix} \tag{76}$$

We now show that seven and only seven of these nine "generators" (of the algebra) are independent. An eighth one emerges from squaring the elements of the first column.

Consider the first two rows of (76). It is easy to find four independent elements among the six of them. There cannot be more than four since all of them are linear combinations of those in (74) (multiplied by $d\mathbf{z}$). The elements of the third row, not containing the factor $d\mathbf{z}$, are independent of the elements of the first two rows. They also are independent among themselves since $d\mathbf{x}d\mathbf{y}\boldsymbol{\varepsilon}^{+}$ (a) contains one term $d\mathbf{x}d\mathbf{y}$ but the other two elements in the third row have it only multiplied by something else, and (b) these two are independent of each other (notice that $d\tau\mathbf{w}$ is present in $\boldsymbol{\tau}^{+}_{xy}d\tau\mathbf{w}$, but not in $d\mathbf{x}d\mathbf{y}d\tau\mathbf{w}$). Together with the unity (which is the square of all terms in the first column), those eight independent elements span the eight dimensional algebra to which we have referred above. Thus the $SU(3)$ idempotents for first generation quarks will be given by

$$\boldsymbol{\chi}_{z}\boldsymbol{\varepsilon}^{*}\boldsymbol{\tau}^{**}_{ij}. \tag{77}$$

There are twelve of them, i.e. 36 $[3_{\chi} \times 2_{\varepsilon} \times (3_{ij} \times 2_{**})]$ between the three generations.

We are leaving for a future paper progress we have made on concepts such as, say, double strangeness, color and its relation to spin and generation jumping. Also, we have assigned actual physical directions to x, y and z. We have not yet started on the stoichiometry.

9. Concluding remarks

In Kähler's scheme of things, the phenomenology associated with the interactions should arise from the expression of a primordial field into a sum of members of ideals defined by idempotents. The range of phenomena classified as electroweak is very small, mainly because of the physical indefiniteness associated with the electroweak interaction (56). The idempotents should contain information speaking of how the different directions in the mathematical sense relate to actual physical direction(s). Incorporating quarks in this scheme of things. is not enough. We leave for a future paper a more close implementation of the inner logic of a preferred frame scenario.

Acknowledgments

Conversations with Prof. Z. Oziewicz, Prof. D. G. Torr and Dr. B. Schmeikal are acknowledged. Funding from PST Associates is deeply appreciated. To Professor Alwyn van der Merwe, former editor of Foundations of Physics, for sustained encouragement while my research meandered towards its goal.

References

[1] Vargas J G Real "Units imaginary" in Kähler's quantum mechanics *Preprint* gen-ph *1207.5781*v1
[2] Vargas J G 2012 *Differential Forms for Cartan-Klein Geometry: Erlangen Program with Moving Frames* (London, Abramis)
[3] Vargas J G and Torr D G 1993 *J. Math. Phys.* **34** 4898
[4] Bohm D 1965 *The Special Theory of Relativity* (New York, Benjamin)
[5] Vargas J G 2011 Opera's neutrinos and the Robertson test theory of the Lorentz transformations *Preprint* gen-ph *1111.2271*v2
[6] Maciel A K A and Tiomno J 1985 *Physical Rev. Lett.* **55**, 143
[7] Vargas J G and Torr D G 1989 *Phys. Rev.* A **39** 2878
[8] Tajmar M, Plesescu F, Seifert B and Marhold K 2007 *Proceedings of the STAIF-2007 Conference* (AIP Conference Proceedings, Vol. 880) pp 1071-1082
[9] Tajmar M, Plesescu F, Marhold K and de Matos C J 2006 Experimental detection of the gravitomagnetic London moment *Preprint* gr-qc/0603/0603033
[10] Tajmar M, Plesescu F, Seifert B, Schnitzer R and I. Vasiljevich 2008 *Proc. of the Second International Conference on Time and Matter* (Nova Gorica, University of Nova Gorica Press) pp 49-74
[11] Vargas J G and Torr D G 1995 *Gen. Rel. Grav.* **27** 629
[12] Vargas J G and Torr D G 1996 *Gen. Rel. Grav.* **28** 451
[13] Vargas J G and Torr D G 1996 *Contemporary Mathematics* **196** 301
[14] Caban P and Rembielinski J 1999 *Phys. Rev.* A **59** 4187
[15] Rembielinski J 1980 *Phys. Lett.* **78A** 33
[16] Rembielinski J and Wlodarczyk M 2012 "Meta relativity: against special relativity? *Preprint* gr-qc/1206.0841v1
[17] Rembielinski J 2000 Superluminal phenomena and the quantum preferred frame *Preprint* quant-ph/0010026v2
[18] Cartan É 1922 *Math. pures et appl.* ***1*** 1
[19] Vargas J G and Torr D G 1997 *Found. Phys.* **27** 533
[20] Rudin W 1976 *Principles of Mathematical Analysis* (New York, Mc-Graw-Hill)
[21] Kähler E 1960 *Abh. Dtsch. Akad. Wiss. Berlin, Kl. Math. Phys. Tech* **4** 1
[22] Kähler E 1962 *Rendiconti di Matematica* **21** 425
[23] Robertson H P 1949 *Rev. Mod. Phys.* **21** *378*
[24] Vargas J G 1984 *Found. Phys.* **14** 625
[25] Vargas J G 1980 *Lett. Nuovo Cimento* **28** 289
[26] Thaller B 1992 *The Dirac Equation* (Berlin, Springer)
[27] Kähler E 1961 *Abh. Dtsch. Akad. Wiss. Berlin, Kl. Math. Phys. Tech.* **1** 1
[28] Vargas J G 2008 *Found. Phys.* **38** 610
[29] Vargas J G 1986 *Found. Phys.* **16** 1231
[30] Vargas J G and Torr D G 1999 *Found. Phys.* **29** 1543
[31] Vargas J G 2007 *Proc. XIII International Scientific Meeting on Physical Interpretations of Relativity Theory*, (Moscow, Baumann State Technical University) pp 70-81
[32] Donaldson S K and Kronheimer B P 1990 *The Geometry of Four-Manifolds* (Oxford, Clarendon Press)
[33] Debever R 1979 *Élie Cartan - Albert Einstein: Letters on Absolute Parallelism* (Princeton, Princeton University)

Hopping in the Phase Model to a Non-Commutative Verlinde Formula for Affine Fusion

M.A. Walton

Department of Physics and Astronomy, University of Lethbridge, Lethbridge, Alberta, T1K 3M4, Canada

E-mail: `walton@uleth.ca`

Abstract. Korff and Stroppel discovered a realization of $su(n)$ affine fusion, the fusion of the $su(n)$ Wess-Zumino-Novikov-Witten (WZNW) conformal field theory, in the phase model, a limit of the q-boson hopping model. This integrable-model realization provides a new perspective on affine fusion, explored in a recent paper by the author. The role of WZNW primary fields is played in it by non-commutative Schur polynomials, and fusion coefficients are thus given by a non-commutative version of the Verlinde formula. We present the extension to all Verlinde dimensions, of arbitrary genus and any number N of points. The level-dependence of affine fusion is also discussed, using the concept of threshold level, and its generalization to threshold weight.

1. Introduction

Affine fusion is found in several mathematical and physical contexts. It is a natural generalization of the tensor product of representations of simple Lie algebras; a simple truncation thereof controlled by a non-negative integer, the level.

One important physical context is conformal field theory, and more specifically, the Wess-Zumino-Novikov-Witten (WZNW) models (see [4], for example). WZNW models realize, at a fixed non-negative integer level k, a non-twisted affine Kac-Moody algebra $g^{(1)}$ based on a simple Lie algebra g, or g_k for short. Their primary fields furnish representations of g_k and their operator products are governed by the corresponding affine fusion algebra.

Recently, Korff and Stroppel [8] found a much simpler physical realization of affine fusion, for the $su(n)_k$ case, at all levels $k \in \mathbb{N}$ together (see also [7, 13]). The phase model [2] is an integrable multi-particle model whose integrals of motion are non-commutative Schur polynomials [8]. The integrals may be diagonalized by the algebraic Bethe ansatz, and their eigenvalues are affine fusion eigenvalues [8]. The non-commutative Schur polynomials play the role in the phase model that the primary fields do in WZNW models [13]. The correlators of non-commutative Schur polynomials equal affine fusion coefficients [8] and more generally, affine Verlinde dimensions of arbitrary genus and numbers of marked points [13]. The Korff-Stroppel formula for these correlators can be thought of as a non-commutative Verlinde formula.

This proceedings contribution is based mostly on the paper [13]. Section 2 reviews the phase model realization of affine fusion found by Korff and Stroppel [8]. Section 3 records the new results of [13]: the extension of the non-commutative Verlinde formula to higher genus and N points, and consideration of the level-dependence of fusion in the phase-model realization. In the latter, the old concept of threshold level [3, 6] plays a prominent role. Generalization to consideration of threshold weights [14] is included here as well. Section 4 is a short conclusion.

2. Phase-Model Realization of Affine $su(n)$ Fusion

In the $su(n)$ phase model, bosons occupy n sites on S^1, or, equivalently, the nodes of the Coxeter-Dynkin diagram of the affine Kac-Moody algebra $\widehat{su}(n) \cong A^{(1)}_{n-1}$. For each $j \in \{1,\ldots,n\}$, the j-th node is associated to an affine fundamental weight Λ^j, and to boson creation and annihilation operators $\varphi^\dagger_j$ and φ_j, respectively. The affine dominant weight $\hat\nu = \sum_{j=1}^n \nu_j \Lambda^j$ can be used to label a basis of states, where $\nu_j \in \mathbb{N}_0$ is the number of bosons at node j, eigenvalue of the number operator N_j. The complete set of possible affine dominant weights is $\hat P_+ := \{\sum_{a=1}^n \nu_a\Lambda^a \,|\, \lambda_a \in \mathbb{N}_0\}$.

The total number of bosons $\sum_{j=1}^n \nu_j$ equals the level of $\hat\nu$, denoted k. So, alternatively, the level k and an $su(n)$ dominant weight $\nu = \sum_{j=1}^{n-1}\nu_j\Lambda^j$ can be used to label states, since $\nu_n = k - \sum_{j=1}^{n-1}\nu_j$. Notice that ν_n is the affine Dynkin label. The 2 notations are useful, so we write $|\hat\nu\rangle = |\nu\rangle_k$. Here $\hat\nu \in \hat P^k_+ := \{\sum_{a=1}^n \nu_a\Lambda^a \,|\, \lambda_a \in \mathbb{N}_0, \sum_{a=1}^n \nu_a = k\} \subset \hat P_+$, and $\nu \in P^k_+ := \left\{\sum_{a=1}^{n-1}\nu_a\Lambda^a \,|\, \lambda_a \in \mathbb{N}_0, \sum_{a=1}^{n-1}\nu_a \le k\right\}$. The basis vectors are orthonormal, so

$$ {}_k\langle\lambda|\mu\rangle_{k'} \;=\; \delta_{\lambda,\mu}\,\delta_{k,k'} \ , \text{ or } \ \langle\hat\lambda|\hat\mu\rangle \;=\; \delta_{\hat\lambda,\hat\mu} \ . \tag{1} $$

It is important to realize, however, that the level, the number of bosons, is not fixed in the phase model.

The algebra of $\{\,\varphi_j, \varphi^\dagger_j, N_j \,|\, j \in \{1,\ldots,n\}\,\}$, subject to the following relations, is known as the *phase algebra*.

$$ \begin{aligned} &[\varphi_i,\varphi_j] \;=\; 0\,, \quad [\varphi^\dagger_i,\varphi^\dagger_j] \;=\; 0\,, \quad [N_i,N_j] \;=\; 0\,, \\ &[N_i,\varphi^\dagger_j] \;=\; \delta_{i,j}\,\varphi^\dagger_i\,, \quad [N_i,\varphi_j] \;=\; -\delta_{i,j}\,\varphi_i\,, \\ &N_i\,(1-\varphi^\dagger_i\varphi_i) \;=\; 0 \;=\; (1-\varphi^\dagger_i\varphi_i)\,N_i\,, \\ &[\varphi_i,\varphi^\dagger_j] \;=\; 0 \ \text{ if } i\ne j\,; \ \text{ but } \ \varphi_i\varphi^\dagger_i \;=\; 1\,. \end{aligned} \tag{2} $$

The last relation follows from the standard actions of φ_i and $\varphi^\dagger_i$ on the basis states:

$$ \varphi_i\,|\hat\nu\rangle \;=\; \begin{cases} |\hat\nu-\Lambda^i\rangle\ , & \hat\nu-\Lambda^i \in \hat P_+\ ; \\ 0\,, & \text{otherwise}\ . \end{cases} $$

$$\varphi_i^\dagger\,|\hat\nu\rangle \;=\; |\hat\nu+\Lambda^i\rangle\ .$$

Similarly, the operator $\varphi_i^\dagger\varphi_i$ can be seen to project onto states of positive i-th boson number $\nu_i>0$, so that $1-\varphi_i^\dagger\varphi_i$ projects onto $\nu_i=0$ states.

The phase model is solved in the standard way; complete details are provided in [8]. The Lax matrices are

$$L_j(u) \;=\; \begin{pmatrix} 1 & u\varphi_j^\dagger \\ \varphi_j & u \end{pmatrix}\ , \tag{3}$$

where u is a spectral parameter. The monodromy matrix is then

$$M(u) \;=\; L_n(u)\,L_{n-1}(u)\,\cdots\,L_1(u) \;=:\; \begin{pmatrix} A(u) & B(u) \\ C(u) & D(u) \end{pmatrix}\ , \tag{4}$$

where the last equality is just standard notation. Integrability results because the fundamental relation

$$R_{12}(u/v)\,M_1(u)\,M_2(v) \;=\; M_2(v)\,M_1(u)\,R_{12}(u/v)\ , \tag{5}$$

is satisfied. The quantum Yang-Baxter equation follows, and (5) defines the so-called Yang-Baxter algebra, that includes the vanishing commutator $[B(u),B(v)]=0$. The algebraic Bethe ansatz yields $B(x_1)\cdots B(x_k)\,|\hat 0\rangle$ for the Bethe vectors. Notice that the latter form is symmetric in the commuting creation operators $B(x_j)$, and so completely symmetric in the variables $x_1,\ldots,x_k$. As a result, it can be expanded in terms of the symmetric Schur polynomials:

$$B(x_1)\cdots B(x_k)\,|\hat 0\rangle \;=\; \sum_{\lambda\in P_+^k} s_{\lambda^t}(x_1,\ldots,x_k)\,|\lambda\rangle_k\ .$$

The Schur symmetric polynomial $s_{\lambda^t}(x_1,\ldots,x_k)$ is also an $su(n)$ character.

The Bethe vectors diagonalize the integrals of motion, provided the Bethe ansatz equations are satisfied, and the latter imply that $x_1,\ldots,x_k$ must equal certain $(k+n)$-th roots of unity. More specifically, one can put the possible sets of values of the x_j in 1-1 correspondence with dominant weights in P_+^k. Adopting the notation $(x_1,\ldots,x_k)=:x_\sigma$ for $\sigma\in P_+^k$, one finds that $s_\lambda(x_\mu)$ is an affine fusion eigenvalue for $\sigma\in P_+^k$. The celebrated Verlinde formula

$$s_\lambda(x_\sigma)\,s_\mu(x_\sigma) \;=\; \sum_{\nu\in P_+^k} {}^{(k)}N_{\lambda,\mu}^{\nu}\,s_\nu(x_\sigma) \tag{6}$$

follows. Here $\lambda,\mu,\nu,\sigma\in P_+^k$ and ${}^{(k)}N_{\lambda,\mu}^{\nu}$ is an affine fusion multiplicity, and the connection with WZNW models is made.

Affine fusion multiplicities count the couplings of primary fields in a WZNW model. Affine fusion is a simple truncation of the tensor product of representations of simple Lie algebras, controlled by the level k. If $T_{\lambda,\mu}^{\nu}$ indicates the $su(n)$ tensor product multiplicity, then

$$ {}^{(k)}N_{\lambda,\mu}^{\nu} \;\le\; T_{\lambda,\mu}^{\nu}\ ; \qquad\qquad {}^{(\infty)}N_{\lambda,\mu}^{\nu} \;=\; T_{\lambda,\mu}^{\nu}\ ,$$

for all $\lambda,\mu,\nu\in P_+^k$.

The direct connection with affine fusion is revealed by examining the integrals of motion [8]. Since fusion has fixed level k, we consider hopping operators $a_j = \varphi_j^\dagger \varphi_{j-1}$, $j \in \{1, 2, \ldots, n\}$, that do not change the number of bosons. For $\hat{\nu} \in \hat{P}_+^k$, their action, depicted in Fig. 1, is

$$a_i\,|\hat{\nu}\rangle \;=\; \begin{cases} |\hat{\nu} - \Lambda^{i-1} + \Lambda^i\rangle\ , & \hat{\nu} - \Lambda^{i-1} + \Lambda^i \in \hat{P}_+^k\ ; \\ 0\ , & \text{otherwise}\ . \end{cases}$$

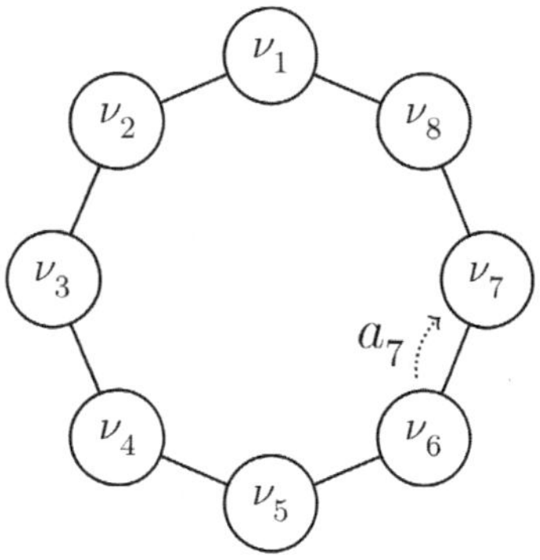

Figure 1. The action of the hopping operator a_7 is indicated in the $su(n) = su(8)$ phase model.

The algebra of hopping operators $\mathcal{A} = \langle a_1, a_2, \ldots, a_n\rangle$ is given by:

$$[a_i, a_j] \;=\; 0\,, \quad \text{if } i \neq j \pm 1 \bmod n\ ;$$

$$a_i a_j^2 \;=\; a_j a_i a_j\ , \quad a_i^2 a_j \;=\; a_i a_j a_i \quad \text{if } i = j+1 \bmod n\ . \tag{7}$$

Here indices are defined mod n. This hopping algebra $\mathcal{A}$ is called the local affine plactic algebra by Korff and Stroppel [8]. The term plactic indicates a connection with Young tableaux (see [5], e.g.). $su(n)$ tensor products can be calculated using Young tableaux in the famous Littlewood-Richardson rule. The integrals of motion of the phase model led Korff and Stroppel to a modified Littlewood-Richardson rule for affine fusion [8].

To see the relation to Young tableaux, note that the hopping operator a_i is associated with the weight $\Lambda^i - \Lambda^{i-1}$. The set of these affine weights have horizontal parts equal to the weights of the basic $su(n)$ irreducible representation $L(\Lambda^1)$, that can be labelled by Young tableau $\boxed{i}$. For example, in the $su(3)$ case, the tableaux for $L(\Lambda^1)$ are

$$\begin{array}{ccc} \boxed{2} & & \boxed{1} \\ & \boxed{3} & \end{array} \quad .$$

They have been arranged here in the pattern of the weight diagram for $L(\Lambda^1)$.

Furthermore, the action of hopping operators on basis states $|\hat{\lambda}\rangle$, $\hat{\lambda} \in \hat{P}_+^k$, is

$$a_i\,|\hat{\lambda}\rangle \;=\; \begin{cases} |\tilde{\lambda} - \Lambda^{i-1} + \Lambda^i\rangle\ , & \hat{\lambda} - \Lambda^{i-1} + \Lambda^i \in \hat{P}_+^k\ ; \\ 0\ , & \text{otherwise}\ . \end{cases}$$

It is precisely reproduced by the product (denoted by $*$ here) of a Young tableaux and a Young diagram that is used in the Littlewood-Richardson rule. In the $su(3)$ example:

$$\begin{array}{|c|}\hline 2\\\hline\end{array} * \begin{array}{|c|c|}\hline & \\\hline\end{array} = \begin{array}{|c|c|}\hline & \\\hline 2 \\\cline{1-1}\end{array}\ , \qquad \begin{array}{|c|}\hline 1\\\hline\end{array} * \begin{array}{|c|c|}\hline & \\\hline\end{array} = \begin{array}{|c|c|c|}\hline & & 1\\\hline\end{array}\ ,$$

$$\begin{array}{|c|}\hline 3\\\hline\end{array} * \begin{array}{|c|c|}\hline & \\\hline\end{array} = \begin{array}{c}\begin{array}{|c|c|}\hline & \\\hline\end{array}\\ \\ \begin{array}{|c|}\hline 3\\\hline\end{array}\end{array} = 0\ .$$

Of course, more complicated Young tableaux must also be considered. But for them, the product with a Young diagram is taken column-by-column in the same way, from the right-to-left, and if no non-dominant shape is encountered, then a contribution results.

For successive tensor products, the algorithm can be repeated. Alternatively, a product $\bullet$ of 2 Young tableaux can be defined. One finds [5] that certain "bumping" relations must be obeyed:

$$\begin{array}{|c|c|}\hline j & k\\\hline\end{array} \bullet \begin{array}{|c|}\hline i\\\hline\end{array} = \begin{array}{|c|c|}\hline i & k\\\hline j \\\cline{1-1}\end{array}\ , \qquad i < j \leq k\ ;$$

$$\begin{array}{|c|c|}\hline i & k\\\hline\end{array} \bullet \begin{array}{|c|}\hline j\\\hline\end{array} = \begin{array}{|c|c|}\hline i & j\\\hline k \\\cline{1-1}\end{array}\ , \qquad i \leq j < k\ . \tag{8}$$

Such computations can be done in plactic algebras, however, instead of with tableaux. First, replace Young tableaux with words. The (column) word of a Young tableau is obtained by listing its entries in the order from bottom to top in the left-most column, then from bottom to top in the next-to-left-most column, continuing until the top entry of the right-most column is listed. Then replace each occurrence of digit j in the word by the corresponding generator of the plactic algebra $\bar{a}_j$.

In this way, the bumping relations (8) are re-written as

$$\bar{a}_{i+1}^2 \bar{a}_i = \bar{a}_{i+1} \bar{a}_i \bar{a}_{i+1} \qquad (j = k = i + 1)\ ;$$

$$\bar{a}_i \bar{a}_{i+1} \bar{a}_i = \bar{a}_{i+1} \bar{a}_i^2 \qquad (j = k = i + 1)\ . \tag{9}$$

Comparing (9) with (7), we see that bumping is compatible with hopping. That is, the relations defining the hopping algebra (7) are very similar to those resulting from the bumping process in (8). The algebra of hopping operators can therefore implement calculations similar to those involved in the Littlewood-Richardson rule for $su(n)$ tensor products. Korff and Stroppel showed that the hopping operators of the phase model realize $su(n)$ affine fusion, a truncation of the $su(n)$ tensor product.

The integrals of motion can be described explicitly. The fundamental relation (5) guarantees that the transfer matrix $T(u) := \mathrm{tr}\, M(u)$ obeys $[T(u), T(v)] = 0$, and so is the generating function $T(u) = \sum_{r=0}^{n} u^r e_r(\mathcal{A})$ of integrals of motion: $[e_r(\mathcal{A}), e_{r'}(\mathcal{A})] = 0$. The form of the integrals can be found from (3,4). $e_r(\mathcal{A})$ indicates the r-th cyclic elementary symmetric polynomial, the sum of all cyclically ordered products of r distinct hopping operators a_i:

$$e_r(\mathcal{A}) \;=\; \sum_{|I|=r} \overset{\circlearrowleft}{\prod_{i\in I}} a_i \;\; .$$

For example, with $n=4$, $e_2(\mathcal{A}) = a_2a_1 + a_3a_1 + a_1a_4 + a_3a_2 + a_4a_2 + a_4a_3$.

Because of the integrability property, one can use the Jacobi-Trudy formula to define the non-commutative Schur polynomial

$$s_\lambda(\mathcal{A}) \;=\; \det\left(e_{\lambda^t_i - i + j}(\mathcal{A}) \right) \;. \tag{10}$$

Here λ^t indicates the partition specified by the transpose of the Young diagram for λ. λ^t_i is the i-th integer in that partition.

For example, the non-commutative Schur polynomial $s_{\Lambda^1+\Lambda^2}(\mathcal{A})$ for the adjoint representation of $su(3)$ equals the sum of:

$$\begin{array}{ccccc} & a_2a_1a_2 & & a_1a_2a_1 & \\ a_2a_3a_2 & & \begin{array}{c}(a_3a_2a_1 + a_1a_3a_2 \\ +a_2a_1a_3 - 1)\end{array} & & a_1a_3a_1 \\ & a_3a_2a_3 & & a_3a_1a_3 & \end{array} \;\; . \tag{11}$$

We already see a couple of important properties. First, (11) contains negative terms, whereas the $su(n)$ analogue contains none. This is general: the vectors of any representation $L(\lambda)$ of $su(n)$ can be put in 1-1 correspondence with Young tableaux. But the corresponding non-commutative Schur polynomial is a sum that includes negative terms, in general. Second, there is a cyclic symmetry present in (11) that is broken in the $su(n)$ case.

The main result of [8] is

$$s_\lambda(\mathcal{A})\,|\,\mu\,\rangle_k \;=\; \sum_{\nu\in P^k_+} {}^{(k)}N^{\nu}_{\lambda,\mu}\,|\,\nu\,\rangle_k \;\; . \tag{12}$$

Inspired by the term non-commutative Schur polynomial, one can call the Korff-Stroppel equation (12) a non-commutative Verlinde formula. Using the orthonormality (1), it can be written as

$${}_k\langle\,\nu\,|\,s_\lambda(\mathcal{A})\,|\,\mu\,\rangle_k \;=\; {}^{(k)}N^{\nu}_{\lambda,\mu} \;\; .$$

To compare with WZNW conformal field theory, consider a special case of (12), $s_\lambda(\mathcal{A})|0\rangle_k = |\lambda\rangle_k$. This is highly reminiscent of the field-state correspondence $\phi_\lambda(0)|0\rangle = |\lambda\rangle$, involving the WZNW primary field $\phi_\lambda(z)$. Furthermore, it can be used to re-write (12) as

$${}^{(k)}N_{\lambda,\mu,\nu} \;=\; {}_k\langle 0|\, s_\lambda(\mathcal{A})\, s_\mu(\mathcal{A})\, s_\nu(\mathcal{A})\, |0\rangle_k \;=$$

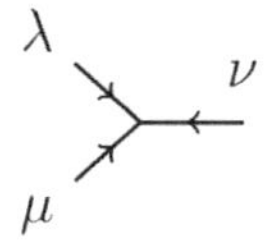

where the trivalent, directed, labeled graph that normally indicates the fusion coefficient has been drawn. Compare the last expression with the notation for the 3-point function in WZNW model: $\langle 0|\,\phi_\lambda(z_1)\,\phi_\mu(z_2)\,\phi_\nu(z_3)\,|0\rangle$. It becomes clear that the non-commutative Schur polynomial $s_\lambda(\mathcal{A})$ plays the role in the phase model of the primary field ϕ_λ in the WZNW conformal field theory.

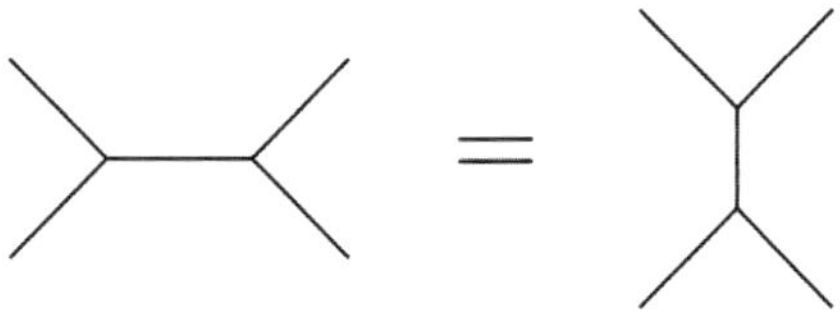

Figure 2. The duality of affine fusion is a consequence of WZNW duality. In the phase model, affine fusion duality is a result of integrability.

Remarkably, it is the integrability of the phase model that gives rise to duality in affine fusion. The non-commutative Schur polynomials are integrals of motion, and so commute: $[s_\lambda, s_\mu] = 0$. Here we have used $s_\lambda := s_\lambda(\mathcal{A})$, for short. Consequently,

$$
{}_k\langle 0|\,s_\lambda\,s_\mu\,s_\nu\,s_\phi\,|0\rangle_k \;=\;
\sum_{\sigma\in P_+^k} {}_k\langle 0|\,s_\lambda\,s_\mu\,|\sigma\rangle_k\;\,{}_k\langle\sigma|\,s_\nu\,s_\phi\,|0\rangle_k \;=\; \sum_{\sigma\in P_+^k} {}_k\langle 0|\,s_\lambda\,s_\phi\,|\sigma\rangle_k\;\,{}_k\langle\sigma|\,s_\mu\,s_\nu\,|0\rangle_k\ .
$$

The graphical representation of this result is shown in Fig. 2.

3. Higher-Genus and Level-Dependence

Let us now turn to the new results reported in [13]. First, consider higher-genus affine fusion. The Verlinde dimension of arbitrary genus and number of points is indicated by its trivalent graph in Fig. 3. It is not difficult to show that the Korff-Stroppel non-commutative Verlinde formula extends to

$$
\begin{aligned}
{}^{(k,g)}N_{\lambda_1,\ldots,\lambda_N} &= \langle\lambda_1^*|\Big(\sum_{\alpha\in P_+^k} s_{\alpha^*}s_\alpha\Big)^g s_{\lambda_2}\cdots s_{\lambda_{N-1}}|\lambda_N\rangle \\
&= {}_k\langle 0|\Big(\sum_{\alpha\in P_+^k} s_{\alpha^*}s_\alpha\Big)^g s_{\lambda_1}\,s_{\lambda_2}\cdots s_{\lambda_N}\,|0\rangle_k
\end{aligned} \tag{13}
$$

for this case. One can then identify the handle operator as $\sum_{\alpha\in P_+^k} s_{\alpha^*}s_\alpha$.

Now consider the simple dependence of affine fusion on the level k, well-described by the concept of threshold level [3, 6]. All possible fusion decompositions can be given simply by treating the level as a variable, and writing multi-sets of threshold levels as subscripts. Consider an example: the decomposition of an $su(3)$ tensor product may be written

$$
\begin{aligned}
L(\Lambda^1+\Lambda^2)^{\otimes 2} \hookrightarrow\ & L(0)_2 \oplus 2\,L(\Lambda^1+\Lambda^2)_{2,3} \oplus L(3\Lambda^1)_3 \\
& \oplus L(3\Lambda^2)_3 \oplus L(2\Lambda^1+2\Lambda^2)_4\ .
\end{aligned} \tag{14}
$$

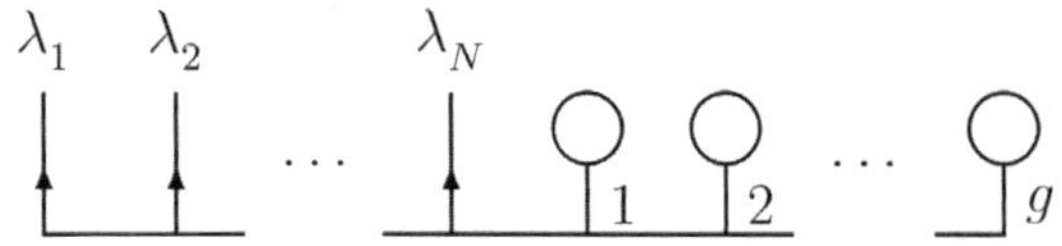

Figure 3. The Verlinde dimension of arbitrary genus g and N points.

The subscripts indicate the threshold levels of the representations in the decomposition, so that

$$ {}^{(2)}N_{\Lambda^1+\Lambda^2,\,\Lambda^1+\Lambda^2}^{\ \Lambda^1+\Lambda^2} = 1\ , \quad {}^{(k\geq 3)}N_{\Lambda^1+\Lambda^2,\,\Lambda^1+\Lambda^2}^{\ \Lambda^1+\Lambda^2} = 2\ , \tag{15} $$

for example.

This threshold-level behaviour in affine fusion is reflected in the Korff-Stroppel result (12) by the striking property of non-commutative Schur polynomials: $s_\lambda(\mathcal{A})$ does not depend on the level. The phase model treats all levels $k\in\mathbb{N}_0$ on an equal footing, and as a consequence, the threshold behaviour of level is clear. The level-dependence can be incorporated into (12) simply by using $|\mu\rangle_k$ with variable level.

In the $su(3)$ example just mentioned, the weight-0 part of the non-commutative Schur polynomial $s_{\Lambda^1+\Lambda^2}$ is $a_3a_2a_1+a_1a_3a_2+a_2a_1a_3-1$. But

$$ \begin{aligned} &(a_3a_2a_1+a_1a_3a_2+a_2a_1a_3-1)\,|\Lambda^1+\Lambda^2+(k-2)\Lambda^3\rangle \\ &\quad = \{\,\theta(k-3)+\theta(k-2)\,\}\,|\Lambda^1+\Lambda^2+(k-2)\Lambda^3\rangle\,, \end{aligned} \tag{16} $$

confirming the presence of $L(\Lambda^1+\Lambda^2)_{2,3}$ in (14). Here we have used $|\Lambda^1+\Lambda^2+(k-2)\Lambda^3\rangle = \theta(k-2)|\Lambda^1+\Lambda^2+(k-2)\Lambda^3\rangle$ and

$$ \theta(x) := \begin{cases} 1\,, & \text{if } x\geq 0\ ; \\ 0\,, & \text{if } x<0\ . \end{cases} \tag{17} $$

As already emphasized, an advantage of the phase-model realization of affine fusion is that, unlike in the WZNW model, the level is not fixed–it is just the total particle number. Changes in level can be described in a simple, algebraic way by the operators $\varphi_i^\dagger$, φ_i of the phase algebra (7).

Let us consider threshold multiplicities [13] in this spirit. The threshold multiplicity ${}^{(t)}n_{\lambda,\mu}^{\nu}$ is the contribution to a fusion multiplicity ${}^{(k)}N_{\lambda,\mu}^{\nu}$ at a fixed threshold level t, so that

$$ {}^{(k)}N_{\lambda,\mu}^{\nu} = \sum_t^k {}^{(t)}n_{\lambda,\mu}^{\nu}\ . \tag{18} $$

We also find

$$ {}^{(k)}n_{\lambda,\mu}^{\nu} = {}^{(k)}N_{\lambda,\mu}^{\nu} - {}^{(k-1)}N_{\lambda,\mu}^{\nu}\ , \tag{19} $$

where we have put ${}^{(k-1)}N_{\lambda,\mu}^{\nu}=0$ if any of λ,μ,ν are not in P_+^{k-1}.

Notice that $\varphi_n^\dagger\,|\mu\rangle_{k-1} = |\mu\rangle_k$. So we calculate

$$ [s_\lambda(\mathcal{A}),\varphi_n^\dagger]\,|\mu\rangle_{k-1} = \sum_{\nu\in P_+^k} {}^{(k)}N_{\lambda,\mu}^{\nu}\,|\nu\rangle_k - \sum_{\nu\in P_+^{k-1}} {}^{(k-1)}N_{\lambda,\mu}^{\nu}\varphi_n\,|\nu\rangle_{k-1}\ . \tag{20} $$

So the phase-model version of (19) is

$$ {}_k\langle\nu|\,[s_\lambda(\mathcal{A}),\varphi_n^\dagger]\,|\mu\rangle_{k-1} \;=\; {}^{(k)}n_{\lambda,\mu}^{\nu}\ . \tag{21}$$

Once a particular non-commutative Schur polynomial $s_\lambda(\mathcal{A})$ is calculated, the interesting operator $[s_\lambda(\mathcal{A}),\varphi_n^\dagger]$ is easy to write down, since

$$ [a_i,\varphi_n^\dagger] \;=\; \delta_{i,1}\,\varphi_1^\dagger\,(1-\varphi_n^\dagger\varphi_n)\ . \tag{22}$$

Recall that $1-\varphi_n^\dagger\varphi_n$ projects onto states with 0 particles at the n-th node.

To conclude this section, let us consider the threshold weight [14], a concept that generalizes threshold level, in the phase model. Instead of treating the level and the n-th Dynkin label as special, as in (12), now let the weight $\hat\mu$ be variable in

$$ s_\lambda(\mathcal{A})\,|\,\hat\mu\,\rangle \;=\; \sum_{\nu\in\hat P_+^k} {}^{(k)}N_{\lambda,\mu}^{\nu}\,|\,\hat\nu\,\rangle\ ,\qquad k=\sum_{c=1}^n \mu_c=\sum_{c=1}^n \nu_c\ . \tag{23}$$

The existence of a threshold level t implies a threshold value for the Dynkin label μ_n, and threshold weight $\hat\theta$ will take account of all the n Dynkin labels of $\hat\mu$. A threshold weight $\hat\theta=\sum_{c=1}^n\theta_c\Lambda^c$ may be defined for each term $a_{j_L}\cdots a_{j_2}a_{j_1}$ in $s_\lambda(\mathcal{A})$. Put $\widehat{\rm wt}(a_j):=\Lambda^j-\Lambda^{j-1}$, $\widehat{\rm wt}(a_ia_j):=\widehat{\rm wt}(a_i)+\widehat{\rm wt}(a_j)$, etc. Then the c-th Dynkin label of the threshold weight is

$$ \theta_c(a_{j_L}\cdots a_{j_2}a_{j_1}) \;=\; -\min\{0,{\rm wt}_c(a_{j_1}),\ldots,{\rm wt}_c(a_{j_L}\cdots a_{j_2}a_{j_1})\,\}\ . \tag{24}$$

That is, $-\theta_c(a_{j_L}\cdots a_{j_2}a_{j_1})$ is the minimum value of the c-th Dynkin label of the weights of the sequence $a_{j_1},a_{j_2}a_{j_1},\ldots,a_{j_L}\cdots a_{j_2}a_{j_1}$ and 0.

For an $su(3)$ example, take $\lambda=\Lambda^1+\Lambda^2$, and consider $a_3a_2a_1$ in $s_\lambda(\mathcal{A})$. The weight sequence associated with $a_3a_2a_1$ is $\{\widehat{\rm wt}(a_1),\widehat{\rm wt}(a_2a_1),\}=\{\Lambda^1-\Lambda^3,\Lambda^2-\Lambda^3,0\}$. Therefore $\hat\theta(a_3a_2a_1)=\Lambda^3$. This threshold weight tells us that $a_3a_2a_1$ contributes to $s_{\Lambda^1+\Lambda^2}(\mathcal{A})\,|\hat\mu\rangle$ iff $\hat\mu-\hat\theta(a_3a_2a_1)=\hat\mu-\Lambda^3\in\hat P_+$.

4. Conclusion

The Korff-Stroppel integrable realization of $su(n)$ affine fusion [8] leads to the non-commutative Verlinde formula (12) for affine fusion coefficients, that can be extended to a non-commutative formula (13) for arbitrary Verlinde dimensions [13]. The realization offers a new perspective on affine fusion, that should deepen our understanding. It already makes clear certain features, like the existence of threshold levels (and weights) [13].

Let us conclude with a brief discussion of possible future work.

Perhaps the most important further development would be to construct the phase-model realizations of affine fusion for all complex, simple Lie algebras. The next-to-simplest case is likely the Lie algebra $sp(2n)\cong C_n$, since tableaux work almost as well in this case as for $su(n)\cong A_{n-1}$. Some progress is reported in [1], where a quantum group approach is used.

From the physical point of view, it would be useful to derive the phase-model realization from affine fusion in another physical context. For example, it is well known that the G/G gauged WZNW model is a topological field theory, with correlation functions equalling the affine Verlinde dimensions [10]. Okuda and Yoshida [9] have already found the Bethe ansatz equations of the phase model in the path-integral formulation of the $U(n)/U(n)$ model, and other indications of the connection. A more direct relation would be helpful, as well as a manifestation of the phase model in the $su(n)$ Chern-Simons theory, which also has Verlinde dimensions as some of its expectation values [12].

More technically, formulas for $s_\lambda(\mathcal{A})$, besides the Jacobi-Trudy formula (10), would likely prove useful. The properties of non-commutative Kostka polynomials should be explored. Of course, similar formulas for non-commutative characters for other Lie algebras are desirable, too.

We are hopeful that progress along these lines can be made in the near future.

Acknowledgments

The author thanks Ali Nassar and Andrew Urichuk for reading the manuscript. This research was supported in part by a Discovery Grant from the Natural Sciences and Engineering Research Council (NSERC) of Canada.

References

[1] H.H. Anderson, C. Stroppel, Fusion rings for quantum groups, preprint(2012) [arXiv:1212.5736]
[2] N.M. Bogoliubov, A.G. Izergin, N.A. Kitanine, Nucl. Phys. B 516 (1998) 501 [arXiv:solv-int/9710002]
[3] C.J. Cummins, P. Mathieu, M.A. Walton, Phys. Lett. B 254 (1991) 386
[4] P. Di Francesco, P. Mathieu, D. Sénéchal, Conformal Field Theory, (Springer-Verlag, 1997)
[5] W. Fulton, Young Tableaux, London Math. Soc. Stud. Texts, vol. 35 (Cambridge U. Press, 1997)
[6] A.N. Kirillov, P. Mathieu, D. Sénéchal, M.A. Walton, Nucl. Phys. B 391 (1993) 651 [arXiv:hep-th/9203004]
[7] C. Korff, RIMS Kokyuroku Bessatsu B28 (2011) 121 [arXiv:1106.5342]
[8] C. Korff, C. Stroppel, Adv. Math. 225 (2010) 200 [arXiv:0909.2347]
[9] S. Okuda, Y. Yoshida, JHEP 1211 (2012) 146 [arXiv:1209.3800]
[10] M. Spiegelglas, S. Yankielowicz, Nucl. Phys. B 393 (1993) 301 [arXiv:hep-th/9201036] ; E. Witten, Commun. Math. Phys. 144 (1992) 189
[11] E. Verlinde, Nucl. Phys. B 300 (1988) 360
[12] E. Witten, Commun. Math. Phys. 121 (1989) 351
[13] M.A. Walton, SIGMA 8 (2012) 086 [arXiv:1208.0809]
[14] M.A. Walton, Can. J. Phys. 72 (1994) 527

Journal of Physics: Conference Series **474** (2013) 012034 doi:10.1088/1742-6596/474/1/012034

Symmetries in Bosonization

Yuan K. Ha

Department of Physics, Temple University, Philadelphia, Pennsylvania 19122 U.S.A.

Email: yuanha@temple.edu

Abstract. Two-dimensional quantum field theories are important in many problems in physics because they contain exact symmetries and are often completely integrable. We demonstrate the power of bosonization in elucidating the structure of a multi-component fermion theory with chiral and internal symmeties. Many remarkable features and quantum symmetries in such a system are most readily uncovered in the bosonized form of the original fermion theory.

1. Introduction

In two-dimensional quantum field theories it is possible to express a fermion theory in terms of a boson theory to uncover hidden structure and new properties. This is known as bosonization. The equivalence has been demonstrated both at the Hamiltonian level and at the Lagrangian level for a number of theories. It is a remarkable occurrence in two-dimensional quantum field theory.

The goal of any symmetry operation is to leave the equations of a system invariant and therefore the physics unchanged. It is a search of self-similarity. Bosonization is a mathematical transformation that leaves the physics of a theory invariant even though the equations appear to be different. It is therefore a search of the self-similarity of the system and the operation can be considered as a quantum symmetry operation.

In two dimensions, Lagrangians have mass dimension 2. Fermions have dimension 1/2; bosons have dimension 0. Derivatives have dimension 1. When a massless fermion becomes a massive boson, the coupling constant in the fermion theory becomes the mass parameter in the boson theory. This is dynamical mass generation via bosonization. Many important symmetries in the fermion theory, such as chiral symmetry, internal symmetry, non-abelian symmetry, duality and triality, are immediately found in the bosonized version of the original theory.

2. Boson Representation

Let ψ be a canonical Fermi field and ϕ a canonical Bose field in two dimensions. The boson representation of a free fermion in terms of a free boson is given by [1]

$$\psi(x) = \begin{bmatrix} \psi_+(x) \\ \psi_-(x) \end{bmatrix} = \frac{1}{(2\pi\alpha)^{1/2}} \begin{bmatrix} \exp[+i\sqrt{4\pi}\phi_+(x)] \\ \exp[-i\sqrt{4\pi}\phi_-(x)] \end{bmatrix}, \tag{1}$$

where $\psi_+(x)$ and $\psi_-(x)$ are the chiral componets of the fermion. Each fermion component obeys the anti-commutation rules

$$\begin{aligned} \{\psi_\pm(x),\psi_\pm{}^\dagger(y)\} &= \delta(x-y) \\ \{\psi_\pm(x),\psi_\mp{}^\dagger(y)\} &= 0. \end{aligned} \tag{2}$$

On the other hand, the components $\phi_+(x), \phi_-(x)$ are the positive and negative frequency parts of the scalar boson $\phi(x)$ satisfying the commutation rules

$$\begin{aligned} [\phi_\pm(x),\phi_\pm(y)] &= \pm\varepsilon(x-y) \\ [\phi_+(x),\phi_-(y)] &= i/4, \end{aligned} \tag{3}$$

with $\phi_+(x)+\phi_-(x)=\phi(x)$, and $\varepsilon(x-y)$ is the odd step function with absolute value of unity. The quantity $1/\alpha$ is an ultraviolet cutoff imposed to guarantee the scalar field to remain finite at high frequencies, and α will be taken to zero at the end of any calculation involving the fermion $\psi(x)$. With the boson representation defined in Eq.(1), all the key properties of a quantum field theory such as correlation functions, cluster properties, Lorentz invariance are all found to be valid under this transformation. In addition, the Hamiltonian for the free fermion can be directly transformed into the Hamiltonian of the free boson. In this respect, we state the formal equivalence of bilinear forms appearing in the Lagrangians of the fermion theory and the boson theory in the following:

$$\begin{aligned} i\bar{\psi}\gamma^\mu\partial_\mu\psi &\leftrightarrow \frac{1}{2}\partial_\mu\phi\partial^\mu\phi \\ \bar{\psi}\psi &\leftrightarrow -\frac{1}{\pi\alpha}\cos\sqrt{4\pi}\phi \\ i\bar{\psi}\gamma^5\psi &\leftrightarrow \frac{1}{\pi\alpha}\sin\sqrt{4\pi}\phi \\ \bar{\psi}\gamma^\mu\psi &\leftrightarrow \frac{1}{\sqrt{\pi}}\varepsilon^{\mu\nu}\partial_\nu\phi \\ \bar{\psi}\gamma^5\gamma^\mu\psi &\leftrightarrow \frac{1}{\sqrt{\pi}}\partial^\mu\phi. \end{aligned} \tag{4}$$

As a demonstration of the instant power of bosonization, we can show directly that the massless Thirring model, a self-interacting fermion theory, is in fact equivalent to a free scalar boson theory [2]. The massless Thirring model, with coupling constant g, is given by the Lagrangian

$$L = i\bar{\psi}\gamma^\mu\partial_\mu\psi - \frac{g}{2}(\bar{\psi}\gamma^\mu\psi)(\bar{\psi}\gamma_\mu\psi). \tag{5}$$

The conservation of the fermion currents in this model shows that the currents can be expressed in terms of a new scalar field $\tilde{\phi}$. The requirement that this new field satisfies the canonical commutation relation for a scalar boson demands that it be rescaled from the free boson ϕ by the factor $\sqrt{1+g/\pi}$, i.e. $\tilde{\phi} = \phi/\sqrt{1+g/\pi}$. Applying the equivalence of the bilinear terms in the Thirring Lagrangian, we find that the fermion kinetic term produces a canonical boson kinetic term in terms of the new scalar field $\tilde{\phi}$, while the vector currents in the Lagrangian contribute an additional anomalous part to the boson kinetic term. The new scalar field and the contribution from the fermion interaction exactly compensate each other and the total effect of the bosonized Lagrangian is simply that of a free scalar boson given by

$$L=\left(1+\frac{g}{\pi}\right)\times\frac{1}{2}\frac{\partial_\mu\phi}{\sqrt{1+g/\pi}}\frac{\partial^\mu\phi}{\sqrt{1+g/\pi}}=\frac{1}{2}\partial_\mu\phi\partial^\mu\phi. \tag{6}$$

If a mass term $m\overline{\psi}\psi$ is included in the Lagrangian of the massless Thirring model, the result is the massive Thirring model. It is well known that the massive Thirring model is equivalent to the quantum sine-Gordon theory [3]. The equivalence is most readily obtained by direct bosonization. The two Lagrangians are respectively,

$$\begin{aligned}L&=i\overline{\psi}\gamma^\mu\partial_\mu\psi-\frac{g}{2}(\overline{\psi}\gamma^\mu\psi)(\overline{\psi}\gamma_\mu\psi)-m\overline{\psi}\psi\\ L&=\frac{1}{2}\partial_\mu\phi\partial^\mu\phi+\frac{m}{\pi\alpha}\cos\sqrt{\frac{4\pi}{1+g/\pi}}\phi.\end{aligned} \tag{7}$$

3. A multi-fermion theory

It is the purpose in this article to demonstrate that the real power of bosonization is in the elucidation of the structure of a multi-component fermion theory, in particular, the uncovering of hidden symmetries inaccessible by other means. Bosonization is uniquely a quantum field theory phenomenon; therefore any symmetry discovered which is not explicitly present in the fermion Lagrangian is a quantum symmetry. It is also shown that the number of fermion components can significantly alter the dynamics.

For our investigation, we consider a multi-component fermion theory with global non-abelian symmetry. This is the $U(N)$ Thirring model constructed with N types of massless fermions ψ^a, $a=1,\dots,N$ [4]. The interaction is through the fermion isospin currents. This model bears similarity to the Kondo problem [5] and is important in the study of many two-dimensional systems. It is of sufficient complexity that it cannot be solved easily. Bosonization can offer rare insights into the structure of this model. By a Lie algebraic generalization of the Fierz identity in two dimensions, the theory is seen to be equivalent to the chiral Gross-Neveu model [6]. The $U(N)$ Thirring Lagrangian is

$$L=\sum_{a=1}^{N}i\overline{\psi}^a\gamma^\mu\partial_\mu\psi^a-g\sum_{A=0}^{N^2-1}\left[\overline{\psi}\gamma^\mu\frac{\lambda^A}{2}\psi\right]\left[\overline{\psi}\gamma_\mu\frac{\lambda^A}{2}\psi\right], \tag{8}$$

where g is the coupling constant and λ^A are the generators of the $U(N)$ group. Upon standard bosonization, the theory becomes a system of N coupled bosons ϕ^a with Lagrangian

$$L=\sum_{a=1}^{N}\frac{1}{2}\partial_\mu\phi^a\partial_\mu\phi^a+\sum_{a\neq b=1}^{N}\frac{g}{4\pi^2\alpha^2}\cos\sqrt{\frac{4\pi}{1+g/\pi}}(\phi^a-\phi^b). \tag{9}$$

We may perform an orthogonal transformation Λ on the ϕ^a fields in the following:

$$\begin{aligned}
B^1 &= (\phi^1 + \phi^2 + ... + \phi^N)/\sqrt{N} \\
B^2 &= (\phi^1 - \phi^2)/\sqrt{2} \\
B^3 &= (\phi^1 + \phi^2 - 2\phi^3)/\sqrt{6} \\
&..... \\
&..... \\
B^N &= (\phi^1 + ... + \phi^{N-1} - (N-1)\phi^N)/\sqrt{N(N-1)}.
\end{aligned} \tag{10}$$

In this case, a massless free boson B^1 will decouple and the effective theory becomes a system of $N-1$ coupled sine-Gordon theories with Lagrangian

$$L = \frac{1}{2}\partial_\mu B^1 \partial^\mu B^1 + \sum_{k=2}^{N} \frac{1}{2}\partial_\mu B^k \partial^\mu B^k + \sum_{i \neq j=2}^{N} \frac{g}{4\pi^2\alpha^2} \cos\sqrt{\frac{4\pi}{1+g/\pi}}(\Lambda^{li}B^l - \Lambda^{mj}B^m). \tag{11}$$

This is a realization of the symmetry $U(N) \approx SU(N) \otimes U(1)$ in the boson theory. Some remarkable equivalences can now be found from the interacting boson Lagrangian from Eq.(11).

3.1. $N = 2$ *Case*. The Lagrangian is that of a free boson and a sine-Gordon theory with symmetry $U(2) \approx SU(2) \otimes U(1)$:

$$L = \frac{1}{2}\partial_\mu B^1 \partial^\mu B^1 + \frac{1}{2}\partial_\mu B^2 \partial^\mu B^2 + \frac{g}{4\pi^2\alpha^2} \cos\sqrt{\frac{8\pi}{1+g/\pi}} B^2. \tag{12}$$

If we take two boson Lagrangians with identical couplings from Eq.(9), one with bosons (ϕ^1, ϕ^2) and the other with bosons (n^1, n^2) , we find the following situation:

$$\begin{aligned}
L = &\left(\frac{1}{2}\partial_\mu \phi^1 \partial^\mu \phi^1 + \frac{1}{2}\partial_\mu \phi^2 \partial^\mu \phi^2\right) + \left(\frac{1}{2}\partial_\mu n^1 \partial^\mu n^1 + \frac{1}{2}\partial_\mu n^2 \partial^\mu n^2\right) \\
&+ \frac{g}{4\pi^2\alpha^2} \cos\sqrt{\frac{4\pi}{1+g/\pi}}(\phi^1 - \phi^2) + \frac{g}{4\pi^2\alpha^2} \cos\sqrt{\frac{4\pi}{1+g/\pi}}(n^1 - n^2).
\end{aligned} \tag{13}$$

This is a system of two independent $U(2)$ models with four bosons. If we further allow the four bosons to mix according to the following orthogonal transformation,

$$\begin{aligned} B^1 &= \frac{1}{2}(\phi^1 + \phi^2 + n^1 + n^2) \\ B^2 &= \frac{1}{2}(\phi^1 + \phi^2 - n^1 - n^2) \\ B^3 &= \frac{1}{2}(\phi^1 - \phi^2 + n^1 - n^2) \\ B^4 &= \frac{1}{2}(\phi^1 - \phi^2 - n^1 + n^2), \end{aligned} \tag{14}$$

we find the fields $(\phi^1 - \phi^2) = (B^3 + B^4)$ and $(n^1 - n^2) = (B^3 - B^4)$. The system now becomes two coupled sine-Gordon theories with two interacting bosons (B^3, B^4), together with two massless free bosons (B^1, B^2). The ensuing Lagrangian then becomes

$$\begin{aligned} L = &\frac{1}{2}\partial_\mu B^1 \partial^\mu B^1 + \frac{1}{2}\partial_\mu B^2 \partial^\mu B^2 + \frac{1}{2}\partial_\mu B^3 \partial^\mu B^3 + \frac{1}{2}\partial_\mu B^4 \partial_\mu B^4 \\ &+ \frac{g}{4\pi^2\alpha^2}\cos\sqrt{\frac{4\pi}{1+g/\pi}}(B^3 + B^4) + \frac{g}{4\pi^2\alpha^2}\cos\sqrt{\frac{4\pi}{1+g/\pi}}(B^3 - B^4). \end{aligned} \tag{15}$$

At this stage, it is auspicious to note that the trigonometric identity

$$\cos(\mathrm{A}+\mathrm{B}) + \cos(\mathrm{A}-\mathrm{B}) = 2\cos\mathrm{A}\cos\mathrm{B} \tag{16}$$

is exclusively applicable to Bose fields in two dimensions. The Lagrangian can be transformed into

$$\begin{aligned} L = &\frac{1}{2}\partial_\mu B^1 \partial^\mu B^1 + \frac{1}{2}\partial_\mu B^2 \partial^\mu B^2 + \frac{1}{2}\partial_\mu B^3 \partial^\mu B^3 + \frac{1}{2}\partial_\mu B^4 \partial_\mu B^4 \\ &+ \frac{g}{2\pi^2\alpha^2}\cos\sqrt{\frac{4\pi}{1+g/\pi}}B^3 \times \cos\sqrt{\frac{4\pi}{1+g/\pi}}B^4. \end{aligned} \tag{17}$$

The interacting part with Bose fields (B^3, B^4) is remarkably the bosonized form of the $O(4)$ Gross-Neveu model [7] constructed with four Majorana fermions $(\chi^1, \chi^2, \chi^3, \chi^4)$, the Lagrangian for which is

$$\begin{aligned} L = &i\bar{\chi}^1\gamma^\mu\partial_\mu\chi^1 + i\bar{\chi}^2\gamma^\mu\partial_\mu\chi + i\bar{\chi}^3\gamma^\mu\partial_\mu\chi^3 + i\bar{\chi}^4\gamma^\mu\partial_\mu\chi^4 \\ &+ \frac{g}{2}\left[\bar{\chi}^1\chi^1 + \bar{\chi}^2\chi^2 + \bar{\chi}^3\chi^3 + \bar{\chi}^4\chi^4\right]^2. \end{aligned} \tag{18}$$

This is a realization of the symmetry $SU(2)\otimes SU(2) \approx O(4)$ through two different fermion theories. The sine-Gordon Lagrangian has a hidden $SU(2)$ symmetry.

Journal of Physics: Conference Series **474** (2013) 012034 doi:10.1088/1742-6596/474/1/012034

3.2. $N = 4$ *Case.* The bosonized Lagrangian from the $U(4)$ Thirring model in Eq.(9) takes the form

$$L = \frac{1}{2}\partial_\mu \phi^1 \partial^\mu \phi^1 + \frac{1}{2}\partial_\mu \phi^2 \partial^\mu \phi^2 + \frac{1}{2}\partial_\mu \phi^3 \partial^\mu \phi^3 + \frac{1}{2}\partial_\mu \phi^4 \partial^\mu \phi^4$$
$$+\frac{g}{4\pi^2\alpha^2}\begin{pmatrix} \cos\sqrt{\frac{4\pi}{1+g/\pi}}(\phi^1-\phi^2)+\cos\sqrt{\frac{4\pi}{1+g/\pi}}(\phi^3-\phi^4)+ \\ \cos\sqrt{\frac{4\pi}{1+g/\pi}}(\phi^1-\phi^3)+\cos\sqrt{\frac{4\pi}{1+g/\pi}}(\phi^2-\phi^4)+ \\ \cos\sqrt{\frac{4\pi}{1+g/\pi}}(\phi^1-\phi^4)+\cos\sqrt{\frac{4\pi}{1+g/\pi}}(\phi^2-\phi^3) \end{pmatrix}. \quad (19)$$

We may again decouple a massless boson B^1 by applying the following transformation

$$\begin{aligned} B^1 &= \frac{1}{2}(\phi^1+\phi^2+\phi^3+\phi^4) \\ B^2 &= \frac{1}{2}(\phi^1+\phi^2-\phi^3-\phi^4) \\ B^3 &= \frac{1}{2}(\phi^1-\phi^2+\phi^3-\phi^4) \\ B^4 &= \frac{1}{2}(\phi^1-\phi^2-\phi^3+\phi^4). \end{aligned} \quad (20)$$

There are six interacting terms in Eq.(19) in the form of cosine fields but after the transformation they may be combined using the previous trigonometric identity into three product terms in the following form:

$$L = \frac{1}{2}\partial_\mu B^1 \partial^\mu B^1 + \frac{1}{2}\partial_\mu B^2 \partial^\mu B^2 + \frac{1}{2}\partial_\mu B^3 \partial^\mu B^3 + \frac{1}{2}\partial_\mu B^4 \partial^\mu B^4$$
$$+\frac{g}{2\pi^2\alpha^2}\begin{pmatrix} \cos\sqrt{\frac{4\pi}{1+g/\pi}}B^2\times\cos\sqrt{\frac{4\pi}{1+g/\pi}}B^3+ \\ \cos\sqrt{\frac{4\pi}{1+g/\pi}}B^3\times\cos\sqrt{\frac{4\pi}{1+g/\pi}}B^4+ \\ \cos\sqrt{\frac{4\pi}{1+g/\pi}}B^4\times\cos\sqrt{\frac{4\pi}{1+g/\pi}}B^2 \end{pmatrix}. \quad (21)$$

The effective theory with three bosons (B^2, B^3, B^4) is remarkably the bosonized form of the $O(6)$ Gross-Neveu model with six Majorano fermions $(\chi^1, \chi^2, \chi^3, \chi^4, \chi^5, \chi^6)$, the Lagrangian for which explicitly is

$$L = i\bar{\chi}^1\gamma^\mu\partial_\mu\chi^1 + i\bar{\chi}^2\gamma^\mu\partial_\mu\chi^2 + i\bar{\chi}^3\gamma^\mu\partial_\mu\chi^3 + i\bar{\chi}^4\gamma^\mu\partial_\mu\chi^4 + i\bar{\chi}^5\gamma^\mu\partial_\mu\chi^5 + i\bar{\chi}^6\gamma^\mu\partial_\mu\chi^6$$
$$+\frac{g}{2}\left[\bar{\chi}^1\chi^1 + \bar{\chi}^2\chi^2 + \bar{\chi}^3\chi^3 + \bar{\chi}^4\chi^4 + \bar{\chi}^5\chi^5 + \bar{\chi}^6\chi^6\right]^2. \tag{22}$$

This is a surprising realization of the symmetry $SU(4) \approx O(6)$ through the fermion theories.

4. A novel symmetry

We return once again to the original bosonized Lagrangian for the $U(N)$ Thirring model in Eq.(9) and this time apply a novel transformation for all cases with $N \geq 3$. This will allow us to uncover a hidden structure which is not evident in the fermion theory. The smallest number for N here is 3 since one of the bosons will later decouple and at least two remaining bosons are necessary in order to form a new linear combination. The transformation is given through the following $N \times N$ matrix,

$$\begin{pmatrix} B^1 \\ B^2 \\ \cdot \\ \cdot \\ \cdot \\ \cdot \\ B^{N-1} \\ B^N \end{pmatrix} = \frac{1}{N}\begin{pmatrix} 2 & 2 & \cdots & 2 & 2-N \\ 2 & 2 & \cdots & 2-N & 2 \\ \cdot & \cdot & \cdots & \cdot & \cdot \\ \cdot & \cdot & \cdots & \cdot & \cdot \\ \cdot & \cdot & \cdots & \cdot & \cdot \\ \cdot & \cdot & \cdots & \cdot & \cdot \\ 2 & 2-N & \cdots & 2 & 2 \\ 2-N & 2 & \cdots & 2 & 2 \end{pmatrix}\begin{pmatrix} \phi^1 \\ \phi^2 \\ \\ \\ \\ \\ \phi^{N-1} \\ \phi^N \end{pmatrix}. \tag{23}$$

The Bose fields $(\phi^1,...,\phi^N)$ are now replaced by the set $(B^1,...,B^N)$. In the above matrix, all the elements have a value equal to 2, except one element in each row and each column has a value equal to $2-N$. The most remarkable property of this transformation is that the difference of any two ϕ fields becomes the difference of two B fields. The total number of possible field differences is the same in both sets. This transformation generates a new symmetry known as boson self-duality. There are two identical boson Lagrangians:

$$L = \sum_{a=1}^{N}\frac{1}{2}\partial_\mu\phi^a\partial_\mu\phi^a + \sum_{a\neq b=1}^{N}\frac{g}{4\pi^2\alpha^2}\cos\sqrt{\frac{4\pi}{1+g/\pi}}(\phi^a - \phi^b)$$
$$L = \sum_{a=1}^{N}\frac{1}{2}\partial_\mu B^a\partial_\mu B^a + \sum_{a\neq b=1}^{N}\frac{g}{4\pi^2\alpha^2}\cos\sqrt{\frac{4\pi}{1+g/\pi}}(B^a - B^b). \tag{24}$$

They correspond to two identical fermion Lagrangians which display fermion self-duality. In terms of the equivalent chiral Gross-Neveu model, they are respectively,

$$L=\sum_{a=1}^{N} i\bar{\psi}^a\gamma^\mu\partial_\mu\psi^a+\frac{g}{2}\left\{[\sum_{a=1}^{N}\bar{\psi}^a\psi^a]^2-[\sum_{a=1}^{N}\bar{\psi}^a\gamma^5\psi^a]^2\right\}$$

$$L=\sum_{a=1}^{N} i\bar{\xi}^a\gamma^\mu\partial_\mu\xi^a+\frac{g}{2}\left\{[\sum_{a=1}^{N}\bar{\xi}^a\xi^a]^2-[\sum_{a=1}^{N}\bar{\xi}^a\gamma^5\xi^a]^2\right\}. \tag{25}$$

The Fermi fields $(\psi^1,...,\psi^N)$ are replaced by the dual set $(\xi^1,...,\xi^N)$ through bosonization. In self-duality, each degree of freedom in one form is a linear or non-linear combination of all the degrees of freedom in the other form. Symmetry dictates interaction. The presence of self-duality symmetry in the dynamics of the $U(N)$ Thirring model is to maintain the equality of masses of the kinks and the particles in the quantum theory. We also see self-triality symmetry in the special case of the O(8) Gross-Neveu model [8]. There are three identical fermion Lagrangians which are connected with one another through their bosons. Duality and triality are discrete symmetries but they are not related to any conservation law. They are important quantum symmetries which are most elegantly found in bosonization.

References

[1] Ha Y K 1984 *Phys. Rev.* D **29** 1744.
[2] Sommerfield C M 1963 *Ann. Phys.* (N.Y.) **26** 1
[3] Coleman S 1975 *Phys. Rev.* D **11** 2088.
Mandelstam S 1975 *Phys. Rev.* D **11** 3026.
[4] Dashen R and Frishman Y 1975 *Phys. Rev.* D **11** 2781.
[5] Andrei N and Lowenstein J H 1979 *Phys. Rev. Lett.* **43** 1698.
[6] Gross D J and Neveu A 1974 *Phys. Rev.* D **10** 3235.
[7] Witten E 1978 *Nucl. Phys.* **B142** 285.
[8] Shankar R 1980 *Phys. Lett.* **92B** 333.

Continuous Emission of A Radiation Quantum

J.X. Zheng-Johansson

Institute of Fundamental Physics Research, 611 93 Nyköping, Sweden

Abstract. It is in accordance with such experiments as single photon self-interference that a photon, conveying one radiation energy quantum "$h\times$ frequency", is spatially extensive and stretches an electromagnetic wave train. A wave train, hence an energy quantum, can only be emitted (or absorbed) by its source (or absorber) gradually. In both two processes the wave and "particle" attributes of the radiation field are simultaneously prominent, where an overall satisfactory theory has been lacking; for the latter process no known theoretical description currently exists. This paper presents a first principles treatment, in a unified framework of the classical and quantum mechanics, of the latter process, the emission (similarly absorption) of a single radiation quantum based on the dynamics of the radiation-emitting source, a charged oscillator, which is itself extensive across the potential well in which it oscillates. During the emission of one single radiation quantum, the extensive charged oscillator undergoes a continuous radiation damping and is non-stationary. This process is in this work treated using a quasi stationary approach, whereby the classical equation of motion, which directly facilitates the correspondence principle for a particle oscillator, and the quantum wave equation are established for each sufficiently brief time interval. As an inevitable consequence of the division of the total time for emitting one single quantum, a fractional Planck constant h is introduced. The solutions to the two simultaneous equations yield for the charged oscillator a continuously exponentially decaying Hamiltonian that is at the same time quantised with respect to the fractional-h at any instant of time; and the radiation wave field emitted over time stretches a wave train of finite length. The total system of the source and radiation field maintains at any time (integer n times) one whole energy quantum, $(n\times)$ $h\times$ frequency, in complete accordance with the standard notion of quantum mechanics and experiment.

1. Introduction

According to the classical electrodynamic theory (J C Maxwell, 1873), the electromagnetic radiation fields are manifestly waves. The theory satisfactorily accounts for such processes as diffraction, interference and superposition. According to quantum theory laid foundation to by M Planck in 1900, electromagnetic radiation fields consist at the scale h of energy quanta, and these are manifestly "particles" (M Planck, 1900; Einstein, 1905). The theory is successful especially for such processes where there presents an energy transfer between the radiation field emitter and an external absorber, in general by one energy quantum (or photon), $\hbar\omega$ or by integral n multiples of $n\hbar\omega$ at a time. A photon, as well as a quantum matter particle, is in the current interpretation of the quantum theory regarded as a statistical point particle. While the classical electrodynamics and the quantum electrodynamics (Dirac, 1927) have proven extremely successful where the wave and the particle attributes are separately prominent, the two essentially parallel theories present apparent clashes where the wave and "particle" attributes are simultaneously prominent[1, 2]. A classical example of this is the single photon self-interference in a double slit as demonstrated in experiments [2]. Here, the clash is a logical one: a photon ($\hbar\omega$), being regarded as a statistical point particle, logically can not pass two slits at the same time. A realistic theoretical representation of the (self) interference would require the radiation quantum to be depicted as a coherent (radial) wave train.

A separate example is the intermediate process of emission (or absorption) of a radiation quantum $\hbar\omega$, or photon; this is closely relevant to the single photon self-interference and will be the central concern of this paper. Based on the classical electrodynamics as well as experiment, the radiation field is gradually and continuously emitted (or absorbed) by the source (or target) charge over a time duration $\sim 10^{-8}$ s [1], and therefore stretches a coherent wave train of a definite phase, a necessary pre-condition for producing single photon self interference. A requisite quantisation of the field however is not facilitated in this theory. According to the quantum theory and experiment, the radiation energy transfer from its emitting charge to an external absorber is by one energy quantum (or less frequently by integral multiples of a quantum) at a time. The radiation energy quantum, when pictured as a statistical point particle and hence a non-dividable unit, on the other hand, can only be emitted (or absorbed) instantaneously (the observational finite elapsing time of a radiation decay instead is attributed entirely to the statistical transitions of many sources, such as atoms, see e.g. H G Kuhn [1]). Similarly here, there presents a pronounced clash between the two theories. The clash, as we have seen from the two examples, arises primarily from the current interpretation of the quantum theory rested on an over simplifying statistical point particle picture, on top of an otherwise rigorous mathematical framework of the quantum mechanics.

The wave and particle pictures may be readily reconciled with one another if we regard the radiation wave field (assuming an angular frequency ω) as being at any one time distributed across a finite distance, whence a wave train of a finite (non-zero) length and yet having one fixed quantum of energy, $\hbar\omega$, as a result of wave amplitude quantisation. This description, in the case of matter wave, is entirely consistent with the corresponding mathematical quantum wave function solution ($\Psi(x,t)$) to, for example, the Schrödinger equation. Namely, $\Psi(x,t)$ is extensive at any one time t and the integration of energy density, being proportional to $|\Psi(x,t)|^2$, across (the space interval enclosing) the wave train leads to one (or integral n times one) energy quantum. For the matter wave, an intrinsically extensive, internally electrodynamic (IED) scheme has been recently proposed by the author (see e.g. a recent review in [3a]), in terms of which the primary outstanding difficulties, including ones associated with wave-particle duality, attributable mainly to the statistical point particle picture in the current interpretation of the quantum theory, may be overcome. The IED description for matter wave can be in principle translated to one for an (existing) photon, although indirectly and for stationary state only in its current form, by imposing together the solutions to the Maxwell's equations for radiation field and to the Schrödinger equation for the radiation-emitting charge. This is mainly because when the continuous emission of one energy quantum $\hbar\omega$ is in question, there requires a treatment which can facilitate both (1) a non stationary process and (2) a differentiable $\hbar\omega$. The two features are as yet not incorporated in the exiting IED representation. Currently, to the author's knowledge a direct theory for the intermediate emission process of single radiation quantum does not exist. The existing time-dependent perturbation quantum theory deals with transitions between sharply defined stationary levels without regarding the intermediate processes from which the stationary levels are finally reached.

In this paper, we give a direct, first principles and relativistic treatment of the intermediate process of the emission of an electromagnetic radiation quantum in an unified framework of the classical and quantum mechanics. The treatment is instrumented by a quasi-stationary quasi-harmonic approach to the radiation-emitting source, a charged oscillator, and the introduction of a fractional Planck constant h which justification is to be provided through its capability of achieving a consistency both within the overall theory representation and between the theory and experiment. We elucidate in Sec. 2 in a unified framework the classical and quantum equations of motion and solutions, in Sec. 3 the unified classical and quantum solutions underlining the continuous emission of a radiation quantum, and in Sec. 4 the transition probability. In the case of a charged matter particle oscillator, the relativistic radiation, in the sense of including both the thermal and rest-mass energy radiations, is facilitated in terms of the IED particle model in Sec. 5.

2. Equations of motion

2.1 Quasi harmonic motion We consider an extensive charged object such as a quantum particle or liquid-like entity (see Sec. 5), of mass $\mathcal{M}$ and charge q, located about an equilibrium position $\mathbf{R}_0(X_{10}, X_{20}, X_{30})$ in the three dimensional (R^3) vacuum. The object was endowed with a mechanical energy in the past time by an external driving force which has ceased action before time $t = 0$. From time $t = 0$ the object is in spontaneous motion under the action of an elastic restoring force $\mathbf{F} = -\nabla V = -\beta\mathscr{U}$ along X_j-direction ($j = 1, 2, 3$), where $V(\mathscr{U}) = \frac{1}{2}\beta\mathscr{U}^2$ and $\mathscr{U} = (X_{j_c}(t) - X_{j_0})\hat{X}_j$ is the displacement of its mass centre X_{j_c} at time t from X_{j_0}. In addition, the object is acted on by a radiation damping force $\mathbf{F}_r = -\alpha\mathcal{M}\frac{d\mathscr{U}}{dt}$ apparently attributable to a viscous (elastic) vacuum medium; $\alpha(> 0)$ is a damping factor or decay rate.

The Hamiltonian of the system acted on by the force $\mathbf{F}_r$ will be time dependent. Supposing that α is small so that its Hamiltonian density is in any brief time interval Δt constant, the extensive object will behave in Δt effectively as if a rigid object (whence facilitating the correspondence principle in the case of a quantum particle). We may thus write down the Newtonian equation of motion for the rigid-like extensive object as a whole, $\mathcal{M}\frac{d^2\mathscr{U}}{dt^2} - (\mathbf{F}_r + \mathbf{F}) = 0$. Or,

$$\frac{d^2\mathscr{U}}{dt^2} + \alpha\frac{d\mathscr{U}}{dt} + \omega^2\mathscr{U} = 0, \quad \omega^2 = \frac{\beta}{\mathcal{M}}. \tag{1}$$

Equation (1) has the general complex damped harmonic oscillation solution,

$$\mathscr{U}^c(t) = \mathcal{A}e^{-(\frac{\alpha}{2}+i\omega^\dagger)t} \tag{2}$$

where $\mathcal{A}$ is the oscillation amplitude at $t = 0$; $\omega^\dagger = \sqrt{\omega^2 - \alpha^2/4}$; $\mathscr{U}(t) = \mathrm{Re}[\mathscr{U}^c] = \mathcal{A}e^{-\frac{\alpha}{2}t}\cos(\omega^\dagger t)$ gives the physical displacement. The initial phase is not relevant here and is in (2) set to zero. Associated with the solution $\mathscr{U}$, the oscillator has at any time t a kinetic energy $\mathscr{E}_{kin}(t) = \frac{1}{2}\mathcal{M}(\frac{d\mathscr{U}(t)}{dt})^2 = \frac{1}{2}\mathcal{M}\omega^2\mathcal{A}^2e^{-\alpha t}(\frac{\omega^\dagger}{\omega}\sin\omega^\dagger t + \frac{\alpha}{2\omega}\cos\omega^\dagger t)^2$, elastic potential energy $V(\mathscr{U}) = \frac{1}{2}\beta\mathscr{U}^2(t) = \frac{1}{2}\mathcal{M}\omega^2\mathcal{A}^2 e^{-\alpha t}\cos^2\omega^\dagger t$ and total mechanical energy or Hamiltonian, assuming $\alpha << \omega$ and thus $\omega \doteq \omega^\dagger$,

$$\mathscr{E}(t) = \mathscr{E}_{kin}(t) + V(\mathscr{U}) = Ee^{-\alpha t}\rho_0(t) \doteq \tfrac{1}{2}\mathcal{M}\omega^2|\mathscr{U}^c|^2, \quad E = \tfrac{1}{2}\mathcal{M}\omega^2\mathcal{A}^2,$$
$$\rho_0(t) = \left[(1+\frac{\alpha^2}{4\omega^2})\cos^2\omega^\dagger t + (1-\frac{\alpha^2}{4\omega^2})\sin^2\omega^\dagger t + \alpha\omega^\dagger \sin\omega^\dagger t\cos\omega^\dagger t\right] \doteq |e^{-i\omega t}|^2 = 1. \tag{3}$$

Here, $|\mathscr{U}^c|^2 \doteq \mathcal{A}^2 e^{-\alpha t}|e^{-i\omega t}|^2 = \mathcal{A}^2 e^{-\alpha t}$, $|e^{-i\omega t}|^2 = \cos^2\omega t + \sin^2\omega t = 1$. The dynamical variables ω, $\mathcal{M}$, β, α, $\mathcal{A}$, $\mathscr{E}$, etc. in general all contain a relativistic effect, which evaluation will be illustrated for the IED oscillator in Sec. 5.3. Identical energy solutions may be obtained from solving Maxwell's equations for the radiation field emitted by charged object (Appendix A).

Under the condition $\alpha << \omega$ which is well fulfilled in the radiation experiments of interest here, about any time t there will in general exist a brief time interval Δt satisfying $\frac{2\pi}{\omega} << \Delta t << \frac{2\pi}{\alpha}$, so that during Δt the displacement (2) effectively has a constant amplitude $\mathcal{A}e^{-\frac{1}{4}\alpha(t+t+\Delta t)}$ and hence is *quasi harmonic*. Accordingly, as further combined with Eq. (3a), the extensive oscillator has in Δt effectively a constant Hamiltonian, and hence is *quasi stationary*, agreeing with our pre-condition for establishing Eq. (1).

2.2 Quasi stationary flow motion Alternately, we may directly describe the extensive oscillator of Sec. 2.1 by a linear probability density $\rho(x_j, t) = |\psi(x_j, t)|^2$ along the X_j direction, where ψ is a complex function for the same reason as $\mathscr{U}^c$ is complex and will serve as a natural independent variable of the Hamiltonian similarly as $\mathscr{U}^c$ in (3a). The coordinate x_j is related to $\mathscr{U}$ as $x_j = X_j - X_{j_0} = \Delta\mathscr{U} + \mathscr{U}$, where $\Delta\mathscr{U} = X_j - X_{j_c}$. With $\mathscr{U}$ given by (2), and necessarily $\Delta\mathscr{U} = \Delta\mathcal{A}e^{-\alpha t}$ for the oscillator being assumed rigid-body like, we have

$$x_j = \mathcal{A}'e^{-\frac{\alpha}{2}t}\cos(\omega t), \quad \mathcal{A}' = (\Delta\mathcal{A} + \mathcal{A}). \tag{4}$$

The $\mathscr{U}$ motion of the oscillator is associated with a ρ- flow motion in the given x_j-direction, with a flow velocity v and flow rate

$$j = \rho v = -D[\psi^*(\nabla\psi) - (\nabla\psi^*)\psi], \quad D = \frac{i\eta}{b\mathcal{M}} \tag{5}$$

Here, D is an imaginary diffusion constant (a general derivation is given in [3b]); $b = 2$ for an oscillator whose $\mathscr{U}$ motion is described by the Newtonian equation (1) as is with ψ_q, ψ_d of Sec. 5.1, and $b = 1$ by the Maxwell's equations as is with the $\psi_{r,q}$, Sec. 5.2. $\eta(t)$ is a new real variable to be determined.

Equation (1) holds, in each brief time interval Δt here, only if ρ satisfies in Δt the continuity equation

$$\frac{\partial\rho}{\partial t} + \nabla j - \mathcal{O}\rho = 0, \tag{6}$$

where $\mathcal{O} = \frac{V(x_j)}{i\eta} - \frac{V(x_j)}{i\eta}$. Substituting the expressions for ρ, j, D and V, Eq. (6) is decomposed into two second order differential equations for ψ, ψ^*, which for ψ and $b = 2$ is given as

$$i\eta\frac{\partial\psi}{\partial t} = H\psi, \quad H = -\frac{\eta^2}{2\mathcal{M}}\nabla^2 + V(x_j), \quad V(x_j) = \frac{1}{2}\mathcal{M}\omega^2 x_j^2; \tag{7}$$

the equation for $\psi_{r,q}$ for the case of $b = 1$ will be given by (37.b), Sec. 5.2. It is easily seen that the Hamiltonian H in (7) is separable into a term (H_0) associated with a potential $V_0(\bar{x}_j)$

dependent only on $\bar{x}_j = \mathcal{A}' \cos(\omega t)$ and independent on radiation, and a term (H_I) with a potential V_I describing the source–radiation interaction, i.e.,

$$H = H_0 + H_I, \quad H_0 = -\frac{\eta^2\nabla^2}{2\mathcal{M}} + V_0(\bar{x}_j), \quad V_0(\bar{x}_j) = \frac{1}{2}\mathcal{M}\omega^2\bar{x}_j^2,$$
$$H_I = V_I = V(x_j) - V_0(\bar{x}_j) = -V_0(\bar{x}_j)(1 - e^{-\alpha t}). \tag{8}$$

If up to the initial time $t = 0$ the oscillator has emitted no (net) radiation and we set for $t = 0$ (as for $t < 0$),

$$2\pi\eta(0) = \eta(0) = h, \tag{9}$$

Eq. (7) identifies then with the usual Schrödinger equation, describing the quasi stationary harmonic oscillator during a brief time interval Δt about $t = 0$ here. And the Eq. (37), Sec. 5, later describes the total wave equation[3c] for the IED particle system. Suppose that upon external perturbation the oscillator begins at $t = 0$ a (net) emission of radiation and is to emit one energy quantum $\hbar\omega$. This is necessarily a gradual process based on experimental indications discussed in Sec. 1. Of the $\hbar\omega$, ω is an intensive quantity and, when as the natural frequency of a specified mass and potential system mainly of interest in this paper, is in general a fixed value —if disregarding possible small broadening due to particle velocity variation, as known in theory and experiment. So during the intermediate process of emitting one radiation quantum $\hbar\omega$, $\eta(t)$ is inevitably the time-dependent counterpart of $\hbar$.

For the extensive oscillator is quasi stationary and hence its H is effectively constant in Δt, $\psi(x_j, t)$ must be factorisable as $\psi(x_j, t) = \phi(x_j)\theta(t)$, $\theta = e^{-i\frac{\mathscr{E}}{\eta}t}$ and $\psi = \phi e^{-i\frac{\mathscr{E}}{\eta}t}$. Placing ψ in (7), we obtain $-i^2\eta\frac{\mathscr{E}}{\eta}\phi\theta = H\phi\theta$, or $\mathscr{E}\phi = H\phi$. The last equation may be rewritten as, with the substitutions of $\xi_j \equiv \sqrt{\frac{\mathcal{M}\omega}{\eta}}\, x_j$ and accordingly $\phi(\xi_j) = e^{-\xi_j^2/2}\varphi(\xi_j)$,

$$\nabla^2\varphi - 2\xi_j\nabla\varphi + 2n\varphi = 0,$$
$$\mathscr{E}(t) \to \mathscr{E}_n(t) = (n + \frac{1}{2})\eta(t)\omega \tag{10}$$

Except that $\eta(t)$ is in place of $\hbar$, Eq. (10a) formally is identical to the time-independent Schrödinger equation for harmonic oscillator and can thus be solved accordingly with respect to $\eta(t)$.

3. Continuous emission of a radiation quantum

$\mathscr{E}$ of (3a), and $\mathscr{E}_n$ of (10b) after subtracting the $n = 0$ term, E_0 (which does not radiate and hence not present in the Newtonian Eq. 3), are the same energy, hence $\mathscr{E}(t) = \mathscr{E}_n(t) - E_0$. This is rewritten as, with (3a) for $\mathscr{E}$ and (10b) for $\mathscr{E}_n$,

$$\mathscr{E}_n(t) - E_0 = E_n\rho_0(t)e^{-\alpha t} \doteq E_n a(t) = n\eta(t)\omega \qquad (a)$$
$$E_n(= \mathscr{E}_n(0)) = \tfrac{1}{2}\mathcal{M}\omega^2\mathcal{A}_n^2 + E_0 = n\hbar\omega + E_0 \qquad (b)$$
$$a(t) = e^{-\alpha t} \qquad (c)$$
$$\mathcal{A}_n = \sqrt{n}\mathcal{A}_1, \quad \mathcal{A}_1 = (\frac{2\hbar}{\mathcal{M}\omega})^{1/2} \qquad (d)$$
$$\eta(t) = \frac{\mathcal{M}\omega\mathcal{A}_n^2}{2n}e^{-\alpha t} = \hbar e^{-\alpha t} \qquad (e) \tag{11}$$

where $n = 1, 2, \ldots$. In (11b), or (3), $\mathcal{M}$ and ω are constants characteristic of the oscillator and potential system; $\mathcal{A}$ only is subject to change upon excitation and thus to quantisation, hence

$\mathcal{A} \to \mathcal{A}_n$. Eqs. (11d) follow from the last two equations of (11b), and Eq. (11e) from the last two equations of (11a), where $\frac{\mathcal{M}\omega \mathcal{A}_n^2}{2n} = \hbar$ from (11d). For the (quasi harmonic) oscillator undergoing electromagnetic radiation here, $E_0 = 0$ based on comparison with the empirical Planck energy equation. (One might also interpret this as the consequence that the ground level energy E_0 is finite but is never emitted as electromagnetic radiation.)

With (11d) in (4) and (2), we obtain $x_j \to x_{jn} = \frac{\Delta \mathcal{A}_n + \mathcal{A}_n}{\mathcal{A}_n} \mathscr{U}_n$ and

$$\mathscr{U} \to \mathscr{U}_n = \mathcal{A}_n e^{-\frac{\alpha}{2}t} \cos(\omega t) = (2\pi B_n \eta)^{1/2} \cos(\omega t), \tag{12}$$

where $B_n = \frac{n}{\pi \mathcal{M}\omega}$. With (11a) for $\mathscr{E}_n(t)$, $x_j \to x_{jn}$, $\xi_j \to \xi_{jn} = \sqrt{\frac{\mathcal{M}\omega}{\eta}} x_{jn}$, and the standard Hermit polynomial solution for $(\varphi \to) \varphi_n$ or the normalised H_n, we obtain the total eigen function solution for (10a)

$$\psi_n(\xi_{jn}, t) = \phi_n(\xi_{jn}) \theta_n(t) = C_n H_n(\xi_{jn}) e^{-\frac{1}{2}\xi_{jn}^2} e^{-in\omega t} \tag{13}$$

Accordingly, $\rho(\xi_{jn}, t) = |\psi_n(\xi_{jn}, t)|^2 = |\phi_n(\xi_{jn})|^2$ which is independent of time, implying that indeed the extensive oscillator moves as a rigid-object in the way that its Hamiltonian density $\mathscr{E}_n |\psi_n(\xi_{j_n}, t)|^2$ during a brief time interval Δt is everywhere constant in time, as presumed in Sec. 2.1. Supposing that $\psi_n(x_{jn}, t)$ is normalised in $[-\frac{L}{2}, \frac{L}{2}]$, where $L \sim \mathcal{A}$, there is then at any time t

$$\int_{-\frac{L}{2}}^{\frac{L}{2}} \mathscr{E}_n(t) |\psi_n(x_{jn}, t)|^2 dx_{jn} = \mathscr{E}_n(t) \int_{-\frac{L}{2}}^{\frac{L}{2}} |\phi_n(x_{jn})|^2 dx_{jn} = \mathscr{E}_n(t) \tag{14}$$

The charged oscillator of a time dependent $\mathscr{E}_n(t)$ begins according to (11a) at time $t = 0$ a (net) radiation-emission, assuming $\alpha > 0$. At a later time t (assuming less than an equilibrium time t_o which may be in question for e.g. an oscillator enclosed between reflection walls), it will have emitted a total amount of radiation energy given as, with $E_{r,n} \equiv E_n$,

$$\begin{aligned} \mathscr{E}_{r,n}(t) &= E_n - \mathscr{E}_n(t) = (1 - e^{-\alpha t}) E_n = a_r(t) E_{r,n} = n \eta_r(t) \omega \qquad (a) \\ a_r(t) &= \frac{\mathscr{E}_{r,n}(t)}{E_{r,n}} = 1 - a(t) \qquad (b) \\ \eta_r(t) &= \hbar(1 - e^{-\alpha t}) = \hbar - \eta(t) \qquad (c) \end{aligned} \tag{15}$$

It follows from Eqs. (11) that during the intermediate process of emission of a radiation quantum, the $\mathscr{E}_n(t)$, being $\propto \mathscr{U}_n^2(t)$ (dotted line in Fig 1a), of the quasi-harmonic oscillator is at any instant of time quantised with respect to an exponentially decaying fractional-h, $\eta(t)$ (solid line in Fig 1a). In this specific way, the oscillator emits electromagnetic radiation gradually and continuously with time. At the same time, as follows from Eqs. (15), the emitted radiation field $\psi_{r,n}$ has a Hamiltonian $\mathscr{E}_{r,n}(t)$ that is quantised with respect to an exponentially increasing fractional-h, $\eta_r(t)$ (dashed line in Fig 1b). The wave field $\psi_{r,n}$ emitted over time, being propagated in the vacuum medium at the speed of light c outward from the source, therefore stretches an extensive wave train, of a length $L_r'(t) = ct$ at time t. The sum of the two fractional-h's of the source and radiation two-component system is at any time equal to

$$\eta(t) + \eta_r(t) = he^{-\alpha t} + h(1 - e^{-\alpha t}) \equiv h \tag{16}$$

(solid horizontal lines $n = 0, 1, 2$ drawn through graphs a and b in Fig 1). The total Hamiltonian is at any time t given as

$$\mathscr{E}_{tot.n}(t) = \mathscr{E}_n(t) + \mathscr{E}_{r,n}(t) = a(t) E_n + a_r(t) E_{r,n} \equiv E_n \equiv E_{r,n} = n\hbar\omega \tag{17}$$

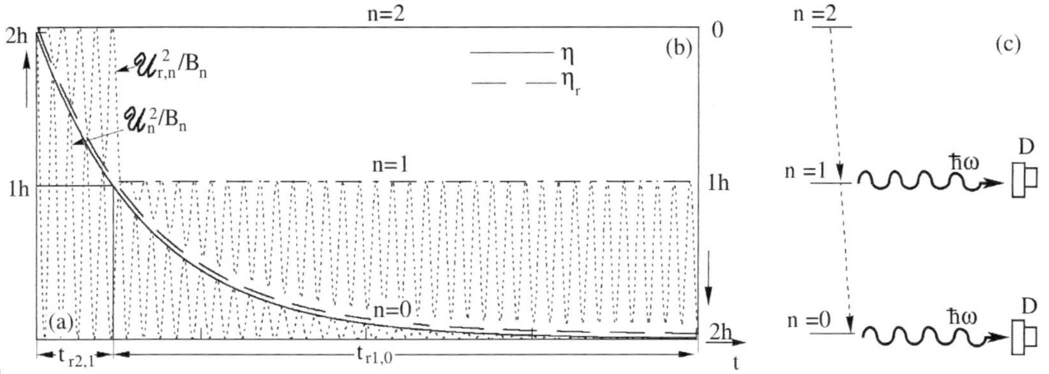

Figure 1. Continuous emission of a radiation quantum $\hbar\omega$ in a $(n=)2 \rightarrow 1$ and $1 \rightarrow 0$ transition of a charged oscillator. The intermediate processes are characterised by an exponentially decreasing and increasing fractional Planck constants of the oscillator and its radiation field as functions of time t, $\eta(t)$ and $\eta_r(t)$ given by Eqs. (11e) and (15c), shown by the solid and long-dashed curves in graphs (a) and (b). The sinusoidal oscillation displacement squares $\mathscr{U}_n^2(t)/B_n$ and $\mathscr{U}_{r,n}^2(t)/B_n$ given after Eq. (12), short-dashed curves in (a) and (b), are modulated accordingly in amplitudes. Used for the plot: $\omega = 2\pi/10$, $\alpha_q = 0.03$, $t_{r2} \doteq 2.3\tau$, $t_{r1} \dot{\geq} 200 - 2.3\tau$. (c) is a schematic illustration of Fig 1a (dotted lines), and of Fig 1b (curly solid lines) as the permanently emitted electromagnetic wave trains of energy quanta $\hbar\omega_q$'s, or photons, being probed by detectors D.

i.e. a constant consisting of n quanta.

The energy difference $(\Delta E_{n,n-1})$ between adjacent stationary levels n and $n' = n-1$ (given by Eq. 18b below) is always one whole energy quantum $\hbar\omega$, in complete accordance with the standard notion of quantum mechanics and with overall experiments. This result may be also stated as that, the difference action $\Delta E_{n,n-1} \times \frac{2\pi}{\omega}$ between two stationary levels is always equal to one full Planck constant h. This is shown in [3d] to be the combined consequence of the least action principle and second law of thermodynamics, i.e. the maximum entropy condition.

If the given charged oscillator (μ) is situated between (fixed) reflection walls, and no other absorbers or perturbing fields present, the radiation wave field will be reflected back to μ, be re-absorbed, and then re-emitted by it, iteratively. After an equilibrium time $t = t_o$, the re-emission and re-absorption of radiation will reach equilibrium, the $\mathscr{E}_{r,n}(t_o), a(t_o), \eta(t_o)$ etc. will be independent of time. We thus have a oscillator (as source) and radiation field two-component system that is as a whole in stationary state and carry n multiples of the one energy quantum $\hbar\omega$, $n\hbar\omega$, a situation as described by Eq. (17). The radiation field, being not charged nor undergoing energy transfer to external absorber (detector), is evidently not observable to an external observer (detector). The above in particular is the scheme by which IED particle maintains as a distinct, stationary quantum system (Sec. 5).

4. Transition time and probability

Of the two-component system above, the radiated fields will fail to be re-absorbed by their emission-oscillator μ if (i) no reflection walls present within distance of reach, (ii) another charged oscillator (as absorber) μ'' nearby begins to absorb the radiation emitted by μ, and/or (iii) the oscillator μ is externally strongly disturbed away from stationary state. The radiation field will then become manifestly permanently emitted by μ.

Consider that the oscillator μ and radiation total system is at initial time $t = t_i$, after having previously undergone an equilibrium time t_o, in the stationary level n. Subjected to any of the circumstances (i)–(iii), the system begins from time $t = t_i$ to undergo a permanent emission of a radiation quantum, transforming at final time $t = t_f = t_i + t_{n.n'}$ to the stationary level n'. Because in between t_o and t_i all the time dependent functions a, a_r etc remain the same, we may generally set $t_i = t_o$; $t_i = t_o = 0$ gives an oscillator with no radiation at initial time. So at

$t_f = t_o + t_{n.n'}$, the energy of the two-component system μ will have reduced by a total amount given based on Eqs. (11a,b), (15a) and (17) in two alternative ways as

$$\Delta\mathscr{E}_{tot.n,n'} = \mathscr{E}_{tot.n}(t_i) - \mathscr{E}_{tot.n}(t_f) = \left[(e^{-\alpha t_o} + a_r(t_o)) - (e^{-\alpha t_f} + a_r(t_f))\right] E_n$$
$$= e^{-\alpha t_o}(1 - e^{-\alpha t_{n,n'}})n\hbar\omega = \Delta\mathscr{E}_{n,n'} \quad (a); \quad \Delta E_{n.n'} = E_n - E_{n'} = (n - n')\hbar\omega \quad (b) \qquad (18)$$

In going from the second to the third of Eqs. (18a) we assumed that during the transition any new radiation emitted by the source contributes to the permanently emitted radiation (i.e. $\Delta\mathscr{E}_{n.n'}$) only and not to $a_r E_n$, so $a_r(t_o) = a_r(t_f) = 1 - e^{-\alpha t_o}$. Further from the identity relation $\Delta\mathscr{E}_{n.n'} = \Delta E_{n.n'}$ we obtain the transition time or lifetime evaluated from a finite $t_i (= t_0)$ and from $t_i = t_0 = 0$ respectively

$$t_{n,n'} = t_f - t_o = \frac{1}{\alpha}\ln\frac{ne^{-\alpha t_o}}{n'[1 - \frac{n}{n'}(1 - e^{-\alpha t_o})]},$$
$$t_{n,n'} = t_f - 0 = \frac{1}{\alpha}\ln\frac{n}{n'} \qquad (19)$$

$t_{n,n'}$, with $n' = n-1$ say, informs the literary time span for the exchange of one photon between two systems μ and μ'' rather than a statistical average, for in accordance to the fundamental quantum principle that one energy quantum or photon can either be exchanged between two systems as a whole, or not exchanged at all. For an oscillator endowed with n energy quanta ($E_n = n\hbar\omega$) at the beginning, the total time t_r required for transiting to final $n' = 0$ level is

$$t_r = \sum_{n'=1}^{n} t_{n',n'-1} = \frac{1}{f\alpha}, \quad \frac{1}{f} = \sum_{n'=1}^{n} \ln\frac{n'}{n'-1} \qquad (20)$$

These have the mean values $\langle t_r\rangle = \frac{\int_0^\infty te^{-\alpha t}dt}{\int_0^\infty e^{-\alpha t}dt} = \frac{1}{\alpha}$ (the mean life time) and $\langle\frac{1}{f}\rangle = 1$. Since for the systems considered $\alpha << \omega$, or $\frac{2\pi}{\alpha} \sim t_{n,n-1} >> \frac{2\pi}{\omega} = \tau$, during a transition time $t_{n,n-1}$ the oscillator in general continuously oscillates a large $\frac{t_{n,n-1}}{\tau} >> 1$ number of oscillation cycles.

Suppose there presenting a large N number of identical oscillators that at initial time all lie at energy level n and will decay to final level n' statistically via $n \to n'$ transitions, elapsing a transition time $t_{n,n'}$ given by (19) each, and emitting a total $\mathfrak{N}_{n,n'}$ energy quanta or photons. The total apparent elapsing time is $T = \sum_{i=1}^{N}(t_{n,n'})_i = Nt_{n,n'}$. So $\mathfrak{N}_{n,n'} = \frac{T}{t_{n,n'}} = N$. The probability per unit time that any oscillator makes a $n \to n'$ transition and emits one photon is thus given as

$$\alpha_{n,n'} = -\frac{d\mathfrak{N}_{n,n'}}{\mathfrak{N}_{n,n'}dt_{n,n'}} = -\frac{1}{\mathfrak{N}_{n,n'}}\frac{T}{t_{n,n'}^2} = \frac{1}{t_{n,n'}} = \frac{\alpha}{\ln\frac{n}{n'}} \qquad (21)$$

Specifically if $n = 1$ and $n' = n - 1 = 0$, Eqs. (19b) or similarly (19a), and (21) give

$$t_{1,0} = \infty, \quad \alpha_{1,0} = \frac{1}{t_{1,0}} = 0; \qquad (22)$$

and for $n = 2$, $n' = 1$, $t_{2,1} = \frac{\ln 2}{\alpha}$, $\alpha_{2,1} = \frac{\alpha}{\ln 2}$; etc. For a fixed α, $t_{n,n-1}$ reduces, and $\alpha_{n,n-1}$ increases with increasing n.

The transition probability may be more generally expressed in the usual terms of a source–radiation interaction Hamiltonian, our H_I given by Eq. (8d) earlier. The total and the unperturbed Hamiltonians $H(= H_0 + H_I)$ and H_0 are given by Eqs. (8a) and (b). The total (or ensemble [3e]) wave function at any time t is of the general form $\psi_{en}(x_j,t) = \sum_m b_m(t)\psi_m(x_j)e^{-i\frac{\mathscr{E}_m(t)}{\eta(t)}t}$, where $b_m(t)$ is the amplitude of state m; ψ_{en} clearly is also a solution to Eq. (7a). Substituting (8a) for H and the equation above for ψ_{en} in the corresponding equation of (7a) we obtain

$$\sum_m (H_0+H_I)b_m(t)\psi_m(x_j)e^{-i\frac{\mathscr{E}_m}{\eta}t} = i\eta\sum_m\left[\dot{b}_m(t)\psi_m(x_j)e^{-i\frac{\mathscr{E}_m}{\eta}t} - \frac{i\mathscr{E}_m}{\eta}b_m(t)\psi_m(x_j)e^{-i\frac{\mathscr{E}_m}{\eta}t}\right]$$

Subtracting $\sum_m H_0\psi_m(x_j,t) = \sum_m \mathscr{E}_m\psi_m(x_j,t)$, multiplying $\psi_k^*(x_j)$ from the left and integrating over all x_j, we obtain, for the eigen functions $\psi_m(x_j)$'s being orthogonal,

$$\sum_m b_m(t)\int\psi_k^*(x_j)H_I\psi_m(x_j)dx_je^{-i\frac{\mathscr{E}_m}{\eta}t} = i\eta\sum_m\dot{b}_m(t)\int\psi_k^*(x_j)\psi_m(x_j)dx_je^{-i\frac{\mathscr{E}_m}{\eta}t}.$$

$$\text{Or}\quad \int b_n(t)\psi_k^*(x_j)H_I\psi_n(x_j)dx_je^{-i\frac{\mathscr{E}_n}{\eta}t} = i\eta(t)\dot{b}_k(t)e^{-i\frac{\mathscr{E}_k}{\eta}t}, \tag{24}$$

assuming the oscillator to be in a definite energy state n at initial time t_i, and thus $b_m = 0$ for all $m \neq n$. Based on the solutions in Sec. 3, throughout the intermediate process of quasi stationary transition from level n to k here across a time duration (t_i,t_f), there present one component wave function which maintains precisely the same as at initial time, $\psi_n(x_j,t) = \psi_n(x_j)e^{-i\frac{\mathscr{E}_n(t)}{\eta(t)}t} = \psi_n(x_j)e^{-i\frac{E_n}{\hbar}t}$; and the other (in a fashion as discussed after Eq. 18) as at the final time t_f, $\psi_k(x_j,t) = \psi_k(x_j)e^{-i\frac{\mathscr{E}_k(t)}{\eta(t)}t} = \psi_k(x_j)e^{-i\frac{E_k}{\hbar}t}$. And the amplitudes of the two quasi stationary component states reduces and increases respectively with time as $a_n(t)$ and $a_{r,n}\to a_k(t) = 1-a_n(t)$. So with $b_n(t) = a_n(t) = e^{-\alpha t}$ (accordingly $b_k(t) = a_k(t) = 1-b_n(t)$; i.e. we are here facilitated with the explicit time dependent functions $b_n(t), b_k(t)$ instead of the usual perturbation approach), and $\eta = \hbar e^{-\alpha t}$ of (11e), denoting $\omega_{kn} = \frac{\mathscr{E}_k(t)-\mathscr{E}_n(t)}{\eta(t)} = \frac{E_k-E_n}{\hbar}$, (24) is rewritten as

$$\dot{b}_k = \frac{1}{i\hbar}H_{I_{kn}}e^{i\omega_{kn}t},$$

$$H_{I_{kn}} = \int\psi_k^*(x_j)H_I\psi_n(x_j)dx_j = -V_{0_0}(1-e^{-\alpha t})\int\left(\frac{x_j}{\mathcal{A}_ne^{-\frac{1}{2}\alpha t}}\right)^2\psi_k^*(x_j)\psi_n(x_j)dx_j = H_{I_0}\mathcal{X}_{kn},$$

$$\mathcal{X}_{kn} = \int\left(\frac{x_j}{\mathcal{A}_ne^{-\frac{1}{2}\alpha t}}\right)^2\psi_k^*(x_j)\psi_n(x_j)dx_j, \tag{25}$$

where $V_{0_0} = \frac{1}{2}\mathfrak{M}\omega^2\mathcal{A}_n'^2$, $H_{I_0} = V_{0_0}(1-e^{-\alpha t})$. Notice that H_{I_0}, the amplitude of source-radiation interaction Hamiltonian, or similarly H_I, is gradually switched on from zero at initial time t_i to maximum V_{0_0} at final time t_f; and this is not necessarily a small or perturbation quantity. For an oscillator with an initial Hamiltonian $E_1 = 2V_{0_0}$ and a corresponding a final time radiation-source interaction Hamiltonian $H_{I_0} = V_{0_0}\cdot(1-0)$, for example, H_{I_0} is equal just to the (maximum) potential or binding energy V_{0_0} of the (quasi) Harmonic oscillator.

The instantaneous transition probability per unit time in a brief time interval say $(t,t+\tau)$, with $\tau << t_{n,k}$, in which $H_{I_0} = V_{0_0}(1-e^{-\alpha t})$ is essentially a constant, is given as

$$\mathcal{P}_{n,k} = \frac{1}{\tau}|b_k|^2, \quad b_k = \int_0^\tau\dot{b}_kdt = -\frac{H_{I_{kn}}(e^{i\omega_{kn}\tau}-1)}{\hbar\omega_{kn}} \tag{27}$$

Suppose that ω_{kn} may have a finite continuous dispersion. The probability per unit time integrated over the entire possible dispersion range $(-\infty, \infty)$ of the transition energy $E_{kn} = E_k - E_n = \hbar\omega_{kn}$, as evaluated at $\frac{1}{2}\tau$ at which $b_k = -\frac{H_{I_{kn}}}{\hbar\omega_{kn}} e^{i\omega_{kn}(\frac{1}{2}\tau)} 2i \sin(\omega_{kn}\frac{1}{2}\tau)$, is

$$\int_{-\infty}^{+\infty} \mathcal{P}_{n,k} dE_{kn} = \frac{1}{\tau}\int_{-\infty}^{+\infty} |b_k|^2 dE_{kn} = \frac{4}{\tau\hbar^2}\int_{-\infty}^{+\infty} |H_{I_{kn}}|^2 \frac{\sin^2 \frac{1}{2}\omega_{kn}\tau}{\omega_{kn}^2} d(\hbar\omega_{kn})$$
$$= \frac{2}{\hbar}|H_{I_{kn}}|^2 \int_{-\infty}^{+\infty} \frac{\sin y^2}{y^2} dy = \frac{2\pi}{\hbar}|H_{I_{kn}}|^2 = \frac{2\pi}{\hbar}|H_{I_{kn}}|^2 \tag{27}$$

where $y = \frac{1}{2}\omega_{kn}\tau$; $\frac{\sin y^2}{y^2} \propto \delta(y)$, and $\int_{-\infty}^{+\infty} \frac{\sin y^2}{y^2} dy = \pi$. (27) after dividing E_{kn} out is by definition equal to the $\alpha_{n,n'}$ earlier for $n' = k$. Except with $H_I(t)$ being here an explicit function of time and describing the instantaneous source-radiation interaction at any time t during the emission of a radiation quantum, the conclusions (25)–(27) are formally as given based on the standard perturbation approach (see e.g. [4]). The source-radiation interaction Hamiltonian at the final completion of emission of one quantum is simply found at time t_f.

5. IED particle oscillator

5.1 The kinetic and the total charge oscillations Consider as two inter-related applications that the (extensive quasi-harmonic) charged oscillator of Secs. 2.1–2.2 firstly represents an usual charged oscillatory quantum particle of a charge q and mass m. The particle has an oscillation displacement $\mathscr{U}_d = X_c - X_0$ at its mass centre along the X direction under the actions of an elastic force $F_d = -\nabla V_d = -\beta_d \mathscr{U}_d$ due to an applied potential $V_d = \frac{1}{2}\beta_d \mathscr{U}_d^2$ (Fig. 2a), and a radiation damping force $F_{rd} = -\alpha_d m \frac{d\mathscr{U}_d}{dt}$. Its kinetic motion is associated with a wave function $\psi_d(x,t)$, where $x = X - X_0$. The equations of motion are given directly by substitutions of $\mathscr{U}_d$, α_d, ω_d, β_d, m, ψ_d, H_d, η_d and x for $\mathscr{U}$, α, ω, β, $\mathcal{M}$, ψ, H, η and x_j in Eqs. (1) and (7),

$$(1.a) \qquad \frac{d^2\mathscr{U}_d}{dt^2} + \alpha_d \frac{d\mathscr{U}_d}{dt} + \omega_d^2 \mathscr{U}_d = 0; \quad \omega_d^2 = \frac{\beta_d}{m} \tag{28}$$

$$(7.a) \qquad i\eta_d \frac{\partial \psi_d}{\partial t} = H_d \psi_d, \quad H_d = -\frac{\eta_d^2}{2m}\nabla_d^2 + \frac{1}{2} m\omega_d^2 x^2 \tag{29}$$

The solutions to (28)–(29) are given directly by substitutions of the corresponding variables n_d, ω_d, m, $\mathcal{A}_{n_d}$, a_d, η_d and α_d for n, ω, $\mathcal{M}$, $\mathcal{A}_n$, a, η and α in Eqs. (2)–(3), (9) and (10)–(20), Sec. 3. Of specific interest here, the Hamiltonian is given upon the substitutions above in Eqs. (11), divided for $t \le 0$ and $t \ge 0$ according to if any of the circumstances (i)–(iii) of Sec. 4 is onsets, as

$$\mathscr{E}_n(t) \to \mathscr{E}_{n_d}(t) = \begin{cases} E_{n_d} = \frac{1}{2} m\omega_d^2 \mathcal{A}_{n_d}^2 = n_d \hbar\omega_d & (t < 0) \\ \mathscr{E}_{n_d}(t) = a_d(t) E_{n_d} = n_d \eta_d(t)\omega_d & (t \ge 0) \end{cases} \tag{30}$$

where $a_d(t) = e^{-\alpha_d t}$, $\eta_d(t) = \hbar e^{-\alpha_d t}$. By (30a), up to time $t = 0$ the particle is in accelerated $(\mathscr{U}_{n_d})$ motion but is in stationary state, with a quantised Hamiltonian E_{n_d}. As such, the accelerated (charged) particle does not radiate. The accelerated $\mathscr{U}_{n_d}$ motion is instead to manifestly augment the frequency of the total wave from Ω_q by an average factor γ (given by Eq. 50b below) to $\omega_q = \gamma\Omega_q$, and accordingly modulate the plane electromagnetic wave Ξ_{r,n_q} into $\psi_{r,n_q} = \Xi_{r,n_q}\psi_{n_d}$ (Sec. 5.2). Under any of the circumstances (i)–(iii) of Sec. 4 only, the charged particle will emit thermal radiation according to (30b).

We shall be interested also in the rest-mass radiation of the particle. As a viable scheme for including the rest-mass radiation here, and for representing the extensive quantum wave of a matter particle in general, we shall represent the particle [assuming for simplicity (effectively) single charged and non-composite, like the electron] in terms of the IED particle model along with a vacuuonic vacuum proposed in [3] based on overall experiments. An IED particle of charge q is composed of a minute oscillatory charge q (as source) and the total electromagnetic radiation field $\mathbf{E}^j(\mathbf{r},t), \mathbf{B}^j(\mathbf{r},t)$ emitted by the charge. So, besides the oscillation of the IED particle itself described by Eqs. (28)–(29) in the X direction here, there is simultaneously an internal oscillation of its generating charge q (the $\mathscr{U}_q$ along Z direction below), and accordingly the dynamical process of its radiation field (the $\psi_{r,q}$ in Sec. 5.2). The IED particle oscillation consists in the oscillation of the charge–radiation field system as a whole (Sec. 5.3).

The vacuum is by construction (on experimental basis[3a,b]) filled of electrically neutral but polarisable vacuuons that are densely and disorderly packed with a mean separation distance $b_v \sim 1 \cdot 10^{-18}$ m. The vacuuons, polarised by the given charge q situated in an interstice i of the vacuuons centred about $\mathbf{R}_{0_i}$, produce at the charge q a vacuum potential

$$V_{vq}(\mathscr{U}_q) = V_{vq}(0) + \sum_n \frac{1}{n!}\nabla^n V_{vq}(\mathbf{R}_{0_i})\mathscr{U}_q^n \doteq V_{vq0} + V_q, \quad V_q = \frac{1}{2}\beta_q \mathscr{U}_q^2, \tag{31}$$

where $V_{vq0} = V_{vq}(0)$ and $\beta_q = \nabla^2 V_{vq}$; $\mathscr{U}_q(t) = \mathbf{R}_{c_i}(t) - \mathbf{R}_{0_i}$ is the displacement of the charge's mass center $\mathbf{R}_{c_i}$ from $\mathbf{R}_{0_i}$ along a direction perpendicular to the maximum intensity of its radiation wave (the $\psi_{r,q}$ later, or the ψ_d above along the X-direction); let this be the Z axis (Fig. 2.b); so $\mathscr{U}_q(t) = Z_{c_i}(t) - Z_{0_i}$. The approximation in (31a) is given for $\mathscr{U}_q$ being relatively small.

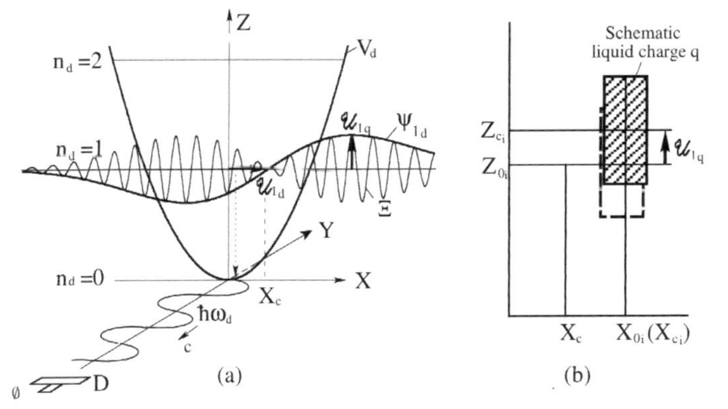

Figure 2: A quasi harmonic IED particle oscillator of oscillation $\mathscr{U}_{n_d}$ (Figure 2a), which contains a simultaneous internal oscillation $\mathscr{U}_{1_q}$ (Figure 2b), transits from initial thermal energy level $n_d = 1$, with an eigen function ψ_{1_d} (solid curves in Figure 2a) plotted after Eq. (13), to a final level $n_d = 0$ (ψ_{0_d} not shown). At the end one energy quantum $\hbar\omega_d$ is emitted.

The charge q is a spinning liquid-like entity (or vortex) of a linear dimension $\sim b_v \sim 1 \times 10^{-18}$ m; a point Z on it has a displacement $z = Z - Z_{0_i}$, and probability density $\psi_q(z,t)$. The charge has a zero rest mass. It however has a total mechanical energy (or Hamiltonian) E_q endowed in the past time, upon the action of an external driving force which has ceased action before time $t = 0$. The charge of the present time is thus an inertial system moving about in the vacuum, spontaneously propelled by its own inertial force F_{ine}. Defining $F_{ine} \equiv \mathfrak{M}_q \frac{d^2\mathscr{U}_q}{dt^2}$ in direct analogy to the Newtonian inertia for the usual matter-particles, we obtain $\mathfrak{M}_q$ as a proportionality constant, or a manifestly dynamical mass upon mapping on to a non-viscous vacuum. Subjected to the vacuum potential V_{vq} of Eq. (31) about the fixed site Z_{0_i} here, the motion of the charge is resisted, by an elastic resistive force $F_{vq} = -\frac{\partial V_{vq}}{\partial \mathscr{U}_q} = -\beta_q \mathscr{U}_q$, and in addition a radiation damping force $F_{rq} = -\mathfrak{M}_q \alpha_q \frac{d\mathscr{U}_q}{dt}$ in the viscous elastic vacuum. The corresponding equations of motion are given directly by substitutions of $\mathscr{U}_q$, α_q, ω_q, β_q, $\mathfrak{M}_q$, ψ_q,

H_q, η_q and z for $\mathscr{U}$, α, ω, β, $\mathcal{M}$, ψ, H, η and x_j in Eqs. (1) and (7) as

$$(1.b) \qquad \frac{d^2\mathscr{U}_q}{dt^2} + \alpha_q \frac{d\mathscr{U}_q}{dt} + \omega_q^2 \mathscr{U}_q = 0, \quad \omega_q^2 = \frac{\beta_q}{\mathfrak{M}_q}; \tag{32}$$

$$(7.b) \qquad i\eta_q \frac{\partial \psi_q}{\partial t} = H_q \psi_q, \quad H_q = -\frac{\eta_q^2}{2\mathfrak{M}_q}\nabla_z^2 + \frac{1}{2}\mathfrak{M}_q \omega_q^2 z^2 \tag{33}$$

The Hamiltonian solution to Eqs. (32)–(33) is given similarly by direct substitutions of n_q, ω_q, $\mathfrak{M}_q$, A_{n_q}, a_q, η_q and α_q for n, ω, $\mathcal{M}$, A_n, a, η and α in Eqs. (11) as

$$\mathscr{E}_n \to \mathscr{E}_{n_q} = \frac{1}{2}\mathfrak{M}_q \omega_q^2 A_{n_q}^2 a_q(t) = n_q \eta_q(t)\omega_q, \quad a_q(t) = e^{-\alpha_q t}, \quad \eta_q(t) = \hbar e^{-\alpha_q t} \tag{34}$$

In sum, for the IED particle oscillator, there present two simultaneous orthogonal (modes of) oscillations $\mathscr{U} = (\mathscr{U}_d, \mathscr{U}_q)$, and two corresponding probablity density $(|\psi_d|^2, |\psi_q|^2)$–flow motions in the $(X_j, X_{j'}) = (X,Z)$ directions, executed by two apparent oscillators, the IED particle oscillator d itself and the IED-particles's source charge q, of the masses $\mathcal{M} = (m, \mathfrak{M}_q)$ and a common charge q. The d and q oscillators each are extensive in their respective potential wells $V = (V_d, V_q)$ of a common quadratic form and are described by the common forms of equations of motion, Eqs. (28)–(29) and (32)–(33). The solutions for the two oscillators are accordingly formally commonly given by Eqs. (2)– (3), (9), and (10)–(20).

5.2 The total wave motion The charge oscillator q, oscillating according to (32)–(33) along the Z direction in a three-dimensional vacuum, generates along any radial $\mathbf{r}$-direction two oppositely travelling, and v_d-motion resultant Doppler-differentiated (Sec. 5.3) electromagnetic wave fields $\mathsf{E}^j(\mathbf{r},t), \mathsf{B}^j(\mathbf{r},t)$ travelling with velocities $\mathbf{c}$ and $-\mathbf{c}$ parallel ($j = \dagger$) and antiparallel ($j = \ddagger$) to $\mathbf{v}_d(t)$. Of direct relevance to the resultant IED particle is the superposed total wave field $\mathsf{E}(\mathbf{r},t) = \mathsf{E}^\dagger(\mathbf{r},t) + \mathsf{E}^\ddagger(-\mathbf{r},t)$; accordingly $\mathsf{B} = -\frac{\mathsf{E}}{c}\hat{\phi}$. The E, B fields in the case of $V_d = 0$ identify with the solutions (A.3)–(A.4) to the Maxwell's equations, Appendix A. In so far as the particle's coherent wave motion, and the associated total radiation power which is a constant independent of r (Eq. A.10, Appendix A), are mainly in question here, the radial radiation waves $\mathsf{E}(\mathbf{r},t)$ may be in effect furthermore (i) represented by the maximum-intensity wave field along the particle's motion X-direction and (ii) solved in regions where $\rho_q = j_q = 0$ only. For the general case of a finite V_d, wave fields are of the general complex forms

$$\mathsf{E}_x^c(x,t) = a_q^{1/2}\mathsf{E}_{0_q}\psi_{r,q}(x,t)\hat{z}, \quad \mathsf{B}_x^c(x,t) = -\frac{\mathsf{E}_{0_q}\psi_{r,q}(x,t)}{c}\hat{y} \tag{35}$$

given similarly as in (A.8), where $\psi_{r,q}(x,t) = \psi_{r,q}{}^\dagger(x,t) + \psi_{r,q}{}^\ddagger(x,t)$ is a complex dimensionless transverse wave and is to be solved in the presence of V_d.

The radiation field has a zero rest mass (similarly as its generating charge in the IED model) but a finite dynamical mass derived based on the following consideration. Suppose that at time t the ratio of the radiation Hamiltonian $(\mathscr{E}_{r,q}(t))$ to the total Hamiltonian of the charge-radiation system $(\mathscr{E}_{tot,q}(t))$ is $a_{r,q}(t)$. Regardless of the applied V_d, the total radiated electromagnetic wave train (of the wave function $\psi_{r,q}$ and total length $L_{r,q}$), propagating at the constant velocity light c in the vacuum, has according to Newtonian mechanics an (intrinsic) total linear momentum $\langle P_{r,q}\rangle (= \frac{\int_0^{\Delta L_{r,q}} \epsilon_0 \mathsf{E}_x^2(x,t)dx}{c}) = mc$ multiplied by $a_{r,q}^{1/2}(t)$, and kinetic energy $a_{r,q}(t)\langle E_{r,q,kin}(t)\rangle (= a_{r,q}(t)\int_0^{\Delta L_{r,q}} \frac{1}{2}\epsilon_0 \mathsf{E}_x^2(x,t)dx) = a_{r,q}(t)\frac{\langle P_{r,q}\rangle^2}{2m} = a_{r,q}(t)\frac{1}{2}mc^2$ (in the electromagnetic energy expression in the brackets the factor $\frac{1}{2}$ in front of $\epsilon_0 \mathsf{E}_x^2$ is because only the E field does work and not the B). m manifestly represents the relativistic dynamical mass of the wave train in a non viscous vacuum (as is assumed in Newtonian mechanics). In the vacuuonic vacuum representation we adopt in this section, m represents a coefficient proportional to the resistive force $F_{\rm vr}$ of the viscous elastic vacuum bulk against the wave train motion, $\langle P_{r,q}\rangle = \int F_{\rm vr} dt = \int_0^c m dc'$. Accordingly, the wave train has in addition an elastic vacuum potential energy $a_{r,q}(t)\langle V_{r,q0}(x)\rangle = a_{r,q}\langle E_{r,q,kin}(t)\rangle = a_{r,q}(t)\frac{1}{2}mc^2$, and therefore a total (intrinsic, i.e. excluding $a_{r,q}(t)V_{d0}$) Hamiltonian

$$\mathscr{E}_{r,q}(t) - a_{r,q}\langle V_{d0}\rangle = a_{r,q}\langle \mathscr{E}_{r,qkin}(t)\rangle + a_{r,q}\langle V_{\rm vr,q0}(x)\rangle = a_{r,q}(t)2\times\frac{1}{2}mc^2 = a_{r,q}(t)mc^2$$
$$\text{Or}\quad \hbar\omega_q - \langle V_{d0}\rangle = mc^2 \tag{36}$$

given after substituting the Eq. (40) below for $\mathscr{E}_{r,q}(t)$ and dividing $a_{r,q}(t)$ out. In the formal sense of the Eq. (41) later, m represents the dynamic mass of the IED particle.

The Maxwell's equations for the fields (35) lead in regions where $\rho_q = j_q = 0$ to the wave equation $\frac{\partial^2\psi_{r,q}}{\partial t^2} = (c^2 + \frac{V_d}{m})\nabla^2\psi_{r,q}$. This further reduces to, by combining with the identity relation $m = \frac{\hbar\omega_q - V_d}{c^2} = \gamma M$ described by Eqs. (36b) (or 41) and (50) and with the procedure described in [3c] except with $\eta_{r,q}$ replacing $\hbar$ in the final result,

$$i\eta_{r,q}\frac{\partial\psi_{r,q}}{\partial t} = H_p\psi_{r,q}, \quad H_p = -\frac{{\eta_{r,q}}^2}{m}\nabla^2 + V_d \tag{37}$$

It is easily seen that Eq. (37) is associated with a continuity equation, $\frac{\partial\rho_{r,q}}{\partial t} + \nabla(-D_p\nabla\rho_{r,q}) - (\frac{V_d}{i\eta_{r,q}} - \frac{V_d}{i\eta_{r,q}})\rho_{r,q} = 0$, given by substitutions of $\rho_{r,q} = |\psi_{r,q}(x,t)|^2$, $b=1$, m, $D_p = \frac{i\eta_{r,q}}{m}$, V_d and x_j for ρ, b, $\mathcal{M}$, D, V and x in (6).

Using the procedure of [3c] except with h being here replaced by the $\eta_{r,q}$, with $\psi_{r,q} = \Xi_{r,q}\psi_d$, Eq. (37) may be decomposed into two separate wave equations for ψ_d and $\Xi_{r,q}$. The wave equation for $\psi_d(x,t)$, with a Hamiltonian $H_d = H_p - H_p^0 = -\frac{\eta_{r,q}^2}{2m}\frac{\partial^2}{\partial x^2} + V_d$, is just the Eq. (29). $\Xi_{r,q}$ is the total radiation field emitted by the charge q when oscillating about a fixed site, i.e. $v_d(t) = 0$, and is described by the wave equation

$$i\eta_{r,q}\frac{\partial\Xi_{r,q}}{\partial t} = H_p^0\Xi_{r,q}, \quad H_p^0 = -\frac{\eta_{r,q}^2}{M}\frac{\partial^2}{\partial x^2}. \tag{38}$$

(38) has the solutions

$$\Xi_{r,n_q}(x,t) = e^{i(\frac{\mathscr{P}^0_{r,n_q}}{\eta_{r,q}}x - \frac{\mathscr{E}^0_{r,n_q}}{\eta_{r,q}}t)}, \quad \mathscr{E}^0_{r,n_q} = \lim_{v_d=0}\mathscr{E}_{r,n_q}(t), \quad \mathscr{P}^0_{r,n_q} = \frac{\mathscr{E}^0_{r,n_q}}{c}, \tag{39}$$

The total Hamiltonian of the radiation field is given after (15) as

$$\mathscr{E}_{r,n}(t) \to \mathscr{E}_{r,n_q}(t) = a_{r,q}(t)n_q\hbar\omega_q = n_q\eta_{r,q}(t)\omega_q,$$
$$a_{r,q}(t) = 1 - a_q(t) = 1 - e^{-\alpha_q t}, \quad \eta_{r,q}(t) = \hbar - \eta_q(t) = \hbar(1 - e^{-\alpha_q t}) \tag{40}$$

5.3 The charge and radiation-wave total system The minute liquid-like charge q and the resulting radiation wave $\psi_{r,q}$ are maintained as one system, the IED particle, by the repeated radiation re-absorption and re-emission scheme commented after Eq. (16). The total Hamiltonian $\mathscr{E}_{tot.n_q}(t)$ of the IED particle thus is at any time t carried a fraction $a_q(t)$ by the charge oscillator q, and $a_{r,q}(t)$ by the total radiation field $\psi_{r,q}$. With the $\mathscr{E}_{n_q}$ of the charge and $\mathscr{E}_{r,n_q}$ of its radiation field given by (34) and (40), we obtain, for $n_q = 1$,

$$\begin{aligned}\mathscr{E}_{tot.1_q}(t) &= \mathscr{E}_{1_q}(t) + \mathscr{E}_{r,1_q}(t) = a_q(t)E_{1_q} + a_{r,q}(t)E_{r,1_q} \\ &\equiv E_{1_q} \equiv E_{r,1_q} = \frac{1}{2}\mathfrak{M}_q\omega_q^2 A_{1_q}^2 = \hbar\omega_q = mc^2 + V_{d0}.\end{aligned} \quad (41)$$

The one energy quantum $\hbar\omega_q$ of the q–$\psi_{r,q}$ system will, applying the general quantum mechanical principle once again, either not be absorbed by an external source at all, or be absorbed as a whole upon an energy exchange.

As elucidated in [3b], an oscillatory charge q (of a zero rest mass) at the stationary level $n_q = 1$ in V_{vq} gives rise to a stationary electron (e) if $q = -e$, and a proton (p) if $q = +e$. A transition of the $-e$ or $+e$ oscillator from initial level $n_q = 1$ to a final $n'_q = 0$ (in V_{vq}) in the vacuum corresponds to a spontaneous decay of the particle e or p, over an infinite transition time $t_{1,0} = \infty$ according to Eqs. (19)–(22), in direct accordance with the empirical fact that the proton and electron have under normal conditions infinite lifetimes. (Nevertheless, the e or p may transit to the ground state $n_q = 0$ if the quadratic V_{vq} condition is strongly distorted, as e.g. would be the case when an anti-particle presents nearby, leading to a pair annihilation.) In a vacuum composed of densely packed vacuuons, the V_{vq} of each site extends only about half way to its neighbouring site and is whereof superseded by V_{vq} of the neighbouring site. So there is no stationary state for $n_q > 1$. The charge of an existing electron, proton or an composite particle of these, carries already one quantum of the relativistic mass energy, $\hbar\omega_q$, of the particle and can not absorb another $\hbar\omega_q$. So, the charge of another such existing particle can only permanently emit its $E_{tot.1_q}$, in a quasi stationary process, through pair annihilation; and the emitted gamma photon can only be absorbed by another "bare" charge (a vaculeon composing a vacuuon) out of the vacuum.

If any of the circumstances (i)–(iii) of Sec. 4 sets in (at time $t = 0$) so that the IED particle is perturbed away from the stationary level n_d, the IED particle oscillator will now (tending to stabilise in a lower and stationary level) manifestly emit thermal radiation (ψ_{r,n_d}) according to Maxwellian electrodynamics, or equivalently the quasi harmonic solution Eq. (30b). The emitted thermal radiation energy is accordingly given as $\mathscr{E}_{r,n_d}(t) = E_{n_d} - \mathscr{E}_{n_d}(t) = E_{n_d}(1 - e^{-\alpha_d t}) = \eta_{r,d}\omega_d$, where $\eta_{r,d}(t) = \hbar - \eta_d(t)$ given after (15) and (30). The maximum intensity, of the thermal radiation is in a direction perpendicular to $\mathscr{U}_{n_d}$, hence lying along a line in the $Y - Z$ plane, passing (X_0, Z_0) in Fig 2.

The sum of the two thermal terms above, $\mathscr{E}_{tot.n_d}(t) = \mathscr{E}_{n_d}(t) + \mathscr{E}_{r,n_d}(t) = E_{n_d}e^{-\alpha_d t} + E_{n_d}(1 - e^{-\alpha_d t})$, represents two briefly co-existing components during the transition only, rather than a distinct system given in (41), for the thermal frequency ω_d is not unique but is one out of a continuous spectrum and the energy quantum $\hbar\omega_d$ can be readily absorbed/emitted by another particle oscillator in the surrounding. At the end of the transition time t_{n_d,n_d-1}, the energy exchanged is given based on (18) as $\Delta\mathscr{E}_{n_d,n_d-1} = E_{n_d}(1 - e^{-\alpha_d t_{n_d,n_d-1}}) = \Delta E_{n_d,n_d-1} = \hbar\omega_d$, i.e. one whole energy quantum. For the thermal electromagnetic radiation here, the result $t_{r.1,0} = \infty$ of Eqs. (19)–(22) implies that the IED particle oscillator will always maintain at least a "zero point" (ground-state) thermal energy, E_{0_d}.

We finally evaluate the relativistic effect due to the v_{n_d} (or $\mathscr{U}_{n_d}$) motion of the particle in n_dth thermal level combining with the thermal and total energy solutions already obtained. The stationary harmonic oscillation (given for $\alpha_d = 0$) of the (IED) particle is an accelerated motion. The instantaneous velocity $v_{n_d}(t)$, accordingly the instantaneous relativistic mass $m(t)$, kinetic energy and linear momentum

$$\mathscr{E}_{kin,n_d}(t) = \frac{p_{n_d}^2(t)}{2m(t)} = \frac{1}{2}m(t)\omega_d^2 A_{n_d}^2 \sin^2(\omega_d t), \quad p_{n_d}(t) = m(t)v_{n_d}(t), \quad (42)$$

and potential energy $V_{n_d}(t)$ (as a short-hand denotation of $V_{n_d}(x(t))$ here) as measured in the laboratory frame (S), each vary with time. In a brief time interval about t, $V_{n_d}(t)$ is effectively constant. Thus the relationship between the instantaneous total (internal) Hamiltonian $\mathscr{E}_{tot.1_q} - V_{n_d}(t)$ (with $\mathscr{E}_{tot.1_q}(t) \equiv E_{tot.1_q} = \hbar\omega_q$ as given in Eq. (41)) and the $p_{n_d}(t)$ may be written according to the solution for a IED particle of constant potential (see e.g. [3c]), or equivalently the Einstein mass energy relation, as

$$[\hbar\omega_q(t) - V_{n_d}(t)]^2 = m^2(t)c^4 = M^2c^4 + p_{n_d}^2(t)c^2, \tag{43}$$

where

$$\omega_q(t) = \gamma(t)\Omega_q, \quad m(t) = \gamma(t)M, \quad V_{n_d}(t) = \gamma(t)V_{n_d}^0(t), \quad p_{n_d}(t) = \gamma(t)p_{n_d}^0(t) = \gamma(t)Mv_{n_d}(t), \tag{44}$$

$$\gamma(t) = \frac{1}{\sqrt{1 - v_{n_d}^2(t)/c^2}}, \quad \gamma^2(t) = 1 + \gamma^2(t)\frac{v_{n_d}^2(t)}{c^2}; \tag{45}$$

Ω_q, M, $V_{n_d}^0(t)$ and $p_{n_d}^0(t)$ are the corresponding rest-mass values measured in S.

The time average of (43) is $\langle[\hbar\omega_q(t) - V_{n_d}(t)]^2\rangle = \langle m^2(t)\rangle c^4 = M^2c^4 + \langle p_{n_d}^2(t)\rangle c^2$. The first two expressions develop as $\langle[\hbar\omega_q(t) - V_{n_d}(t)]^2\rangle = \langle\gamma^2(t)\rangle \left[\hbar\Omega_q - V_{n_d}^0\right]^2$ and $\langle m^2(t)\rangle = \langle\gamma^2(t)\rangle M^2$, where

$$\langle\gamma^2(t)\rangle = \left\langle\frac{1}{1 - v_{n_d}^2(t)/c^2}\right\rangle = 1 + \frac{\langle v_{n_d}^2(t)\rangle}{c^2} + \frac{\langle v_{n_d}^4(t)\rangle}{c^4} + \ldots \tag{46}$$

The $\langle v_{n_d}^2(t)\rangle$, $\langle v_{n_d}^4(t)\rangle$, etc. may be readily individually evaluated as

$$\langle v_{n_d}^2(t)\rangle = \frac{\frac{1}{2}m(t)\omega_d^2 A_{n_d}^2 \frac{1}{\tau_d}\int_0^{\tau_d}\sin^2\omega_d t dt}{\frac{1}{2}m(t)} = \frac{1}{2}\omega_d^2 A_{n_d}^2, \tag{47}$$

and $\langle v_{n_d}^4(t)\rangle = \frac{1\cdot3}{2\cdot4}\omega_d^4 A_{n_d}^4$, etc., where $\int_0^{\tau_d}\sin^2\omega_d t dt = \frac{\tau_d}{2}$, $\int_0^{\tau_d}\sin^4\omega_d t dt = \frac{1\cdot3}{2\cdot4}\tau_d$, etc. Multiplying (46) by $\frac{\langle v_{n_d}^2(t)\rangle}{c^2}$ and in turn adding 1 on each side, we obtain $\langle\gamma^2(t)\rangle\frac{\langle v_{n_d}^2(t)\rangle}{c^2} + 1 = 1 + \frac{\langle v_{n_d}^2(t)\rangle}{c^2}\frac{\langle v_{n_d}^2(t)\rangle}{c^2} + \frac{\langle v_{n_d}^4(t)\rangle}{c^4}\frac{\langle v_{n_d}^2(t)\rangle}{c^2} + \ldots$, or

$$\langle\gamma^2(t)\rangle = 1 + \langle\gamma^2(t)\rangle\frac{\langle v_{n_d}^2(t)\rangle}{c^2} = 1 + \langle\gamma^2(t)\frac{v_{n_d}^2(t)}{c^2}\rangle \tag{48}$$

The last of Eqs. (48) follows from the equality of the left side of Eqs. (48) with that of the time average of Eq. (45b), $\langle\gamma^2(t)\rangle = 1 + \langle\gamma^2(t)\frac{v_{n_d}^2(t)}{c^2}\rangle$. With Eqs. (48), $\langle p_{n_d}^2(t)\rangle = \langle\gamma^2(t)M^2v_{n_d}^2(t)\rangle$ is written as

$$\langle p_{n_d}^2(t)\rangle = \langle\gamma^2(t)\rangle M\langle v_{n_d}^2(t)\rangle \tag{49}$$

Combining with (49), the root mean of Eqs. (43) is developed as $\hbar\omega_q - V_{n_d} = mc^2 = \sqrt{M^2c^4 + \gamma^2M^2v_{n_d}^2c^2}$, where

$$m = \sqrt{\langle m^2(t)\rangle} = \gamma M, \quad \gamma = \sqrt{\langle\gamma^2(t)\rangle}, \quad v_{n_d} = \sqrt{\langle v_{n_d}^2(t)\rangle}. \tag{50}$$

Accordingly the time averages of $\omega_q(t)$, $p_{n_d}(t)$, etc. are given by $\omega_q = \sqrt{\langle\omega_q^2(t)\rangle} = \gamma\Omega_q$, $p_{n_d} = \sqrt{\langle p_{n_d}^2(t)\rangle} = mv_{n_d}$, etc. Hence $E_{1_q} = \hbar\omega_q$, where $E_{1_q}^0 = \hbar\Omega_q$; $E_{n_d} = n\hbar\omega_d = \frac{1}{2}mv_{n_d}^2 + V_{n_d} = \gamma E_{n_d}^0$, where $E_{n_d}^0 = n\hbar\Omega_d = \frac{1}{2}Mv_{n_d}^2 + V_{n_d}^0$.

In sum, the dynamical variables m, ω_q etc of the particle in harmonic, hence accelerated motion in the in X direction here relative to the laboratory frame S are each on average augmented by the factor γ as measured in S. Accordingly, as measured in S, the space and time variables x, $\mathscr{U}_{n_d}$, $\mathcal{A}_{n_d}$ and t of the particle in the particle's motion X direction are each contracted by the average factor γ from their rest values (indicated by the superscript 0 as elsewhere if not specified otherwise) x^0, $\mathscr{U}^0_{n_d}$, $\mathcal{A}^0_{n_d}$ and t^0 as (Lorentz-Einstein transformation) $\frac{x}{x^0} = \frac{\mathscr{U}_{n_d}}{\mathscr{U}^0_{n_d}} = \frac{\mathcal{A}_{n_d}}{\mathcal{A}^0_{n_d}} = \frac{t}{t^0} = \frac{1}{\gamma}$. The space variables $z, \mathscr{U}_q$ etc. in the transverse direction are particularly here the projections of the longitudinally contracted (by γ) elastic deformation of the vacuum. These are therefore proportionally contracted by the same factor γ from their rest values: $\frac{z}{z^0} = \frac{\mathscr{U}_{1_q}}{\mathscr{U}^0_{1_q}} = \frac{\mathcal{A}_{1_q}}{\mathcal{A}^0_{1_q}} = \frac{1}{\gamma}$. Finally, $\mathfrak{M}_q = \gamma \mathfrak{M}^0_q$, $(\beta_d, \beta_q) = \gamma^3(\beta^0_d, \beta^0_q)$, $(\alpha_d, \alpha_q) = \gamma(\alpha^0_d, \alpha^0_q)$ given as derivative relations based on Eqs. (28b), (32b), (41) and (30) combined with the foregoing transformation relations.

The author thanks Professor C Burdik for kindly inviting the author for presenting a contribution at the XXIst Int Conf on Integrable Systems and Quantum Symmetries in Prague, June, 2013, during which stay the author had very much enjoined extensive scientific discussions with several of the participants. Several Swedish professors have kindly communicated with the author regarding academic funding possibilities for the research. The author's this research is privately financed by P-I Johansson (emeritus scientist, Uppsala Univ.).

Appendix A. Radiation damping based on solution to Maxwell equations

A.1 Electromagnetic Hamiltonian quantisation and damping For a self-contained illustration of the subject we derive in this appendix the electromagnetic equations for $\mathscr{E}_r$ (the radiation Hamiltonian), α, F_r, and H_1 based on solutions to the Maxwell's equations. The results are formally mostly well known; although, the following solutions will be obtained in terms of the radiation Hamiltonian, complex Poynting vector, and in the end, the quantised radiation fields.

Consider the quasi-stationary charged oscillator as specified in Sec. 2.1, oscillating about position $\mathbf{r} = 0$ at time $t = 0$ along a specified X_j direction, which we set as the Z axis here. In the case where the charged oscillator has also an (instantaneous) linear motion, and thus its radiated electromagnetic waves are Doppler -differentiated as in Sec. 5.2, we shall be concerned mainly with the superposed radiation electric field $\mathsf{E}(\mathbf{r}, t)(= \mathsf{E}^\dagger(\mathbf{r}, t) + \mathsf{E}^\ddagger(\mathbf{r}, t))$ and $\mathsf{B}(\mathbf{r}, t)$ similarly as in Sec. 5.2. Similarly as the $\mathsf{E}^j, \mathsf{B}^j$, the E, B fields are governed by the Maxwell's equations $\nabla \cdot \mathsf{E} = \frac{\rho_q}{\epsilon_0}$, $\nabla \times \mathsf{B} = -\mu_0 \mathbf{j}_q + \frac{1}{c'^2}\frac{\partial \mathsf{E}}{\partial t}$, $\nabla \cdot \mathsf{B} = 0$, $\nabla \times \mathsf{E} = -\frac{\partial \mathsf{B}}{\partial t}$, where $c'^2 = c^2 + \frac{V}{\mathbb{M}}$; ρ_q and j_q are the charge and current densities of the radiation-emitting charged oscillator. In regions where $\rho_q = j_q = 0$, choosing here the radiation gauge $f = -\int \Phi dt$, Φ being the Coulomb potential of q, such that $\nabla \cdot \mathsf{A} = 0$ and (the vector potential) A will be a transverse field while E, B maintain unaltered, furthermore setting for simplicity $V = 0$, the Maxwell's equations lead to the wave equations

$$\nabla^2 \mathsf{Y} - \frac{1}{c^2}\frac{\partial^2 \mathsf{Y}}{\partial t^2} = 0, \quad \mathsf{Y} = \mathsf{A}, \mathsf{E}, \mathsf{B}. \tag{A.1}$$

Eqs. (A.1) have the solutions given in the spherical polar coordinates as, under the easily satisfied condition $\mathcal{A}/r << 1$ for the systems of interest here,

$$\mathsf{A}(\mathbf{r}, t) = -\frac{a^{1/2} \mathsf{A}_0 r_0 \sin\theta \sin(\mathbf{k}\cdot\mathbf{r} - \omega t)\hat{\theta}}{r}, \quad \mathsf{A}_0 = \frac{q\mathcal{A}\omega}{4\pi\epsilon_0 r_0 c^2}, \tag{A.2}$$

$$\mathsf{E}(\mathbf{r}, t) = -\nabla(\Phi + \frac{\partial f}{\partial t}) - \frac{\partial(\mathsf{A} - \nabla f)}{\partial t} = -\frac{\partial \mathsf{A}}{\partial t} = \frac{a^{1/2}\mathsf{E}_0 r_0 \sin\theta \cos[\omega(\frac{r}{c} - t)]}{r}\hat{\theta} = \frac{\mathsf{E}_{r\theta}\boldsymbol{\mathscr{U}}_r}{\mathcal{A}}\hat{\theta},$$

$$\mathsf{E}_0 = \mathsf{A}_0\omega = \frac{q\mathcal{A}\omega^2}{4\pi\epsilon_0 r_0 c^2}, \quad \mathsf{E}_{r\theta} = \frac{\mathsf{E}_0 r_0 \sin\theta}{r},$$

$$\mathscr{U}_r(\mathbf{r},t) = a^{1/2}\mathcal{A}\cos(\mathbf{k}\cdot\mathbf{r}-\omega t)\hat{\theta}; \tag{A.3}$$

$$\mathsf{B} = \nabla\times(\mathsf{A}-\nabla f) = -\frac{\mathsf{E}}{c}\hat{\phi}; \tag{A.4}$$

Here, $\mathbf{k} = k\hat{\mathbf{r}}$, $k = \omega/c$. r_0 represents an (effective) radius of the oscillator. $a^{1/2}(t)\mathsf{E}_0$ is the radiation-damped amplitude of a spherical wave front at radius r_0 and angle $\theta = \frac{\pi}{2}$ at time t, and E_0 is the amplitude without damping. a is to be (separately) determined. For either of the two oscillators of Sec. 5.2 or any other compatible system, $\omega = \gamma\Omega$, $\mathcal{A} = \mathcal{A}^0/\gamma$, $\mathcal{M} = \gamma\mathcal{M}^0$ and $r = (x^{0^2}/\gamma^2 + y^{0^2}/\gamma^2 + z^{0^2}/\gamma^2)^{1/2}$ are the relativistic values of Ω, $\mathcal{A}^0$, $\mathcal{M}^0$ and r^0 as in Sec. 5.2. The transverse coordinates y, z, representing the projections of longitudinally contracted vacuum for the same (IED) particle here, are contracted similarly as discussed in Sec. 5.3. Accordingly as derivative relations, $\mathsf{E}_0 = \gamma^2\mathsf{E}_0^0$ and $\mathsf{E} = \gamma^2\mathsf{E}^0$ in the above, and $\mathscr{E} = \gamma\mathscr{E}^0$, $E = \gamma E^0$, $t_r = t_r^0/\gamma$ and $L_r = L_r^0/\gamma$ later, are the relativistic values of E_0^0, E^0, $\mathscr{E}^0$, E^0, t_r^0 and L_r^0.

$\mathscr{U}_r(\mathbf{r},t)$ of Eq. (A.3) represents the (quasi harmonic) oscillation displacement of a constituent of the vacuum, a coupled polarised vacuuon according to [3], along Z direction at position $\mathbf{r}$ at time t, generated in response to the perturbation by $\mathscr{U}(0, t-\frac{\mathbf{r}}{c})$ of the charged oscillator at $\mathbf{r} = 0$ and $t - \frac{\mathbf{r}}{c}$. Accordingly, as follows directly from the last relation of Eqs. (A.3a), $\mathsf{E}(\mathbf{r},t)$ is an electromagnetic representation of this vacuum deformation $\mathscr{U}_r(\mathbf{r},t)$. The radiated time-dependent electromagnetic Hamiltonian $\mathscr{E}_r(r,\theta,t)$ therefore necessarily consists of a kinetic oscillation term $\mathscr{E}_{r,kin}(r,\theta,t)$ and in addition, an elastic vacuum potential term $V_{\rm vr}(r,\theta,t)$. The two terms and hence their sum may be written down by observing both the usual form of electromagnetic energy equation and the underlying dynamics, given for per volume within a volume element dV about position $\mathbf{r}$ at time t, as

$$\frac{d\mathscr{E}_{r,kin}}{dV} = \frac{1}{2}\left[\epsilon_0\mathsf{E}^2(r,\theta,t,-\tfrac{\pi}{2}) + \frac{1}{\mu_0}\mathsf{B}^2(r,\theta,t,-\tfrac{\pi}{2})\right] = \frac{1}{2}\left[\epsilon_0\frac{\dot{\mathsf{E}}^2(r,\theta,t)}{\omega^2} + \frac{1}{\mu_0}\frac{\dot{\mathsf{B}}^2(r,\theta,t)}{\omega^2}\right]$$

$$= \epsilon_0\mathsf{E}_{r\theta}^2 a\sin^2[\mathbf{k}\cdot\mathbf{r}-\omega t] = \mathfrak{M}^{em}\omega^2\mathsf{E}_{r\theta}^2 a\sin^2[\mathbf{k}\cdot\mathbf{r}-\omega t], \quad \mathfrak{M}^{em} = \frac{\epsilon_0}{\omega^2} \tag{A.5}$$

$$\frac{dV_{\rm vr}}{dV} = \frac{1}{2}[\epsilon_0\mathsf{E}^2(r,\theta,t) + \frac{1}{\mu_0}\mathsf{B}^2(r,\theta,t)] = \epsilon_0\mathsf{E}^2(r,\theta,t) = \epsilon_0\mathsf{E}_{r\theta}^2 a\cos^2[\mathbf{k}\cdot\mathbf{r}-\omega t] \tag{A.6}$$

$$\frac{d\mathscr{E}_r}{dV} = \frac{d(\mathscr{E}_{r,kin}+V_{\rm vr})}{dV} = \epsilon_0\mathsf{E}_{r\theta}^2 a_1[\sin^2(\mathbf{k}\cdot\mathbf{r}-\omega t) + \cos^2(\mathbf{k}\cdot\mathbf{r}-\omega t)] = \frac{\epsilon_0 r_0^2\sin^2\theta\,|\mathsf{E}_x^c(\mathbf{r},t)|^2}{r^2}, \tag{A.7}$$

$$\mathsf{E}_x^c(\mathbf{r},t) = a^{1/2}\mathsf{E}_0 e^{-i(\mathbf{k}\cdot\mathbf{r}-\omega t)}\hat{\theta} = a^{1/2}\mathsf{E}_0\psi_r(\mathbf{r},t)\hat{\theta} = \frac{\mathsf{E}_0}{\mathcal{A}}\mathscr{U}_r^c(\mathbf{r},t)\hat{\theta},$$

$$\psi_r(\mathbf{r},t) = e^{-i(\mathbf{k}\cdot\mathbf{r}-\omega t)}, \quad \mathscr{U}_r^c(\mathbf{r},t) = a^{1/2}\mathcal{A}\psi_r(\mathbf{r},t) \tag{A.8}$$

In the above, the first of Eqs. (A.5a) preserves the usual form of electromagnetic energy equation except with the cosine function being shifted by a phase $-\frac{\pi}{2}$. The second of Eqs. (A.5a) is alternatively and equivalently expressed as the time rate of E; that is, $\mathscr{E}_{kin}$ is now proportional to $\dot{\mathsf{E}}^2$ and a corresponding "electromagnetic inertial mass" $\mathfrak{M}^{em}$, by making analogy to the kinetic energy of an oscillator. It is rather the elastic potential term (A.6), which does not present in the usual empty vacuum representation, that formally directly corresponds to the usual (phenomenological) electromagnetic energy equation.

The "complex Poynting vector", defined as the vector intensity of the electromagnetic radiation Hamiltonian $d\mathscr{E}_r(r,\theta,t)$ passing per unit time per unit area through a differential cross-section area $d\sigma = \frac{dV}{dr} = r^2 \sin\theta d\theta d\phi$ along $\mathbf{r}(r, \theta, \phi)$ direction, is given as

$$\mathbf{I}^c(r, \theta, t) = \frac{d\mathscr{E}_r(r, \theta, t)}{dt d\sigma}\,\hat{r} = \frac{d\mathscr{E}_r(r, \theta, t)}{dV} c\,\hat{r} = \frac{\epsilon_0 r_0^2 \sin^2\theta\, |\mathsf{E}_x^c(r, t)|^2 c}{r^2}\,\hat{r} \quad (A.9)$$

The negative time rate of the Hamiltonian of the oscillator, $-\frac{d\mathscr{E}(t)}{dt}$, is equal to the total radiation power passing a sphere of radius r, i.e., $\frac{d\mathscr{E}_r(r,\theta,t)}{dt d\sigma} d\sigma = I(r, \theta, t) d\sigma$ integrated over 4π solid angle:

$$\begin{aligned} -\frac{d\mathscr{E}(t)}{dt} = P(t) \equiv \frac{d\mathscr{E}_r(t)}{dt} &= \int_0^{\pi}\int_0^{2\pi} I(r, \theta, t) r^2 \sin\theta d\theta d\phi = s_0 \epsilon_0 |\mathsf{E}_x^c(r, t)|^2 c \quad (a) \\ &= \frac{q^2\omega^4 \mathcal{A}^2 a}{6\pi\epsilon_0 c^3} = \frac{q^2|\ddot{\mathscr{U}}_r^c|^2}{6\pi\epsilon_0 c^3} \quad (b) \\ &= \frac{q^2\omega^4\mathcal{A}^2 a \frac{1}{2}\mathcal{M}}{6\pi\epsilon_0 c^3 \frac{1}{2}\mathcal{M}} = \frac{q^2\omega^2\mathscr{E}}{3\pi\epsilon_0 c^3 \mathcal{M}} = \alpha\mathscr{E}, \quad (c) \end{aligned} \quad (A.10)$$

$$\alpha = \frac{q^2\omega^2}{3\pi\epsilon_0 c^3\mathcal{M}} = \frac{q^2\omega^3 \mathcal{A}_1^2}{6\pi\epsilon_0 c^3\hbar} = \frac{4qr_0\mathsf{E}_0}{3\mathcal{A}\mathcal{M}c} \quad (A.11)$$

where $s_0 = \frac{8\pi}{3} r_0^2$ represents the apparent area passed by the spherical wave E at $r = r_0$. Eqs. (A.10c) are given after multiplying and then dividing the right sides of (A.10.b) by $\frac{1}{2}\mathcal{M}$, and substituting into it by $\mathscr{E} = \frac{1}{2}\mathcal{M}\omega^2\mathcal{A}^2 a(t)$ given by (3a). The second expression of Eqs. (A.11) is given by substituting Eq. (11b) for $\mathcal{M}$; and the third by substituting (A.3a) for E_0. Eqs. (A.10b) and (A.11) give the well-known Larmor formula in complex form here and formula for damping factor. Integrating (A.10c,a) over time $(0, t)$ gives $\ln \mathscr{E}(t)|_0^t = -\alpha t|_0^t$, $-\mathscr{E}(t)|_0^t = \mathscr{E}_r(t)|_0^t$. Combining with the quantisation results of (11a)–(b) for the same energies $\mathscr{E}$, E as here, we obtain

$$\mathscr{E}(t) \to \mathscr{E}_n(t) = E_n e^{-\alpha t} = E_n a(t), \quad E_n = \mathscr{E}_n(0) = \frac{1}{2}\mathcal{M}\omega^2\mathcal{A}_n^2, \quad a(t) = e^{-\alpha t} \quad (A.12)$$

$$\mathscr{E}_r(t) \to \mathscr{E}_{r,n}(t) = E_{r,n}(1 - e^{-\alpha t}) = E_{r,n} a_r(t) \quad E_{r,n} = \mathscr{E}_{r,n}(\infty) = L_r s_0 \epsilon_0 \mathsf{E}_{0n}^2, \quad a_r(t) = 1 - e^{-\alpha t},$$

$$\text{where} \quad L_r = \frac{1}{\alpha} c = t_r c = \frac{3\pi\epsilon_0 c^4 \mathcal{M}}{q^2\omega^2}, \quad t_r = \frac{1}{\alpha}, \quad \mathsf{E}_{0n} = \frac{q\mathcal{A}_n\omega^2}{4\pi\epsilon_0 r_0 c^2}. \quad (A.13)$$

The equations for a, a_r above are the same as Eqs. (11c), (15b). t_r of (A.13) is identical to the mean transition time $\langle t \rangle = \frac{\int_0^\infty t e^{-\alpha t} dt}{\int_0^\infty e^{-\alpha t} dt} = \frac{1}{\alpha}$. So L_r represents an average length of the radiated electromagnetic wave train.

Specifically for a charge oscillator $q = +e$ or $-e$ which oscillates in the vacuum potential field V_{vq} and generates an electron or proton based on Eqs. 31–33, Sec. 5, accordingly with the substitutions of $|\pm e|$, $\mathfrak{M}_q = \frac{2\hbar}{\mathcal{A}_{1q}^2 \omega_q} = \frac{2\hbar^2}{\mathcal{A}_{1_q}^2 mc^2}$ (given after Eq. 11d), $\omega_q = \frac{mc^2}{\hbar}$ for q, $\mathcal{M}$, ω, Eqs. (A.11), (A.13b), (a) and (c) are written as

$$\alpha_q = \frac{e^2\omega_q^2}{3\pi\epsilon_0 c^3 \mathfrak{M}_q} = \frac{e^2(\frac{mc^2}{\hbar})^2}{3\pi\epsilon_0 c^3 (\frac{2\hbar^2}{\mathcal{A}_{1_q}^2 mc^2})} = \frac{e^2 m^3 c^3 \mathcal{A}_{1_q}^2}{6\pi\epsilon_0\hbar^4}, \quad t_{r,q} = \frac{1}{\alpha_q}, \quad L_{r,q} = ct_{r,q} = \frac{6\pi\epsilon_0\hbar^4}{e^2 m^3 c^2 \mathcal{A}_{1_q}^2},$$

$$\mathsf{E}_{01_q} = \frac{e\mathcal{A}_{1q}\omega_q^2}{4\pi\epsilon_0 r_0 c^2} = \frac{e\mathcal{A}_{1_q} m^2 c^2}{4\pi\epsilon_0 r_0 \hbar^2} \quad (A.14)$$

Calculated values of α_q, $t_{r,q}$, $L_{r,q}$ and $E_{0,1_q}$ based on Eqs. (A.14) for the charge oscillators $-e$ and $+e$, which generate an IED electron and proton, are given in Table A1, where the $\mathfrak{M}_q$ and

Table A1.

Oscillator q:	$\mathfrak{M}_q^{(a)}$ (kg)	$\omega_q(=\frac{mc^2}{\hbar})$ (r/s)	α_q (r/s)	$t_{r,q}$ (s)	$L_{r,q}$ (m)	$\mathsf{E}_{01_q}^{(b)}$ (N/C)
$-e$	1.083×10^{-18}	7.763×10^{20}	6.354×10^{6}	1.574×10^{-7}	47.18	9.656×10^{15}
$+e$	5.896×10^{-22}	1.425×10^{24}	3.935×10^{16}	2.541×10^{-17}	7.619×10^{-9}	3.256×10^{22}

(a): Values from [3b], with the free parameter f_1 set to 1.
(b): Based on Eq. (A.14b), where we have set $r_0 = \mathcal{A}_{1q}$; clearly $\mathcal{A}_{1q} \sim \frac{b_V}{2}$, $b_V \sim 1 \times 10^{-18}$ m being the inter-vacuuon distance estimated based on experiment [3b].

$\omega_q(=\frac{mc^2}{\hbar})$ values (second and third columns in the table) are as input data. In comparison, for an electron oscillator described by Eqs. (28)–(29), Sec. 5, with an oscillation frequency $\omega_d = 2\pi \cdot 10^{14}$ r/s, corresponding evaluations give $t_{r,e} = 2.02 \times 10^{-7}$ s, $L_{r,e} = 60.6$; and for a proton oscillator with the same oscillation frequency, $t_{r,p} = 3.71 \times 10^{-7}$ s, $L_{r,p} = 1.11 \times 10^5$ m. Common with all of the examples, the condition $\alpha << \omega$ is well satisfied.

A.2 Radiation–charge electromagnetic interaction force and work In the mechanical representation of Sec. 2.1 the radiation damping force F_r acts on a charged oscillator through the deformation of a viscous elastic vacuum medium, $\mathscr{U}_r \propto \mathscr{U} \propto \mathcal{A}$ whose electromagnetic counterpart is E (Eq. A.3). We shall below derive the electromagnetic counterpart of the F_r, F_r^{em}. Consider that a charged oscillator μ of charge q is emitting radiation of radiation electric field $\mathsf{E}(r,\theta,t)$ at position r at time t; our specific attention is the r lying in the region $[-r_0, r_0]$ occupied by the oscillator here. μ must inevitably in turn be acted by E a Coulomb force given for per unit length along the X axis and per unit cross-sectional area as, with (A.2) for E,

$$\mathbf{F}_e(x,t) = -q\mathsf{E}(r,\frac{\pi}{2},t) = -\frac{a^{1/2}q^2\mathcal{A}\omega^2\cos[kx-wt]\hat{z}}{4\pi\epsilon_0 rc^2}$$
$$= -\mathcal{M}\left(\frac{q^2\omega^2}{3\pi\epsilon_0 c^3\mathcal{M}}\right)\frac{3a^{1/2}c\mathcal{A}\cos[\omega(\frac{x}{c}-t)]\hat{z}}{4r} = -\frac{3\alpha\mathcal{M}c\boldsymbol{\mathscr{U}}(x,\frac{\pi}{2},t)}{4r} \qquad (A.15)$$

For obtaining the last of Eqs. (A.15) we used (A.11) and $\mathscr{U}(t)$ for $\mathscr{U}_r(x,\frac{\pi}{2},t) = a^{1/2}\mathcal{A}\cos[\omega(\frac{x}{c}-t)]$.

Substituting into the last of Eqs. (A.15) the identity relation $\mathscr{U}(t) = \frac{d\mathscr{U}(t')}{\omega dt}$ (where $t' = t - \frac{\pi}{2\omega}$), and $r = r_0$ and $\omega = 2\pi c/\lambda$, we obtain

$$\frac{s_0}{r_0\lambda}\mathbf{F}_e = -\alpha\mathcal{M}\frac{d\boldsymbol{\mathscr{U}}}{dt} \qquad (A.16)$$

where $s_0 = \frac{8\pi}{3}r_0^2$ as before. The right side of (A.16) is just the expression for $\mathbf{F}_r(= -\alpha\mathcal{M}\frac{d\mathscr{U}}{dt})$ of Sec. 2.1, hence

$$\mathbf{F}_r^{em}(=\mathbf{F}_r) = \frac{s_0\mathbf{F}_e}{r_0\lambda} = -\frac{s_0 q\mathsf{E}}{r_0\lambda} = -\frac{s_0 q\omega\mathsf{A}}{r_0\lambda} \qquad (A.17)$$

We shall next derive the source-radiation interaction potential (V_I) from the work done by F_r, or F_r^{em}. F_r^{em} does to the oscillator μ a dissipative work W_I; and F_e, hence F_r^{em} or F_r, is a conservative force (i.e., it depends on the position of the charge in the E field only). So W_I amounts to a negative source-radiation interaction potential, V_I. Suppose that at time $t = 0$, the oscillator has a potential energy $V(t=0) = V_0 = \frac{1}{2}\mathcal{M}\omega^2\bar{\mathscr{U}}^2$ ($V(t)$ is a short hand notion of $V(\mathscr{U}(t))$ used in this and the next sub-section), with $\bar{\mathscr{U}} = \frac{\mathscr{U}}{a^{1/2}}$, which is undamped. The corresponding work done during one oscillation cycle $2\pi\tau/\tau = 2\pi$ int $0 \leq t' \leq \tau$, thus is given as, with $d\mathscr{U}/dt = -\omega\mathscr{U}(x, t - \pi/2\omega)$,

$$\Delta V_I(0) = -4\times\frac{2\pi(\tau/4)}{\tau}\int_{\mathscr{U}(0)=0}^{\mathscr{U}(\frac{\tau}{4})=\mathcal{A}_n} F_r(t')d\mathscr{U}(t') = 2\pi\int_{\mathscr{U}(0)=0}^{\mathscr{U}(\frac{\tau}{4})=\mathcal{A}_n} \mathcal{M}\alpha\omega\mathscr{U}(t')d\mathscr{U}(t')$$
$$= -(\frac{1}{2}\mathcal{M}\omega^2\mathscr{U}^2(0))2\pi\frac{\alpha\tau}{2\pi} = -V_0\alpha\tau \doteq -V_0(1-e^{-\alpha\tau}) \quad (A.18)$$

The last equality holds for $\alpha\tau << 1$, a condition characteristic with the quasi harmonic oscillation here. The total potential at time $t = \tau$ thus is, $V(\tau) = V(0) + \Delta V_I(\tau) = V_0 - V_0(1 - e^{-\alpha_q\tau}) = V_0e^{-\alpha\tau}$. Similarly, for $\tau \leq t' \leq 2\tau$, $\Delta V_I(2\tau) = -V(\tau)(1 - e^{-\alpha\tau})$, $V(2\tau) = V(\tau) + \Delta V_I(\tau) = V(\tau)e^{-\alpha\tau} = (V_0e^{-\alpha\tau})e^{-\alpha\tau} = V_0e^{-\alpha 2\tau}$; and so forth. Finally, for $t - \tau \leq t' \leq t$, $\Delta V_I(t) = -V(t-\tau)(1 - e^{-\alpha\tau})$,

$$V(t) = V(t-\tau) + \Delta V_I(t) = (V_0e^{-\alpha(t-\tau)})e^{-\alpha\tau} = V_0e^{-\alpha t} = V_0 + V_I(t)$$
$$H_I = V_I(t) = -V_0(1-e^{-\alpha t}) = -\frac{1}{2}\mathcal{M}\omega^2x_j^2(1-e^{-\alpha t}) = -\hbar\omega\cos^2(\omega t)(1-e^{-\alpha t}) \quad (A.19)$$

Eqs. (A.19) and $H_0 = H - H_I$ are the same as Eqs. (7.b), (8c), and (8a-b), Sec. 2.2.

A.3 Radiation-charge interaction Hamiltonian expressed in **A** Suppose that due to the action of the $V_I(t)$ above, a photon is emitted by its source during a particular transition time $t_r = \tau = 2\pi/\omega$, and we want to express $H_I = V_I$ using **A**. For $\alpha t_r = \alpha\tau << 1$, Eq. (A.19b) reduces to

$$H_I = V_I(\tau) = -\hbar\omega\cos^2(\omega\tau)(1-(1-\alpha\tau+\frac{1}{2!}\alpha^2\tau^2-\ldots)) \doteq -2\pi\hbar\alpha$$
$$\text{or} \quad H_I = -2\pi\hbar(\frac{4qr_0\mathsf{A}\cdot(\upsilon k)}{3\mathcal{A}\mathcal{M}c}) = -\frac{\hbar q(\frac{8\pi r_0\upsilon}{3\mathcal{A}c})\mathsf{A}(-i\nabla_{x_j})}{\mathcal{M}} \quad (A.20)$$

The first of Eqs. (A.20) is given after substituting (A.11) for α, $\mathsf{E}_0 = \mathsf{A}_0\omega \approx \mathsf{A}\omega$ from (A.3a) and $\omega = \upsilon k$; and the second $k = -i\nabla_{x_j}$. If the radiation is emitted by the charge oscillator q along the Z direction as specified in Sec. 5, substituting the corresponding variables z, $\mathfrak{M}_q$, $\mathcal{A}_q$, $\omega_q = kc$ and c for x_j, $\mathcal{M}$, $\mathcal{A}$, ω and υ, (A.20) is written as

$$H_I = \frac{i\hbar q(\frac{8\pi r_0}{3\mathcal{A}_q})\mathsf{A}\cdot\nabla_z}{\mathfrak{M}_q} = \frac{i\hbar q\mathsf{A}'\cdot\nabla_z}{\mathfrak{M}_q}, \quad \mathsf{A}' = (\frac{8\pi r_0}{3\mathcal{A}_q})\mathsf{A} \quad (A.21)$$

If alternatively the radiation is emitted by the charged particle oscillator d along the X direction, substituting accordingly x, m, $\mathcal{A}_d$, $\omega_d = \frac{1}{2}\upsilon_{n_d}k_d$, $\frac{1}{2}\upsilon_{n_d}$, and k_d for x_j, $\mathcal{M}$, $\mathcal{A}$, ω, υ, and k, (A.20) is written as

$$H_I = \frac{i\hbar q(\frac{8\pi r_0\frac{1}{2}\upsilon_{n_d}\hat{y}}{3\mathcal{A}_dc}\times\mathsf{A})\cdot\nabla_x}{m} = \frac{i\hbar q(\mathbf{P}_r\times\mathsf{A})\cdot\nabla_x}{m} = \frac{i\hbar q\mathsf{A}''\cdot\nabla_x}{m},$$
$$\mathsf{A}'' = \mathbf{P}_r\times\mathsf{A}, \quad \mathbf{P}_r = \frac{8\pi(r_0/c)}{6(\mathcal{A}_d/\upsilon_{n_d})}\hat{y} \quad (A.22)$$

The electromagnetic radiation fields, the vector field **A** in the above, emitted (or absorbed) by either the oscillator q or d is oriented along the transverse Z direction and propagated in the X direction. So when expressed into the dot product form in the second to third of Eqs. (A.22), an apparent A'' is involved given after the projection of **A** by $\mathbf{P}_r$.

The same result (A.22) for a charged particle oscillator d is in the usual derivation given by considering the effect of $\mathbf{A}$ and Φ on the particle's linear momentum. Assume that $\mathbf{A}$ is along the transverse Z direction as in the above and is projected on to the X direction into $\mathbf{A}''$ according to Eq. (A.22). Under the action of $\mathbf{A}''$, the linear momentum of the charged oscillator is reduced from $\mathbf{p}_d$ by an amount $\Delta \mathbf{p}_d = -q\mathbf{A}''$ to $\mathbf{p}'_d = \mathbf{p}_d + \Delta \mathbf{p}_d = \mathbf{p}_d - q\mathbf{A}''$. The Hamiltonian operator H_d, subtracted by V_{d0}, is thus given as, with the substitutions $\mathbf{p}_d \cdot (q\mathbf{A}'') = \mathbf{A}'' \cdot \mathbf{p}_d - i\hbar \nabla \cdot \mathbf{A}''$ and $\mathbf{p}_d = \frac{\hbar}{i}\nabla$,

$$H_d - V_{d0} = \frac{1}{2m}(\mathbf{p}_d - q\mathbf{A}'')^2 + q\Phi = \frac{1}{2m}[\mathbf{p}_d \cdot \mathbf{p}_d - 2q\mathbf{A}'' \cdot \mathbf{p}_d + i\hbar \nabla \cdot \mathbf{A}'' + q^2\mathbf{A}''^2] + q\Phi$$
$$= -\frac{\hbar^2}{2m}\nabla^2 + \frac{i\hbar q\mathbf{A}'' \cdot \nabla}{m} + \frac{i\hbar q \nabla \cdot \mathbf{A}''}{2m} + \frac{q^2\mathbf{A}''^2}{2m} + q\Phi \doteq -\frac{\hbar^2}{2m}\nabla^2 + \frac{i\hbar q\mathbf{A}'' \cdot \nabla}{m} \quad (A.23)$$

The last of (A.23) is given by considering the situation that $\nabla \cdot \mathbf{A}'' = 0$ for $\mathbf{A}''$ is essentially of uniform amplitude along the radiation path; $\frac{q^2\mathbf{A}''^2}{2m}$ is a higher order term and thus omitted; and the static Φ field produced by the charge q does not act on q itself, thus $q\Phi = 0$. The remaining last term in the last of Eqs. (A.23) gives just the H_{dI}.

[1] Bekefi G and A H Barrette 1977 *Electromagnetic vibrations, waves, and radiation* (The MIT Press, Massachusetts); H G Kuhn 1961 *Atomic spectra* (Lonmans).

[2] Taylor G I 1909, Interference fringes with feeble light, *Proc Cam Philos Soc* **15** 114-5; A J Dempster and H F Batho 1927 Light quanta and interference *Phys Rev* **30** 644-8; P Grangier, G Roger and A Aspect 1986 Experimental evidence for a photon anticorrelation effect on a beam splitter: a new light on single-photon interferences *Europhys Lett* **1** 173-9.

[3] Zheng-Johansson J X (a) 2010 Internally electrodynamic particle model: its experimental basis and its predictions *Phys Atom Nucl* **73** 571-581 (*Preprint* arxiv:0812.3951), Proc Int 27th Int Colloq Group Theory in Math Phys ed G Pogosyan (Ireven, 2008); (b) 2012 Vacuum potentials for the two only permanent free particles, proton and electron. Pair productions, *J. Phys: Conf Series* **343** 012135; *Proc 7th Int Symp Quantum Theory & Symm* (Prog, Aug, 2011) ed. C. Burdik; (*Preprint*arxiv:1111.3123); (c) 2006 Inference of Schrödinger equation from classical mechanics solution *Suppl Blug J Phys* **33**, 763, *Quantum Theory and Symmetries* **IV.2**, ed V K Dobrev (Heron Press, Sofia), p763 (*Preprint* arxiv:physics/0411134v5), with P-I Johansson; (d) 2012 Inference of the Universal constancy of Planck constant based on first principles, *Preprint* arxiv:1302.2572; (e) *Preprint* arxiv:1011.1344.

[4] Schiff L I 1968 *Quantum Mechanics*, 3rd ed (McGraw-Hill, London).

Quantum Lie algebra solitons

Alexander Zuevsky
School of Mathematics, National University of Ireland, Galway, Ireland
E-mail: zuevsky@yahoo.com

Abstract. We construct a special type of quantum soliton solutions for quantized affine Toda models. The elements of the principal Heisenberg subalgebra in the affinised quantum Lie algebra are found. Their eigenoperators inside the quantized universal enveloping algebra for an affine Lie algebra are constructed to generate quantum soliton solutions.

1. Conformal and affine abelian Toda systems: classical region

1.1. Lie-algebraic setting

Let M be a manifold, $\mathbb{R}^2$ or $\mathbb{C}^1$, with the standard coordinates $z_\pm$. For $\mathbb{C}^1$ we suppose that $z_- = z_+^*$. Let G be a complex simple Lie group [25] of rank r with the Lie algebra $\mathcal{G}$ endowed with the principal grading. In the decomposition $\mathcal{G} = \oplus_{m\in\mathbb{Z}}\mathcal{G}_m$, the subspace $\mathcal{G}_0$ is abelian. Denote by G_0 and $G_\pm$ the subspaces corresponding to $\mathcal{G}_0$ and $\oplus_{m>1}\mathcal{G}_{\pm m}$ respectively. Denote by h_i and $x_{\pm i}$ Cartan and Chevalley generators of $\mathcal{G}$. In the principal grading, the elements of $\mathcal{G}_0$ and $\mathcal{G}_{\pm 1}$, respectively, satisfy the defining relations

$$[h_i, h_j] = 0,\ [h_i, x_{\pm j}] = \pm k_{ij} x_{\pm j},\ [x_{+i}, x_{-j}] = \delta_{ij} h_i,\ 1 \le i, j \le r, \tag{1}$$

where k is Cartan matrix of $\mathcal{G}$.

1.2. Conformal Toda systems

Conformal abelian Toda fields $\phi = \sum_{i=1}^r h_i\phi_i$ satisfy the equations [26, 18]

$$\partial_+\partial_-\phi + \frac{4\eta^2}{\beta}\sum_{i=1}^r m_i \frac{\alpha_i}{\alpha_i^2} e^{\beta\alpha_i\cdot\phi} = 0, \tag{2}$$

with some coupling constant β and a length scale factor η. The formal general solution [18, 17] to (2), found in holomorphically factorisable form, is given by

$$e^{-\beta\lambda_j\cdot\phi} = \langle\Lambda_j|\gamma_+^{-1}\mu_+^{-1}\mu_-\gamma_-|\Lambda_j\rangle, \tag{3}$$

where $\gamma_\pm(z_\pm)$ and $\mu_\pm(z_\pm)$ are holomorphic and antiholomorphic mappings $M \to G_0$, $M \to G_\pm$ to the Gauss decomposition subgroups G_0, $G_\pm$, respectively; $|\Lambda_i\rangle$ is the highest vector of the i-th fundamental representation of $\mathcal{G}$. Mappings $\mu_\pm(z_\pm)$ satisfy the initial value problem

$$\partial_\pm\mu_\pm = \mu_\pm\kappa_\pm, \quad \kappa_\pm(z_\pm) = \sum_{i=1}^r \phi_i^0\, x_{\pm i}, \tag{4}$$

where $\kappa_\pm$ are mappings $M \to \mathcal{G}_{\pm 1}$ that can be also represented as

$$\kappa_\pm(z_\pm) = \gamma_\pm^{-1} E_{\pm 1} \gamma_\pm, \quad E_{\pm 1} = \sum_{i=1}^{r} m_i x_{\pm i}, \tag{5}$$

with some nonzero constants m_i. Note that to obtain parametric solutions from (3), in particular instantons [17], it is pretty enough to take the screening functions $\phi_i^0(z_\pm) = c_{\pm i} z_\pm^{n_i}$ parametrized by constants $c_{\pm i}$.

1.3. Affine Toda systems

Affine Toda fields satisfy the equations

$$\partial_+ \partial_- \phi + \frac{4\eta^2}{\beta} \left(\sum_{i=1}^{r} m_i \frac{\alpha_i}{\alpha_i^2} e^{\beta \alpha_i \cdot \phi} - \frac{\psi}{\psi^2} e^{-\beta \psi \cdot \phi} \right) = 0. \tag{6}$$

Here $\partial_\pm$ stand for the partial derivatives with respect to $z^\pm$, η conventionally denotes a real inverse length scale, and β is an imaginary coupling constant. The coefficients are arranged in such a way that $\phi = 0$ is a constant solution. The formal general solution to (6) was found in [21]

$$e^{-\beta \lambda_j \cdot \phi} = e^{-\beta \lambda_j \cdot \phi^0} \frac{\langle \Lambda_j | \gamma_+^{-1} \mu_+^{-1} \mu_- \gamma_- | \Lambda_j \rangle}{\langle \Lambda_0 | \gamma_+^{-1} \mu_+^{-1} \mu_- \gamma_- | \Lambda_0 \rangle^{m_j}}, \quad 1 \le i \le r, \tag{7}$$

with some quite clear different meaning of the ingredients here. Namely, $\widehat{\mathcal{G}}$ is an affine Kac–Moody Lie algebra [14] of rank $r+1$, k (in the relations of type (1)) is an affine (degenerated with the single zero eigenvalue) matrix. For more details on affine Lie algebras see [14]. Moreover, it is convenient to enlarge Cartan subalgebra of $\widehat{\mathcal{G}}$ by the derivative element d, such that $[d, h_i] = 0$, $[d, x_{\pm i}] = \pm x_{\pm i}$, and then completed Cartan subalgebra has dimension $r+2$. Positive integers m_i in (7) are defined as the lowest for which $\sum_i k_{ji} m_i = 0$. In terms of these integers the dual Coxeter numbers of $\widehat{\mathcal{G}}$ are written as $\tilde{h} = \sum_{i=0}^{r} m_i$, while the centre of $\widehat{\mathcal{G}}$ is $c = \sum_{i=0}^{r} m_i h_i$. Finally, $|\Lambda_i\rangle$ is the highest vector of the i-th fundamental representation of affine algebra $\widehat{\mathcal{G}}$. Note that, contrarily to those for finite dimensional systems, the general solution (7) has a rather complicated structure. In particular, it can be represented as infinite series, though absolutely convergent ones.

1.4. Solitonic specialisation on classical level

In [22] a remarkable specialisation of the general solution (7) which led to soliton solutions was suggested. Before we explain it, let us introduce some useful information and notations. The subspaces $\widehat{\mathcal{G}}_{\pm 1}$ of an affine Lie algebra $\widehat{\mathcal{G}}$ endowed with the principal grading contain elements $\widehat{E}_{\pm 1} = \sum_{i=0}^{r} \sqrt{m_i}\, x_{\pm i}$ of the principal Heisenberg subalgebra of $\widehat{\mathcal{G}}$. These elements, thanks to the defining relations, satisfy a very important condition

$$[\widehat{E}_{+1}, \widehat{E}_{-1}] = c.$$

The whole principal Heisenberg subalgebra of $\widehat{\mathcal{G}}$ commuting with $\widehat{E}_{+1}$, modulo the central term c, is spanned by $\{\widehat{E}_m\}$ with m being exponents of $\widehat{\mathcal{G}}$. One can normalise the invariant bilinear form on $\widehat{\mathcal{G}}$ in such a way that

$$[\widehat{E}_m, \widehat{E}_{-n}] = \delta_{mn} m c.$$

The subspace of $\widehat{\mathcal{G}}$, $\widehat{\mathcal{G}}_m$, $m \neq 0$ contains, in addition to $\widehat{E}_m$, the elements $\widehat{F}_m^i$, $1 \le i \le r$, satisfying the relations

$$[\widehat{E}_m, \widehat{F}_n^i] = \chi_i \widehat{F}_{m+n}^i,$$

with some constants χ_i that belong to the root space of $\widehat{\mathcal{G}}$. Then the generating elements for $\widehat{F}^i_n$, $\widehat{F}^i := \widehat{F}^i(t) = \sum_{n\in\mathbb{Z}} t^{-n}\widehat{F}^i_n$ with a complex parameter t, is an eigenvector of $\widehat{E}_m$,

$$[\widehat{E}_m, \widehat{F}^i(t)] = \chi_i^{(m)} t^m \widehat{F}^i(t). \tag{8}$$

The simplest solitonic specialisation found in [22] consists in the following. Let $\widehat{G}$ be an infinite-dimensional group with the Lie algebra $\widehat{\mathcal{G}}$. One chooses mappings $\gamma_\pm$ in (5) to be unit elements of the subgroup $\widehat{G}_0$ of $\widehat{G}$, and writes solutions to (4) as

$$\mu_\pm = \mu^0_\pm e^{\eta^\pm z_\pm \widehat{E}_{\pm 1}}.$$

with some constants $\eta^\pm$, and constant mapping $\mu^0 \equiv (\mu^0_+)^{-1}\mu^0_- : M \to \mathcal{G}$ independent of the coordinates $z_\pm$. The elements $\widehat{E}_{\pm 1}$ can be removed from (7) if one chooses μ^0 in the form

$$\mu^0 = \prod_{i=1}^{N} Q_i e^{\widehat{F}^i},$$

where $Q_i \in \mathbb{R}$ are some constants, and uses relations (8), namely that $[\widehat{E}_{\pm 1}, \widehat{F}^i] = \chi_i^{(\pm 1)} t^m \widehat{F}^i(t)$. Then (7) delivers an N-soliton solutions [5] to the affine Toda model characterized by the parameters Q_i and $\chi_i^\pm$, $i = 1, \ldots, N$.

2. Quantum Lie algebras

2.1. The quantised universal enveloping algebra $U_q(sl_2)$

In the spirit of [4, 13], the quantised enveloping algebra $U_q(sl_2)$ is an associative algebra generated by X^+, X^-, H with q-deformed commutation relations

$$X^+X^- - X^-X^+ = \left(q^H - q^{-H}\right)\left(q - q^{-1}\right)^{-1}, \qquad HX^\pm - X^\pm H = \pm 2X^\pm.$$

It possesses a Hopf algebra structure with the deformed adjoint action

$$(ad_{X^\pm})_q a = X^\pm a q^{H/2} - q^{\mp 1} q^{H/2} a X^\pm, \qquad (ad_H)_q a = Ha - aH,$$

for all $a \in U_q(sl_2)$.

2.2. The quantum algebra $(sl_2)_q$

In [3, 8] a new deformed basis for the generators in the quantised universal enveloping algebra $U_q(sl_2)$ was introduced

$$X_h^\pm = \sqrt{2(q+q^{-1})^{-1}}\, q^{-H/2} X^\pm, \qquad H_h = 2(q+q^{-1})^{-1}\left(qX^+X^- - q^{-1}X^-X^+\right), \tag{9}$$

which form a three dimensional subspace $(sl_2)_q$ generated by $\{X_h^+, X_h^-, H_h\}$ in $U_q(sl_2)$ closed under the quantum Lie bracket

$$[a, b]_h := (ad_a)_q b, \qquad a, b \in (sl_2)_q. \tag{10}$$

The generators (9) possess the following commutation relations:

$$[H_h, X_h^\pm]_h = \pm 2q^{\pm 1} X_h^\pm, \quad [X_h^\pm, H_h]_h = \mp 2q^{\mp 1} X_h^\pm, \quad [X_h^+, X_h^-]_h = H_h,$$

$$[X_h^-, X_h^+]_h = -H_h, \quad [H_h, H_h]_h = 2(q - q^{-1})H_h, \quad [X_h^\pm, X_h^\pm]_h = 0.$$

The algebra $(sl_2)_q$ is not a Lie algebra in the standard sense. The generators (9) do not satisfy Jacobi identity and deformed Lie bracket (10) is not skew-symmetric. Although the q-analogue of Jacobi identity for $(sl_2)_q$ is missing, nevertheless (10) is q-skew-symmetric in accordance with [8]. By q-skew-symmetry we mean a symmetry under q-conjugation (which we will denote with a tilde), the automophism of $(sl_2)_q$ defined by $q \mapsto 1/q$. Then for an element in $(sl_2)_q$ $(aX_h^+ + bX_h^- + cH_h)\tilde{} = \tilde{a}X_h^+ + \tilde{b}X_h^- + \tilde{c}H_h$, $a, b, c \in \mathbb{C}$, the q-deformed Lie bracket satisfies $\widetilde{[a, b]}_h = [\tilde{b}, \tilde{a}]_h$.

2.3. The affinisation of $(sl_2)_q$

In this subsection we introduce the affinisation of the algebra $(sl_2)_q$. Denote $\mathcal{G} = (sl_2)_q$. Let $\mathcal{L} = \mathbb{C}[t, t^{-1}]$ be the algebra of Laurent polynomials in t and $\mathcal{L}(\mathcal{G}) = \mathcal{L} \otimes_{\mathbb{C}} \mathcal{G}$. Introduce the complex vector space $(sl_2)_q^t$: $\widetilde{\mathcal{L}}(\mathcal{G}) = \mathcal{L}(\mathcal{G}) \oplus \mathbb{C}c \oplus \mathbb{C}d$. This is a loop algebra $\mathcal{L}(\mathcal{G})$ completed with the derivation d (acting as $t\frac{d}{dt}$ in $\mathcal{L}$ and trivially on c) extended by one dimensional center c corresponding to $\mathbb{C}$–valued q-deformed 2-cocycle $\Psi_q(a,b) = (x|y)_h \Phi(P,Q)$, $\Phi(P,Q) = Res_t \frac{dP}{dt} Q$ on $\mathcal{L}(\mathcal{G})$. Here $(x|y)_h$ is a non-degenerate bilinear form on $(sl_2)_q$ and P, Q are polynomials in t. We define the q-deformed Lie bracket in this algebra as

$$\begin{aligned}&[t^m \otimes x \oplus \omega c \oplus \nu d, \quad t^n \otimes y \oplus \omega_1 c \oplus \nu_1 d]_h \\ &\quad = (t^{m+n} \otimes [x,y]_h + \nu n t^n \otimes y - n\nu_1 m t^m \otimes x) \oplus m\delta_{m+n,0}(x|y)_h c,\end{aligned} \tag{11}$$

where x, $y \in \mathcal{G}$; ν, ω, ν_1, $\omega_1 \in \mathbb{C}$. Now we introduce generators that form the affinisation of the quantum algebra $(sl_2)_q$

$$H_1 = 1 \otimes H_h,\ H_0 = 1 \otimes (c - H_h),\ e_1 = 1 \otimes X_h^+,\ e_0 = t \otimes X_h^-,\ f_1 = 1 \otimes X_h^-,\ f_0 = t^{-1} \otimes X_h^+.$$

Then we derive the adjoint action:

$$\begin{aligned}&[H_0, e_0]_h = 2e_0 q^{-1},\ [H_0, f_0]_h = -2f_0 q^{-1},\ [H_0, e_1]_h = -2e_1 q,\ [H_0, f_1]_h = 2f_1 q^{-1},\\ &[H_1, e_1]_h = 2e_1 q,\ [H_1, f_1]_h = -2f_1 q^{-1},\ [H_1, e_0]_h = 2e_0 q^{-1},\ [H_1, f_0]_h = 2q f_0,\\ &[H_0, H_1]_h = -2(q - q^{-1})H_1,\ [H_0, H_0]_h = 2(q - q^{-1})H_1,\\ &[H_1, H_1]_h = 2(q - q^{-1})H_1,\ [H_1, H_0]_h = -2(q - q^{-1})H_1,\\ &[e_0, f_0]_h = H_0,\ [e_1, f_1]_h = H_1,\ [f_0, e_0]_h = -H_0,\ [f_1, e_1]_h = -H_1,\\ &[e_0, H_0]_h = -2qe_0,\ [e_0, H_1]_h = 2qe_0,\ [e_1, H_0]_h = 2e_1 q,\ [e_1, H_1]_h = -2e_1 q,\\ &[f_0, H_0]_h = 2q^{-1} f_0,\ [f_0, H_1]_h = -2q^{-1} f_0,\ [f_1, H_1]_h = 2q f_1,\ [f_1, H_0]_h = -2q f_1.\end{aligned}$$

2.4. The Heisenberg subalgebra and eigenvectors of the q-deformed adjoint action

Recall that the basic point in the construction of solitonic solutions [22] to the affine Toda equations (6) is the existence of eigenvectors with respect to elements of the Heisenberg subalgebra of underlying affine algebra. Here we define the elements of the principal Heiseberg subalgebra

$$\widehat{E}_h^+ = 1 \otimes X_h^+ + t \otimes X_h^-, \qquad \widehat{E}_h^- = 1 \otimes X_h^- + t^{-1} \otimes X_h^+,$$

so that the following generating series for $\zeta \in \mathbb{C}$,

$$\begin{aligned}\mathcal{F}_q = \sum_{k=-\infty}^{+\infty} \zeta^k A_k, \quad A_{2m} = (q + q^{-1})^m t^m \otimes (-H_h),\\ A_{2m+1} = (q + q^{-1})^m t^m \otimes q^{-1} X_h^+ - t^{m+1} \otimes q X_h^- + \left(X_h^- | X_h^+\right)_h,\ m \in \mathbb{Z},\end{aligned} \tag{12}$$

is an eigenvector of $\widehat{E}_h^{\pm}$ with respect to the bracket (11) with eigenvalies $\zeta^{\pm 1}$. Using $\mathcal{F}_q$ one can find a quantum analogue of solitonic solutions corresponding to the affinisation of the quantum algebra $(sl_2)_q$.

3. Quantum group solutions

3.1. Formal quantum group solutions

There are a few ways to quantize the conformal affine Toda models [6, 9, 10, 11, 12, 15, 16, 20]. In [20] the light cone quantization was performed, appropriate quantum equations and Lax pairs found, and formal quantum solutions [20, 6, 15, 16] constructed. Previously, in [15, 16] it was shown that the formal quantum solution in terms of Heisenberg field operators and in the Yang–Feldman perturbative formalism for the quantized conformal affine Toda equations can be written in the form (7), (3) but with ingredients replaced by their quantum analogues. namely the group-like elements ${}_q\gamma_\pm$, ${}_q\mu_\pm$ of the quantized universal enveloping algebras $U(\mathcal{G})$ in the Gauss decomposition [2], and vectors $|\Lambda\rangle_q$ in highest weight fundamental representations for corresponding quantum group [1].

The quantised affine Toda Heisenberg field operators (see [20] for the conformal Toda model) associated to an affine Lie algebra $\widehat{\mathcal{G}}$ satisfy the equations

$$\partial_+\partial_-\phi^{(q)} + \frac{4\eta^2}{\beta}\left(\sum_{i=1}^{r} m_i \frac{\alpha_i}{\alpha_i^2} : e^{\beta\alpha_i\cdot\phi^{(q)}} : - \frac{\psi}{\psi^2} : e^{-\beta\psi\cdot\phi^{(q)}} :\right) = 0.$$

We can get quantum group deformed solutions to the affine Toda equations corresponding to $U_q(\widehat{\mathcal{G}}))$. Starting from the general solution (7) to the affine Toda equations for $\widehat{\mathcal{G}}$ we replace the vectors of fundamental highest weight $\widehat{\mathcal{G}}$-representations of by the fundamental highest weight representation vectors of $U_q(\widehat{\mathcal{G}})$. One shows [20] that the quantum Toda system possess the Heisenberg field operator solution of the form (dots denoting normal ordering) for $1 \leq i \leq r$,

$$: e^{-\beta\lambda_j^{(q)}\cdot\phi^{(q)}} :=: e^{-\beta\lambda_j^{(q)}\cdot\phi_0^{(q)}} : {}_q\langle\Lambda_j|{}_q\gamma_+^{-1}\, {}_q\mu_+^{-1}\, {}_q\mu_-\, {}_q\gamma_-|\Lambda_j\rangle_q \left[{}_q\langle\Lambda_0|{}_q\gamma_+^{-1}\, {}_q\mu_+^{-1}\, {}_q\mu_-\, {}_q\gamma_-|\Lambda_0\rangle_q\right]^{-m_j}, \tag{13}$$

where $\phi_0^{(q)}$ are free field operators, and $\partial_\pm {}_q\mu_\pm = {}_q\mu_\pm\, {}_q\kappa_\pm$, ${}_q\gamma_\pm$, ${}_q\mu_\pm$, and ${}_q\kappa_\pm$ are mappings $M \to U(\widehat{\mathcal{G}})$.

3.2. Quantum group soliton solutions

Quantum solutions [9, 10, 11] generated by quantum soliton operators [24] and corresponding to classical soliton solutions [22] can be obtained from the quantum formal solution (13). For that purpose we put ${}_q\gamma_\pm = Id$, i.e., the unit mapping, and let ${}_q\kappa_\pm(z_\pm) = {}_q\gamma_\pm^{-1}\, {}_q\widehat{E}_{\pm 1}\, {}_q\gamma_\pm$, where ${}_q\widehat{E}_{\pm 1}$ are supposed to be generators of a Heisenberg subalgebra in $U_q(\widehat{\mathcal{G}})$. Then ${}_q\mu_\pm$ are of the form

$${}_q\mu_\pm = {}_q\mu_\pm^0\, e^{\eta^\pm z_\pm\, {}_q\widehat{E}_{\pm 1}}.$$

By letting ${}_q\mu^0 = ({}_q\mu_+^0)^{-1}\, {}_q\mu_-^0 = e^{Q\mathcal{F}}$ for $Q \in \mathbb{R}$, we therefore obtain from (13)

$$: e^{-\beta\Lambda_i\cdot\phi^{(q)}} :=: e^{-\beta\lambda_j\cdot\phi_0^{(q)}} : {}_q\langle\Lambda_j|e^{-{}_q\widehat{E}^+z_+}e^{Q_q\mathcal{F}}e^{{}_q\widehat{E}^-z_-}|\Lambda_j\rangle_q \left[{}_q\langle\Lambda_0|e^{-{}_q\widehat{E}^+z_+}e^{Q\mathcal{F}}e^{{}_q\widehat{E}^-z_-}|\Lambda_0\rangle_q\right]^{-m_j}. \tag{14}$$

Here $|\Lambda_i\rangle_q$ denotes the highest vector in the $i-$th fundamental representation of $U_q(\widehat{\mathcal{G}})$. Though the main fundamental problem remains unsolved, i.e., a suitable analogue for the principal Heisenberg subalgebra of $U(\widehat{\mathcal{G}})$ is unknown, we are still able to deduce certain soliton solutions from (14) by means of the algebraic considerations in subsection 2. To illustrate that we construct an example (for the case of $(\widehat{sl}_2)_q^t$) in the next subsection.

3.3. Example: soliton solutions from quantum Lie algebra $(sl_2)_q^t$

The case $\widehat{\mathcal{G}} = \widehat{sl}_2$ of the affine Toda system corresponds to the sine–Gordon equation [17]. We use the quantum Lie algebra constructions given in section 2 to generate quantum group soliton solutions based on $(sl_2)_q^t$. Starting from (13), we put ${}_q^t\mu_\pm = {}_q^t\mu_\pm^0\, e^{\eta^\pm z_\pm\, \widehat{E}_h^\pm}$, ${}_q^t\mu^0 = ({}_q^t\mu_+^0)^{-1}\, {}_q^t\mu_-^0 = e^{Q\mathcal{F}_q}$ for the group-like mappings associated with $(sl_2)_q^t$, and include the fundamental highest weight vectors $|\Lambda_j\rangle^t_q$, $j = 0, \dots, r$ (some examples of finite-dimensional representations for $(sl_2)^t_q$ are given in Appendix 4). Then we obtain for $j = 0,\ 1$

$$: e^{-\beta\lambda_j\cdot\phi^{(q)}} :=: e^{-\beta\lambda_j\cdot\phi_0^{(q)}} : {}^t_q\langle\Lambda_j|e^{-\widehat{E}^+_h z_+} e^{Q\mathcal{F}_q} e^{\widehat{E}^-_h z_-}|\Lambda_j\rangle^t_q \left[{}^t_q\langle\Lambda_0|e^{-\widehat{E}^+_h z_+} e^{Q\mathcal{F}_q} e^{\widehat{E}^-_h z_-}|\Lambda_0\rangle^t_q\right]^{-m_j}$$
$$=: e^{-\beta\lambda_j\cdot\phi_0^{(q)}} : \ {}^t_q\langle\Lambda_j|F|\Lambda_j\rangle^t_q \left[\langle\Lambda_0|F|\Lambda_0\rangle^t_q\right]^{-m_j},$$

where we have denoted $F = exp\left(Qe^{-2z_+\zeta-2\frac{1}{t}z_-\zeta^{-1}}\mathcal{F}_q\right)$. Thus, for $j = 0{,}1$ we have

$$: e^{-\beta\lambda_0\cdot\phi^{(q)}} :=: e^{-\beta\lambda_0\cdot\phi_0^{(q)}} :,$$
$$: e^{-\beta\lambda_1\cdot\phi^{(q)}} :=: e^{-\beta\lambda_1\cdot\phi_0^{(q)}} : exp\left(-Qe^{-2z_+\zeta-2\frac{1}{t}z_-\zeta^{-1}}\Xi(\zeta,t,q)\right),$$
$$\Xi(\zeta,t,q) \equiv \sum_{m=-\infty}^{+\infty} \zeta^{2m}t^m(q+q^{-1})^m.$$

Finally, we would like to make several remarks. The construction of a quantum Lie algebra $(\mathcal{G})_h$ find its extensions to all semisimple Lie algebras $\mathcal{G}$ in [8]. For cases (other than sl_2) of quantum Lie algebras assosiated to $\mathcal{G}$, one can also derive formulae for generators of their Heisenberg subalgebras. In these notes we only discussed the case of the principal grading [17] of $\mathcal{G}$. Similar constructions of Toda systems associated to the homogeneous and non-abelian [23] gradings of corresponding Lie algebras can be found. Conformal affine Toda systems have natural generalizations [7] which involve in the construction generators of higher grading subspaces of Lie algebras. Multi-solitonic solutions for the generalized Toda systems were found in [7]. Generalizations of the constructions mentioned in this paragraph for quantum Lie algebras will be given in a forthcoming publication.

4. Appendix: representations for $(sl_2)_h$

4.1. A two-dimensional representation

In [3, 8] a two-dimensional q-representation of $(sl_2)_h$ was given in the form

$$\pi(X^+_h) = \sqrt{(q+q^{-1})/2}\begin{pmatrix} 0 & 1 \\ 0 & 0 \end{pmatrix}, \quad \pi(H_h) = \begin{pmatrix} q & 0 \\ 0 & -q^{-1} \end{pmatrix}, \quad \pi(X^-_h) = \pi(X^+_h)^t,$$

satisfying $\pi([a,b]_h) = \pi(a)\pi(b) - (\tilde{\pi}(b)\tilde{\pi}(a))$, $\pi(\tilde{a}) = \tilde{\pi}(a)$, with $\tilde{h} = -h$, $h \in sl_2$, $\tilde{q} = q^{-1}$. In [8] it is mentioned that there exist similar construction of $(sl_2)_h$-q-representations of any dimension.

4.2. Fundamental highest weight representations of $(sl_2)_q$

The highest weight vector v_0 of the fundamental representation of the quantum Lie algebra $(sl_2)_q$ satisfies the conditions

$$H_h v_0 = v_0, \quad X^+_h v_0 = 0, \quad X^-_h v_0 = v_1.$$

Thanks to definition (9) of generators $(sl_2)_h$ we have

$$[H, X^+_h]_h = [H, X^+_h] = 2X^+_h, \ [H, X^-_h]_h = [H, X^-_h] = -2X^-_h, \ [H, H_h]_h = [H, H_h] = 0.$$

The element H lies in Cartan subalgebra of $(sl_2)_q$. Therefore

$$H_h(Hv_0) = (Hv_0), \quad Hv_0 = \lambda v_0.$$

Using the definition (9) we find that

$$H_h = 2(q+q^{-1})^{-1}\left(qX^+X^- - q^{-1}X^-X^+\right) = q^H(X_h^+X_h^- - X_h^-X_h^+),$$

and the action of H_h on v_0 in such a form gives

$$q^H(X_h^+X_h^- - X_h^-X_h^+)v_0 = v_0.$$

Finally $X_h^{\mp}v_1 = q^{-\xi}v_0$, where ξ can be written in the following fashion

$$2(q+q^{-1})^{-1}\left(qX^+X^- - q^{-1}X^-X^+\right)v_0 = 2(q+q^{-1})^{-1}[H]v_0 = v_0,$$

$$2q(q^{\xi}-q^{-\xi})(q+q^{-1})^{-1}(q-q^{-1})^{-1} = 1.$$

This procedure can be recurrently continued for all v_n of the basis of the representation.

4.3. Fundamental highest weight representation for the affinised $(sl_2)_q$

Above we introduced a quantum affine algebra $(sl_2)_q^t$ as an affinisation of $(sl_2)_q$. The highest weight vector of the i-th $(i=1,2)$ fundamental representation of $(sl_2)_q^t$ possesses the properties similar to the properties of the highest weight vector of fundamental representation of $(sl_2)_q$. The actions of $h_{1,2}$, $e_{1,2}$ and $f_{1,2}$ generators on highest weight vectors $v_0^{(1)}$ and $v_0^{(0)}$ are given by

$$\begin{array}{l} h_1v_0^{(1)} = v_0^{(1)},\ h_1v_0^{(0)} = 0,\ h_0v_0^{(1)} = 0,\ h_0v_0^{(0)} = v_0^{(0)},\ e_1v_0^{(0)} = e_1v_0^{(1)} = e_0v_0^{(0)} = e_0v_0^{(1)} = 0, \\ f_0v_0^{(1)} = f_1v_0^{(0)} = 0,\ f_0v_0^{(0)} = v_1^{(0)},\ f_1v_0^{(0)} = v_1^{(0)}, \end{array}$$

where superscripts correspond to representation and subscripts label the vectors of the basis. In the same way as for $(sl_2)_q$, we have

$$\begin{array}{l} e_1v_1^{(1)} = q^{-H}v_0^{(1)},\ e_1v_1^{(1)} = q^{-\lambda^{(1)}}v_0^{(1)},\ hv_0^{(1)} = \lambda^{(1)}v_0^{(1)},\ e_1v_1^{(0)} = 0,\ h = 1\otimes H, \\ e_0v_1^{(0)} = q^{-H}v_0^{(0)},\ e_0v_1^{(0)} = q^{-\lambda^{(0)}}v_0^{(0)},\ hv_0^{(0)} = \lambda^{(0)}v_0^{(0)},\ e_0v_1^{(1)} = 0, \\ H_1v_1^{(1)} = (1-2qq^{-(\lambda^{(1)}-2)})v_1^{(1)},\ H_1v_1^{(0)} = 2q^{-1}q^{(2+\lambda_0^{(0)})}v_1^{(0)}. \end{array}$$

References

[1] Chari V and Pressley A 1994 *A guide for quantum groups* (Cambridge University Press)
[2] Damaskinski E V, Kulish P P, Lyakhovsky V V and Sokolov M A 1996 *Symmetry methods in physics*, 1 (Dubna, Joint Inst. Nuclear Res., Dubna)
[3] Delius G W and Hueffmann A 1996 J. Phys.A29
[4] Drinfeld V G 1985 Sov. Math. Dokl. 32
[5] Faddeev L D and Takhtajan L A 1987 *Hamiltonian methods in the theory of solitons* (Springer–Verlag)
[6] Fedoseev I A and Leznov A N 1984 Phys. Lett. 141B, 1-2
[7] Ferreira L A, Gervais J-L, Guillen, J S and Saveliev M V 1996 Nuclear Phys. B 470 1-2
[8] Gould M D and Delius G W 1997 Commun. Math. Phys. 185
[9] Hollowood T 1991 PUPT-1286, hep-th/9110010
[10] Hollowood T 1993 Int. J. Mod. Phys. A8
[11] Hollowood T 1993 Phys. Lett. B300
[12] Hollowood T and Mansfield P 1990 Nucl. Phys. B 330 2-3
[13] Jimbo M 1985 Lett. Math. Phys. 10, 63
[14] Kac V G 1990 *Infinite dimensional Lie algebras.* Third edition (Cambridge university Press, Cambridge)
[15] Leznov A N and Fedoseev I A 1982 Theor. Math. Phys. 53 3
[16] Leznov A N and Mukhtarov M A 1987 Theor. Math. Phys. 71 1
[17] Leznov A N and Saveliev M V 1992 *Group-Theoretical Methods for Integration of Non–Linear Dynamical Systems.* Progress in Physics Series, v. 15, (Birkhauser–Verlag, Basel)

[18] Leznov A N and Saveliev M V 1979 Lett. Math. Phys. 3
[19] Lyubashenko V and Sudbery 1995 arXiv:q-alg/9510004
[20] Mansfield P 1983 Nucl. Phys. B 222
[21] Olive D I Turok N and Underwood J W R 1993 Nucl. Phys. B401
[22] Olive D I, Turok N and Underwood J W R 1993 Nucl. Phys. B 409
[23] Razumov A V, Saveliev M V and Zuevsky A 1999 *In the Memorial volume dedicated to M. V. Saveliev, Dubna*
[24] Saveliev M V and Zuevsky A B 2000 Internat. J. Modern Phys. A 15 24
[25] Serr J-P 1992 *Lie Algebras and Lie Groups, 2nd ed* (Lecture Notes in Mathematics), Vol. 1500
[26] Toda M 1975 Phys. Rep. 18C 1

AUTHOR INDEX

Institute of Physics
Dirac House, Temple Back
Bristol BS1 6BE UK

CURRAN ASSOCIATES INC
proceedings
.com